FRONTS, INTERFACES, AND PATTERNS

Fronts, Interfaces and Patterns

Proceedings of the
Third Annual International Conference
of the Center for Nonlinear Studies
Los Alamos, New Mexico 87545, USA
May 2-6, 1983

Editors:

Alan R. Bishop
Laurence J. Campbell
Paul J. Channell

Center for Nonlinear Studies
Los Alamos, National Laboratory
Los Alamos, NM 87545, USA

N·H
P·C

1984
NORTH-HOLLAND PHYSICS PUBLISHING

Amsterdam · Oxford · New York · Tokyo

ISBN: 0 444 86906 9

PUBLISHED BY:

North-Holland Physics Publishing
a division of
Elsevier Science Publishers B.V.
P.O. Box 103
1000 AC Amsterdam
The Netherlands

SOLE DISTRIBUTORS FOR THE USA AND CANADA:

Elsevier Science Publishing Company Inc.
52, Vanderbilt Avenue
New York, N.Y. 10017
U.S.A.

REPRINTED FROM PHYSICA D VOLUME 12D (1984)

Library of Congress Cataloging in Publication Data

International Conference of the Center for Nonlinear Studies on Fronts, Interfaces, and Patterns (3rd : 1983 : Los Alamos, N.M.)
Fronts, interfaces, and patterns.

"Reprinted from Physica, D volume 12D (1984), nos. 1-3"--T.p. verso.
Includes bibliographies and index.
1. Nonlinear theories--Congresses. 2. Matter--Properties--Congresses. 3. Solidificaton--Congresses. 4. Dynamics--Congresses. I. Bishop, A. R. (Alan R.) II. Campbell, Lawrence J. (Lawrence James), 1931- . III. Channell, Paul J. IV. Center for Nonlinear Studies (Los Alamos National Laboratory) V. Title.

QC20.7.N6I53 1983 530.1'5 84-10282
ISBN 0-444-86906-9

Printed in The Netherlands

PREFACE

In many fields of science, the march of progress has confronted, and not infrequently been halted by, the difficulties of understanding, describing, calculating, measuring, and otherwise dealing with dynamic interfaces between homogeneous media. Evidence that these difficulties can be overcome is becoming clear in a number of cases, and the third annual international conference of the Los Alamos Center for Nonlinear Studies (CNLS) was convened during May 2–6, 1983 to acknowledge and advertise this hopeful message. Much of the recent progress can be attributed to the creative use of vastly more powerful computers, not merely as solution finders for specific problems, but, more importantly, as searchlights on the hitherto dark terrain of nonlinear dynamical phenomena. This point was stressed at the conference by Norman Zabusky in his keynote address, which will be published in more popular form in "Physics Today".

The conference title, "Fronts, Interfaces, and Patterns", is even more general than the wide variety of topics contained in these proceedings, so additional specification may be helpful. The focus here is on the dynamic behavior of interfaces, such as their motion, emergence, collective properties, or internal processes. Topics were excluded for one or more of the following reasons: thorough coverage at other CNLS conferences, lack of either a distinct or a moving or a non-trivial boundary, lack of wider relevance to other topics of this conference, lack of space, and editorial arbitrariness. Thus, we excluded convection patterns, random interfaces and roughening transitions, boundary layers, and domain walls. The topics of geophysical fronts in the ocean and atmosphere were discussed at the conference, but are absent from the proceedings for lack of manuscripts. Although the theory and method of tracking free boundaries, in a computational sense, are necessarily important ingredients of many papers here, and indeed the principal ingredients of some, they were not the exclusive theme of the conference. A useful review of moving boundary problems in this more restricted sense can be found in recent conference proceedings [1].

The papers have been classified primarily according to the source of the interfacial dynamics. *Section 1,* Boundaries of Demarcation, deals with interfaces that are not physically or chemically active but are essentially passive markers of the boundary between two media whose internal dynamical constraints are expressed through interfacial deformation. There is no significant transport of material across these interfaces except in the case of shock fronts, which are included in this category because material crossing the shock is only transformed to another state of the same phase. Section 1 is subdivided into liquids, gases, plasmas, and solids. The case of liquid interfaces is generally familiar but the recent progress in this classical subject may be less so. We should point out that Aref and Tryggvason's passive and active interfaces are all passive by our definition. A marvelous variety of interfacial phenomena occur between gaseous regions if their relative velocity is supersonic. Plasmas exhibit significant interfaces at the boundary with a vacuum or in neutral sheets between regions of opposite magnetic field. Solids contain shock fronts and also cracks, which in the continuum limit form an unusual medium for wave propagation.

In *Section 2,* Boundaries of Reaction and Transformation, are found those interfaces that can be established only by the transport of material across the interfacial region and that are sites of chemical reactions, diffusion gradients, or phase transformations. The subsections are combustion, solidification, and "other", a very rich class touching on colloids, metamorphic layering, cardiology, embryology, and basic surface physics. For completeness it seemed desirable to have some survey of the important earlier work on solidification and on spatial structures in chemical reactions, and we are pleased that Robert Sekerka and John Ross consented to provide bibliographic abstracts on these well-documented subjects. The current high interest in solidification patterns is reflected in the large number of papers on that subject.

Most of the papers discuss a specific physical system and so contain a large measure of physics or chemistry, and consequently, direct real world relevance. While one cannot expect to learn the subject from a single article the references will, at least, give an entry to the varied literatures. Mathematical development and/or numerical techniques are discussed in almost every paper, and some of the more general discussions are collected separately in *Section 3* for emphasis.

This is the third interdisciplinary conference organized by CNLS, and represents the commitment to broader scientific exchanges that CNLS pursues in addition to its sponsorship of specialized workshops [2]. The rewards of interdisciplinary publication can be meager, and the authors of these papers deserve the thanks of all who believe scientists have some responsibility to risk communication across disciplinary boundaries in homage to the unity of science.

Because the range of subjects at this conference outstripped the collective expertise of the organizers/editors, we sought, and here gratefully acknowledge, the sound advice of our Los Alamos colleagues, particularly Stirling Colgate, Arthur Cox, Darryl Holm, Mac Hyman, and Tetsuji Yamada. Alas, they cannot be held responsible for any editorial oversights. We also thank session chairmen Martin Kruskal, Jerry Marsden, Philip Morrison, Tony Maxworthy, Charles Conley, Mark Bitensky, Bruce West, Michael Nauenberg, and Geoffrey Ludford. We especially thank our banquet speaker, Milton Van Dyke. The experienced office staff at the CNLS, Frankie Gomez, Marian Martinez, and Laurie Lauer, carried through the complicated mechanics of conference organization with professionalism and panache. We are most grateful to the Applied Mathematical Sciences Program of the USDOE for a grant that made the conference possible and to Donald Kerr, Director of Los Alamos National Laboratory, for his overall support and for making available the Laboratory's excellent conference facilities.

A. R. Bishop
L. J. Campbell
P. J. Channell

Editors

[1] Moving Boundary Problems, D. G. Wilson, A. D. Solomon and P. T. Boggs, eds. (Academic Press, New York, 1978).

[2] The proceedings of the earlier CNLS conferences are Nonlinear Problems: Present and Future, A. R. Bishop, D. K. Campbell, and B. Nicolaenko, eds. (North-Holland, Amsterdam, 1982) and Order in Chaos, D. K. Campbell and H. A. Rose, eds. (North-Holland, Amsterdam, 1983).

CONTENTS

Section 2 Boundaries of reaction and transformation

2.1. *Combustion*

2.2 *Solidification*

2.3. *Diffusion, wetting, biological differentiation and excitation*

Section 3 Mathematical structure and computational methods

SECTION 1

BOUNDARIES OF DEMARCATION

Physica 12D (1984) 3–18
North-Holland, Amsterdam

1.1. LIQUIDS

AN OVERVIEW OF RAYLEIGH–TAYLOR INSTABILITY*

D.H. SHARP
Theoretical Division, Los Alamos National Laboratory, Los Alamos, New Mexico 87545, USA

The aim of this talk is to survey Rayleigh–Taylor instability, describing the phenomenology that occurs at a Taylor unstable interface, and reviewing attempts to understand these phenomena quantitatively.

1. Introduction

The Rayleigh–Taylor instability is a fingering instability of an interface between two fluids of different densities, which occurs when the light fluid is pushing the heavy fluid [1, 2]. The aim of this talk is to survey Rayleigh–Taylor instability, describing the phenomenology that occurs at a Taylor unstable interface, and reviewing attempts to understand these phenomena quantitatively. I will also emphasize some critical questions which require further study.

2. Simplest explanation of the occurrence of Rayleigh–Taylor instability

This conference affords the pleasure of learning about a great variety of topics from speakers with the most diverse backgrounds. In view of this diversity, I hope the experts will forgive me if I begin with the simplest possible description of Rayleigh–Taylor instability.

Imagine the ceiling of a room plastered uniformly with water to a depth of 1 meter (fig. 1). The layer of water will fall. However, it is not through lack of support from the air that the water will fall. The pressure of the atmosphere is equivalent to that of a column of water 10 meters thick, quite sufficient to hold the water against the ceiling. But in one respect the atmosphere fails as a supporting

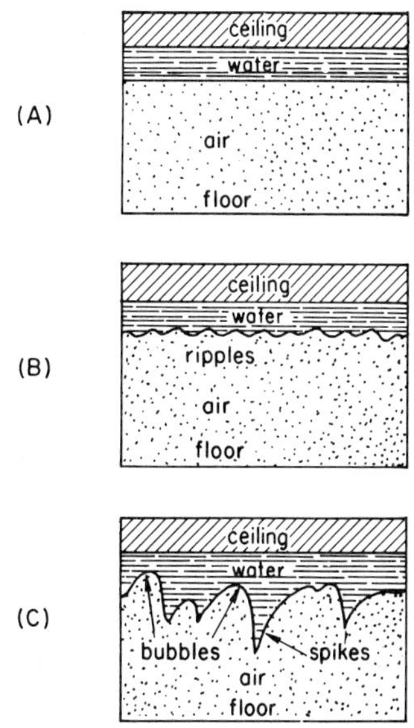

Fig. 1. (A) The pressure of the air is quite sufficient to support a perfectly uniform layer of water 1 meter thick against the ceiling. (B) But the air pressure can not constrain the air–water interface to flatness. Ripples or irregularities will inevitably be present at the interface. (C) The irregularities grow, forming "bubbles" and "spikes." The water falls to the floor.

medium. It fails to constrain the air–water interface to flatness. No matter how carefully the water layer was prepared to begin with, it will deviate from planarity by some small amount. Those portions of the fluid which lie higher than the average experience more pressure than is needed for their support. They begin to rise, pushing aside water as they do so. A neighboring portion of the fluid, where the surface hangs a little lower than average, will require more than average pressure

* Work supported by the U.S. Department of Energy.

0167-2789/84/$03.00 © Elsevier Science Publishers B.V.
(North-Holland Physics Publishing Division)

for its support. It begins to fall. The air cannot supply the variations in pressure from place to place necessary to prevent the interface irregularities from growing. The initial irregularities therefore increase in magnitude, exponentially with time at the beginning. The water which is moving downward concentrates in spikes. The air moves upward through the water in round topped columns. The water falls to the floor.

The same layer of water lying on the floor would have been perfectly stable. Irregularities die out. Thus we can infer a simple criterion for the onset of Taylor instability at the interface between two fluids of different densities: *If the heavy fluid pushes the light fluid, the interface is stable. If the light fluid pushes the heavy fluid, the interface is unstable.* A criterion of comparable simplicity governs the onset of Kelvin–Helmholtz instability: *The interface between two fluids is unstable if there is a jump in the tangential component of the velocity across the interface.* These two criteria are among the most basic principles in the subject of interface instability.

3. Examples of Rayleigh–Taylor instability

Taylor instability occurs in diverse situations. As examples, we mention:

A. *Natural phenomena*

i) Overturn of the outer portion of the collapsed core of a massive star [3];

ii) the formation of high luminosity twin-exhaust jets in rotating gas clouds in an external gravitational potential [4].

B. *Technological applications*

i) Laser implosion of deuterium–tritium fusion targets [5];

ii) electromagnetic implosion of a metal liner [6], and several others.

Let us take a quick look at one of these examples. A highly schematic picture of the implosion of a deuterium–tritium (DT) pellet is as follows (fig. 2). A sphere of DT is surrounded by a glass or metal tamper. This tamper is irradiated with

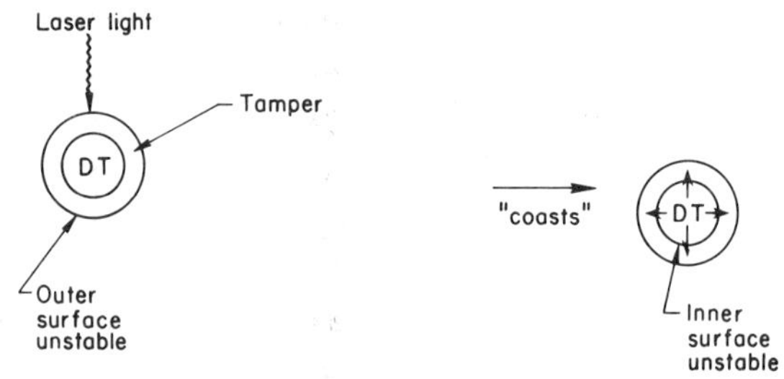

Fig. 2. Schematic diagram of the implosion of DT pellet.

intense laser light, which causes it to accelerate inward. The outer surface of the tamper is the interface between a heavy fluid (metal) and a light fluid (vaporized metal) and is unstable during the initial phase of the implosion. As the pellet is compressed to perhaps 1,000 times its normal density, its pressure increases until it is sufficient to slow the inward motion of the tamper. This phase of the implosion is also Taylor unstable; here it is the DT which is pushing the tamper. Although this picture of the implosion of a DT pellet is so oversimplified as to be almost ridiculous, it nevertheless suggests quite clearly that Taylor instability is a factor to be dealt with in evaluating pellet performance [7].

4. Phenomenology of Rayleigh–Taylor instability

There is a complex phenomenology associated with the evolution of a Taylor unstable interface. This includes the formation of spikes, curtains and bubbles, the development of Helmholtz instability on the side of the spikes, competition among bubbles leading to their amalgamation, formation of droplets, entrainment and turbulent mixing, and a possible chaotic limit with a fractalized interface.

It is helpful to organize a description of the growth of the instability into a number of stages. This can be done as follows [8, 9].

Stage 1. If the initial perturbations in the interface or velocity field are extremely small, the early stages in the growth of the instability can be analyzed using the linearized form of the dynamical equations for the fluid. The result is that small amplitude perturbations of wavelength λ grow exponentially with time. When the amplitude of the initial perturbation grows to a size of order 0.1λ to 0.4λ, substantial deviations from the linear theory are observed.

Stage 2. During the second stage, while the amplitude of the perturbation grows nonlinearly to a size of order λ, the development is strongly influenced by three-dimensional effects and the value of the density ratio, or Atwood number $A = (\rho_H - \rho_L)/(\rho_H + \rho_L)$, where ρ_H = density of heavy fluid, ρ_L = density of light fluid.

If $A \lesssim 1$, the light fluid moves into the heavy fluid in the form of round topped bubbles with circular cross sections. Note that "two-dimensional" plane bubbles are unstable to perturbations along the axis perpendicular to the plane of the bubble, and a trough having a plane bubble as a cross section will break up into three-dimensional bubbles. The heavy fluid will form spikes and walls or curtains between the bubbles, so that a horizontal section would show a honeycomb pattern. If $A \gtrsim 0$, one will instead find a pattern more like two sets of interpenetrating bubbles.

Note that these pictures are rather different from the simpler patterns of bubbles and spikes that we think of in two dimensions. I will nevertheless often refer to bubbles and spikes for the sake of simplicity.

Stage 3. The next stage is characterized by the development of structure on the spikes and interactions among the bubbles. These phenomena can originate from several sources. There is a nonlinear interaction among initial perturbations of different frequency. Also Helmholtz instability along the side of the spike can cause it to mushroom, increasing the effect of drag forces on the spike. This effect is more pronounced at low density ratios. There is some experimental evidence for bubble amalgamation, a process in which large bubbles absorb smaller ones, with the result that large bubbles grow larger and move faster.

The presence of heterogeneities in various physical quantities can modify the shape and speed of bubbles and spikes to a degree which depends on the strength and length scale of the heterogeneity.

Stage 4. In the final stage, we encounter the breakup of the spike by various mechanisms, the penetration of a bubble through a slab of fluid of finite thickness and other complicated behavior

Table I
Some factors influencing the development of Rayleigh–Taylor instability

Factor	Relative size of effect (dimensionless parameter)	Effect on growth of instability
Density ratio	ρ_H/ρ_L or $A = (\rho_H - \rho_L)/(\rho_H + \rho_L)$	A key factor governing the growth rate of Rayleigh–Taylor or Kelvin–Helmholtz instability for small amplitude perturbations of wavelength λ.
Surface tension	Weber number $= 2\sigma/(\rho_H - \rho_L)g\lambda^2$	In linear theory, stabilizes wavelengths shorter than a critical wavelength $\lambda = \sqrt{\sigma/g(\rho_H - \rho_L)}$. Establishes a most unstable wavelength, hence probably makes problem well posed mathematically.
Viscosity	$R = \nu t/\lambda^2$	Reduces growth rate; regularizes fluid flow.
Compressibility	$G = g/kc^2 = \frac{(\text{phase velocity of gravity waves})^2}{(\text{sound speed})^2}$	Reduces growth rate of long wavelength perturbations; decreases active volume of fluid.
Heterogeneity	$\Delta L/\lambda$, $\Delta v/v \ldots$	Can excite secondary, tertiary, ... instabilities of various wavelengths.

that leads to a regime of turbulent or chaotic mixing of the two fluids.

5. Factors influencing the development of Rayleigh–Taylor instability

Numerous factors influence the development of Taylor instability in a simple fluid. These include surface tension, viscosity, compressibility, effects of converging geometry, three-dimensional effects, the time dependence of the driving acceleration, shocks, and a variety of forms of heterogeneity. An assessment of some of these factors is given in table I.

6. Other factors which may be operative in realistic problems

In natural phenomena and technological applications where Taylor instability occurs, there are many other factors that can play an important role. For example, material properties and the equation of state of the fluids may be important. The fluids may conduct heat or diffuse mass. The material may change phase or consist of several components. Radiation often couples to hydrodynamics.

It is not easy for me to imagine dealing scientifically with the whole range of factors that can influence Rayleigh–Taylor instability, so in this talk I will restrict myself to a few of the factors which effect the behavior of simple fluids. It is a little bit humbling to recall that engineers must deal with Taylor instability in its full complexity.

7. Analytic and quasi-analytic modeling

The purpose of analytic modeling is to identify the effects which are dominant during a given stage in the development of the instability.

7.1. *Linear analysis*

There is a considerable body of literature which analyzes the initial stage in the growth of small amplitude Taylor instability, where the linearized form of the equations of fluid dynamics can be used.

7.1.1. *Plane geometry*

As an example, consider two infinitely extended inviscid fluids which meet at a plane interface (fig. 3). For definiteness, we suppose the upper fluid is heavier, $\rho_H > \rho_L$. The fluids are subjected to a constant acceleration in a direction normal to the interface. We write the total effective acceleration as $\boldsymbol{G} = (\boldsymbol{a} - \boldsymbol{g}) = (a + g)\hat{z} = G\hat{z}$, with $g > 0$ and $\hat{z}$ a unit vector normal to the interface, pointing into the heavy fluid. The gravitational acceleration is $\boldsymbol{g} = -g\hat{z}$ and $\boldsymbol{a} = a\hat{z}$ is a uniform external acceleration applied to the system as a whole. Thus, when $G > 0$, the effective acceleration acts vertically upward, the light fluid accelerates the heavy fluid, and the configuration is unstable according to the criterion we discussed above.

One works in a noninertial frame comoving with the unperturbed interface. In this frame, the unperturbed fluids are at rest and the unperturbed interface is defined by $z = 0$. The pressure fields in the fluids vary with the vertical coordinate z to balance the total acceleration and permit static equilibrium (in the comoving frame).

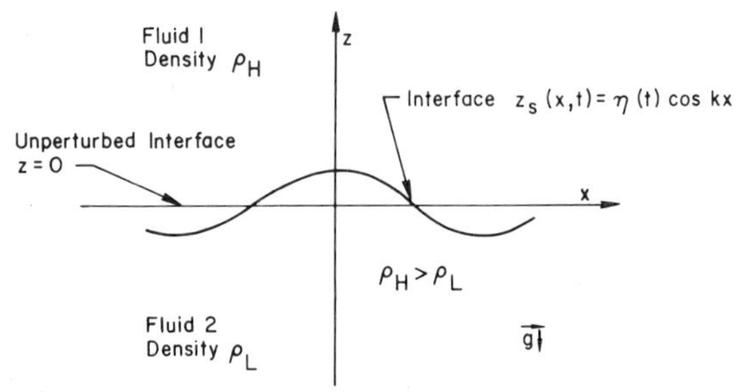

Fig. 3. Two incompressible fluids of infinite depth, having densities ρ_H, ρ_L, meet at an interface. For $t < 0$, the interface is the plane $z = 0$. For times $t \geqslant 0$, the interface has a perturbed shape. The simple case $Z_s = \eta(t)\cos kx$ is illustrated in the figure.

This is the picture for $t < 0$. At $t = 0$, we perturb the configuration in some way. We might suppose that the fluid is initially at rest but that the interface is perturbed so as to have the form

$$Z_s = \eta(t)\cos kx, \quad [\eta(t)/\lambda \ll 1]. \tag{1}$$

Thus the interface now consists of a set of crests and troughs parallel to the y-axis.

One may then show, using either a potential theory argument via Bernoulli's equation or a simple energy analysis, that the amplitude of the perturbation is determined by the equation [10]

$$\ddot{\eta}(t) = \alpha^2(k)\eta(t), \tag{2}$$

with

$$\alpha^2(k) = G\left(\frac{\rho_H - \rho_L}{\rho_H + \rho_L}\right)k - \left(\frac{\sigma}{\rho_H + \rho_L}\right)k^3. \tag{3}$$

Here σ is the coefficient of interfacial tension.

The solution to (2) for fluids initially at rest is

$$\eta(t) = \eta(0)\cosh \alpha t. \tag{4}$$

Several simple but useful conclusions can be drawn from (3):

If $\sigma = 0$, $G > 0$ and $\rho_H > \rho_L$, α is real and the interface is unstable. The growth rate for short wavelengths is unbounded, so on the basis of linear analysis the Rayleigh–Taylor problem would appear to be ill-posed in the absence of surface tension. However, we note that there is no rigorous theorem available which says either that the Rayleigh–Taylor problem is ill-posed in the absence of surface tension or that it is well-posed when surface tension is included. For $G < 0$, α is imaginary and one has stable gravity waves.

Surface tension stabilizes perturbations shorter than a critical wavelength

$$\lambda_c = [\sigma/G(\rho_H - \rho_L)]^{1/2}. \tag{5}$$

The shape of the dispersion curve makes it plain (fig. 4) that there is a fastest growing, or most unstable, wavelength λ_M. This is given by

$$\lambda_M = \sqrt{3}\lambda_c. \tag{6}$$

The above analysis can be generalized in several ways. For example, in the linear approximation we can superpose harmonic interface perturbations in the x and y directions to give a three-dimensional treatment of the instability.

Also, the linear treatment can be generalized to include other physical effects such as compressibility [11, 12], nonuniform accelerations [13], shocks [14], density gradients [10, 15], slab geometry [16], and so forth. A thorough analysis of the role of viscosity is available [10, 17, 18].

Finally, one can treat general initial conditions [18, 19]. Solutions of the linearlized equations satisfying general initial conditions can be expressed in terms of Fourier–Laplace transforms of the hydrodynamic variables, although the results can get quite complicated.

7.1.2. *Spherical geometry*

Taylor instability at a spherical interface has been studied by several authors [20–25]. Results of generality comparable to those obtained in plane geometry are not available, owing to the greater mathematical complexity of the equations encountered in curved geometries. There is, moreover, a new effect at work in curved geometries – a convergent geometry can itself be destabilizing.

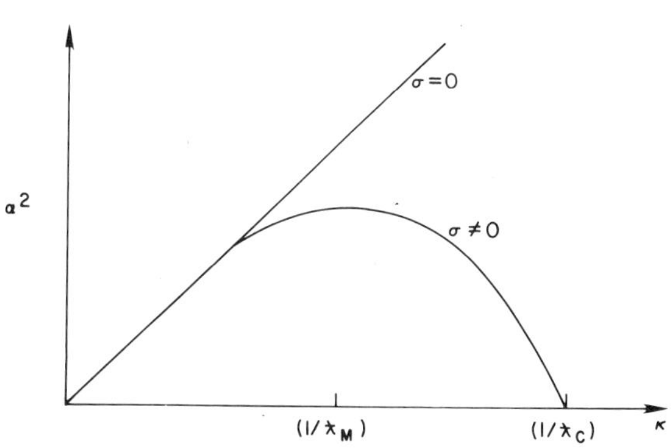

Fig. 4. Schematic plot of α^2 vs. k, eqn (3).

This can be seen readily in the case where the fluids are incompressible [25]. If $R(t)$ is the radius of the unperturbed spherical interface, and if the perturbations δR are expanded in Legendre functions, $\delta R = \Sigma_n a_n(t) P_n(\cos\phi)$, then the amplitudes $a_n(t)$ can be shown to satisfy

$$\ddot{a}_n(t) + (3\dot{R}/R)\dot{a}_n(t) - \alpha^2(n)(\ddot{R}/R)a_n(t) = 0\,, \tag{7}$$

with

$$\alpha^2(n) = \frac{n(n-1)\rho_2 - (n+1)(n+2)\rho_1}{n\rho_2 + (n+1)\rho_1}\,. \tag{8}$$

Here ρ_2 = the density outside the interface, ρ_1 = the density inside the interface, and n = the mode number of the spherical harmonic describing the perturbation.

For large R, (7) goes over to the usual Rayleigh–Taylor result. However, if $\rho_1 \to \rho_2$, (7) takes the form

$$\ddot{a}_n(t) + (3\dot{R}/R)\dot{a}_n(t) + (2\ddot{R}/R)a_n(t) = 0\,. \tag{9}$$

This equation can have unstable solutions, depending on the behavior of $R(t)$. Thus a spherical interface can be unstable even if there is no discontinuity in density across it. The relative importance of classical Rayleigh–Taylor effects and geometric or convergence effects depends strongly on compressibility and the acceleration history of the interface. This dependence can best be studied numerically.

7.2. *Nonlinear modeling*

Next I will briefly discuss some attempts to model the nonlinear growth of spikes and bubbles.

First, an exact closed form solution valid for a finite time T has been derived by Ott [26] for the Taylor unstable motion of a fluid layer idealized as having infinitesimal thickness. An initial sinusoidal perturbation evolves into a cycloid. After time T adjacent segments of the fluid collide and the development cannot be followed analytically.

There have been a number of attempts to model the growth of spikes and bubbles in more general settings. These models are based on one or another set of drastic simplifying assumptions which it is hoped capture the zeroth order physics correctly and permit the dynamics to be described by ordinary differential equations.

The easiest model to describe is that of Fermi [27], who considered incompressible fluids in the limit of infinite density ratio. He considered an initial sinusoidal perturbation, which was then approximated by a polygon, i.e., by a square wave profile (fig. 5). The interface is thus described by 3 parameters; the heights of the spike and bubble (a, b), and the width of the spike (x). The condition of incompressibility provides one relation between these parameters. Fermi next estimates the kinetic energies associated with the horizontal and vertical motions of the spike, and with the motion of the fluid above the spike and bubble, as well as the potential energy of the fluid, in terms of the parameters a and x. This yields a set of coupled nonlinear ordinary differential equations for $a(t)$ and $x(t)$. The results were that the asymptotic speed of the spike was roughly correct, but the speed of the bubble was not correct.

More recent attempts along these lines are due to Baker and Freeman [28] and Crowley and Levermore [29]. Baker and Freeman [28] derive (uncoupled) ODE's for the time evolution of spike

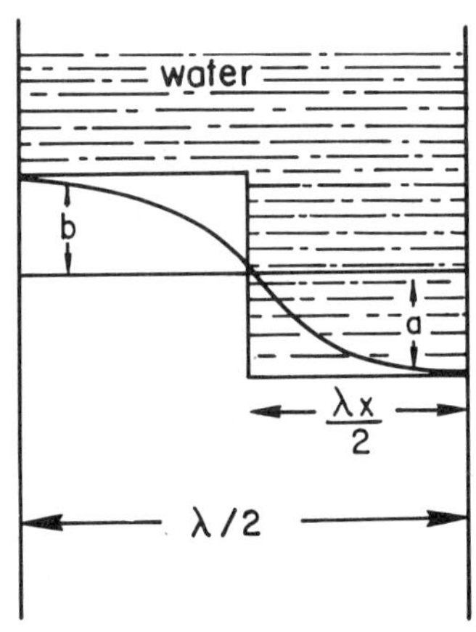

Fig. 5. Fluid configuration analyzed in Fermi's model.

and bubble peak amplitudes by devising functions which interpolate regimes of small and large amplitude behavior. The model produces results for the motion of the tips of spikes and bubbles which appear to be in reasonable agreement with code and experimental determination of these quantities.

Crowley and Levermore [29] have devised an ODE for the time development of an amplitude which is in some sense an envelope of an ensemble of initial perturbations.

My appraisal of this kind of work is as follows: If such models should turn out to be sufficiently accurate for their intended application, they would enjoy the advantage of any good model: They would express in relatively simple form the governing physics of the problem.

It is not possible to assess a priori the validity of the approximations made in deriving the models. This means that they must be developed in close conjunction with numerical codes and, where possible, experiment. This is no bad thing – indeed, one would hope that high quality codes might help one to develop more refined and better validated models.

A weak point of such models is that they do not seem to help with the most difficult aspects of Rayleigh–Taylor instability: interactions of perturbations of different frequency, the effects of statistically distributed heterogeneities, break-up of spikes and bubble amalgamation.

7.3 *Rising bubbles*

There are a number of studies which analyze the motion of symmetrical bubbles rising under gravity. The steady state motion of cylindrically symmetric bubbles rising in a tube of circular cross section has been treated by Davies and Taylor [30], who give approximate expressions for the speed of the tip of the bubble and for the profile of the bubble near the tip. Layzer [31] has given an approximate solution to the nonlinear equations determining this flow which interpolates between the initial small amplitude motion and the steady state motion.

Birkhoff and Carter [32] and Garabedian [33] discuss plane bubbles rising between parallel walls. They formulate this problem rigorously in terms of nonlinear integral equations, and their work includes an investigation of the existence and uniqueness of solutions to these equations.

Although bubble rise in a gravitational field is not identical to the Rayleigh–Taylor problem, it is closely related and is of course of interest in its own right. Moreover, the results of Davies and Taylor and of Layzer have been incorporated into certain phenomenological models of the late stage of Taylor instability, while those of Birkhoff and Carter have proven useful in validating numerical codes which compute Taylor instability. Also, the papers of Birkhoff and Carter and of Garabedian are examples of a rather sparse body of mathematically rigorous work on problems closely related to Taylor instability.

8. Numerical computation

In the simplest formulation of Rayleigh–Taylor instability, the governing equations are the two-fluid, two-dimensional incompressible Euler equations. As mentioned above, these equations appear to be ill-posed in that the growth rate of short wavelength perturbations is unbounded. We have also pointed to many factors which can influence the development of Taylor instability. These result in modifications to the simple form of the Euler equations, possibly so as to render them well-posed. However, in many cases of physical interest, the extra physical factors occur multiplied by a small parameter. In others words, they occur on small length scales, inaccessible to feasible calculations.

There are two obstacles to correct calculation in such circumstances. The small parameter, small length scale effects must somehow be included by mathematical or computational modeling. At the same time, it is necessary to avoid the incorrect simulation of physical effects by numerical artifacts. In particular convergence under mesh refinement, while a necessary aspect of validation for such problems, is not sufficient. Validation requires quantitative agreement with an independently correct calculation, analytic solution or laboratory experiment.

The computational strategies which have been developed for this problem fall into the two classes: special purpose codes and general purpose codes. The special purpose codes are ones not readily adaptable to include the variety of physical factors mentioned above. This narrower scope permits analytic simplifications which are utilized to a maximum degree.

Two notable examples of such codes have been developed by Menikoff and Zemach [34, 35] and by Baker, Meiron, and Orszag [36]. The Menikoff–Zemach code is based on conformal mappings. In the case of an infinite density ratio and two fluids in an infinite strip, a time dependent conformal mapping takes the region occupied by the heavy fluid into an infinite half strip. In this half strip, the known Green's function for Laplace's equation is used to express the interface velocity as a quadrature. The code of Baker et al. [36] is based on boundary integral techniques, in which the velocity potential is expressed as an integral over a dipole sheet distributed over the fluid interface. Coupled Fredholm integral equations can then be derived which determine the strength of the dipole sheet, and its time development.

The strong points of the special purpose codes are accuracy and speed. Thus they can be used to validate general purpose codes. Of course, these codes have also been applied in their own right to study several interesting questions. For example, the code of Baker et al. [36] has been used to study Taylor instability of a thin fluid layer [37] (fig. 6), and both codes have been used to confirm results on rising plane bubbles originally obtained by Birkhoff and Carter [32].

There have been numerous calculations of Rayleigh–Taylor instability using codes which solve the full (two-dimensional) Euler or Navier–Stokes equations. Notable examples include the work of F. Harlow and J. Welch [38], B.J. Daly [39, 40], W.P. Crowley [29], J.R. Freeman, M.J. Clauser and S.L. Thompson [41], K.A. Meyer and P. J. Blewett [42, 43] as well as the work cited in refs. 5 and 37.

Time does not permit a systematic review of this work, nor would it be easy to carry out such a review, even if time were not a constraint, because the published work often lacks the detail necessary to evaluate just what is done.

I will instead merely summarize a few of my general impressions:

The codes run up to an aspect ratio of ≈ 1 or 2,

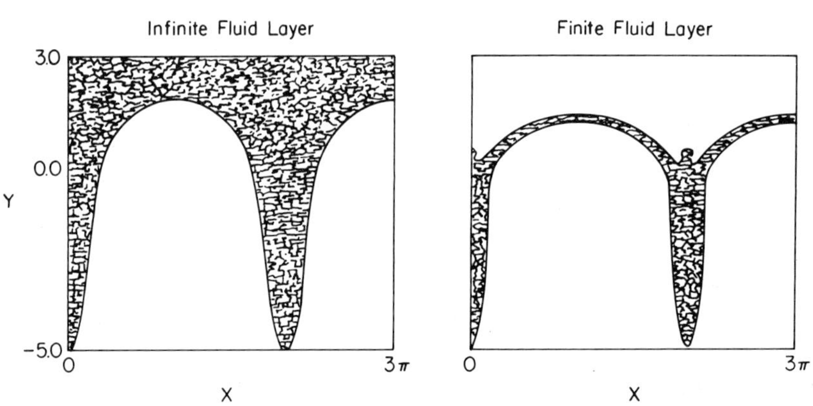

Fig. 6. Plot of interfaces for Rayleigh–Taylor instability of layered flow of an ideal fluid. The Atwood number is unity. Left: Case of a semi-infinite fluid. Right: A finite fluid layer. Figure is adapted from Verdon et al., ref. 37.

where they break down for reasons that are often not thoroughly understood.

The codes have been used to study the effect of factors such as the Atwood number, surface tension, viscosity, compressibility, thermal conductivity and others on the development of the instability. The case of single frequency perturbations is the one commonly treated;

The results are often compared to those of the linear theory. In general, they lack sufficient validation in the nonlinear regime. This is a situation which could be improved substantially if cross comparisons of general purpose codes were more common and more precise.

I now want to turn briefly to a code which I have been working on with Glimm, McBryan, and Menikoff [44]. It is based on the method of front tracking, whose goal is to achieve the accuracy of a special purpose calculation within the context of a general purpose method. The main idea of front tracking is to introduce as a computational degree of freedom an interface consisting of a (codimension one) set of curves, composed for example of piecewise linear bonds joining vertices. The front is propogated using the velocity or acceleration fields of the fluid in the case of a fluid interface discontinuity. Thus, in the Rayleigh–Taylor problem front tracking is a mixed Euler–Lagrange approach, with the front being a Lagrangian degree of freedom and all other grid points being Eulerian. In the incompressible case, it is necessary to solve elliptic PDE's at each time step for the pressure and stream function. The density discontinuity leads to elliptic equations which are singular exactly on the interface, either in their coefficients or their source terms or both.

I will not go into the computational strategies we employ to deal with these problems, or into the structure of the code and related computer science questions, or into a discussion of our current results since these points will be covered in McBryan's talk [45].

I would like to emphasize that in most cases of practical interest, one does not know in detail the nature of the many disturbances perturbing the fluid motion. In general, statistically distributed heterogeneities on various length scales will be present in the driving forces, the velocity fields and in other physical quantities. In some cases, these heterogeneities can strongly influence the fluid motion. Thus, an adequate treatment of Rayleigh–Taylor instability will require analysis not only of the growth of perturbations of fixed wavelength, but of statistical perturbations as well.

The input to our code presently allows one to add a statistically distributed component on one or several length scales to the x and y components of the velocity or acceleration, to the vorticity in the interior of the fluid, and to other physical quantities [44]. This is done by using a random number generator to define a randomly distributed variable on a lattice of points in the (x, y)-plane. The computed functions are modified at each time step by the addition of a random function.

There are many points which need to be studied in connection with statistical perturbations. First, there are now two competing effects at work: heterogeneity and nonlinear growth of fixed perturbations. Introduction of heterogeneity directly affects the physics. The driving forces associated with heterogeneity may set a length scale of their own for the formation of bubbles and spikes. These will compete with and modify the development of deterministically specified perturbations, which are channeled into their own length scales by the nonlinear dynamics.

These points are illustrated in figs. 7–12. Fig. 7 shows the growth of a small amplitude, fixed wavelength perturbation at an air–water interface (density ratio $\approx$ 500:1). In fig. 8, we show how the results in fig. 7 are modified when a strong statistical heterogeneity is added to the vorticity in the interior of the fluids. In particular, two spikes are present now where there was one before. The velocity fields are displayed in figs. 9 and 10, to show the nature of the vorticity which has been added.

In fig. 11, we show the evolution of a larger amplitude, fixed wavelength perturbation at an interface between fluids having a 4:1 density ratio.

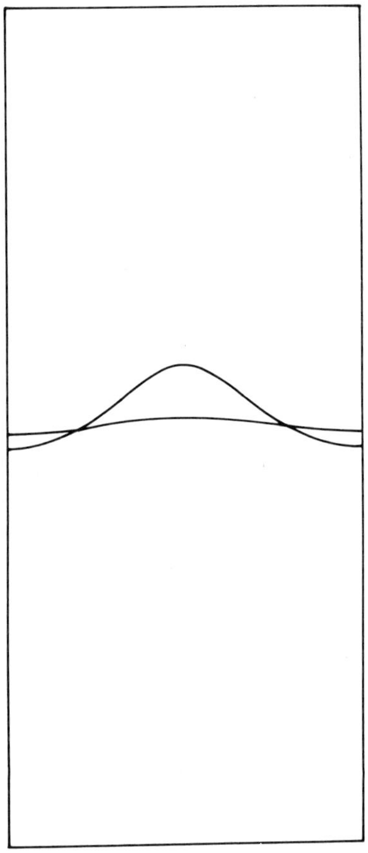

Fig. 7. Nonlinear growth of a fixed wavelength, small amplitude perturbation. In this run the density ratio was $\rho_H/\rho_L \approx 500$, the initial amplitude (in units of wave number × amplitude) was 0.1, and the grid size was 48 × 48.

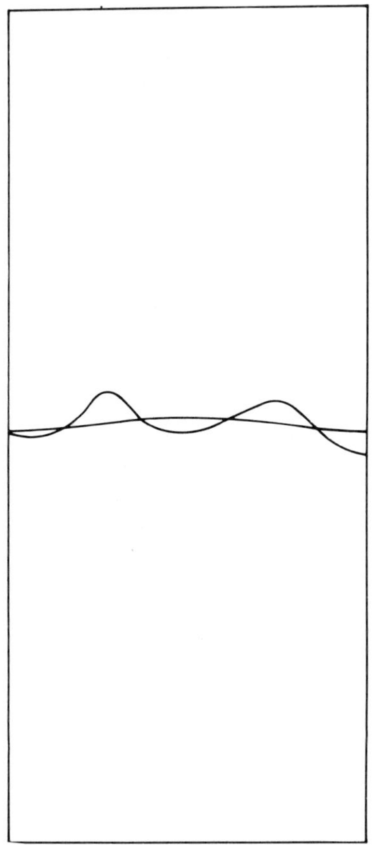

Fig. 8. Growth of a fixed wavelength, small amplitude perturbation in the presence of strong statistical heterogeneity in the vorticity in the interior of the fluid. Other parameters are as in fig. 7.

Fig. 12 shows the effect of adding a strong statistical heterogeneity to the vorticity in this problem, which is to modify the speed and shape of both bubble and spike.

To obtain useful results, one will have to identify aspects of the nonlinear growth which tend to be independent of the details of how the instability is excited. In other words, we need to seek functionals of the solution which are statistically stable. These will be the quantities on which it is appropriate to base design considerations. It is my hope that our work can be joined to the very interesting results recently obtained by Youngs [46] in the case of statistical perturbations of an interface to learn more about these questions in the near future.

9. The late stage of Rayleigh–Taylor instability

I next want to discuss some issues relating to the very late stage of Rayleigh–Taylor instability, during which processes such as bubble amalgamation, spike break-up, and turbulent mixing occur.

9.1. *Bubble amalgamation*

A very primitive statistical model for the process of bubble amalgamation was proposed quite some time ago by Wheeler and myself [47].

We suppose that after an initial period of linear growth, one may consider the Taylor unstable interface to consist of a collection of bubbles of

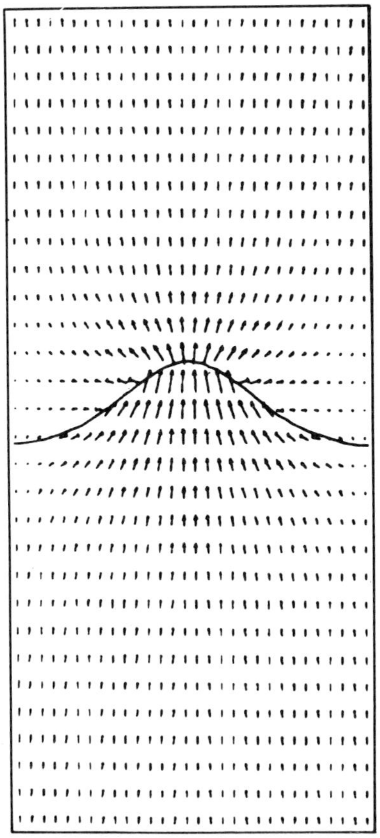

Fig. 9. This figure shows the velocity field associated with the interface motion displayed in fig. 7.

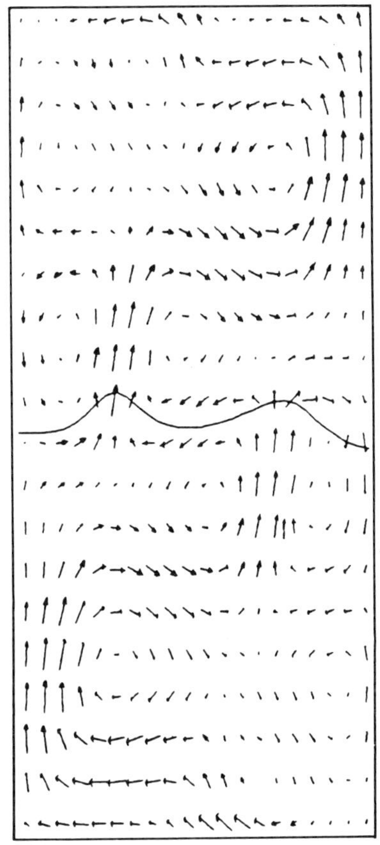

Fig. 10. This figure shows the velocity field associated with the interface motion displayed in fig. 8.

light fluid rising through a slab of heavy fluid. The cross-sectional area of the ith bubble is assumed to be circular, of radius R_i, so that it rises with a velocity given by the Davies–Taylor formula [30]

$$v_i \approx \tfrac{1}{2}\sqrt{gR_i}\,. \tag{10}$$

We also suppose that the sizes of the bubbles are not all the same, but are statistically distributed about an average size R_{av}. As a result, the bubbles will move at different rates and, in particular, a large bubble will move ahead of a smaller neighbor. The experimental evidence suggests that two such neighboring bubbles will merge, forming a single larger bubble.

Our first step was to write down a simple but crude set of rules governing bubble merger. These were:

Rule # 1. Two contiguous bubbles merge when;

$$z_+ - z_- \geqslant R_-\,. \tag{11}$$

Here $z_{+(-)}$ = (volume of bubble)/(cross sectional area of bubble) = effective height of large (small) bubble, R_- = radius of smaller bubble. This rule expresses the idea that the lead of the larger bubble over the smaller bubble necessary for merger will increase as the fraction R_-/R_+ increases.

Rule #2. Conservation of cross-sectional area;

$$\pi R_m^2 = \pi R_+^2 + \pi R_-^2\,. \tag{12}$$

This rule provides a way to calculate the radius of the merged bubble.

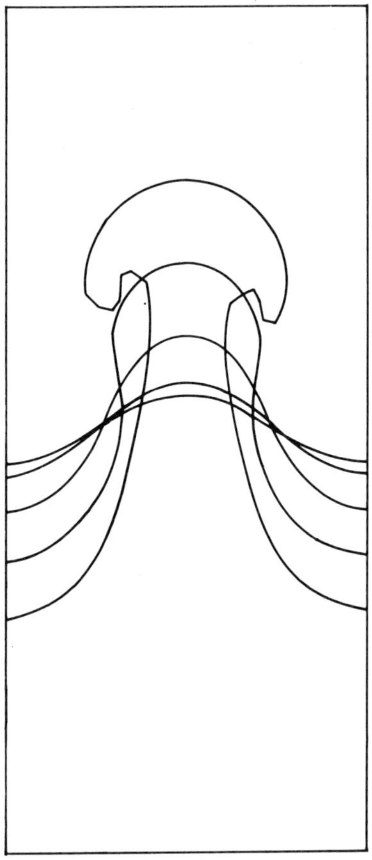

Fig. 11. Nonlinear growth of a fixed wavelength perturbation. In this run the density ratio was $\rho_H/\rho_L = 4$, the initial amplitude (in units of wave number × amplitude) was 0.5, and the grid size was 20 × 20.

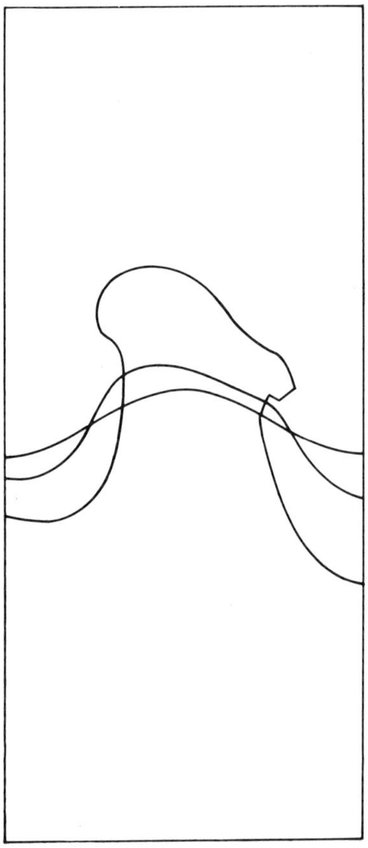

Fig. 12. Growth of a fixed wavelength perturbation in the presence of strong statistical heterogeneity in the vorticity in the interior of the fluid. Other physical parameters are as in fig. 11, except that the grid size here was 40 × 40.

Rule # 3. Conservation of volume;

$$\pi R_m^2 z_m = \pi R_+^2 z_+ + \pi R_-^2 z_- . \tag{13}$$

This rule provides a way to calculate the height of the merged bubble.

Thus as initial conditions one gives a statistically distributed set of values R_i and z_i for the collection of bubbles, and the connectivity of the bubble diagram. One can then follow the process of bubble amalgamation, generation by generation.

Analysis of the model has led to two main results:

(i) Since smaller bubbles get continually absorbed into larger ones, and the opposite process of break-up of large bubbles into small ones seems not to occur, the average size of the bubbles, and hence the average velocity of bubble rise, increases with time.

Numerically, we found

$$v_{av} = k_1 g t, \quad k_1 \approx \frac{1}{20} \text{ to } \frac{1}{100}, \tag{14}$$

where g is the bulk acceleration of the slab of fluid. This leads to a simple prediction. The slab will travel a distance $x = \frac{1}{2}gt^2$ in time t. In this time, the bubble will advance a distance $\delta = \frac{1}{2}(k_1 g)t^2$ into the slab. The bubble breaks through a slab of thickness H when $\delta = H$. Hence the slab can only travel a distance $L = (g/k_1 g)H$ before breakup, i.e., a slab can only be pushed a distance 20–100 times its thickness.

(ii) A second prediction is that a bubble anomalously larger than average will eventually grow to the point where a single large bubble will dominate the flow, provided the slab is thick enough for the merger process to go to completion before breakthrough. It does not seem to me that available computational and experimental information suffice to distinguish between the following possibilities:

(a) one bubble always quickly runs away;

(b) depending on the degree of anomaly in the original distribution of sizes, either one bubble runs away, *or* a standard, stable statistical distribution in sizes is attained, with a corresponding standard law of growth;

(c) a standard distribution of size and rate of rise is almost always reached.

Nor does available information suffice to validate the rules of bubble merger which we have adopted.

9.2. *Break-up of spikes*

At this time, little can be said with confidence about the details of the processes whereby Rayleigh–Taylor spikes break up – into droplets or otherwise. In particular, droplet formation is a three-dimensional effect which occurs on a small length scale, and as such is likely to fall outside the scope of numerical computation for some time. Available experiments on Taylor instability are inadequate as a guide for modeling these effects.

An objective in modeling spike break-up would be to derive semi-empirical formulae, valid in specified parameter ranges, for the spectra of droplet sizes and velocities. This information could then provide the input for models of entrainment, transport, and mixing.

In thinking about these questions, it may be helpful to regard the heavy falling spike as being somewhat analogous to a liquid jet, with the hope that some of the ideas that have been developed to analyze the stability and atomization of jets can be adapted to spikes.

There are a variety of mechanisms leading to jet disintegration [48, 49]. One of the first studies was by Rayleigh [50], who showed that an idealized cylindrical jet is unstable to varicose perturbations if $L > \pi D$ (L = length of the jet, D = jet diameter). This analysis ceases to apply if the jet is too thick or moves too fast, or if surface tension is negligible.

Kevin–Helmholtz instability on the side of a jet is a frequent cause of jet break-up. In this case, the nonlinear phase of Kelvin–Helmholtz instability results in the formation of a turbulent mixing zone, which spreads into the jet. The size and shape of the mixing zone depends on the shape of the nozzle, the presence of turbulence and other factors.

The formation of a mixing zone in the case of plane two-dimensional Taylor instability has recently been studied by Youngs [46, 51]. A very interesting, very important step has been taken towards understanding the mixing zone approach to spike break-up. I am glad Youngs was able to describe his work at this conference [52], because I do not have time to do it justice in this talk.

Numerous patterns of drop sizes and velocities are observed to occur in the atomization regime of jet disintegration [48, 53, 54]. Factors affecting the character of these patterns include surface tension, viscosity and the nature of "aerodynamic" forces that may be present. The variability of the observed patterns also bespeaks the frequent presence of statistical heterogeneities. Hence in the engineering literature one finds several different drop size distributions used to correlate data. For example, one distribution for the probability $p(d)$ for a drop to have a diameter between d and $d + \delta d$ is the Nukiyama–Tanasawa law [54]

$$p(d) = A d^m \mathrm{e}^{-bd^n}, \tag{15}$$

where A, b, m, and n are parameters. This law is interesting in that its general form can be derived from statistical considerations together with simple (but not necessarily correct!) assumptions about droplet formation.

10. The need for experiments

I believe that to carry out informative experiments on the late stage of Rayleigh–Taylor instability is as difficult a challenge for the experimenter as the calcuation of this phase of the instability is for the theoretician.

The classic experiments on Rayleigh–Taylor instability are those of Lewis [8] and Allred and Blount [55]. Additional experiments have been carried out by Duff, Harlow and Hirt [56], Emmons, Chang and Watson [57], Ratafia [58], Cole and Tankin [59], Popil and Curzon [60], Read and Youngs [51], and J.F. Barnes et al. [61, 62].

The available experiments are adequate to confirm semi-quantitatively, or perhaps even quantitatively, several predictions of the linear analysis for the initial growth of the instability. The experiments also provide pictures of the development of bubbles and spikes and the work of Ratafia [58] shows Kelvin–Helmholtz roll-up on the spike. Furthermore, the work of Lewis [8], Allred and Blount [58] and Read and Youngs [51] provides some information on bubble amalgamation.

There is a clear need for more and better experiments. First, available experiments are still inadequate for modeling the very late stage of Taylor instability, although the recent experiments of Read and Youngs [51] provide interesting new information on the mixing zone question. Second, experiments are needed to benchmark codes which compute Rayleigh–Taylor instability in circumstances where accurate special purpose codes do not exist for comparison.

To be of most use, the experiments should be designed with two criteria in mind:

They should be analyzable to produce quantitative data on the time history of the unstable interface. The quantitative data may well refer to appropriately chosen statistical quantities rather than to the detailed properties of a specific interface.

The experiments should be performed in an environment where the equation of state of the fluids can be regarded as known. Compounding an analysis of fluid dynamics with uncertainties about material properties will result in confusion about both.

11. Summary – critical issues

I will close by summarizing what I consider to be some critical issues concerning Rayleigh–Taylor instability.

First, it is very important to carry out three-dimensional calculations of Taylor instability [63]. There are several reasons for this: (a) Several features of Rayleigh–Taylor instability are intrinsically three-dimensional, e.g., bubble merger, processes leading to the break-up of spikes, and turbulent mixing; (b) Experiments are likely to pertain to three-dimensional flows; (c) There may be some surprises.

Second, it is important to assess the role of statistically distributed heterogeneities on the growth of the instability. Such heterogeneties will frequently be present in practical situations, and in some cases can modify the flow substantially.

As a final remark, I think it is interesting to ask whether geometric structures having fractal properties can be helpful in understanding a possible chaotic limit of Taylor instability [64].

The background for this idea is roughly as follows. It has been suggested [65] that in fully developed three-dimensional turbulence, in the limit as the Reynolds number approaches infinity, geometric structures are formed having the properties of fractal sets. The formation of objects on smaller and smaller length scales is believed to be the result of repeated generations of Kelvin–Helmholtz instability. The new idea is that the eddies formed in this way may not be space-filling. One consequence of this picture is so-called intermittancy in turbulence; another is corrections to the self-similar behavior described by Kolmogorov's Law.

We believe that a Taylor unstable interface is also subject to Kelvin–Helmholtz instability. Particularly in the case where surface tension is negligible, could successive generations of Kelvin–Helmholtz instability, possibly initiated by small scale heterogeneities, lead to a fractalized interface evolving in a self-similar manner? What would be some observable consequences of this behavior?

Acknowledgments

The author would like to thank J. Glimm, D. Holm, O. McBryan, R. Menikoff, R. Mjolsness, and H. Rose for helpful advice on the preparation of this article.

References

[1] Lord Rayleigh, Scientific Papers, Vol. II (Cambridge Univ. Press, Cambridge, England, 1900), p. 200.
[2] G.I. Taylor, Proc. R. Soc. London Ser. A 201 (1950) 192.
[3] L. Smarr, J.R. Wilson, R.T. Barton and R.L. Bowers, Ap. J. 246 (1981) 515.
[4] M.L. Norman, L. Smarr, J.R. Wilson and M.D. Smith, Ap. J. 247 (1981) 52.
[5] Taylor instability of laser driven fusion targets has been the subject of numerous papers. Some examples are: J.D. Lindl and W.C. Mead, Phys. Rev. Lett. 34 (1975) 1273; R.L. McCrory, L. Montierth, R.L. Morse and C.P. Verdon, Phys. Rev. Lett. 46 (1981) 336; R.G. Evans, A.J. Bennett and G.J. Pert, Phys. Rev. Lett. 49 (1982) 1639; M.H. Emery, J.H. Gardner and J.P. Boris, Phys. Rev. Lett. 48 (1982) 677.
[6] R.A. Gerwin and R.C. Malone, Nucl. Fusion 19 (1979) 155.
[7] A more realistic description of the interaction of laser light with the surface of a fusion target is given in D.W. Forslund, "The Importance of Surface Physics in Laser-Plasma Interactions".
[8] D.J. Lewis, Proc. R. Soc. London Ser. A 202 (1950) 81.
[9] G. Birkhoff, "Taylor Instability and Laminar Mixing", Los Alamos National Laboratory report LA-1862 (1955); Appendices A–H issued as report LA-1927 (1956).
[10] S. Chandrasekhar, Hydrodynamic and Hydromagnetic Stability (Oxford Univ. Press, Oxford, 1961), Chap. X.
[11] M. Mitchner and R.K.M. Landshoff, Phys. Fluids 7 (1964) 862.
[12] M.S. Plesset and D-Y. Hsieh, Phys. Fluids 7 (1964) 1099.
[13] G.H. Wolf, Z. Physik 227 (1969) 291.
[14] R.D. Richtmyer, Comm. Pure and Appl. Math. **13** (1960) 297.
[15] R. LeLevier, G.J. Lasher and F. Bjorklund, "Effect of a Density Gradient on Taylor Instability", Lawrence Livermore Laboratory report UCRL-4459 (1955).
[16] J.N. Hunt, Appl. Sci. Res. A 10 (1961) 45.
[17] R. Menikoff, R.C. Mjolsness, D.H. Sharp and C. Zemach, Phys. Fluids 20 (1977) 2000.
[18] R. Menikoff, R.C. Mjolsness, D.H. Sharp, C. Zemach and B.J. Doyle, Phys. Fluids 21 (1978) 1674.
[19] R.A. Axford, "Initial Value Problems of the Rayleigh–Taylor Instability Type", Los Alamos National Laboratory report LA-5378 (1974).
[20] G.I. Bell, "Taylor Instability on Cylinders and Spheres in the Small Amplitude Approximation", Los Alamos National Laboratory report LA-1321 (1951).
[21] G. Birkhoff, Quart. Appl. Math. 12 (1954) 306.
[22] M.S. Plesset, J. Appl. Phys. 25 (1954) 96.
[23] M.S. Plesset and T.P. Mitchell, Quart. Appl. Math. 13 (1956) 419.
[24] G. Birkhoff, Quart. Appl. Math. 13 (1956) 451.
[25] H.N. Fisher, "Instabilities in Converging Compressible Systems", unpublished, Los Alamos (1982).
[26] E. Ott, Phys. Rev. Lett. 29 (1972) 1429.
[27] E. Fermi, "Taylor Instability of an Incompressible Fluid", Document AECU-2979, Part 1 (1951). Also published in *The Collected Papers of Enrico Fermi*, vol. 2, E. Segré, editor-in-chief (Univ. of Chicago, Press, Chicago, 1965), p. 816. The case of finite density ratio was treated by E. Fermi and J. von Neumann, "Taylor Instability at the Boundary of Two Incompressible Liquids", Document AECU-2979, Part 2 (1953). This article is also published in Fermi's collected papers, vol. 2, p. 821.
[28] L. Baker and J.R. Freeman, "Heuristic Model of the Nonlinear Rayleigh–Taylor Instability", Sandia National Laboratories report Sand80-0700J (1980).
[29] W.P. Crowley, "An Empirical Theory for Large Amplitude Rayleigh–Taylor Instability", Lawrence Livermore Laboratory report UCRL-72650 (1970); D. Levermore, private communication, 1983.
[30] R.M. Davies and G.I. Taylor, Proc. R. Soc. London Ser. A 200 (1950) 375.
[31] D. Layzer, Ap. J. 122 (1955) 1.
[32] G. Birkhoff and D. Carter, J. Math. Mech. 6 (1957) 769.
[33] P.R. Garabedian, Proc. R. Soc. London Ser. A 241 (1957) 423.
[34] R. Menikoff and C. Zemach, J. Comp. Phys. 36 (1980) 366.
[35] R. Menikoff and C. Zemach, "Rayleigh–Taylor Instability and the Use of Conformal Maps for Ideal Fluid Flow", J. Comp. Phys., in press, 1983.
[36] G.R. Baker, D.I. Meiron and S.A. Orszag, Phys. Fluids 23 (1980) 1485.
[37] C.P. Verdon, R.L. McCrory, R.L. Morse, G.R. Baker, D.I. Meiron and S.A. Orszag, Phys. Fluids 25 (1982) 1653.
[38] F.H. Harlow and J.E. Welch, Phys. Fluids **9** (1966) 842.
[39] B.J. Daly, Phys. Fluids 10 (1967) 297.

[40] B.J. Daly, Phys. Fluids 12 (1969) 1340.
[41] J.R. Freeman, M.J. Clauser and S.L. Thompson, Nucl. Fusion **17** (1977) 223.
[42] K.A. Meyer and P.J. Blewett, "Some Preliminary Numerical Studies of Taylor Instability which Include Effects of Material Strength", Los Alamos National Laboratory report LA-4754-MS (1971).
[43] K.A. Meyer and P.J. Blewett, Phys. Fluids 15 (1972) 753.
[44] J. Glimm, O. McBryan, R. Menikoff and D.H. Sharp, "Front Tracking Applied to Rayleigh–Taylor Instability", in preparation, 1983.
[45] O. McBryan, "Computing Discontinuous Flows", these proceedings.
[46] D.L. Youngs, private communication, 1983.
[47] D.H. Sharp and J.A. Wheeler, "Late Stage of Rayleigh–Taylor Instability", Institute for Defense Analyses, Unpublished report (1961).
[48] G. Birkhoff and E.H. Zarantonello, Jets, Wakes and Cavities (Academic Press, New York, 1957), chap. XV.
[49] S. Chandrasekhar, ref. 10, chap. XII.
[50] Lord Rayleigh, Scientific Papers, vol. I (Cambridge Univ. Press, Cambridge, England, 1899), p. 361.
[51] K.I. Read and D.L. Youngs, "Experimental Investigation of Turbulent Mixing by Rayleigh–Taylor Instability", AWRE report No. 011/83, (1983).
[52] D.L. Youngs, "Turbulent Mixing by Rayleigh–Taylor Instability", Physica 12D (1984) 32, these Proceedings.
[53] R.D. Reitz and F.V. Bracco, Phys. Fluids 25 (1982) 1730.
[54] G.B. Wallis, One-Dimensional Two-phase Flow (McGraw-Hill, New York, 1969), chap. 12.
[55] J.C. Allred and G.H. Blount, "Experimental Studies of Taylor Instability", Los Alamos National Laboratory report LA-1600 (1953).
[56] R.E. Duff, F.H. Harlow and C.W. Hirt, Phys. Fluids 5 (1962) 417.
[57] H.W. Emmons, C.T. Chang and B.C. Watson, J. Fluid Mech. 7 (1960) 177.
[58] M. Ratafia, Phys. Fluids 16 (1973) 1207.
[59] R.L. Cole and R.S. Tankin, Phys. Fluids 16 (1973) 1810.
[60] R. Popil and F.L. Curzon, Rev. Sci. Instr. 50 (1979) 1291.
[61] J.F. Barnes, P.J. Blewett, R.G. McQueen, K.A. Meyer and D. Venable, J. Appl. Phys. 45 (1974) 727.
[62] J.F. Barnes, D.H. Janney, R.K. London, K.A. Meyer and D.H. Sharp, J. Appl. Phys. 51 (1980) 4678.
[63] Preliminary work on the calculation of three-dimensional Taylor instability is reported in S.A. Orszag, "Generalized Vortex Methods in 3-Dimensional Rayleigh–Taylor Instability", Physica 12D (1984) 19, these Proceedings.
[64] This suggestion was formulated in conversations with J. Glimm and H.A. Rose.
[65] B. Mandelbrot, J. Fluid Mech. 62 (1974) 331.

Physica 12D (1984) 19–31
North-Holland, Amsterdam

BOUNDARY INTEGRAL METHODS FOR AXISYMMETRIC AND THREE-DIMENSIONAL RAYLEIGH–TAYLOR INSTABILITY PROBLEMS

Gregory R. BAKER and Daniel I. MEIRON
Department of Mathematics, University of Arizona, Tucson, Arizona 85721, USA

and

Steven A. ORSZAG
Department of Mathematics, Massachusetts Institute of Technology, Cambridge, Massachusetts 02139, USA

1. Introduction

The large-scale motion of free surfaces can often be understood by means of a relatively simple model. The fluid is assumed inviscid and incompressible and its density is constant in regions separated by the free surface. In the absence of vorticity, except possibly at the free surface, the fluid flow equations reduce to Poisson's equation for the velocity potential, but kinematic and dynamic conditions must be imposed at a moving boundary, introducing severe non-linearities.

In general, solutions to free-surface flow problems are best found by numerical techniques. Experience has shown that it is important to resolve the free surface with a high degree of precision. Boundary integral techniques are a natural choice since they are adaptive in that information only at the free surface is used to predict its motion. Thus the dimension of the problem is reduced by one. Moreover, the discretisation of the free surface can be done to resolve areas of high curvature, a much more difficult task for other methods. In addition, boundary integral techniques automatically account for far field asymptotic behaviour.

Boundary integral techniques have been used successfully to study a variety of free surface flows. For example, their application to studies of the steady motion of free surfaces is straightforward and has led to several results; Bryatt-Smith and Longuet-Higgins [1], Meiron and Saffman [2], and Miksis et al. [3], among others.

Only recently have boundary integral techniques been used to study the unsteady motion of free surfaces. There are several cases where such studies are important, including the breaking of surface waves, studied by Longuet-Higgins and Cokelet [4] and Baker et al. [5], and the non-linear development of the Rayleigh–Taylor instability, studied by Baker et al. [6] in its classical form and by Verdon et al. [7] in the case of an accelerating thin layer of fluid. The results are so accurate that they have become benchmarks for calculations based on other techniques.

Previously, unsteady calculations have been restricted to two-dimensional flows. There are several important simplifications to the numerical procedures in two dimensions. In particular, the numerical quadrature for the integrals is simple and highly accurate. In this paper we shall discuss the difficulties in obtaining a reliable numerical quadrature for surface integrals in three dimensions. There are two main cases to consider: closed, spherically-shaped surfaces considered in section 3 and open, doubly-periodic surfaces, considered in section 4. For completeness, we include a review of the boundary integral equations in section 2, following Baker [8].

2. Boundary integral equations

We review the formulation of boundary integral equations with the following application in mind. Suppose that a bubble lies in a homogeneous, inviscid, incompressible fluid of constant density ρ. The bubble is composed of a gas whose density is so slight compared to that of the fluid that the motion of the gas is unimportant. Nevertheless it is capable of sustaining a pressure comparable to that of the fluid. Let P_0 be the pressure of the fluid in the far field, while P_i is the uniform pressure in the bubble. The pressure difference $P_0 - P_i$ acts to collapse or expand the bubble depending on its sign. With a suitable choice of $P_0 - P_i$, which may in general be time-dependent, it is possible to accelerate outwards the free surface between the gas and fluid, hence resulting in Rayleigh–Taylor instability. The appearance of Rayleigh–Taylor instability in such circumstances is of interest to designers of inertial confinement devices.

Suppose that at a given time the location of the free surface x and ϕ are known. Evolution equations are needed to advance the location of the free surface. The velocity potential may be expressed in terms of a dipole distribution of strength μ along the surface and a sink of strength $4\pi\sigma$ located inside the bubble at x_0 to accommodate the change in mean radius as the bubble collapses. The dipole sheet strength μ is determined by a Fredholm integral equation of the second kind,

$$\frac{\mu(x)}{2} + \frac{1}{4\pi} \oint \mu(x') \frac{\partial}{\partial n'}\left(\frac{1}{|x - x'|}\right) \mathrm{d}x' = \frac{\sigma}{|x - x_0|} - \phi(x), \quad x \in S, \tag{2.1}$$

where the principal value of the integral is taken over the surface S, n is the outward normal, and normal derivatives are taken with respect to x'. In order for (2.1) to have a solution, σ must satisfy

$$\sigma \int_S \frac{\tau(x)}{|x - x_0|} \, \mathrm{d}x = \int_S \phi(x) \tau(x) \, \mathrm{d}x, \tag{2.2}$$

where τ satisfies the adjoint integral equation

$$\frac{\tau(x)}{2} + \frac{1}{4\pi} \oint \tau(x') \frac{\partial}{\partial n}\left(\frac{1}{|x - x'|}\right) \mathrm{d}x' = 0, \tag{2.3}$$

where the normal derivative is taken with respect to x. As pointed out by Baker [8], both (2.1) and (2.3) may be solved iteratively and since a good first guess is always available from a previous time-step, only a few iterations are usually required. Once μ is known, the associated vector velocity potential A is determined by

$$A(x) = -\frac{1}{4\pi} \int \mu(x') n(x) \times \nabla\left(\frac{1}{|x - x'|}\right) \mathrm{d}x', \tag{2.4}$$

where a cross product has been taken between the normal and the gradient. The velocity u of the surface now follows from tangential derivatives of ϕ and A. The tangential velocity is computed from $t \cdot \nabla\phi$ where t is a tangent vector and the normal velocity is computed from $n \cdot \nabla \times A$ plus the contribution from the sink. The kinematic condition at the surface is satisfied by

$$\frac{\mathrm{d}x}{\mathrm{d}t} = u. \tag{2.5}$$

The motion prescribed by (2.5) is Lagrangian. The rate of change of ϕ, following (2.5), is determined by Bernoulli's equation evaluated at the surface:

$$\frac{\mathrm{d}\phi}{\mathrm{d}t} = \tfrac{1}{2}(u \cdot u) + \frac{P_0 - P_\mathrm{i}}{\rho}. \tag{2.6}$$

Thus (2.5) and (2.6) constitute the evolution equations for x and ϕ.

The numerical approximation to the above equations may seem straight-forward. If N points are introduced to represent the surface, collocation may be used to determine their velocities. The problem then becomes one of approximating the integrals in (2.1), (2.3) and (2.4). Note that at each collocation point, these integrals must be approximated using information at all the points. The difficulty lies in designing a quadrature that does not promote numerical instabilities. In general, the truncation error should not vary rapidly among neighbouring collocation points. In the next sections we shall discuss our attempts to find reliable quadratures for specific cases.

3. Quadrature for closed surfaces

Axisymmetric free surface flow provides a simple example of three-dimensional effects in that the boundary integral approach remains essentially one dimensional, but the integrals contain the complexities of the three-dimensional formulation. In a full three-dimensional flow the integrands behave singularly like an inverse square, while in two dimensions the singularity is of the nature of a pole singularity and is much easier to treat. For the axisymmetric case, the integrals involve complete elliptic functions of the first and second kind and so involve complicated expressions containing logarithmic and pole singularities. Thus there is comparable difficulty in designing a reliable numerical quadrature for both the axisymmetric and full three-dimensional cases.

Before discussing possible quadratures, it is important to note that the integrals in (2.1), (2.3) and (2.4) may be modified by using the following identities:

$$\frac{1}{4\pi} \oint_S \frac{\partial}{\partial n'} \left(\frac{1}{|x - x'|} \right) \mathrm{d}x' = \begin{cases} -\tfrac{1}{2}, & S \text{ closed}, \\ 0, & S \text{ open}. \end{cases} \tag{3.1}$$

$$\frac{1}{4\pi} \oint_S n(x) \times \nabla \left(\frac{1}{|x - x'|} \right) \mathrm{d}x' = 0. \tag{3.2}$$

In particular, the integral in (2.1) may be written as

$$\frac{1}{4\pi} \oint_S (\mu(x') - \mu(x)) \frac{\partial}{\partial n'} \left(\frac{1}{|x - x'|} \right) \mathrm{d}x' - \frac{\mu(x)}{2}. \tag{3.3}$$

In this way, the singularities in the integrals are reduced; the integrals are no longer singular for the axisymmetric case, but the integrands have discontinuous derivatives. The other integrals in (2.3) and (2.4) are treated similarly.

An axisymmetric surface may be parametrized as $(R(e), Z(e))$ where R and Z are radial and axial locations respectively in a cylindrical coordinate system. Define

$$f=\frac{1}{\left[(Z(e)-Z(e'))^2+(R(e)+R(e'))^2\right]^{1/2}}, \tag{3.4a}$$

$$g=\frac{Z_e(e')\left[(Z(e)-Z(e'))^2+R^2(e)-R^2(e')\right]-2R(e')R_e(e')(Z(e)-Z(e'))}{(Z(e)-Z(e'))^2+(R(e)-R(e'))^2}, \tag{3.4b}$$

$$F(A,B)=\frac{1}{4\pi}(\mu(e')-\mu(e))\cdot f\cdot\left[Z_e(e')A-gB\right], \tag{3.5}$$

where the subscript refers to differentiation. The integral in (3.3) may now be written explicitly for an axisymmetric surface as

$$I(e)=\int F(K(k),E(k))\,\mathrm{d}e', \tag{3.6}$$

where $K(k)$ and $E(k)$ are complete elliptic functions of the first and second kind respectively and

$$k^2=\frac{4R(e)R(e')}{(Z(e)-Z(e'))^2+(R(e)-R(e'))^2}. \tag{3.7}$$

The elliptic function $K(k)$ and the derivative of $E(k)$ have logarithmic singularities at $k=1$. Note that despite appearances g is not singular.

For a closed surface, the parameter e can be chosen so that $e=0$ or π corresponds to the two poles where the surface intersects the axis of symmetry. The integrand $F(K, E)$ in (3.5) for $e\neq 0$ or π, varies from zero at the pole $e'=0$ to an asymptotic behaviour

$$-\frac{\mu_e(e)Z_e(e)}{8\pi R(e)}(e'-e)\ln|e'-e| \tag{3.8}$$

when e' is near e and finally vanishes again at the other pole $e'=\pi$. A simple illustration is given in fig. 1; the surface is chosen to be a unit sphere, $R=\sin(e)$, $Z=\cos(e)$ and $\mu=\sin(e)$. The integrand $F(K, E)$ is plotted as a function of e' for $e=\pi/8$ and $e=\pi/2$. Note that for $e=\pi/2$, the asymptotic behaviour of F for e' near e is $(e'-e)^2\ln|e'-e|$. However, as e approaches a pole, the asymptotic form (3.8) shows that there is a strong variation in F near that pole.

For computational purposes, $K(k)$ and $E(k)$ may be approximated as

$$K(k)=A_k(m)-B_k(m)\ln|m|, \tag{3.9a}$$

$$E(k)=A_E(m)-B_E(m)\ln|m|, \tag{3.9b}$$

where A_K, A_F, B_K and B_E are polynomials in $m=1-k^2$ (see Abramowitz and Stegun (1970)). Thus we

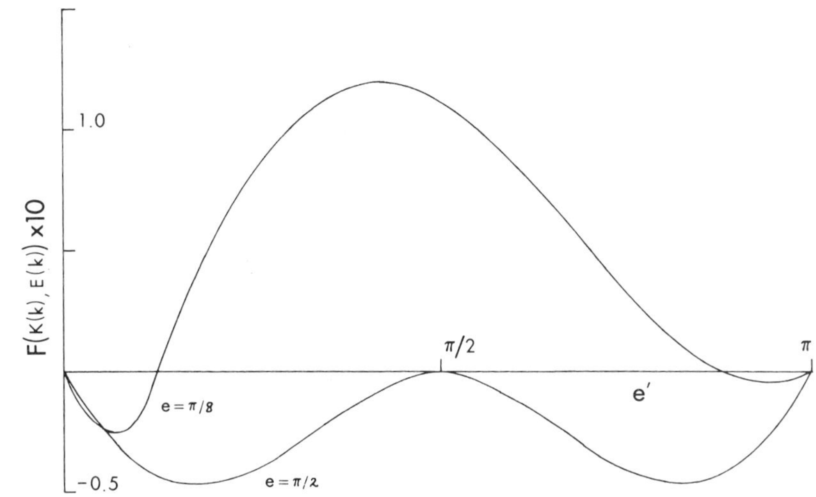

Fig. 1. The integrand $F(K(k), E(k))$ as a function of e' for $e = \pi/8$ and $e = \pi/2$.

may split off the part of the integrals that contain logarithmic singularities. If

$$I_1 = \int F(A_K, A_E)\, \mathrm{d}e', \tag{3.10}$$

$$I_2 = -\int F(B_K, B_E) \ln|m|\, \mathrm{d}e', \tag{3.11}$$

then $I = I_1 + I_2$. The integrand in I_1 is a smooth function of e' for all choices of e, and so is amenable to standard quadrature. The difficulty lies in computing I_2 numerically.

By a simple trick, one may simplify the problem as follows. Let

$$I_{2a} = -\int F(B_K, B_E)(\ln|m| - \ln|\overline{m}|)\, \mathrm{d}e', \tag{3.12}$$

$$I_{2b} = -\int F(B_K, B_E) \ln|\overline{m}|\, \mathrm{d}e', \tag{3.13}$$

then $I_2 = I_{2a} + I_{2b}$. The function $\overline{m}$ is chosen to model the behaviour of m. In particular, $\overline{m}$ must have a zero at the zero of m so that I_{2a} is not a singular integral. In addition, $\overline{m}$ is chosen so that I_{2b} may be approximated by product integration techniques, that is, $F(B_K, B_E)$ is approximated by a polynomial and the resulting integral performed analytically. Unfortunately, the simple choice $\overline{m} = (e' - e)^2$ leads to inaccuracies, since, as $e' \to e$, $m/\overline{m} \simeq (Z_e^2(e) + R_e^2(e))/4R^2(e)$, which takes on large values near the axis of symmetry. A more suitable choice is

$$\overline{m} = \frac{\sigma(e' - e)^2}{e'(\pi - e') + \sigma(e' - e)^2}, \tag{3.14}$$

where $\sigma = 1/\pi^2 + \pi^2/[4e(\pi - e)]$. Now $m/\overline{m} \simeq (Z_e^2(e) + R_e^2))e(\pi - e)/4R^2(e)\sigma$ as $e' \to e$ which approaches 1 near the axis of symmetry.

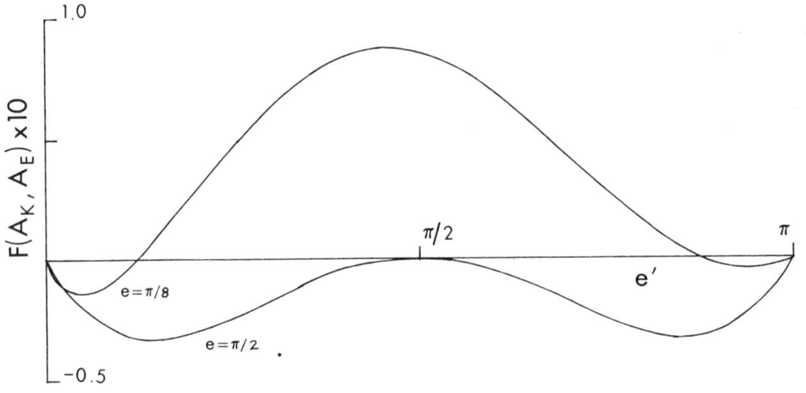

Fig. 2. The integrand $F(A_K, A_E)$ as a function of e' for $e = \pi/8$ and $e = \pi/2$.

The trapezoidal rule is a convenient choice for the quadrature in approximating I_1 and I_{2a}. For I_{2b}, $F(B_K, B_E)$ is approximated by a linear function and the resulting integral integrated analytically. The details are tedious, but a reliable, special quadrature can be obtained this way. Thus a uniform mesh $\{e_j\}$ may be introduced and the integrals computed at each e_j. Unfortunately the results still lack high accuracy. The reason can be found by examining plots of $F(A_K, A_E)$ and $-F(B_K, B_E)$ as shown in figs. 2 and 3, respectively. As before, the free surface is a sphere. The factors $F(A_K, A_E)$ and $-F(B_K, B_E)$ are plotted for $e = \pi/8$ and $e = \pi/2$. Imagine that only 9 points are used to discretise the surface. Then $e = \pi/8$ corresponds to the point adjacent to a pole. Clearly, $F(A_K, A_E)$ and $F(B_K, B_E)$ are not well resolved in the vicinity of the pole and the quadrature is correspondingly inaccurate. Note that even if more points are used in the discretization, $F(A_K, A_E)$ and $F(B_K, B_E)$ will have sharp troughs in the interval between the pole and the first point adjacent to it. To overcome this difficulty, a more accurate quadrature must be designed. The simplest way to achieve this is to interpolate r, Z, and μ at additional points and compute the integrals numerically with the straightforward quadrature but including the information at the additional points. Using 9 points to discretize the surface, the maximum absolute error is reduced from 0.00323 to 0.00014 when three additional interpolated points are used per interval.

The other integrals in (2.3) and (2.4) may be treated similarly. The vector potential A has only one non-vanishing component, $\hat{\theta}\psi/R$, where $\hat{\theta}$ is a unit vector in the azimuthal direction and ψ is a pseudo-streamfunction for axisymmetric flow. The integral that determines ψ has a similar form to (3.6) except that the factor g now has a pole singularity and so the integral must be split into three parts which are, respectively, regular, logarithmically singular, and pole singular. Finally, a special quadrature to treat the pole singularity can be developed following the approach adopted to treat the logarithmic singularity.

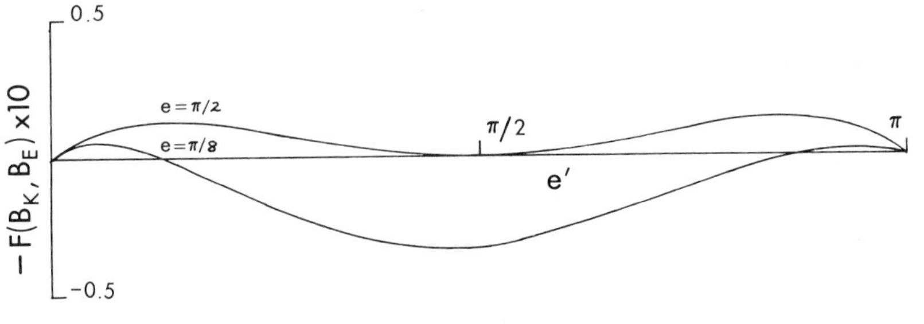

Fig. 3. The integrand $-F(B_K, B_E)$ as a function of e' for $e = \pi/8$ and $e = \pi/2$.

Now consider a general surface in three-dimensions. It may be parametrized by $(x(r, s), y(r, s), z(r, s))$ where x, y, z are Cartesian coordinates and r, s surface variables. Local unit tangent vectors are given by

$$t_1 = \frac{(x_r, y_r, z_r)}{(x_r^2 + y_r^2 + z_r^2)^{1/2}}, \tag{3.15a}$$

$$t_2 = \frac{(x_s, y_s, z_s)}{(x_s^2 + y_s^2 + z_s^2)^{1/2}}, \tag{3.15b}$$

where the subscripts refer to partial derivatives. For suitable choices of r, s, the outward normal is $\boldsymbol{n} = \boldsymbol{t}_1 \times \boldsymbol{t}_2$.

The integral in (3.3) takes the form

$$I(r, s) = \frac{1}{4\pi} \iint (\mu(r', s') - \mu(r, s)) \frac{J(y', z')(x - x') + J(z', x')(y - y') + J(x', y')(z - z')\,\mathrm{d}r'\,\mathrm{d}s'}{\left[(x - x')^2 + (y - y')^2 + (z - z')^2\right]^{3/2}}, \tag{3.16}$$

where the primes indicate that the variable is evaluated at r', s', and

$$J(A, B) = A_r B_s - A_s B_r. \tag{3.17}$$

Note that the integral (3.16) is not singular and may be treated numerically in a straight-forward fashion. However, this is not the case for the vector potential A. For the tangential components of A, the integrand has a singularity of the form

$$H = \frac{D(r - r')^2 + E(r - r')(s - s') + F(s - s')^2}{\left[A(r - r')^2 + B(r - r')(s - s') + C(s - s')^2\right]^{3/2}}. \tag{3.18}$$

where A, B, C, D, E and F are coefficients that depend on r, s. A similar approach to the development of a special quadrature for axisymmetric surfaces may be followed here. The same difficulty arises, too; the coefficients may show strong variations for certain values of r and s. Interpolation or some other means must then be used to improve the accuracy.

4. Quadrature for open, periodic surfaces

In an unbounded or closed domain the difficulties encountered in a numerical implementation of three-dimensional boundary integral methods lie strictly in the accurate evaluation of the integrals. In a periodic domain, there is the extra difficulty of the accurate evaluation of the Green's function G. In a periodic domain, G is a lattice sum of periodic extensions of the dipole kernel;

$$G = \sum_{m=-\infty}^{\infty} \sum_{n=-\infty}^{\infty} \left[(x - x' - 2n\pi)^2 + (y - y' - 2m\pi)^2 + (z - z')^2\right]^{-1/2}. \tag{4.1}$$

When the sum is written as in (4.1), it is strictly divergent but it may be given meaning as the integral of the sums

$$\frac{\partial G}{\partial x'}=\sum_{m=-\infty}^{\infty}\sum_{n=-\infty}^{\infty}\frac{(x-x'-2n\pi)}{\left[(x-x'-2n\pi)^2+(y-y'-2m\pi)^2+(z-z')^2\right]^{3/2}}, \tag{4.2}$$

$$\frac{\partial G}{\partial y'}=\sum_{m=-\infty}^{\infty}\sum_{n=-\infty}^{\infty}\frac{(y-y'-2m\pi)}{\left[(x-x'-2m\pi)^2+(y-y'-2m\pi)^2+(z-z')^2\right]^{3/2}}, \tag{4.3}$$

$$\frac{\partial G}{\partial z'}=\sum_{m=-\infty}^{\infty}\sum_{n=-\infty}^{\infty}\frac{(z-z')}{\left[(x-x'-2n\pi)+(y-y'-2m\pi)^2+(z-z')^2\right]^{3/2}}. \tag{4.4}$$

These sums may be evaluated as principal values. In two dimensions, the corresponding periodic Green's functions may be summed in closed form in terms of simple trigonometric functions. Using the identity

$$\frac{2}{\sqrt{\pi}}\int_0^{\infty}\mathrm{d}q\exp(-Aq^2)=\frac{1}{\sqrt{A}},$$

it is possible to rewrite (4.1) as

$$G=\frac{1}{\sqrt{\pi}}\sum_{m=-\infty}^{\infty}\sum_{n=-\infty}^{\infty}\int_0^{\infty}\mathrm{d}q\exp\left(-q^2\left[\left(\frac{x-x'}{2}-n\pi\right)^2+\left(\frac{y-y'}{2}m\pi\right)^2+\left(\frac{z-z'}{2}\right)^2\right]\right). \tag{4.5}$$

The sums then factor and (4.5) becomes

$$G=\frac{1}{\sqrt{\pi}}\int_0^{\infty}\mathrm{d}q\exp\left(-q^2\left(\frac{z-z'}{2}\right)^2\right)\theta_3\left(\frac{x-x'}{2}\middle|q\right)\theta_3\left(\frac{y-y'}{2}\middle|q\right), \tag{4.6}$$

where θ_3 is the 3rd Jacobi theta function.

We have been unable to evaluate (4.6) in terms of simpler functions. We may however rewrite G as a rapidly convergent sum of error functions. This is accomplished by splitting the integral into two parts,

$$G_1=\sum_{m=-\infty}^{\infty}\sum_{n=-\infty}^{\infty}\frac{1}{\sqrt{\pi}}\int_0^{1}\mathrm{d}q\exp\left(-q^2\left[\left(\frac{z-z'}{2}\right)^2+\left(\frac{x-x'-n\pi}{2}\right)^2+\left(\frac{y-y'-m\pi}{2}\right)^2\right]\right),$$

$$G_2=\sum_{m=-\infty}^{\infty}\sum_{n=-\infty}^{\infty}\frac{1}{\sqrt{\pi}}\int_1^{\infty}\mathrm{d}q\exp\left(-q^2\left[\left(\frac{z-z'}{2}\right)^2+\left(\frac{x-x'-n\pi}{2}\right)^2+\left(\frac{y-y'-m\pi}{2}\right)^2\right]\right).$$

Clearly, as $q>1$ in G_2, the sum converges very rapidly and the integrals may be performed to give

$$G_2=\frac{1}{2}\sum_{m=-\infty}^{\infty}\sum_{n=-\infty}^{\infty}\frac{\operatorname{erfc}\left\{[(x-x')/2-n\pi]^2+[(y-y')/2-m\pi]^2+(z-z')^2\right\}}{\left\{[(x-x')/2-n\pi]^2+[(y-y')/2-m\pi]^2+(z-z')^2\right\}^{1/2}} \tag{4.7}$$

for G_1 convergence slows markedly as $q \to 0$. Rapid convergence of G_1 is achieved by applying the Jacobi identity

$$\frac{1}{\sqrt{\pi q}} \sum_{n=-\infty}^{\infty} \exp\left(-n^2/q^2 + 2nib\right) = \sum_{n=-\infty}^{\infty} \exp\left(-(b-n\pi)^2 q^2\right).$$

The resulting integrals may also be performed in terms of error functions to give

$$G_2 = \frac{2}{\pi} \sum_{m=0}^{\infty} \sum_{n=0}^{\infty} w_{nm} R(n,m) \cos(n(x-x')) \cos(m(y-y')),$$

where

$$w_{nm} = \begin{cases} 1, & \text{for } n \neq 0, m \neq 0, \\ 1/2, & \text{for } n \text{ or } m = 0, \\ 1/4, & \text{for } n = m = 0, \end{cases}$$

and

$$R(n,m) = \cosh\frac{\left((z-z')\sqrt{n^2+m^2}\right)}{\sqrt{n^2+m^2}} - \frac{1}{2\sqrt{n^2+m^2}}\left[\exp\left[-(z-z')\sqrt{n^2+m^2}\right] \operatorname{erf}\left[\sqrt{n^2+m^2} - \left(\frac{z-z'}{2}\right)\right]\right.$$
$$\left. + \exp\left[(z-z')\sqrt{n^2+m^2}\right] \operatorname{erf}\left[\sqrt{n^2+m^2} + \left(\frac{z-z'}{2}\right)\right]\right]. \tag{4.8}$$

Note that the divergence of G is reflected in the infinite DC component of G_2. The derivatives G_x, G_y, G_z are however all well defined. The above derivation is a simple application of the Ewald summation technique used in solid state physics.

The numerical evaluation of G is still relatively time consuming as we have converted a slowly convergent sum of standard algebraic functions into a rapidly convergent sum of transcendental functions. Thus computation on a periodic domain is at present much slower than on closed surfaces.

5. Rayleigh–Taylor instability of axisymmetric interfaces

In this section, we consider some applications of the methods described above. To begin, let us consider an incompressible, inviscid fluid of unit density exterior to a gas bubble that is subjected to an external pressure p_0 at infinity. Suppose that the internal gas is so slight in density that its dynamics is completely specified by the pressure law, $p_i = 0.1[(v_i/v)^{5/3} - 1]$ where $v(v_i)$ is the (initial) volume of the bubble. Initially the fluid is at rest and the free surface is parametrized by $R = [1 + \varepsilon P_4(\cos\theta)]\sin\theta$, $z = [1 + \varepsilon P_4(\cos\theta)]\cos\theta$, where P_4 is the 4th order Legendre polynomial and θ is the polar angle. The difference in pressure between p_0, the pressure at infinity, and p_i, the pressure at the free surface, accelerates the fluid inwards, collapsing the bubble. As the bubble shrinks, p_i increases until it exceeds p_0 whence the fluid

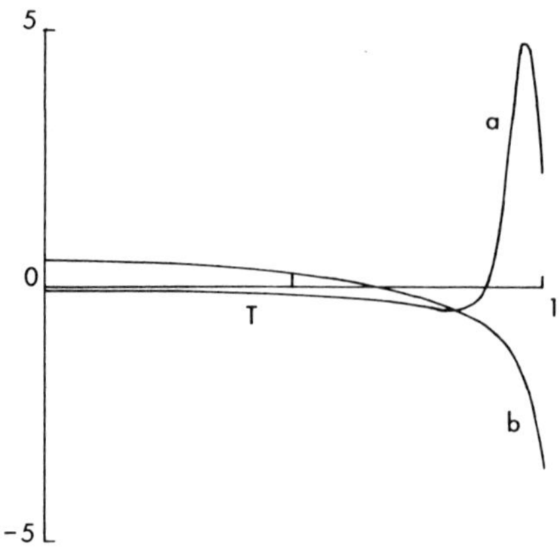

Fig. 4. Temporal variation of an axisymmetric interface using the linear perturbation theory described in section 5: (a) 0.1 × maximal free surface acceleration; (b) $10r(t)$.

decelerates. The free surface becomes Rayleigh–Taylor unstable as is evident from the behavior of r plotted in fig. 4 as determined using the linear analysis described below. The initial value of ε is 0.05. The outward acceleration is also plotted in fig. 4. The parameters in the gas law were chosen so that the initial inward acceleration would be 1 and there would be a strong outward acceleration after the surface perturbation had changed phase. Thus the surface is rather dramatically unstable.

The linear theory used to obtain the results plotted in fig. 4 is derived by simply perturbaing around the mean motion of the interface. Denote the mean radius of the interface by R, the mean velocity potential by ϕ/R and let the perturbation radius and velocity potential at the surface be, respectively,

$$R(t)+r(t)P_n(\cos\theta), \phi(t)/R + a(t)P_n(\cos\theta)/R^{n+1}.$$

Then the equations for $R(t)$, $\phi(t)$, $r(t)$, and $a(t)$ are

$$\begin{aligned} &\mathrm{d}R/\mathrm{d}t = -\phi/R^2, \\ &\mathrm{d}\phi/\mathrm{d}t = -\phi^2/2R^3 + RD, \\ &\mathrm{d}r/\mathrm{d}t = R^{n+1}\left[r\mathrm{d}\phi/\mathrm{d}t/R^2 + 2\phi^2 r/R^5 - (n+1)\phi a/R^{n+4}\right], \\ &\mathrm{d}a/\mathrm{d}t = 2\phi r/R^3 - (n+1)a/R^{+2}, \end{aligned}$$

where $D = p_0 - p_\mathrm{i}$ is the pressure difference. In fig. 4, we assume that $p_0 = 1$ and $p_\mathrm{i} = 0.1(1/R^5 - 1)$.

The nonlinear boundary integral code was also run under the same conditions used to obtain the linear results plotted in fig. 4 except that $\varepsilon = -0.05$ initially. In fig. 5, the location of the free surface is shown at several times, a) $t = 0$, b) $t = 0.7$, when the surface perturbation is changing phase $(r(t) \simeq 0)$ and c) $t = 0.98$, when the instability is developing so rapidly that the 65 points used to describe the surface will no longer adequately resolve the spike. Also shown is the surface location as predicted by above linear theory at d) $t = 0.98$. A 4th-order Adams–Bashforth predictor–corrector scheme with time-step 0.005 was used to advance in time the results of the perturbation theory and the numerical code.

Next, the behavior of an axisymmetric, thin shell of incompressible, inviscid fluid of unit density is studied as it is accelerated by an externally imposed pressure difference. The formulation and solution of boundary integral equations for this problem parallels the previous case of a single interface; details are

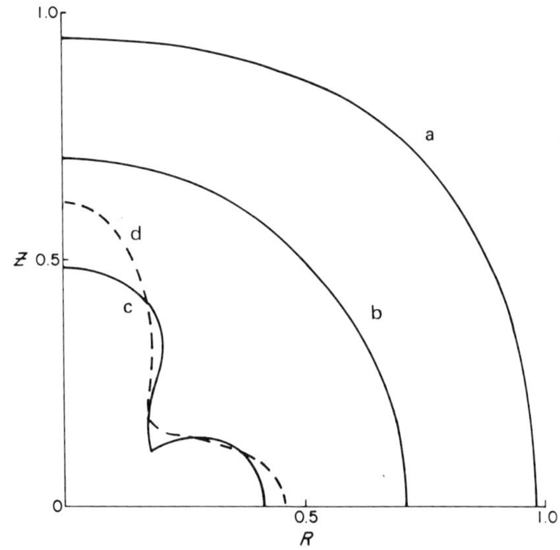

Fig. 5. Location of the axisymmetric free surface at $t=0$ (a), $t=0.7$ (b), and $t=0.98$ (c) obtained using the nonlinear boundary integral code. The location of the interface on the basis of the linear theory is plotted in (d) at $t=0.98$. All the profiles are symmetric about the R axis.

given by Baker (1983). The shell has an external free surface located initially at $R=[1.2+P_2(\cos\theta)]\sin\theta$, $Z=[1.2+P_2(\cos\theta)]\cos\theta$ and an internal free surface initially at $R=\sin\theta$, $Z=\cos\theta$. The outside pressure is 1 and the internal pressure is $(v_i/v)^{2/3}-1$. As the shell is accelerated inwards, the perturbation on the outer surface induces a perturbation on the inner surface which becomes Rayleigh–Taylor unstable when the internal pressure exceeds the external pressure and the inward motion of the shell is decelerated. In fig. 6, the location of the free surfaces is shown at $t=0$ (a), 0.5 (b), and, finally, 0.58 (c) [including 'spikes' at

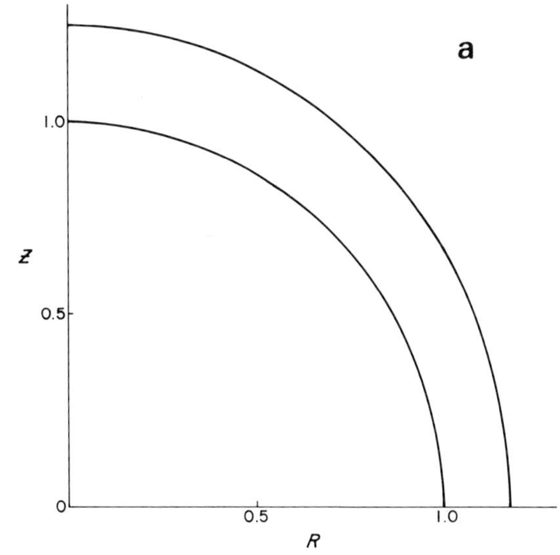

Fig. 6a.

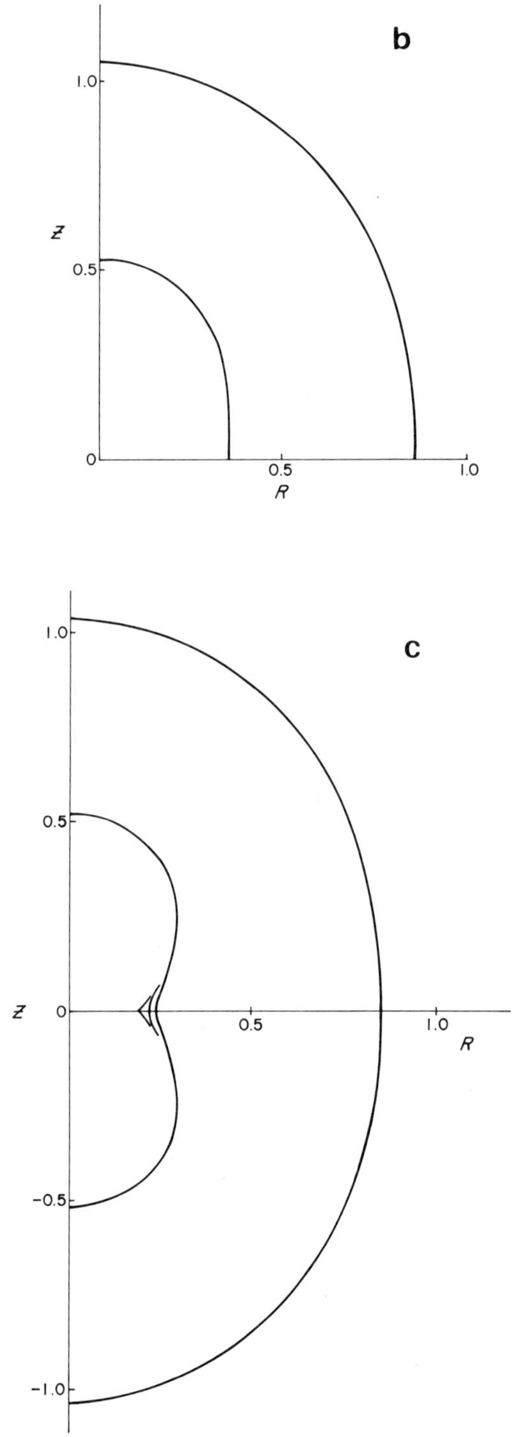

Fig. 6. Location of the boundaries of the axisymmetric thin shell at (a) $t = 0$; (b) $t = 0.5$; and (c) $t = 0.57$. In (c) the additional spikes on the inner interface are the results at $t = 0.58$ and $t = 0.59$. All the profiles are symmetric about the R axis.

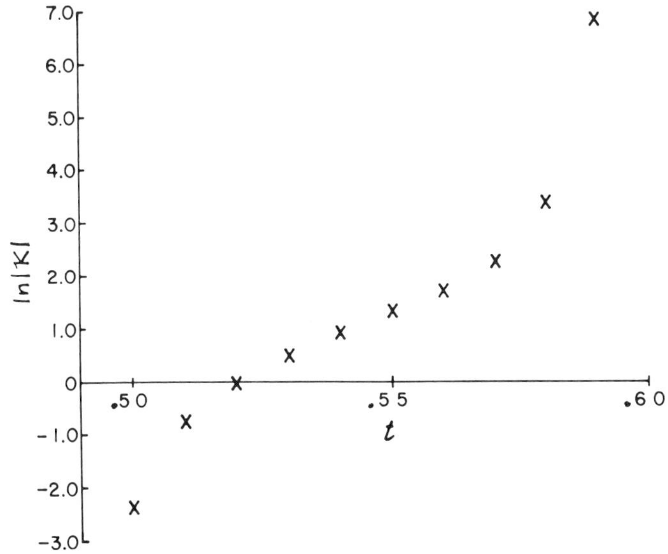

Fig. 7. A plot of the maximal meridional curvature (curvature at fixed azimuthal angle) as a function of time for the run plotted in fig. 6. The last data point at $t = 0.59$ is spectacular, but may be suspect because of the lack of numerical resolution.

$t = 0.58,\ 0.59]$ when the instability results in spikes of such high curvature on the inner surface that 65 points no longer adequately resolve the region of high curvature. Nevertheless, we plot in fig. 7 the behaviour of the curvature of the spike as a function of time. The growth seems to be faster than exponential in time, but this result may be a spurious feature of our lack of numerical resolution at these late times in the evolution of this flow. This leads naturally to the question whether these Rayleigh–Taylor unstable inviscid flows have smooth solutions for all time or whether they inevitably lead to curvature singularities in a finite time.

Acknowledgements

We would like to thank Drs. R.L. McCrory and C.P. Verdon for helpful discussions and suggestions. One of us (DIM) would like to thank the Bantrell Foundation for support. This work was supported by the Los Alamos National Laboratory and the Lawrence Livermore National Laboratory, which are supported by the Department of Energy and by the Office of Naval Research under Contract N00014-83-C-0048.

References

[1] J.G.B. Bryatt-Smith and M.S. Longuet-Higgins, Proc. Roy. Soc. London A350 (1976) 175.
[2] D.I. Meiron and P.G. Saffman, J. Fluid Mech. 129 (1983) 213.
[3] M. Miksis, J.M. Vanden-Broeck and J.B. Keller, J. Fluid Mech. 108 (1981) 89.
[4] M.S. Longuet-Higgins and E.D. Cokelet, Proc. Roy. Soc. London A350 (1976) 1.
[5] G.R. Baker, D.I. Meiron and S.A. Orszag, J. Fluid Mech. 123 (1982) 47.
[6] G.R. Baker, D.I. Meiron and S.A. Orszag, Phys. Fluids 23 (1980) 1485.
[7] C.P. Verdon, R.L. McCrory, R.L. Morse, G.R. Baker, D.I. Meiron and S.A. Orszag, Phys. Fluids 25 (1982) 1653.
[8] G.R. Baker (1983), to appear in Proc. on Waves on Fluid Interfaces, MRC, University of Wisconsin.

Physica 12D (1984) 32–44
North-Holland, Amsterdam

NUMERICAL SIMULATION OF TURBULENT MIXING BY RAYLEIGH–TAYLOR INSTABILITY

David L. YOUNGS
Atomic Weapons Research Establishment, Aldermaston, Berkshire, England

Two-dimensional hydrodynamic codes are used to simulate the growth of perturbations at an interface between two fluids of different density due to Rayleigh–Taylor instability. Problems where the interface is subjected to a constant acceleration and where it is accelerated and decelerated by shock waves are considered. Emphasis is placed on the case when the initial perturbation consists of many different wavelength modes. Results are compared with the experimental data of Andronov et al. (1976) and Read (1983). The use of a simple empirical model of the mixing process based on the equations of two-phase flow is discussed.

1. Introduction

Rayleigh–Taylor instability occurs when the interface between two fluids of different density is subjected to a normal pressure gradient with direction such that the pressure is higher in the light fluid than in the dense fluid, see, for example ref. 1. A related process occurs when shock waves pass through an interface. The small amplitude theory for this case was investigated by Richtmyer [2]; shock tube instability experiments were performed by Meshkov, see [3] and the references therein. As a result this process is often referred to as the Richtmyer–Meshkov instability.

Such instabilities are of current interest because of their deleterious effect on the performance of Inertially Confined Fusion (ICF) capsules, see [4] for example. Much of the previous theoretical work described in the review article [5] has concentrated on the growth of single wavelength perturbations. The main purpose of the present investigation is to study the growth of instabilities from random initial perturbations where many different wavelength modes are present. This situation is considered to be more relevant to practical applications. The complex flow that evolves should be regarded as an example of turbulent mixing and analysis of both computational and experimental results should involve measurement of the average properties of the mixed region.

Two-dimensional computer simulation using Eulerian hydrodynamics codes has been used to investigate the Rayleigh–Taylor and Richtmyer–Meshkov processes. This gives considerable insight into the behaviour of the large scale features of the mixed region. However, the real phenomena are three dimensional. For example the dissipation of turbulence kinetic energy by the transfer of energy to successively smaller and smaller eddies until viscous damping occurs, will certainly not be treated correctly in the two dimensional simulations. Three-dimensional simulation of the problems of most interest, where multiple wavelength initial perturbations are present was thought to be impractical using the computer facilities presently available at AWRE (a CRAY-1). Experimental studies, which are of course three dimensional were therefore considered to be essential. Rayleigh–Taylor instability experiments performed at AWRE are described by Read [6] and Read and Youngs [7]. Computational results for the Richtmyer–Meshkov process have been compared with the experimental data of Andronov et al. [3].

The understanding of the mixing processes achieved from computer simulation and experimental studies needs to be applied to practical prob-

0167-2789/84/$03.00 © Elsevier Science Publishers B.V.
(North-Holland Physics Publishing Division)

lems. This article concludes with some thoughts on how this might be carried out. It is suggested that the equations of two-phase flow may be used to model the mixing process in a one-dimensional hydrocode. Such a technique could be used, for example, to predict the break up due to hydrodynamic instabilities of interfaces in ICF capsules.

2. Rayleigh–Taylor instability

2.1. *Basic picture of the mixing process*

Two incompressible fluids of densities ρ_1 and ρ_2 ($\rho_1 > \rho_2$), separted by a plane boundary, are subjected to a constant acceleration g normal to the interface with direction from fluid 2 to fluid 1.

In the absence of viscosity the amplitude, a, of a small sinusoidal perturbation of wavelength λ grows according to the formula, see Chandrasekhar [1]

$$a = a_0^+ \exp(n_\lambda t) + a_0^- \exp(-n_\lambda t) . \tag{1}$$

The growth rate n_λ is given by

$$n_\lambda^2 = \frac{2\pi g}{\lambda} \frac{\rho_1 - \rho_2}{\rho_1 + \rho_2} . \tag{2}$$

Consider the case when many different wavelength components are present. According to eqn (2) short wavelength small perturbations grow more rapidly than long wavelength small perturbations. When the amplitude of the perturbation becomes significant compared to its wavelength the rate at which the instability penetrates the denser fluid approaches a limiting value proportional to $\sqrt{g\lambda}$, see Lewis [8]. Short wavelength perturbations then grow more slowly than long wavelength perturbations. The result of this behaviour is that one expects small structures to appear first. As time proceeds larger and larger structures should dominate the flow. The way in which the dominant wavelength increases, by bubble competition, was obseved in the experiments of Lewis [8] and Emmons et al. [9]. If a bubble of the lighter fluid penetrating the denser fluid is slightly larger than its neighbours it grows more rapidly and eventually crowds out the surrounding bubbles.

In the absence of viscosity or any other mechanism which stabilises the short wavelength perturbations, the growth rate, n_λ, increases without limit as the wavelength is decreased. However, if viscosity is present, n_λ reaches a maximum value ($= n_m$) corresponding to the most unstable wavelength λ_m. According to [1]

$$\lambda_m \approx 4\pi \left\{ \frac{\nu^2}{g} \frac{\rho_1 + \rho_2}{\rho_1 - \rho_2} \right\}^{1/3}$$

and

$$n_m = \left\{ \frac{\pi g}{\lambda_m} \frac{\rho_1 - \rho_2}{\rho_1 + \rho_2} \right\}^{1/2} ,$$

where

$$\nu = \frac{\mu_1 + \mu_2}{\rho_1 + \rho_2} \tag{3}$$

is the mean kinematic viscosity.

The evolution of larger and larger structures should begin with the appearance of perturbations with wavelength of order λ_m.

The phenomenon appears to be analogous to that seen in a number of turbulent shear flows such as the plane mixing layer, see ref. 10 for example. Here, larger and larger structures appear as fluid is carried downstream. An important feature of this flow is that memory of the initial conditions tends to be lost as the downstream distance increases. Then, as expected from dimensional analysis, the time averaged mixing profile at a given downstream station is described by a similarity solution with length scale equal to the distance downstream from the point of initiation of the mixing process.

These arguments suggest that the evolution of Rayleigh–Taylor instability at a plane boundary from a low initial perturbation level should give rise to the following three stages of mixing:

Stage 1. A perturbation corresponding to the most unstable wavelength λ_m appears. This should happen after a time of a few times τ_m where $\tau_m = 1/n_m$.

Stage 2. When the perturbation of wavelength λ_m reaches a height of the order of $\frac{1}{2}\lambda_m$ its exponential growth rate slows down and longer wavelength perturbations begin to grow more rapidly. In stage 2 larger and larger structures appear. If the initial perturbation level is low enough, these larger structures will evolve from the non-linear interaction between smaller structures rather than directly from an initial perturbation of the corresponding wavelength. Then by the end of stage 2 memory of the initial conditions will be lost. This should occur when the dominant wavelength λ_d reaches about $10\lambda_m$.

Stage 3. In stage 3, λ_d, grows from about $10\lambda_m$ to infinity. Viscosity now has little effect on the growth of the large-scale features. Since memory of the initial conditions has been lost, gt^2 is the only length scale of importance. For a given density ratio the mixing process is described by a similarity solution with length scale proportional to gt^2. The width of the mixed region is given by

$$\delta = F\left(\frac{\rho_1}{\rho_2}\right) gt^2 . \tag{4}$$

If large amplitude long wavelength perturbations are present initially the growth of the mixed region will exceed that indicated by eq. (4).

The width δ could be defined in a variety of ways. One possibility is as follows. Let $f_1(x, y, z)$ denote the volume fraction of the dense fluid 1 at the point (x, y, z), z is measured in the direction normal to the interface. The plane average of f_1 at height z is

$$\bar{f}_1(z) = \iint f_1(x, y, z)\, \mathrm{d}x\, \mathrm{d}y \Big/ \iint \mathrm{d}x\, \mathrm{d}y .$$

The mixed region width, δ = difference in height between points where $\bar{f}_1 = 0.01$ and $\bar{f}_1 = 0.99$; the penetration of dense fluid, h_1 = difference in height between position undisturbed interface would have reached and point where $\bar{f}_1 = 0.99$; the penetration of light fluid, h_2 = difference in height between position undisturbed interface would have reached and point where $\bar{f}_1 = 0.01$. By definition $\delta = h_1 + h_2$.

A simple argument may be used to indicate the dependence on density ratio in eq. (4). Suppose that throughout stage 3 of the mixing process the dominant wavelength takes N exponential growth periods to appear, so that

$$n_{\lambda_d} t = N = \text{a constant.}$$

Then

$$\delta \propto \lambda_d = \frac{2\pi}{N^2} \frac{\rho_1 - \rho_2}{\rho_1 + \rho_2} gt^2 \tag{5}$$

by using eq. (2).

Computer simulation and experimental results both show that the density variation in eq. (5) is more applicable to h_1, the penetration of the denser fluid, than to δ the total width.

2.2. *Computer simulation of the growth of the instability from a single wavelength initial perturbation*

Although multiple wavelength initial perturbations are of most interest, it is instructive to begin by considering the growth of the instability from a single sinusoidal perturbation. In the computer simulation, fluid of density ρ_1 rests initially on top of fluid of density ρ_2. Gravity acts downwards. The interface between the two fluids is initially horizontal but a velocity perturbation is present. This is a combination of the eigenfunctions (at $t = 0$) obtained from the linear theory, corresponding to the case $a_0^+ = -a_0^- = 0.5a_0$ in eq. (1). The stream function is $\psi = -(a_0 n/k) \sin kx \exp(-k|z|) \cosh nt$, where x and z are the distances in the horizontal and vertical directions. The displacement of the interface, according to linear theory is then

$$\zeta(x) = \int -\left.\frac{\partial \psi}{\partial x}\right|_{z=0} \mathrm{d}t = a_0 \cos kx \sinh nt. \qquad (6)$$

Simulations have been performed using the 2-D Eulerian compressible flow code of Youngs [11]. Equations of state were chosen to give a low Mach number flow. The effect of compressibility on the results is then negligible. An interface tracking method is used to follow the boundary between the two fluids. In each mixed cell, fluid volume fractions for the cell under consideration and its eight neighbours are used to construct a straight-line interface. This type of interface tracking method was first used by DeBar [12].

Fig. 1 shows calculations at density ratios of $\rho_1/\rho_2 = 3$ and 20. Gravity is chosen so that $g(\rho_1 - \rho_2)/(\rho_1 + \rho_2)$ is the same in each case. The penetration of the denser fluid, h_1, then grows at a similar rate for the two density ratios (growth is in fact a little more rapid at $\rho_1/\rho_2 = 3$). At $\rho_1/\rho_2 = 3$, as the spike of denser fluid falls, the interface soon rolls up. This phenomenon was first computed by Daly [13]. The present calculations show that a similar process happens at $\rho_1/\rho_2 = 20$ but at a larger value of nt. At the larger density ratio, the asymmetry of mixed region, h_2/h_1, is greatly increased. The spike penetration, h_2, does, however fall a long way short of the free-fall line $h_2 = \frac{1}{2}gt^2$. At the latest time shown in fig. 1, h_2 is only $0.35 \times \frac{1}{2}gt^2$.

2.3. *Computer simulation of the growth of the instability from a multiple wavelength initial perturbation*

The multiple wavelength calculations require a large number of computational meshes. In order to carry out such calculations a numerical technique was devised which was simpler than that used for the single wavelength instability calculations. An incompressible hydrocode somewhat similar to the MAC code, [14] was written. As a high degree of fine scale mixing was expected, the interface tracking method was abandoned. Instead, the fluid volume fraction, f_1, is treated as a continuous function of position. Changes in f_1 are calculated by solving the advection equation

$$\frac{\partial f_1}{\partial t} + \mathrm{div}(f_1 \boldsymbol{u}) = 0.$$

The monotonic advection method of Van Leer [15] is used. The method (which is also used in the advection phase of the compressible code [11]) gives minimal numerical diffusion whilst ensuring that the values of f_1 remain in the interval [0, 1].

For the Rayleigh–Taylor instability calculations

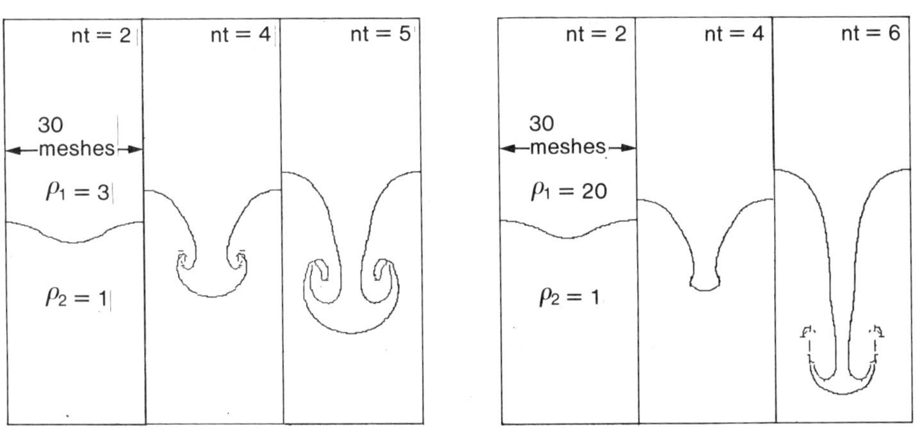

Fig. 1. Rayleigh–Taylor instability, single wavelength initial perturbation.

shown here, where flow is confined to the region $0 \leqslant x \leqslant L$, the initial velocity perturbation is derived from the stream function

$$\psi = -\sum_r a_r \frac{n_r}{k_r} \cos k_r x \exp(-k_r|z|), \quad k_r = \frac{r\pi}{L}.$$

Examples of two such calculations are shown in fig. 2. The initial perturbation, the same for each case consists of 12 modes with $r = 49$ to 60. The values of a_r were chosen at random from a Gaussian distribution. Viscosity is present in these calculations. Fluid viscosities are chosen so that the most unstable wavelength λ_m is equal to $L/100$. Viscosity then has little effect on the large scale features that are formed by the end of the calculation. In the calculations shown here $L = 1$ unit and $g(\rho_1 - \rho_2)/(\rho_1 + \rho_2)$ has the same value at the two density ratios; g is in fact chosen so that the exponential growth rate, n, is unity for wavelength L. It is then found that the penetration of the dense fluid, h_1, grows at a similar rate for the two density ratios. In these calculations the dominant wavelength increases from $\approx L/30$ at the beginning of the calculation to $\approx L/3$ at the end of the calculation. Perturbations of wavelength $L/3$ are not present initially. Hence these structures must have been formed from the non-linear interaction between smaller structures. In these circumstances it is expected that the average mixing rate, i.e. that deduced from the profile of $\bar{f}_1$ versus z, should be insensitive to the initial conditions. This is found to be the case. For the calculations shown here and

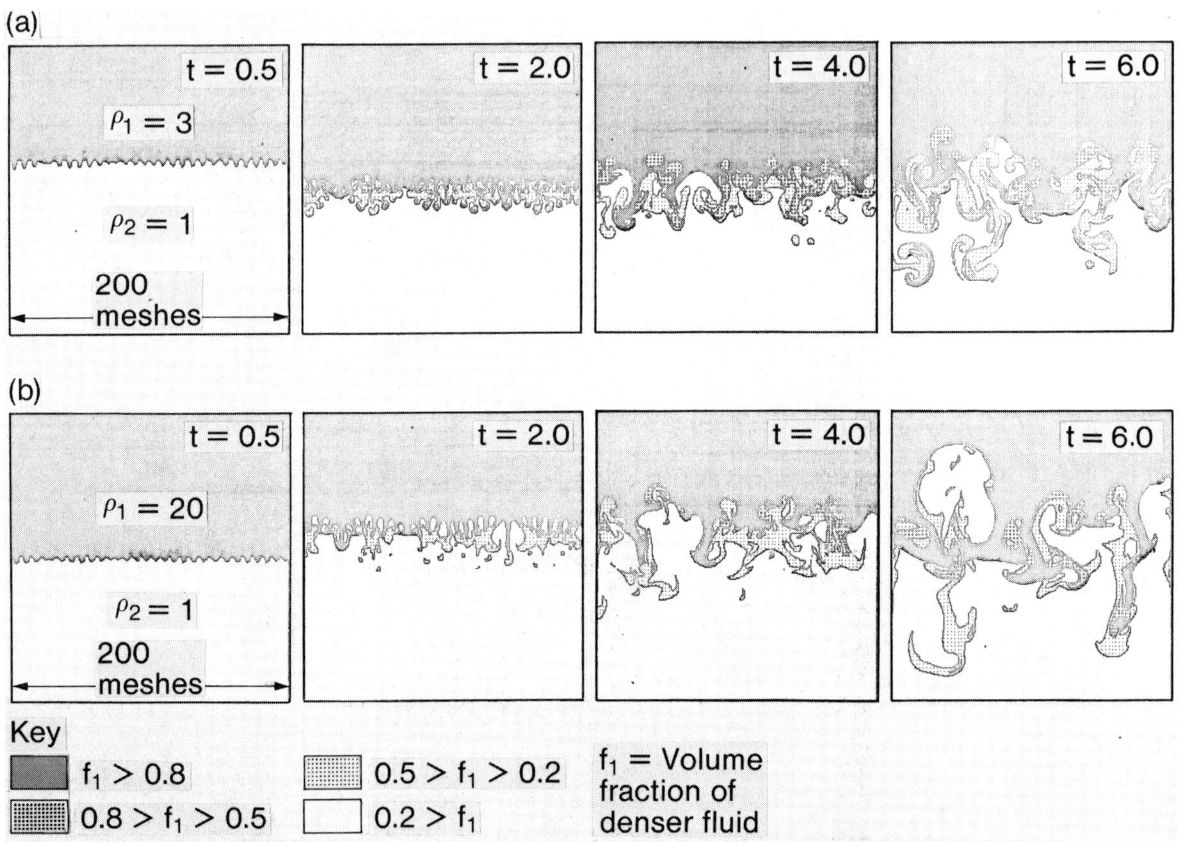

Fig. 2. Rayleigh–Taylor instability, multiple wavelength initial perturbation.

others with different initial conditions:

$$h_1 = \alpha \frac{\rho_1 - \rho_2}{\rho_1 + \rho_2} g t^2 , \tag{7}$$

where

$\alpha \approx 0.04$ to 0.05.

Eq. (7) was confirmed by experimental investigation [6, 7]. For experiments in which the instability evolved from small random perturbations and developed uniformly over the interface, good linear correlations were obtained between h_1 and gt^2. Values of α for the three dimensional experiments were in the range 0.063 to 0.077. The experimental estimates of α tend to be higher than the values from the two-dimensional calculations. It is argued in [6] that two-dimensional constraints inhibit the growth of the large scale structures.

Fig. 3 shows plots of the average volume fraction $\bar{f}_1$ versus z obtained from the calculations illustrated in fig. 2. For each density ratio the shape of the volume fraction profile remains similar as the mixed region grows. The ratio h_2/h_1 increases as the density ratio increases:

$$\rho_1/\rho_2 = 3, \quad h_2/h_1 \approx 1\tfrac{1}{2},$$

$$\rho_1/\rho_2 = 20, \quad h_2/h_1 \approx 2\tfrac{1}{2}.$$

The calculation shown in fig. 2a was repeated with an additional perturbation of wavelength $L/2$. This long wavelength initial velocity perturbation was chosen to make the interface displacement, as given by eq. (6)

$$\zeta(x) = 0.01L \cos kx \sinh nt .$$

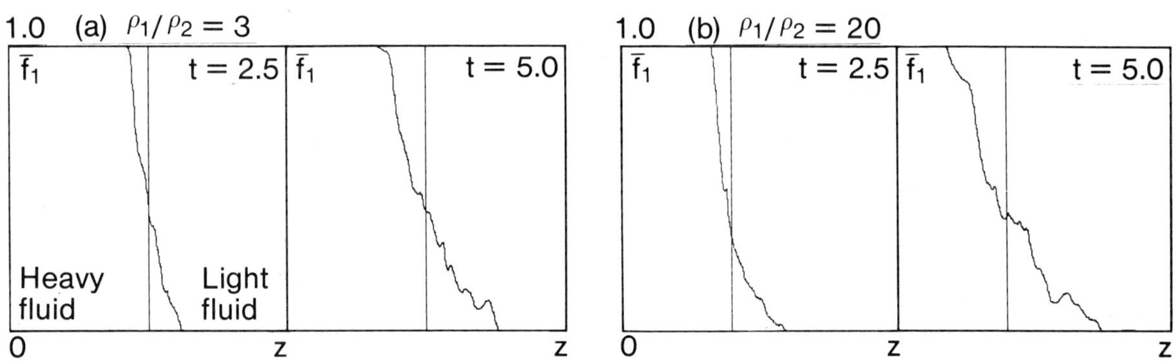

Fig. 3. Dense fluid volume fraction averaged over horizontal layers, $\bar{f}_1$, plotted against depth z.

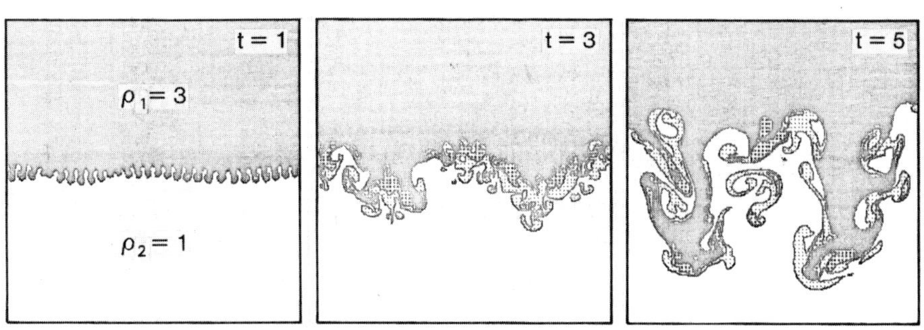

Fig. 4. Multiple wavelength calculation shown in fig. 2(a) repeated with an additional long wavelength initial perturbation.

The effect is very similar to that of introducing an amplitude perturbation, with zero initial velocity, for which the initial growth is given by

$$\zeta(x) = 0.01L \cos kx \cosh nt\,.$$

In this case the initial amplitude is 2% of the wavelength.

The addition of this long wavelength perturbation increases the instability penetration, h_1, by about 50% towards the end of the calculation, see fig. 4. On the other hand, if the size of the perturbation is reduced by a factor 10, so that it corresponds to an initial amplitude of 0.2% of the wavelength, the enhancement of the instability is negligible.

The quadratic law, eq. (7) will therefore cease to be valid if large amplitude long wavelength perturbations are initially present.

3. Richtmyer–Meshkov instability

3.1. *Small amplitude theory*

Mixing induced by the transit of shocks normal to the interface is now considered. Firstly, the small amplitude results of Richtmyer [2] are noted. Consider the growth of an initial amplitude perturbation at an interface between two incompressible fluids, i.e. $a_0^+ = a_0^- = \frac{1}{2}a_0$ in eq. (1). After time t the amplitude and velocity are

$$a = a_0 \cosh nt\,,$$

$$v = \dot{a} = na_0 \sinh nt\,.$$

If $t \to 0$ and $gt \to \Delta U$ in these equations the following results are obtained, for an impulsive acceleration with velocity change ΔU:

$$a = a_0, \quad v = \frac{2\pi \Delta U}{\lambda} a_0 \frac{\rho_1 - \rho_2}{\rho_1 + \rho_2}\,. \tag{8}$$

Richtmyer showed that eq. (8) gave a good approximation for the effect of a shock passing through the interface between compressible fluids provided that a_0, ρ_1 and ρ_2 were measured after the shock transit.

3.2. *The shock tube experiment*

Computer simulation has been used to investigate the shock tube experiment of Andronov et al. [3]. The experimental geometry is illustrated in fig. 5. The helium/air interface is first accelerated by a shock wave passing from the air to the less dense helium. This shock transit amplifies the perturbation introduced when the thin film which separates the two gases is ruptured. Subsequently the interface is decelerated by reflected shocks. Most of the mixing occurs in this deceleration phase. A region of turbulent mixing was observed and a visual estimate of its width as a function of time was obained [3]. The experiment was of course three dimensional; the shock tube had cross-section 40 × 120 mm. The two-dimensional computer calculations were confined to a region with width 50 mm normal to the direction of shock propagation.

3.3. *Computer simulation with a single wavelength initial perturbation*

The 2-D compressible Eulerian code [11] was adapted to simulate the shock tube experiment. Calculations were performed in plane geometry with x measured in the shock direction and y measured in the normal direction. A rectangular mesh was used. The y-direction mesh was fixed. The x-direction mesh moved with velocity $\bar{u} = \int u\mathrm{d}y/\int \mathrm{d}y$, so that in the absence of interface

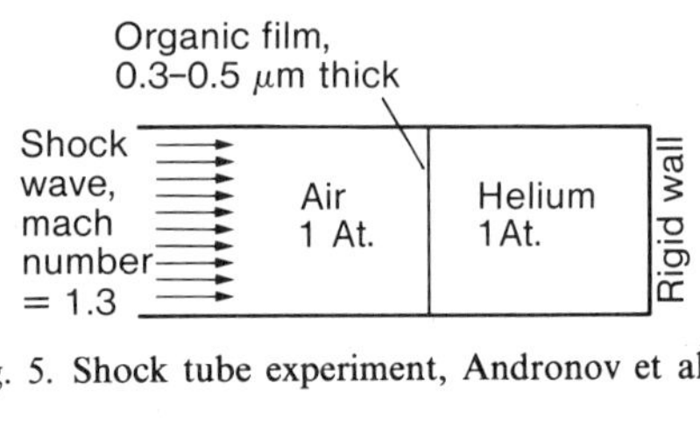

Fig. 5. Shock tube experiment, Andronov et al. [3].

perturbations, when the flow was one dimensional, the mesh moved in a Lagrangian manner. This maintained the resolution in the helium region as it compressed. Moving rectangular mesh methods have been used before, by Emery et al. [16], to improve the accuracy of Rayleigh–Taylor instability calculations. They are also discussed by Pert [17].

Fig. 6 shows the variation of the width of the helium reigon with time for a calculation with no interface perturbation. The graph shows the changes in the velocity at the interface due to shock transits. 80 meshes were used in the x-direction in the helium region. The helium and the air were treated as perfect gases with adiabatic constants $\gamma = 1.4$ and $\gamma = 1.63$, respectively. The density ratio, ρ_1/ρ_2, at the interface remained in the interval 7 to 8.

A two-dimensional calculation was carried out with the following initial amplitude perturbation at the interface:

$$\zeta(y) = a_0 \cos \frac{2\pi y}{\lambda},$$

with

$$a_0 = 1\ \text{mm} \quad \text{and} \quad \lambda = 50\ \text{mm}.$$

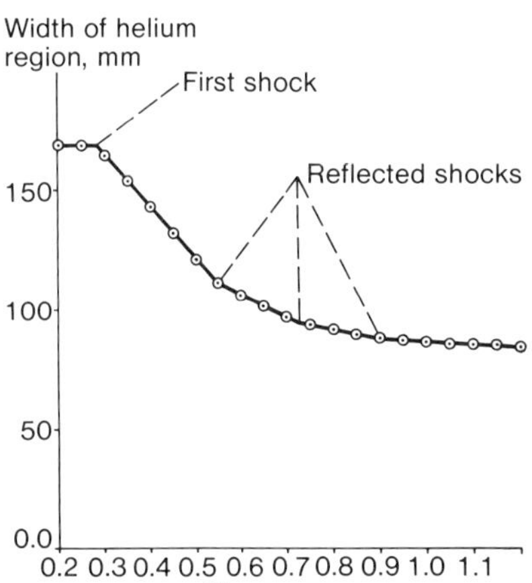

Fig. 6. Shock tube experiment; computer simulation with no interface perturbation.

Plots of the helium/air interface and vectors of the perturbed velocity $(u - \bar{u}, v)$ are shown in fig. 7. A total of 200×40 meshes are used. A uniform 80×40 mesh is initially present in the helium region. Then, at peak compression the mesh in the helium region is approximately square.

The first shock reaches the interface at 0.283 ms. As expected from eq. (8), since $\Delta U(\rho_1 - \rho_2)$ is negative, the sign of the perturbation is reversed. The computer simulation shows that for this shock $\Delta U = -220$ mm/ms. After the shock has passed

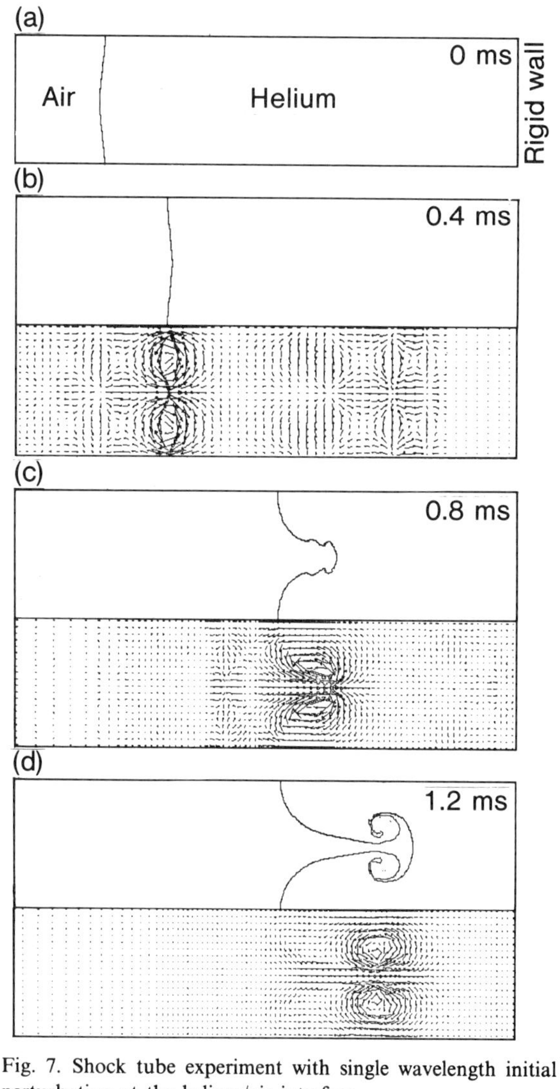

Fig. 7. Shock tube experiment with single wavelength initial perturbation at the helium/air interface.

$\rho_1/\rho_2 = 7.4$ and the fluid is compressed by a factor 1.3. Eq. (8) then gives

$$v = -\frac{2\pi \times 220}{50}\frac{1}{1.3}\frac{7.4-1}{7.4+1} = -16\,\text{mm/ms}.$$

The perturbation amplitude changes from +1 mm to −1 mm in about 0.12 ms (see fig. 8). This gives velocity $v \approx -2/0.12 = -17$ mm/ms in good agreement with Richtmyer's formula.

After the first shock has passed through the interface, perturbations are present on the shock front. This is demonstrated by the perturbation velocity vector plot in fig. 7 at 0.4 ms.

For the subsequent reflected shocks $\Delta U(\rho_1 - \rho_2)$ is positive. These shocks enhance the growth of the instability. Towards the end of the calculation, as the spike of air penetrates the less dense helium, the interface rolls up. The shape of the interface is in fact very similar to that seen in section 2.2 for Rayleigh–Taylor instability at a constant acceleration.

Fig. 8 shows the variation with time of the width of the instability for this calculation. The growth agrees reasonably well with the observed data. This agreement is somewhat fortuitous as the flow is not, in reality, dominated by the growth of a single wavelength perturbation.

3.4. *Computer simulation with a multiple wavelength initial perturbation*

A more realistic simulation of the shock tube experiment is obtained if the initial perturbation consists of a mixture of different wavelength modes. In order to achieve this the following interface perturbation was used:

$$\zeta(y) = \sum_{r=10}^{40} a_r \cos\frac{r\pi y}{W}.$$

The coefficients a_r were chosen from a Gaussian distribution. The width of the computational region in the y-direction, W, was 50 mm. The range of wavelengths present in the initial perturbation was $2\frac{1}{2}$ to 10 mm. The magnitude of the perturbation was scaled to make the standard deviation $\overline{(\zeta^2)}^{1/2} = 0.2$ mm.

The calculation was performed with 370 meshes in the x-direction and 100 meshes in the y-direction. As it was anticipated that fine scale mixing would occur in some regions, the interface

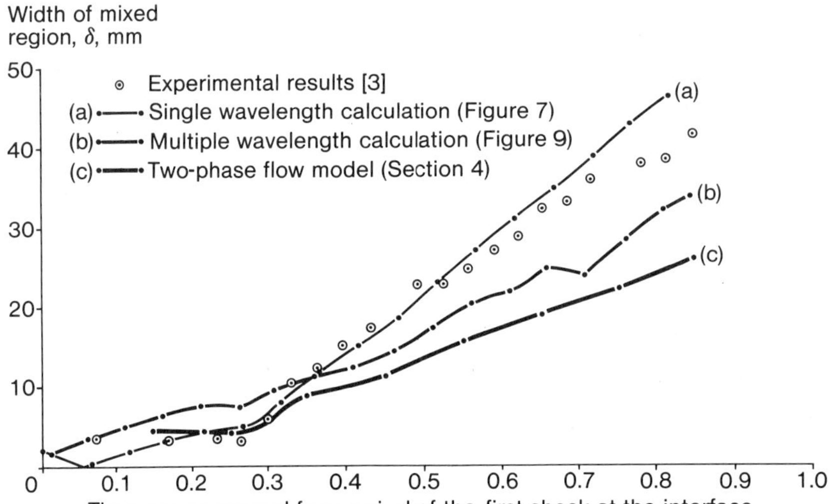

Fig. 8. Shock tube experiment, growth of the mixed region.

tracking method was modified. A straight line interface was only defined in a mixed cell if one of its eight neighbours was pure i.e. all helium or all air. If all the surrounding eight cells were mixed, it was assumed that the volume fractions varied continuously with position and the Van Leer method was used to calculate transport across cell sides as in the calculations described in section 2.3.

Fig. 9 shows plots of the helium/air interface, perturbation velocity vectors $(u - \bar{u}, v)$ and air volume fraction contours. Cells where the interface tracking method has not been used are marked with a + on the interface plot. The results show that, as in the Rayleigh–Taylor calculations of section 2.3, the dominant wavelength associated with the instability increases as the mixed region grows. Initially λ_d is about $3\frac{1}{2}$ mm. By the end of the calculation λ_d has increased to about 15 mm.

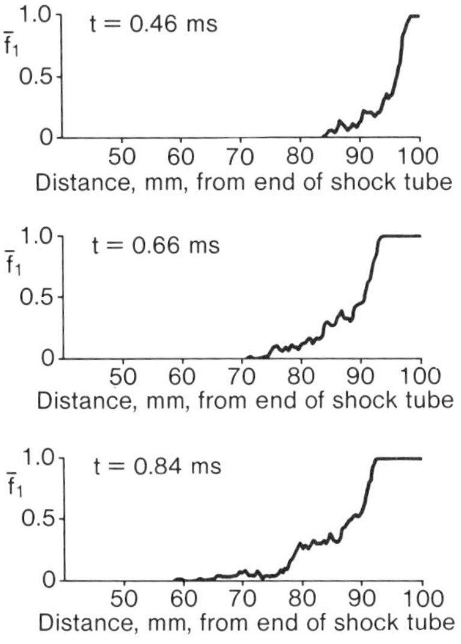

Fig. 10. Mean air volume fraction profiles for the shock tube experiment. Multiple wavelength instability calculation. Times are measured from the arrival of the first shock at the interface.

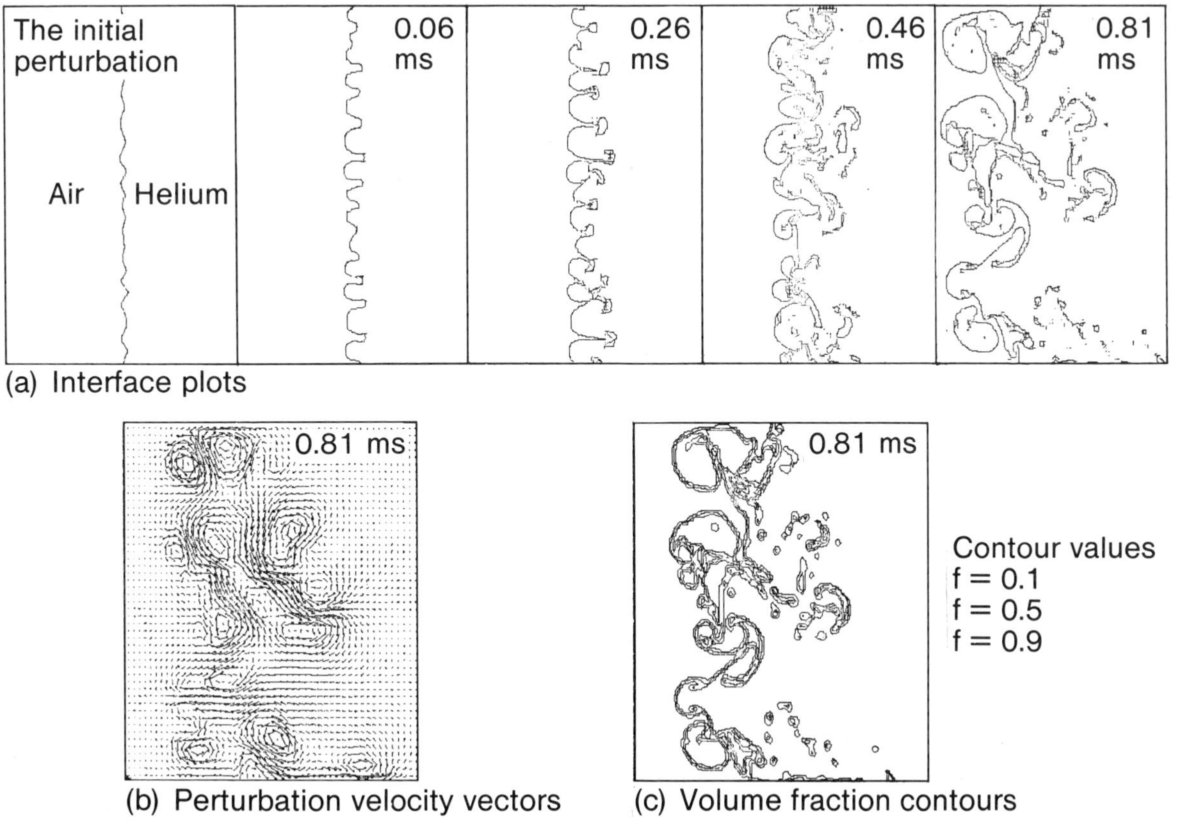

Fig. 9. The multiple wavelength instability calculation for the shock tube experiment. Times are measured from the arrival of the first shock at the interface.

Profiles of the mean air volume fraction, $\bar{f}_1 = \int f_1\, dy / \int dy$ are shown in fig. 10. The shapes obtained are similar to those for Rayleigh–Taylor instability with constant acceleration. The width, δ, of the mixed region i.e. the distance between the points where $\bar{f}_1 = 0.99$ and $\bar{f}_1 = 0.01$ is plotted in fig. 8. Towards the end of the calculation the growth of the mixed region is less than observed. This may be due to the fact that the simulation is two-dimensional whereas the experiment was three dimensional.

4. Empirical models

The two dimensional calculations of multiple wavelength instability growth described in sections 2.3 and 3.4 each took about two hours on a CRAY-1 computer. For the more complex geometries present in real applications, such calculations would be impractical in two dimensions, out of the question in three dimensions. Simple empirical models are therefore needed to assess the effects of interface break up in these situations. Such a model would not attempt to predict the mixing processes from first principles. Instead it should be thought of as a way applying the results of experiments and computer simulation in simple situations, to real problems.

Andronov et al. [3] described a one-dimensional model for the mixing process in the shock tube experiment. Mixing of the two gases was represented by turbulent diffusion. The mass flux of fluid 1 was

$$F_1 = -\rho D_t \frac{\partial c_1}{\partial x},$$

where c_1 is the concentration, per unit mass, of fluid 1.

The coefficient of turbulent diffusion was

$$D_t = \text{a constant} \times k^{1/2} L,$$

where k is the turbulence kinetic energy and L is the width of the mixed region.

An alternative approach is suggested here. The mixed region grows because, within the mixed region, the pressure gradient induces different accelerations in the two fluids which have different densities. The difference in the velocities of the two fluids is limited by drag, as in the case of a single gas bubble rising in a liquid when gravity is present. This behaviour suggests that the two-phase flow equations, see for example [18], should be used to model the process. Some preliminary calculations have been carried out with the following simple one dimensional two-phase flow model:

mass conservation:

$$\frac{\partial}{\partial t}(f_r \rho_r) + \frac{\partial}{\partial x}(f_r \rho_r u_r) = 0, \quad r = 1 \quad \text{and} \quad 2;$$

momentum conservation:

$$\frac{\partial}{\partial t}(f_r \rho_r u_r) + \frac{\partial}{\partial x}(f_r \rho_r u_r^2)$$

$$= -f_r \frac{\partial p}{\partial x} + K(u_s - u_r) + f_r \rho_r g,$$

$$r = 1 \quad \text{and} \quad 2, \quad s = 2 \quad \text{and} \quad 1.$$

f_r, ρ_r and u_r denote the volume fraction, density and velocity of fluid r averaged over planes perpendicular to the x direction. For compressible flows, internal energy equations for each fluid, see [18], are also solved. The interphase drag coefficient used here has the following form:

$$K = C \frac{f_1 f_2 \bar{\rho}}{L} |u_1 - u_2|, \tag{9}$$

where

$$\bar{\rho} = f_1 \rho_1 + f_2 \rho_2$$

and

$$C = \text{a constant}.$$

The length scale L, which in this model depends only on t, is proportional to the dominant bubble or drop radius at time t. It is assumed to grow in proportion to the width of the mixed region. This is appropriate to the case when the instability evolves from small random perturbations.

The equation used for L was

$$\frac{\mathrm{d}L}{\mathrm{d}t} = V_0 + L\left(\frac{\partial \bar{u}}{\partial x}\right)_0,$$

where V_0 is the value of $|u_1 - u_2|$ at the centre of the mixed region. The velocity gradient term $(\partial \bar{u}/\partial x)_0$ ensures that, in the absence of velocity separation, the length scale is compressed when the fluid is compressed.

Further work needs to be done to perfect this model. It could be extended to deal with two dimensional problems where the average values $f_r, \rho_r, u_r \ldots$ are functions of two space variables. A more sophisticated length scale equation would then be needed. Although it is suggested here that gradient diffusion by itself does not represent the mixing process well, such effects may need to be included in the model. The interphase drag is a source of turbulence kinetic energy. In periods where the pressure gradient falls to zero this may cause the fluids to mix in a manner similar to gradient diffusion. For the transient flows considered here it is likely that added mass terms, as described by Ishii and Chawla [9] for example, need to be included in the momentum equations.

Results obtained with the simple model described here are now discussed. It has been applied to mixing at the interface between two incompressible fluids subject to constant acceleration g. If the constant in the drag formula (9) is chosen to be $C = 4.5$, then the penetration of the dense fluid given by

$$h_1 = \beta \frac{\rho_1 - \rho_2}{\rho_1 + \rho_2} g t^2,$$

where β varies only slightly with density ratio. At $\rho_1/\rho_2 = 8$, $\beta = 0.07$ in agreement with the experimental results [6].

Volume fraction profiles obtained from calculations at $\rho_1/\rho_2 = 3$ and 20 are shown in fig. 11. The shapes obtained agree well with those indicated by the numerical simulation of section 2.3.

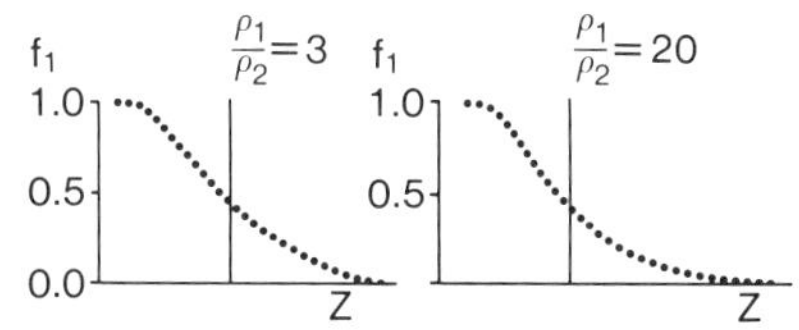

Fig. 11. Volume fraction profiles predicted by the two phase flow model for the mixing of two incompressible fluids.

The two-phase flow model has also been applied to the shock tube experiment. The present model does not cope with the reversal and amplification of the initial perturbation by the first shock. The mixing calculation was started just after the first shock had passed through the interface, with a mixed region 4 mm wide as observed in the experiment. The length scale L was initially set equal to 2 mm. A one-dimensional Lagrangian code was used; the Lagrangian mesh moved with the mean fluid velocity $\bar{u} = f_1 u_1 + f_2 u_2$. Profiles of the air volume fraction at a series of times are shown in fig. 12. Again, it is found that the shapes predicted by the model agree with those obtained from the

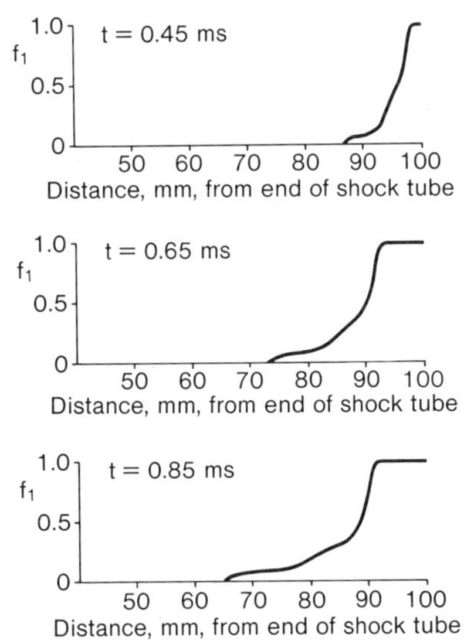

Fig. 12. Air volume fraction profiles for the shock tube experiment predicted by the two-phase flow model. Times are measured from the arrival of the first shock at the interface.

two-dimensional instability calculation, fig. 10. The width, δ, of the mixed region predicted by the model does not grow as fast as observed, see fig. 8. It does, however, grow at a similar rate to that in the two-dimensional numerical simulation of the shock tube experiment.

The tie-up between application of the model to the constant acceleration case and the shock tube experiment is not as good as one would like. It is not clear at present if this is due to inadequacies in the model or difficulties in interpreting the experiments. More precise measurements of the amount of mixing in the experiments, for example measurement of the mean volume fraction profiles would be invaluable.

5. Conclusions

As already noted, most of the past research into Rayleigh–Taylor instability, both theoretical and experimental, has investigated the growth of a single wavelength perturbation. At first sight, extending the research to cope with multiple wavelength initial perturbations seems a daunting task. However, there are compensations. If the mixing process evolves from small random perturbations then the growth of the instability is soon controlled by the nonlinear interaction between bubbles of different sizes. It is then found that the mixing rate is insensitive to the initial conditions, which are rarely known in a real situation. Also, numerical simulation of mixing at a plane boundary has shown that the average properties of the mixed region, such as the volume fraction ($\bar{f}_1$) profiles, then fit into a simple pattern.

Much further work needs to be done. In the author's opinion the most effective way to proceed is to extend the understanding of the mixing process by the use of two-dimensional computer simulation and three-dimensional experiment. Simple empirical models should then be used to apply the results obtained to real situations.

References

[1] S. Chandrasekhar, Hydrodynamic and Hydromagnetic Stability (Oxford Univ. Press, Oxford, 1961).

[2] R.D. Richtmyer, "Taylor Instability in Shock Acceleration of Compressible Fluids", Comm. in Pure and Appl. Math. 13 (1960) 297.

[3] V. Andronov, S.M. Bakhrakh, E.E. Meshkov, V.N. Mokhov, V.V. Nikiforov, A.V. Pevnitskii and A.I. Tolshmyakov, Sov. Phys. JETP 44 (1976) 424–427.

[4] P.D. Roberts, S.J. Rose, P.C. Thompson and R.J. Wright, "The Stability of Multiple Shell ICF Targets", J. Phys D 13 (1980) 1957.

[5] D.H. Sharp, "An Overview of Rayleigh–Taylor Instability", Physica 12D (1984) 3 (these proceedings).

[6] K.I. Read, "Experimental Investigation of Turbulent mixing by Rayleigh–Taylor Instability", Physica 12D (1984) 45 (these proceedings).

[7] K.I. Read and D.L. Youngs, "Experimental Investigation of Turbulent Mixing by Rayleigh–Taylor Instability", AWRE Report 011/83 (1983).

[8] D.J. Lewis, The Instability of Liquid Surfaces when Accelerated in a Direction Perpendicular to their Plane", Proc. Roy. Soc. 202A (1950) 81.

[9] H.W. Emmons, C.T. Chang and B.C. Watson, "Taylor Instability of Finite Surface Waves", J. Fluid Mech. 7 (1963) 177.

[10] G.L. Brown and A. Roshko, "On Density Effects and Large-Scale Structure in Turbulent Mixing Layers", J. Fluid Mech. 64 (1974) 775.

[11] D.L. Youngs, "Time-Dependent Multi-Material Flow with Large Fluid Distortion", in: Numerical Methods for Fluid Dynamics, K.W. Morton and M.J. Baines, eds. (Academic Press, New York, 1982).

[12] R. DeBar, "A Method in Two-D Eulerian Hydrodynamics", UCID-19683 (1974).

[13] B.J. Daly, "Numerical Study of Two Fluid Rayleigh–Taylor Instability", Phys. Fluids 10 (1967) 297.

[14] J.E. Welch, F.H. Harlow, J.P. Shannon and B.J. Daly, "The MAC Method: a Computing Technique for Solving Viscous Incompressible Transient Fluid Flow Problems Involving Free Surfaces", LA-3425 (1966).

[15] B. van Leer, "Towards the Ultimate Conservative Difference Scheme: IV, A New Approach to Numerical Convection", J. Comp. Phys. 23 (1977) 276.

[16] M.H. Emery, J.H. Gardner and J.P. Boris, "Rayleigh–Taylor and Kelvin–Helmholtz Instabilities in Targets Accelerated by Laser Ablation", Phys. Rev. Lett. 48 (1982) 677.

[17] G.J. Pert, "Quasi-Lagrangian Rezoning of Fluid Codes Maintaining an Orthogonal Mesh", J. Comp. Phys. 49 (1983) 1.

[18] F.H. Harlow and A.A. Amsden, "Numerical Calculation of Multiphase Fluid Flow", J. Comp. Phys. 17 (1975) 19.

[19] M. Ishii and T.C. Chawla, "Local Drag Laws in Dispersed Two-phase Flow", NUREG/CR-1230 (1979).

Physica 12D (1984) 45–58
North-Holland, Amsterdam

EXPERIMENTAL INVESTIGATION OF TURBULENT MIXING BY RAYLEIGH–TAYLOR INSTABILITY

K.I. READ
Atomic Weapons Research Establishment, Foulness, England

A novel experimental technique using solid fuel rocket motors, has been developed at AWRE Foulness to study the growth of Rayleigh–Taylor instabilities in fluids.

The technique, which achieves near constant accelerations up to 750 m s^{-2} over distances of 1.25 m, is currently providing results for comparison with the theoretical studies on interfacial mixing by Youngs.

1. Introduction

This presentation describes an investigation of turbulent mixing at a plane boundary between two fluids due to Rayleigh–Taylor instability. The fluids, of densities ρ_1 and ρ_2 ($\rho_1 > \rho_2$) are near incompressible and are subject to an approximately constant acceleration (g) normal to the interface, directed from the lighter to the denser fluid.

The experiments show that if the instability arises from small random perturbations then the width of the mixed region grows in proportion to t^2. The depth h_1 to which the instability penetrates the denser fluid is given by

$$h_1 = \alpha \frac{\rho_1 - \rho_2}{\rho_1 + \rho_2} g t^2 \quad \text{(Refs. 1, 2)}.$$

The coefficient α is found to be insensitive to the density ratio and varies little between experiments. The best experimental estimate for α in a truly three-dimensional situation is 0.07.

Previous experimental work [3–8] in this field has concentrated on the study of the growth of a single wavelength perturbation. The major purpose of the present investigation is to study the evolution of the instability from small random perturbations which, it is believed, is more relevant to many practical applications.

The experiments described were performed in a region where surface tension and viscosity should have negligible effects on the overall growth rate of the mixed region. This has been achieved by ensuring that the acceleration of the interface was sufficiently high.

A novel technique was used to provide the desired acceleration. The two fluids were enclosed in a rectangular tank, the less dense fluid initially resting on top of the denser fluid. One or more rocket motors, attached to the top of the tank, were used to drive the tank vertically downwards. Accelerations in the range $25g_0$ to $75g_0$ have been achieved. ($g_0 = 9.81$ m s^{-2} denotes the acceleration due to gravity).

In several experiments a tank with square cross section, i.e. $D = W$, was used. This enables a truly three-dimensional instability to develop. Other experiments used a tank with $D \ll W$, referred to here as the two-dimensional tank. Visualization of the mixing process was easier in the latter case, but the flow was to some extent constrained to follow a two-dimensional pattern.

0167-2789/84/$03.00 © Elsevier Science Publishers B.V.
(North-Holland Physics Publishing Division)

2. The experimental method

2.1. *Alternatives*

Consideration was first given to the use of either stretched rubber cord or compressed air as the accelerating force, both methods having been successfully employed by earlier workers [3, 4, 6, 7, 8]. For the current investigation however, they were thought to have disqualifying limitations. The former was ruled out due to the fast decay of acceleration after an initial peak, while the latter method would have been expensive both in money and manufacturing time.

The decision was made therefore, that the requirements could best be met by the use of solid fuel rocket motors having an appropriate thrust profile [1]. This method has the advantage of a high and approximately constant acceleration which would be maintained over a long useful distance (≈ 1.5 m). The necessary apparatus could be easily and quickly constructed at low cost.

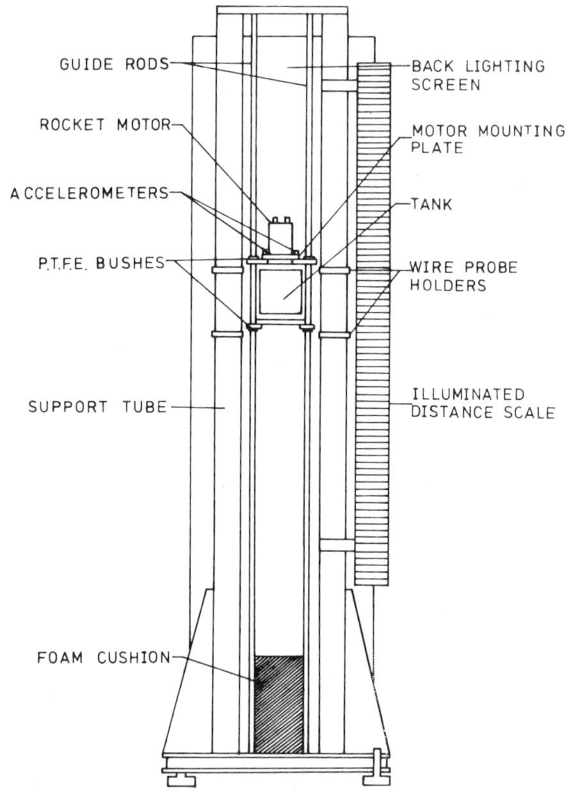

Fig. 1. Experimental apparatus.

2.2 *Apparatus*

2.2.1. *Description*

The apparatus is shown in fig. 1. The tank, driven by the rocket motors, is constrained to move vertically downwards between two parallel steel rods. Lateral motion is restricted by means of four accurately machined PTFE bushes, two on either side of the tank.

Motion is stopped at the end of useful travel by means of an expanded polystyrene cushion, thus causing minimal damage to the moving parts.

2.2.2. *Initial conditions*

The tank is held in position at the top of the apparatus by cotton thread, the breaking strength of which is very much less than the motor thrust, and both the initial position of the tank relative to the frame, and the fluid interface relative to the tank, are clearly marked. The majority of the experiments had no introduced interface perturbation, but methods of creating initial disturbances have been investigated [1].

2.2.3. *Tank detail*

Experiments have been conducted with both "two-dimensional" and "three-dimensional" tanks. These tanks are essentially glass walls within a metal framework which also supports the rocket motor mounting plate. The internal dimensions of the tanks for the two cases were $150 \times 150 \times 25$ mm and $150 \times 150 \times 150$ mm. It has been found helpful to have the inner top surface of the tanks ridged to assist in the removal of bubbles introduced during filling.

2.2.4. *Fluids*

In order to vary the density ratio, various combinations have been used:

e.g.: Ethyl alcohol/air, $\rho_1/\rho_2 = 600$;

Aqueous solution of sodium iodide/pentane, $\rho_1/\rho_2 = 3$;

Water/pentane, $\rho_1/\rho_2 = 1.6$.

A surface active agent, Vulcastab LW, was used in some cases to reduce interfacial surface tension.

Since all fluids were essentially clear, early experiments used a weak (0.0025%) solution of Nigrosine dye in the denser fluid in order to determine the mix region. It was found, however, that when mixing began, refraction within the mixed region enables this region to be easily identified without the inclusion of dye.

2.2.5. *Rocket motors*

Suitable rocket motors are manufactured by the Propellants, Explosives and Rocket Motor Establishment, England. Two types of motor with similar near-constant thrust profiles (mean thrust $\approx$ 1.9 kN), but with different burn times, were used.

To obtain the highest accelerations with the larger three-dimensional tank it was necessary to use two rocket motors, fired simultaneously, for propulsion.

Since the only change in mass of the system during the period of the experiment will be that due to the burning of the propellant, at most 1.2% of the total mass, the acceleration profile of the tank will have essentially the same shape as the rocket motor thrust curve and will remain almost constant throughout most of the burn.

2.2.6. *Instrumentation*

Acceleration is measured by two accelerometers attached via insulated studs to the motor mounting plate. The output from the accelerometers is recorded on magnetic tape together with the pulses from thin wire probes placed in the path of the tank. These probes which are at known distances from the initial position of the tank, are activated by the passage of the tank and are used to confirm the mean acceleration over a given distance. Since the output of the probes are recorded on the same time base as the accelerometer signals, accurate values of acceleration at known positions may be determined.

A distance scale is attached to the support tubes of the apparatus in the plane of the front of the tank so that direct measurements are also obtainable from photography.

Photographs of the tank motion are obtained using two Vinten 35 mm framing cameras operating at a nominal 200 frames per second and with a frame exposure time of 0.14 ms. The vertical separation of the camera is such as to give an overlap of their images so that similar positions of the tank are photographed on the films from both cameras.

Time markers are recorded on the edges of the films at millisecond intervals, and back lighting is provided by a bank of diffused photoflood bulbs. Since the earlier experiments were performed, available films and processing agents have been compared using the same photographic conditions as the experiments [9]. As a result we are now using a much finer grain film (Pan F) and a higher accutance developer (Qualitol).

3. Results and analysis

Space permits the presentation of only a few of the results so far obtained (figs. 2–7). Further results are given in ref. 1.

For each experiment h_1, the depth to which the instability penetrates the denser fluid, is measured on the photographic prints. However, h_2, the depth to which the instability penetrates the less dense fluid, is ill defined for the alcohol/air experiments but has been measured for the liquid/liquid combinations. In some cases edge effects due to the meniscus give enhanced penetration in the tank corners. Then h_1 and h_2 are measured in the central region of the tank (fig. 8).

For the experiments using the three-dimensional tank parallax corrections are applied as explained in fig. 9. These corrections have been ignored for the experiments using the two-dimensional tank.

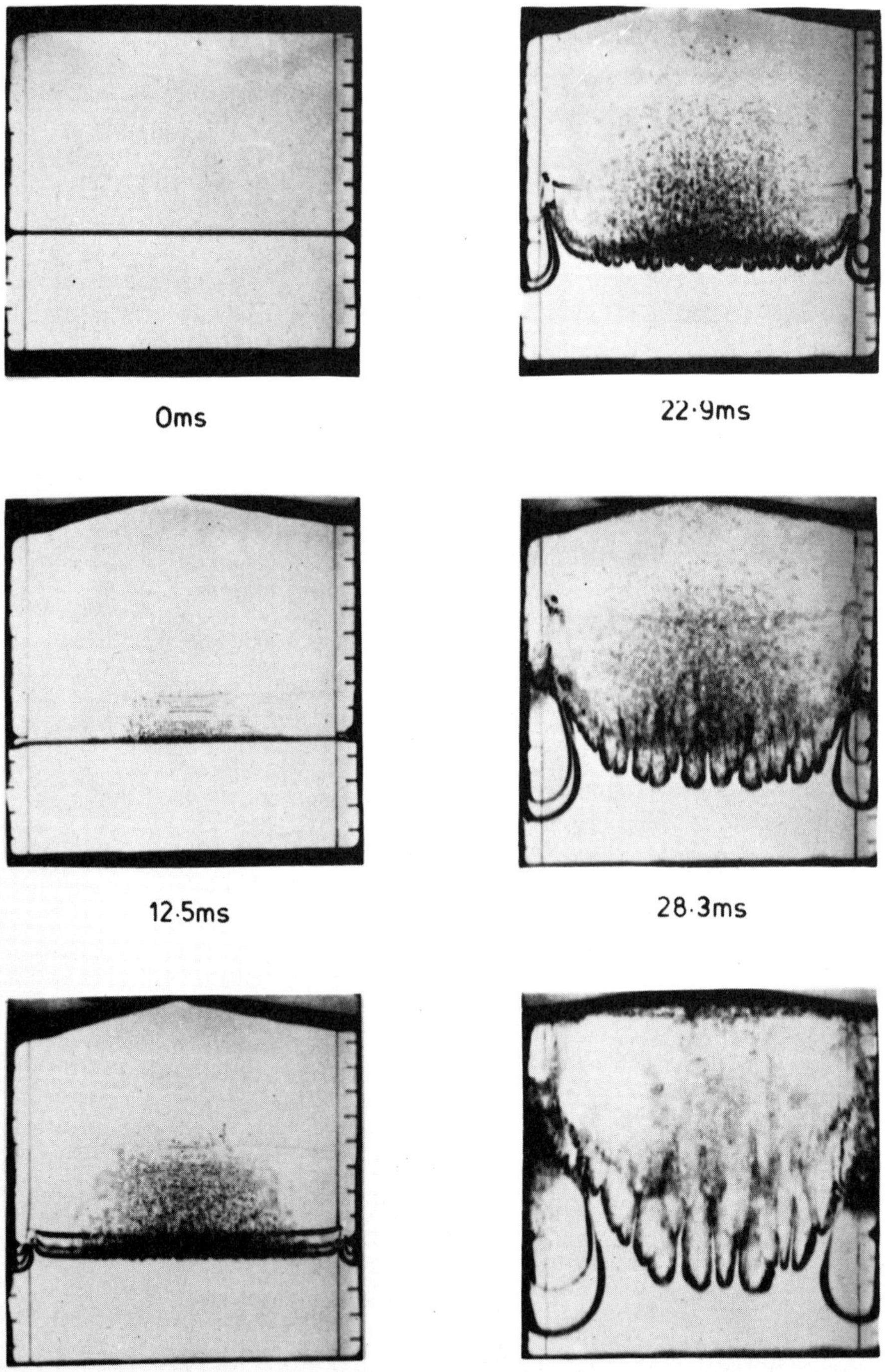

Fig. 2. Experiment 56 – air/ethanol. 2-D.

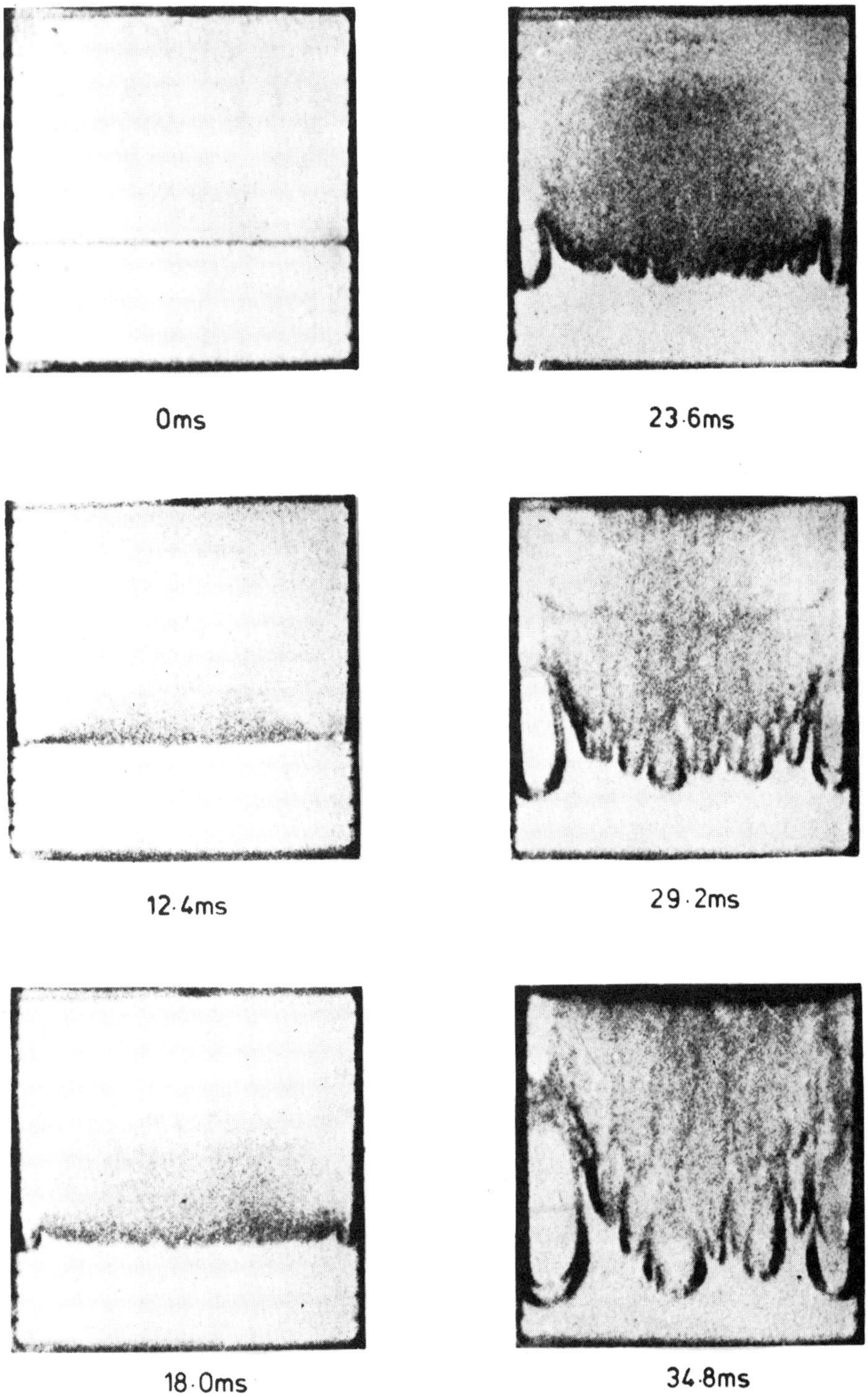

Fig. 3. Experiment 34 – air/ethanol. 2-D. Random noise.

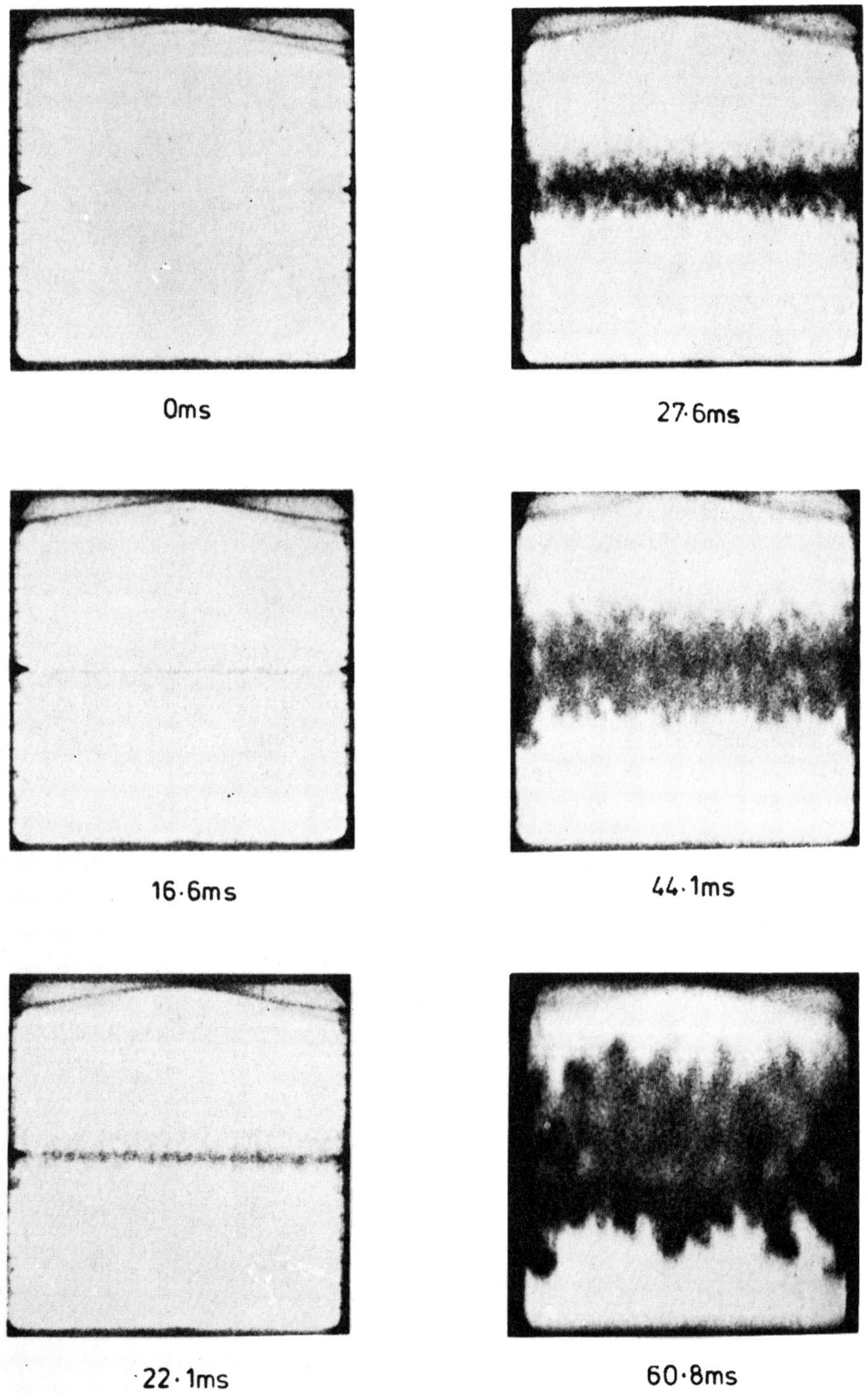

Fig. 4. Experiment 36 – pentane/NaI solution. 2-D.

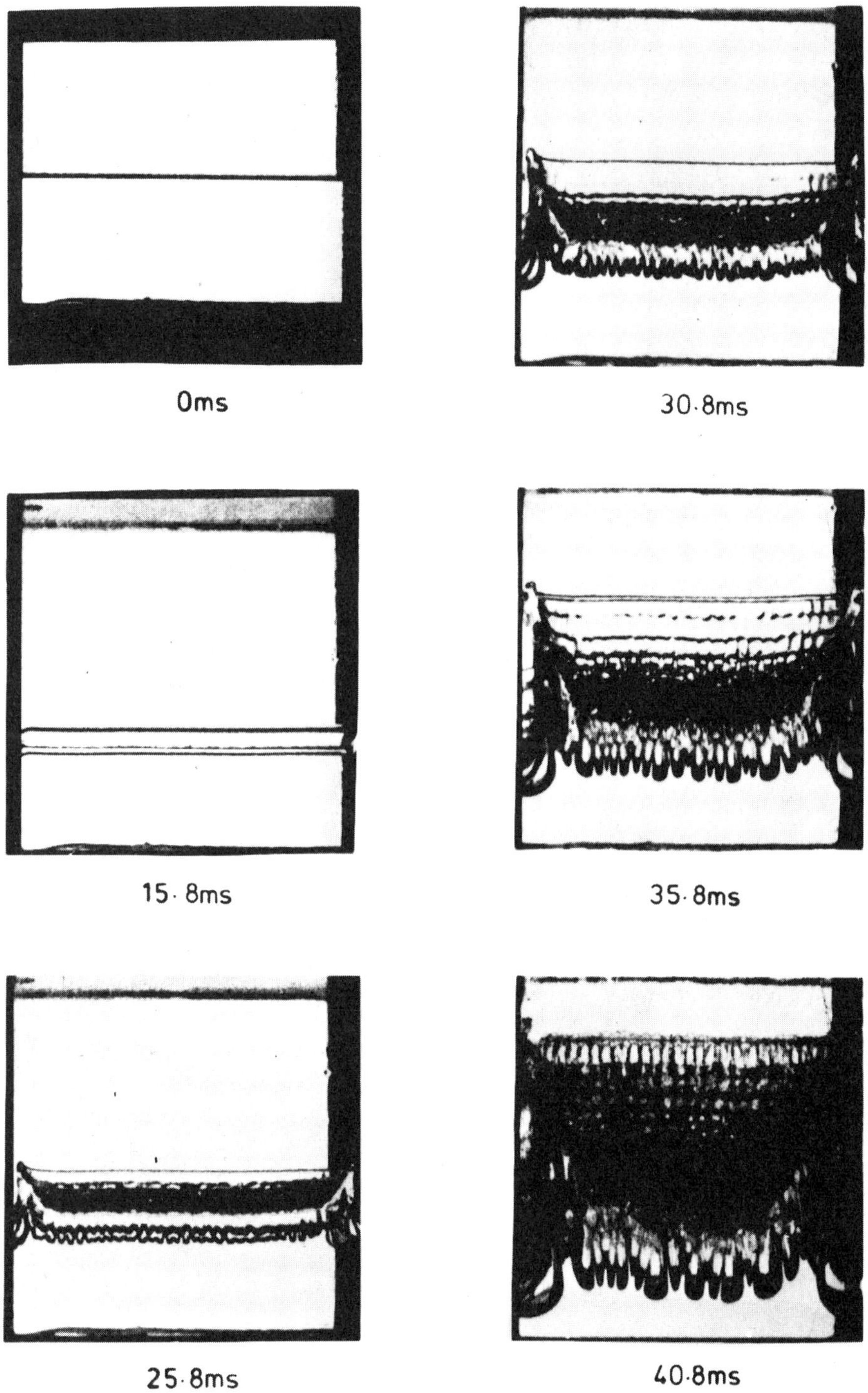

Fig. 5. Experiment 29 – air/ethanol. 3-D.

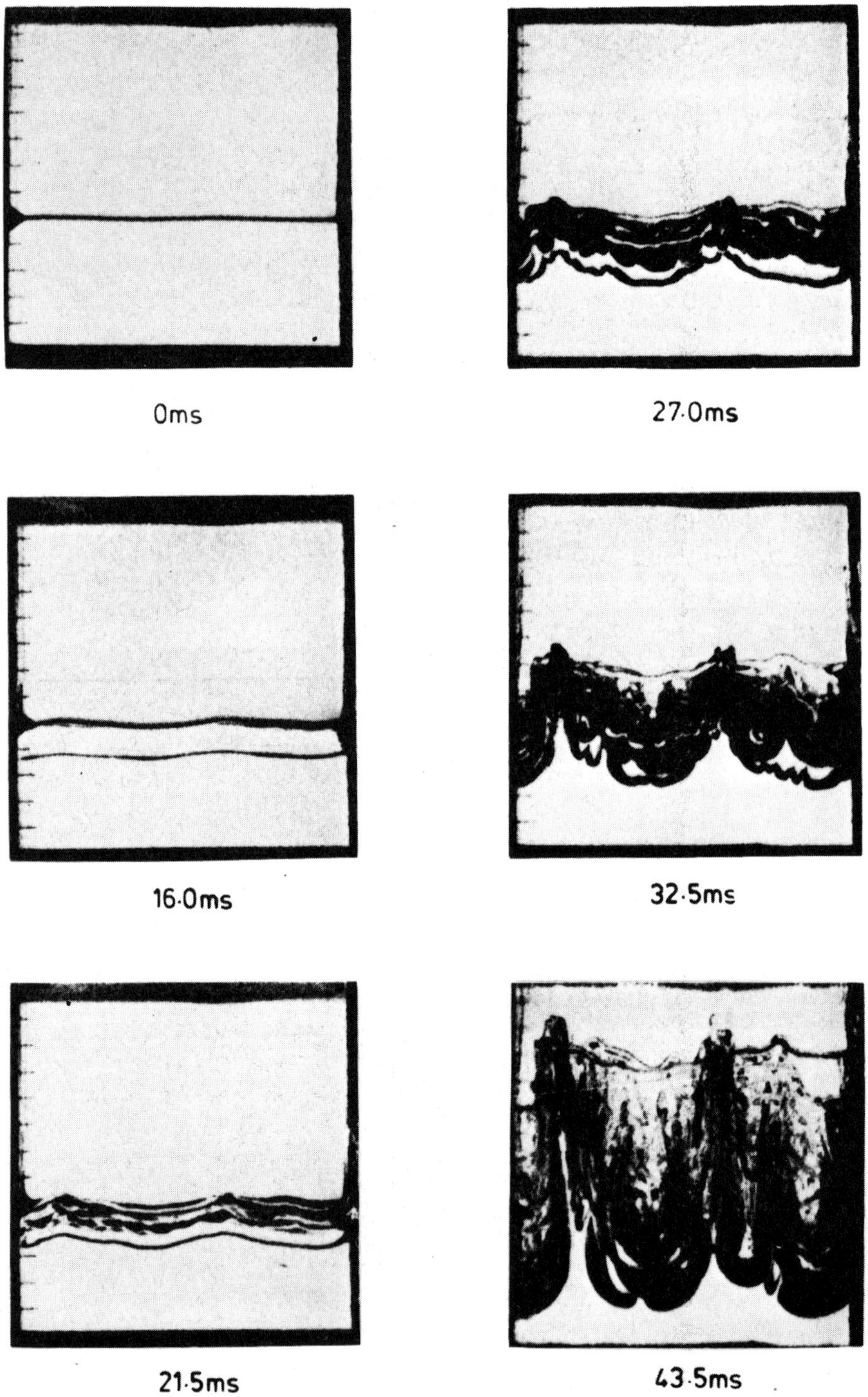

Fig. 6. Experiment 37 – air/ethanol. 3-D. Long wavelength disturbance.

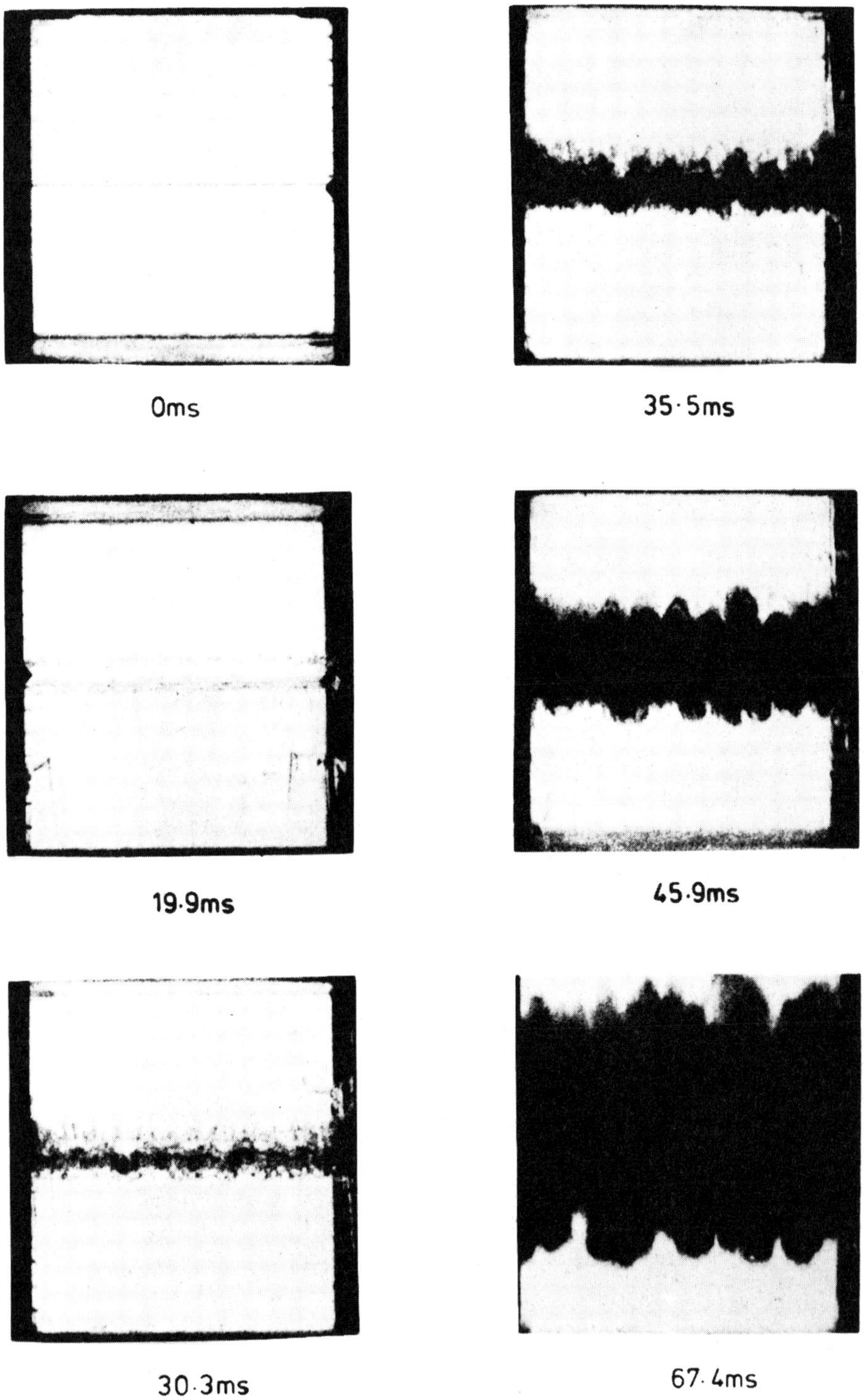

Fig. 7. Experiment 35 – pentane/NaI solution. 3-D.

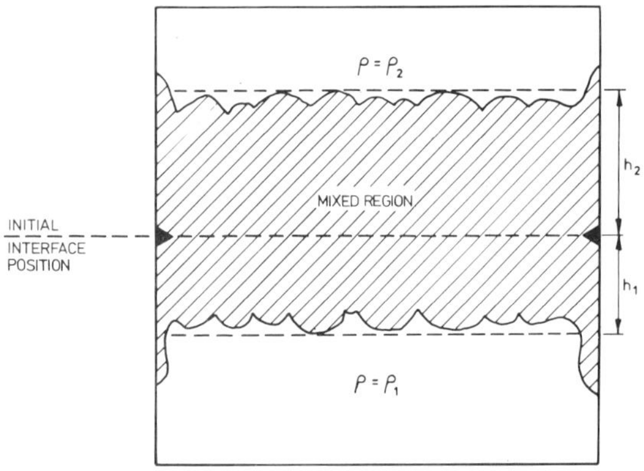

Fig. 8. Analysis of experiments: definition of h_1 and h_2.

The experiments are analysed by plotting h_1 against X, the scaled acceleration distance (figs. 10–12) where

$$X = \frac{\rho_1 - \rho_2}{\rho_1 + \rho_2}\left\{\int_0^t (g[t^1])^{1/2}\,\mathrm{d}t^1\right\}^2$$

$$= \frac{\rho_1 - \rho_2}{\rho_1 + \rho_2}\phi S\,.$$

S is the distance moved by the tank and the correction factor ϕ is calculated from the accelerometer record of g_1 versus time [1]. The coefficient α is obtained from the straight line fit

$$h_1 = \alpha(X - X_0)\,.$$

Values of α are quoted in table I for those experiments in which the instability developed uniformly across the width of the tank and the slope of the h_1 versus X graph settled down to an approximate limiting value. The ratio h_2/h_1 has been estimated for the liquid/liquid experiments. At $\rho_1/\rho_2 = 3$, h_2/h_1 is approximately 1.3. For the experiment at $\rho_1/\rho_2 = 1.6$ there was no detectable asymmetry, i.e. h_2/h_1 was close to unity.

Values of α for the two-dimensional tank appear to be higher than those obtained from the two-dimensional numerical simulation [2], with the three-dimensional tank giving even higher values

Table I
Values of coefficient

Type	Exp't	ρ_1/ρ_2	α
2D numerical simulation		20	0.04–0.05
ref. 2		3.0	0.04–0.05
2D alcohol/air	56	600	0.058
	34	600	0.062
	40	600	0.063
	26	600	0.065
2D NaI soln./pentane	23	3.0	0.060
	42	3.0	0.061
	36	3.0	0.064
2D NaI soln. + SAA/pentane	61	2.74	0.057
	57	2.74	0.061
2D water + SAA/pentane	50	1.6	0.058
2D water + $CaCl_2$/hexane	54	1.74	0.065
3D alcohol/air	29	600	0.073
	39	600	0.076
	58	600	0.077
3D NaI soln./pentane	33	3.0	0.066
	35	3.0	0.071
3D NaI soln./hexane	62	2.77	0.063
	60	2.77	0.073

on average. These results indicate that two-dimensional constraints tend to inhibit the growth rate. The front and rear walls of the two-dimensional tank should slow down the growth rate slightly when $h_1 > D$, the distance between the walls. This reduces the average value of α for the experiment. The plots of h_1 versus X for the two-dimensional tank should be slightly curved (fig. 11).

Experiment 37 (figs. 6 and 10) illustrates the effect of an additional long wavelength initial perturbation of the form

$$\zeta = a_0 \sin\frac{2\pi X}{\lambda}$$

on the evolution of the instability.

For experiment 37: $a_0 \approx 0.5$ mm, $\lambda = 65$ mm.

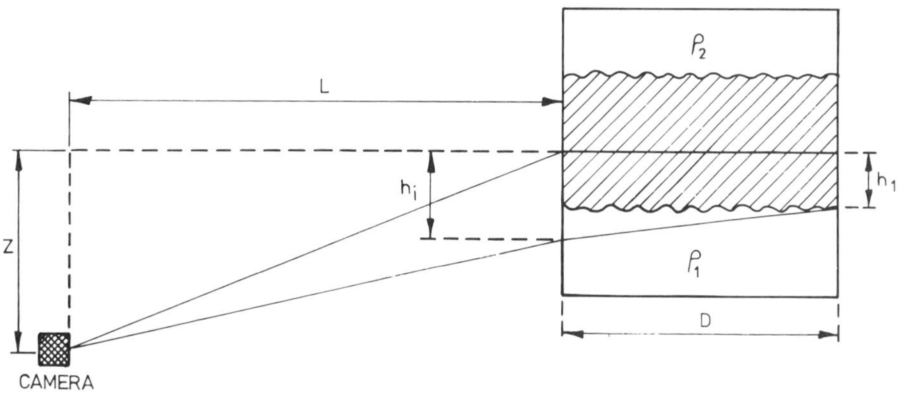

h_i = APPARENT PENETRATION OF DENSE FLUID.

h_1 = TRUE PENETRATION OF DENSE FLUID.

$$= \frac{D + n_1 L}{n_1 L} h_i - \frac{D}{n_1 L} Z$$

n_1 = REFRACTIVE INDEX AT AIR/LIQUID 1 INTERFACE.

(a) CAMERA BELOW LOWER BOUNDARY OF MIXED REGICN

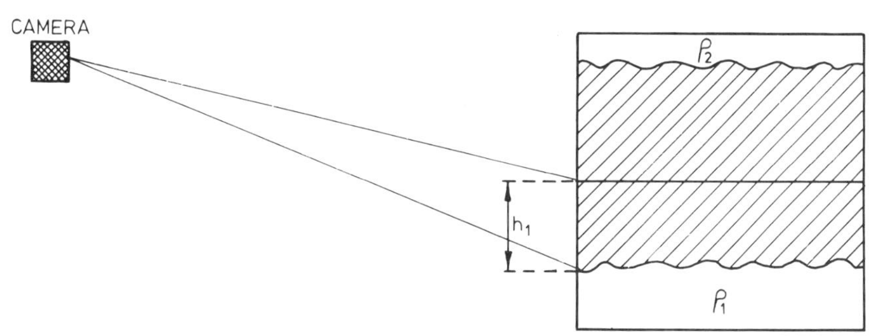

h = TRUE PENETRATION OF DENSE FLUID.

(b) CAMERA ABOVE LOWER BOUNDARY OF MIXED REGION

Fig. 9. Analysis of experiments using 3-D tank.

At early times an approximately two-dimensional disturbance was present (see early frames fig. 6), but towards the end of the experiment the flow became three-dimensional. Several bubbles developed throughout the depth of the tank as shown in the final frame of fig. 6.

As the wavelength of the instability remains constant throughout this experiment, at late stages the bubble velocity should be proportional to $\sqrt{g\lambda}$. An estimate for this limiting velocity was obtained by Layzer [10] who derived a theory for the case $\rho_1/\rho_2 = \infty$.

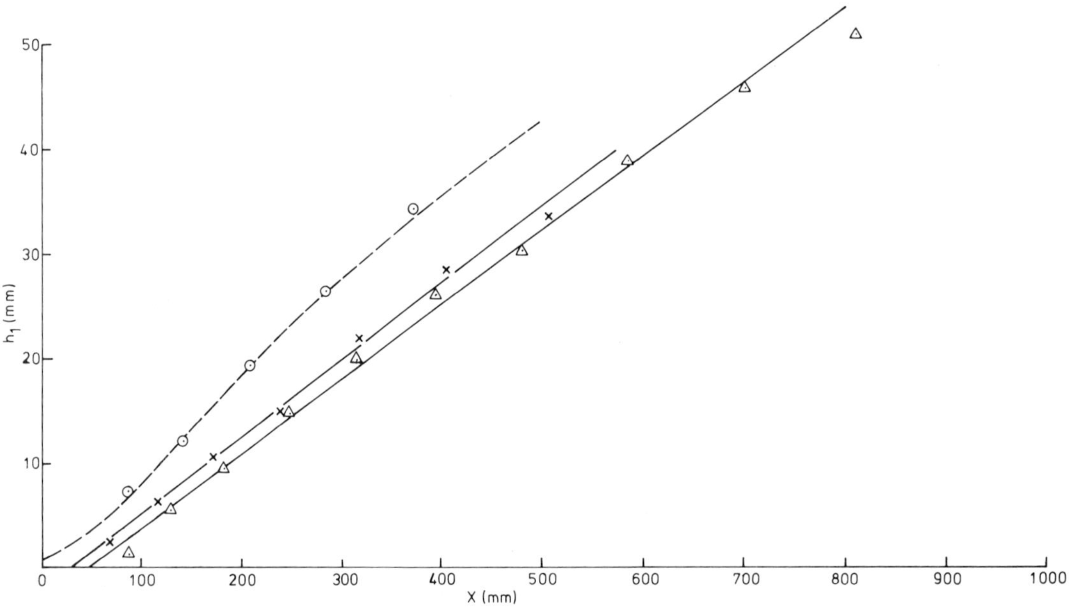

Fig. 10. Graph h_1 versus X. 3-D tank.

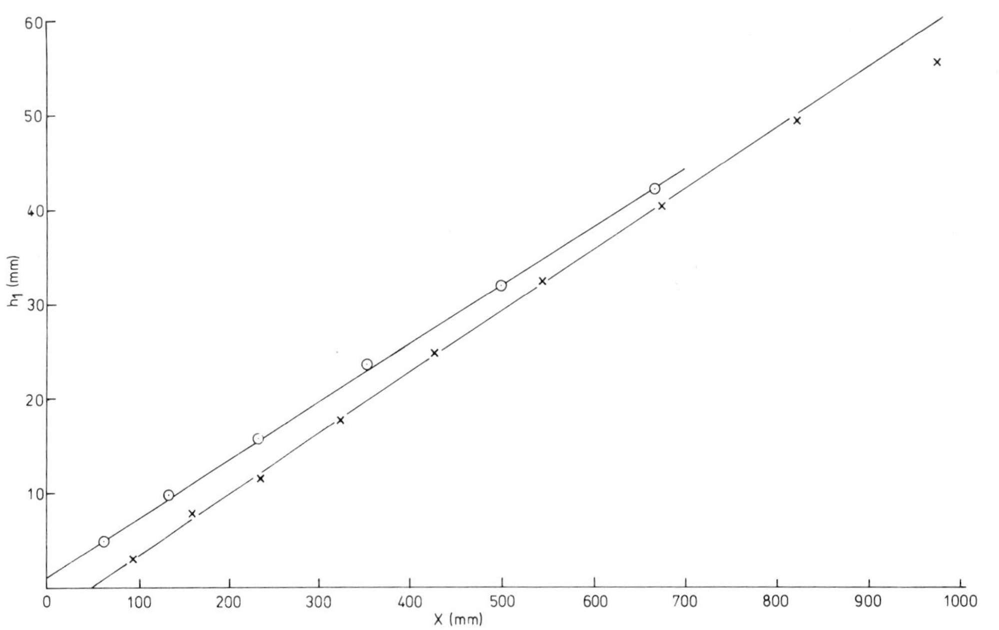

Fig. 11. Graph h_1 versus X. 2-D tank.

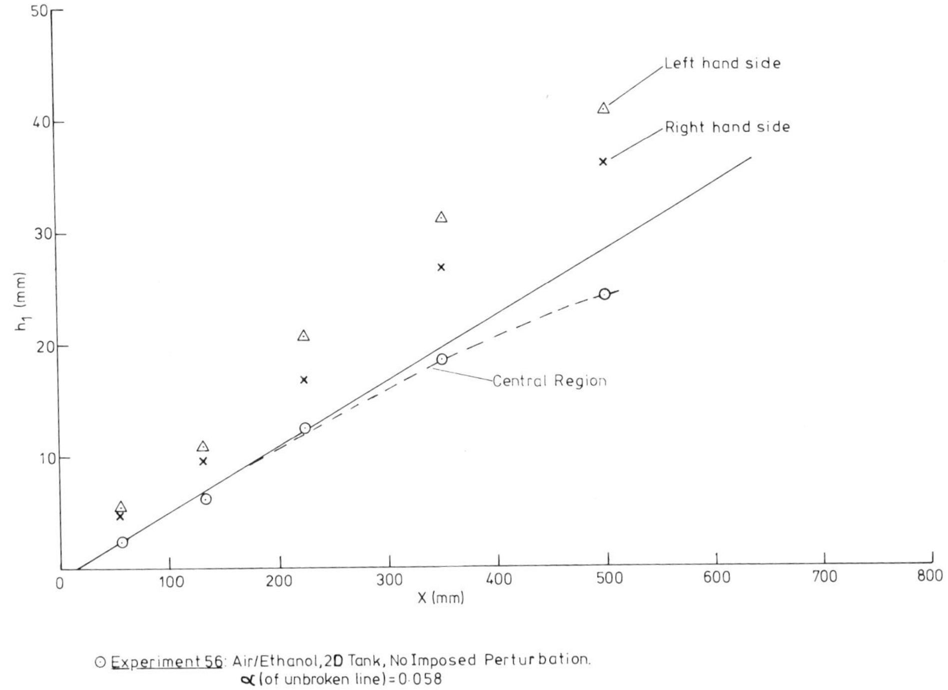

Fig. 12. Graph h_1 versus X. Experiment 56. 2-D tank.

For "bubbles" in the form of archways evolving from an initial perturbation,

$$\zeta_0(x) = a_0 \sin \frac{2\pi x}{\lambda},$$

the limiting bubble velocity is

$$V_B = \left(\frac{g\lambda}{6\pi}\right)^{1/2} = 0.23\sqrt{g\lambda}.$$

Layzer also considered bubbles rising in a cylindrical tube evolving from an initial perturbation of the form

$$\zeta_0(r) = a_0 J_0\left(\frac{2\beta_1 r}{\lambda}\right),$$

where λ is now interpreted as the diameter of the cylinder and $\beta_1 = 3.83171$ is the first zero of the Bessel function J_1. The limiting bubble velocity is now

$$V_B = \left(\frac{g\lambda}{2\beta_1}\right)^{1/2} = 0.36\sqrt{g\lambda}.$$

Analysis of the later stages of experiment 37 gave the results shown in table II.

For this experiment, Layzer's formula for cylindrical bubbles gives

$$V_B = 0.36 \times (21 \times 9.81 \times 65 \times 10^{-3})^{1/2}$$
$$= 1.32 \text{ m/s}.$$

Table II

Time (ms)	h_1 (mm)	$V_B = \frac{\Delta h_1}{\Delta t}$ (m/s)
27.0	12.2	
		1.33
32.5	19.5	
		1.29
38.0	26.6	
		1.42
43.5	34.4	

This agrees well with the observed value; the theory for archways would give too low a value.

We see that the development of a three-dimensional instability consisting of cylindrical bubbles, from the initial approximately two-dimensional disturbance, enables the instability to grow more rapidly. It is suggested that this is the reason for the growth rate (i.e. α-value) in the three-dimensional tank being about 40% larger than that seen in the two-dimensional computer simulation.

Experiment 37 may be compared in more detail with Layzer's theory for which the bubble penetration, h_1, as a function of time is given by

$$\frac{\mathrm{d}h_1}{\mathrm{d}t} = V\,,$$

$$\frac{\mathrm{d}V}{\mathrm{d}t} = g\left(\frac{1-E-(V/V_\mathrm{B})^2}{1+E}\right),$$

where

$$V_\mathrm{B} = \left(\frac{g\lambda}{2\beta_1}\right)^{1/2}$$

is the limiting bubble velocity and

$$E = \exp\left(-\frac{4\beta_1 h_1}{\lambda}\right).$$

These equations have been integrated for constant g and $\lambda = 65$ mm, starting with $V = 0$ and $h_1 = a_0 = 0.65$ mm (which is close to the estimated initial amplitude) at $t = 0$. Agreement between Layzer's theory and experiment is good, see fig. 10.

4. Conclusions

Experiment has shown that in the absence of large amplitude, long wavelength, initial perturbations and where the flow is not dominated by edge effects, the mix rate at a given density ratio varies little from experiment to experiment. The penetration, h_1, of the denser fluid is described well by the linear law

$$h_1 = \alpha(X - X_0)\,,$$

where

$$X = \frac{\rho_1 - \rho_2}{\rho_1 + \rho_2} g t^2\,.$$

The coefficient α has been found to vary little with the density ratio. The best estimate for α is 0.07 in a three-dimensional situation.

Additional long wavelength perturbations enhance the rate of growth of the instability and further work is planned to more precisely quantify this effect.

Acknowledgements

The contributions of Mr. V. Smeeton, Mr. M. Berry and Mr. B. Farr are gratefully acknowledged. The later experiments in the series were conducted under the direction of Mr. J.D. Herbert.

References

[1] K.I. Read and D.L. Youngs, "Experimental Investigation of Turbulent Mixing by Rayleigh–Taylor Instability", AWRE Report 011/83 (1983).

[2] D.L. Youngs, "Numerical Simulation of Turbulent Mixing by Rayleigh–Taylor Instability", Physica 12D (1984) 32 (these proceedings).

[3] D.J. Lewis, "The Instability of Liquid Surfaces When Accelerated in a Direction Perpendicular to their Planes, II", Proc. Roy. Soc. 202A (1950) 81.

[4] H.W. Emmons, C.T. Chang and B.C. Watson, "Taylor Instability of Finite Surface Waves", J. Fluid Mech. 7 (1960) 177.

[5] R.E. Duff, F.H. Harlow and C.W. Hirt, "Effects of Diffusion on Interface Instability Between Gases", Physics of Fluids 5 (1962) 417.

[6] R.L. Cole and R.S. Tankin, "Experimental Study of Taylor Instability", Physics of Fluids 16 (1973) 1810.

[7] M. Ratafia, "Experimental Study of Taylor Instability", Physics of Fluids 16 (1973) 1207.

[8] R. Popil and F.L. Curzon, "Production of Reproducible Rayleigh–Taylor Instabilities", Rev. Sci. Instr. 50 (1979) 1291.

[9] V. Smeeton and J.D. Herbert, private communication (1983).

[10] D. Layzer, "On the Instability of Superposed Fluids in a Gravitational Field", Astrophysical J. 122 (1955) 1.

Physica 12D (1984) 59–70
North-Holland, Amsterdam

VORTEX DYNAMICS OF PASSIVE AND ACTIVE INTERFACES

Hassan AREF and Grétar TRYGGVASON
Division of Engineering, Brown University, Providence, RI 02912, USA

Two classes of interface motion are exemplified. In the first example the advection of a line of passive marker particles by the unsteady flow due to a pair of vortex agitators is considered. The qualitative nature of the subsequent motion of the line depends sensitively on the parameters governing the unsteady flow. The advection of the particles determines either an integrable or a non-integrable dynamical system. The consequences of the transition from one regime to the other are explored by numerical experiments. In the second example a calculation of piecewise potential flow with a sharp interface is presented. The flow of two immiscible fluids of different but constant density and viscosity in a Hele–Shaw cell is considered. The interface separating the fluids is formally a tangential discontinuity of velocity and, thus, may be modelled as a vortex sheet. The evolution of fingers on the interface and the variability of the finger topology with the control parameters in the problem are accessible to numerical computation using a variant of the vortex-in-cell method. Results of numerical experiments employing such a code are presented and discussed. The paper concludes with some general speculations on the dynamics of interfaces.

1. Passive and active interfaces

We report on two sets of numerical experiments involving sharp interfaces moving in two-dimensional flows of a rather idealized nature. We shall want to classify interfaces into "passive" and "active" in order to bring out more clearly the differences between our two calculations. By a passive interface we shall mean a line of marker particles in the flow which is simply advected without in any way affecting the flow field itself. The flow field would be exactly the same without the passive interface. With it we have a device for visualizing the flow. Passive interfaces are used extensively in flow visualization studies. Essentially all the pictures in the recent collection edited by van Dyke [1] (the banquet speaker at this conference) are of passive interfaces or, at least, of interfaces that the experimenter who took the pictures would have us believe are passive! Advection of passive particles, whether initially arranged on a line or not, is of interest to the chemical engineer in connection with stirring and mixing devices, and to the meteorologist, oceanographer and environmentalist in connection with the dispersal of pollutants in the atmosphere and oceans. Our example in section 2 is oriented towards stirring devices.

Most interfaces in fluid mechanics are of the other, "active" type, in the sense that the configuration of the interface "feeds back" into the flow field and alters it. Formally, the paradigm for this situation is a free boundary problem. Examples include several of the topics discussed at this conference such as waves on the free surface of a fluid in a gravitational field (Yuen [2]), shock waves (Shaner [3]) and atmospheric and ocean fronts (Hoskins [4], Simpson [5]). There are obvious qualitative analogies between idealized problems of this type, such as the finite amplitude Rayleigh–Taylor instability (see Sharp [6]) or the fingering in stratified Hele–Shaw flow (Tryggvason and Aref [7] and section 3), and the problem of dendritic growth (Langer [8], Sekerka [9]).

A particularly simple case arises if the flow field consists of regions individually in potential flow. Then the flow in the interior of a region is uniquely and explicitly determined by what goes on at the boundaries. This classical observation has led to the development of boundary integral techniques for the numerical computation of piecewise poten-

0167-2789/84/$03.00 © Elsevier Science Publishers B.V.
(North-Holland Physics Publishing Division)

tial flow. In two dimensions a very convenient representation arises by using vortex sheets at all interfaces, since one may then take full advantage of the techniques developed, mainly in the context of high-speed aerodynamics, to evolve vortex sheets in time. In section 3 we discuss the results of applying such a method to the problem of sharply stratified flow in a Hele–Shaw cell.

Lamb [10] ascribes the mathematical description of an advected interface to Kelvin. Consider

$$F(x, y, z, t) = 0 \tag{1}$$

to be the equation of the interface at time t. Then in a velocity field V the motion of the interface is given by

$$\frac{\partial F}{\partial t} + V \cdot \nabla F = 0 . \tag{2}$$

Within this framework the interface is "active" if V depends on the location of $F = 0$. If V is independent of this data, in particular if V can be prescribed, the interface is "passive".

2. Passive interface: chaotic advection and the efficiency of stirring

The problem of advection of passive particles lends itself immediately to the methods and tools of the theory of dynamical systems. In the Lagrangian representation of the problem the advection of a passive particle amounts to the solution of d simultaneous, ordinary differential equations, where d is the dimensionality of the physical space. If a three-dimensional velocity field, V, has Cartesian components (u, v, w), the equations to be solved are

$$\begin{aligned} \dot{x} &= u(x, y, z, t) , \\ \dot{y} &= v(x, y, z, t) , \\ \dot{z} &= w(x, y, z, t) . \end{aligned} \tag{3}$$

This is just a dynamical system with three independent variables. For a steady flow the system (3) is autonomous. For a laminar flow, as we have in mind here, the system should be deterministic. Turbulent advection can be accomodated by using a system of stochastic equations.

This connection between advection problems and dynamical systems has been apparent for many years. In 1966 Hénon [11], at the suggestion of Arnold, studied the advection problem

$$\begin{aligned} \dot{x} &= A \sin z + C \cos y , \\ \dot{y} &= B \sin x + A \cos z , \\ \dot{z} &= C \sin y + B \cos x , \end{aligned} \tag{4}$$

where A, B and C are constants. The flow given by the right-hand sides of eqs. (4) is an incompressible, steady solution of the three-dimensional Euler equation. The solution has distributed vorticity. In fact, the flow field is of a particular type, sometimes associated with the name of Beltrami, in which the vorticity and velocity vector are everywhere parallel. Now, it can be shown by computing a Poincaré section for the dynamical system (4) that when $A = \sqrt{3}$, $B = \sqrt{2}$, $C = 1$, the phase space for this system is decomposed into regular and chaotic regions. The volume-preserving transformation that this flow induces in a blob of particles will clearly be very different in the regular and in the chaotic regions. The consequences for stirring must be that one achieves more efficient stirring from chaotic advection than from regular motion. Note that the transition from one regime to the other is available by changing parameters for an incompressible, steady flow.

We have pursued this application of chaos in a slightly modified model format. Instead of a steady three-dimensional flow we employ an unsteady two-dimensional flow, which arises, in the models considered so far, by alternately switching point vortex agitators on and off [12]. We recall that the steady flow about a spinning cylinder in a viscous fluid is of exactly the same form as that due to a rectilinear vortex in an ideal fluid.

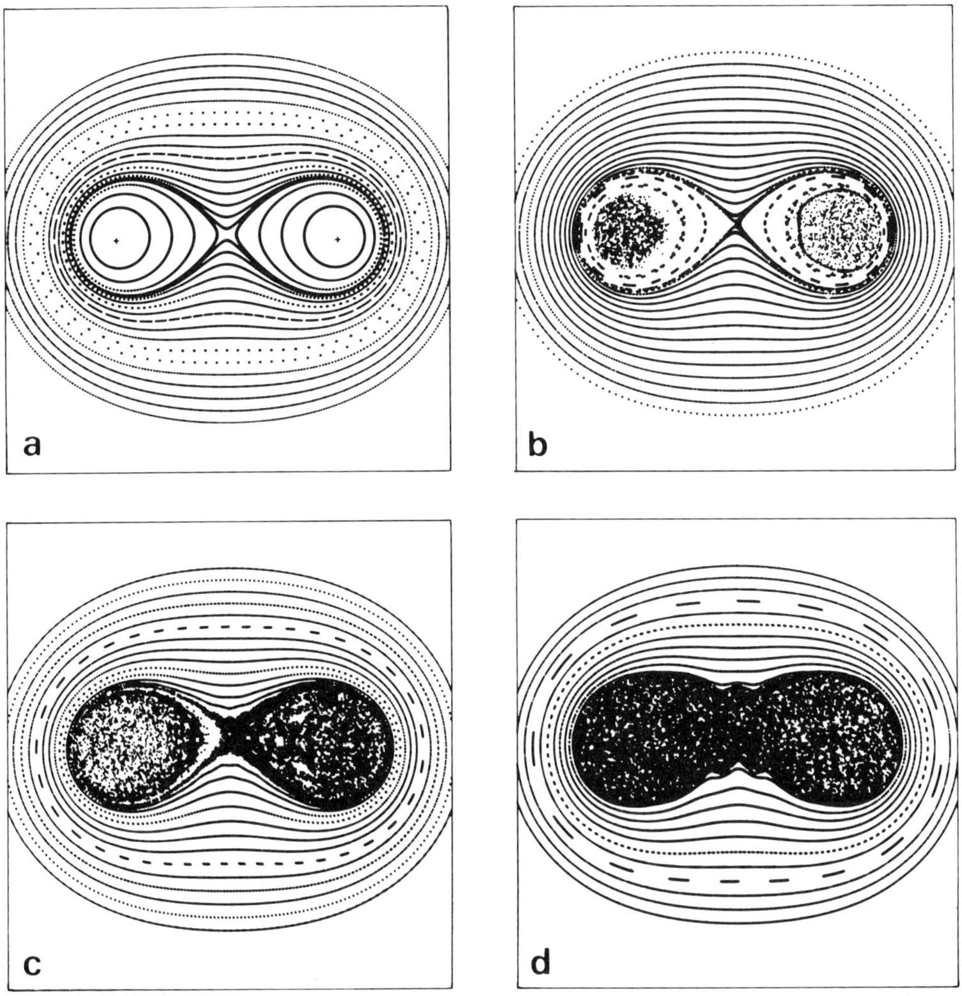

Fig. 1. Poincaré sections for the "alternating agitator model" of section 2. The square frames are not physical boundaries. Values of μ, eq. (8), are: (a) $\mu = 0.1$; (b) $\mu = 0.25$; (c) $\mu = 0.35$; (d) $\mu = 0.5$.

To present the simplest example of this kind of model, let the flow plane be parametrized as the complex z-plane and let two vortex agitators be situated at $x = \pm b$. Let their common circulation by Γ and assume that they are "on" (alternately) for a time interval $T/2$ (followed by an equal time interval where they are "off".) Consider now a particle at $z = z_0$ at time $t = 0$. At time $t = T/2$ this particle will be at

$$z_1 = b + (z_0 - b)\exp(\mathrm{i}\Gamma T/4\pi|z_0 - b|^2) \tag{5}$$

(assuming that the vortex agitator at $+b$ was turned on first.) A similar formula may be derived for the position $z = z_2$ at time $t = T$ (i.e. after advecting by the agitator at $-b$) in terms of z_1. In all, then, we have reduced this model advection problem to an explicitly given, area-preserving mapping, $z_0 \to z_2$. This mapping may be shown to be the second iterate of the mapping $z_0 \to z_1'$, where $z_1' = -z_1$. In non-dimensional form we have

$$z' = -1 + (1 - z)\exp(\mathrm{i}\mu/|1 - z|^2), \tag{6}$$

where

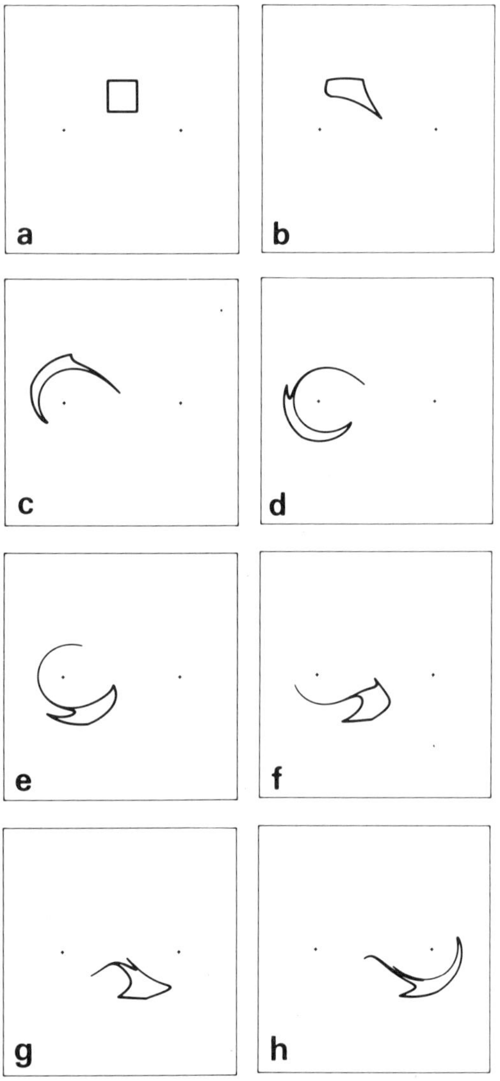

Fig. 2. A sequence of stages in the stirring of an initially square region by the model of section 2. The parameter $\mu = 0.1$. The frames shown correspond to "physical time" $n\mu$ where (a) $n=0$; (b) $n=5$; (c) $n=10$; (d) $n=15$; (e) $n=20$; (f) $n=25$; (g) $n=30$; (h) $n=35$.

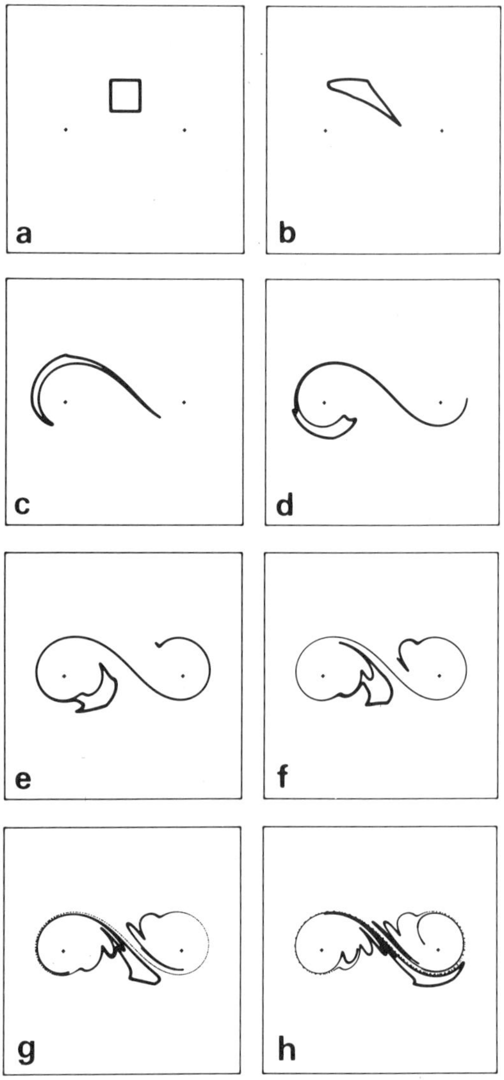

Fig. 3. A sequence of stages in the stirring of an initially square region by the model of section 2. The parameter $\mu = 0.5$. The frames shown correspond to "physical time" $n\mu$ where (a) $n=0$; (b) $n=1$; (c) $n=2$; (d) $n=3$; (e) $n=4$; (f) $n=5$; (g) $n=6$; (h) $n=7$. These are the same physical times as shown in fig. 2.

$$z = z_0/b, \quad z' = z_1'/b\,, \tag{7}$$

and the parameter

$$\mu = \Gamma T/4\pi b^2\,. \tag{8}$$

Numerical experiments show that the mapping (6) displays a transition to chaotic behavior as μ is increased with "large scale chaos" setting in at $\mu = 0.35$, approximately. This result has been discussed in detail elsewhere [12], and here we simply show four Poincaré sections at different values of μ (see fig. 1). In this paper we are particularly concerned with the consequences of the transition to chaos for advecting interfaces. Intuitively one would expect the interface to be wrapped around

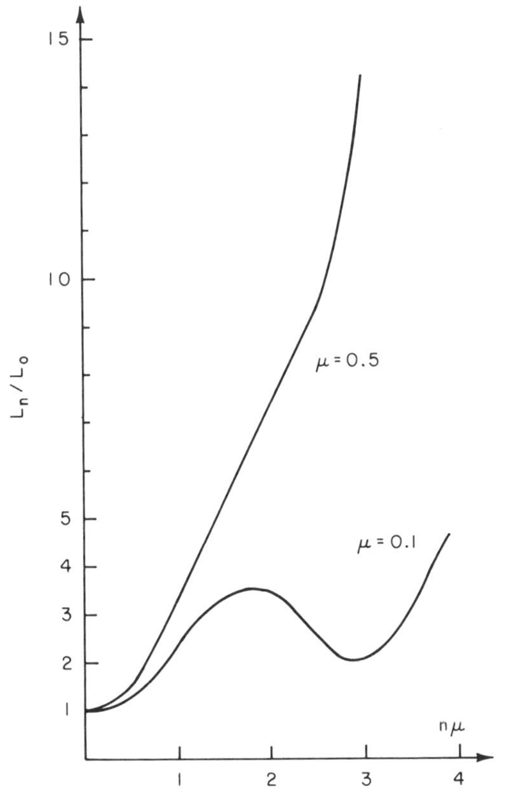

Fig. 4. The chord length L_n of the perimeter of the advected region (normalized by its initial value L_0) at "time" $n\mu$ for the sequences shown in figs. 2 and 3. The curves are labelled with the value of μ.

in a smooth and regular fashion for a predominantly integrable mapping. For a chaotic mapping, on the other hand, highly irregular interfacial structure should evolve.

In fact, in their discussion of "quantum maps" Berry et al. [13] digress to comment on the fate of a line of particles as it is advected chaotically. They focus on the fixed points of the flow, which in two dimensions are of two types, elliptic and hyperbolic, and they discuss the kind of characteristic "signal" that each kind of fixed point will leave on the advected line or curve. For an elliptic fixed point the main feature is the difference in winding number as one recedes from the fixed point. This leads to a spiral structure that Berry et al. [13] propose to call a "whorl". Hyperbolic fixed points, on the other hand, are generally surrounded by an intricate system of homoclinic (or heteroclinic) points, at which the stable and unstable manifolds intersect. The large undulations set up on these manifolds, as they weave in and out, lead to similar wiggles in the advected line, which Berry et al. [13] propose to call "tendrils". An advected interface, then, should evolve into a highly complicated shape consisting of whorls and tendrils, sometimes separated by regions of the curve that are reasonably undistorted, but frequently intermingled so as to form the very complex patterns that are familiar from our everyday experiences (oil spreading on a puddle of water, cream spreading on coffee etc.).

Tendrils are readily apparent in the advection of an interface by our model stirring device. Figs. 2 and 3 compare regular and chaotic advection of an initially square region. In the sequence of fig. 2 the parameter $\mu = 0.1$ and the mapping is predominantly regular (cf. fig. 1a). Although the square is obviously deformed and its sides are stretched, this happens in a very regular fashion compared to the sequence in fig. 3. For the latter $\mu = 0.5$, and now we see, clearly, the evolution of tendrils, i.e. secondary oscillations in the advected sides on scales that are not obvious from the steady streamline pattern.

It is clear that when a closed boundary is subjected to advection of the kind in fig. 3, the combination of area-preservation and the violent stretching of the boundary will lead to a tremendous increase in the ratio of circumference to area. This is likely to be of considerable importance in applications to mixing devices, combustors, chemical reactors etc. For the cases in Figs. 2 and 3 we continually computed the total chord length of the polygon line formed by successsively connecting the 10,000 particles that make up the boundary. This data is shown in fig. 4. (When $\mu = 0.5$ the boundary stretches so violently that the points on it are no longer recognizable as a curve shortly after fig. 3h. This explains the restriction in fig. 4 to "times" less than 4 for $\mu = 0.5$. It would in principle be possible to redistribute the points along the boundary as it deforms. However, any such redistribution will lead to small numerical errors in the exact location of the boundary, and, since we are investigating a phenomenon that amplifies such errors, it was decided not to attempt any redistribution and to simply terminate the calculation when the boundary "disappeared".) There is an obvious difference in the rate at which the perimeter stretches for the two cases. When $\mu = 0.1$ there is manifestly a net increase in the perimeter during the duration of the run. However, the increase is much less than in the comparison run for $\mu = 0.5$. Indeed, for $\mu = 0.1$, the perimeter actually oscillates with time, as may also be seen directly from the sequence in fig. 2. The results in fig. 4 can in principle be understood quantitatively by computing the Liapunov exponents [14] for the mapping (6). The degree of stetching of the various portions of the square can then be found by averaging the Liapunov exponents over each portion. Although in all but the simplest cases this promises to be a nontrivial piece of analysis, the point is that tools and methods for predicting the degree of stretching as a function of control parameters for a given model flow do exist. It is encouraging to see an entity as "sophisticated" as a Liapunov exponent emerge as a quantity of practical significance for the chemical engineer interested in stirring devices.

3. Active interface: simulation of stratified Hele–Shaw flow by a vortex method

If a passive interface can be understood in terms of the theory of dynamical systems with a few degrees of freedom, active interfaces present considerably greater challenges. Now, even for the case of a flow determined by boundary integrals, every element of the interface acquires dynamical degrees of freedom so that, even in two dimensions, where we arrive at a one-dimensional continuum for the interface, we have an infinity of strongly coupled degrees of freedom. We do not understand the morphology of these interfaces in terms of whorls and tendrils. However, even a brief sampling leads one to realize that these interfaces also are dominated by structural features of various kinds. And, indeed, there has evolved a separate vocabulary to describe the multitude of shapes that active interfaces can present. We speak of, and intuitively understand what is meant by, crests and troughs (of waves), by spikes and bubbles (in the Rayleigh–Taylor problem) and by fingers ("digitations"!) in the context of the application to stratified Hele–Shaw flow below. We may think of passive interfaces as representing a kind of "weak-coupling limit" of the general dynamics of a curve, and thus being limited to displaying, at worst, the effects of the flow singularities topologically available in the underlying space. Active interfaces, on the other hand, occupy the entire spectrum of interactions, including the "strong-coupling limit", and their potential for generating "shapes" seems essentially unlimited. Additional comments along these lines appear in section 4.

Interfaces separating two or more immiscible fluids, individually in potential flow, are, as already mentioned, ideally suited for vortex methods. The reason is that the interface in this situation is a tangential discontinuity of the velocity and thus has an immediate representation as a vortex sheet. The vortex sheet (shear layer) in ordinary hydrodynamics may be seen as the simplest example. (Note, by the way, that for late stages of vortex sheet roll-up the outer portion of the sheet can, to some extent, be thought of as being "passive".) In this case there is no discontinuity of physical properties, such as density or viscosity, across the interface, and the circulation is preserved in the limit of inviscid flow (Kelvin's circulation theorem.) Another well-known example is the Rayleigh–Taylor problem in which the interface marks a discontinuity surface of density. In this case the sheet circulation is not constant in time and Kelvin's theorem is replaced by the circulation theorem due to Bjerknes. The vortex sheet strength must now be evolved independently by solving a certain integro-differential equation due originally to Birkhoff [15]. Recent implementations of this method may be found in the work of Baker, Meiron and Orszag [16].

The Taylor–Saffman instability, realized in the flow of two immiscible fluids confined in a Hele–Shaw cell [17], provides an example that in many ways is intermediate between the two just mentioned, yet is of considerable practical and methodological interest. For this case the interfacial circulation is given by an integral equation due to de Josselin de Jong [18]. We have taken this representation of the problem and coupled it with the well-established vortex-in-cell method [19–22] for evolving vorticity configurations to produce a code with the capability of following highly contorted interfaces in a physically consistent way [7].

The basic equations to be solved in the Hele–Shaw problem are as follows: There is a vortex sheet strength or circulation per unit arclength, γ, which must satisfy the following equation:

$$\gamma = \frac{\Delta\mu}{\bar{\mu}} \boldsymbol{U}\cdot\hat{\boldsymbol{s}} + \frac{\Delta\rho g b^2}{12\bar{\mu}} \hat{\boldsymbol{j}}\cdot\hat{\boldsymbol{s}} + \frac{\alpha b^2}{12\bar{\mu}} \frac{\partial}{\partial s}\left(\frac{1}{R}\right). \tag{9}$$

In this equation $\bar{\mu}$ is the average of the two fluid viscosities μ_1 and μ_2, b is the distance between the parallel plates of the Hele–Shaw cell, g is the acceleration of gravity, $\Delta\rho$ is the density difference and $\Delta\mu$ the viscosity difference across the interface and α is the surface tension coefficient at the interface. Vector $\hat{\boldsymbol{s}}$ is the unit tangent, vector $\hat{\boldsymbol{j}}$ a unit vector opposite to the direction of gravity, and R is the radius of curvature of the interface in the plane of flow. We have in mind a gravity driven cell so that the top fluid (fluid 2) has larger density than the bottom fluid (fluid 1). This situation is obviously always unstable, but the nature of the instability at finite amplitude depends on the viscosity contrast between the two fluids. Eq. (9) is an integral equation since $\boldsymbol{U}$, the interface velocity, is given by the Birkhoff integral formula [15]

$$\boldsymbol{U} = \boldsymbol{U}(s, t) = \frac{1}{2\pi} \mathscr{P}\int \frac{\hat{\boldsymbol{z}}\times(\boldsymbol{x}(s,t) - \boldsymbol{x}(s',t))}{|\boldsymbol{x}(s,t) - \boldsymbol{x}(s',t)|^2} \gamma(s',t)\,\mathrm{d}s' \tag{10}$$

where the integral is on arclength along the interface, $\boldsymbol{x} = \boldsymbol{x}(s,t)$, at time t, and $\hat{\boldsymbol{z}}$ is a unit vector perpendicular to the flow plane. As discussed in our paper [7] it is possible to cast (9) in the dimensionless form

$$\tilde{\gamma} = 2A(\tilde{\boldsymbol{U}}\cdot\hat{\boldsymbol{s}}) + 2\hat{\boldsymbol{j}}\cdot\hat{\boldsymbol{s}} + 2B\frac{\partial}{\partial\tilde{s}}\left(\frac{1}{\tilde{R}}\right), \tag{11}$$

where the "tilded" variables are dimensionless versions of the same quantities without the tilde,

$$A = \frac{\mu_2 - \mu_1}{\mu_2 + \mu_1} \tag{12}$$

is an "Atwood ratio" of viscosities, and B is a nondimensional version of the surface tension coefficient. It is clear from what has been said so far that this problem can be solved by tracking a vortex sheet with the one additional complication of continually updating the sheet strength. As mentioned above, we have performed numerical simulations that follow the evolution to very large amplitudes by using the vortex-in-cell method [19–22] to calculate $\boldsymbol{U}$ from γ. In vortex sheet calculations the vortex particles are frequently not redistributed along the sheet, but for interface tracking that is absolutely essential. Hence, to convert a vortex-in-cell code to an interface code for piecewise potential flows it is also necessary to include algorithms for particle redistribution. These can be kept quite primitive when using the vortex-in-cell scheme [7].

In fig. 5 we show stages in the evolution of an initially flat interface perturbed by a combination of very small amplitude waves with arbitrarily selected wavenumbers. The calculation in fig. 5 was performed using the method mentioned above, and discussed in more detail in our paper [7], with 8192 particles on a 256^2 grid. It is included here mainly to illustrate the degree of convolution that our code can simulate, and the number of competing interfacial structures that can be accomodated. Note, in particular, the physically plausible results obtained for the very thin necks that connect droplets at the ends of fingers to the main finger

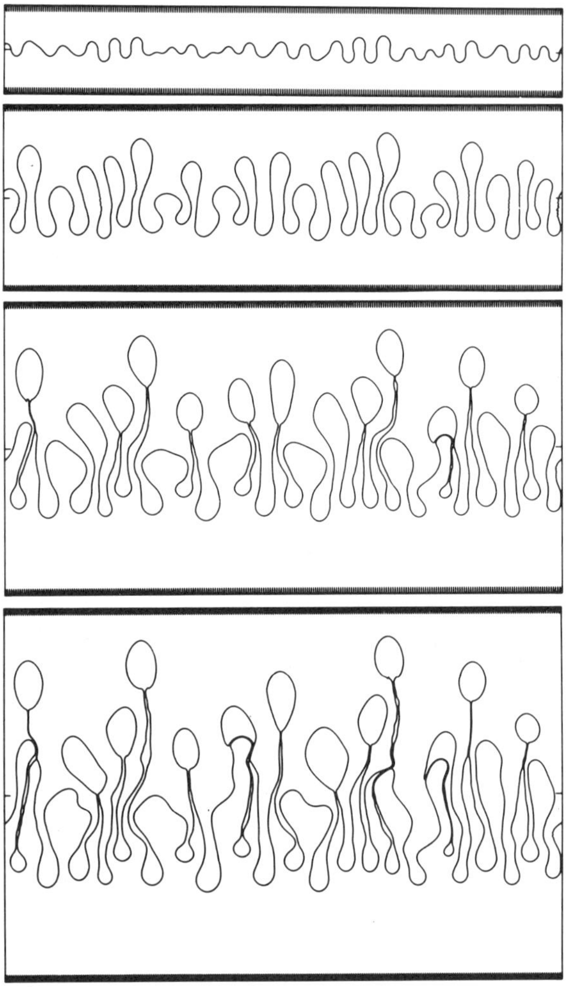

Fig. 5. Stages in the evolution of sharply stratified Hele–Shaw flow initially perturbed by a collection of small amplitude waves of arbitrarily selected wave lengths. Parameters are $A = 0.75$, $B = 1.25 \times 10^{-5}$. Time increases from top to bottom. Only a band centered about the interface is shown. The full computational domain is square. For this calculation 8192 vortex elements on a 256^2 grid were used.

itself. Note also the physically consistent representation of the double interfaces, which arise when two fingers collide.

As discussed above, the global dynamics of this type of interface depends on the two dimensionless control parameters A and B. For the run in fig. 5, $A = 0.75$ and $B = 1.25 \times 10^{-5}$. (Results of similar runs with $A = 0$, $A = 0.5$ and $A = 1$ appear in ref. 7.) Increasing B corresponds to increasing the surface tension (with other parameters held fixed) and leads to a reduction in the ratio of largest to smallest unstable scale size on the interface. Thus the most interesting regime arises when B is kept as small as possible. On the other hand, as B tends to zero, the wavelength of the shortest unstable wave (on a flat interface) also tends to zero [17, 23]. Hence, B must be kept large enough for this wavelength to exceed a grid spacing. With B fixed at (or close to) its physically acceptable lower limit, the main parameter is A, eq. (12), which excercises most of the control over the interface evolution. This is not the place to repeat all our conclusions on the role of parameter A, for which the reader should consult ref. 7, however it is essential to recapitulate one or two main points.

It is obvious from eq. (11) that $A = 0$ corresponds to a limiting case of the problem for which the self-induced velocity of the vortex sheet interface does not "feed back" into the vortex sheet strength. In this limit the interface will remain up-down symmetric on average. At the opposite extreme is the case $A = 1$ (where, taken literally, the Hele–Shaw equations break down, since one fluid now has zero viscosity). In this case, with which the celebrated experiments and theory of Saffman and Taylor [17] are concerned, there is a powerful "competition" between fingers and as the flow evolves some large fingers emerge at the expense of smaller ones. The interface is now completely asymmetrical with the less viscous fluid penetrating much more readily into the more viscous one than conversely. These simple trends, which are in fact consistent with experimental results [17, 24], have interesting consequences for the evolution of a system with several layers of fluid.

We have recently begun considering the case of three layers with the middle layer being of small thickness compared to the other two. Let the fluids be labelled 1, 2 and 3 from "top" to "bottom" with densities ρ_1, ρ_2 and ρ_3 respectively. Three cases may be distinguished: I. $\rho_1 > \rho_3 > \rho_2$; II. $\rho_1 > \rho_2 > \rho_3$; III. $\rho_2 > \rho_1 > \rho_3$. In cases I and III one interface is linearly unstable (if taken by itself) and one is stable. In case II both layers are unstable. These cases can now be combined in an almost unlimited number of ways with values of A at the two interfaces. So far we have restricted ourselves to having $A = 0$ or $A = \pm 1$ at the interfaces and to investigate all the physically allowed combinations. There has recently been a renewed interest in stratified flows of this type stimulated in part by a linearized stability analysis for layered fluids in the Rayleigh–Taylor problem [25]. Stratified Hele–Shaw flow is similar to Rayleigh–Taylor flow at infinitesimal amplitudes where A, in fact, does not enter. At finite amplitudes, however, the parameter A takes over and analogy with the Rayleigh–Taylor problem ceases.

Fig. 6 shows two examples. The top and bottom interfaces are perturbed by different, arbitrary collections of waves but the perturbations in figs. 6a and 6b are the same on each interface. In fig. 6a the density stratification is also the same at each interface, i.e. $\rho_1 - \rho_2 = \rho_2 - \rho_3$, and $A = 0$ at both. In fig. 6b, however, $A = 0$ at the upper interface only. At the lower interface $A = 1$. The densities are such that $\rho_1 - \rho_2 = 2(\rho_2 - \rho_3)$ which ensures that the coefficient of $\hat{\boldsymbol{\jmath}} \cdot \hat{\boldsymbol{s}}$ in eq. (9) has the same numerical value (unity) on each interface.

We notice in both cases in fig. 6 that at the tips of fingers very steep gradients of density ("density shocks") are produced. For a thin layer, such as we are considering, there is clearly considerable correlation between the fingers that occur on the two interfaces. However, in fig. 6b we see that the top interface (where $A = 0$) generates additional structures on the section where the $A = 1$ fingers are not growing rapidly. This result suggests that by interjecting a layer of fluid between two given fluids one can modulate the "net" interface distortion. Such mechanical modulation of fingering instabilities may be technologically important.

Another view of the three layer results comes from asking to what degree the middle fluid is mixed. This question is similar to the one posed in section 2. Again the middle region is being mapped in an area-preserving way (by incompressibility) as the flow evolves and the degree of stretching of the

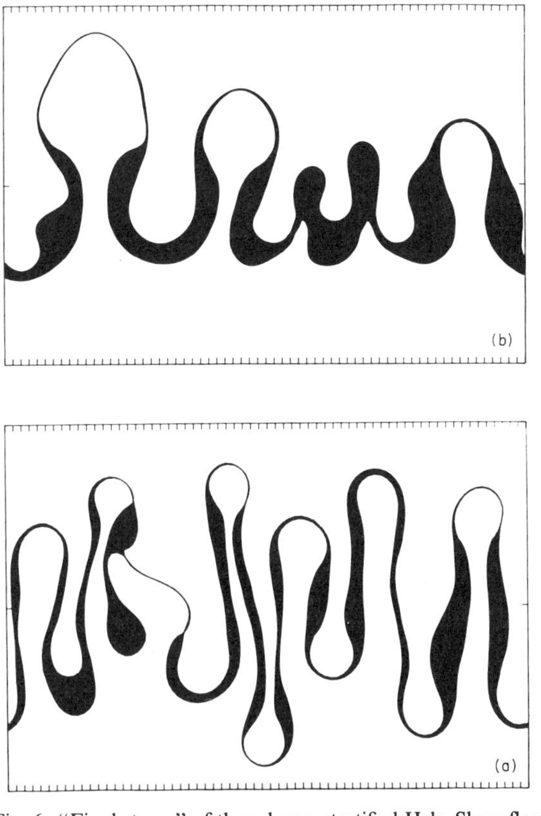

Fig. 6. "Final stages" of three-layer, stratified Hele–Shaw flow with sharp interfaces. Both interfaces are gravitationally unstable (and have the same non-dimensional surface tension). In (a) $A = 0$ at both interfaces. In (b) $A = 0(1)$ at the upper (lower) interface. The same initial perturbations were applied to the interfaces in (a) and (b).

interfaces becomes a key quantity. Fig. 7 displays data of this type and in this sense is the analog of fig. 4 (note: the total interface length, appropriately normalized, is plotted versus the maximum width of the middle layer). Again we see that by varying the control parameter A (at the two interfaces) different degrees of stretching can be obtained. However, the changes are not nearly as dramatic as those in fig. 4.

We conclude this section with the remark that although Hele–Shaw interfaces are "active" for any value of A some are formally more active than others. In particular, when $A = 0$ the integral equation (11) reduces to a simple algebraic formula for the sheet strength. A similar reduction occurs

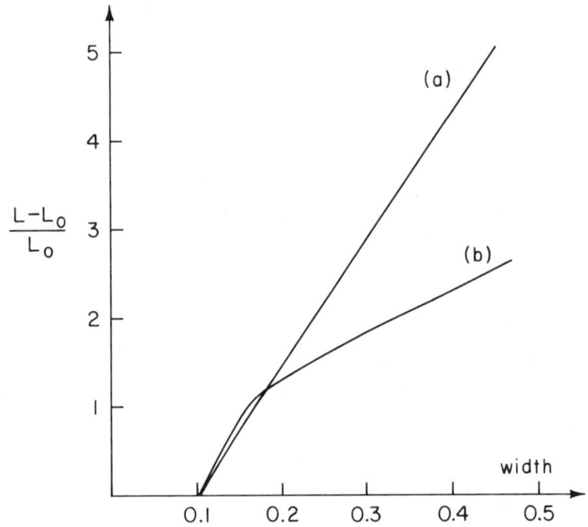

Fig. 7. The total interface length L, normalized by its initial value L_0 as indicated, was plotted vs. the maximum width of the middle layer as the calculations in fig. 6 proceeded. This gives an indication of the degree of stretching similar to the diagnostic in fig. 4. Labels (a) and (b) refer to fig. 6.

in the Rayleigh–Taylor problem for weak density stratification [26]. In both these cases the vortex sheet strength depends on the geometry of the interface. Similarly for an ordinary vortex sheet in two-dimensional hydrodynamics, the sheet strength depends on the degree of stretching. These three examples, then, may formally correspond to a weak-coupling or "minimally active" limit of the two-dimensional interface problem.

4. Concluding remarks

We have considered two examples of interface motion that in a sense lie at opposite extremes of the subject: On one hand the self-propelling "active" interfaces between immiscible fluids in creeping flow through a very narrow space (section 3), on the other, the deformations of a completely inert, "passive", line of marker particles, which can evolve very differently according to the nature of the underlying flow (section 2).

As we have remarked, "active" interfaces seem to know few bounds when it comes to producing shapes. Indeed, in his well known treatise D'Arcy Thompson [27] has analyzed a large variety of shapes that occur in nature, and has pointed out again and again examples from dynamically evolving (primarily hydromechanical) systems that lead to similar shapes, albeit just for an instant during their evolution. For example, he finds analogies between the shapes of a hydroid polyp and phases of a splash, or between the configuration of a falling drop of ink in water and various medusoids, or between the soap film suspended from a polyhedral frame and the skeletons of certain radiolaria. His list of analogies goes on and on. To the extent that these dynamical problems are concerned just with interfaces, the interfaces are definitely "active" in our terminology. Thus, if we accept Thompson's [27] conclusions, we may say that the problem of shapes of active interfaces is about as rich as the morphology of all biological organisms! This conference may have contributed to the possibility of a rough classification but we feel that with regard to active interfaces there will be open questions for a very long time to come. (A more up-to-date discussion of mechanistic models for pattern formation in biological systems may be found in ref. 28.)

Another way of stating this somewhat pessimistic conclusion is to say that if there are general results about classes of active interfaces, these still remain to be discovered. In our discussion we have adopted one classification, viz. the terminology of "active" and "passive", but there are clearly other descriptive terms that deserve to be introduced. For example, one may focus on whether the dynamics of an interface is local or non-local [29]. The motion of a passive interface would be classified as local. The dynamics of unstable Hele–Shaw interfaces, on the other hand, would never be local, since the velocity is given by the non-local Birkhoff integral (10). Note, however, that when $A = 0$ the formula for the vortex sheet strength, eq. (9), becomes local. Ben-Jacob [30] discusses a local dynamics model for the evolution in dendritic growth. The work of Mandelbrot [31] on fractals contains numerous models of what we would now classify as active, local evolution, although the physical processes that might lead to these models are frequently obscure. It seems to us that local dynamics leads more easily to fractal structures than does non-local dynamics. Indeed, we expect a passively advected interface as in fig. 3 to become fractal as the whorls and tendrils evolve, a statement that presumably can be made quantitative on the basis of the general correspondence between chaos in dynamical systems and fractal sets. On the other hand, the highly convoluted interface in fig. 5 is probably chaotic (in the sense of exponential growth of small differences in initial configuration) but it does not seem to have the hierarchy of scales necessary for a fractal. We conjecture that for the Hele–Shaw problem the non-local dynamics of the interface biases the resulting shape away from a fractal.

For passive interfaces the outlook may be more optimistic. We have already argued that such interfaces can at worst display the singularities of the underlying flow. And, although the signal produced may be more or less complicated, these singularities are in turn constrained by the dimensionality and topology of the physical space in which the flow occurs. Hence, in two dimensions we arrived at whorls and tendrils (section 2) and that, basically, exhausted the possibilities. The word "whorls" brings to mind the poem that Richardson transcribed from Swift about the hierarchy of eddies that makes up a turbulent flow [32]. This self-similar, physical space hierarchy leads to exponents in the more conventional Fourier space power laws, of which Kolmogoroff's [33] is the best known, that describe the spectral properties of homogeneous, isotropic turbulence. There are similar spectral laws for the advection of a passive scalar by turbulence [33]. Now, in the interpretation of section 2 the advection by a deterministic flow could become "stochastic" by tuning parameters in such a way that the advection equations become non-integrable. Furthermore, it is known that in the non-integrable regime a chaotic Hamiltonian system, such as advection by a time-dependent, incompressible flow in two dimensions (!), produces an essentially infinite hierarchy of "eddies" of varying scale sizes [14]. Recently, it has been shown that certain aspects of this hierarchy are associated with scaling properties [34]. The obvious conjectures, then, are (1) that the spectral properties of an advected scalar, even for very simple flows, may be governed by power laws, (2) that the exponents of these powers may be related to the hierarchy of fixed points for chaotic Hamiltonian systems, and, thus, may in some sense be "universal", and, possibly, (3) that there is a connection between these power laws and the earlier, phenomenological results for turbulent advection. This seems to be a very exciting possibility that is worth pursuing*.

Acknowledgments

Both authors are indebted to the Center for Nonlinear Studies at Los Alamos National Laboratory for its hospitality. The research reported on here was supported by National Science Foundation grant MEA81-16910 to Brown University. The allocation of computer resources by the Scientific Computing Division at the National Center for Atmospheric Research is gratefully acknowledged. NCAR is sponsored by the NSF.

*The ideas and conjectures presented here were in large measure stimulated by discussions with U. Frisch.

References

[1] M. van Dyke, An Album of Fluid Motion (Parabolic Press, Stanford, California, 1982).
[2] H.C. Yuen, "Nonlinear dynamics of interfacial waves". Physica 12D (1984) 71 (these proceedings).
[3] J.W. Shaner, "Pattern formation by shock processes", Physica 12D (1984) 154 (these proceedings).
[4] B.J. Hoskins, Ann. Rev. Fluid Mech. 14 (1982) 131.
[5] J.E. Simpson, Ann. Rev. Fluid Mech. 14 (1982) 213.
[6] D.H. Sharp, "Overview of Rayleigh-Taylor instability", Physica 12D (1984) 3 (these proceedings).
[7] G. Tryggvason and H. Aref, "Numerical experiments on Hele-Shaw flow with a sharp interface", J. Fluid Mech. 136 (1983) 1.
[8] J.S. Langer, Rev. Mod. Phys. 52 (1980) 1.
[9] R.F. Sekerka, "Morphological and hydrodynamic instabilities during phase transformations", Physica 12D (1984) 212 (these proceedings).
[10] H. Lamb, Hydrodynamics (Dover, New York, 1945), art. 9.
[11] M. Hénon, C. R. Acad. Sc. Paris Ser. A 262 (1966) 312.
[12] H. Aref, "Stirring by chaotic advection", J. Fluid Mech., in press.
[13] M.V. Berry, N.L. Balazs, M. Tabor and A. Voros, Ann. Phys. 122 (1979) 26.
[14] A.J. Lichtenberg and M.A. Lieberman, Regular and Stochastic Motion (Springer, New York, 1983), sec. 5.2b.
[15] G. Birkhoff, Los Alamos Sci. Lab. Rep. LA-1862, LA-1927 (1954).
[16] G.R. Baker, D.I. Meiron and S.A. Orszag, Phys. Fluids 23 (1980) 1485.
[17] P.G. Saffman and G.I. Taylor, Proc. Roy. Soc. London Ser. A 245 (1958) 312.
[18] G. de Josselin de Jong, J. Geophys. Res. 65 (1960) 3739.
[19] J.P. Christiansen, J. Comp. Phys. 13 (1973) 363.
[20] J.P. Christiansen and N.J. Zabusky, J. Fluid Mech. 61 (1973) 219.
[21] H. Aref and E.D. Siggia, J. Fluid Mech. 100 (1980) 705.
[22] H. Aref and E.D. Siggia, J. Fluid Mech. 109 (1981) 435.
[23] R.L. Chuoke, P. van Meurs and C. van der Poel, J. Petrol. Tech. 11 (1959) 64.
[24] R.A. Wooding, J. Fluid Mech. 39 (1969) 477.
[25] K.O. Mikaelian, Phys. Rev. Lett. 48 (1982) 1365.
[26] J.C.S. Meng and J. A. L. Thomson, J. Fluid Mech. 84 (1978) 433.
[27] D.W. Thompson, On Growth and Form, J.T. Bonner, ed. (Cambridge Univ. Press, London, 1961).
[28] G.F. Oster, "The Mechanochemistry of Cytogels", Physica 12D (1984) 333 (these proceedings).
[29] R.C. Brower, D.A. Kessler, J. Koplik and H. Levine, "Simple Models of Interface Growth", Physica 12D (1984) 241 (these proceedings).
[30] E. Ben-Jacob, N. Goldenfeld, J.S. Langer and G. Schön, "String models of interfacial pattern formation", Physica 12D (1984) 245 (these proceedings).
[31] B.B. Mandelbrot, Fractals, Form, Chance, and Dimension (Freeman, San Francisco, 1977).
[32] L.F. Richardson, Weather Prediction by Numerical Process (Cambridge Univ. Press, Cambridge 1922) p. 66.
[33] H. Tennekes and J.L. Lumley, A First Course in Turbulence (MIT Press, Cambridge, MA, 1972), Chap. 8.
[34] L.P. Kadanoff, Phys. Rev. Lett. 47 (1981) 1641.

Physica 12D (1984) 71–82
North-Holland, Amsterdam

NONLINEAR DYNAMICS OF INTERFACIAL WAVES

Henry C. YUEN
Fluid Mechanics Department, TRW Space & Technology Group, One Space Park, Redondo Beach, CA 90278, USA

Results on the stability of a steady two-dimensional interfacial gravity wave of finite amplitude to infinitesimal three-dimensional disturbances in the presence of a current jump are presented. It is found that the waves are unstable, and the instability is predominantly two-dimensional for small amplitudes, but becomes predominantly three-dimensional for larger amplitudes. In addition, for sufficiently large amplitude, a new long wave instability develops due to subharmonic distortions of the eigenfunctions of the Kelvin–Helmholtz short wave instability.

1. Introduction

In this paper, we consider gravity waves on the interface between two infinite fluids which are assumed to be incompressible, inviscid and irrotational. Specifically, we report results on the stability of a steady, two-dimensional, interfacial gravity wave of finite amplitude to infinitesimal three-dimensional disturbances. These steady waves have been calculated by Saffman and Yuen [7] and Saffman [5] for cases in which there is a jump in horizontal velocities between the two fluids.

When the heavier fluid is beneath the lighter fluid and the two fluids are at rest relative to each other, the flat interface is stable and supports gravity waves. However, in the presence of a current jump, the flat interface becomes unstable to disturbances of sufficiently short wavelengths, which is the well-known Kelvin–Helmholtz instability. This instability can be suppressed by the inclusion of surface tension (Lamb [1]). However, in this work, we shall merely confine ourselves to long-wave phenomena, and rely on the fact that surface tension will suppress short wave instabilities, so that we can concentrate on long-wave instabilities without the actual inclusion of surface tension.

A special case of interfacial waves is the water waves, for which the upper fluid is assumed to have zero density. There has been significant progress in the understanding of nonlinear water wave dynamics in the past two decades. Among the most important results are those by McLean et al. [4] (see also McLean [2, 3]) on the stability of finite amplitude waves to two- and three-dimensional perturbations, and the experimental observation of these instabilities by Su [8].

This paper presents results which are extensions of those by McLean et al. [4] to the case of interfacial waves with current jump. In order to illustrate the approach and to introduce the concept, the results of McLean et al. are summarized first.

2. Governing equations

The method of analysis follows that used by McLean et al. [4]. In a frame of reference moving with phase speed C of the steady wave, the unperturbed interface can be expressed as

$$Z(x)=\sum_{n=1}^{\infty} A_n \cos nx. \tag{1}$$

The scale is chosen so that the wavelength of the undisturbed wave is 2π. We impose a three-dimensional infinitesimal disturbance η of the form

$$\eta=\mathrm{e}^{-\mathrm{i}\sigma t}\mathrm{e}^{\mathrm{i}(px+qy)}\sum_{-\infty}^{\infty} a_n \mathrm{e}^{\mathrm{i}nx}. \tag{2}$$

0167-2789/84/$03.00 © Elsevier Science Publishers B.V.
(North-Holland Physics Publishing Division)

It is a consequence of Floquet theory that the normal modes of the disturbed system are of this form where p and q are arbitrary real numbers. The equations of motion will determine σ as a function of p and q and the amplitude of the undisturbed wave. σ is a complex number. The real part of σ gives the speed of the disturbance relative to the undisturbed wave, and the imaginary part of σ gives the growth rate of an unstable mode. Imaginary part of σ being zero indicates stability. We shall now describe the procedure to calculate σ as a function of p, q and the amplitude of the unperturbed wave as given by (1).

The velocity potentials of the perturbed motion below and above the interface are given respectively by

$$\begin{aligned} &\Phi(x,y)+\phi(x,y,z,t) \quad \text{Above,}\\ &\Phi'(x,y)+\phi'(x,y,z,t) \quad \text{Below.}\end{aligned} \tag{3}$$

We shall denote by U the magnitude of the current jump, and by C the phase speed of the unperturbed wave relative to the fluid at great depths. Without loss of generality, we shall take $g=1$, the density of the lower fluid also to be unity, and denote the density of the upper fluid by R $(R<1)$. The height of the undisturbed wave (i.e., vertical distance between crest and trough) is denoted by H.

We assume that the disturbances are small compared with the unperturbed quantities:

$$|\eta| \ll Z, \quad |\phi| \ll |\Phi| \tag{4}$$

so that the governing equations can be linearized about the unperturbed state. We then obtain

$$\begin{aligned} &\nabla^2\phi' = 0, \quad \text{Above,}\\ &\nabla^2\phi = 0, \quad \text{Below.}\end{aligned} \tag{5}$$

In the fluids above and below the interface, with the dynamical condition

$$\begin{aligned}&\phi_t + \eta + \nabla\Phi\cdot\nabla\phi + (\nabla\Phi\cdot\nabla\Phi_z)\eta\\ &\quad = R\{\phi'_t + \eta + \nabla\Phi'\cdot\nabla\phi' + (\nabla\Phi'\cdot\nabla\Phi'_z)\eta\}\end{aligned} \tag{6}$$

and the kinematic conditions

$$\begin{aligned}&\nabla\Phi'\cdot\nabla\eta + \nabla Z\cdot\nabla\phi'\\ &\quad + (\nabla\Phi'_z\cdot Z_x - \Phi'_{zz})\eta - \phi'_z = 0,\end{aligned} \tag{7}$$

$$\begin{aligned}&\nabla\Phi\cdot\nabla\eta + \nabla Z\cdot\nabla\phi\\ &\quad + (\nabla\Phi_z\cdot Z_x - \Phi_{zz})\eta - \phi_z = 0,\end{aligned} \tag{8}$$

to be applied at the unperturbed surface $z = Z(x)$. In addition, the disturbances are assumed to vanish far from the interface, so that we have

$$\phi' \simeq 0, \quad z \to \infty, \tag{9}$$

$$\phi \simeq 0, \quad z \to -\infty. \tag{10}$$

Consistent with the assumed form (2) of the perturbation to the surface, the disturbed velocity potentials take the forms

$$\begin{aligned}\phi' &= e^{-i\sigma t} e^{i(px+qy)}\\ &\quad \times \sum_{-\infty}^{\infty} b'_n e^{inx} \exp\left\{-\sqrt{(p+n)^2+q^2}\,(z-z_t)\right\},\end{aligned} \tag{11}$$

$$\begin{aligned}\phi &= e^{-i\sigma t} e^{i(px+qy)}\\ &\quad \times \sum_{-\infty}^{\infty} b_n e^{inx} \exp\left\{\sqrt{(p+n)^2+q^2}\,(z-z_c)\right\},\end{aligned} \tag{12}$$

where z_c is the height of the crest and z_t is the height of the trough. Substitution of the expressions (2), (11) and (12) into the governing equations (6), (7) and (8) gives the following system of equations for σ, a_n, b_n and b'_n:

$$\begin{aligned}&\left[(1+\Phi_x\Phi_{xz}+\Phi_z\Phi_{zz}) - R(1+\Phi'_x\Phi'_{xz}+\Phi'_z\Phi'_{zz})\right]\\ &\qquad \times \sum_{-\infty}^{\infty} a_n e^{inx}\\ &\quad + \sum\left\{i(p+n)\Phi_x + \left[(p+n)^2+q^2\right]^{1/2}\Phi_z\right\}\\ &\quad \times b_n e^{inx} \exp\left\{\left[(p+n)^2+q^2\right]^{1/2}\left[Z(x)-z_c\right]\right\}\\ &\quad - R\sum\left\{i(p+n)\Phi'_x - \left[(p+n)^2+q^2\right]^{1/2}\Phi'_z\right\} b'_n e^{inx}\\ &\qquad \times \exp - \left\{\left[(p+n)^2+q^2\right]^{1/2}\left[Z(x)-z_t\right]\right\}\\ &= i\sigma\left\{\sum b_n e^{inx} \exp\left\{\left[(p+n)^2+q^2\right]^{1/2}\left[Z(x)-z_c\right]\right\}\right.\\ &\quad \left. + \sum b'_n e^{inx} \exp - \left\{\left[(p+n)^2+q^2\right]^{1/2}\left[Z(x)-z_c\right]\right\}\right\},\end{aligned} \tag{13}$$

$$\sum_{-\infty}^{\infty} [\Phi_{xz} Z_x - \Phi_{zz} + \mathrm{i}(p+n)\Phi_x] a_n \mathrm{e}^{\mathrm{i}nx}$$
$$+ \sum_{-\infty}^{\infty} \left\{ \mathrm{i}(p+n) Z_x - \left[(p+n)^2 + q^2 \right]^{1/2} \right\}$$
$$\times b_n \mathrm{e}^{\mathrm{i}nx} \exp\left\{ \left[(p+n)^2 + q^2 \right]^{1/2} [Z(x) - z_c] \right\}$$
$$= \mathrm{i}\sigma \sum_{-\infty}^{\infty} a_n \mathrm{e}^{\mathrm{i}nx}, \tag{14}$$

$$\sum_{-\infty}^{\infty} [\Phi'_{xz} Z_x - \Phi'_{zz} + \mathrm{i}(p+n)\Phi'_x] a_n \mathrm{e}^{\mathrm{i}nx}$$
$$+ \sum_{-\infty}^{\infty} \left\{ \mathrm{i}(p+n) Z_x + \left[(p+n)^2 + q^2 \right]^{1/2} \right\}$$
$$\times b'_n \mathrm{e}^{\mathrm{i}nx} \exp - \left\{ \left[(p+n)^2 + q^2 \right]^{1/2} [Z(x) - z_t] \right\}$$
$$= \mathrm{i}\sigma \sum_{-\infty}^{\infty} a_n \mathrm{e}^{\mathrm{i}nx}. \tag{15}$$

This can be interpreted as an eigenvalue problem for the normal modes of the disturbance with σ as the eigenvalue and a_n, b_n and b'_n as the eigenfunctions. From the structure of the Euler equations, which are time reversible (since there is no dissipation) and real, we know that whenever σ is an eigenvalue, so are $-\sigma$, σ^* and $-\sigma^*$. Stable eigenvalues appear in pairs, whereas unstable eigenvalues appear in quartets. Consequently, the transition from stability to instability is marked by the coalescence of pairs of eigenvalues which can also be interpreted as a resonance between two modes which have the same frequency in the frame of reference moving with the unperturbed wave.

It is important to note that there is a degeneracy in the dependence of the eigenmodes on p, since the value of p is arbitrary to the addition of an integer as can be seen in the representations (2), (11) and (12). This means that when p is changed to $p + n$, where n is an integer, we obtain the same eigenmode provided that a_n, b_n and b'_n are changed to a_{n-p}, b_{n-p} and b'_{n-p}, accordingly. In principle, there would be no loss of generality in confining ourselves to the range $0 < p < 1$, but it turns out to be convenient not to do this and allow for some arbitrariness in the labeling of the eigenmodes.

3. The wave of zero height

For the limiting case where the undisturbed wave has zero wave height (the flat surface), the solutions of the eigenvalue problem give

$$\sigma_{\pm} = \frac{kUR}{1+R} \pm \sqrt{\frac{(1-R^2)k - Rk^2U^2}{(1+R)^2}}$$
$$-(p+n)C, \tag{16}$$
$$k = \left[(p+n)^2 + q^2 \right]^{1/2} \tag{17}$$

and the eigenvectors are such that only those components with suffices $p + n$ are nonzero. Since the flat surface here is regarded as the limit of steady waves with the wave height approaching zero, the eigenvalue contains the wave speed C. It follows from (16) that the flat surface is unstable if

$$U^2 > \frac{(1-R^2)}{Rk}, \quad \text{for given } k, \tag{18}$$

or

$$k > \frac{(1-R^2)}{U^2R}, \quad \text{for given } U, \tag{19}$$

which is the classical Kelvin–Helmholtz instability result. Note that (19) confines the Kelvin–Helmholtz instability to sufficiently short waves for a given value of U. We are concerned with properties of waves and disturbances with wavelengths longer than this critical value and are hence Kelvin–Helmholtz stable.

As pointed out by McLean et al. [4], instability modes of finite amplitude waves can be identified by resonance among normal modes associated with the flat surface. This resonance will occur whenever there exist p, q and integers n and n' such that

$$\sigma_+(p+n, q) = \sigma_-(p+n', q). \tag{20}$$

Making use of the degeneracy associated with p, we can separate the resonance conditions into two classes

$$n = m, n' = -m \quad \text{Class I}, \tag{21}$$
$$n = m, n' = -(m+1) \quad \text{Class II}. \tag{22}$$

The results for interfacial waves with current are rather complicated, and it will be helpful to consider first a special case.

4. Water waves: a special case

An important and interesting special case is water waves. In this case, the density ratio R is assumed to be zero, and the current jump U drops out of the equations. The wave height H then becomes the only parameter describing the unperturbed wave. This case was studied and reported in detail by McLean et al. [4] and McLean [2]. The results are summarized here since they permit a clearer insight into the effects of finite amplitude and provide a framework for the discussion of the more complicated effects when density ratio and current strength are considered.

The condition (20) gives the two classes of resonance

$$[(p+m)^2+q^2]^{1/4} + [(p-m)^2+q^2]^{1/4} = 2m \quad \text{Class I,} \tag{23}$$

$$[(p+m)^2+q^2]^{1/4} + [(p-m-1)^2+q^2]^{1/4} = 2m+1 \quad \text{Class II,} \tag{24}$$

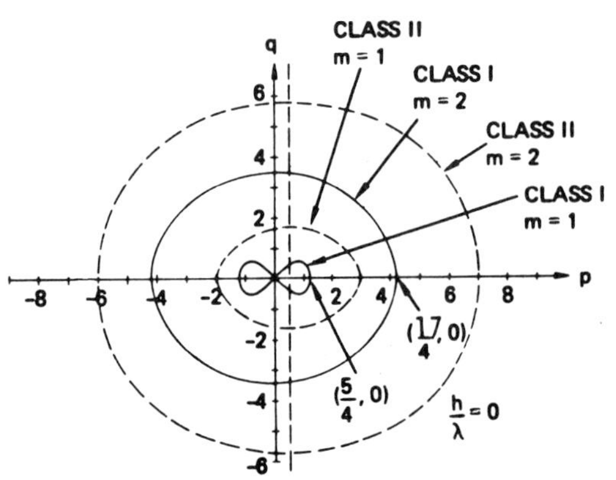

Fig. 1. Resonance diagram for water waves ($R=0, U=0, H=2\pi h/\lambda=0$) in p, q plane. Solid lines correspond to class I resonance (eq. (23)). Dashed lines correspond to class II resonance (eq. (24)).

where m is a positive integer. A plot of the two sets of resonance curves for $m=1$ and $m=2$ is given in fig. 1. These curves describe the loci of perturbation wave-vectors which have two pairs of eigenvalues that coincide in the limit $H=0$. The effect of finite H is to merge these pairs of real eigenvalues into a quartet of complex eigenvalues representing instability. The result is a band of instability developing from the resonance curves, with the width of the band proportional to some power of H. The stability diagrams for the two classes of instabilities with $m=1$ are shown in fig. 2 for different values of H, up to nearly the limiting value of $H=0.892$. The location and growth rate of the most unstable disturbance in each class are noted. It can be seen that for values of $H<0.55$, the class I instability dominates, and the strongest instability is two dimensional ($q=0$). For larger values of H, the class II instability overtakes the class I, with the most unstable mode being three dimensional. The most unstable class II modes always have $p=0.5$, which represents modulations in which every other wave crest is enhanced. The value of q for the most unstable class II mode decreases with increasing H but remains nonzero even for the highest value of H calculated, indicating that the instability is three dimensional. Another interesting property of the class II waves is that along the line $p=0.5$, the unstable disturbances (including the most unstable disturbances) have zero real parts, so that they copropagate with the unperturbed wave. It has been pointed out by Saffman and Yuen [6] that this copropagating instability can lead to the bifurcation of the uniform wave train to a steady three-dimensional modulated wave train, a phenomenon which has been observed by Su [8].

5. Results for interfacial waves with current jump

In this general case, the undisturbed system is described by three parameters, namely, H, U and R. For a given density ratio, it proves convenient to scale U against U_{cr}, which is defined as the value of the current at which the unperturbed

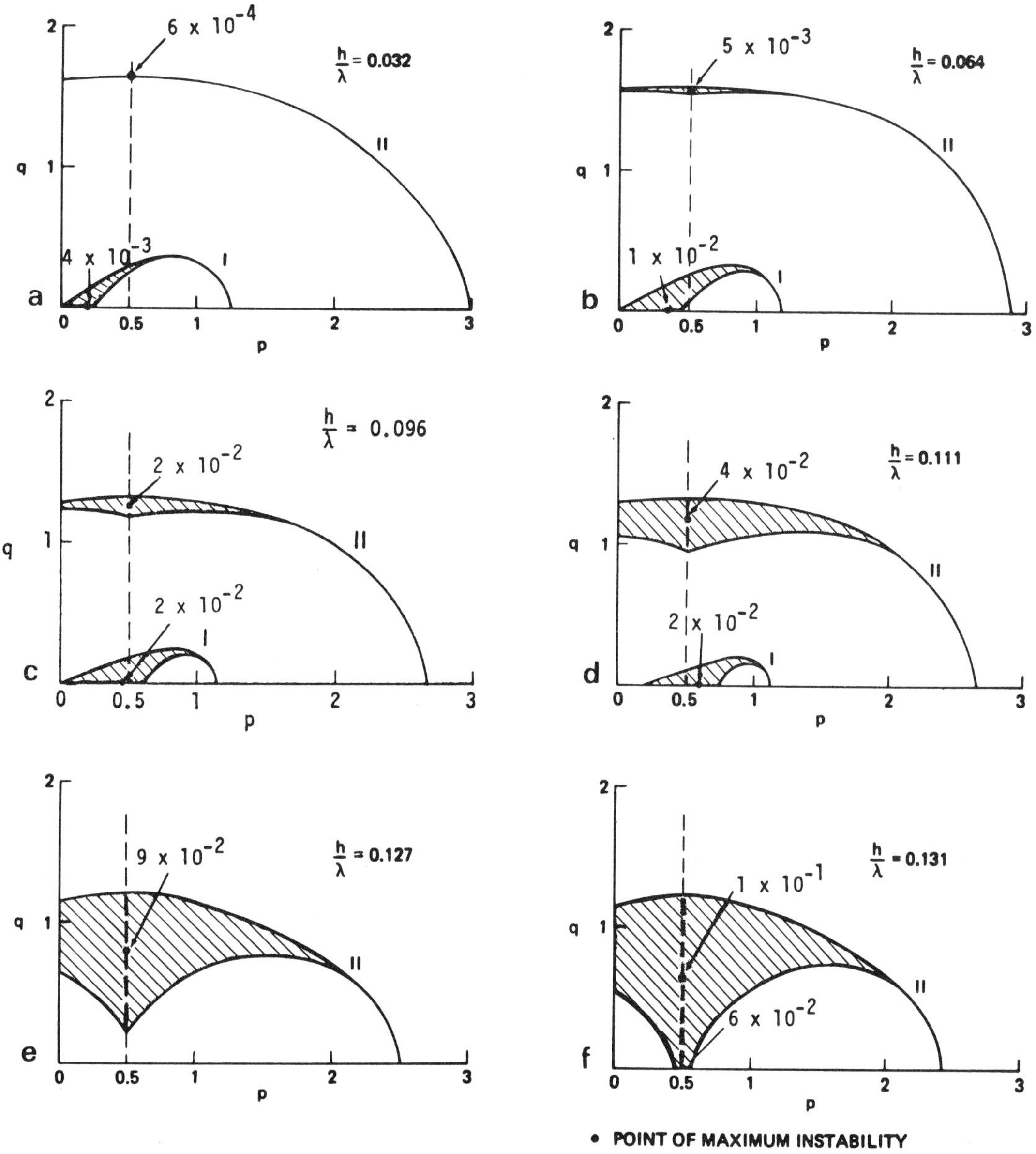

Fig. 2. Instability diagram for water waves ($R=0, U=0$) in p, q plane for various values of $H=2\pi h/\lambda$. The location and magnitude of the maximum instability for each class of instability are indicated. (a) $H=0.2$; (b) $H=0.4$; (c) $H=0.6$; (d) $H=0.7$; (e) $H=0.8$; (f) $H=0.823$.

wave of wavelength 2π in the zero amplitude limit is Kelvin–Helmholtz unstable,

$$U_{\mathrm{cr}}=\left[\frac{(1-R)(1+R)}{R}\right]^{1/2}. \tag{25}$$

We define $U'=U/U_{\mathrm{cr}}$.

When U' is zero, the resonance curves are not altered from the water wave case since the effect of density ratio is merely to scale the value of gravity. In fig. 3, we show the resonance curves for $U'=0.0$, 0.5 and 0.9, for class I and class II with $m=1$. It can be seen that the $U'=0.0$ curves are identical with those in the water wave case. When U' is nonzero, a Kelvin–Helmholtz instability boundary exists, and disturbances lying in the unstable regime are themselves Kelvin–Helmholtz unstable in

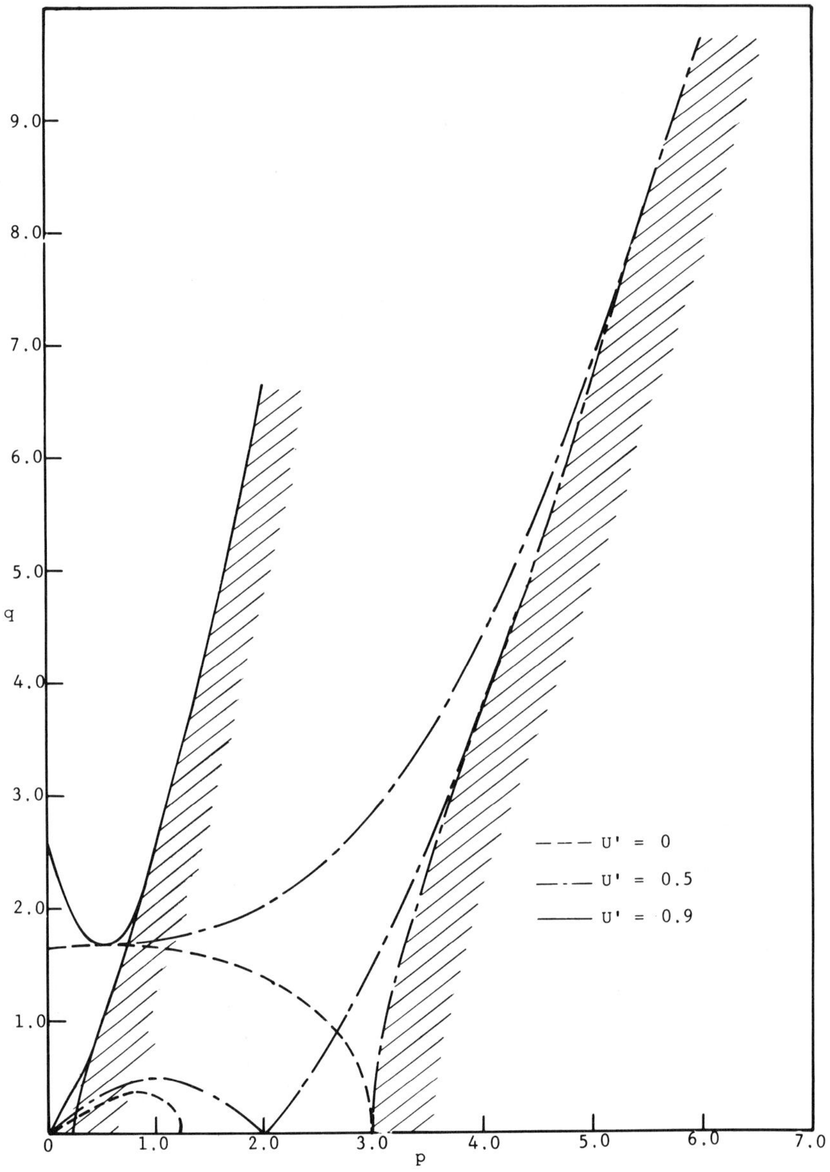

Fig. 3. Resonance diagram for interfacial waves with a current jump. Dashed lines correspond to $U' = 0$ case. Dash–dot lines correspond to $U' = 0.5$ case. Solid lines correspond to $U' = 0.9$ case. Shaded regions indicate the boundaries of Kelvin–Helmholtz instability, which is present only for nonzero U'. Note that the $U' = 0$ case is similar to that for water waves.

the limit when the undisturbed waves have zero amplitude. The class I resonance takes on a new branch which merges into the Kelvin–Helmholtz boundary. The class II resonance curve, instead of closing on itself as in the $U' = 0.0$ case, also turns up and merges into the Kelvin–Helmholtz boundary. For values of U' greater than 0.5, the two branches of the class I resonance curve become one and merge into the Kelvin–Helmholtz boundary as seen in the $U' = 0.9$ example shown. Based on the water wave results, we expect bands of instability to develop along these resonance curves when the unperturbed wave has finite amplitude.

Table I
Finite amplitude cases calculated. Results for each case are shown in figures as indicated

R	H	U'	Figure
0.1	0.2	0.0	4a
		0.5	4b
		0.9	4c
0.9	0.2	0.0	4d
		0.5	4e
		0.9	4f
0.9	0.5	0.5	4g
		0.9	4h

We present the results of a number of selected cases listed in table I. At the present time, cases with larger values of H have not been calculated. This justifies the use of the Fourier series representation (1) for the unperturbed wave. As discussed in Saffman [5], large amplitude overhanging waves exist when H is greater than values around one. For such cases, other formulations of the stability analysis would be required. For the values of H in table I, it was found that the undisturbed waves can be adequately represented by 20 Fourier modes. We calculated numerically the eigenvalues for finite values of H by truncating the infinite

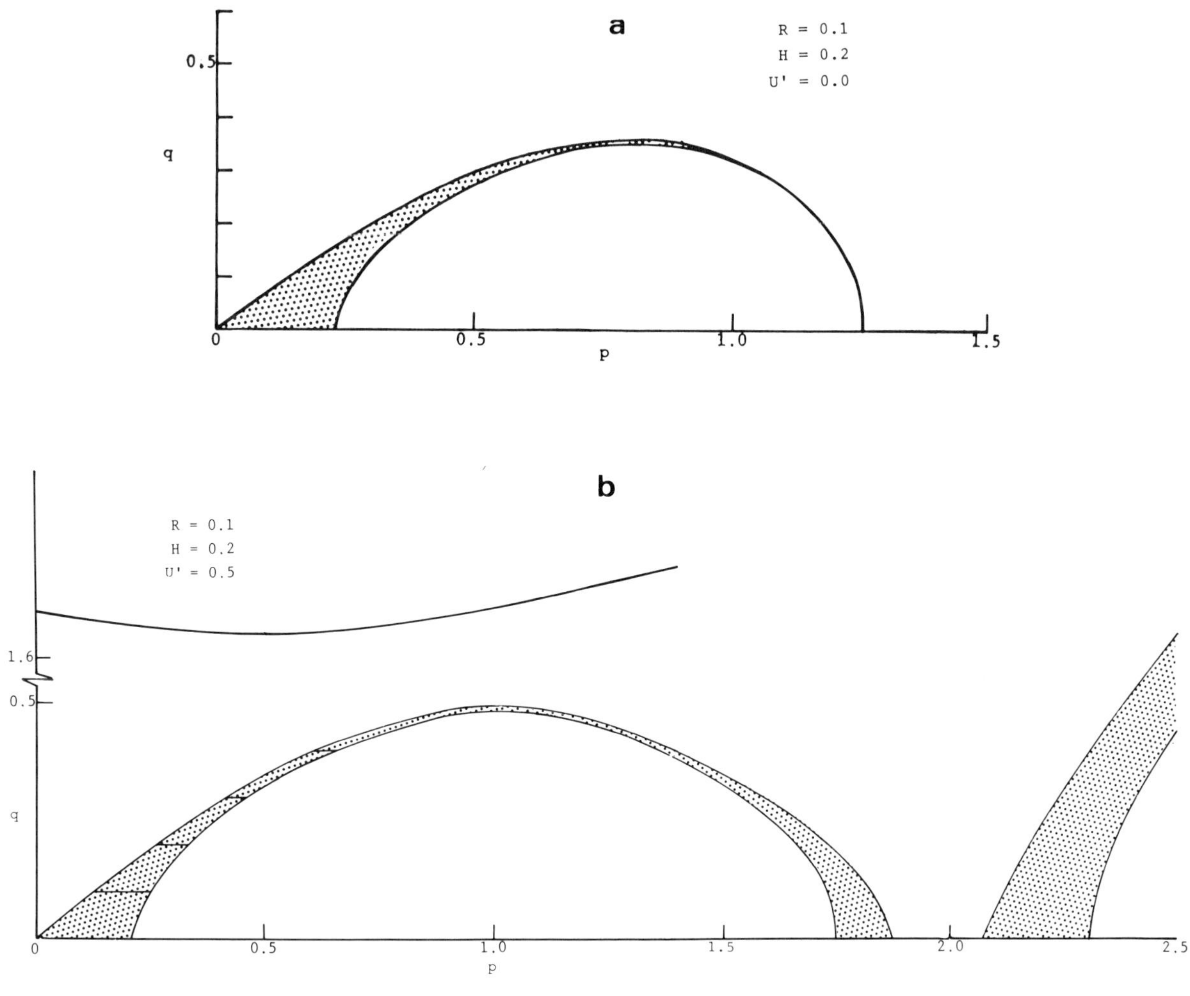

Fig. 4a, b. Instability diagrams for interfacial waves with a current jump. Different values of R, H and U' correspond to cases indicated in table I.

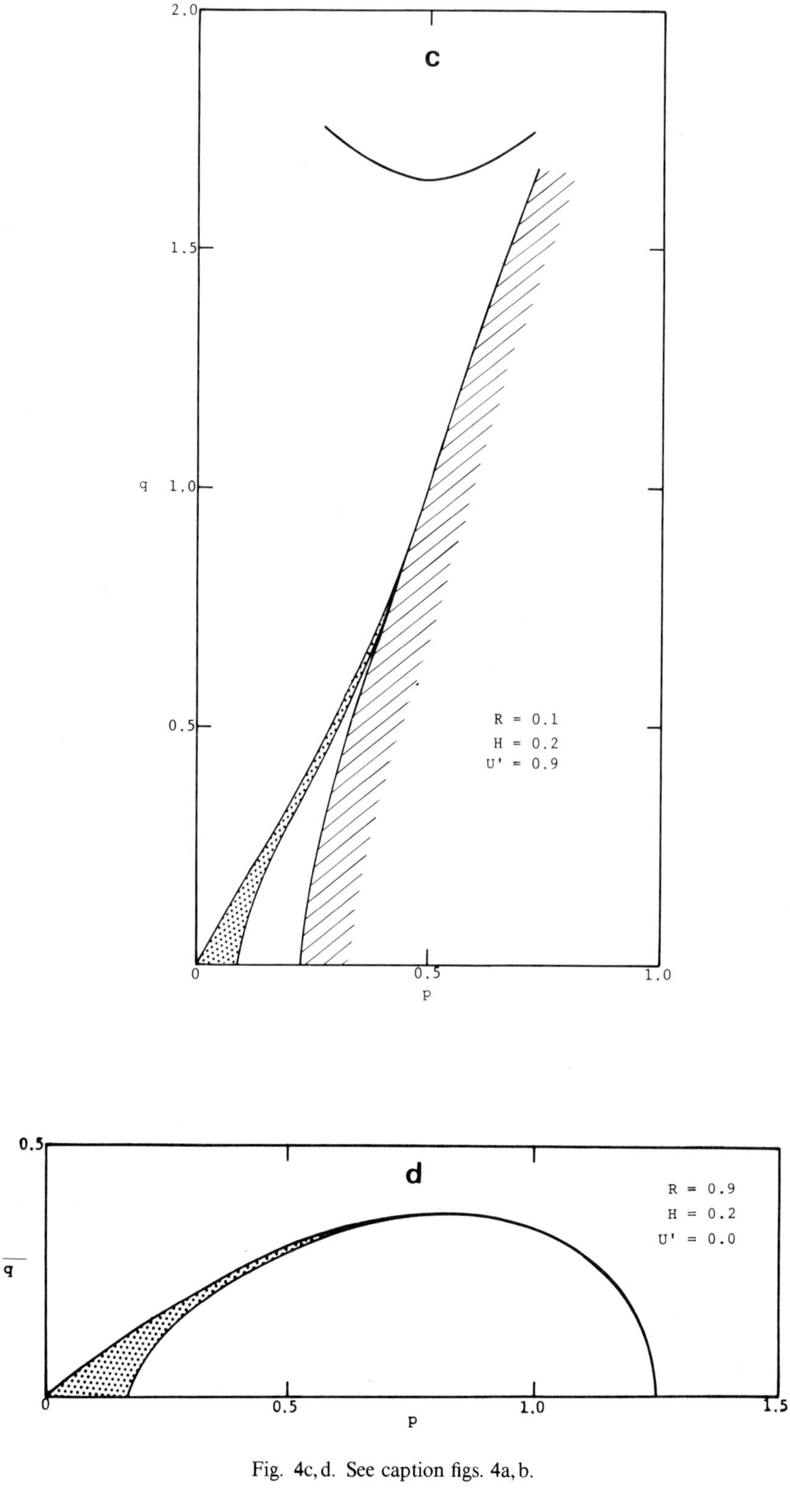

Fig. 4c, d. See caption figs. 4a, b.

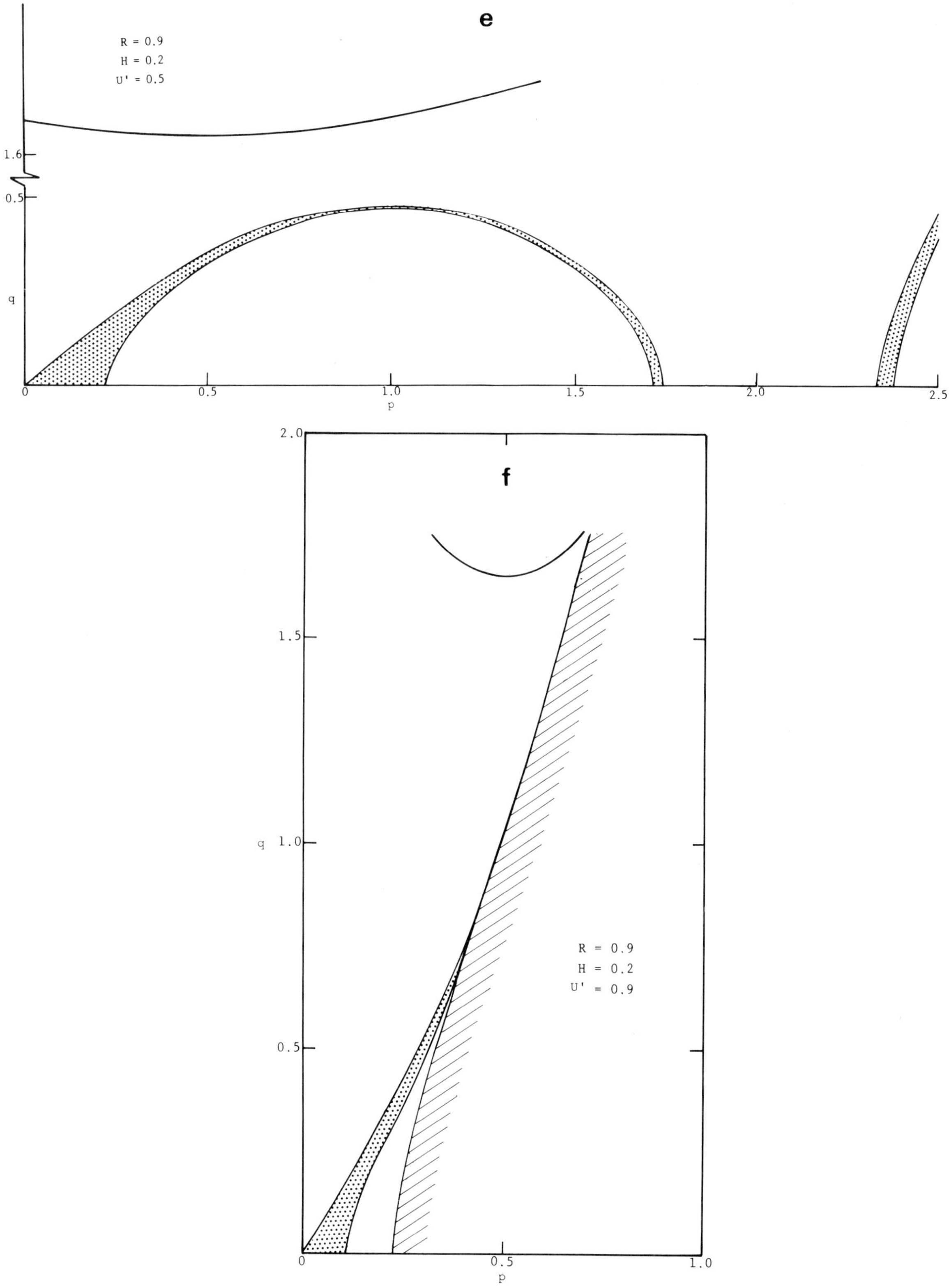

Fig. 4e, f. See caption figs. 4a, b.

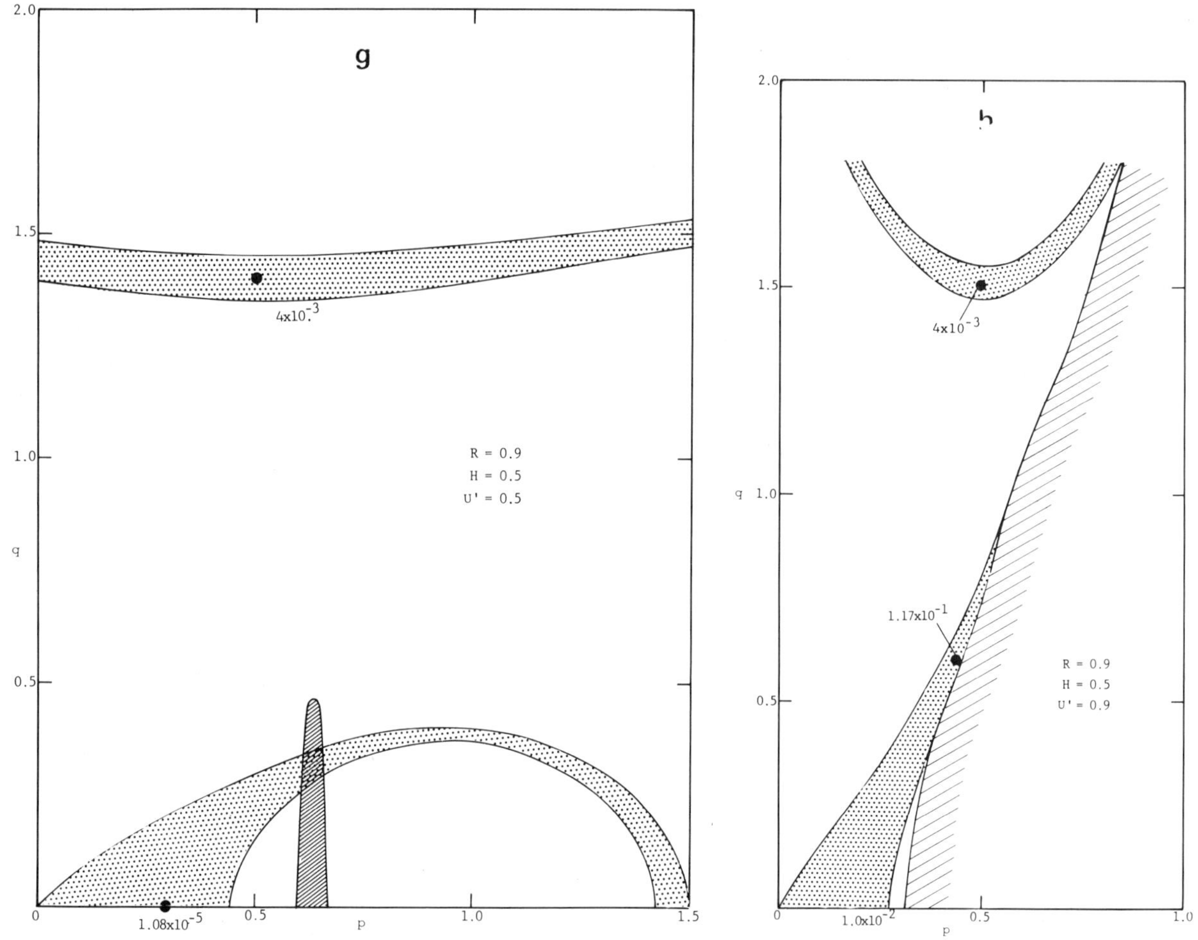

Fig. 4g, h. See caption figs. 4a, b.

system (13)–(15) to 5 modes and using library routines for the complex generalized eigenvalue problem. The calculations were carried out on a PRIME 750 minicomputer. Accuracy of the calculations was spot checked by repeating with a larger number of modes, and it was determined that the accuracy of the stability boundary results was within plotting resolution. It is expected that large values of H will require considerably more modes for the description of the unperturbed wave and the perturbation.

The stability diagrams for the selected cases are shown in fig. 4.

It can be seen that the cases where $U' = 0$ (figs. 4a and 4d) closely resemble each other and the water wave results. The cases with $H = 0.2$ and $U' = 0.5$ (figs. 4b and 4e) show that the two branches of the class I resonance curves develop into two distinct regions of instability, but that the general trend is consistent with the expectation that instability bands lie near the resonance curves. The cases with $H = 0.2$ and $U' = 0.9$ (figs. 4c and 4f) show the merging of the class I instability into the Kelvin–Helmholtz instability. For this value of H, the class II instability has a very narrow bandwidth.

The large amplitude cases (figs. 4g and 4h) show significantly broadened instability bands for class II, and a strong increase in the maximum growth rate for the class II instability, even though it is still smaller than that for the class I. Fig. 4h shows that the maximum growth rate associated with the class I instability is no longer two dimensional as in the case of $U' = 0$ (and the water wave case).

An interesting phenomenon is encountered in the results shown in fig. 4g. It is seen that a band of instability, not identified with any resonance curves and absent for smaller values of H, appears around $p = 0.7$. Detailed examination of this instability reveals that it is associated with the Kelvin–Helmholtz instability in the following way. We recall that the value of p is arbitrary to the addition of an integer. In the results shown in fig. 4, the value of p was fixed by the requirement that the Fourier component of the perturbation with largest magnitude has wavenumber $p + 1$. In other words, we identify a disturbance by its dominant component. The effect of finite amplitude is usually to increase the relative magnitude of the other components. It can then happen that at some wave height a different component becomes dominant and "nonlinear mode jumping" occurs. This is what has happened in fig. 4g. The effect of finite amplitude on the eigenfunctions of the Kelvin–Helmholtz unstable waves is to introduce a low wavenumber (or long-wave) component into their structures, and the appearance of the apparently new longer wave instability is when the low wavenumber component becomes dominant. The appearance of this type of instability is abrupt as one increases H. To our knowledge, it is the first reported example of a nonlinear mode jumping phenomenon caused by finite amplitude effects.

6. Discussion

Apart from the mode jumping phenomenon just referred to, the results more or less conform to expectations and resemble the phenomena found in surface waves. Mode resonance found for the flat surface translates into narrow bands of instability, which become wider and are associated with larger growth rates, as the unperturbed wave height increases. In addition to the Kelvin–Helmholtz instability, which is present for short waves in the presence of current, there are long wave instabilities of finite amplitude waves. We have neglected surface tension in our discussion. Since the stabilizing effect of surface tension is more effective the shorter the wave, it is expected that its presence will make the long wave finite amplitude instabilities more important, especially when the Kelvin–Helmholtz is suppressed.

It should be noted that the present results are all for C_+ waves. It is expected that similar results will hold for C_- waves, but this remains to be confirmed. The cases considered are also all for undisturbed waves with positive energy. It was mentioned in the discussion of steady waves of permanent form [5] that such waves exist with negative energy (relative to that of the flat surface). It remains to investigate these solutions, but there is not reason to believe that their stability properties are in any way significantly different. Indeed, it should be noted that there is not necessarily any significance to negative energy, because energy is not Galilean invariant. In the case of waves, it is dangerous to compare the energies of states which do not have the same momentum, and calculation indicates that the interfacial waves of the type considered here always have nonzero momentum compared to the flat surface case when the relative energies are negative. There is clearly a pressing need for good experiments on interfacial waves in the presence of current in a stably stratified situation.

Acknowledgement

This work was done in collaboration with Professor Philip G. Saffman and was partially supported by the National Science Foundation Grant OCE8100517 and the Internal Research and Development Funds of TRW. The results have been previously presented in the University of Wisconsin Mathematics Research Center Symposium on "Waves on Fluid Interfaces" and its proceedings (edited by R.E. Meyer).

References

[1] Sir Horace Lamb, Hydrodynamics (Dover, London, 1932).
[2] J.W. McLean, Instabilities of finite-amplitude water waves, J. Fluid Mech. 114 (1982) 315.
[3] J.W. McLean, Instabilities of finite-amplitude gravity waves on water of finite depth, J. Fluid Mech. 114 (1982) 331.
[4] J.W. McLean, Y.C. Ma, D.U. Martin, P.G. Saffman and H.C. Yuen, Three-dimensional instability of finite amplitude water waves, Phys. Rev. Letts. 46 (1981) 817.
[5] P.G. Saffman, Finite amplitude interfacial waves, in: Waves on Fluid Interfaces, Meyer, ed. (Academic Press, New York, 1983), p. 1.
[6] P.G. Saffman and H.C. Yuen, A new type of three-dimensional deep-water wave of permanent form, J. Fluid Mech. 101 (1981) 797.
[7] P.G. Saffman and H.C. Yuen, Finite amplitude interfacial waves in presence of a current, J. Fluid Mech. 123 (1982) 459.
[8] M.Y. Su, Three-dimensional deep-water waves. Part I. Experimental measurements of skew and symmetric wave patterns, J. Fluid Mech. 124 (1982) 73.

Physica 12D (1984) 83–106
North-Holland, Amsterdam

1.2. GASES

SHOCKS, INTERFACES, AND PATTERNS IN SUPERSONIC JETS

Larry L. SMARR†*
Departments of Astronomy and Physics, University of Illinois, Urbana, Illinois 61801, USA

and

Michael L. NORMAN and Karl-Heinz A. WINKLER
Max-Planck-Institut für Physik und Astrophysik, Institut für Astrophysik, 8046 Garching bei München, Fed. Rep. Germany

Supersonic gaseous jets exhibit coherent nonlinear dynamic structures. We use a supercomputer to obtain solutions to the equations of two-dimensional inviscid hydrodynamics, representing both axisymmetric and planar jets boring their way through a uniform medium. We use color images to display these high resolution computations. Several basic morphologies of interfaces and shock structures arise which we discuss in simple terms. The global structure of the two-parameter solution space is analyzed. A number of the basic structures, which appear in the supercomputer simulations, seem to occur in both terrestrial and astrophysical supersonic jets.

1. Introduction

Supersonic gaseous jets have been studied experimentally and theoretically for a century. Their scales vary from centimeters in the laboratory to millions of light years in radio galaxies. We briefly review their observed properties in section 2. A two-dimensional Eulerian hydrodynamics code is used to solve the dynamical evolution of a supersonic jet as it bores through a pressure confining medium. The numerical program is outlined in section 3. We explore in some detail the morphological patterns which occur in the axisymmetric case of matched jets, i.e. those in which the inlet pressure of the jet equals the exterior pressure of the confining medium. Color computer images are used to exhibit the physical properties of our solutions. In section 4, the existence of a cocoon of gas returning from the advancing head of the supersonic beam is shown to depend dramatically on the ratio of the beam density to the density of the surrounding medium: light hot beams have cocoons, while dense cold beams are naked. The Mach number of the beam, the other dimensionless parameter of our study, determines how mixed the cocoon becomes. Our numerical survey of solution space covers a factor of 1000 in density ratio and in beam luminosity. Shock systems are found to be ubiquitous in these jets. In section 5 we describe the various types of shocks which appear in the flow and show how they arise from unstable perturbations. We find that the nature of the shock systems is intimately coupled to the patterns of cocoon and beam discussed in the previous section. The stability of the interfaces in axisymmetric supersonic jets is examined in section 6. Again, the dynamics are dominated by nonlinear waves whose development depends on the density ratio and the Mach number. In section 7, we show how the shocks, interfaces, and patterns differ in slab jets. Finally, we conclude with some speculations about how fully three-dimensional jets will resemble our two limits of axisymmetry and planar symmetry. Throughout

†Visitor, Max-Planck-Institut für Astrophysik.
*Alfred P. Sloan Fellow.

0167-2789/84/$03.00 © Elsevier Science Publishers B.V.
(North-Holland Physics Publishing Division)

our study, our focus is on using the computer simulations to reveal the basic nonlinear structures which occur in the solutions of a fairly simple system of laws of physics. These nonlinear structures appear similar to many of the observed structures of terrestrial supersonic jets and are very suggestive as explanations of the knots and hot spots seen in astrophysical jets.

2. Real supersonic jets

The laboratory study of supersonic gaseous jets was begun a century ago by E. Mach and Salacher (1889) [1]. This work was extended by L. Mach, Emden, and Prandtl. What they discovered is illustrated in fig. 1, taken from the textbook on gas dynamics by Prandtl (1952) [2]. Here gas is shot through a small nozzle into the air. The nozzle pressure can be greater than (underexpanded jet), equal to (matched jet), or less than (overexpanded jet) the atmospheric pressure. The nozzle opening can be a long thin rectangle (slab jet) or a circle

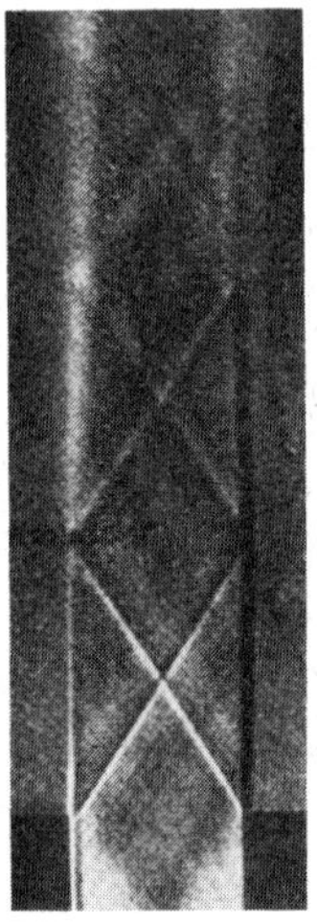

Fig. 1. A Schlieren photograph of the supersonic air flow emerging from a narrow slit. The long axis of the slit is into the page. The pressure in the inlet has been matched to the outside ambient pressure. There are a series of three crisscrossing shock waves standing in the flow. Notice that the jet/atmosphere interface is almost straight, even though there are large variations in the internal pressure caused by the shock waves. (From Prandtl, p. 285, 1952.)

(axial jet). In any of these cases, a nearly periodic array of shock waves are found to be embedded in the flow. Fig. 1 is a matched slab jet photographed by Schlieren techniques, which rely on the rarefactions and compressions in the gas to modulate the illuminating light. If the Mach number is high enough, the intersecting shocks form a transverse Mach stem (in slabs) or Mach disk (in axial jets). For examples, see Landenburg et al. (1949) [3].

In aerodynamics, the existence of these periodic internal shocks has been known since the beginnings of supersonic flight. Fig. 2 shows the supersonic exhaust plume from the Bell X-1 rocket plane, the first manned aircraft to break the sound barrier (taken from Anderson [4]). One sees the long axisymmetric jet with periodic bright and dark spots. The bright spots can be caused by either secondary recombustion of fuel in the high temperature regions following the internal shocks or by chemiluminescence [5].

In our galaxy, a growing number of astrophysical objects of roughly solar mass are being found to produce jets. Examples are: protostars (e.g. Mundt and Fried [6]), old binaries in supernova remnants (e.g. SS433, Margon [7]) and in planetary nebulae (e.g. R Aqr, Tapia et al. [8]). A number of these jets are knotty, see fig. 1 of Norman et al. [9], with the emission coming from high excitation atomic lines.

Finally, an enormous number of jets have been discovered in the last few years emerging from the centers of galaxies and quasars. Here the primary energy source is assumed to be accretion onto a supermassive black hole of 100 million solar masses. The focusing of that energy into two oppositely directed jets occurs either near the hole (see the review by Blandford [10]) or in the galactic atmosphere (Smith et al. [34]). The jets range in size from thousands to millions of light years in length, with a typical length of 50–100 jet diameters. Many of these jets exhibit knots of emission internal and external to the beams. Fig. 3a shows a well-studied example of internal knots, NGC 6251 (Perley, Bridle, and Willis [12]). In several cases, e.g. M87 and Centaurus A, these internal knots are

Fig. 2. The supersonic exhaust plume from the rocket plane Bell X-1, the first manned plane to break the sound barrier. Note the periodic bright emission knots in the exhaust beam, with spacing about two beam diameters. There are at least six of these knots, which are caused by secondary combustion of fuel as the exhaust gas moves through the embedded series of crisscross shocks. Again note that the jet diameter is essentially constant even though there are strong internal shocks. (From Anderson, 1981).

seen in all wavelengths from radio to X-ray. The *internal* knots are caused by synchrotron emission from electrons, shocked up to high energy, spiraling around a local magnetic field. The *external* knots (e.g. Miley [13]) are similar to the stellar jets in showing high excitation atomic lines.

In addition to the knotty beam, the advancing head of the jet can be seen in many extragalactic jets. The beam ends in a characteristic "hot spot", usually embedded in an amorphous lobe of emission. Again, the emission is synchrotron radiation. Fig. 3b shows a typical example (3C111 taken from an observation of Perley, see Fomalent [14]). There one can see another feature which often occurs. Although the jet is remarkably straight, the jet begins to bend toward the end, making the knots and hot spots appear slightly off the center line of the jet.

In summary, the observations of supersonic jets point to a very similar underlying hydrodynamics flow. The beams appear to preserve their initial symmetry (planar or axial) for many beam diameters downstream. The opening angles are usually small, indicating that the supersonic flow preserves collimation unless the external pressure is suddenly altered (Sanders [15]). In most cases, the external gas pressure seems to be the confining agent, although in some cases magnetic fields may be necessary. Knots of higher pressure, density, and temperature, caused by internal shocks, seem

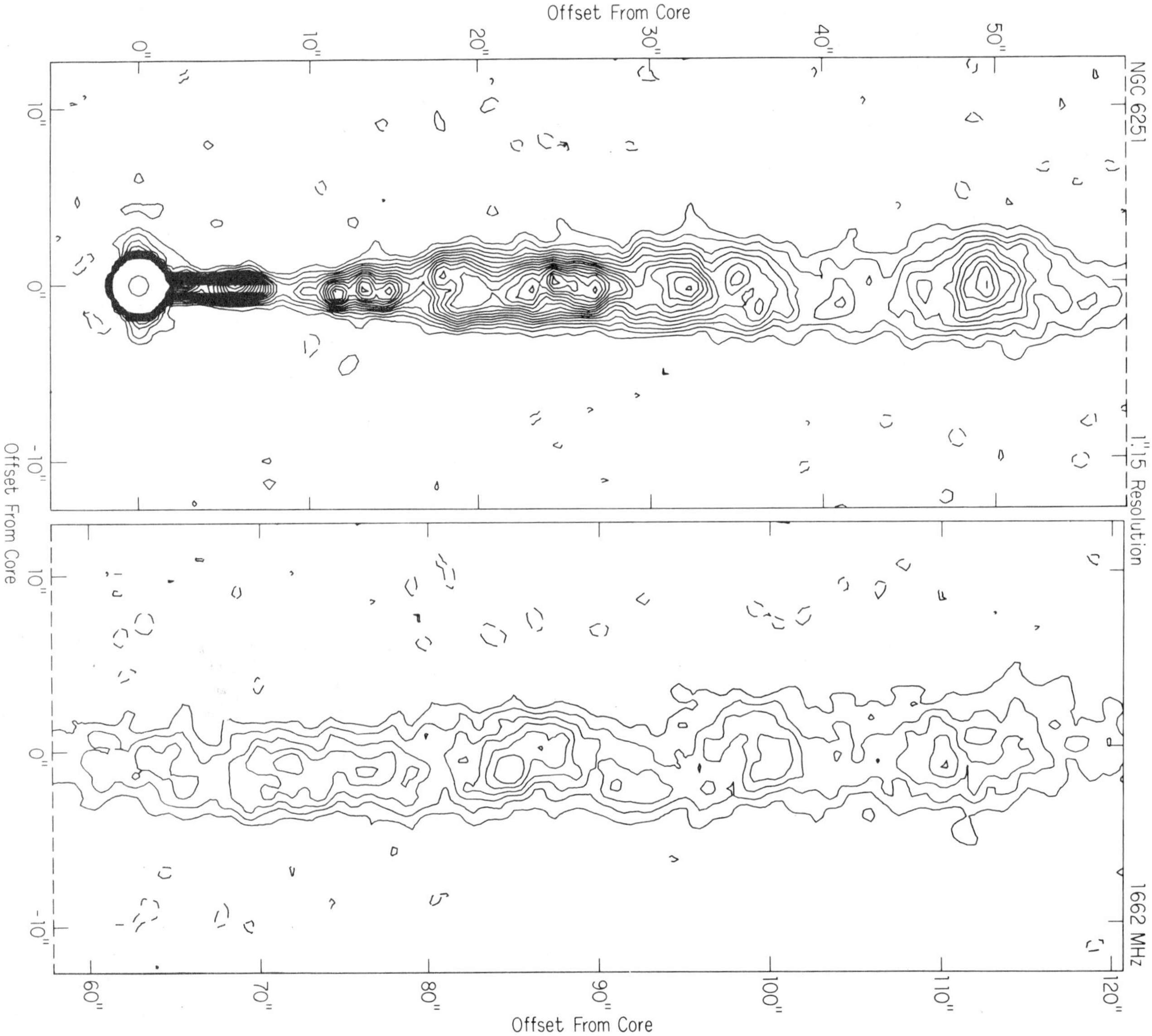

Fig. 3. Examples of astrophysical supersonic jets. Fig. 3a. A high resolution radio interferometric map made with the Very Large Array of radio telescopes in Socorro, New Mexico. This map details a portion of the inner jet of the radio source associated with elliptical galaxy NGC 6251. One can see that the emission is knotty along the beam. The circular feature at 0″ is the center of the radio galaxy. The map has been broken into two sections (0″–58″, 58″–120″). The distance scale is that 10″ = 22 light years. The piece of the jet shown is therefore 260 light years = 2×10^{18} meters in length (reprinted by permission of A. Bridle and R. Perley).

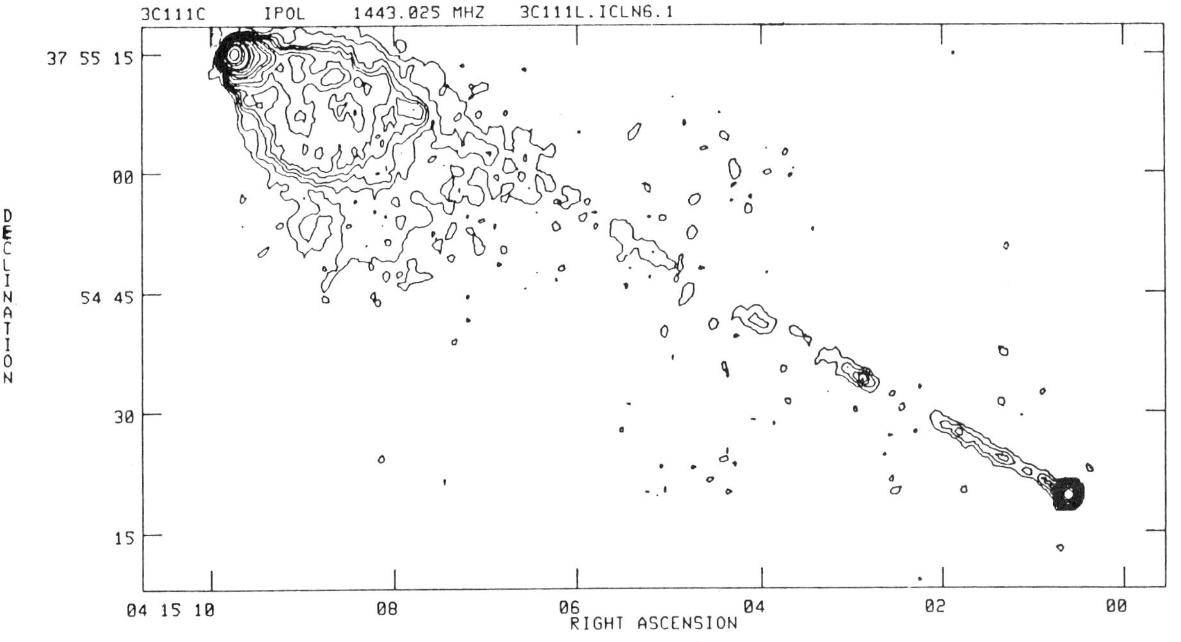

Fig. 3b. A VLA map of radio source 3C311. The contours are of total radio intensity at 1.4 GHz. Note the long knotty linear jet ending in a hot spot at the end of a lobe of amorphous emission. The circular feature at the lower right-hand of the map is the center of the radio galaxy (reprinted by permission of R. Perley).

to be ubiquitous in the supersonic beams. These internal shocks typically do not disrupt the beam, however the very strong shock at the head of the beam (the "hot spot") does terminate the flow. The head of the beam is often embedded in an amorphous lobe and indeed the beam itself is often surrounded by a "cocoon" of low surface brightness emission. Finally, the jets seem remarkably stable. The shear layers on the surface of the jets are subject to Kelvin–Helmholtz instabilities, which can grow to nonlinear amplitude. The large wavelength instabilities (pinching and kinking modes) threaten global beam stability, while small wavelength instabilities should lead to turbulent mixing. In spite of this, supersonic jets are able to propagate to large distances.

To supplement the data obtained from laboratory experiments and astrophysical observations, we have performed high resolution supercomputer simulations of the interior of the supersonic beam, its surroundings, and the entire length of the jet from the near-inlet region out to the advancing head. This provides an appropriate set of solutions to the compressible gas dynamics equations from which we can study whether observed properties of jets naturally develop as nonlinear phenomena in supersonic flow. The enormous volume of "data" produced by the supercomputer is first reduced graphically. Then simple models are devised which can explain many of the morphologies seen in the computer images. This approach thus provides a new tool to the century old research effort to understand supersonic jets.

3. Computational supersonic jets

Our use of computers to solve a basic physics problem follows the strategy foreseen many years ago by John von Neumann. This approach was recently paraphrased by the mathematician Garett Birkoff:

> ... it seems clear that von Neumann was envisioning fluid dynamics as a *mathematical science*, as had Euler, Lagrange, Stokes, Riemann, and Poincaré before him. His main point was that mathematicians had nearly exhausted *analytical* methods, which apply mainly to *linear* differential equations and *special geometries*... In short, von Neumann's proposal was that, with large scale, high speed digital computers, one could substitute *numerical* for analytical methods, tackling *nonlinear* problems in *general* geometries.
>
> SIAM Review 25 (1983) 1.

Our program is to solve the basic equations (mass continuity, Euler equation, energy equation) of inviscid compressible hydrodynamics for a well-posed physics problem and see what features of the solutions can be understood a posteriori in simple terms. This approach follows in a direct way the synergetic method, described by Zabusky [17], which was used in the discovery of the properties of solitons which arise in nonlinear dispersive equations. Even though our solutions are much more complicated, they will turn out to possess features in common with many of the solutions of the other nonlinear equations discussed in these proceedings.

There are a number of important advantages to using a computer to obtain solutions to a complicated physics problem. First, one can control precisely which laws of physics are operating. In both observation and experimentation, one is dealing with a variety of physical effects which may be difficult to sort out. An example is the difficulty in producing a truly inviscid flow in the laboratory. With the computer one can see which phenomena emerge from the inviscid hydrodynamics, then carefully add in viscosity, magnetic fields, etc. and see how that alters the solutions. Second, one has complete control over the environment of the solution. It is simple to produce gradients in the background pressure or change from atmospheric to interstellar pressures. Third, one can explore how the solution space phenomonology changes as one varies the fundamental parameters of the problem. Fourth, one can compute over a much greater dynamical range of parameters than can be done in the laboratory. To go from a dense jet to a light jet requires only the change of an input statement, not building a new laboratory apparatus. Finally, modern color computer image techniques allow one to read out in great detail any physical variable of interest. In many cases, the only observables in experiments or observations are complicated combinations of the basic physical variables.

Unfortunately, computers are still not fast enough with large enough memories to handle general 3-dimensional dynamic flows at sufficient resolution to describe jet physics. However, they are quite capable of resolving two spatial dimensions. The solutions we present here use a large fraction of the central memory of a CRAY-1 supercomputer and require between one and thirty hours of CPU time per evolution. We compute both axisymmetric and planar symmetric jets. The computations and graphics were done at the Max-Planck-Institut für Physik und Astrophysik, Institut für Astrophysik, in Garching b. Munchen, West Germany.

The code we use for solving the equations of gas dynamics is described in Norman and Winkler [18]. It is a time explicit, Eulerian, finite differenced code with interface trackers, which has been extensively calibrated on many problems. The advection is by van Leer's monotonic method. There were two sets of axisymmetric evolutions which we performed. The first set (completed in 1981) used a grid which was 30 jet radii long and 7.5 wide. The supersonic beam had 8 zones across its radius. The second set of runs (made in 1982–83) used the same resolution but extended the grid to 80 jet radii in length and used a ratioed zoning from 7.5 out to 15 jet radii in width. The planar jets were run on a Cartesian grid with 20 equidistant zones across the jet and 70 ratioed zones on either side of the jet. The grid was 300 zones long. In all cases, gas of a specified Mach number (initial velocity of beam gas in the lab frame divided by the internal sound speed of the beam gas) and density ratio (of beam gas divided by ambient density) was shot into the grid and allowed to propagate through a uniform gaseous ambient medium. In most cases the inlet pressure in the jet equaled the ambient gas pressure so that a matched jet resulted. We have assumed a perfect gas law for both beam and ambient gas, with the ratio of specific heats in each gas equal to 5/3.

The results of the first set of runs were reported in paper I (Norman et al. [19]). The propagation and morphology of the second run of jets was graphically presented (in grey tones) in paper II (Norman et al. [20]). An in-depth discussion of the internal shocks and the resulting knots was given in paper III (Norman et al. [9]). The comparison of the terminal strong shock with astrophysical hot spots was made in paper IV (Smith et al. [11]). As stated above, the code itself was documented in paper V (Norman and Winkler [18]), while the new color image algorithm was given in paper VI (Winkler [21]). This paper can be considered a summary paper in which we focus on the generic nonlinear structures we discovered in this project. It will also serve as a comprehensive documentation of the solution space of the axisymmetric jet problem.

4. Patterns in axisymmetric jets

As the supersonic gas emerges onto the grid, it drives a bow shock ahead of itself. The boundary of the jet gas with the ambient gas is a contact discontinuity. Since in general this contact will propagate more slowly than the gas in the supersonic jet beam, a shock will form behind the contact which both stops the beam gas before it hits the contact and deflects the gas out of the beam. This much of the structure of the advancing head of the jet was reasoned out in the landmark paper on astrophysical jets by Blandford and Rees [22].

In fig. 4, one can see all these features in four snapshots of the evolution of a Mach 6, density ratio 0.01 jet. The color at any point represents the intensity level of the quantity being plotted, with dark blue being the lowest value and bright red being the highest value. The order of the colors is that of the spectrum (see color bar in fig. 6). There are 73 increments in color space, each representing an equal logarithmic increment in the quantity plotted. Thus, our color bar spans the entire dynamic range of the physical quantity plotted. Since our grid has 640 grid points along the axis of symmetry, while the color image has only 512 pixels, the image slightly underresolves our solution. The physical values are those in a plane of symmetry which contains the axis of rotation. Because of the assumed axisymmetry, the calculation need only be done in the upper half plane, but we have plotted the image with both half planes.

The physical quantity represented in fig. 4 is entropy. As the jet propagates, the inner beam maintains a fairly constant entropy (yellow) which is much higher than the cold ambient gas (dark blue). The beam is terminated in a strong shock where the entropy jumps from yellow to bright red. This termination process is geometrically complex and time dependent. This dynamics is intimately related to the development of the regularly spaced surface waves, which one can see generated in fig. 4. These waves are on the interface between the return gas flow (the cocoon) and the ambient medium. These breaking waves, combined with the swirling red color in the entropy under the waves point to the existence of large scale vortices in the cocoon flow. These vortices are responsible for mixing the low entropy gas from the ambient medium into the high entropy shocked gas in the cocoon leading to the well mixed yellow and then green of the entropy further downstream. One can repeat this process of unraveling the physics of jet propagation, by studying the evolution of each of our axisymmetric jets pictured in grey scale images in paper II.

The evolutions we ran are located in the Mach number–density ratio plane as shown in fig. 5. In this paper we concentrate on four different density ratios (0.01, 0.1, 1, and 10) and three different Mach numbers (3, 6, and 12). We do not study in any detail the Mach 1.5 jets which we have also computed, because we believe that such low Mach number jets are so unstable to nonaxisymmetric modes that our axisymmetric computations are not relevant to the real world. The dynamic range of our sample is then 1000 in density ratio. Since the dimensionless kinetic energy luminosity scales as (density ratio)$^{-1/2}$ (Mach number)^3, then (see fig. 1 of paper II) a Mach 12, density 0.01 beam is 2000 times as luminous as a Mach 3, density 10 beam.

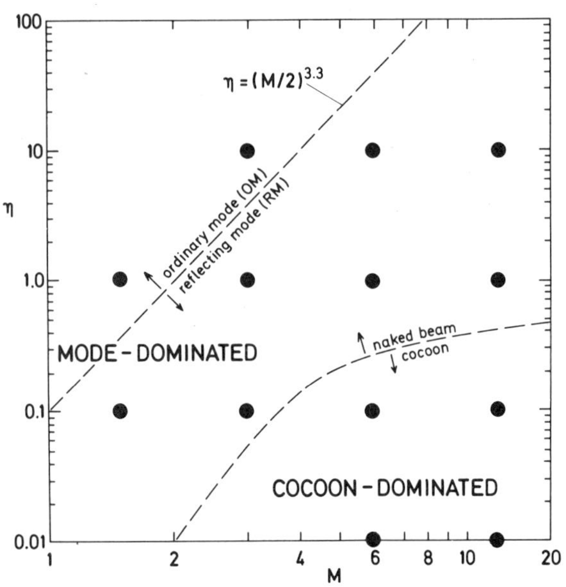

Fig. 5. The two-parameter space of pressure matched supersonic jets. The vertical axis is the log of the density ratio between the inlet gas and the ambient medium. The horizontal axis is the log of the internal Mach number of the inlet gas. The heavy black dots are the jets we have calculated. The jets have a cocoon of return flow below the curved dashed line; they have only a naked supersonic beam above it. These jets are called mode-dominated if the jet is naked and cocoon-dominated if not. All of the jets develop internal shocks. The shocks are planar and disruptive if the jet lies above the straight dashed line; they are biconical and nondisruptive if they are below the line. The equation for this dashed line comes from Cohn's 1983 study of linear perturbations of axisymmetric jets.

We present in figs. 6, 7, and 8 a picture album of the morphology of the solution space for these two parameters. The color graphics scale is as above. Each page represents jets of the same Mach number, with the densest jet (density ratio 10) at the top, and the lightest jet (density ratio 0.01) at the bottom. The images are snapshots of each evolution taken at a time (table I) in which the jet has reached a length of ~ 75 jet radii. Facing pages are the same jets with their density field plotted on one page and its pressure field plotted on the other page. The numerical values of the maximum and minimum density and pressure for each jet shown are given in table I.

The reason for showing both pressure and density is twofold. First, the density plots show very

Table I
The quantitative data for the jets shown in figs. 6, 7, and 8. The units of time, density ratio, density, and pressure are fixed by the ambient medium values. The radius of the jet inlet is taken to be 1, the sound speed in the ambient medium is 1, the time unit is the time it takes the external sound speed to cross the jet radius, the external density is 1, the external pressure is 0.6 (since the ratio of specific heat is 5/3). The Mach number is the ratio of the inlet velocity to the internal sound speed of the supersonic beam. The density ratio is the inlet beam density in units of the external medium density

Mach Number	Density Ratio	time	ρ_{min}	ρ_{max}	p_{min}	p_{max}
3	10.0	95	4×10^{-1}	18	0.13	2.0
	1.0	100	5.3×10^{-2}	1.4	0.03	1.2
	0.1	60	9.7×10^{-3}	1.7	0.06	3.2
6	10.0	60	3.1×10^{-1}	19.0	0.08	2.5
	1	34	7.6×10^{-2}	2.4	0.03	3.3
	0.1	27	6.7×10^{-3}	3.3	0.01	7.8
	0.01	26	9.6×10^{-4}	3.4	0.04	13
12	10.0	28	1.4×10^{-1}	25	0.03	6.0
	1.0	14	2.3×10^{-2}	5.7	.03	32
	0.1	10	4.0×10^{-3}	5.4	0.02	64.
	0.01	7.6	4×10^{-4}	4.0	0.04	50.

clearly the location of contact discontinuities, which are interfaces in pressure equilibrium. These discontinuities often have large jumps in density across them, but they are invisible in the pressure plots. Second, the pressure plots are much more sensitive to the strength of shocks than density plots. The jump in density across a strong planar shock saturates at four, while the jump in pressure across a shock continues to rise with the shock strength. By comparing the two plots for each jet, one can infer most of the important physics of the flow.

The most obvious pattern that one sees in these plots is that the cocoon rapidly disappears as one goes from light to dense jets. Futhermore, if one keeps the density ratio fixed, one gets more cocoon as one increases the Mach number. As discussed in detail in papers I and II, there are several effects which cause these trends. First, energy and momentum balance, applied in a frame moving with the bow shock, tells one that a light, low inertia, jet will bounce back from its collision with the dense ambient medium, thus producing a large backflow in light beam jets. A similar argument predicts that when dense beam gas is halted in the frame of the head of the jet, its inertia will keep it moving with the bow shock, thus creating no appreciable backflow for dense jets. Second, as the Mach number of the beam increases, so does the Mach number of the backflow. This raises the relative Mach number of cocoon/ambient medium interface, which stabilizes it against the disruptive Kelvin–Helmholtz instability. Finally, we are comparing our jets at fixed length, not at fixed time. For those jets with lower Mach number, not only is this interface more unstable, but there have been more eddy turnover timescales in which to disrupt the cocoon, than in the faster moving high Mach number jets.

Another pattern one sees is that the location of the pressure maximum varies considerably with the parameters. It can be either located between the bow shock and the shock which terminates the flow ("hot spot dominated" jets), or it can be in the beam itself, often appearing as a quasiperiodic set of maxima ("knot dominated" jet). The jets become more hot spot dominated as the Mach number increases, since the strength of the bow shock, and therefore the pressure jump across it, increases with the Mach number. In the following sections, we will study this set of solutions in more detail. Our aim will be to draw out general properties of such flows and explain them in simple terms.

5. Shocks in axisymmetric jets

By solving the full nonlinear dynamic equations, one can determine the global structure of shock systems in these jets. We have discovered that there are three major classes of shocks: those which are internal to the beam, those which are external to the beam, and those which terminate the beam. Since each class will have observational signatures, particularly in astrophysical jets, we will discuss the patterns which we have found in each class.

We illustrate the different shock structures by considering in more detail two of the Mach 6 jets, the density ratio 10 and 0.01 cases (top and bottom jets in fig. 7). In figs. 9a and 11a, we exhibit $\nabla \cdot V$, the scalar convergence (dark grey) and divergence (light grey) of the velocity field for the two jets at the same times as shown in fig. 7. These dark and light grey regions plotted represent the most extreme volumes of compression and rarefaction, respectively. In both the dense and light jet cases (figs. 9a, 11a), the supersonic beam has alternating regions of rarefaction and compression. These show up as low (blue/green) and high (yellow/red) pressure regions, respectively, in fig. 7. The compression zones occur across shock fronts, where the velocity suffers a discontinuity. The sharpness of the compression zones indicates the resolution of the shock waves in our code. The strength of each shock or rarefaction can be judged by looking at the color pressure plot to see what jump in pressure occurs across each feature.

5.1. *Mode dominated internal shocks*

Fig. 9a illustrates a mode dominated beam; here a regularly spaced pattern of shocks occurs in the beam. This induces a quasiperiodic string of high pressure knots along the beam as seen in fig. 7. We have explored this regime in depth in paper III. We find that these jets do not have extensive cocoons around the beam. As a result, the unstable

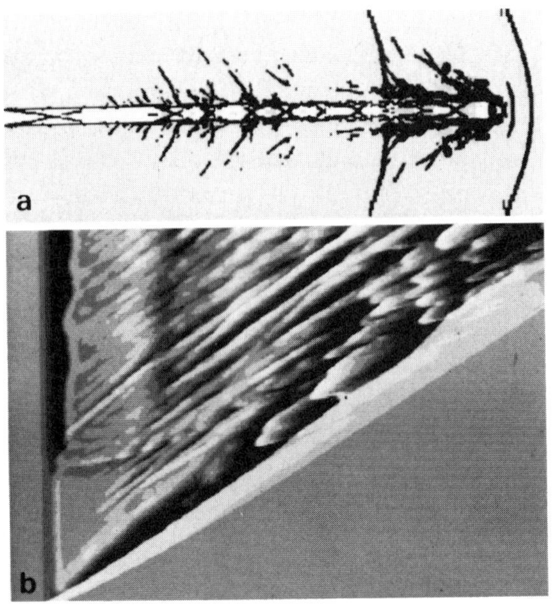

Fig. 9. Shock structure of the Mach 6, density ratio 10 jet. The top image (a) shows regions of extreme compression in black and extreme rarefaction in gray. The sharply defined compression regions delineate the shock waves in the flow. There are external shocks in ambient gas and internal shocks in the beam. One can see the characteristic "x" shape of the internal biconical shocks. The lower figure (b) is a spacetime diagram of the pressure along the symmetry axis. The horizontal scale is space and the vertical axis is time. The maximum value of pressure is white, the minimum is black, and the intermediate values are 15 shades of grey representing equal logarithmic intervals of pressure between the maximum and the minimum. The shocks are located where the tone abruptly changes from dark to light. One can see that the periodic internal shocks propagate at the same velocity as the bow shock. The shocks preserve themselves as they propagate, with new shocks appearing behind them as they move downstream.

axisymmetric modes predicted by the linear theory of the Kelvin–Helmholtz instability (Hardee [23–25]; Ferrari, Trussoni, and Zaninetti [26, 27]; Ray [28]; Cohn [29]), as applied to the naked beam/ambient medium interface, can grow freely. The linear theory can accurately estimate the growth length of the instability, i.e. how far down the beam the instability is advected before it has grown to nonlinear amplitude. This length scales as the Mach number. This is why the dense Mach 12 jets in fig. 8 without cocoons do not show knots – the beam moves so fast that the instability has not had time to become nonlinear. If the Mach 12, density ratio 10 jet is evolved further (fig. 11 of paper III), the knots appear looking much like the Mach 6, density ratio 10 jet in fig. 7.

What the linear theory cannot predict at all is the nature of the final saturated nonlinear state. As can be seen by viewing Figs. 6 and 7, the unstable surface perturbations saturate by creating shock waves across the beam. The shocks are planar if the Mach number is low enough and the density high enough (above the straight dashed line in fig. 5). Of the jets studied in this paper, only the Mach 3, density ratio 10 jet (fig. 6) is in this regime. Note that these shocks disrupt the beam, causing it to neck down severely. Below this line in fig. 5, the shocks appear as "x's" (biconical in our axisymmetric geometry) and they do not disrupt or pinch off the flow.

The dividing line between planar and biconical shocks, discovered in our numerical survey, coincides closely with the line discovered in the linear theory (Cohn, [29]) which divides the unstable jets into those in which the ordinary mode (OM) is most unstable and those in which the reflecting mode (RM) is most unstable. The difference between these two modes is that the OM is homogeneous across the beam, whereas the RM has internal nodes in the beam. We hypothesize that the homogeneous OM's lead to planar shocks and the RM's with one node lead to a biconical shock. We have begun making high resolution computations on "clean" beams to verify this hypothesis (see, e.g. fig. 9 of paper III).

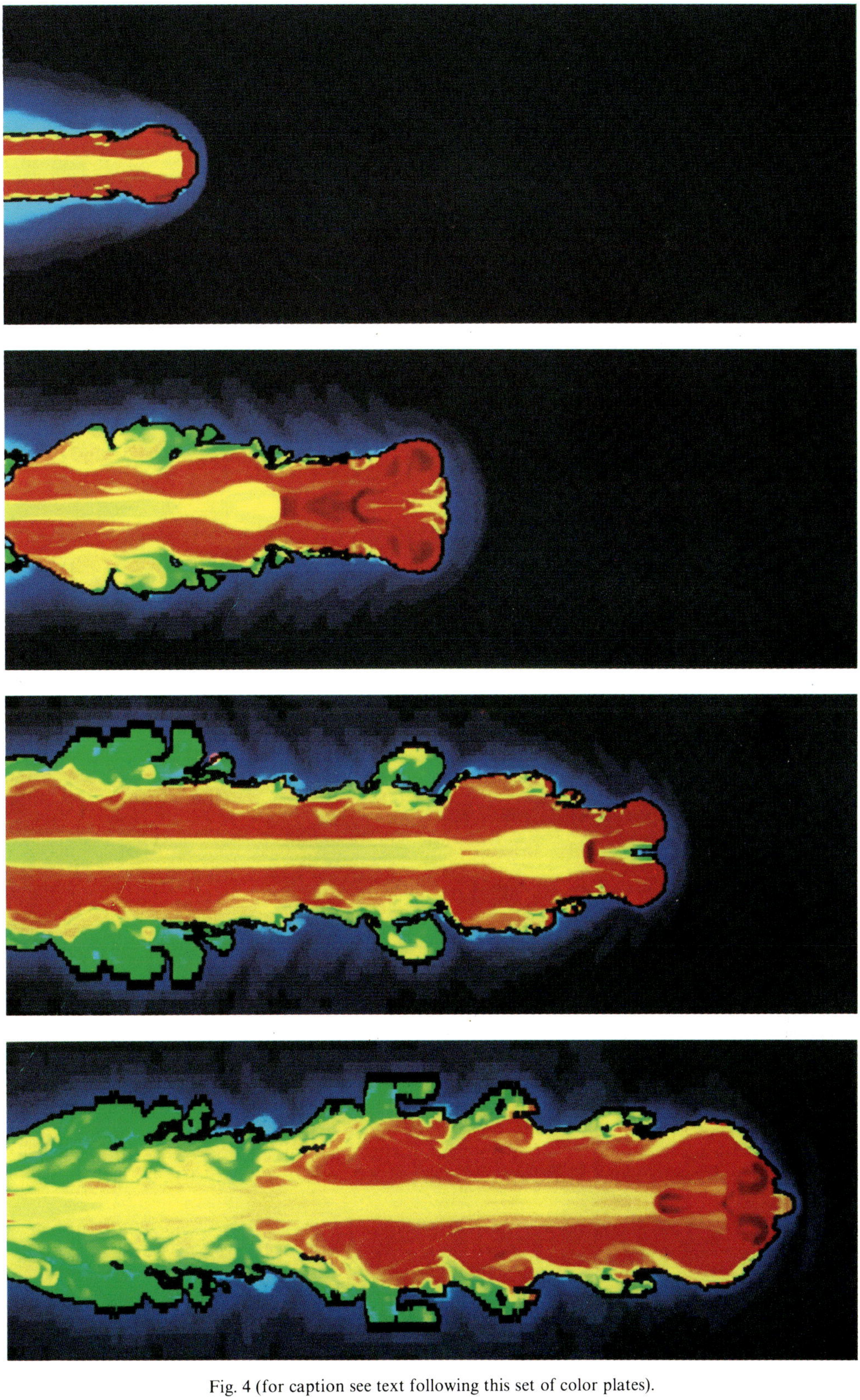

Fig. 4 (for caption see text following this set of color plates).

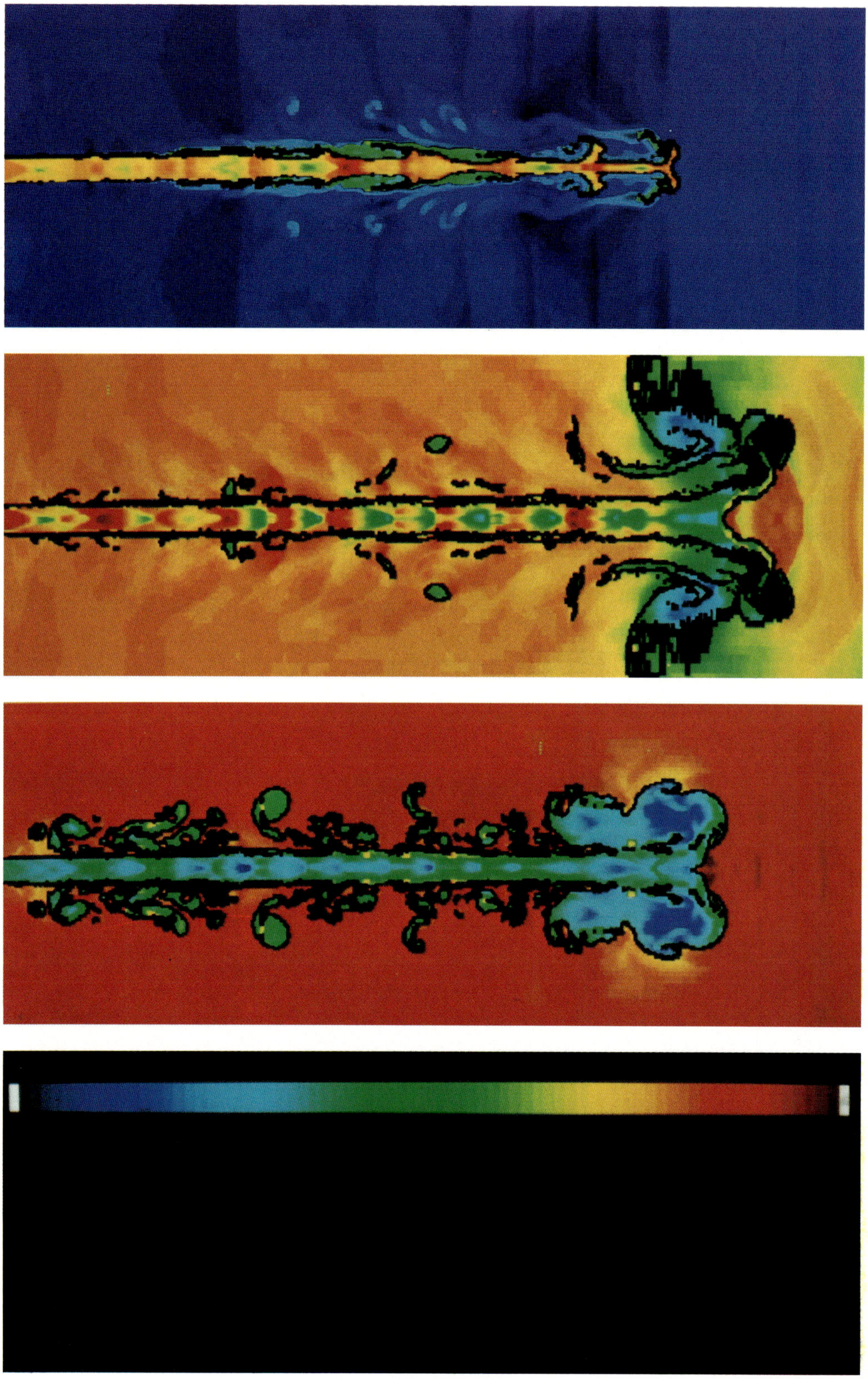

Fig. 6a (for caption see text following this set of color plates).

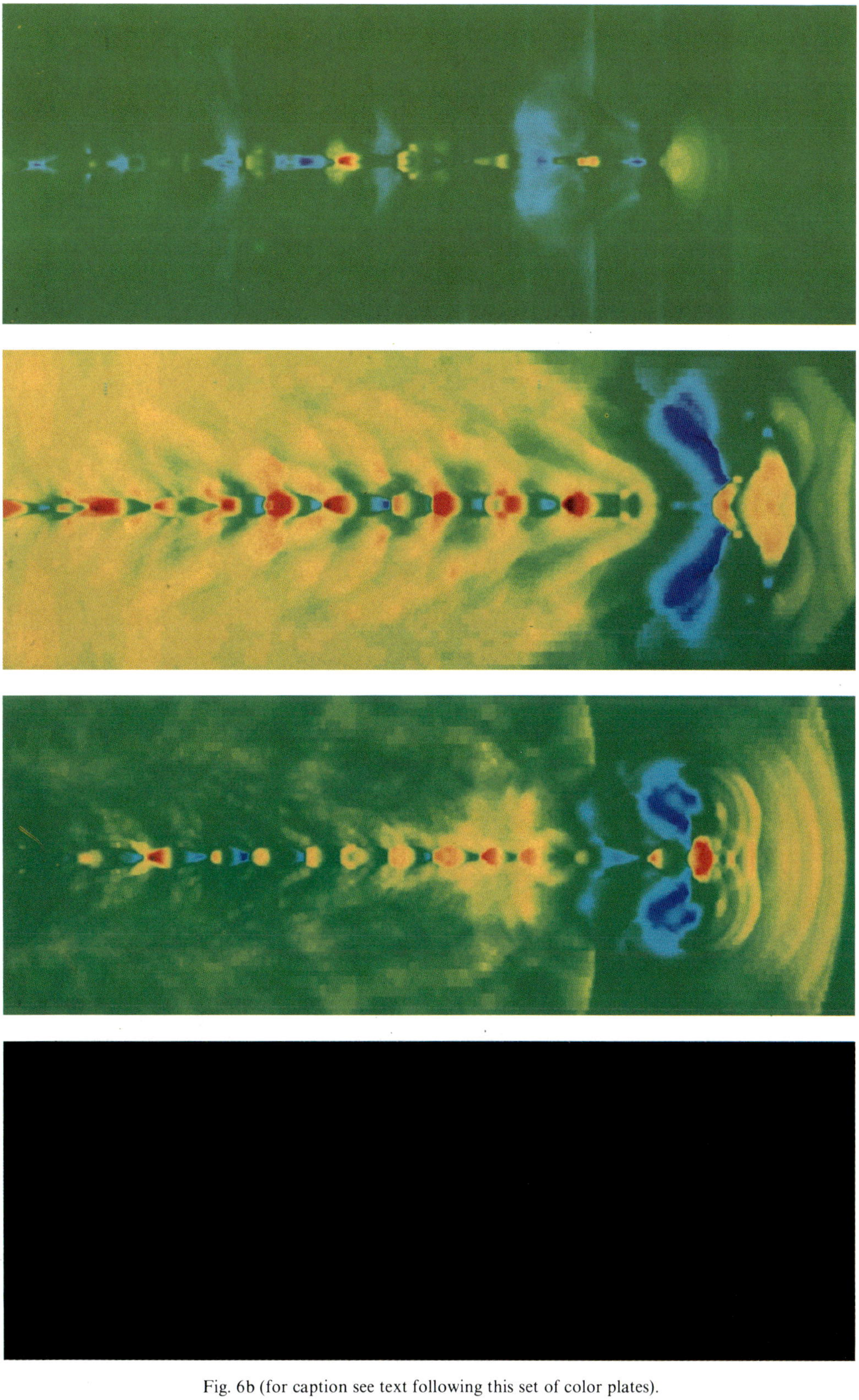

Fig. 6b (for caption see text following this set of color plates).

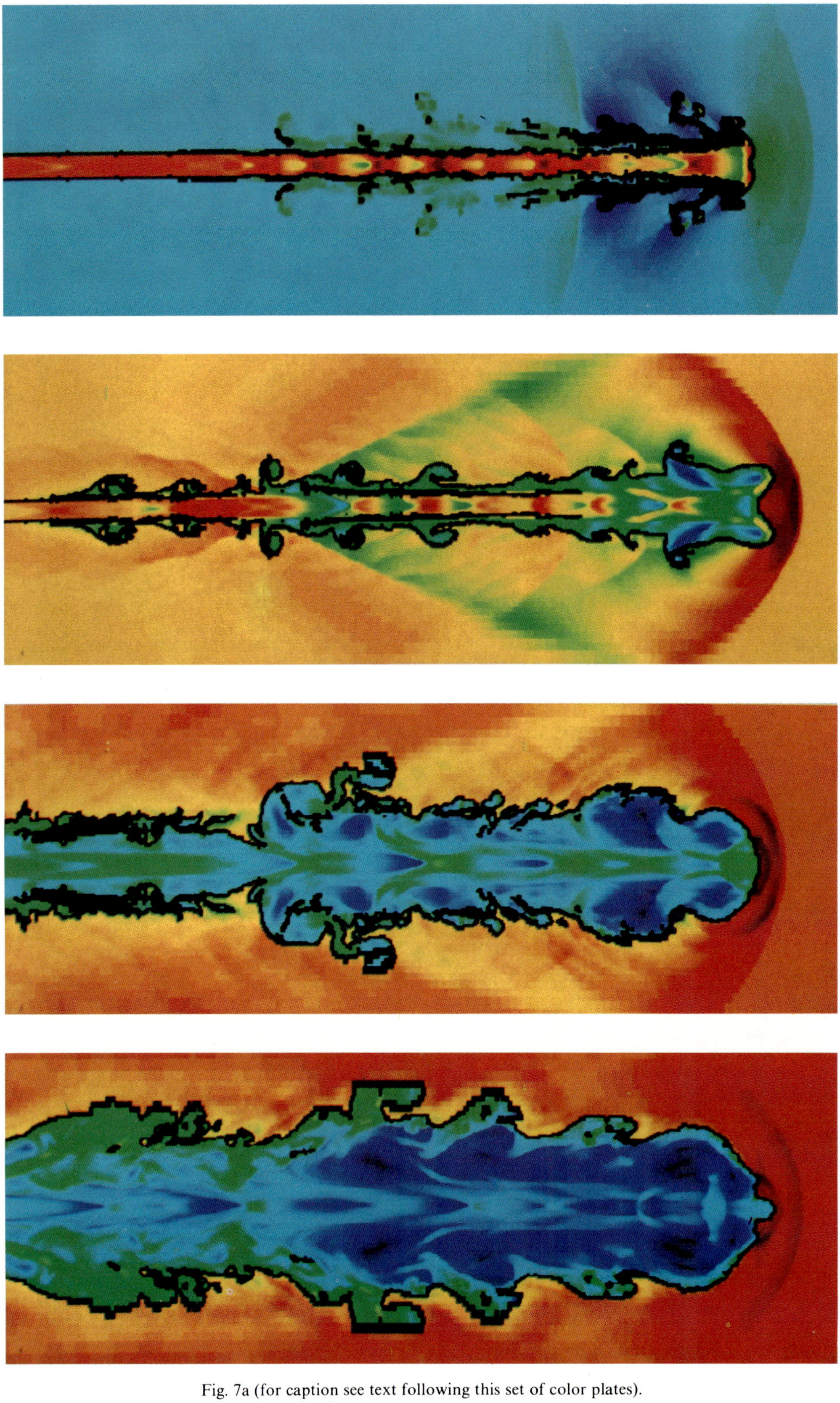

Fig. 7a (for caption see text following this set of color plates).

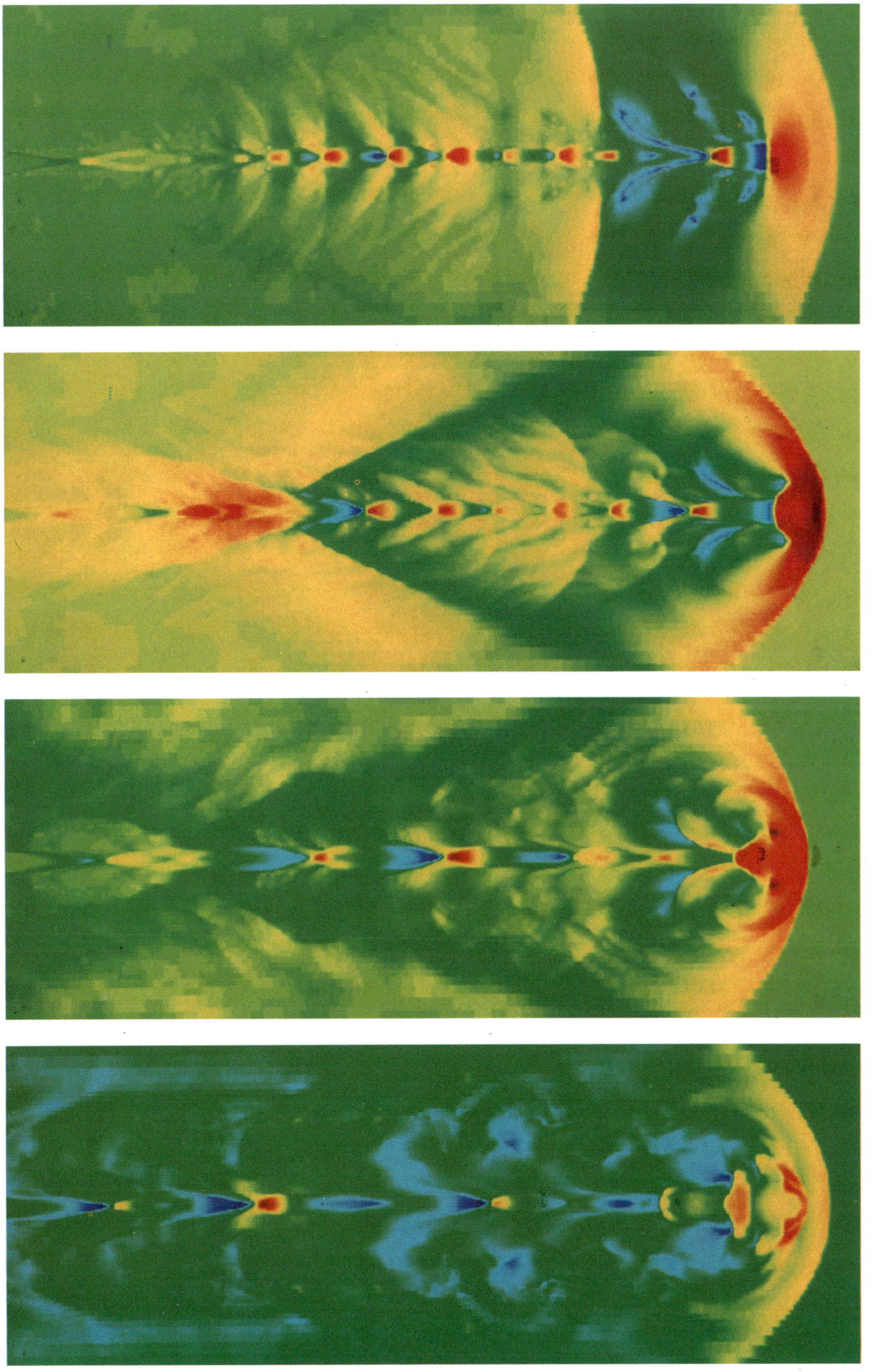

Fig. 7b (for caption see text following this set of color plates).

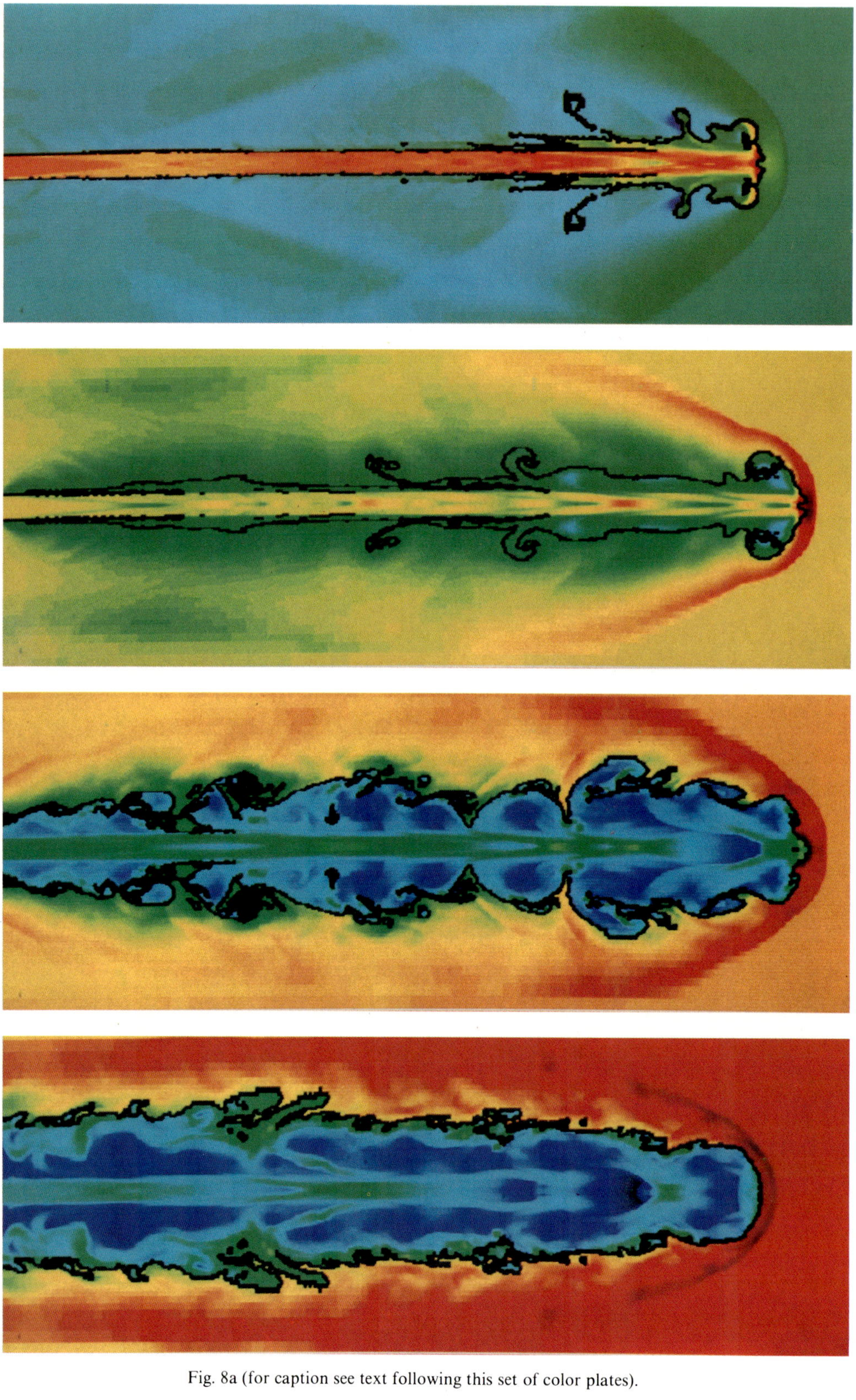

Fig. 8a (for caption see text following this set of color plates).

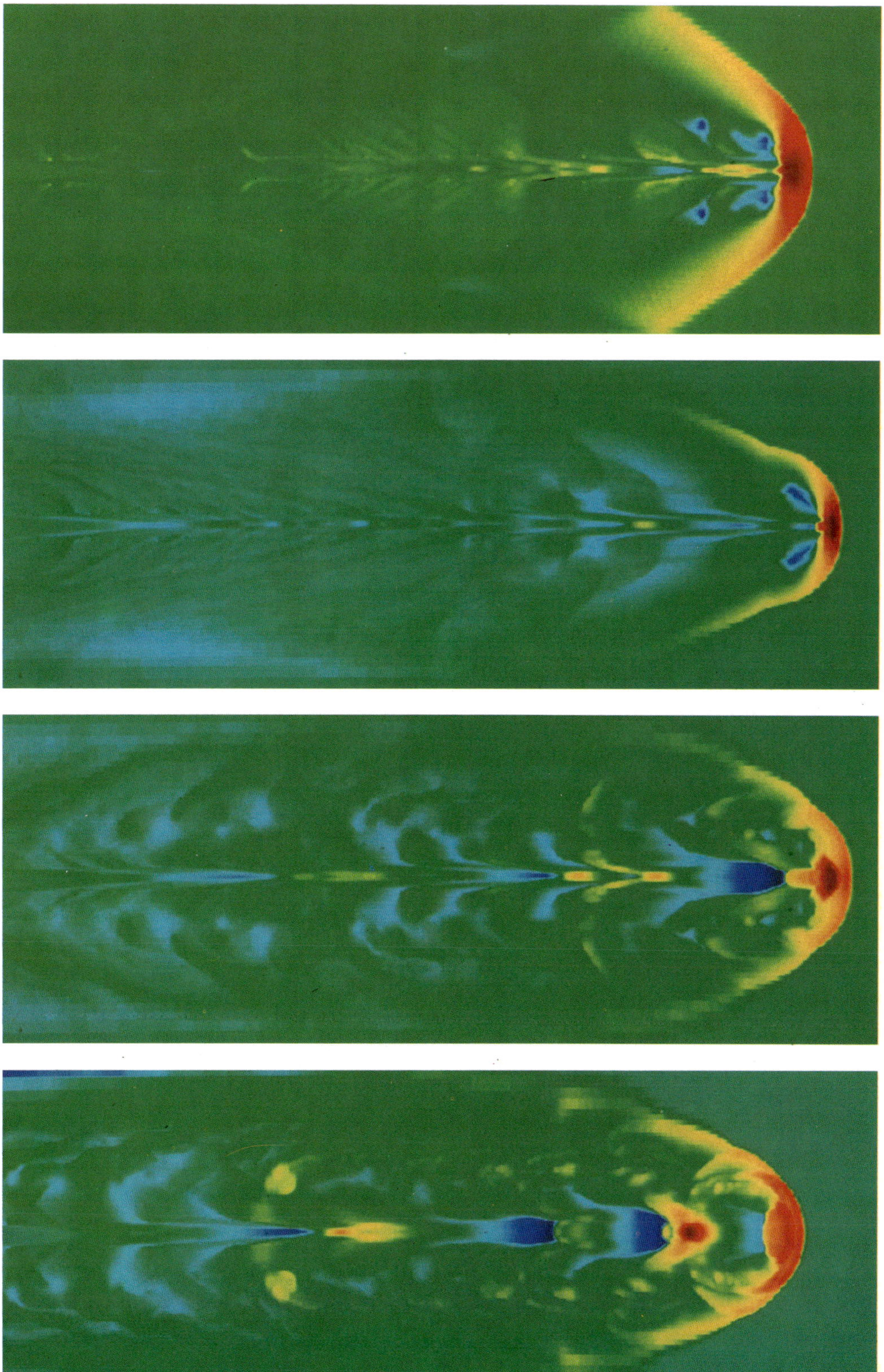

Fig. 8b (for caption see text following this set of color plates).

Fig. 14 (for caption see text on opposite page).

Captions to color plates

Fig. 4. The time evolution of an axisymmetric jet. The plane of the picture contains the symmetry axis of the jet. Gas is continuously introduced from a circular inlet at the left, with an internal Mach number of 6, a density 0.01 times the ambient density, and a pressure equal to the ambient density. The quantity plotted is entropy. It is represented by 73 shades of color, each color representing an equal logarithmic interval between the maximum and minimum value of the entropy. The order of the colors is the same as the spectrum, see color bar in fig. 6. Dark blue is the minimum and bright red is the maximum. The jet is divided into a forward moving beam and a surrounding backward moving cocoon. The interface between the cocoon and the ambient medium has nonlinear Kelvin–Helmholtz waves on it. Note that these are roughly stationary in space as the beam advances. The swirls in the red region of the cocoon reveal large scale vortices which mix low entropy material from the ambient medium into the cocoon, changing its color from red to green.

Fig. 6. The Mach 3 jets. Density fields are shown on the left page (fig. 6a) for beam ambient medium density ratios of 10.0, 1.0, and 0.1 (from top to bottom, respectively). Corresponding pressure fields are given on the right page (fig. 6b). The heavy dark line in the density plot represents the interface between the jet gas and the ambient gas. The standard color bar is given at the bottom of the density plots. Note that the top jet almost pinches itself off. The lower two jets have periodic internal shocks where the pressure color changes abruptly from blue to red.

Fig. 7 The Mach 6 jets. Same as in fig. 6, except that a density ratio 0.01 jet is also included at the bottom of the page. Note that the lower two jets have extensive cocoons, which causes the internal shocks to be much more widely spaced. The pressure plots show, in the middle two jets, that the bow shock reflects off the constant pressure boundary at the edge of the grid. The interface of the density ratio 1.0 jet has a periodic set of nonlinear Kelvin–Helmholtz waves on it.

Fig. 8. The Mach 12 jets. Same as in fig. 7. Those jets with naked beams (the top two) do not yet show strong internal shocks because they are shorter than the growth length for the reflecting mode perturbation to go nonlinear. If they are evolved to twice this length, they will develop internal shocks similar to fig. 7. The bottom two do have some internal shocks because they are driven by finite amplitude pressure variations in the cocoon.

Fig. 14. A Mach 6, density ratio 10.0 slab jet. The two images on the left are from a run where no perturbation was applied to the inlet flow. The two on the right are the same inlet conditions except that a periodic wiggle is applied to the inlet velocity. The bottom figures are density plots, the top two are pressure plots. Note the similarity of the left jet with axisymmetric Mach 12, density ratio 10 jet in fig. 8. No internal shocks have developed because the jet is shorter than its growth length. The asymmetric alternating shocks which develop in the right run will soon cause the jet to kink sharply and break up.

The spacing of the knots depends only slightly on Mach number. We find that it varies between 2.3 and 2.6 jet diameters. These knots are not stationary in space, but propagate with a well-defined pattern velocity. This can be graphically seen in fig. 9b, which is a spacetime diagram of the pressure along the symmetry axis. The maximum value of pressure is white, the minimum is black, and the intermediate values are 15 shades of gray representing equal logarithmic intervals of pressure between the maximum and the minimum. The shocks are located at the places where the shade goes abruptly from black to white. The vertical axis is time, while a horizontal slice through the diagram gives the pressure profile on the spatial axis at that instant. The line at about 45 degrees is the bow shock advancing through the ambient medium.

One sees that the pressure knots have a remarkably parallel spacetime flow. New shocks appear behind the older ones as the older ones move downstream. Thus the nonlinear knot structure has a well-defined spacing and pattern velocity. This figure illustrates the power of computer graphics to highlight the physics of interest in dynamic nonlinear systems, just as Zabusky [17] has argued in the much simpler problem of the one-dimensional nonlinear dispersive equations.

To quantify the knot structure, we present graphs of the key physical variables on a cut along the axis of symmetry (fig. 10a). One can see that although the velocity remains fairly constant along the jet, the other properties of the jet are very inhomogeneous. For instance, the Mach number varies from 3 to 10 and the density from 2 to 15. These large variations are caused by the internal shocks in the beam. Note that because these are not planar shocks, the density jump across the shock can be greater than 4. Although these internal shocks are strong, they do not cause the jet to become subsonic, in the ambient medium frame, until the terminal shock.

This set of graphs illustrates the conceptual difficulty in applying perturbation theory to supersonic phenomena. The perturbation theory must assume a smooth stationary background to perturb around. However, precisely because the flow is unstable, the self-consistent background is unsmooth and unstationary. Therefore, the interpretation of perturbation theory, in the absence of fully nonlinear calculations, must be treated with caution.

In fig. 10b, we make a cut at right angles to the beam at a distance of 45 units down the symmetry axis. The boundary of the beam is sharply defined by the kinetic luminosity plot. Note that there is no appreciable backflow (negative V_z) surrounding the beam. The dip or peak at the center of the beam occurs because we have chosen a cut through one of the "x" shocks. The cut is made just forward of the center of the "x" so that the axis has been shocked, while the off-axis flow has not been. Thus, the axis is temporarily at high pressure compared to the edge of the beam. These graphs show that very large *transverse* pressure gradients occur in supersonic jets, in addition to the large jumps in pressure *along* the beam. Nonetheless, the beam diameter stays almost constant along the jet.

5.2. *Cocoon dominated jets*

The internal shocks in the supersonic beam divide the jets into two broad subcategories: the mode dominated (fig. 9) and the cocoon dominated (fig. 11) jets. The boundary in parameter space between these two domains is not sharp, but is estimated by the curved dashed line in fig. 5; it represents the division between those jets which have cocoons and those which do not.

We now turn to the cocoon dominated regime of low density, high Mach number jets, which possess strong backflows. A representative example is the Mach 6, density ratio 0.01 jet shown in figs. 7, 11, and 12. The spacetime diagram (fig. 11b) shows that the mode dominated shocks do not exist here. Instead, there is a slow backward drift of large regions of low and high pressure separated by shocks; (note that the vertical timescale is not the same as in fig. 9b). The beam

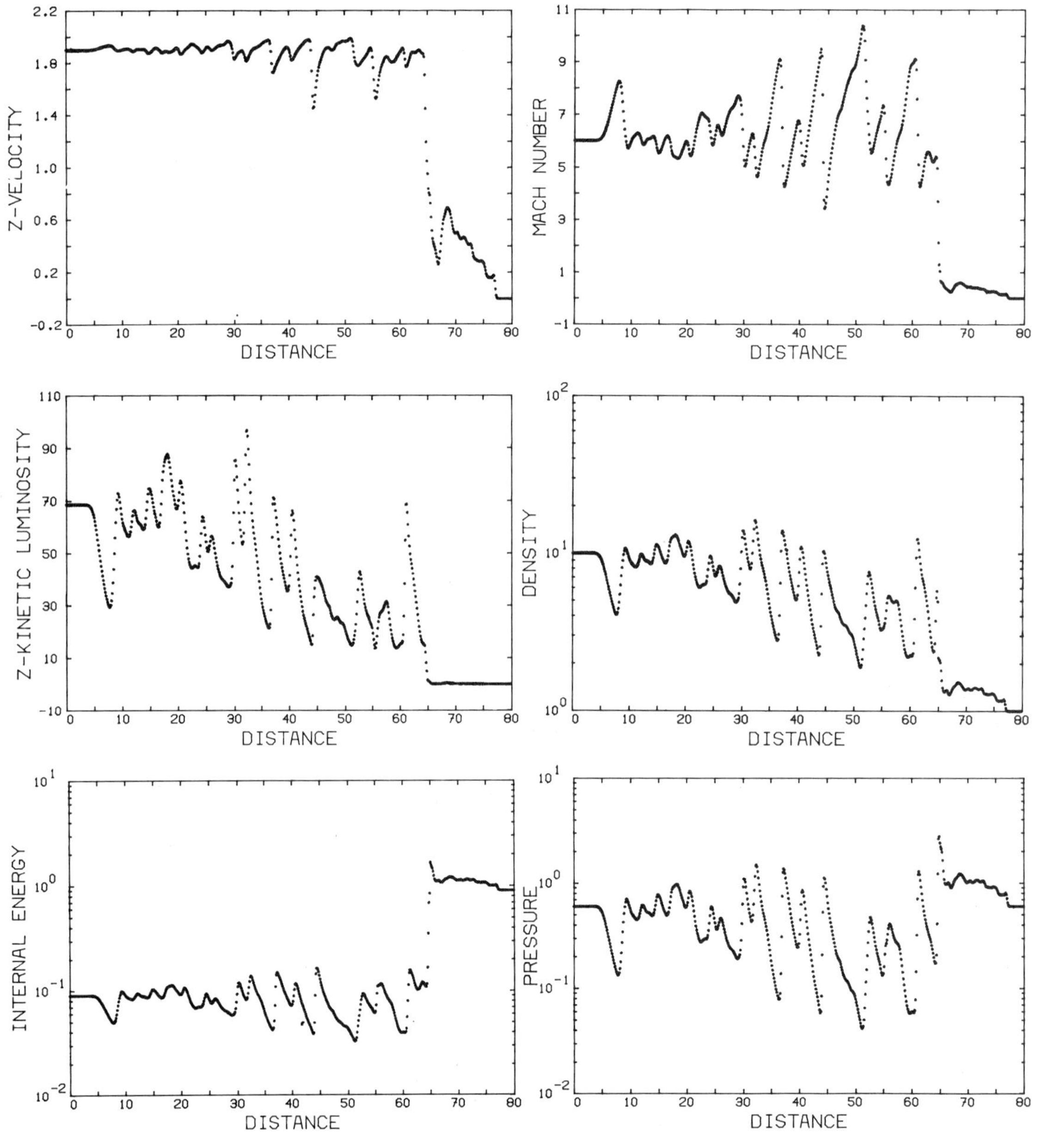

Fig. 10a. Physical variables along the symmetry axis (Mach 6, density ratio 10). Shown are the velocity (V_z), the internal Mach number (M), the kinetic luminosity (ρV_z^3), the density (ρ), the specific internal energy or temperature, and the pressure. Note how inhomogeneous the beam is. The quasiperiodic internal shocks cause the abrupt jumps. The dots are the grid points used in the computer simulations. The distance is in units of the beam radius.

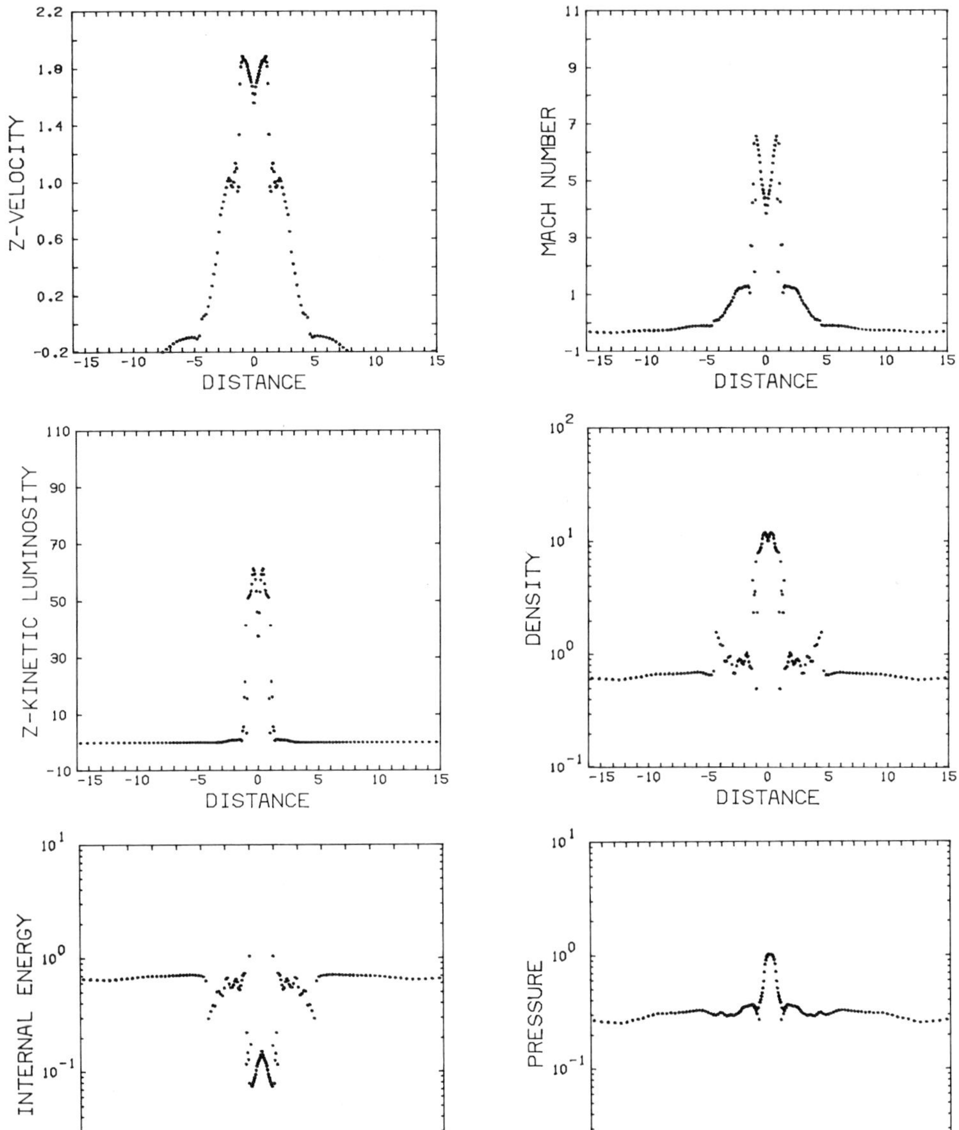

Fig. 10b. Same as 10a, except the cut is across the jet at $Z = 45$ units. The location of the beam is seen by the abrupt falloff in the kinetic luminosity. Note that the beam is still the same diameter as when it left the inlet. The peak or valley at the center of the beam occurs because the cut goes across one of the "x" shocks embedded in the beam.

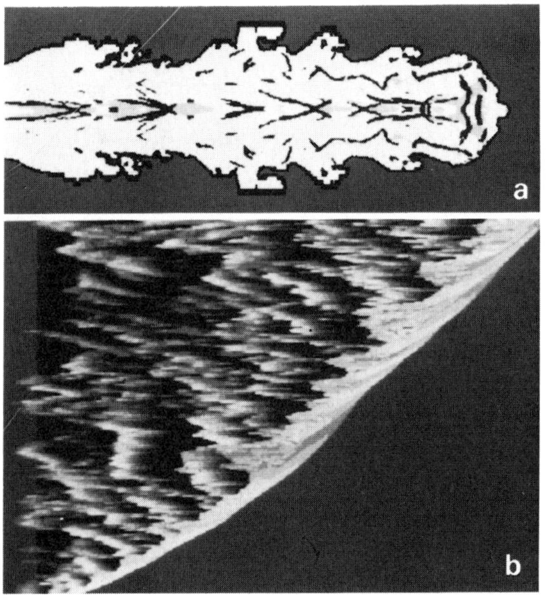

Fig. 11. Same as Fig. 9, except for the Mach 6, density ratio 0.01 jet. The shocks external to the beam are in the cocoon rather than the external medium. Note that they sometimes smoothly connect with the internal shocks. Even though the internal shocks are forced by the cocoon, they are biconical just as the free shocks in fig. 9. The terminal shock system is complex, but part of it is a strong Mach disk which stops the beam. The pressure spacetime is totally different than in the mode-dominated jet in fig. 9. The shocks which were part of the terminal shock system are left behind when the vortices are shed. These shocks gradually drift backward with the vortices. This shows that the shocks are pinned by cocoon physics and are not the free modes of a naked beam. The time-scale is about one half of that in fig. 9, since the light jet advances at about twice the speed of the dense one.

shocks are still biconical, but they start outside of the beam in the cocoon (fig. 11a). The spacing between shocks is much longer than in the mode dominated case, being roughly equal to the spacing of the large "breaking waves" seen on the interface between the cocoon and the ambient medium. These vortices sit in the cocoon, imposing pressure variations along the surface of the beam, which in turn, forces shocks into the beam.

In fig. 12a, we give the physical variables along the axis. We notice the fundamental effect, that the beam is broken up by biconical internal shocks, is present here as in fig. 10a. However, the spacing, while still fairly regular, is much larger than in the mode-dominated case. This is because the spacing in fig. 12a is set by the spacing of the forcing vortices in the cocoon, whereas the spacing in the mode-dominated case is set by the most unstable reflecting mode of a "free beam". The spacing of the cocoon vortices is ultimately set by the vortex shedding instability of the head of the jet.

Even though both jets (figs. 10a and 12a) start with Mach 6 inflow, the velocity of the light jet is $(1000)^{1/2} \sim 31.6$ times higher than the heavy jet. This is because both jets are in pressure equilibrium with the same ambient medium. As the density of the heavy jet is 1000 times that of the light jet, the temperature (see internal energy graphs in figs. 10a and 12a) of the light jet is 1000 times that of the heavy jet. The sound speed goes as the square root of the temperature, so the same Mach number in the two jets leads to the difference of ~ 31.6 in the actual velocities. This also means that the kinetic luminosity ρv_z^3, has a net increase of 31.6 in going from the heavy jet to the light jet.

The final point we make is simply to reemphasize the existence of enormous jumps in pressure along the jet axis. Note that while the shocks are very strong, leading to jumps in pressure as high as 500, the rarefactions immediately following the shocks are equally strong. This means that the knots are very isolated, with high contrast against the mean pressure background of the jet. This is particularly important for the astrophysical jets, in which the synchrotron emission scales roughly as p^2 (see paper IV).

We turn now to the cut across the beam at $Z = 47$, fig. 12b. The major contrast with the heavy beam in fig. 10b is that the light beam has a low density cocoon with negative velocity surrounding it. This means (compare density plots in figs. 10b and 12b) that the supersonic beam is always surrounded by lighter material, whether the inlet density ratio of the beam to the ambient material is high or low. This contributes strongly to the beam stability. Once again we see that a linearized stability analysis of a *naked* light beam would not have had the correct self-consistent background,

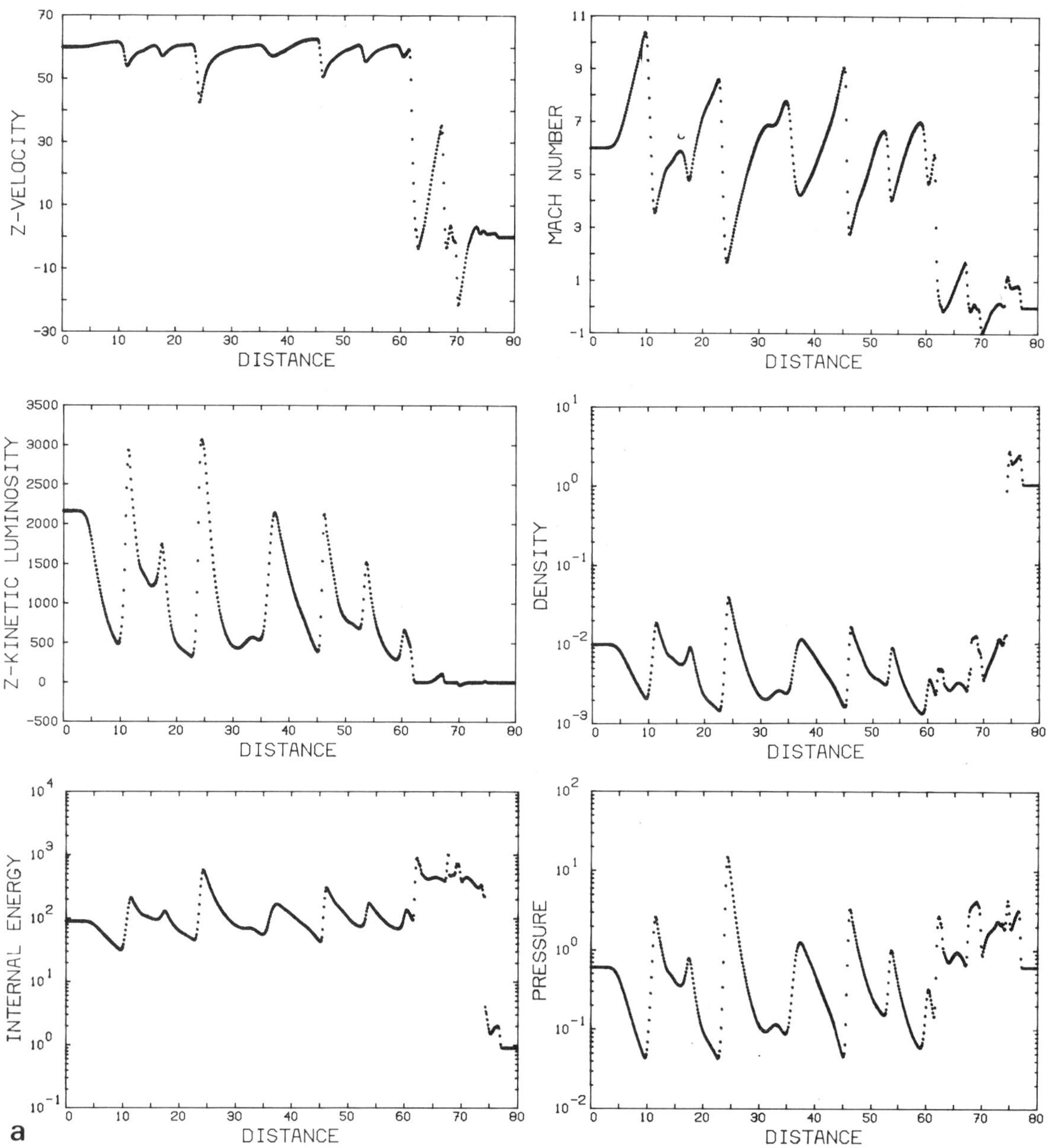

Fig. 12a.

i.e. one could never have guessed that one should surround it by an even lighter cocoon. This is another illustration of the importance of using supercomputers to solve the full problem.

In summary, the cocoon makes the jet structure much wider than in the dense beam case. However, because the cocoon is so light, the kinetic luminosity profile is even sharper and narrower than in the dense beam case. That is, although there is energy transport backward along the cocoon, it is negligible compared to that energy carried forward by the beam. The gas in the cocoon

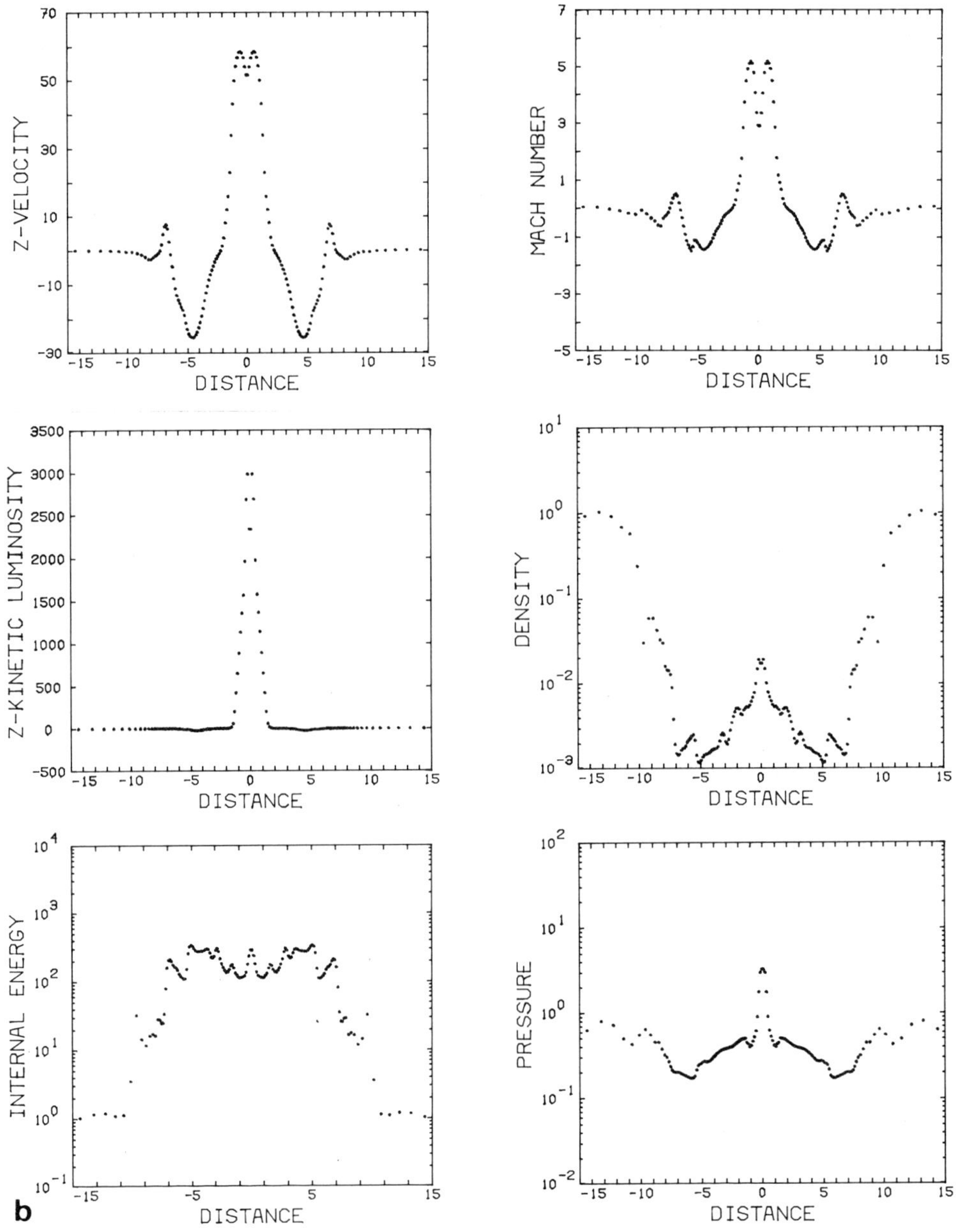

Fig. 12. Same as fig. 10, except for the Mach 6, density ratio 0.01 jet. Note that the spacing between the cocoon-dominated shocks is much larger than for the mode-dominated shocks. Here the spacing represents the distance between shed vortices in the cocoon. The jumps in pressure across shocks along the axis can be as great as 300. Even so, the beam is not disrupted. The beam diameter is roughly constant as can be seen in the kinetic luminosity plot across the jet. The wide return flow (negative velocity) in the cocoon can be seen in the V plot across the jet.

is shocked to higher temperatures by the terminal shock, but then it is adiabatically expanded and cooled in the larger volume of the cocoon. Thus, the resulting temperature profile is fairly constant across both the beam and cocoon (internal energy plot in fig. 12b), and much higher than in the ambient medium. Again, large transverse pressure gradients exist in both the beam and in the cocoon. Finally, note that the interface tracer (solid line in the density plot in fig. 7) locates the boundary between the jet gas in the cocoon and the ambient medium gas. It is much sharper than the density or temperature profiles across this boundary in fig. 12b.

5.3. *External shocks*

In both the dense (fig. 9) and light (fig. 10) jets, we see that there are shocks *external* to the supersonic beam. However, the dense jet drives the shocks directly into the *ambient medium*, whereas the light jet has its external shocks *inside of the cocoon*. In astrophysical jets, these will have very different observational characteristics. The radio emission from astrophysical jets is caused by the synchrotron process of shocked electrons spiraling around local magnetic fields. The magnetic field is carried along the jet from inside the galaxy. This means that in the cocoon the field will be present where the shocks are, thus producing a diffuse emission from large portions of the cocoon. The dense jets, on the other hand, do not possess a cocoon, and therefore the magnetic field will not be present where the external shocks are. These shocks will thus be invisible, unless the shocks compress the ambient gas enough to ionize it and produce emission lines. This provides a means, in principle, of distinguishing between dense and light jets observationally.

There is a spurious external shock in several of our runs (the Mach number 6, density ratios 1.0 and 0.1 and Mach number 12, density ratio 10.0). This is caused by the reflection of the bow shock off the outer constant pressure boundary condition. Only in the Mach 6, density ratio 1.0 case does it have a significant impact on the solution. There the effect is to give a finite amplitude pressure perturbation to the beam, thereby guaranteeing the development of internal shocks. However, since the beam was already long enough to have the mode-dominated shocks anyway (see density ratio 10.0 jet above it), this is not a major contribution. These unwanted external shocks are not a feature solely of computational aerodynamics. In wind tunnels, such reflected shocks can occur from apparatus in the tunnel (see, e.g., fig. 271 in Van Dyke [35]). The important lesson here is that with computer graphics, such numerical features can be easily identified in the solution and be taken into account.

5.4. *The terminal shock system*

Finally, we come to the most important shock system in the problem, the one which terminates the supersonic beam at the head of the jet. As mentioned in section 4, Blandford and Rees [22] concluded on the basis of simple reasoning that there must be bow shock in the ambient medium, a contact discontinuity between the ambient gas and the beam gas, and a terminal shock in the beam gas. However, figs. 6, 7, and 8 demonstrate that the nature of these three discontinuities is much more complicated than this simple picture suggests.

An important discovery of our computer simulation is that the structure and shape of the terminal shock and the contact discontinuity are highly time dependent. Fig. 4 shows how the head of the jet changes shape as the jet advances. Fig. 2 of paper IV shows the detailed dynamics of the contact for the Mach 6 jet with a density ratio of 0.1. As part of this time dependence, the terminal shock changes its pattern dramatically. Figs. 3–6 in paper IV give high resolution views of the terminal shock at four times in the history of this single jet evolution.

We will describe one example of a terminal shock configuration to illustrate the complexity of this dynamic structure. Consider the Mach 6, density ratio 0.01 jet. From the entropy plot of the final state in fig. 4, one sees that the beam bifurcates as it approaches the head of the jet. The red color indicates high entropy gas, so we can tell that the center part of the beam was shocked more strongly (dark red) than the outer part of the beam (orange). This is confirmed by noting that the kinetic luminosity contours in fig. 13 also bifurcate in the same way. The velocity vectors in fig. 13 show that the beam gas is almost brought to rest near the axis, but continue at high speed in the outer part of the beam. From fig. 11a, one can see a characteristic triple point shock configuration of incident shock, Mach disk shock and reflected shock is responsible for this phenomenon. The strong Mach disk stops the axis gas, while the off-axis beam flow goes through the weaker and oblique incident and reflected shocks. The reflected shock also deflects the flow upward, causing the bifurcation.

The axis flow then accelerates until its speed is the same as the off-axis flow (fig. 13). At this point, a strong transverse shock stops all forward flow. In fig. 4, this causes the yellow in the bifurcation to jump to red, the kinetic luminosity contours stop abruptly (fig. 13), and a high pressure disk (yellow/red in fig. 7) forms. The flow now diverts upward with most of the gas being turned into the cocoon (fig. 13). However, a portion of the gas is trapped in the beam "cap" at the head of the jet in a clockwise rotating vortex ring. Note that this causes the gas to flow *backward* on and near the axis (see also the V_z plot in fig. 12a). Thus, the beam gas is actually *colliding* from both sides (in

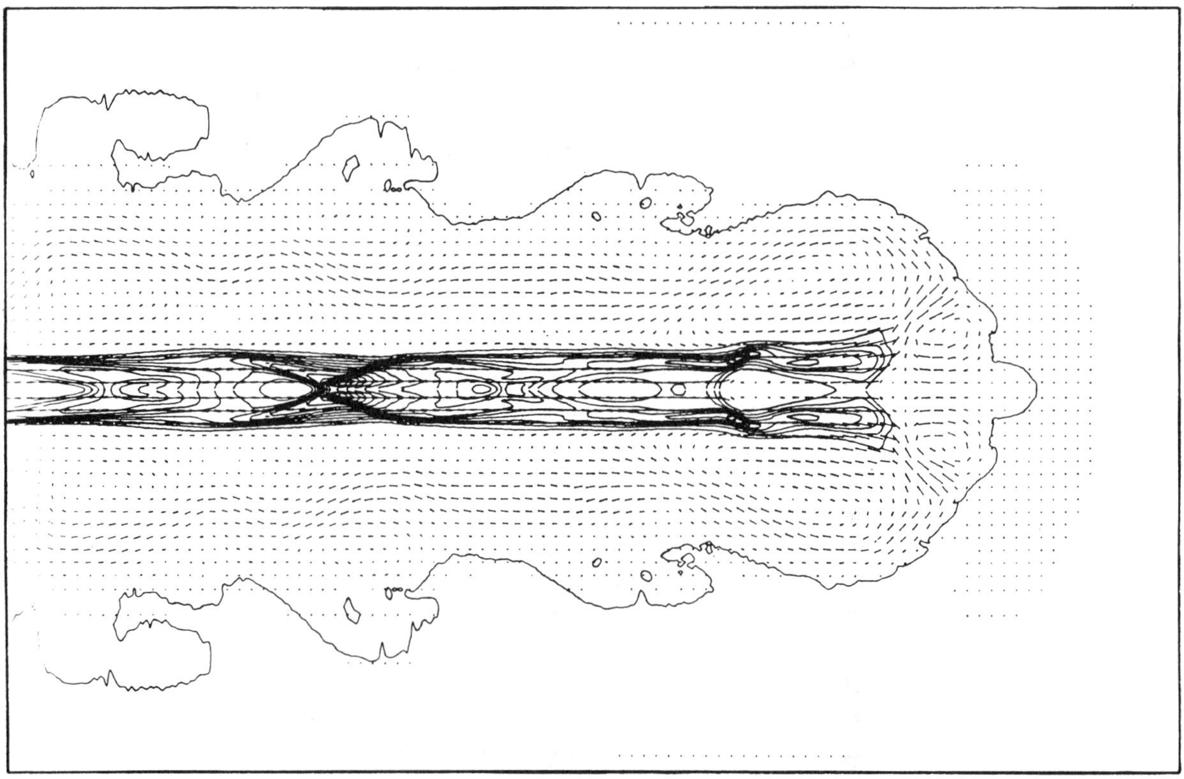

Fig. 13. The velocity field of the head of the Mach 6, density ratio 0.01 jet. The short line segments give the direction of the local velocity, with their length proportional to the speed. The contour lines are of kinetic luminosity (ρV_z^3). One can see a number of vortices embedded in the flow.

the ambient medium frame) of the strong terminal shock.

Finally, the gas which flows into the cocoon froms a large counterclockwise rotating vortex directly over the bifurcation. This flow produces a low pressure (blue in fig. 7) region which is responsible for creating the underexpanded triple point Mach disk configuration in the first place (see detailed discussion in paper I). This vortex is just about to be shed in place, while the beam head advances forward (see shedding sequence in fig. 4 of paper I). The vortex will then swirl the cocoon interface around, creating another large surface wave like the three already present in fig. 13. As the vortex separates from the head of the beam and slowly moves backward, it causes the underlying shock structure to also move backward, thus explaining the backward moving pressure features seen in the spacetime of fig. 11b.

Since the nature of the terminal shock system is discussed in great detail in papers I and IV, we will not go into any more depth here. We can summarize by saying that the patterns at the end of the jet, seen in the pressure plots in figs. 6, 7, and 8, are caused by this time varying terminal shock system. Note that this can cause either single "hot spots" as in the Mach 12, density ratio 0.1 case, or multiple "hot spots" as in the Mach 12, density ratio 0.01 case. Furthermore, these multiple hot spots can be off-axis as seen in fig. 6, paper IV.

6. Interfaces in axisymmetric jets

There are three major interfaces in an axisymmetric jet. The primary one is the interface between the supersonic beam and its surrounding medium. The next one is the contact discontinuity between the jet gas and the ambient medium at the head of the jet. Finally, there is the interface between the cocoon and the ambient medium, in those cases where the jet has a major return flow. All these interfaces are unstable to either Kelvin–Helmholtz or Rayleigh–Taylor instabilities. It has been a matter of great concern to astrophysicists as to whether these instabilities, in their nonlinear limit, can pinch off or otherwise destroy the jet. These are essentially nonlinear coupled processes, whose elucidation requires detailed numerical simulation.

Our computer code includes an interface tracker (described in paper V) which follows the position of the interface between jet gas and ambient gas. In a naked beam, this is the interface along the supersonic beam, because there is no significant backflow of shocked beam gas. In the jets with large backflow, the interface is between the cocoon gas and the ambient gas. In this latter case, the boundary between the beam and cocoon is not followed by the interface tracker, but must be determined self-consistently by the hydrodynamics. The sharpness of the interfaces can be seen in the density plots of figs. 6, 7, and 8 for the cases of naked beams and cocoon-dominated jets.

6.1. *The beam/cocoon interface*

If the supersonic beam were to pinch off then the jet could not be used to transport the energy, mass and momentum to the great distances required by the observations of astrophysical jets. As mentioned in the last section, pinching modes in supersonic beams saturate in internal shocks. On the upper side of the dashed straight line in the parameter space plot (fig. 5), we find that the shocks are planar and the beam disrupts by severe necking down. On the lower side, the shocks are biconical and the beam is in general not disrupted. In this section, we will examine what the existence of these shocks do to the interface between the beam and its surroundings.

The location of that interface is given directly by our interface tracker in the case of the high density jets. In the color density plot for the Mach 6, density ratio 10 jet, fig. 7, one can see that with the biconical shocks fully developed, the beam interface is modulated by a periodic oscillation.

That oscillation has saturated at a small enough amplitude that the beam does not pinch off. This behavior of the interface is typical of naked beams in the portion of parameter space (fig. 5) which is dominated by biconical internal shocks. It is verified by our higher resolution "clean" runs in fig. 9c of paper III.

The behavior of the interface is radically different in the ordinary mode region *above* the dotted line in fig. 5. An example of such a jet which we have studied in some detail is the Mach 1.5, density ratio 1 jet. As the head of the jet moves along (fig. 10b of paper I), a large vortex above the head efficiently entrains material across the equal density interface. After the jet has moved off the right side of our grid, the beam interface develops large "breaking waves" (fig. 14 of paper I), which not only entrain material in the beam, but also breaks it up into a series of subsonic/supersonic pieces. A high resolution run (figs. 9a,b of paper III) graphically shows how the Kelvin–Helmholtz waves grow as they propagate downstream, leading to greater and greater necking down of the beam. The planar shocks interior to the beam are clearly associated with these necking down regions. Another example is the Mach 3, density ratio 10 jet shown in fig. 6. Again the necking down is quite clear. We expect these pinched jets to rapidly become 3-D unstable after the first wave or two on the interface, much like the round turbulent subsonic jets pictured in laboratory studies (Dimotakis, et al. [30]).

Finally, we come to those low density, high Mach number jets which have extensive cocoons. Here an effective way to visualize the beam interface is to plot the kinetic luminosity, ρv_z^3, since there is an abrupt jump in both density and velocity at the beam/cocoon interface. We did this for all our jets in paper II. There one can see that the beam has the same sort of small amplitude oscillations as the naked beams exhibit. There is no tendency for the beam to pinch off. In fig. 13, we plot this same quantity for a blow-up of the front end of the Mach 6, density ratio 0.01 jet. Again, one can see the oscillations in the beam/cocoon interface.

6.2. *The working surface interface*

We used the term "working surface", coined by Blandford and Rees [22], for the head of the jet where the beam gas is brought to rest, relative to the contact discontinuity which separates the beam gas from the ambient gas. Ahead of this contact sits the bow shock in the ambient medium. As mentioned above, we have found the shape of this contact to be time dependent. We make a cautionary warning that this time dependence may be accentuated by the assumed axisymmetry of the calculation. This forces the intersection of the symmetry axis and the contact to be a stagnant point and therefore be more unstable than it would be in a 3-dimensional jet. However, there is some evidence from laboratory studies of the dynamics of supersonic fluid jets, that the head of the jet does suffer oscillations (Brad Sturtevant, Cal. Tech., private communication). Although the working surface accelerates and decelerates in time, as can be seen by its oscillating trajectory in spacetime in fig. 11, the interface does not develop a strong Rayleigh–Taylor instability.

As described in section 5.4., the head of the jet exhibits a highly nonlinear coherent self-excited structure which combines shocks, interfaces and patterns in an intimate way. Our studies have opened up a panorama of complicated dynamic phenomena.

6.3. *The cocoon interface*

After the supersonic beam gas is stopped in the rest frame of the terminal shock, it is pushed to the side. If the beam gas is dense enough (density ratio 10), it can maintain its forward momentum in the rest frame of the ambient gas. Thus, the waste gas rides along with the head of the jet. If the beam gas is just as dense as the ambient gas (density ratio 1), then it is brought to rest in the frame of the ambient gas. This leads to a thin sheath which is laid down around the beam as the beam propagates. As the beam density is lowered below the ambient density, the return flow begins to move backward relative to the ambient medium, until at a density ratio of 0.01 it has become supersonic in the ambient medium rest frame. This causes a well inflated cocoon to surround the beam. All three of these cases can be clearly seen in figs. 6, 7, and 8.

In the case of the low velocity backflow, e.g. the Mach 3, density ratio 0.1 jet, the overturn time of the vortices in the cocoon is comparable to the timescale for the head of the jet to advance a vortex width. This is accompanied by efficient entrainment of the denser ambient medium gas into the cocoon. This process is shown at early times in the pair of pictures in fig. 9 of paper I. The long term result is that the cocoon can only stay inflated near the head of the jet, forming a lobe as seen in fig. 6. Those cocoons which have a supersonic velocity relative to the ambient medium, have a much more stable interface. However, even in the Mach 6, density ratio 0.01 case, the embedded vortices in the cocoon ultimately cause entrainment of ambient medium gas. This can be seen very clearly in the entropy sequence for this jet in fig. 5. Note that the entropy decreases (color goes from red to yellow to green), due to mixing, the further one gets from the head of the jet.

The embedded vortices, clearly seen in the velocity vector field in fig. 13, are caused by the periodic vortex shedding mentioned in the last section. Note the string of regularly spaced vortices which distort the cocoon interface in the Mach 6, density ratio 1 plot in fig. 7. This brings to mind the regularly spaced vortices which are often seen along shear layers in incompressible flows.

7. Nonaxisymmetric jets

It is clear that jets in the real world do not maintain exact axisymmetry forever, even if they start with axisymmetric initial conditions. There are two reasons for this. First, the environment in which the jet propagates may impose a finite nonaxisymmetric pressure gradient on the jet. Second, axisymmetric jets are unstable to nonaxisymmetric perturbations and these can exponentiate infinitesimal disturbances to finite ones. In an axisymmetric code, neither of these effects can occur. To investigate them we ran several slab symmetric jets. These are jets which are infinite in extent in one of the directions perpendicular to propagation, but finite in the other perpendicular direction. This symmetry is a good approximation to jets which issue from a thin slit orifice, such as the laboratory jet shown in fig. 1.

In figs. 14a, b, we show the density and pressure plots for a Mach 12, density ratio 10 slab jet with no perturbation applied. It has propagated about 2/3 as far (in units of the inlet diameter) as the same parameter axisymmetric jet shown in fig. 8. The two jets are remarkably similar in structure. In particular, both jets have internal and external shocks, with the highest pressure in the region between the terminal shock and the bow shock. However, as mentioned in section 5.1., the Mach 12 jet has not had time enough for the instabilities to grow to nonlinear strength. When we evolved further the axisymmetric jet (fig. 11 in paper III), we found that the reflecting mode biconical shocks developed.

In the slab symmetric case, there are two competing instabilities, the ones which are symmetric about the midline of the slab (Xs) and the ones which are not symmetric (kinks). Which of these wins depends on the strength of the initial perturbations in both modes. In the laboratory jet shown in fig. 1, the symmetric mode clearly wins. This is because, although the jet pressure was fairly closely matched to the ambient pressure, a mismatch of a few percent is likely. Since the mismatch is the same above and below the slab, it exerts a symmetric disturbance on the jet causing the beautiful crisscrossed shocks seen in fig. 1. Just as with the biconical axisymmetric internal shocks, these symmetric shocks do not seem to interfere with the jet propagation.

To illustrate the other extreme, we did a numerical experiment in which we applied a few percent nonsymmetric perturbation to the slab jet inlet. In detail, we wiggled the incoming flow (Mach 12, density ratio 10 as above) up and down through

0.05 radians, conserving speed, at a frequency corresponding to twice the most unstable nonsymmetric mode. In figs. 14c, d, one sees the resulting density and pressure plots at the same time as in figs. 14a, b. As in the symmetric laboratory case (fig. 1), the perturbations immediately saturate into shocks, because of the finite amplitude of the perturbation. However, here the flow is severely affected, with the beam kinking more and more with time, ultimately disrupting the beam. The high pressure and density "knots" after the shocks are no longer collinear with the beam, but instead alternate from side to side. As in the late time axisymmetric Mach 12 jet, the internal and external shocks are intimately coupled. The oscillations in the jet interface, caused by the internal shock systems, in turn cause shock systems in the external medium. Other numerical investigations of kink instability, mainly in periodic flow boundary conditions, have been reported (see e.g. Tajima and Leboeuf [31], Nepveu [32], and Woodward [33]), which show in more detail the structure of these shocks.

8. Conclusions

We started this paper by observing the real world of supersonic jets. There is evidence that in the laboratory, in aerodynamics, and in astrophysics, supersonic jets exist that are nearly axisymmetric or slab symmetric for many jet diameters. These jets range in length from centimeters to millions of light years. Almost all of these jets exhibit knots of emission inside of their beams. There was clear evidence in the laboratory case that these knots are caused by periodic shock wave systems. Although in the laboratory and in aerodynamics one is usually not interested in the transient behavior of the jet head boring through the ambient medium, it seems clear that one is seeing the astrophysical jets in just such an evolution. These cosmic jets seem to require strong terminal shocks to explain their hot spots near the head of the jet.

These patterns, reoccuring over such enormous length scales, indicate that some fundamental scale-free nonlinear hydrodynamics is at work here. To investigate this possibility we undertook to use a supercomputer to simulate these flows. The equations integrated were the basic inviscid equations of hydrodynamics (continuity equation, adiabatic energy equation, and Euler equation). Because these equations have no other physics in them, they are scale-free and their solutions should apply to laboratory or astrophysical jets equally well. We fixed a uniform ambient medium and then ran a two-parameter set of jets through it. We realize that other physics (heating and cooling, phase changes, magnetic fields, etc.) may break the scale invariance in some particular jets and we are extending our basic analysis to incorporate such effects.

We found that all jets develop internal shock waves if they propagate to a distance long enough for perturbations to grow to nonlinear amplitude. In the axisymmetric case, there are two regimes of shock physics. One in which the shock waves are planar and disrupt the beam and one in which the shocks crisscross the beam and do not disrupt it. The boundary between the two regimes coincides well with the boundary found (Cohn, [29]) in perturbation theory between ordinary Kelvin–Helmholtz pinching modes and the higher order reflecting modes.

These shocks occur at regular intervals along the beam, causing enormous jumps in pressure internal to the beam. A spacetime diagram of the knots reveals two qualitatively different dynamical patterns. The one is that of closely spaced knots propagating downstream with the velocity of the leading bow shock. The other is that of the slow backward drift of widely spaced knots with a more chaotic appearance. The difference between these two behaviors was traced to the other major pattern in supersonic jets: the existence of a cocoon of return gas flow around fast light jets. If there is such a cocoon, the vortices in the return flow drive the instabilities in the beam and produce the second type of spacetime diagram. If the beam is naked, then the knots seem to develop from the free growth of perturbations to nonlinear strength, producing the first type of spacetime diagram.

The terminal shock physics was found to be highly time dependent and fairly complex. However, almost all features of the flow can be understood in terms of simple concepts from textbooks on supersonic flow. The strong shock at the end of the jet is likely to explain the hot spot phenomenon in astrophysical jets. There is a shedding of vortices and the development of large breaking waves on the cocoon/ambient medium interface.

Our slab jet calculations indicated that such jets can propagate for many jet diameters if they are not disturbed. However, if hit with a nonsymmetric finite amplitude perturbation, they go into a kinking instability, which ultimately destroys the beam. This produces alternating shock waves along the beam, more widely spaced than in the free axisymmetric jet case.

In the real world, the ability of jets to propagate to large distances without destroying themselves is observed. However, ultimately many such jets begin to wiggle and soon end. Thus, we expect that fully three-dimensional computer simulations will be necessary to model these jets. These calculations await larger and faster computers. Nonetheless, much of the basic physics of the shocks, patterns, and interfaces of supersonic jets seems to come out of our two-dimensional models. A crucial tool of analysis for computer simulations are the color images that we displayed herein. One of the great challenges for tomorrow's calculations is how to visualize the basic physics of time dependent three-dimensional flows.

Acknowledgements

We wish to thank the Los Alamos Center for Nonlinear Physics for financial support to attend the conference. The National Science Foundation (grants number PHY 80-01496 and PHY 83-08826) and the Alfred P. Sloan Foundation provided partial support for one of us (LS). We are particularly grateful to Norman Zabusky for his clarifying and encouraging remarks.

References

[1] E. Mach and P. Salcher, Wien. Ber. 98 (1889) 1303.
[2] L. Prandtl, Essentials of Fluid Dynamics (Blackie and Son, London, 1952).
[3] R. Landenburg, C.C. Van Voorhis and J. Winckler, Phys. Rev. 76 (1949) 662.
[4] F.W. Anderson, Orders of Magnitude, NASA Special Pub. (1981) 4403.
[5] J. Wormhoudt, and V. Yousefian, J. Spacecraft and Rockets 19 (1982) 4, 382.
[6] R. Mundt, and J.W. Fried, Submitted to Ap. J. Letters (1983).
[7] B. Margon, Science 215 (1982) 247.
[8] S. Tapia, G. Jacoby, H.R. Butcher, E.R. Craine and H.S. Stockman, submitted to the Astrophysical Journal (1983).
[9] M.L. Norman, L. Smarr and K.-H.A. Winkler, To appear in Numerical Astrophysics: A Festschrift in Honor of James R. Wilson, J. Centrella, R. Bowers, J. LeBlanc and M. LeBlanc, eds., (1983b), (paper III).
[10] R.D. Blandford, to appear in Numerical Astrophysics: A Festschrift in Honor of James R. Wilson, J. Centrella, R. Bowers, J. LeBlanc and M. LeBlanc, eds., (1983).
[11] M.D. Smith, M.L. Norman, K.-H.A. Winkler and L. Smarr, "Hot Spots in Radio Galaxies: A Comparison with Hydrodynamics Beam Cap Simulations", submitted to Monthly Notices Roy. Astron. Soc. (1983), (paper IV).
[12] R.A. Perley, A.H. Bridle and A.G. Willis, Submitted to the Astrophysical Journal, (1983).
[13] G. Miley, in "Astrophysical Jets", A. Ferrari and A.G. Pacholczyk, eds., (Reidel, Dordrecht, 1983).
[14] E.B. Fomalont, in Astrophysical Jets, A. Ferrari and A.G. Pacholczyk, eds., (Reidel, Dordrecht, 1983).
[15] R.H. Sanders, Ap. J. 266 (1983) 73.
[16] G. Birkoff, SIAM Review 25 (1983) 1.
[17] N. Zabusky, J. Comp. Phys. 43 (1981) 195.
[18] M.L. Norman and K.-H.A. Winkler, in Astrophysical Radiation Hydrodynamics, M.L. Norman and K.-H.A. Winkler, eds., (1984), (paper V).
[19] M.L. Norman, L. Smarr, K.-H.A. Winkler and M.D. Smith, Astron. Astrophys. 113 (1982) 285 (paper I).
[20] M.L. Norman, K.-H.A. Winkler and L. Smarr, in Astrophysical Jets, A. Ferrari and A.G. Pacholczyk, eds., (Reidel, Dordrecht, 1982). (paper II).
[21] K.-H.A. Winkler, in Astrophysical Radiation Hydrodynamics, M.L. Norman and K.-H.A. Winkler, eds., (1984) (paper VI).
[22] R.D. Blandford and M. Rees, Mon. Not. Roy. Astr. Soc. 169 (1974) 395.
[23] P.E. Hardee, Ap. J. 234 (1979) 47.
[24] P.E. Hardee, Ap. J. 257 (1982) 509.
[25] P.E. Hardee, Ap. J. 269 (1983) 94.

[26] A. Ferrari, E. Trussoni and L. Zaninetti, Astron. Astrophys. 64 (1978) 43.
[27] A. Ferrari, E. Trussoni, and L. Zaninetti, Mon. Not. Roy. Astr. Soc. 196 (1981) 1054.
[28] T.P. Ray, Mon. Not. Roy. Astr. Soc. 196 (1981) 195.
[29] H. Cohn, Ap. J. 269 (1983) 500.
[30] P.E. Dimotakis, R.C. Miake-Lye and D.A. Papantoniou, preprint GALCIT FM82-01 of the Graduate Aeronautical Laboratories of Cal. Tech. (1982).
[31] T. Tajima, and J.N. Leboeuf, Phys. Fluids 23 (1980) 884.
[32] M. Nepveu, Astron. Astrophys. (1982).
[33] P.R. Woodward, Bull. Amer. Astron. Soc. 14 (1982) 4, 810.
[34] M.D. Smith, L. Smarr, M.L. Norman and J.R. Wilson, Ap. J. 264 (1983) 432.
[35] M. Van Dyke, An Album of Fluid Motion (Parabolic Press, Stanford, CA, 1982).

Physica 12D (1984) 107–132
North-Holland, Amsterdam

1.3. PLASMAS

PLASMA–VACUUM INTERFACE PROBLEMS IN MAGNETOHYDRODYNAMICS

J.P. GOEDBLOED*
Massachusetts Institute of Technology, Plasma Fusion Center, Cambridge, MA 02139, USA

A review of ideal MHD theory pertaining to thermonuclear confinement problems is presented, starting with fundamental questions concerning the spectrum and ending with optimization studies of high-β tokamaks.

1. Introduction

In magnetic fusion research interface problems arise naturally from the requirement that a thermonuclear plasma be confined and isolated from the outside world by a vacuum magnetic field (fig. 1). In the early investigations this goal was thought to be realizable by the technique of pinch implosion, leaving behind a hot plasma of high density and separated from the vacuum region by a sharp boundary layer in which skin-currents are flowing. Ideally, this interface represents a surface of discontinuity where the description of the state of the system changes abruptly. The plasma would be described as an infinitely conducting fluid, whereas the little material that is left behind in the outer region would be assigned zero conductivity.

Historically, this ideal picture has played an important role in identifying and analyzing the most global instabilities, like external kink modes, resulting in a violent destruction of the plasma as a whole. With the success of tokamak performance and the concomitant decline of θ-pinch research, the stress has shifted more to the analysis of diffuse plasma configurations where there is hardly any space left for regions that would qualify as a vacuum. As is always the case with changing fashions, this has tended to obscure some of the previous insights obtained from the sharp-boundary model and, also, it has led away from an area of research where there might have been more benefit from developments in other parts of fluid dynamics. [The latter insight dawned on the author while attending the International Conference on Fronts, Interfaces, and Patterns.]

This paper is an attempt at reviewing some of the developments in magnetohydrodynamics (MHD) as they are relevant to the subject of interfaces, both in a narrow sense as pertaining to the phenomena described by skin-current models and in a wider sense as pertaining to a description of the global plasma motion of confined, i.e., inhomogeneous plasmas. It should be added at this point that this review is strongly biased toward some aspects of spectral theory of 1D systems and methods of analyzing 2D problems where certain personal preferences are obvious [1]. Some of the older literature pertinent to the subject of ideal magnetohydrodynamic equilibrium and stability is found in refs. 2 and 3. For more up to date reviews of the subject matter the reader is referred to refs. 4–6.

In section 2 ideal MHD is introduced as a relevant model for the study of toroidal plasma confinement properties. Interface problems and characteristics are discussed in section 3 to illustrate the complementarity of global and local approaches. Section 4 presents some of the basic issues of spectral theory of 1D MHD systems, in particular the gravitating slab and the diffuse linear pinch. In section 5 some 2D toroidal confinement problems are outlined and a method of

*On leave of absence from FOM Institute for Plasma Physics, Nieuwegein, The Netherlands.

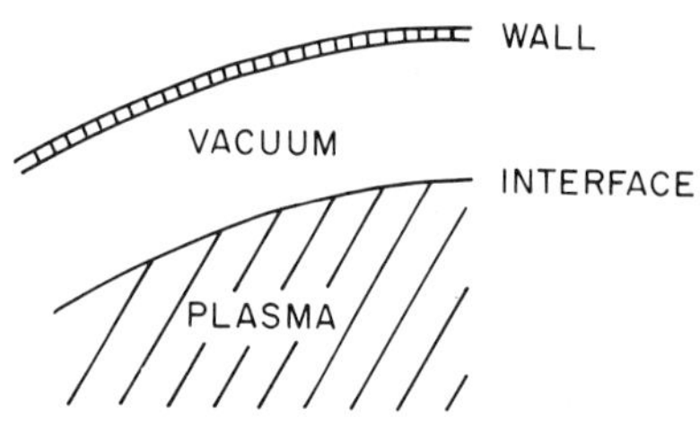

Fig. 1. Plasma–vacuum interface system.

solving them by means of conformal mapping is discussed. Explicit results with respect to the stability of high-β tokamaks are presented in section 6. Finally, some possibilities for future work are indicated in section 7.

2. The model of ideal MHD

2.1. *Experimental starting point: toroidal plasma confinement*

Our starting point is the task of properly describing the global properties of confinement of plasmas in toroidal geometry (fig. 2), as considered for the purpose of CTR (controlled thermonuclear reactions).

First, let us introduce some of the usual terminology and parameters. Plasma confinement in tokamaks is effected by an externally created toroidal magnetic field B_φ, which is modified by the poloidal plasma current I_p, and a poloidal magnetic field B_p, which is created by the toroidal plasma current I_φ. The resulting configuration is characterized by helical magnetic field lines which are wound on toroidal magnetic surfaces nested around a single magnetic field line which is purely toroidal, the magnetic axis. For a toroidal tube of circular cross-section the geometry is fixed by the value of the inverse aspect ratio $\epsilon \equiv b/R$, which is extensively exploited as a small parameter in many theoretical investigations.

An important problem presented by this approach to plasma confinement is to find the spatial distribution of the magnetic fields, currents, energies, pressure, density, etc., and, in particular, to find the critical value of the parameter β,

$$\beta_{\text{crit}} \equiv 2\langle p\rangle_{\max}/\langle B_\varphi^2\rangle_{\min}, \tag{1}$$

in dependence of the safety factor

$$q \equiv \frac{1}{2\pi}\oint \frac{B_\varphi}{RB_{\text{p}}}\,\mathrm{d}l \approx \frac{2\pi r^2 B_\varphi}{RI_\varphi}. \tag{2}$$

Here, $\langle p\rangle$ is the average plasma pressure and $\langle B_\varphi^2\rangle$ is the average squared toroidal field so that β presents a figure of merit for a particular plasma confinement machine determining the amount of thermonuclear power output that is eventually to be expected.

The parameter q has a simple geometric interpretation (fig. 3). It measures the average pitch of the magnetic field lines or the inverse rotational transform, which varies from one magnetic surface to the next. Hence, $q = q(\psi)$, where ψ is the poloidal flux which is a convenient radial coordinate to label the magnetic surfaces in toroidal

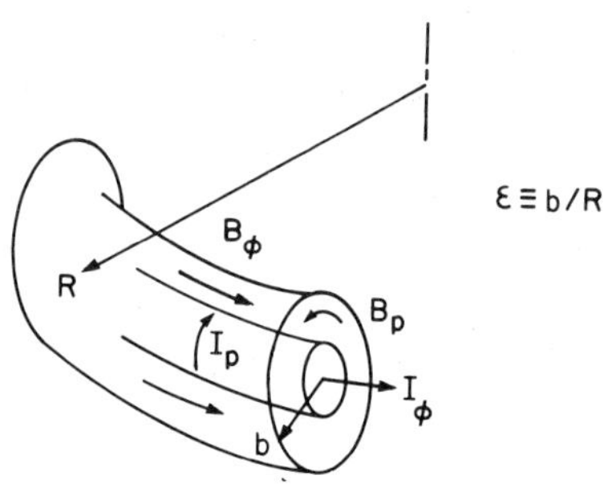

Fig. 2. Toroidal confinement geometry.

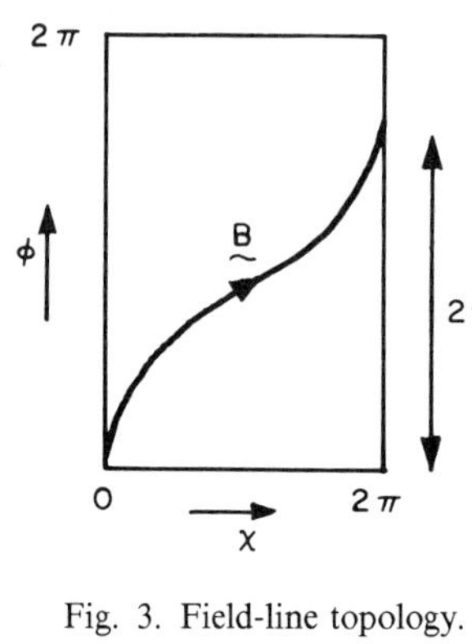

Fig. 3. Field-line topology.

geometry. In fig. 3 we have depicted a particular magnetic surface on which $q<1$ although usually $q>1$ is required on all magnetic surfaces for stability of tokamaks.

An alternative interpretation of q is indicated in the approximate equality of eq. (2), which is valid for $\beta \ll \epsilon \ll 1$ (so-called low-β tokamaks) and magnetic surfaces with circular cross-section. Here, the toroidal current I_φ appears in the denominator, so that low-q devices (e.g., reversed field pinches) correspond to large toroidal current and high q corresponds to small toroidal current. For tokamaks $q \sim 1$. The important point to notice is that the expression on the right of eq. (2) provides a more convenient parameter than q itself, when interpreted as a global parameter evaluated at the outermost magnetic surface, i.e., at the plasma–vacuum interface:

$$q^* \equiv \frac{2\pi a^2 B_\varphi}{R I_\varphi}. \tag{3}$$

Here, a is a measure for the radial dimension of the plasma column, usually taken to be the half-width in the equatorial plane. For so-called "high-β" tokamaks, where $\beta \sim \epsilon$, q^* may deviate appreciably from the value of q at the plasma surface.

We may now formulate one of the central questions of CTR: Find the distribution of the pressure p and of the magnetic field components B_p and B_φ, and of the associated current densities j_φ and j_p, such that the quantity β_crit defined in eq. (1) is maximum. Concerning the parameter q^*, this optimization leads to conflicting requirements with respect to Ohmic heating and equilibrium of the plasma, where one would like to have q^* as small as possible (i.e., I_φ maximum) and stability of the plasma, where high q^* operation (i.e., I_φ minimum) is more desirable. A surprisingly large part of present day plasma theory is devoted to precisely this single question. Accordingly, the present paper will address a number of interesting theoretical issues, but always aiming at a contribution to the solution of this question.

2.2. *Requirements for a theoretical model*

In order to solve the mentioned problems one needs a theoretical model that meets the following requirements [1]:

- It should respect the main physical conservation laws;
- it should permit a genuine treatment of the global toroidal geometry;
- it should have a decent mathematical structure.

Ideal MHD is presently the only model that satisfactorily combines these features. This model treats the plasma as a perfectly conducting fluid interacting with a magnetic field. It is interesting to notice that all modern physics seems to have been eliminated from this theory: No quantum effects are taken into account, neither are relativistic corrections considered, all kinetic effects have been removed by averaging over the velocity distributions of the individual particles, even electromagnetic waves have been eliminated by the neglect of the displacement current, and, finally, the model is considered to be ideal, i.e., without dissipation. Nevertheless, ideal MHD represents the simplest physical theory that still makes sense in the context of plasma confinement for CTR.

2.3. *Equations of ideal MHD*

The equations of ideal MHD may be written as follows:

$$\rho \frac{\mathrm{d}\boldsymbol{v}}{\mathrm{d}t} = -\nabla p + (\nabla \times \boldsymbol{B}) \times \boldsymbol{B}, \tag{4}$$

$$\frac{\partial \boldsymbol{B}}{\partial t} = \nabla \times (\boldsymbol{v} \times \boldsymbol{B}), \tag{5}$$

$$\frac{\partial p}{\partial t} = -\boldsymbol{v} \cdot \nabla p - \gamma p \nabla \cdot \boldsymbol{v}, \tag{6}$$

$$\frac{\partial \rho}{\partial t} = -\nabla \cdot (\rho \boldsymbol{v}). \tag{7}$$

Here, ρ is the density, $\boldsymbol{v}$ is the velocity, p is the pressure, and $\boldsymbol{B}$ is the magnetic field. These four equations with associated initial data and boundary conditions, together with a specification of the geometry (as in fig. 2), constitute the most global approach to the description of confinement problems in CTR.

Concerning the three requirements for an acceptable model mentioned above, we notice that eqs. (4)–(7) express the conservation of momentum, magnetic flux, entropy, and mass, respectively. It has been established by Friedrichs [7] that these equations constitute a system of nonlinear symmetric hyperbolic partial differential equations, where the nonlinearity is only of a quasilinear nature. Consequently, this set of equations is well-suited for numerical solution as an initial value problem for toroidal geometries [8–10].

If we wish to study plasma–vacuum systems with an interface (fig. 1) we need to supplement eqs. (4)–(7) with the equations describing the vacuum magnetic field $\hat{\boldsymbol{B}}$:

$$\nabla \times \hat{\boldsymbol{B}} = 0, \tag{8}$$

$$\nabla \cdot \hat{\boldsymbol{B}} = 0. \tag{9}$$

These equations are all that is left from Maxwell's equations when the displacement current is negligible. The plasma variables $\rho, p, \boldsymbol{v}, \boldsymbol{B}$ are connected with the vacuum variable $\hat{\boldsymbol{B}}$ through two boundary conditions:

$$\boldsymbol{n} \cdot \boldsymbol{B} = \boldsymbol{n} \cdot \hat{\boldsymbol{B}} = 0, \tag{10}$$

$$[\![p + \tfrac{1}{2}B^2]\!] = 0, \tag{11}$$

which guarantee that $\boldsymbol{B}$ and $\hat{\boldsymbol{B}}$ remain tangential to the plasma vacuum interface and that the total pressure is continuous across this interface. The discontinuity of the tangential component of $\boldsymbol{B}$, which is allowed by eqs. (10) and (11), is thought to originate from a surface current flowing at the plasma surface:

$$\boldsymbol{j}^* = \boldsymbol{n} \times [\![\boldsymbol{B}]\!]. \tag{12}$$

Finally, the whole configuration is either closed off from the outside world by the presence of a perfectly conducting wall where $\boldsymbol{n} \cdot \hat{\boldsymbol{B}} = 0$ or it is open with respect to a set of coils which feed in energy and flux. In the present paper we will be exclusively concerned about the first possibility.

3. Two approaches

3.1. *Global interface problems: Rayleigh–Taylor and kink instabilities*

Two typical interface problems that are relevant to the stability of plasmas confined by magnetic fields are shown in fig. 4: The Rayleigh–Taylor instability of a gravitating plasma slab embedded in a magnetic field which abruptly changes direction at the plasma–vacuum interface (fig. 4a) and the kink instability of a straight plasma cylinder with a longitudinal magnetic field and surrounded by a helical magnetic field in the vacuum (fig. 4b). The two situations are quite similar and represent just a minor modification of the two examples that were first investigated by Kruskal and Schwarzschild [11].

In both cases the change of direction of the magnetic field at the plasma–vacuum interface is caused by perpendicular surface currents. These currents also create a jump in the magnitude of $\boldsymbol{B}$ which, according to eq. (11), allows for a finite value of β of the confined plasma: $\beta \equiv 2p/\hat{B}^2 = [\![B^2]\!]/\hat{B}^2$.

To analyze the stability of these configurations the static equilibrium is subjected to a small perturbation of the form

$$\boldsymbol{\xi}(x)\mathrm{e}^{\mathrm{i}(\boldsymbol{k}\cdot\boldsymbol{r}-\omega t)}, \tag{13}$$

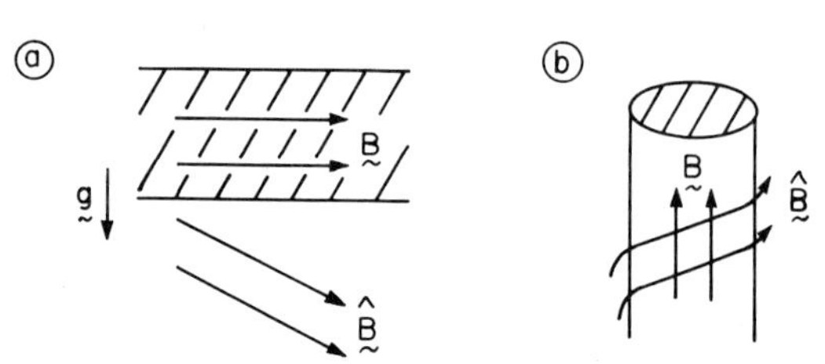

Fig. 4. Geometries for Rayleigh–Taylor and kink instabilities: (a) gravitating slab; (b) linear pinch.

where $\boldsymbol{x}$ is the direction of inhomogeneity (r for the cylindrical case) and $\boldsymbol{k}$ is the wavevector in the two homogeneous directions: $(0, k_y, k_z)$ for the slab and $(m/r, 0, k)$ for the cylinder. Exploiting the equations of section 2, a straightforward analysis leads to ordinary second order differential equations of the normal components of $\boldsymbol{v} = \partial \boldsymbol{\xi}/\partial t$ in the plasma and of the perturbation $\hat{\boldsymbol{Q}}$ of the vacuum magnetic field. Satisfying the appropriate boundary conditions then leads to the following dispersion equation for long-wavelength perturbations:

$$\rho\omega^2 = (\boldsymbol{k}\cdot\boldsymbol{B})^2 + (\boldsymbol{k}\cdot\hat{\boldsymbol{B}})^2 - \begin{cases} k^2\rho g a, & \text{for the slab,} \\ |m|\hat{B}_\theta^2/a, & \text{for the cylinder.} \end{cases} \tag{14}$$

Eq. (14) is a typical example of the kind of results one obtains from a global analysis like the one we have just described. There are three terms which may be associated with the perturbation of the plasma, of the vacuum, and of the plasma–vacuum interface, respectively. Here, the first two terms, which are positive, are the stabilizing contributions due to the bending of the plasma and vacuum magnetic field lines. The third term, which is negative, is the term responsible for the Rayleigh–Taylor instability of the slab or the kink instability of the cylinder.

Although the driving mechanisms of the two instabilities are different (gravity in the first and curvature of the magnetic file lines in the second case), the magnitude of the unstable term is determined in both cases by the amount of surface currents flowing at the plasma–vacuum interface. Thus, the parameter β enters into the calculation as a parameter associated with instability. On the other hand, it is clear from the form of eq. (14) that the instability is counteracted by shear of the magnetic field, i.e., by the change of direction of the field lines across the plasma–vacuum interface. This effect prevents the occurrence of perturbations for which both stabilizing terms would vanish simultaneously. Consequently, perturbations for

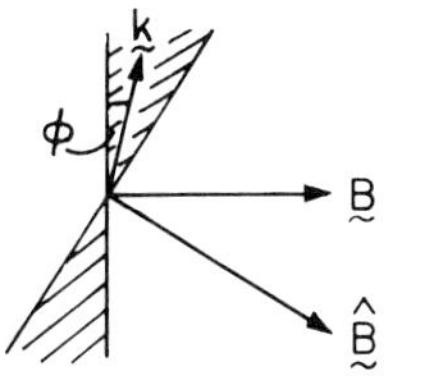

Fig. 5. Direction of $\boldsymbol{k}$-vector for the most unstable perturbations of the configurations of fig. 4.

which $\boldsymbol{k}\cdot\boldsymbol{B}$ is minimal ($\boldsymbol{k}$ lying in the shaded area of fig. 5) still create a finite amount of stabilizing energy.

In fig. 6 the frequencies of the waves (or the growth rates of the instabilities when $\omega^2 < 0$) are plotted versus the component of the wavevector parallel to the magnetic field in the plasma. Also

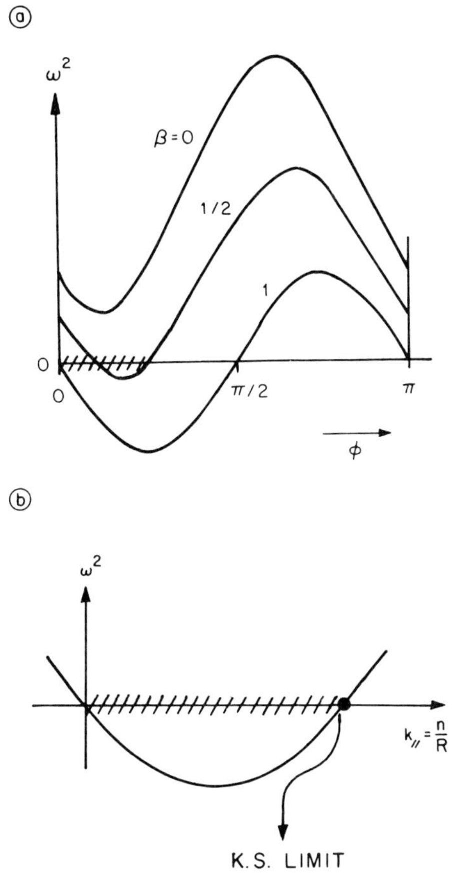

Fig. 6. Stability diagrams for the Rayleigh–Taylor (a) and kink instability (b).

indicated are the regions corresponding to fig. 5, where the influence of field-line bending is minimal. From fig. 6a the destabilizing influence of β is obvious. In fig. 6b the growth rate of the kink instability is plotted for $\beta = 0$. Clearly, already in the absence of plasma pressure the plasma is violently unstable for a wide range of values of $k_{\parallel}$. Typically, growth rates of the order of $10^6\ \mathrm{s}^{-1}$ are obtained from eq. (14), so that one cannot tolerate these kinds of instabilities in a thermonuclear confinement experiment.

Considering the straight cylinder as the leading order approximation of a toroidal geometry, a very simple stabilization method of the kink instability is found to reside in the quantization condition $k_{\parallel} = n/R$, where n is the toroidal wave number. Clearly, when the first mode ($n = 1$) has a longitudinal wavelength that does not fit into the torus, one does not have to worry about the kink mode any more. A simple derivation by means of eq. (14) shows that this is equivalent to the celebrated Kruskal–Shafranov limit,

$$q^* = q(a) > 1, \tag{15}$$

which is a limit on the toroidal plasma current according to eq. (3).

3.2. *Local phenomena*: *characteristics*

An entirely different approach to the solution of the equations of section 2 is obtained when one considers the problem from a local point of view. Rather than trying to catch the global structure of the solutions at once, as in section 3.1, one pays more attention to the internal inhomogeneities of the plasma and tries to construct solutions in the large from local constituents. This is the method of characteristics, which is very well described in refs. 7 and 12.

Along these lines the phase and group diagrams of the MHD waves are obtained from eqs. (4)–(7) by perturbation methods. The phase diagram (fig. 7a) is obtained by computing the distance which a plane wave front travels along the wavevector $\boldsymbol{k}$

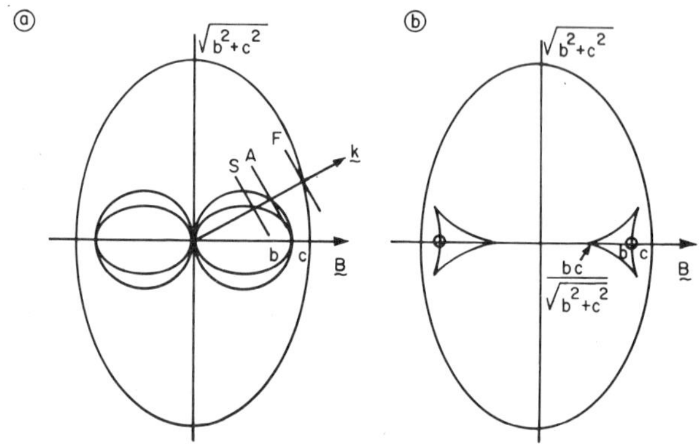

Fig. 7. Phase (a) and group (b) diagrams for the three MHD waves: fast magneto acoustic (F), Alfvén (A), and slow magneto acoustic (S). The Alfvén speed b is taken smaller than the sound speed c.

after having passed the origin at $t = 0$. Since the MHD equations comprise both magnetic and acoustic phenomena, two distinctive speeds enter the description, viz. the Alfvén speed b and the sound speed c:

$$b \equiv B/\sqrt{\rho}, \quad c \equiv \sqrt{\gamma p/\rho}. \tag{16}$$

The interplay of these phenomena constitutes part of the beauty of ideal MHD. Three waves are obtained: (1) The fast magnetoacoustic wave, which transforms into an ordinary sound wave if the magnetic field vanishes ($b \to 0$), (2) The Alfvén wave, which propagates preferentially along the magnetic field; (3) The slow magnetoacoustic wave, which also exhibits a strong anisotropy with respect to the magnetic field.

The actual characteristic surfaces, or rather the spatial parts of them, are obtained by considering the response to a point disturbance emitted at $t = 0$ from the origin, i.e., by constructing the envelopes of all plane waves passing through the origin of fig. 7a. This gives the group diagram, or ray surface, depicted in fig. 7b. Notice that the anisotropy is even more pronounced now, with the Alfvén waves moving as two points ($\pm b$) and the slow waves as two cusped figures in opposite directions along the magnetic field.

The three MHD waves are most easily distinguished on the basis of their polarizations, which are shown in fig. 8. The Alfvén waves are characterized by a displacement $\boldsymbol{\xi}_A$ and a magnetic perturbation $\boldsymbol{Q}_A$ which are both perpendicular to the plane through $\boldsymbol{k}$ and $\boldsymbol{B}$, whereas the polarizations $\boldsymbol{\xi}_S$ and $\boldsymbol{\xi}_F$ and the magnetic perturbations $\boldsymbol{Q}_S$ and $\boldsymbol{Q}_F$ of the magneto acoustic waves are situated in this plane. Of particular importance is the fact that the three vector $\boldsymbol{\xi}_S$, $\boldsymbol{\xi}_A$ and $\boldsymbol{\xi}_F$ form an orthogonal triad so that arbitrary initial conditions can be decomposed in terms of the three underlying waves.

The method of characteristics consists of solving the nonlinear MHD equations by integrating along the characteristics. The overall inhomogeneity of the medium (e.g., due to the toroidal geometry) only enters indirectly through the fact that we propagate the solution from one space-time point to the next. The advantage of this method is that it is general, not restricted to certain specific geometries. An obvious disadvantage as compared to the global approach sketched in section 3.1 is that it is difficult to obtain global solutions this way and that one is quickly led to a large computational effort.

In the next section we will describe a kind of merging of the two approaches, where both the global and the local aspect of the theory are taken seriously, albeit to the cost of losing the nonlinearities.

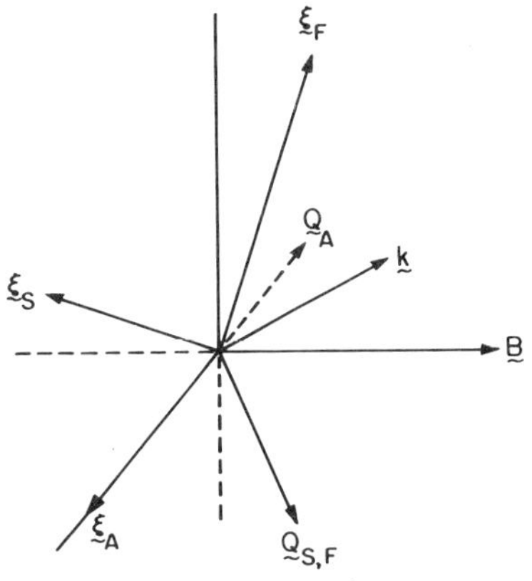

Fig. 8. Displacements $\boldsymbol{\xi}$ and magnetic perturbations $\boldsymbol{Q}$ for the three MHD waves.

4. Spectral theory of ideal MHD

4.1. *Outline*

The global analysis of plasma–vacuum systems as exemplified by the two cases treated in section 3.1 has the obvious shortcoming of doing little justice to the internal structure of the plasma equilibrium, as expressed, e.g., by the profiles $\rho(r)$, $p(r)$, $B_\theta(r)$, $B_z(r)$ of the linear pinch. In toroidal geometry these profiles even depend on two coordinates $(r, \theta$ or $\psi, \chi)$. On the other hand, the method of characteristics indicated in section 3.2 does not permit us to gain much analytical insight into the overall solutions for confined plasmas. Here, the spectral theory of ideal MHD, which has been extensively investigated over the past ten years, suggests itself as the most logical approach for the solution of these problems.

The following questions present themselves: What is the fate of the three ideal MHD waves shown in fig. 8 when the locally homogeneous background is replaced by the globally inhomogeneous and curved environment of confined plasmas? What is their role in determining the stability of toroidal systems? Significant progress in the solution of these questions has been obtained along three lines: (1) For 1D inhomogeneous plasmas (slab, cylinder) the spectrum has been partly disentangled [13–20]. (2) For 2D inhomogeneous plasmas (axisymmetric torus) analytical progress has been obtained by means of the so-called ballooning representation [21–24]. (3) Numerical results on a wide variety of 2D problems have been produced by means of perfected equilibrium and stability codes, notably PEST and ERATO [25–34]. We will discuss some of these items in the following, while misusing the occasion by adding some new results obtained along the first line (section 4.3) and by turning to a discussion of the program HBT when discussing 2D numerical techniques (section 5).

Spectral theory of ideal MHD starts with the assumption that the plasma motion may be described as a departure from a static equilibrium state, where the plasma is at rest: $\boldsymbol{v}_0 = 0$. While the density, ρ_0, may be chosen arbitrarily, the two other variables p_0 and $\boldsymbol{B}_0$ according to eq. (4) then have to satisfy the equilibrium equations

$$\nabla p_0 = (\nabla \times \boldsymbol{B}_0) \times \boldsymbol{B}_0, \quad \nabla \cdot \boldsymbol{B}_0 = 0. \tag{17}$$

This leads to trivial equations in one dimension, whereas for 2D systems a nonlinear partial differential equation is obtained, the Grad–Shafranov equation.

Perturbations of this equilibrium state are described by the plasma displacement vector $\boldsymbol{\xi}(\boldsymbol{r}, t)$, connected with the velocity by the relation $\boldsymbol{v}_1 = \partial \boldsymbol{\xi}/\partial t$. Inserting this expression into the eqs. (4)–(7) directly leads to the famous linearized equation of motion for ideal MHD [35]:

$$\rho_0 \frac{\partial^2 \boldsymbol{\xi}}{\partial t^2} = \mathscr{F}(\boldsymbol{\xi}), \tag{18}$$

where

$$\begin{aligned} \mathscr{F}(\boldsymbol{\xi}) &\equiv -\nabla p_1 + (\nabla \times \boldsymbol{B}_0) \times \boldsymbol{B}_1 \\ &\quad + (\nabla \times \boldsymbol{B}_1) \times \boldsymbol{B}_0, \\ \boldsymbol{B}_1 &= \nabla \times (\boldsymbol{\xi} \times \boldsymbol{B}_0), \\ p_1 &= -\boldsymbol{\xi} \cdot \nabla p_0 - \gamma p_0 \nabla \cdot \boldsymbol{\xi}. \end{aligned} \tag{19}$$

Turning to a study of normal mode solutions $e^{-i\omega t}$, the basic equation for ideal MHD spectral theory is obtained,

$$\mathscr{F}(\boldsymbol{\xi}) = -\rho \omega^2 \boldsymbol{\xi}, \tag{20}$$

where $\mathscr{F}$ is a Hermitian operator. Inserting a particular equilibrium solution of eq. (17), we may obtain the waves ($\omega^2 > 0$) and instabilities ($\omega^2 < 0$) of the system by solving eq. (20) subject to certain boundary conditions which may be found in the literature [35].

4.2. *1D inhomogeneous plasmas: gravitating slab*

Let us now reconsider the Rayleigh–Taylor problem of section 3.1, where we eliminate the vacuum region and concentrate the study on the diffuse plasma. According to eq. (17), supplemented with a gravitational term, the variables $\rho_0(x)$, $p_0(x)$, and $\boldsymbol{B}_0(x)$ should satisfy the equilibrium condition

$$\left(p_0 + \tfrac{1}{2}B_0^2\right)' + \rho_0 g = 0, \tag{21}$$

which leaves great freedom in the choice of equilibrium profiles. In particular, we may choose a field $\boldsymbol{B}_0 = B_y(x)\boldsymbol{e}_y + B_z(x)\boldsymbol{e}_z$ which arbitrarily changes direction when we move in the direction of inhomogeneity, i.e., along the x-axis.

Because inhomogeneities are restricted to x, we may write

$$\boldsymbol{\xi}(\boldsymbol{r}, t) = \boldsymbol{\xi}(x) \exp\left[i(k_y y + k_z z - \omega t)\right]. \tag{22}$$

It is now expedient to introduce a projection for the perturbations and the gradient operator which distinguishes between the directions (1) normal to the magnetic surfaces in the direction of the inhomogeneity (unit vector $\boldsymbol{e}_n$); (2) tangential to the magnetic surfaces and perpendicular to $\boldsymbol{B}_0$ (unit vector $\boldsymbol{e}_\perp$), and (3) parallel to $\boldsymbol{B}_0$ (unit vector $\boldsymbol{e}_\parallel$) [see fig. 9]. This produces the following decomposition:

$$\boldsymbol{\xi}(x) = (X, Y, Z), \quad \nabla = (\hat{D}, \hat{G}, \hat{F}), \tag{23}$$

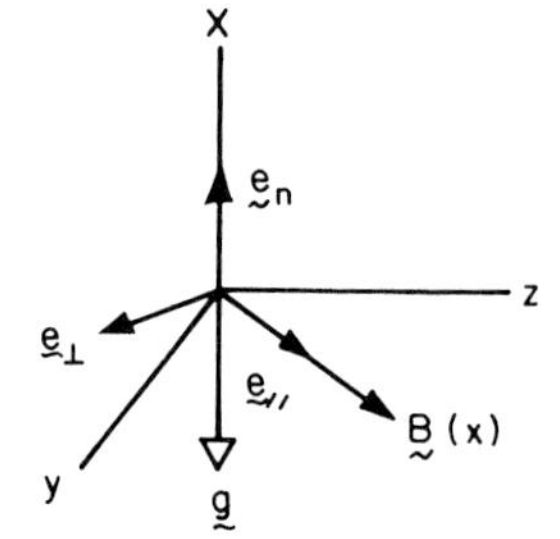

Fig. 9. Projection for inhomogeneous plasmas.

where $\hat{D} \equiv \mathrm{d}/\mathrm{d}x$, $\hat{G} \equiv k_\perp(x)$, $\hat{F} \equiv k_\parallel(x)$. Using this projection, eq. (20) becomes

$$\begin{pmatrix} \hat{D}(\gamma p + B^2)\hat{D} - \hat{F}B^2\hat{F} & [\hat{D}(\gamma p + B^2) + \rho g]\hat{G} & [\hat{D}\gamma p + \rho g]\hat{F} \\ -\hat{G}[(\gamma p + B^2)\hat{D} - \rho g] & -\hat{G}(\gamma p + B^2)\hat{G} - \hat{F}B^2\hat{F} & -\hat{G}\gamma p\hat{F} \\ -\hat{F}[\gamma p\hat{D} - \rho g] & -\hat{F}\gamma p\hat{G} & -\hat{F}\gamma p\hat{F} \end{pmatrix} \begin{pmatrix} X \\ Y \\ Z \end{pmatrix} = -\rho\omega^2 \begin{pmatrix} X \\ Y \\ Z \end{pmatrix}, \tag{24}$$

which is the basic equation for the study of the spectrum of a 1D gravitational slab. This system of equations turns out to be equivalent to one ordinary second order differential equation for X [eq. (28) below], which may be solved subject to the boundary conditions $X = 0$ at $x = 0$ and $x = a$, say. For $g = 0$ and $\mathrm{d}/\mathrm{d}x = \mathrm{i}k_x$, we recover the three waves of fig. 8.

The first effect of the inhomogeneity is the appearance of two continuous spectra, corresponding to the existence of singular perturbations that are localized on particular magnetic surfaces. These may be found from eq. (24) by taking the limit $\mathrm{d}/\mathrm{d}x \to \infty$. In this limit the first component of eq. (24) may be integrated once to give

$$(\gamma p + B^2)\,\mathrm{d}X/\mathrm{d}x \approx -\hat{G}(\gamma p + B^2)Y - \hat{F}\gamma p Z. \tag{25}$$

Redefine: $F \equiv B\hat{F}$ and $G = B\hat{G}$. Inserting expression (25) into the second component of eq. (24) then yields

$$(\rho\omega^2 - F^2)Y \approx 0, \tag{26}$$

having the solution $Y \approx \delta(x - x_\mathrm{A})$, where $x = x_\mathrm{A}$ is the singular point where $\omega^2 = \omega_\mathrm{A}^2 \equiv F^2/\rho$. The collection of improper eigenvalues $\{\omega_\mathrm{A}^2\}$ for the interval $0 \le x \le a$ constitutes the continuous Alfvén spectrum. Similarly, inserting the expression (25) into the third component of eq. (24) gives

$$\left(\rho\omega^2 - \frac{\gamma p}{\gamma p + B^2}F^2\right)Z \approx 0, \tag{27}$$

having the solution $Z \approx \delta(x - x_\mathrm{S})$, where $x = x_\mathrm{S}$ is the singular point where $\omega^2 = \omega_\mathrm{S}^2 \equiv [\gamma p/(\gamma p + B^2)]F^2/\rho$. The collection of improper eigenvalues $\{\omega_\mathrm{S}^2\}$ constitutes the continuous slow wave spectrum.

In order to go beyond these singular continuum modes, which are extremely localized perturbations, we need to solve the differential equations (24) on a global scale. This is most conveniently done by means of the single second order differential equation for X, which is obtained from eq. (24) by eliminating Y and Z:

$$\frac{\mathrm{d}}{\mathrm{d}x}\frac{N}{D}\frac{\mathrm{d}X}{\mathrm{d}x} + \left[\rho\omega^2 - F^2 + \rho' g - \frac{k^2\rho^2 g^2(\rho\omega^2 - F^2)}{D} + \left\{\rho g\frac{\rho\omega^2(\rho\omega^2 - F^2)}{D}\right\}'\right]X = 0, \tag{28}$$

where

$$N \equiv \rho^2(\gamma p + B^2)(\omega^2 - \omega_\mathrm{A}^2)(\omega^2 - \omega_\mathrm{S}^2), \tag{29}$$

$$D \equiv \rho^2(\omega^2 - \omega_\mathrm{I}^2)(\omega^2 - \omega_\mathrm{II}^2),$$

$$\rho\omega_{\mathrm{I,II}}^2 \equiv \frac{1}{2}k^2(\gamma p + B^2) \times \left[1 \pm \sqrt{1 - 4\frac{\gamma p F^2}{k^2(\gamma p + B^2)^2}}\right]. \tag{30}$$

Notice that the Alfvén and slow mode singularities of eqs. (26) and (27) now appear as the zeros of the coefficient N in eq. (28). In addition, two more special frequencies ω_I^2 and ω_II^2 appear, which were originally thought to be associated with two more continua [13], subsequently proved to be apparent singularities [14], and finally shown to represent regions of non-monotonicity of the discrete spectrum [15]. For a more extensive derivation of eqs. (28), see ref. 36.

One, admittedly somewhat primitive, motive behind this explicit derivation is to impress the reader

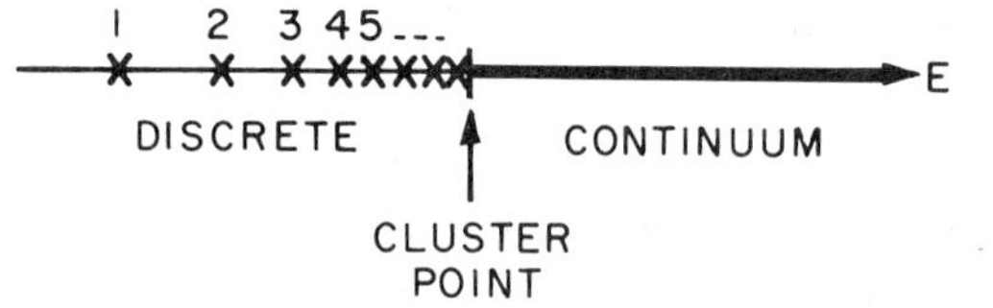

Fig. 10. Structure of the spectrum of a 1D quantum mechanical system.

with the complexity of the basic MHD equations for even the simplest cases. Consider, e.g., the comparable equation for 1D quantum mechanical systems,

$$\left[-\frac{\hbar^2}{2m}\nabla^2 + V(r)\right]\psi = E\psi, \tag{31}$$

where the gradient operator may still be reduced with respect to the two ignorable coordinates, corresponding to quantum numbers l and m, say. For given l and m, the eigenvalue E of the Hamiltonian operator is then a function of n, the number of nodes of $\psi(r)$. Since eq. (31) is of a classical Sturm–Liouville type, this dependence is monotonic, so that the schematic representation of fig. 10 is obtained for the spectrum.

In contrast, the presence of the factor N/D in eq. (28) points to the fact that the ideal MHD problem is not of a classical Sturm–Liouville type. Consequently, for fixed k_y and k_z, the eigenvalue ω^2 of the MHD force operator $\mathscr{F}$ is not a monotonic function of the number of nodes of $X(x)$. Instead, partial monotonicity is obtained for the discrete subspectra separately, as schematically shown in fig. 11 for the case of weak inhomogeneity. It can be shown [15, 18] that the three ideal MHD waves split into five discrete subspectra which cluster at the tips of the slow and Alfvén continua $\{\omega_S^2\}$ and $\{\omega_A^2\}$ and at $\omega^2 = \infty$ for the fast subspectra. Sturmian regions, where $\omega^2(n)$ monotonically increases (indicated by the upward arrows in fig. 11), and anti-Sturmian regions, where $\omega^2(n)$ monotonically decreases (downward arrows), alternate, where it depends on the sign of N/D which of the two options is realized.

The clear separation of subspectra only obtains when the frequency ranges $\{\omega_S^2\}, \{\omega_I^2\}, \{\omega_A^2\}$, $\{\omega_{II}^2\}$ do not overlap, i.e., when the inhomogeneities are not too strong. It is not difficult to design situations where the whole positive ω^2-axis is covered with continua so that there are no discrete modes at all. On the other hand, if the inhomogeneities are weak, not all discrete subspectra need to be present. This depends on whether certain conditions on the equilibrium variables are satisfied. E.g., expansion in the neighborhood of the Alfvén continuum provides the following expression for the cluster spectrum caused by gravity:

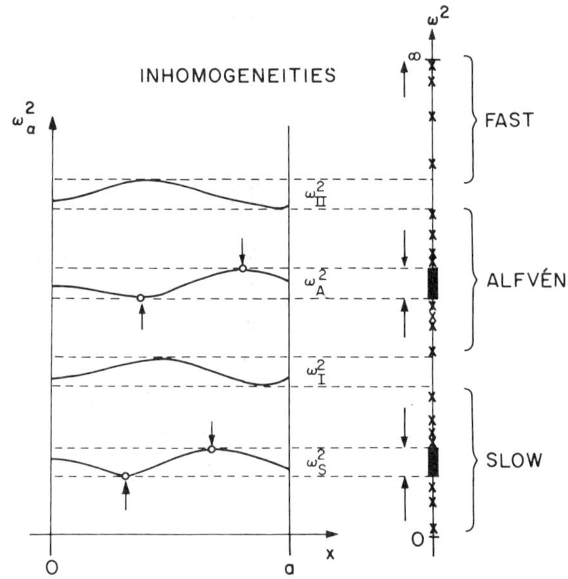

Fig. 11. Structure of the spectrum of a 1D magnetohydrodynamic system. Accumulation of point eigenvalues both from above and below.

$$\omega^2 \approx \omega_A^2 - \frac{G^2}{q^2 B^2}\frac{\rho' g}{\rho}, \tag{32}$$

where q is the local wave number in the x-direction. Hence, if $\rho' g < 0$ an anti-Sturmian Alfvén spectrum is obtained, whereas $\rho' g > 0$ provides a Sturmian Alfvén spectrum. It is of interest to notice that only in the latter case the discrete Alfvén spectrum may spill over to the negative side of the ω^2-axis. This is associated with the interchange instability of a gravitating slab.

To complete this study would require a systematic numerical investigation of eq. (28) for all kinds of representative equilibria. Such a study has not yet been undertaken. It is embarrassing to notice that ten years of active research in spectral theory of ideal MHD has not yet resulted in a final classification of the spectrum of even the simplest systems. In particular, the first paper initiating a systematic analysis of MHD spectra from a group-theoretical point of view is yet to be written. [Challenge!]

4.3. 1*D inhomogeneous plasmas*: *diffuse linear pinch*

For plasma confinement problems the geometry of the diffuse linear pinch is of more intrinsic interest than that of the gravitating slab, although the analysis proceeds along the same lines. The equation analogous to eq. (28) was first derived by Hain and Lüst [37]. Writing $r\xi_r = \chi(r)\exp\mathrm{i}(m\theta + kz - \omega t)$ it takes the form

$$(f\chi')' - g\chi = 0, \tag{33}$$

where

$$\begin{aligned} f &= N/(rD), \\ g &= -\frac{1}{r}(\rho\omega^2 - F^2) + \left(\frac{B_\theta^2}{r^2}\right)' \\ &\quad + \frac{4k^2B_\theta^2}{r^2D}(B^2\rho\omega^2 - \gamma pF^2) \\ &\quad - \left\{\frac{2kB_\theta G}{r^2D}[(\gamma p + B^2)\rho\omega^2 - \gamma pF^2]\right\}', \end{aligned} \tag{34}$$

and

$$N = (\rho\omega^2 - F^2)[(\gamma p + B^2)\rho\omega^2 - \gamma pF^2], \tag{35}$$

$$\begin{aligned} D &= \rho^2\omega^4 - (m^2/r^2 + k^2)(\gamma p + B^2)\rho\omega^2 \\ &\quad + (m^2/r^2 + k^2)\gamma pF^2, \\ F &= mB_\theta/r + kB_z, \\ G &= mB_z/r - kB_\theta. \end{aligned} \tag{36}$$

Notice immediately: Although the structure of the equations (28) and (33) is very similar, there is no simple recipe to translate gravity into field-line curvature.

A similar picture as fig. 11 may be obtained for the spectrum of a weakly inhomogeneous linear pinch, where the frequencies ω_{I}^2 and ω_{II}^2 occupy a larger portion of the ω^2-axis though because of the singularity m^2/r^2 in the expression for D.

It is of interest to analyze the cluster spectra in a little more detail for this case because of the renewed interest in Alfvén wave heating [17, 38, 39] and the discovery of global Alfvén modes just below the lower edge of the continuum [40, 41]. To that end, consider the case of weak inhomogeneities where eq. (33) may be solved by means of a WKB solution. Writing

$$\chi(r) = p(r)\exp\left[\mathrm{i}\int q(r)\,\mathrm{d}r\right], \tag{37}$$

the expressions $p(r)$ and $q(r)$ are determined by requiring that the solution (37) be correct to leading order in the inhomogeneity. Substitution into eq. (33) yields

$$p \approx (-fg)^{-1/4}, \quad q \approx (-g/f)^{1/2}, \tag{38}$$

where we have to demand that $q^2L^2 \gg 1$ in order for the WKB approximation to be valid. Here, L is the scale length for the inhomogeneities.

The relation (38) gives a local dispersion equation relating ω^2 and the local radial wave number q:

$$q^2 = -(rD/N)g, \tag{39}$$

which is a quintic in ω^2. [No, this has nothing to do with the fact that the discrete spectrum consists of five subspectra]. This equation may be solved in the neighborhood of the Alfvén and slow continua, when these are sufficiently far apart. This gives

$$\rho\omega^2 \approx F^2 - \frac{1}{q^2}A, \tag{40}$$

$$A \equiv \frac{4k^2B_\theta^2(B^2 - \gamma p)}{r^2B^2} - \frac{rG^2}{B^2}\left(\frac{B_\theta^2}{r^2} + \frac{2kB_\theta B^2}{r^2G}\right)'$$

and

$$\rho\omega^2 \approx \frac{\gamma p}{\gamma p + B^2} F^2 - \frac{1}{q^2} S,$$

$$S \equiv -\left(\frac{\gamma p}{\gamma p + B^2}\right)^2 \left[\frac{F^4}{\gamma p + B^2} + \frac{rF^2}{B^2}\left(\frac{B_\theta^2}{r^2}\right)' - \frac{4k^2 B_\theta^2}{r^2 B^2}(\gamma p + B^2)\right]. \tag{41}$$

Consequently, for weakly inhomogeneous plasmas discrete Alfvén modes may be found either above the continuum when $A < 0$ or below the continuum when $A > 0$. Likewise, anti-Sturmian slow modes are found above the slow continuum when $S < 0$ and Sturmian slow modes are found below the slow continuum when $S > 0$.

Very similar results are obtained from a Frobenius expansion around extrema of the functions $\omega_A^2(r)$ and $\omega_S^2(r)$, as indicated in fig. 11. Here, there is no restriction on the strength of the inhomogeneities, except that $\{\omega_A^2\}$ and $\{\omega_S^2\}$ should not overlap. We find the following condition for the Alfvén continuum to have a clusterpoint from below:

$$\omega_A^{2\prime} = 0, \quad 0 < (1/8)\rho\omega_A^{2\prime\prime} < A, \tag{42}$$

where A is the same expression as defined in eq. (40b). Likewise, the condition for the slow continuum to have a clusterpoint from below is found to be

$$\omega_S^{2\prime} = 0, \quad 0 < (1/8)\rho\omega_S^{2\prime\prime} < S, \tag{43}$$

where S is defined in eq. (41b). The conditions for the existence of clusterpoints from above are found from the equations (42) and (43) by just reversing both inequality signs.

Since the expressions for A and S involve many free parameters and profile functions it is always possible to find equilibria which satisfy any of the four conditions at some point in the plasma. For such equilibria the general structure of a fully developed spectrum as shown in fig. 11 is to be expected. In particular, it turns out to be quite easy to satisfy condition (42) for tokamaks, so that Alfvén wave heating by resonant excitation of global Alfvén modes below the continuum is a viable possibility.

[To my knowledge the conditions (40)–(43), which were announced in ref. 18 appear here for the first time in the literature.]

4.4. *Interface problems*

Returning to the study of plasma–vacuum interface systems it should be clear that the internal structure of the plasma may have a significant influence on the behavior of the global eigenfunctions. On the other hand, the very presence of a discontinuity in the plasma density at the plasma edge also produces a very large effect on the internal modes, even on the continuum modes. This is clear from the expression for the Alfvén continuum frequencies $\omega_A^2 \equiv F^2/\rho$, which blows up when $\rho \to 0$. Consequently, a continuum as shown in fig. 12 is obtained. We have seen that such a continuum with a minimum allows for the excitation of global Alfvén modes below the continuum when the condition (42) is satisfied.

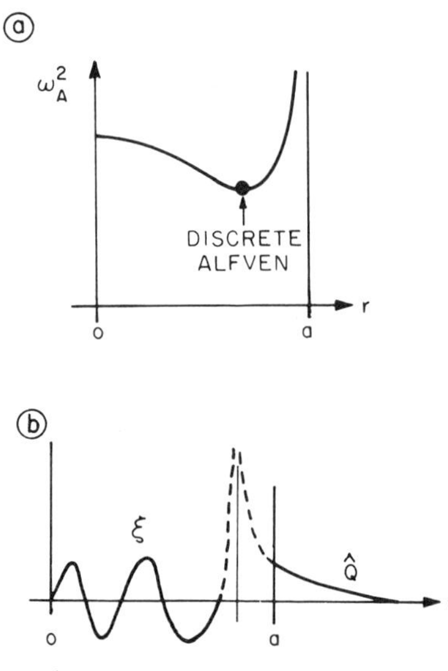

Fig. 12. Typical continuous spectrum of Alfvén waves (a) producing a clusterpoint of discrete global Alfvén modes and transforming the fast magneto acoustic waves into damped quasi-modes (b).

Another interesting result of such a radial dependence of $\omega_A^2(r)$ is the fact that the discrete fast wave spectrum is swallowed by the Alfvén continuum. Consequently, discrete fast modes do not exist anymore. Instead, one finds modes with eigenfunctions as schematically indicated in fig. 12b. The interior part of the mode behaves much like an ordinary fast wave, but an Alfvén singularity always interferes before it can be joined to the exterior solution, expressed by the vacuum field perturbation $\hat{Q}$. At the singularity the mode necessarily picks up a non-square integrable contribution, both with respect to the potential energy and with respect to the proper norm $I = \frac{1}{2}\int \rho \boldsymbol{\xi}^2 \, dV$ to be used in MHD spectral studies. The improper part of the eigenfunction may consist both of large contributions, that exhibit a logarithmic behavior of the normal component and a $1/x$ dependence of the tangential component of $\boldsymbol{\xi}$, and of "small" contributions, that exhibit a jump of the normal component and a δ-function behavior of the tangential component.

Different kinds of improper fast magnetoacoustic modes are obtained, depending on the relative amplitude of the "small" and large components. An interesting discrete set of special improper fast modes is obtained when the small contribution does not jump. Such modes are to be considered as the diffuse analogs of the fast surface modes of a sharp-boundary plasma–vacuum system where ρ does not go to zero continuously, but jumps. For the similar case of Alfvén surface waves this situation has been analyzed by Tataronis and Grossmann [42] and by Hasegawa and Chen [43] who have shown, by an analysis adapted from the analogous one for the electrostatic oscillations of an inhomogeneous cold plasma [44, 45], that these quasi-modes exhibit damping due to phase mixing.

With respect to radio frequent heating by means of excitation in the Alfvén frequency range, we thus come to the conclusion that both discrete global Alfvén waves below the continuum and damped quasi-modes of the fast magnetoacoustic spectrum may be excited. A problem to realize here is the very sensitive profile dependence of these two kinds of modes, where the manner in which $\rho \to 0$ at the plasma edge enters in a critical fashion. With respect to the slow continuum modes the situation is even worse since $\omega_S^2 \sim p/\rho$ so that the radial dependence of $\omega_S^2(r)$ at the plasma edge can do almost anything. In particular, one can easily envisage situations where all stable low-frequency discrete modes are swallowed by the slow continuum.

We have derived in section 4.3 conditions for the existence of clusterpoints of the spectrum of the diffuse linear pinch. From the expressions (40) and (41) it is clear that clustering of both Alfvén modes (Sturmian and anti-Sturmian) and of the Sturmian slow modes is caused by field line curvature ($B_\theta \neq 0$). So far, we have not mentioned a particularly important kind of clusterpoint, also associated with field line curvature, viz. the one that occurs at marginal stability when we have a point $F \equiv \boldsymbol{k} \cdot \boldsymbol{B} = 0$ in the plasma so that the Alfvén and slow continuum have a common minimum at that point: fig. 13. The condition for such a point to be a clusterpoint is the violation of Suydam's criterion [46] when unstable discrete modes accumulate at $\omega^2 = 0$, i.e., a sufficient condition for instability.

Incidentally, it should be noticed that the mere existence of a point $F = 0$ in a plasma–vacuum system where $\rho \to 0$ continuously at the plasma edge, guarantees that the whole positive ω^2-axis is covered with continua so that stable purely discrete modes no longer exist. Also notice that in this case the continua may fold over onto themselves several times, so that a particularly complicated situation arises.

Let us now return to our discussion of the stability of confined plasmas for CTR purposes.

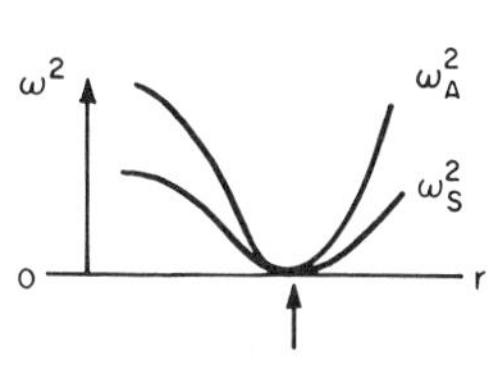

Fig. 13. Clusterpoint of slow magneto acoustic and Alfvén modes when Suydam's criterion is violated.

The generalization of Suydam's criterion for a torus is the well-known Mercier criterion [47], which refers to modes that are localized with respect to the field lines at a particular magnetic surface. In the theory of ballooning modes [21–24] a more general kind of localization is considered where also the poloidally localized regions of unfavorable field line curvature can be made to dominate. In the context of the study of the ideal MHD spectrum the discussion of these conditions tends to become quite involved [48]. However, for $\omega^2 < 0$ the spectrum is purely discrete, at least for finite mode numbers. Consequently, if we restrict the study to the question of (in)stability only, we may now move on to the discussion of 2D systems, even though many questions about the spectrum of 1D systems still remain to be solved.

5. 2D toroidal confinement problems

5.1. *Outline*

In toroidal confinement geometry (see fig. 2) the solution of the equilibrium itself [eqs. (17)] becomes a problem. Note that the equations are nonlinear. A very effective method has proven to be the introduction of the poloidal flux function ψ. Since $\nabla \cdot \boldsymbol{B} = 0$, ψ enters as the stream function for the poloidal field:

$$\boldsymbol{B}_{\mathrm{p}} = -\frac{1}{R} \nabla_{\perp}^{*} \psi, \quad \text{where } \nabla_{\perp}^{*} \equiv (\partial/\partial Z, -\partial/\partial R). \tag{44}$$

Here, R and Z are the coordinates in the poloidal plane: R is the distance from the major axis of the torus and Z is the distance above the equatorial plane. Likewise, since $\nabla \cdot \boldsymbol{j} = 0$, we may introduce a stream function I for the poloidal current:

$$\boldsymbol{j}_{\mathrm{p}} = \frac{1}{R} \nabla_{\perp}^{*} I, \quad \text{where } I \equiv RB_{\varphi} = I(\psi). \tag{45}$$

One also easily shows that $p = p(\psi)$. Consequently, ψ labels the magnetic surfaces, which are tangent to the vectors $\boldsymbol{j}$ and $\boldsymbol{B}$ and orthogonal to the vector ∇p.

The equilibrium problem may be considered as solved if we know $\psi(R, Z)$, since all physical variables are then known in terms of the geometric coordinates R and Z of the poloidal cross-section of the torus. Inserting the expressions (44) and (45) into the force-balance equation $\boldsymbol{j} \times \boldsymbol{B} = \nabla p$ leads to the famous Grad–Shafranov equation:

$$R \frac{\partial}{\partial R} \frac{1}{R} \frac{\partial \psi}{\partial R} + \frac{\partial^2 \psi}{\partial Z^2} = -R^2 p' - II' \quad (= Rj_{\varphi}), \tag{46}$$

which should be solved subject to the boundary condition that $\psi = \text{const.}$ at the plasma edge. The physically relevant solutions of eq. (46) are those that represent a set of nested closed flux contours around a point, the position of the magnetic axis, where $\partial\psi/\partial R = \partial\psi/\partial Z = 0$.

For the theoretical analysis it is also convenient to introduce a poloidal angle $\chi(R, Z)$, orthogonal to $\psi(R, Z)$, so that we obtain an orthogonal coordinate system ψ, χ, φ based on the magnetic surfaces. Here, φ is the toroidal angle, which is an ignorable coordinate for the equilibrium. [It should be mentioned that χ is never used in numerical calculations since it misbehaves at the origin when the magnetic surfaces are elliptical. Therefore, other angles are exploited, but we will not go into these details, important as they may be for numerical accuracy.]

The theoretical task posed by the CTR program is to optimize the equilibrium configuration with respect to MHD stability and also to provide a description of the global geometry, including the interaction with the external coils that are needed to create the optimum configuration. Since this problem is too complex to be solved at once, a two-stage procedure is generally followed.

First, the internal problem is considered by itself. One prescribes a certain shape of the cross-section of the plasma and proceeds to solve eq. (46) and the stability equations. This involves many free parameters and functions: β, q^*, ϵ, shape of the cross-section, and the profiles $p(\chi)$ and $I(\chi)$ [or, equivalently, $q(\chi)$ from eq. (2)], so that a very efficient solution of eq. (46) is required to scan parameter space. The internal problem is complete in itself if the plasma is bounded by a perfectly conducting wall of the desired shape. However, if we wish to consider plasma–vacuum interface systems where a wall or a set of coils is situated at a distance from the plasma, the problem becomes ill-posed since small variations in the desired shape of the plasma cross-section will produce large variations of the coil or wall parameters.

The second stage consists of a self-consistent determination of the plasma equilibrium, starting with a prescribed position of the wall or of the coils with prescribed currents. This is the free-boundary problem, which is mathematically well-posed, but significantly harder to solve. Except for the solution of eq. (46), one now also needs a solution of the vacuum field equations, which is usually obtained by means of Green's function methods, and one needs to match these solutions [through eqs. (10) and (11)] at an interface whose location is unknown.

From a purely physical point of view the latter procedure might appear to be the correct one, identifying well-posed mathematical problems with physically relevant ones. However, it should be noticed that the actual technological problem is intrinsically ill-posed: Eventually one is not interested in the purely physical question of a description of the equilibrium and stability of a certain plasma created by certain coils. One is interested in finding out how to produce a plasma with desirable properties (low q^*, high β) by action at a distance, i.e., by external manipulation of the coils and currents. [Accordingly, one should expect the problem of reproducibility and reliability to become a major one in future fusion reactor studies.]

Both fixed-boundary and free-boundary methods have been used extensively in the study of tokamak equilibrium problems [26, 27, 49–54], where eq. (46) is usually solved by means of some finite difference scheme. Recently, an entirely different approach has presented itself. Starting with the observation that all stability theories are formulated in terms of flux coordinates ψ, χ, φ, one tries to solve the nonlinear equations for the inverse coordinates $R(\psi, \chi)$, $Z(\psi, \chi)$ directly. [Remember: χ should be read here as any poloidal angle]. This eliminates the necessity of having to invert the coordinates in between the equilibrium and stability parts of the numerical calculation. In this method of inverse coordinates or method of moments [55–58] the solution of the equilibrium problem is reduced to that of solving a set of nonlinear ordinary differential equations for the moments of R and Z. In section 5.2 we will present a third approach to the solution of the equilibrium equations which is based on conformal mapping. In this method the flux function is determined as a function of polar coordinates s, t in a computational plane obtained from the physical plane by conformal mapping. By construction, these coordinates are globally adapted to the geometry of the magnetic surfaces, so that there is no need for coordinate inversion (with one exception, to be mentioned later).

After the equilibrium has been determined with sufficient accuracy, it makes sense to consider the stability problem. The stability of toroidal equilibrium configurations is usually carried out by means of some form of the Rayleigh–Ritz variational principle. Eigenvalues ω^2 of the basic eigenvalue problem (20) are obtained as stationary values of a functional $\Omega^2[\boldsymbol{\xi}]$, the ratio of the potential energy W and the norm I:

$$\delta\Omega^2 = 0, \quad \Omega^2 \equiv W[\boldsymbol{\xi}]/I[\boldsymbol{\xi}]. \tag{47}$$

Here, $W[\boldsymbol{\xi}]$ is the well-known quadratic form [35] involving the ideal MHD force operator defined in eq. (19):

$$\begin{aligned} W &= -\tfrac{1}{2}\int \boldsymbol{\xi}^* \cdot \mathscr{F}(\boldsymbol{\xi})\,\mathrm{d}V^{\mathrm{p}} \\ &= W^{\mathrm{p}}[\boldsymbol{\xi}] + W^{\mathrm{s}}[\xi_n] + W^{\mathrm{v}}[\hat{\boldsymbol{Q}}], \end{aligned} \tag{48}$$

where the three expressions to the right refer to the separate contributions of the plasma, of the plasma–vacuum interface, and of the vacuum, respectively. The plasma variable $\boldsymbol{\xi}$ and the vacuum variable $\hat{\boldsymbol{Q}}$ are connected through the boundary condition

$$\boldsymbol{n}\cdot\nabla\times(\boldsymbol{\xi}\times\hat{\boldsymbol{B}})=\boldsymbol{n}\cdot\hat{\boldsymbol{Q}} \tag{49}$$

at the plasma–vacuum interface, whereas $\boldsymbol{n}\cdot\hat{\boldsymbol{Q}}=0$ should be posed at the conducting wall surrounding the configuration. The proper norm corresponding to a complete solution of the eigenvalue problem (20) reads

$$I=\tfrac{1}{2}\int\rho\boldsymbol{\xi}^{*}\cdot\boldsymbol{\xi}\,\mathrm{d}V^{\mathrm{P}}. \tag{50}$$

However, if one is interested in the stability boundaries only, any other convenient norm may be chosen for I.

The global stability problem given by eqs. (47)–(50) corresponds to a set of linear partial differential equations in the two dimensions ψ and χ when single Fourier components $e^{in\varphi}$ with respect to the ignorable coordinate are considered. This is a problem involving many terms and many parameters. For ballooning modes, localized with respect to a single magnetic surface, a major simplification results since the equations then become ordinary differential equations involving the poloidal angle χ only.

5.2. *Conformal mapping as a tool for solving 2D MHD problems*

The global 2D geometry of a toroidal plasma equilibrium is characterized by two important features: (1) the shape of the poloidal cross-section of the plasma–vacuum interface, which we may either prescribe arbitrarily (fixed-boundary problem) or which has to be determined self-consistently (free-boundary problem); (2) the outward shift of the magnetic axis due to finite β.

On the other hand, a fair amount of theoretical insight in ideal MHD stability properties has been gained from 1D straight cylinder theory (cf. section 4.3), i.e., a geometry where the poloidal cross-section of the plasma is a circle with the magnetic axis in the center. The question arises: Is it possible to establish a simple connection between the two pictures?

The Riemann mapping theorem states that a conformal mapping exists which brings about the desired effect: See fig. 14, which depicts the mapping of an arbitray curve C (the poloidal cross-section of the plasma) in the z-plane onto a circle C' in the w-plane while mapping the position $z=\delta$ of the magnetic axis onto the origin of the w-plane. A simple example of such a mapping is given by the Moebius transformation:

$$z(w)=(w+\delta)/(1+\delta w), \tag{51}$$

which maps a circular region onto itself while transforming a point $z=\delta$ into $w=0$. This simple mapping already provides a transformation which is quite useful in the study of high-β tokamaks with a circular cross-section [59]. Clearly, we need a numerical generalization of such kinds of mappings for arbitrary regions.

The required tool is provided by the recently developed numerical conformal mapping techniques exploiting the fast Hilbert transformation [60]. This very effective method has been successfully applied to the study of the equilibrium and stability of high-β tokamaks [61, 62], which will be discussed in section 6. Since all techniques are extensively described in the mentioned references we will only indicate here the essence of the ideas.

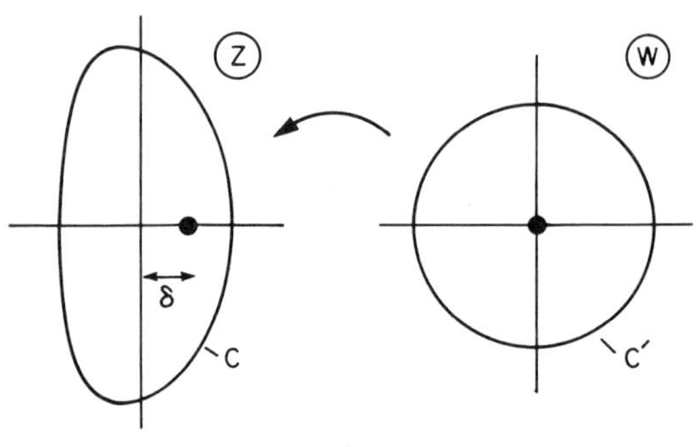

Fig. 14. Conformal mapping of the poloidal cross-section of a toroidal plasma onto a circle with the image of the magnetic axis in the center.

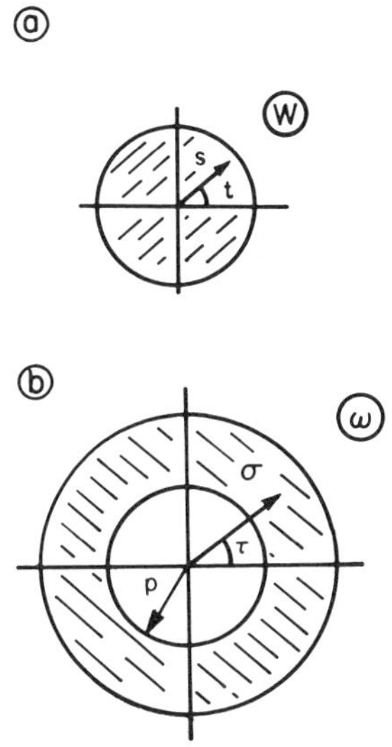

Fig. 15. Domains for the Fast Hilbert Transformation of analytic functions in simply-connected (a) and doubly-connected (b) regions.

Consider an analytic function $\Omega(w)$ on the unit disk (fig. 15a),

$$\Omega(w)=\Phi(s,t)+\mathrm{i}\Psi(s,t), \tag{52}$$

which attains the values $\varphi(t)+i\psi(t)$ at the boundary. The conjugate periodic functions $\varphi(t)$ and $\psi(t)$ then satisfy the Hilbert transformation

$$\varphi(t)=A-\frac{1}{2\pi}\mathscr{P}\oint \operatorname{cotg}\frac{1}{2}(t-t')\psi(t')\,\mathrm{d}t'. \tag{53}$$

Indicating the Fourier components of $\varphi(t)$ and $\psi(t)$ as a_m and γ_m, respectively, a discrete representation of this equation is obtained by taking twice the fast Fourier transformation (FFT):

$$\{\varphi(t_i)\}\xrightarrow{\mathrm{FFT}^-}\{a_m\}\xrightarrow{\cdot sg(m)}\{\gamma_m\}\xrightarrow{\mathrm{FFT}^+}\{\psi(t_i)\}. \tag{54}$$

We have termed this algorithm the fast Hilbert transformation (FHT).

The conformal mapping $z(w)$ of a region in the z-plane bounded by a curve C, given by $r=f(\vartheta)$, to the unit disk $w\leq 1$ in the w-plane is obtained by considering the following choice for the analytic function:

$$\Omega(w)\equiv\ln[z(w)/w], \tag{55}$$

which attains the values $\ln f[\vartheta(t)]+i[\vartheta(t)-t]$ at the boundary. Here, $f(\vartheta)$ is known (the curve C), but $\vartheta(t)$ represents the unknown relationship between the angles at the boundary curves in the z- and w-plane, respectively. It follows that $f[\vartheta(t)]$ and $\vartheta(t)-t$ are conjugate periodic functions with respect to each other, satisfying the Hilbert transformation (53). This nonlinear integral equation for the so-called boundary correspondence function $\vartheta(t)$ has been derived by Theodorsen, as long ago as 1931, to study the problem of potential flow around a wing of arbitrary cross-section (fig. 16a). Its solution becomes nearly trivial by the iterative application of the fast Hilbert transformation given by eq. (54).

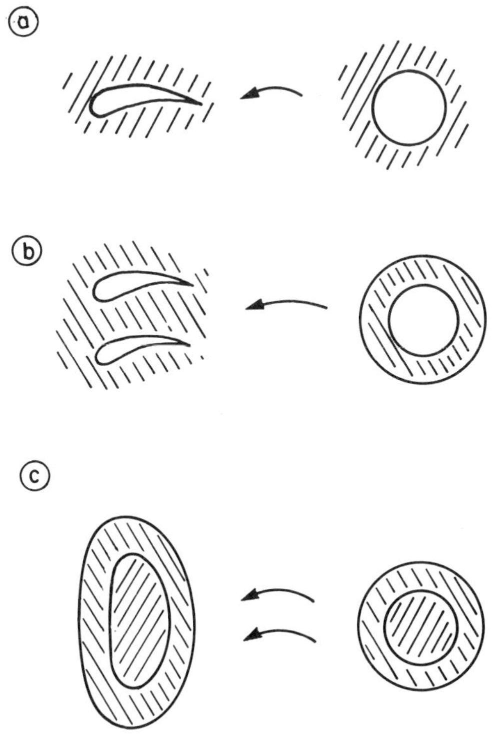

Fig. 16. (a) Theodorsen's mapping of flow around a wing, (b) Garrick's mapping of flow around a biplane, and (c) superposition of the two to describe the stability of tokamaks.

In this manner, an arbitrary shape of the plasma cross-section may be mapped onto a circle. This numerical mapping leaves the position of the origin invariant. A subsequent analytic Moebius transformation of the form (51) them provides the requires final situation where the image of the magnetic axis is situated at the center of the circle.

For the study of discontinuous plasma–vacuum systems we need to be able to treat the vacuum region surrounding the plasma in a similar fashion. This requires the construction of a mapping $z(\omega)$ of a doubly connected region, bounded by two given curves C_0 [given by $r=f_0(\vartheta)$] and C_1 [given by $r=f_1(\vartheta)$] in the z-plane, onto an annulus $p \le \sigma \le 1$ in the ω-plane (fig. 15b). This involves the determination of the two boundary correspondence functions $\vartheta_0(\tau)$ and $\vartheta_1(\tau)$ and of the modulus p, which are obtained from the solution of a pair of nonlinear integral equations. These equations were derived by Garrick, in 1936, to study the problem of potential flow around biplanes (fig. 16b). The solution of Garrick's integral equations becomes equally simple as that of Theodorsen's equation by the use of the corresponding pair of fast Hilbert transformations, where the number of operations just doubles as compared to that of eq. (54).

In conclusion: We have effected the simplification of the geometry of a plasma–vacuum system to that of concentric circles (fig. 16c). The plasma region is mapped onto a disk by means of the internal mapping $z(w)$, which is found in terms of the solution $\vartheta=\vartheta^{in}(t)$ of Theodorsen's integral equation. The vacuum region is mapped onto an annulus by means of the external mapping $z(\omega)$, which is found in terms of the solution $\vartheta=\vartheta^{ex}_{0,1}(\tau)$ of Garrick's integral equation. In order to connect the physical variables across the plasma–vacuum interface we will have to connect the internal grid with the external one by computing $t=t(\tau)$ or vice versa by computing $\tau=\tau(t)$.

Of course, the construction of the conformal mappings does not solve the physical problem of equilibrium and stability, unless the pertinent equations happen to be the 2D Laplace equation. This is the case for high-β tokamaks with a skin-current when the appropriate ordering is applied (section 6.1). For diffuse high-β tokamaks the equations are more general elliptic partial differential equations so that the mappings do not solve the problem but just facilitate its solution. We will see how in section 6.2.

In closing this section it is appropriate to point out a number of papers of related interest [63–66] concerning the numerical construction of conformal mappings and applications to fluid dynamics. In these papers the problem of severe distortion of boundaries is considered, a problem that becomes urgent in time-dependent studies of fluid interfaces in the nonlinear domain. Here, we have restricted the study to linear perturbations of equilibria that are only moderately distorted, so that these further complications could be avoided.

6. High-β tokamaks

6.1. *Free-boundary high-β tokamaks with a skin current*

In fig. 17a a particular free-boundary equilibrium of a tokamak with skin currents is shown. Equilibrium results here when the outward shift of the plasma column, caused by the finite value of β/ϵ, is balanced by the squeeze of the poloidal flux against the conducting wall. For simplicity the cross-section of the conducting wall has been taken circular and the equilibrium shape $r=f_0(\theta)$ of the plasma cross-sections has been determined self-consistently by solving the free-boundary equilibrium equations in the high-β tokamak ordering. The latter is formally given by

$$\beta \sim \varepsilon \ll 1, \quad q^* \sim 1, \quad a\varepsilon\beta_p \equiv (\beta/a\varepsilon)q^{*2} \sim 1, \tag{56}$$

where

$$\beta_p \equiv 2p/\langle \hat{B}_p \rangle^2$$

[$\sim \varepsilon^{-1}$: "high-β_p" would have been a more descriptive term than "high-β"]. The physical

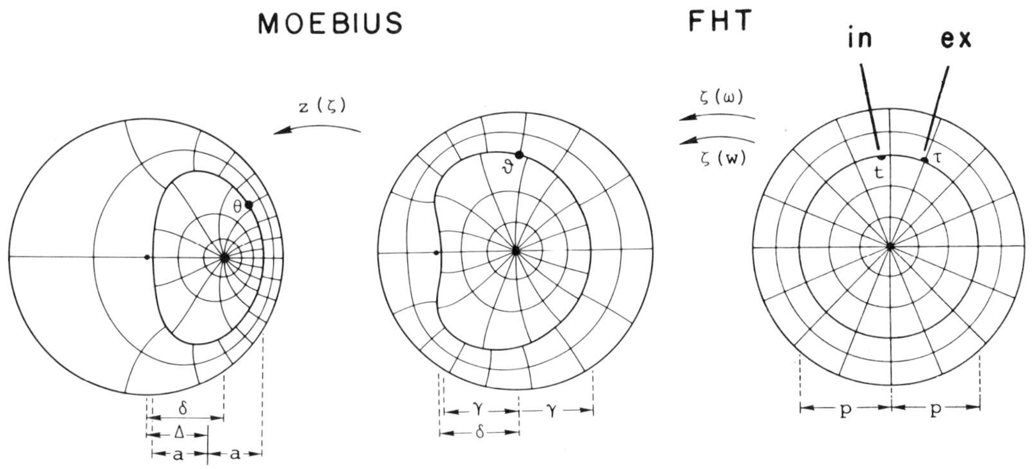

Fig. 17. Complete sequence of conformal mappings for the solution of the stability of free-boundary high-β tokamaks with a skin-current.

meaning of this ordering is that it forces the above-mentioned equilibrium mechanism to appear in the leading order. The method of solution of the free-boundary equilibrium problem may be found in ref. 62a. Here, we want to continue our discussion of the previous section, concerning the way in which conformal mapping leads to a complete solution of the stability problem.

Figs. 17b and c show the two mappings that transform the plasma–vacuum geometry in the physical z-plane into the geometry of two concentric circles in the superposed w- and ω-planes. First, a Moebius transformation $z(\zeta)$ is used to map the whole domain, plasma and vacuum region, onto itself. Next, the two numerical mappings $\zeta(w)$ and $\zeta(\omega)$ are used to separately map the internal plasma region onto the disc $s \le p$ and the external vacuum region onto the annulus $p \le \sigma \le 1$. The intermediate Moebius transformation $z(\zeta)$ is somewhat arbitrary since the skin-current plasma has no magnetic axis. We have chosen this mapping for numerical convenience such that the image of the plasma becomes centered with respect to the image of the wall in the auxiliary ζ-plane.

Turning now to the stability analysis, it may be shown [62c] that the high-β tokamak ordering (56) turns both the interior problem for the plasma perturbation $\boldsymbol{\xi}_\perp$ and the exterior problem for the magnetic field perturbation $\hat{\boldsymbol{Q}}_\perp$ into a potential flow problem:

$$\begin{aligned} \boldsymbol{\xi}_\perp &= \nabla_\perp^* \chi, \quad \Delta_\perp \chi = 0, \\ \hat{\boldsymbol{Q}}_\perp &= \nabla_\perp^* \hat{\psi}, \quad \Delta_\perp \hat{\psi} = 0. \end{aligned} \tag{57}$$

Consequently, the construction of the two conformal mappings $z(\zeta(w))$ and $z(\zeta(\omega))$ is equivalent to solving these problems. The stream functions χ and $\hat{\psi}$ then need to be connected across the plasma–vacuum interface, which involves the conversion of the internal angle t into the external angle τ.

Through all these simplifications the variational principle (47) may be reduced to the following one-dimensional form for the growth rates of kink modes:

$$\omega^2 = \frac{\oint\oint \xi^*(\tau) N(\tau,\tau') \xi(\tau')\, \mathrm{d}\tau \mathrm{d}\tau'}{\oint\oint \xi^*(\tau) D(\tau,\tau') \xi(\tau')\, \mathrm{d}\tau \mathrm{d}\tau'}, \tag{58}$$

Here, the kernels N and D are explicit algebraic expressions involving the connection $t(\tau)$ between the internal angle t and the external angle τ, the modulus p of the annulus in the ω-plane, the scale factor $H = |\mathrm{d}z/\mathrm{d}\omega|_0$ of the complete external mapping, and the physical variables $a\epsilon\beta_p$, nq^*, and a. Because the whole problem depends on only these three parameters a complete scan of parameter space can be performed. The resulting stability diagram is shown in fig. 18.

The whole diagram shown in fig. 18 may be considered as the high-β generalization of the simple results shown in fig. 6b, which lead to the Kruskal–Shafranov criterion. Along the nq^* axis two marginal points are found, viz. $nq^* = a^2$ and $nq^* = 1$, i.e., just the usual low-β marginal points in the presence of a conducting wall. [Notice that in fig. 6b the left marginal point is situated at $nq^* = 0$ since there is no wall assumed in that calculation]. If we increase β the two branches of the marginal curve move toward each other until they close at a certain value of $\beta/a\epsilon n^2$. Above this value complete stability is obtained. Consequently, at sufficiently high β, when the plasma is squeezed against the wall, the Kruskal–Shafranov limit may be surpassed and complete stability is obtained for $\beta > \beta_{min}$. The particular value of β_{min} depends on the magnitude of a, the width of the plasma relative to the radius of the tube. In fig. 18 the

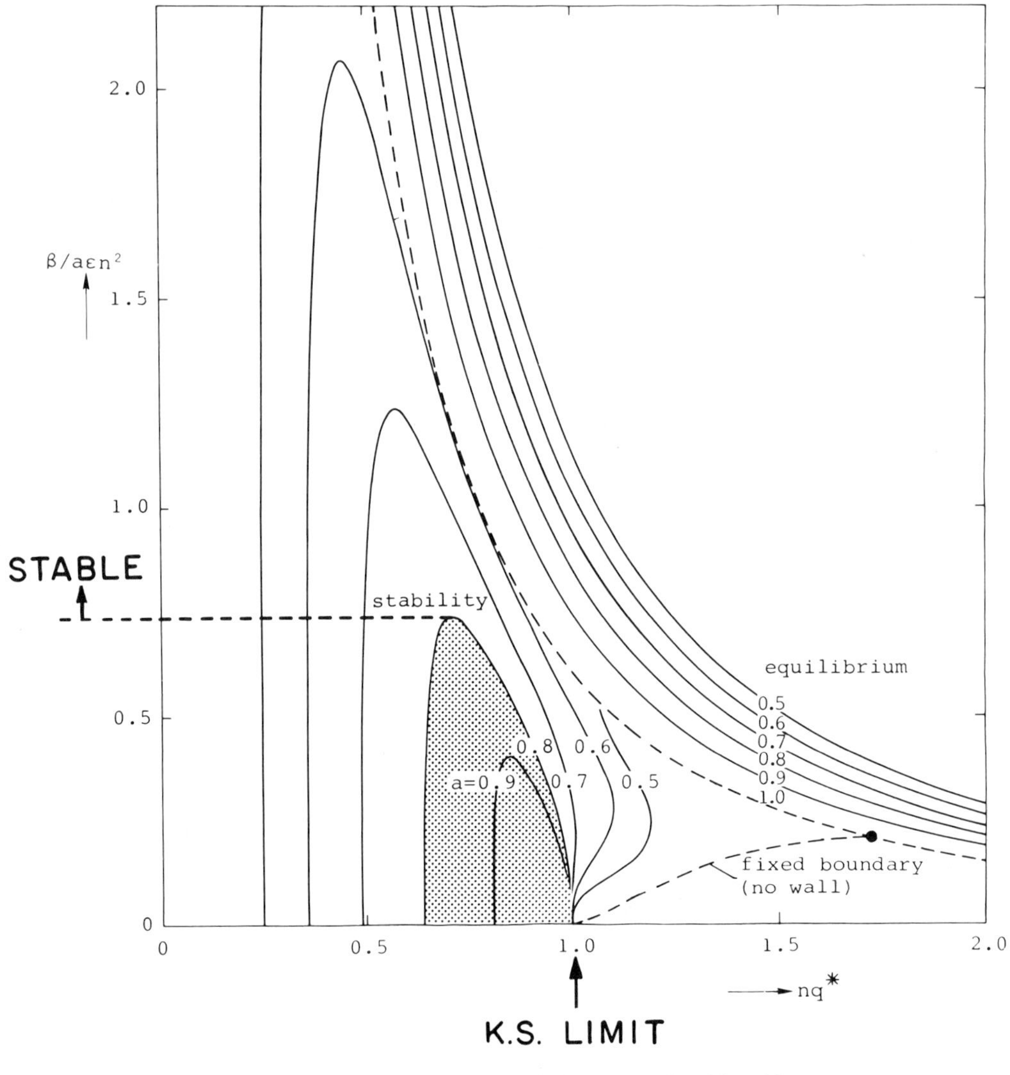

Fig. 18. Stability diagram of free-boundary high-β tokamaks with a skin current.

shaded area represents the unstable region for $a = 0.8$. For that case one obtains $\beta_{\min} = 0.59\epsilon n^2$.

Numerous investigations [67–74] of the fixed-boundary skin-current model with prescribed plasma cross-section have been carried out. A comparison with the present free-boundary calculation show that the self-consistent shaping of the plasma column by compression against the wall increases the stability significantly, and qualitatively.

6.2. *Diffuse high-β tokamaks*

For the skin-current model of a high-β tokamak conformal mappings solve the problem completely. What of more realistic toroidal systems, where the pressure and current profiles are diffuse and, consequently, the equations are no longer the 2D Laplace equations? In this case conformal mappings no longer solve the equations, but they do provide coordinates that are globally adapted to the geometry of the magnetic surfaces. This idea has been implemented in a diffuse toroidal equilibrium and stability code HBT [75].

Although most of the subsequent discussion generalizes to arbitrary diffuse toroidal equilibria, for the purpose of the present review it is most illustrative to restrict the discussion again to the high-β tokamak ordering (56). In that approximation, the Grad–Shafranov equation (46) may be simplified as follows [61]:

$$\Delta_{\perp}\Psi = A[\Gamma(\Psi) + Bx\Pi(\Psi)] \quad (= j_{\varphi}). \tag{59}$$

Here, we have introduced (1) dimensionless coordinates x and y centered at the midpoint of the plasma column, (2) a new dimensionless flux function $\Psi = \psi/\psi_{\text{edge}}$ of unit range: $\Psi = 0$ at the magnetic axis ($x = \delta$), $\Psi = 1$ at the plasma edge, (3) unit profiles $\Gamma(\Psi)$ and $\Pi(\Psi)$, related to the profiles $I^{2\prime}$ and p' occurring in eq. (46): $\Gamma(0) = \Pi(0) = 1, \Gamma(1) = \Pi(1) = 0$. Under these conditions A and B become eigenvalues to be determined along with $\Psi(x, y)$.

If we now prescribe the plasma cross-section C and the position of the magnetic axis $x = \delta$ (so that β is to determined from the solution of eq. (59) and not the other way around), a conformal mapping $z = z(w)$ as shown in fig. 14 is determined. This mapping turns eq. (59) into an equation for $\Psi(s, t)$ where the transformation of the LHS is given by $\Delta_{\perp}^{(z)} = h^{-2}\Delta_{\perp}^{(w)}, h \equiv |\mathrm{d}z/\mathrm{d}w|$ and the RHS only changes through the introduction of the known function $x = x(s, t)$. Thus, the same structure as eq. (59) is obtained, but we now have to solve this equation subject to the boundary conditions $\Psi = 0$ at $s = 0$ and $\Psi = 1$ at $s = 1$. Consequently, the flux contours in the w-plane will be near circular, which represents a considerable advantage for the numerics: it speeds up the convergence of the iterative solution of the Grad–Shafranov equation.

In fig. 19a we show the s, t grids obtained from the solution of Theodorsen's equation and a Moebius transformation for a number of representative plasma cross-sections. The corresponding equilibrium solutions are shown in fig. 19b for a linear choice of the profiles $\Gamma(\Psi)$ and $\Pi(\Psi)$. The essential point of this figure is the demonstration that the s, t coordinates of the w-plane are very well adapted to the description of the geometry of the magnetic surfaces. Therefore, these coordinates are used throughout the code HBT to compute the equilibrium as well as the stability of diffuse high-β tokamaks.

Once a solution $\Psi(s, t)$ of eq. (59) is found, a number of different equilibria may be computed from it through a peculiar scaling of high-β tokamaks. A dimensionless parameter α may be introduced:

$$\alpha \equiv a^2 B_0/\psi_{\text{edge}}, \tag{60}$$

where a is the halfwidth of the plasma column and B_0 is the toroidal magnetic field. This parameter is directly related to $q^* = q^*(\alpha)$ so that we get the equilibria for free if we scan the parameter q^* in a search for the marginally stable points in the $\epsilon\beta_{\mathrm{p}} - nq^*$ diagram (figs. 20–22).

Through the use of the parameters α, ε, a, and B_0 all physical quantities can be made dimension-

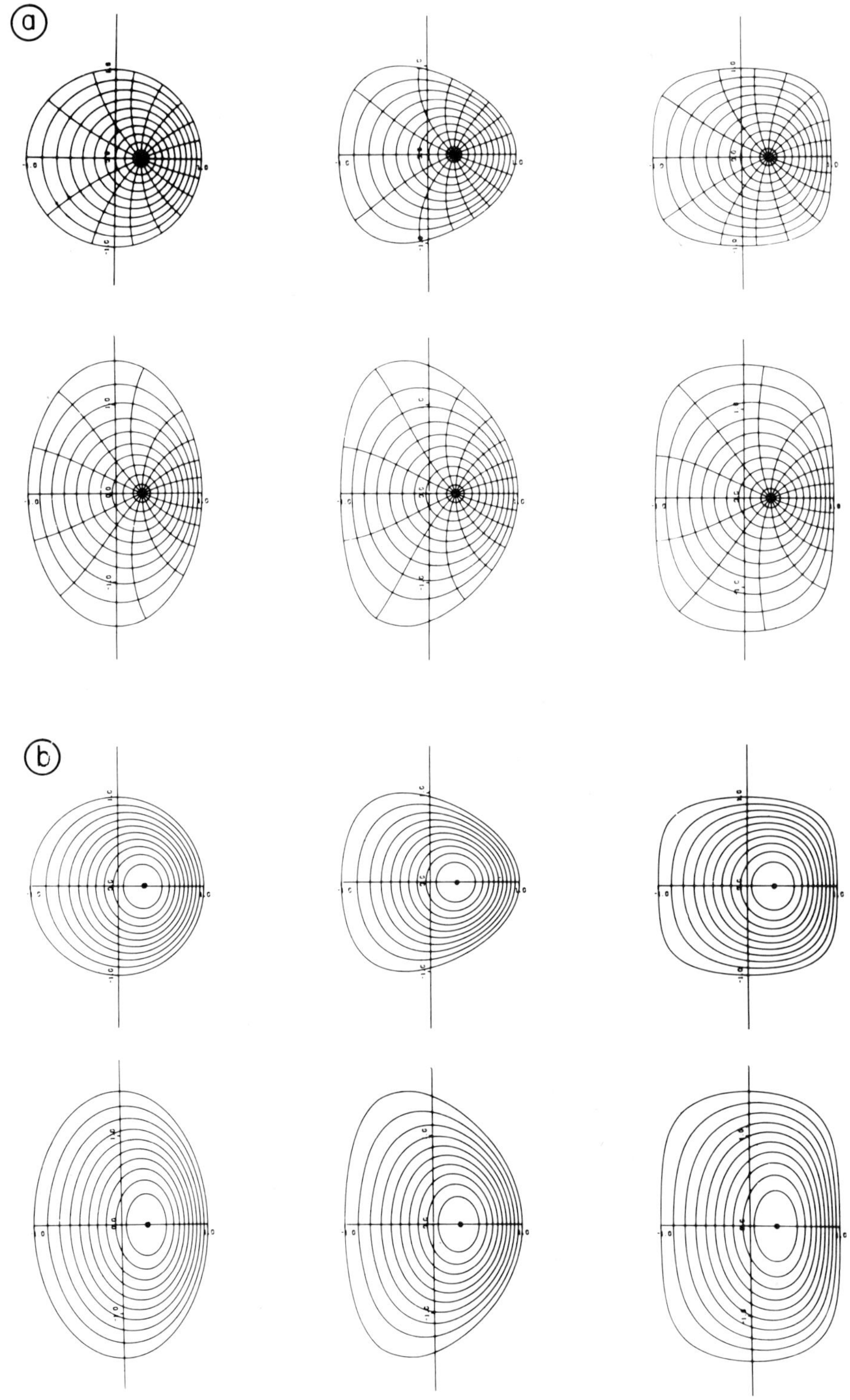

Fig. 19. Grids of the conformal mapping (a) and equilibrium solutions (b) for diffuse high-β tokamaks obtained from program HBT.

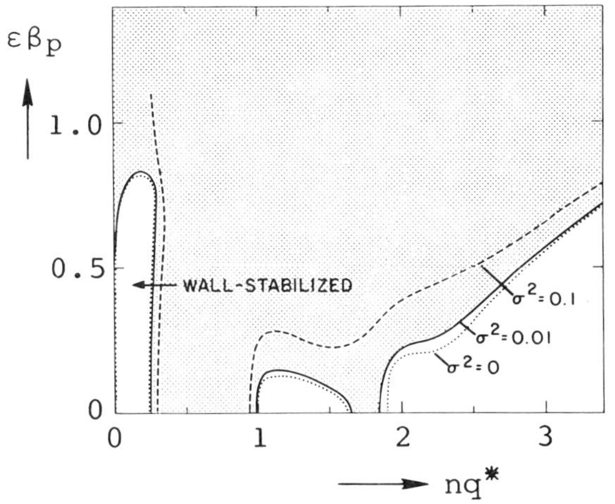

Fig. 20. σ-stability diagram for diffuse high-β tokamak with circular cross-section, $b/a = 2$.

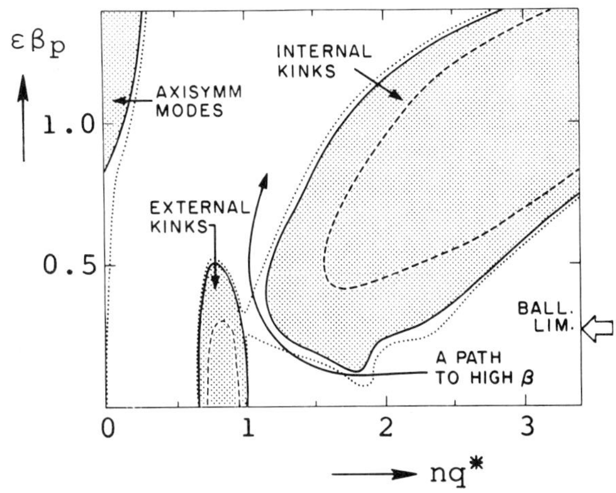

Fig. 21. σ-stability diagram for diffuse high-β tokamak with circular cross-section, $b/a = 1.25$.

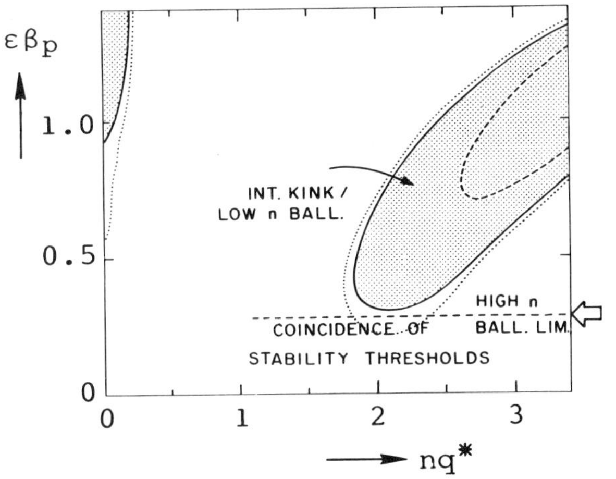

Fig. 22. σ-stability diagram for diffuse high-β tokamak with circular cross-section, $b/a = 1$.

less and of the order unity. From eqs. (48) and (50) we then get the following simple expressions for the potential energy W^{p} of the plasma and the norm I in the high-β tokamak approximation,

$$W = \pi \int\int \left[|B_{\mathrm{p}}^{-1} F X|^2 + |B_{\mathrm{p}} F Y + 2\kappa_{\mathrm{p}} X|^2 - 2 B_{\mathrm{p}}^{-1} (j_\varphi \kappa_{\mathrm{p}} - p' \kappa_{\mathrm{t}}) |X|^2 \right] J \,\mathrm{d}\Psi \,\mathrm{d}\chi,$$

$$I = \pi \int\int \left[|B_{\mathrm{p}}^{-1} X|^2 + |B_{\mathrm{p}} Y|^2 \right] J \,\mathrm{d}\Psi \,\mathrm{d}\chi. \qquad (61)$$

Here, J is the Jacobian of the Ψ, χ, φ coordinates, κ_{p} and κ_{t} are the poloidal and toroidal curvatures of the magnetic surfaces:

$$\kappa_{\mathrm{p}} \equiv J^{-1}(J B_{\mathrm{p}})', \quad \kappa_{\mathrm{t}} \equiv B_{\mathrm{p}} x', \qquad (62)$$

F is the parallel gradient operator:

$$F \equiv -\mathrm{i} J^{-1} \partial/\partial\chi + n\alpha, \qquad (63)$$

and X and Y are the normal and perpendicular

components of $\boldsymbol{\xi}$:

$$X \equiv B_p \xi_\psi, \quad Y \equiv \mathrm{i} B_p^{-1} \xi_p, \tag{64}$$

which are related through a compressibility relation in the poloidal plane:

$$\partial(JX)/\partial\Psi = \mathrm{i}\,\partial Y/\partial\chi. \tag{65}$$

This latter condition permits us to introduce a stream function S for the perturbations. This, obviously, does not satisfy Laplace's equation in 2D (as in the skin-current case), but it does permit us to express the variational principle for the stability of high-β tokamaks in the form

$$\delta\Omega^2 = 0, \quad \Omega^2 \equiv W[S]/I[S]. \tag{66}$$

As compared to eq. (47) this is a significant simplification.

This is the essence of the analysis which has been incorporated in the code HBT, where all functions of Ψ and χ are computed in terms of the coordinates s and t. The formulation also contains the driving forces which are responsible for ballooning modes. Thus, a separate local analysis as in ref. 23 was carried out and incorporated in HBT [75]. At this point it was necessary to introduce a coordinate inversion in order to be able to localize the modes with respect to a particular magnetic surface. The use of the nonorthogonal coordinates Ψ and t turned out to be the most convenient choice for this purpose.

Finally, we present a number of stability diagrams for high-β tokamaks [75] with a fixed-boundary circular cross-section of the equilibrium in which the wall is moved in from $b/a = 2$ (fig. 20), to $b/a = 1.25$ (fig. 21), to $b/a = 1$ (fig. 22). The curves labeled with different values of σ^2 refer to different σ-stability boundaries [15, 76]. Recall that an equilibrium is called σ-stable if it does not manifest growth of instabilities faster than $\exp(\sigma t)$.

There is a marked difference between stability at low β and at high β, depending on where the wall is situated. When the wall is far away (fig. 20) increasing β leads to decreased stability for all values of nq^*. When the wall is moved in (fig. 21) different unstable regions start to separate. In particular if a σ-stability criterion is used it is easier to distinguish different stability regions. E.g., in fig. 21 a gap between the external and internal kink mode regions appears which is not present if a strict marginal stability criterion ($\sigma^2 = 0$) is exploited. This creates room for a path in parameter space connecting the stable regions at low β with those at high β.

When the wall is close (fig. 21, 22) several stabilization effects at high β become visible. First, we notice stabilization of the external $m = 1$ kink mode, which we already encountered when discussing the stability of the free-boundary skin-current model (fig. 18). Next, we encounter a large unstable region of the internal kink mode having both a lower and an upper stability boundary. The lower stability boundary has been known for some time [77, 78], but the upper boundary is of more recent interest [79, 80]. The latter boundary is very similar to the second stability region of localized high-n ballooning modes [81], which has not been indicated in fig. 21.

With the wall at the plasma (fig. 22) we found [75] a very striking coincidence of the stability thresholds for internal kink modes and high-n ballooning modes (indicated by the arrow) for a variety of different pressure and current profiles. This suggests that the instability mechanism is quite similar for the two modes. It also lends support for the use of the high-β tokamak ordering leading to the variational principle (66).

7. Perspectives

In the last section we have finally returned to the central questions formulated in section 2.1 concerning the optimization of plasma configurations with respect to the value of β_{crit} for equilibrium and stability. We have purposely taken a long excursion to arrive at that point because this gave us the opportunity to discover some omissions in the development of ideal MHD. One of them was the lack of clarity in the classification of spectral results for simple 1D systems, like the plane gravitating slab and the diffuse linear pinch. If progress would be made there, this would undoubtedly have important implications for the understanding of the more complex 2D systems. One of the problems with a strongly mission-oriented field of research is precisely that these basic questions are never fully solved, due to lack of motivation.

Another omission concerns the development of a nonlinear theory of simple plasma–vacuum interface systems. Again, a number of more sophisticated approaches exist, incorporated in large numerical codes solving the initial-value problem for diffuse 2D and 3D systems, but nothing is known about the nonlinear fate of the gross ideal MHD instabilities in genuine plasma-vacuum interface systems. This may well be explained by the bias of researchers in CTR for plasmas that are diffuse and relatively quiescent, so that the usual split in a static equilibrium state and a small linear perturbation is an adequate picture. Nevertheless, it should be clear that a number of interesting possibilities is eliminated in this manner. E.g., the free-boundary high-β tokamak model discussed in section 6.1 should allow an extension in the nonlinear domain by means of the contour dynamical methods discussed in ref. 82.

Acknowledgements

The author is indebted to Daniel A. D'Ippolito, Jeffrey Freidberg, Ricardo Galvão, Jan Rem, and Paulo Sakanaka for many years of fruitful interactions.

This work was supported by the U.S. Department of Energy under Contract #DE-AC02-78ET-51013 and by the "Stichting voor Fundamenteel Onderzoek der Materie" (FOM), the "Nederlandse Organisatie voor Zuiver-Wetenschappelijk Onderzoek" (ZWO), and Euratom.

References

[1] J.P. Goedbloed, Lecture Notes on Ideal Magnetohydrodynamics, [Lectures given at Universidade Estadual de Campinas, Brasil, 1978], Rijnhuizen Report 83-145 (1983).

[2] V.D. Shafranov, in: Rev. of Plasma Physics 2, M.A. Leontovich, ed. (Consultants Bureau, New York, 1966) p. 103.

[3] B.B. Kadomtsev, in: Rev. of Plasma Physics 2, M.A. Leontovich, ed. (Consultants Bureau, New York, 1966) p. 153.

[4] J.A. Wesson, Nuclear Fusion 18 (1978) 87.

[5] G. Bateman, MHD Instabilities (MIT Press, Cambridge, 1978).

[6] J.P. Freidberg, Rev. Mod. Phys. 54 (1982) 801.

[7] K.O. Friedrichs and H. Kranzer, Notes on Magnetohydrodynamics VIII. Non-Linear Wave Motion (New York University, 1958) NYO-6486.

[8] J.U. Brackbill, in: Methods in Computational Physics, vol. 16, J. Killeen, ed. (Academic Press, New York, 1976) p. 1.

[9] F. Bauer, O. Betancourt and P. Garabedian, A Computational Method in Plasma Physics (Springer, New York, 1978).

[10] A. Schlüter and U. Schwenn, Comput. Phys Comm. 24 (1981) 263.

[11] M. Kruskal and M. Schwarzschild, Proc. Roy. Soc. London A223 (1954) 348.

[12] A.I. Akhiezer, I.A. Akhiezer, R.V. Polovin, A.G. Sitenko and K.N. Stepanov, Plasma Electrodynamics, vol. I, Linear Theory (Pergamon, Oxford, 1975).

[13] H. Grad, Proc. Nat. Acad. Sci. 70 (1973) 3277.

[14] K. Appert, D. Berger and J. Vaclavik, Phys. Fluids 17 (1974) 1471.

[15] J.P. Goedbloed and P.H. Sakanaka, Phys. Fluids 17 (1974) 908.

[16] R.L. Dewar, R.C. Grimm, J.L. Johnson, E.A. Frieman, J.M. Greene and P.H. Rutherford, Phys. Fluids 17 (1974) 930.

[17] J. Tataronis, J. Plasma Physics 13 (1975) 87.

[18] J.P. Goedbloed, Phys. Fluids 15 (1975) 1090.

[19] Y.P. Pao, Nucl. Fusion 18 (1975) 631.

[20] M.S. Chance, J.M. Greene, R.C. Grimm and J.L. Johnson, Nucl. Fusion 17 (1977) 65.

[21] B. Coppi, Phys. Rev. Lett. 39 (1977) 939.

[22] D. Dobrott, D.B. Nelson, J.M. Greene, A.H. Glasser, M.S. Chance and E.A. Frieman, Phys. Rev. Lett. 39 (1977) 943.

[23] J.W. Connor, R.J. Hastie and J.B. Taylor, Phys. Rev. Lett. 40 (1978) 396; Proc. Roy. Soc. London A365 (1979) 1.

[24] F. Pegoraro and T.J. Schep, Phys. Fluids 24 (1981) 478.

[25] R.L. Dewar, J.M. Greene, R.C. Grimm and J.L. Johnson, J. Comput. Phys. 18 (1975) 132.

[26] R.C. Grimm, J.M. Greene and J.L. Johnson, in: Methods in Computational Physics, vol. 16, J. Killeen, ed. (Academic Press, New York, 1976), 253.

[27] J.L. Johnson, H.E. Dalhed, J.M. Greene, R.C. Grimm, Y.Y. Hsieh, S.C. Jardin, J. Manickam, M. Okabayashi, R.G. Storer, A.M.M. Todd, D.E. Voss and K.E. Weimer, J. Comp. Phys. 32 (1979) 212.

[28] A.M.M. Todd, J. Manickam, M. Okabayashi, M.S. Chance, R.C. Grimm, J.M. Greene and J.L. Johnson, Nucl. Fusion 19 (1979) 743.
[29] L.A. Charlton, R.A. Dory, Y.-K.M. Peng, D.J. Strickler, S.J. Lynch and D.K. Lee, Phys. Rev. Lett. 43 (1979) 1395.
[30] L.C. Bernard, D. Dobrott, F.J. Helton and R.W. Moore, Nucl. Fusion 20 (1980) 1199.
[31] D. Berger, L.C. Bernard, R. Gruber and F. Troyon, J. Appl. Math. Phys. (ZAMP) 31 (1980) 113.
[32] W. Kerner, P. Gautier, K. Lackner, W. Schneider, R. Gruber and F. Troyon, Nucl. Fusion 21 (1981) 1383.
[33] R. Gruber, F. Troyon, D. Berger, L.C. Bernard, S. Rousset, R. Schreiber, W. Kerner, W. Schneider and K.V. Roberts, Comput. Phys. Commun. 21 (1981) 323.
[34] R.C. Grimm, R.L. Dewar and J. Manickam, J. Comp. Phys. 49 (1983) 94.
[35] I.B. Bernstein, E.A. Frieman, M.D. Kruskal and R.M. Kulsrud, Proc. Roy. Soc. London A224 (1958) 1.
[36] J.P. Goedbloed, Physica 53 (1971) 412, 501, 535; 100C (1980) 273.
[37] K. Hain and R. Lüst, Z. Naturforsch 13a (1958) 936.
[38] J.A. Tataronis and W. Grossmann, New York University, Courant Institute of Mathematical Sciences, Report C00-3077-102, MF-84 (1977).
[39] A. Hasegawa and C. Uberoi, The Alfvén Wave (Technical Information Center, U.S. Dept. of Energy, 1982).
[40] K. Appert, R. Gruber, F. Troyon and J. Vaclavik, Plasma Phys. 24 (1982) 1147.
[41] S. Mahajan, D.W. Ross and G.-L. Chen, University of Texas, Report FRCR249 (1982).
[42] J. Tataronis and W. Grossmann, Z. Phys. 261 (1973) 203.
[43] A. Hasegawa and L. Chen, Phys. Rev. Lett. 32 (1974) 454.
[44] E.M. Barston, Ann. Phys. (N.Y.) 29 (1964) 282.
[45] Z. Sedlaček, J. Plasma Phys. 5 (1971) 239.
[46] B.R. Suydam, Proc. of the Second U.N. Intern. Conf. on the Peaceful Uses of Atomic Energy, Geneva, 1958 (Columbia Univ. Press, New York, 1959), vol. 31, p. 157.
[47] C. Mercier, Nucl. Fusion 1 (1960) 47.
[48] T.M. Antonsen, A. Ferreira and J.J. Ramos, Plasma Phys. 4 (1982) 197.
[49] Y. Suzuki, Nucl. Fusion 14 (1974) 345.
[50] K. Lackner, Comput. Phys. Comm. 12 (1976) 33.
[51] R. Teman, Commun. Part. Diff. Eq. 2 (1977) 563.
[52] J. DeLucia, S.C. Jardin and A.M.M. Todd, J. Comput. Phys. 37 (1980) 183.
[53] J. Blum, J. Le Foll and B. Thooris, Comput. Phys. Comm. 24 (1981) 235.
[54] T.S. Wang and F.J. Helton, Comput. Phys. Comm. 24 (1981) 255.
[55] P.N. Vabishchevich and L.M. Degtyarev, Dokl. Akad. Nauk 247 (1979) 1342 [Sov. Phys.-Dokl. 24 (1979) 612].
[56] L.L. Lao, S.P. Hirschman and R.M. Wieland, Phys. Fluids 24 (1981) 1431.
[57] L.E. Zakharov and V.D. Shafranov, in: Reviews of Plasma Physics 11 (Energoisdat, Moscow, 1982) p. 118. [Russian, translation not yet available].
[58] A. Bhattacharjee, J.C. Wiley and R.L. Dewar, to be published.
[59] J.P. Freidberg and J.P. Goedbloed, in: Pulsed High-Beta Plasmas, D.E. Evans, ed. (Pergamon, Oxford, 1976), p. 117.
[60] P. Henrici, SIAM Rev. 21 (1979) 481.
[61] J.P. Goedbloed, Comput. Phys. Comm. 24 (1981) 311.
[62] J.P. Goedbloed, Phys. Fluids 25 (1982) 852; 2062; 2073.
[63] B. Fornberg, SIAM J. Sci. Stat. Comput. 1 (1980) 386.
[64] R. Menikoff and C. Zemach, J. Comput. Phys. 36 (1980) 366.
[65] D.I. Meiron and S.A. Orszag, J. Comput. Phys. 40 (1981) 345.
[66] R. Menikoff and C. Zemach, submitted to J. Comput. Phys.
[67] G. Bateman, Phys. Fluids 14 (1971) 1506.
[68] J.P. Freidberg and F.A. Haas, Phys. Fluids 16 (1973) 1909; 17 (1974) 440.
[69] B.M. Marder, Phys. Fluids 17 (1974) 447; 639.
[70] J.P. Freidberg and W. Grossmann, Phys. Fluids 18 (1975) 1494.
[71] E. Rebhan and A. Salat, Nucl. Fusion 16 (1976) 805; 17 (1977) 251; 18 (1978) 1639; 20 (1980) 839.
[72] M. Kito and T. Honma, J. Phys. Soc. Jpn. 41 (1976) 1749; Phys. Fluids 21 (1978) 272.
[73] D.A. D'Ippolito, J.P. Freidberg, J.P. Goedbloed and J. Rem, Phys. Fluids 21 (1978) 1600.
[74] M. Kito, Phys. Fluids 24 (1981) 1142.
[75] R.M.O. Galvão, J.P. Goedbloed, J. Rem, P.H. Sakanaka, T.J. Schep and M. Venema, Proc. 9th Intern. Conf. on Plasma Physics and Controlled Nuclear Fusion Research, Baltimore (1982) paper P-1-1.
[76] P.H. Sakanaka and J.P. Goedbloed, Phys. Fluids 17 (1974) 919.
[77] M.N. Bussac, R. Pellat, D. Edery and J.L. Soule, Phys. Rev. Lett. 35 (1975) 1638.
[78] L.E. Zakharov, Fiz. Plasmy 4 (1977) 898 [Sov. J. Plasma Phys. 4 (1978) 503].
[79] S. Tokuda, T. Tsunematsu, M. Azumi, T. Takizuka and T. Takeda, Nucl. Fusion 22 (1982) 661.
[80] G.B. Crew and J.J. Ramos, Phys. Rev. A26 (1982) 1149.
[81] B. Coppi, A. Ferreira and J.J. Ramos, Phys. Rev. Lett. 44 (1980) 990.
[82] N.J. Zabusky, Physica 12D (1984) 145 (these proceedings).

Physica 12D (1984) 133–144
North-Holland, Amsterdam

NONLINEAR INTERACTIONS DURING MAGNETIC FIELD-LINE RECONNECTION IN PLASMAS

R.L. STENZEL and W. GEKELMAN
Department of Physics, University of California, Los Angeles, CA 90024, USA

Stimulated by important dynamic processes in space and fusion plasmas, the reconnection of magnetic fields in plasmas is investigated. A well-controlled large laboratory plasma is established, and direct observations of magnetic fields and particle distributions are performed. High speed computers are interfaced with the experiment so as to permit the analysis of multidimensional functions. These include the average magnetic field topology $\langle \boldsymbol{B}(\boldsymbol{r}, t)\rangle$ and the tensor correlation function $\langle \boldsymbol{B}_1\boldsymbol{B}_2\rangle_{\omega,\boldsymbol{k}}$ of magnetic fluctuations, as well as the electron phase space distribution function $f(\boldsymbol{v}, \boldsymbol{r}, t)$. The observations show that the characteristic field topology of a neutral sheet is subject to a high level of magnetic turbulence consisting of low frequency whistler waves. These are excited by anisotropies in the electron distribution functions inside the current sheet, in particular by energetic electrons which run away in the electric field along the separator. The observed microinstabilities and kinetic processes cannot be described by classical fluid theories or MHD simulations but are important in determining the effective resistivity and reconnection rate.

1. Introduction

Solar flares [1] and magnetic substorms [2] present examples from space plasma physics where a strong nonlinear interaction of magnetic fields with plasmas takes place. These dynamic processes are both characterized by a rapid energy transfer from stored magnetic field energy into plasma kinetic energy [3]. The events appear to violate the classical MHD picture which predicts that magnetic fields are essentially "frozen in" [4] highly conducting fluids and cannot be dissipated except in small regions near magnetic null points where finite resistivity permits diffusion, hence reconnection of field lines and dissipation [5]. The discrepancy between the observed impulsive energy exchange and the classical steady-state theories has long been recognized, and models for various plasma instabilities near null points have been proposed [6]. These involve either macroscopic changes in the magnetic field topology by tearing instabilities [7] or microscopic instabilities which generate small scale fluctuations and effectively enhance the plasma resistivity [8]. Either case leads to nonlinear phenomena since field and plasma properties are highly coupled. In spite of many years of theoretical work, numerical simulation effort and observations from spacecraft, the complex reconnection process is still an open fundamental problem in plasma physics.

Its significance is not restricted to space plasmas. In various fusion devices such as tokamaks, reverse field pinches, spheromaks, and laser fusion, magnetic null point geometries arise [9]. In most confinement devices there is evidence that tearing instabilities lead to a loss of confinement of particles and energy even though the details are not known.

A common problem in fusion and space plasmas is the difficulty of making direct detailed observations of the reconnection processes near magnetic null points. The basic physical processes are best observed in a well-controlled laboratory plasma designed for optimizing diagnostic access. A large diameter discharge plasma has been used at UCLA for reconnection studies for several years [10–15].

Although one cannot model in a laboratory device global configurations of cosmic plasmas, the characteristic magnetic null point geometry can be reproduced [16–18]. Here we will show the formation of a classical magnetic neutral sheet [5]. It presents the interface between two oppositely directed magnetic field regions, and it is maintained by a plasma current sheet. Its stability has been investigated and is found to contain a high level of magnetic turbulence. It consists of a broad spectrum of unstable whistler waves. The source of these instabilities lies in the non-equilibrium electron distribution function in the current sheet. Anisotropies in the form of runaway electron tails are observed from direct measurements of $f(\boldsymbol{v}, \boldsymbol{r}, t)$. The significance of these findings to the reconnection physics is twofold: (i) Non-Maxwellian distribution functions modify the classical transport parameters; for example, the effective resistivity is determined by the runaway electrons rather than the bulk electron "temperature". (ii) Velocity space instabilities create a high level of turbulence which interacts with resonant particles. These wave–particle interactions are outside the models described by MHD theory, yet may explain "anomalous" resistivities, enhanced reconnection, and energy transfer.

Most of the physical quantities measured in this experiment involve higher dimensions and multiple variables. In order to be able to investigate a wide parameter space the experiment has been interfaced with high speed computers. This approach opens a new dimension in experimental work. Without digital data processing it would be essentially impossible to investigate the seven-variable distribution function or the Fourier transformed cross-correlation tensor of magnetic fluctuations. We will display some of these functions in stereoscopic form; the details for viewing such figures are described in the appendix.

2. Experimental arrangement

The present reconnection experiment (see fig. 1) employs a linear discharge plasma column produced with a large diameter (1 meter) cathode [19]. Typical densities are $n_e \approx 10^{12}\,\mathrm{cm}^{-3}$, temperature $kT_e \approx 10\,\mathrm{eV}$, axial magnetic fields $B_y \approx 10\,\mathrm{G}$. The plasma is uniform, quiescent, nearly collisionless, and highly reproducible in pulses of duration $t_p \approx 5\,\mathrm{ms}$ repeated every $t_r \approx 2\,\mathrm{s}$. After establishing the plasma a time varying transverse magnetic field ($\langle B_\perp \rangle \approx 10\,\mathrm{G}$) is applied. Its topology in vacuum contains an X-type neutral point on the axis of the device (see fig. 1a). In the presence of the plasma electron currents ($I_p \approx 1000\,\mathrm{A}$) are induced which flow preferentially in regions of $B_\perp \approx 0$ but are sufficiently large to modify the applied field topology. During the period of rising magnetic fields ($0 < t < 100\,\mu\mathrm{s}$) the self-consistent field distribution is that of a classical neutral sheet [5]. The plasma current, driven by inductive and space charge electric fields, $\boldsymbol{E} = -\dot{\boldsymbol{A}} - \nabla\phi_p$, is mainly carried by electrons which drift from the cathode toward the end anode. The current closes via the grounded chamber wall to the cathode. No net currents flow to the insulated plates or to the radial chamber walls which are separated from the plasma by vacuum.

The plasma processes are analyzed with various probes which provide time and space resolved data. A digital data acquisition system consisting of 8 bit, 100 MHz analog-to-digital (A/D) converters, minicomputer (LSI 11/23), array processor and direct data link to large size computers (CDC 7600, Cray 1) is employed. Probes include magnetic loops, Langmuir probes, high-frequency rf probes, and a novel directional velocity analyzer [20]. In a typical magnetic field measurement the signals of orthogonal loops are recorded vs. time at 300 spatial locations. At each position the experiment is repeated up to 80 times so as to obtain statistical averages, i.e., mean, standard deviation, and correlation functions. Extensive space-time maps may involve > 100 million numbers which can only be handled by high speed computers.

3. Magnetic field topologies

A function of multiple variables such as the vector field $\boldsymbol{B}(x, y, z, t)$ cannot be visualized at

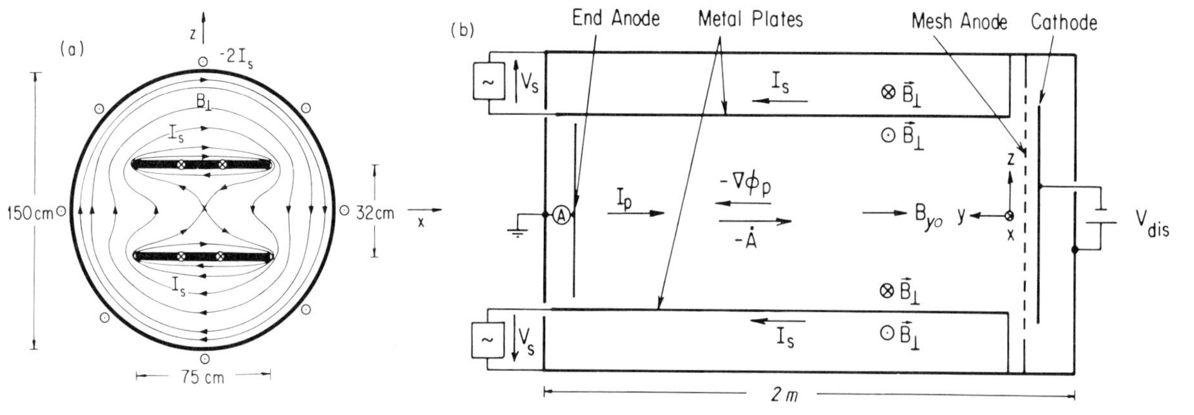

Fig. 1. Schematic picture of the experimental arrangement. (a) Cross-sectional view showing parallel plate electrodes with pulsed currents I_s and magnetic field lines $\boldsymbol{B}_\perp$ without plasma. (b) Side view of the device with main electrodes, currents (I_p, I_s) electric fields ($\boldsymbol{E} = -\dot{\boldsymbol{A}} - \nabla\phi_p$), and magnetic fields ($\boldsymbol{B} = \boldsymbol{B}_\perp + \boldsymbol{B}_{y0}$). The coordinate system common in magnetospheric physics has been adopted where y is along the neutral line (device axis), x is along the horizontal neutral sheet, and z is normal to the sheet.

once but only in a reduced parameter space. We start to show the two-dimensional field $\boldsymbol{B}_\perp = (B_x, B_z)$ in the transverse plane (x, z) at a given time t, ensemble averaged at each point over $N = 25$ repeated experiments. Fig. 2a depicts the field topology of a neutral sheet, the classical configuration treated in most steady-state theories of reconnection [21]. We observe this field pattern during the rise of the applied currents. The induced plasma current tends to slow down the penetration of the external field and its reconnection in the null region. The current sheet $\boldsymbol{J} = \nabla \times \boldsymbol{B}/\mu_0$ has a half width of $\Delta z \approx 3$ cm which is on the order of the collisionless skin depth $c/\omega_{pe} \approx 0.7$ cm. Current sheets investigated in space are often broader than an ion Larmor radius so that, in comparison, we are resolving the fine structure of current sheets. The field topology is macroscopically stable for many Alfvén transit times across the plasma ($t_A \approx 7\ \mu$s in He, 20 μs in Ar). However, in the case of very large current densities ($J \approx nev_e$) disruptive instabilities are observed [22]. These will not be considered in this paper.

In response to the induced currents and $\boldsymbol{J} \times \boldsymbol{B}$ forces, the plasma is vertically compressed and horizontally expelled. We have measured the fluid motion with orthogonal differential ion velocity analyzers [12]. Fig. 2b depicts the characteristic ion

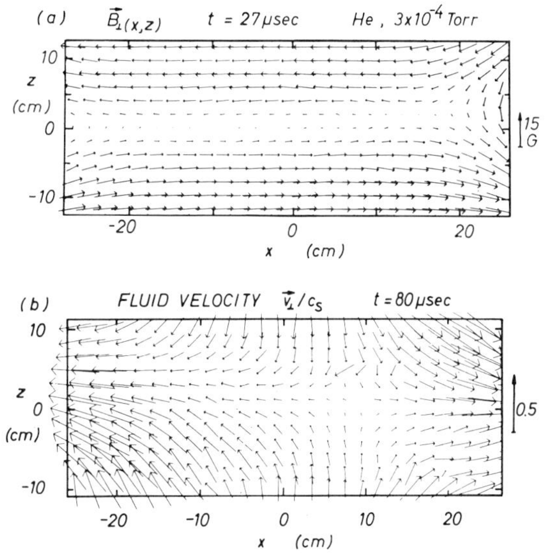

Fig. 2. Measured magnetic field and flow field during reconnection. (a) Transverse field $\boldsymbol{B}_\perp(x, z)$ at a fixed time t, position $y = 137$ cm, showing the classical neutral sheet topology. (b) Transverse ion flow velocity, $\boldsymbol{v}_\perp(x, z)$, normalized to the local sound or Alfvén speed, $C_S = (kT_e/m_i)^{1/2} \approx C_A = (B^2/2\mu_0 nm_i)^{1/2}$. Vertical compression and horizontal jetting are in qualitative agreement with reconnection models.

flow pattern which is observed to develop after an initial phase of turbulence. The measured flow field clearly shows the fluid jetting at Alfvénic speeds, which is predicted by theory and computer simulations [23] and thought to explain the generation of high energy particles in solar flares and the magnetotail.

We now turn to the time dependence of the magnetic field topology. Two time scales can be distinguished: a slow one corresponding to the rise time of the external fields ($t \lesssim 100\ \mu$s), and a fast time scale associated with fluctuations and instabilities. The slowly varying processes are well reproducible over repeated experiments. Thus, ensemble averages such as shown in fig. 2 are generated in typically 1024 successive, small time increments. The time evolution of the field topologies is then viewed in sequence in the form of a computer processed movie. Here, we can display only a few frames depicting the major features.

Fig. 3 shows magnetic field lines in the transverse x–z plane at three different time steps, $t = 57$, 58, and 59 μs. Instead of the vector field of fig. 2a, we have integrated the $\boldsymbol{B}$ field and display contours of constant vector potential, $A_y = \text{const.}$, where $\boldsymbol{B} = \nabla \times \boldsymbol{A}$. The motion of magnetic field lines through the plasma is best followed by the two inner field lines drawn heavier. With increasing time the opposing field lines move vertically ($\pm z$) towards the center, merge at the null point, and reconnect to form a new field line pair which moves horizontally ($\pm x$) outward. This process of field line reconnection occurs continuously during the external current rise. Magnetic flux generated at the two current carrying plates is transferred across the separatrix into regions of common flux where field lines encircle both current plates. During the current decrease the process reverses but the resultant vertical current sheet usually tears to form a large magnetic island [10]. The motion of field lines through plasmas is highly descriptive but becomes physically significant only through the coupling of plasma and field lines. However, near the null region, diffusion due to finite resistivity dominates. The field-line "velocity" is then a measure of the axial inductive electric field which is often taken as another measure for the reconnection rate [21]. The effective plasma resistivity is an important parameter since it determines the size of the diffusion region, the reconnection rate, and

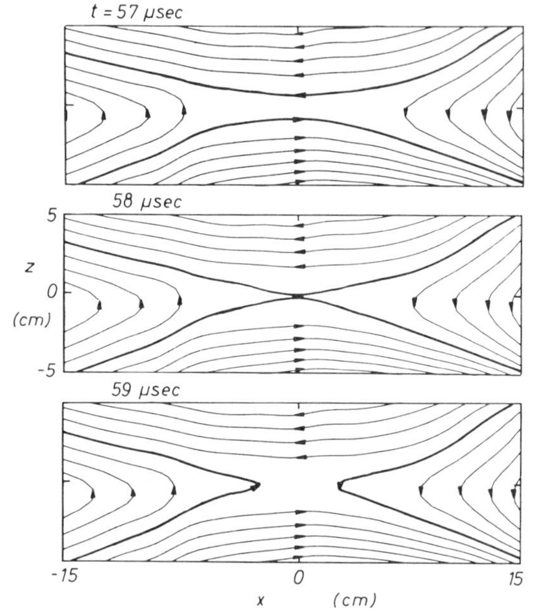

Fig. 3. Snapshots of magnetic field lines ($A_y = \text{const.}$) during reconnection in a plasma. The two inner field lines merge and reconnect to form a pair of horizontally outward moving field lines. The entire time evolution has been displayed in a motion picture.

the plasma heating. Direct measurements of the resistivity [13] show a level well above the classical one, and an inhomogeneous pattern in the neutral sheet. These features may be understood after having discussed wave and particle measurements.

The present discussion was so far limited to the two-dimensional field topology. Because of the axial component B_{y0} the total magnetic field is three-dimensional, $\boldsymbol{B} = (B_x, B_{y0}, B_z)$. Fig. 4 is a display of the three-dimensional vector field in the transverse plane $\boldsymbol{B}_{(x,z)}$. It is presented as a picture pair for stereoscopic viewing, the techniques for which are explained in the appendix. The field lines are straight only in the current sheet, and exhibit increasing shear both above and below it. The neutral sheet is a null region only with respect to the transverse field ($B_x = B_z = 0$) but not the total field ($|B| = B_{y0} \approx 10$ G). Thus, the electrons in the current sheet always remain magnetized. In a different investigation we have varied the axial field strength from $B_{y0} = 0$ to $B_{y0} \gg |\boldsymbol{B}_\perp|$ but here we limit ourselves to the case $B_{y0} \approx |\boldsymbol{B}_\perp|$. Both in

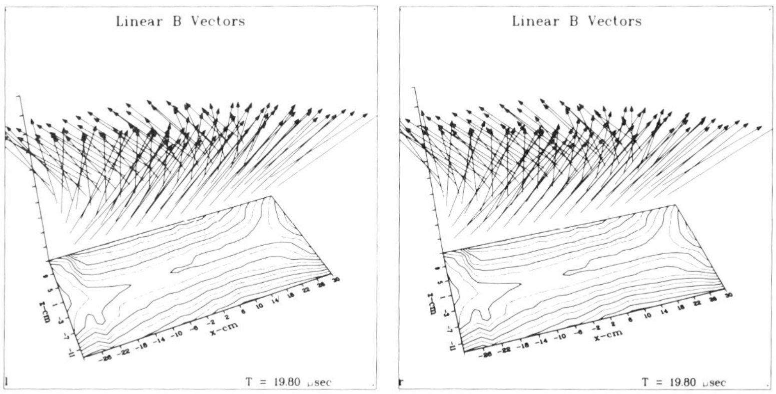

Fig. 4. Stereoscopic display of the three-dimensional vector magnetic field $\boldsymbol{B} = (B_x, B_y, B_z)$ in the transverse $x - z$ plane. For reference, bottom plane gives a projection of the transverse field lines $A_y(x, z) = \text{const.}$ ($y = 137$ cm). See appendix for viewing stereo pictures.

tokamaks and in the magnetosphere the fields are, in general, three-dimensional, while most theoretical models assume $B_y = 0$, for simplicity. When the axial field component is absent the magnetic field gradients are largest. Fig. 5 presents a stereoscopic view of the field strength $|\boldsymbol{B}_\perp|_{(x,z)}$ along the neutral sheet which forms a deep long valley. In such inhomogeneous fields the particle orbits are, in general, nonadiabatic and the distributions are anisotropic [24].

We now turn to the rapidly varying magnetic field fluctuations ($\delta B/B \approx 10\%$). These result in rippled and distorted field lines. Although with a single probe we cannot map the instantaneous field lines, we can study the nature of the fluctuations statistically, e.g., by two-point cross-correlations. The first step in this investigation is to determine the fluctuation spectrum in terms of characteristic modes. Once the waves are identified the instability mechanism and the saturation processes have to be identified, which requires the knowledge of particle distributions. The final step is to assess the effect of the instability on the macroscopic plasma parameters, i.e., the transport properties.

Cross correlation studies are performed with two identical pairs of magnetic probes, one movable (# 1) and one stationary (# 2). At a given position an ensemble of $N = 80$ repeated mea-

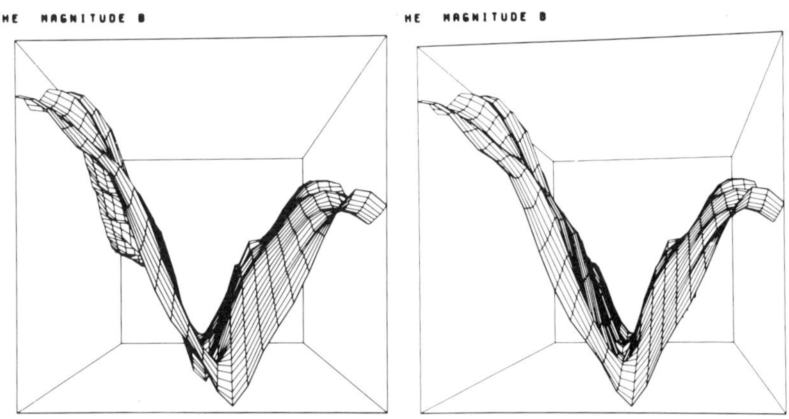

Fig. 5. Stereoscopic view of the transverse magnetic field strength $|\boldsymbol{B}_\perp(x, z)|$ showing steep gradients adjacent to neutral sheet ($0 < B < 15$ G, $-25 < x < 25$ cm, $-12 < z < 12$ cm, $t = 25\,\mu$s). See appendix for viewing stereo pictures.

surements of the vector components B_x, B_z is recorded vs. time. Due to the slow time variation the fluctuation processes are weakly nonstationary so that statistical averages must be obtained by ensemble averaging rather than time averaging [25]. From the stored data we calculate on-line the ensemble mean

$$\langle \boldsymbol{B} \rangle = \frac{1}{N} \sum_{i=1}^{N} \boldsymbol{B}_i , \tag{1}$$

the root mean square value of the fluctuations,

$$\boldsymbol{B}_{\mathrm{rms}} = \left(\frac{1}{N} \sum_{i=1}^{N} (\boldsymbol{B}_i - \langle \boldsymbol{B} \rangle)^2 \right)^{1/2} , \tag{2}$$

and the cross correlation function

$$\langle \boldsymbol{B}_1 \boldsymbol{B}_2 \rangle = \frac{1}{N} \sum_{i=1}^{N} (\boldsymbol{B}_{i1} - \langle \boldsymbol{B}_1 \rangle)(\boldsymbol{B}_{i2} - \langle \boldsymbol{B}_2 \rangle) \tag{3}$$

We note that the correlation of two vectors results in a tensor of, at least, rank $R = 2$. Here, its components are given by

$$\langle \boldsymbol{B}_1 \boldsymbol{B}_2 \rangle = \begin{pmatrix} \langle B_{x1} B_{x2} \rangle & \langle B_{x1} B_{z2} \rangle \\ \langle B_{z1} B_{x2} \rangle & \langle B_{z1} B_{z2} \rangle \end{pmatrix} \tag{4}$$

In order to identify the mode spectrum we Fourier transform the time dependence of the fluctuations, select a narrow spectral component, and analyze its spatial properties. The resultant cross spectral function is a function of probe separation $\Delta \boldsymbol{r} = (\Delta x, \Delta y, \Delta z)$, delay time τ between the two probes, and a weaker function of the time t during the reconnection process, $\langle \boldsymbol{B}_1 \boldsymbol{B}_2 \rangle_\omega (\Delta \boldsymbol{r}, \tau, t)$.

In fig. 6 we display an example of a diagonal tensor component $\langle B_{x1} B_{x2} \rangle$ in the axial x–y plane at a given frequency ($\omega / 2\pi = 1$ MHz), time ($t = 21.4\ \mu$s), and delay time ($\tau = 1.3\ \mu$s). The upper part of the figure shows a topological view of the cross spectral function while the lower part presents a corresponding contour plot. The displays on the right-and left-hand side are rotated by $\approx 6°$ for purpose of stereoscopic viewing (see appendix A).

The measurement indicates that even at a selected frequency a broad spectrum of spatial modes is excited. This behavior is also found at

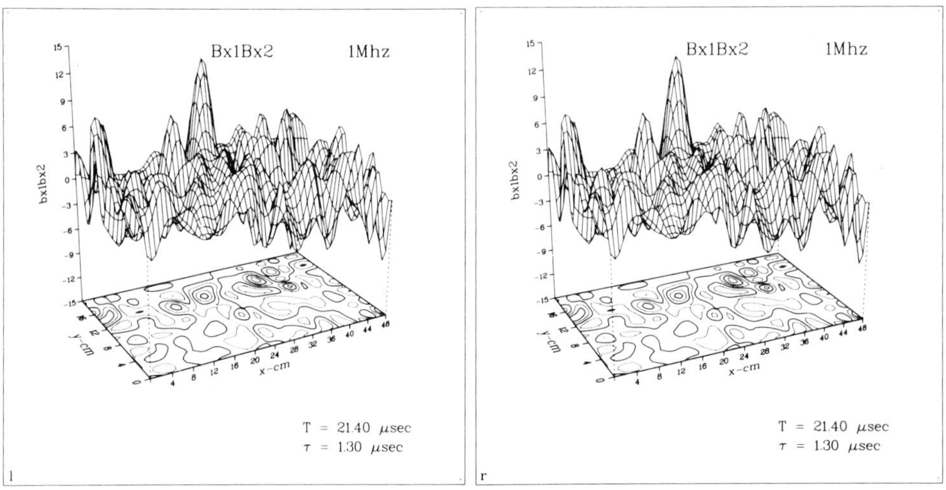

Fig. 6. Stereoscopic view of the cross-spectral tensor components $\langle B_{x1} B_{x2} \rangle_{(\Delta x, \Delta y)}$ of magnetic field fluctuations. The data are taken at $t = 21.4\ \mu$s during reconnection with $\tau = 1.3\ \mu$s time delay between the two probe signals, each filtered at $\omega / 2\pi = 1$ MHz. A two-dimensional spatial mode spectrum is observed. See appendix for viewing stereo pictures.

other frequencies in the range investigated (100 kHz $< f <$ 1 MHz). The characteristic electromagnetic modes in this frequency band are the whistler waves [26]. These modes are not only dispersive ($\omega/k \neq$ const.) but also anisotropic, i.e., at a given ω the wave number k depends on the direction of wave propagation with respect to the magnetic field [$k = k(\theta)$]. From our cross-spectral measurements we can obtain the wavenumber spectrum by a spatial Fourier transformation from $\boldsymbol{r}$-space into $\boldsymbol{k}$-space. Since the time and space dependence is recorded the complex fast Fourier transform yields both magnitude and direction of $\boldsymbol{k}$. Using the data from two orthogonal planes (x–y, x–z) we have constructed the three-dimensional wave vector spectrum from the two projections in the k_x–k_y and k_x–k_z planes.

An example of such data analysis is shown in fig. 7. It is a stereoscopic display of the dominant modes of $\langle B_{x1} B_{x2} \rangle_\omega$ in the full three-dimensional $\boldsymbol{k}$ space. The relative intensity of the modes is proportional to the dot size. A more detailed amplitude profile is provided by contour plots in the two orthogonal planes. In order to compare the measurements with theory we also display the wave vector surfaces $\pm k_{(\theta)}$ calculated from the dispersion relation for whistlers [26]

$$\left(\frac{kc}{\omega}\right)^2 = 1 - \frac{\omega_p^2}{\omega(\omega - \omega_c \cos\theta)}, \tag{5}$$

where $\omega/2\pi = 1$ MHz, electron plasma frequency $\omega_p/2\pi = 20$ GHz, and cyclotron frequency $\omega_c/2\pi = 28$ MHz are typical experimental parameters. One can see that the unstable modes fall close so the theoretical refractive index (kc/ω) surface for whistlers. An exact fit cannot be expected due to magnetic field and density inhomogeneities.

Besides the dispersion properties we have also investigated the polarization of the unstable modes. Theory predicts that the magnetic field of whistler waves has a right hand circular polarization [26]. By applying a digital filter to the measured k vector spectrum we select one mode at a time, inverse Fourier transform it, and inspect its space-time dependence ($\Delta \boldsymbol{r}, \tau$). This procedure is applied to the two tensor components $\langle B_{x1} B_{x2} \rangle$ and $\langle B_{z1} B_{x2} \rangle$ which have two orthogonal components of vector $\boldsymbol{B}_1$ correlated with the same reference signal B_{x2}. A display of $\langle \boldsymbol{B}_1 B_{x2} \rangle$ versus τ at a given $\Delta \boldsymbol{r}$ (or vice versa) provides a hodogram which reveals the polarization.

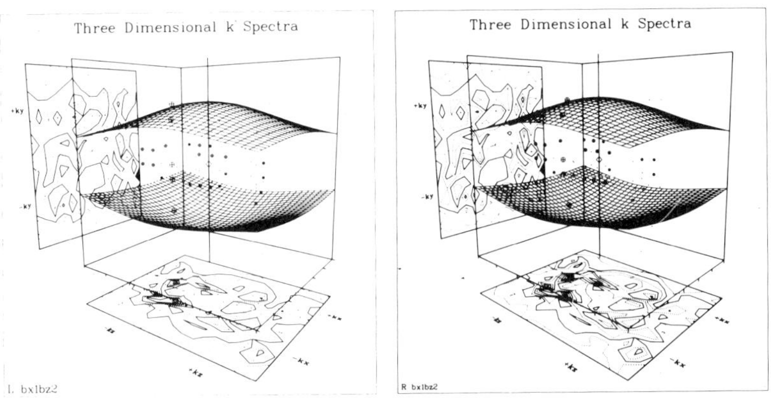

Fig. 7. Three-dimensional wavevector space (k_x, k_y, k_z) with measured modes of the cross-spectral function $\langle B_{x1} B_{z2} \rangle$ and the theoretical dispersion surfaces $\pm k_{(\theta)}$ for whistler waves ($\omega/2\pi = 1$ MHz, $\omega_c/2\pi = 28$ MHz, $\omega_p/2\pi = 20$ GHz). Vertical axis k_y is along average $\boldsymbol{B}_{y0}$, origin $k = 0$ is indicated by square dot. The mode intensity is proportional to the dot size, the mode location $\boldsymbol{k}$ is constructed from measurements of $\langle B_{x1} B_{z2} \rangle$ in two orthogonal spatial planes, Fourier transformed into k planes and presented as contour plots adjacent to the $\boldsymbol{k}$ space. The observations indicate that the magnetic turbulence consists of oblique whistler modes. See appendix for viewing stereo pictures.

Fig. 8 is a stereoscopic view of the hodogram $\langle \boldsymbol{B}_1 B_{x2} \rangle$ vs. τ for a mode propagating nearly along the average magnetic field. The helical trajectory indicates that the mode is essentially right-hand circularly polarized. However, for long delay times ($\tau \approx 3$ μs) phase mixing due to the finite filter bandwidth ($\Delta\omega/\omega \approx 10^{-1}$) tends to randomize the polarizations. For modes propagating at an angle to the magnetic field we first transform the wave $\boldsymbol{B}$-field into a plane normal to $\boldsymbol{k}$ ($\boldsymbol{k} \cdot \boldsymbol{B} = 0$) and then inspect the polarization. Most modes which fall close to the dispersion surface are right-hand polarized.

Thus, from both the dispersion and polarization properties the eigenmodes which make up the magnetic field fluctuations are found to be whistler waves.

We will next investigate the cause for the instability of whistler waves. According to theoretical models, anisotropies in the electron velocity distribution can lead to whistler instabilities [27].

4. Distribution functions

The distribution of particles in velocity space $\boldsymbol{v}$, real space $\boldsymbol{r}$, and time t, $f(\boldsymbol{v}, \boldsymbol{r}, t)$, provides the fundamental description of plasmas on a microscopic level. All macroscopic plasma properties can be derived from it by forming moments, i.e., averages over the velocity distribution. Kinetic theories and numerical particle simulations all depend on the knowledge of the distribution function. In spite of the theoretical significance of this function there are, surprisingly, few experimental observations of, e.g., electron distributions in laboratory plasmas. In order to measure $f(\boldsymbol{v})$ a velocity analyzer with directional sensitivity is required. Such detectors are available for low density, high energy particles in space [28] but have only recently been developed for denser laboratory plasmas [20, 29].

Our electron detector consists of a small (3 mm radius) retarding potential velocity analyzer whose directional sensitivity ($\Delta\Omega \approx 4\pi \times 10^{-3}$ sterad) arises from filtering electrons through a microchannel plate. It can be rotated through the two spherical angles ($0 < \theta < \pi$, $0 < \phi < 2\pi$) so as to collect electrons from any direction in velocity space. The three-dimensional distribution function $f(v, \theta, \phi)$ is assembled from individual flux measurements at typically 200 different directions θ, ϕ. Furthermore, at each angular position the experiment is repeated N times so as to form the ensemble average $\langle f \rangle$ and standard deviation f_{rms}, analogous to the case of the magnetic field

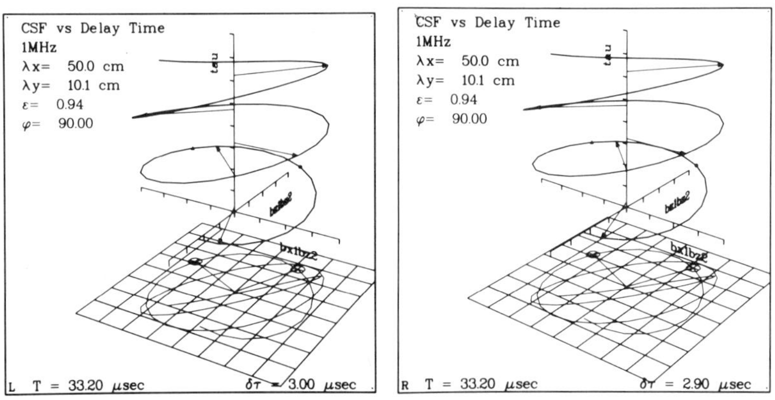

Fig. 8. Polarization properties of a single (ω, k) mode of the magnetic turbulence spectrum. With increasing delay time τ the correlation vector $\langle B_1, B_{z2} \rangle$ rotates initially in a right-hand circular sense as expected for whistler waves ($\omega/2\pi = 1$ MHz). See appendix for viewing stereo pictures.

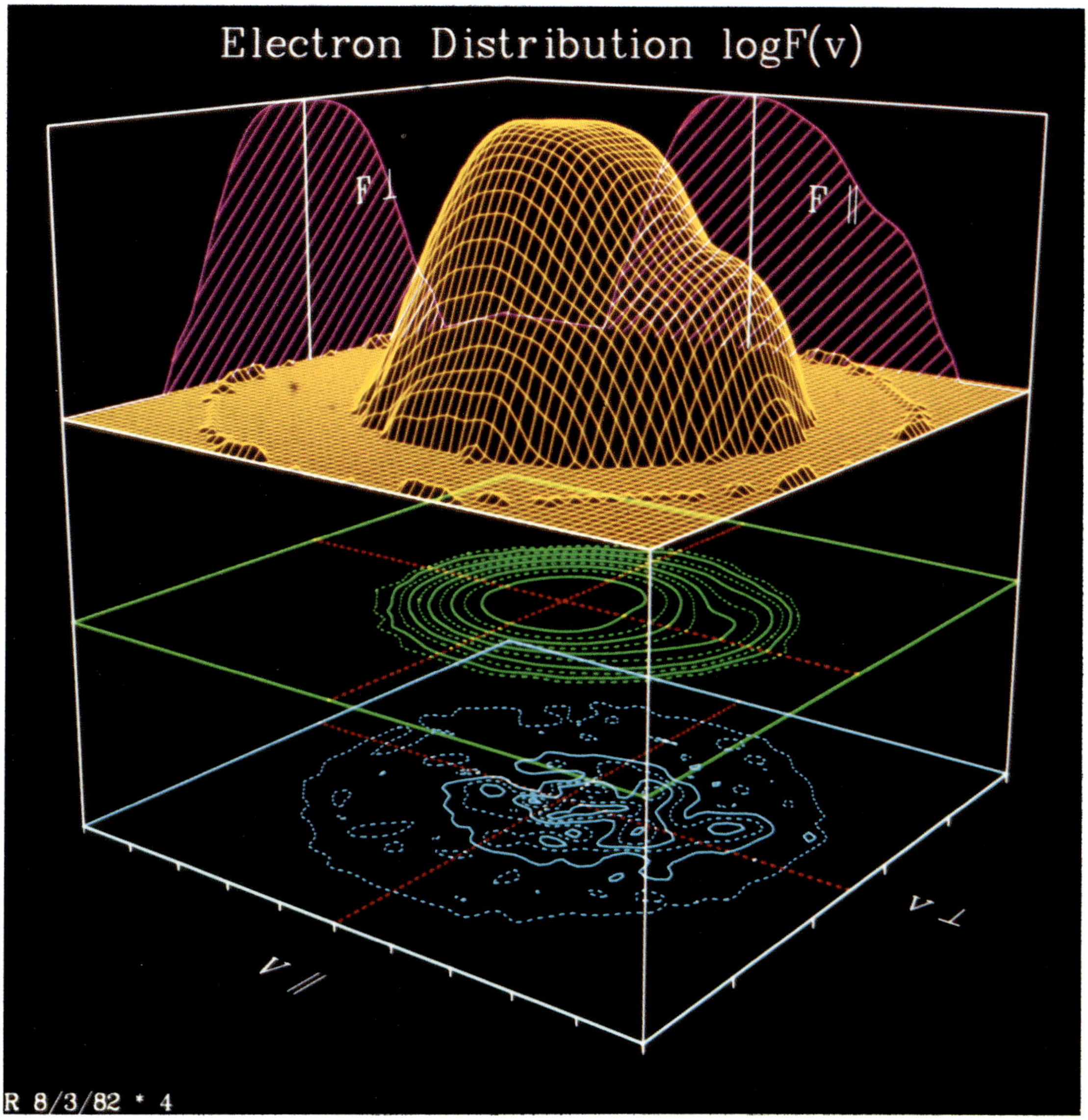

Fig. 12. Comparison between the average distribution function and its fluctuations in the neutral sheet.
Upper panel: Logarithmic display of the average value $\langle f(v_\parallel, v_\perp)\rangle$ and two one-dimensional cuts along $v_\parallel = v_y$ axis and $v_\perp = v_z$ axis. Note tail of runaway electrons. [$f(0,0) \cong 10^{-14}$ cm^{-6} s^3; reference plane at $f = 10^{-17}$ cm^{-6} s^3.]
Middle level: Contours of constant average $\log\langle f(v_\parallel, v_\perp)\rangle$ (3 dB contour spacing).
Bottom level: Contours of root-mean-square fluctuations $\log f_{rms}(v_\parallel, v_\perp) = $ const. (3 dB per contour, $f_{rms\ max}/\langle f_{max}\rangle \cong 20\%$). Note large fluctuations in regions of runaway electron tails. [$0 \leqslant |v_\parallel|, |v_\perp| \leqslant 6 \times 10^8$ cm/s.]

measurements [see eqs. (1), (2)]. Besides the velocity space dependence we also measure the real space dependence by moving the analyzer to various positions in the neutral sheet. The measurements are also time resolved to within $\Delta t \approx 3\ \mu$s, the time required to sweep out the current–voltage characteristics. The data processing for a function of up to seven variables is made possible only through the use of high speed computers with large memory.

Fig. 9 presents measurements of the electron distribution function in velocity space. In two dimensions the distribution $f(v_x, v_y)$ can either be displayed topographically (fig. 9a) or as a contour plot (fig. 9b). In three-dimensional velocity space a display analogous to fig. 9a cannot be visualized but we can extend fig. 9b and plot surfaces of constant $f(\boldsymbol{v})$ as done in fig. 9c. The maximum value $f_{\max} \approx 10^{-14}\ \mathrm{cm}^{-6}\ \mathrm{s}^3$ generates a surface near the origin. Decreasing values of $f(\boldsymbol{v})$ produce nested surfaces of expanding radius in velocity space, three of which are shown in fig. 9c where the outer surfaces are cut open for purpose of display. Using time as the variable f we have displayed the continuous variation $0 < f < f_{\max}$ in a computer generated movie.

Anisotropic distribution functions exhibit nonspherical surfaces of $f(\boldsymbol{v}) = \text{const.}$ Fig. 10 shows a stereoscopic view of a surface near the tail of the distribution. We choose to display $v^2 f(\boldsymbol{v}) = \text{const.}$ rather than $f(\boldsymbol{v}) = \text{const.}$ to stress the importance of a seemingly small tail (see fig. 9) on the moments of $f(\boldsymbol{v})$. The quantity displayed is the kernel for the zeroth moment, i.e., the density

$$n_e = \int f(\boldsymbol{v})\, d\boldsymbol{v} = \int_0^{2\pi}\int_0^{\pi}\int_0^{\infty} v^2 f(v, \theta, \phi)\, dv \sin\theta\, d\theta\, d\phi\,. \tag{6}$$

Higher moments such as those leading to the current density $\boldsymbol{j} = \int \boldsymbol{v} f(\boldsymbol{v})\, d\boldsymbol{v}$, mean energy $E = \int \frac{1}{2} m v^2 f(\boldsymbol{v})\, d\boldsymbol{v}$, and heat flow $\boldsymbol{q} = \int \frac{1}{2} m v^2 \boldsymbol{v} f(\boldsymbol{v})\, d\boldsymbol{v}$ involve progressively higher powers of v which puts a strong weight on the fast tail electrons. For example, integration over the distri-

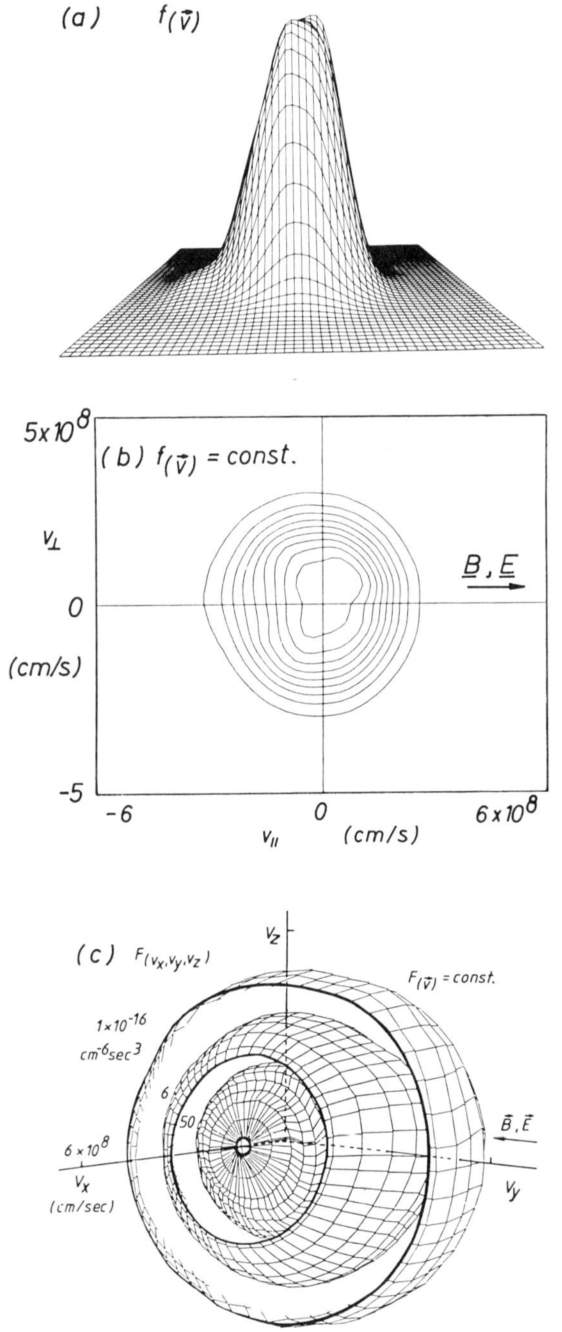

Fig. 9. Measured electron distribution function in the neutral sheet. (a) Linear display of $f(v_\parallel, v_\perp)$; $f_{(0,0)} \approx 10^{-14}\ \mathrm{cm}^{-6}\ \mathrm{s}^3$. (b) Contour plot $f(v_\parallel, v_\perp) = \text{const.}$; $\Delta f / f_{\max} = 10\%$ per contour. (c) Distribution function in three-dimensional velocity space displayed as surfaces of $f(v_x, v_y, v_z) = \text{const.}$ Outer surfaces are cut for purpose of display. Continuous variation of $0 < f < f_{\max}$ has been displayed in a motion picture.

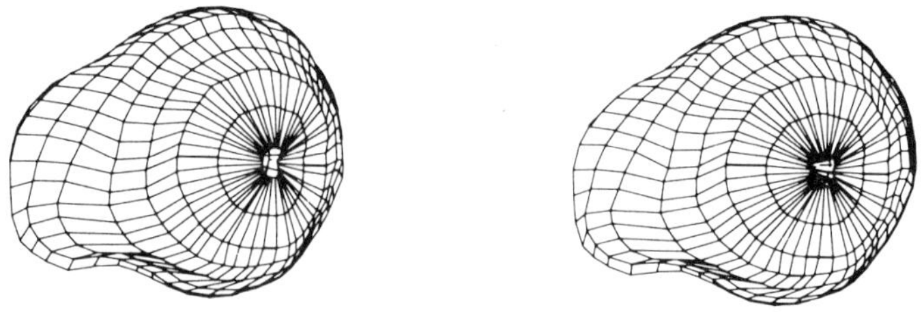

Fig. 10. Stereoscopic display of a surface $v^2 f(v) = \text{const}$ (50 cm^{-4} s) in three-dimensional velocity space. The distortion on the left-hand side reflects an anisotropy caused by runaway electrons in the electric field $E_y \| B_{y0}$ of the current sheet. v_z-axis along maximum protrusion to the left (180° rotation with respect to fig. 9c). See appendix for viewing stereo pictures.

bution function of fig. 9 reveals that the asymmetric tail electrons have a relative density $n_t/n_e \approx 10\%$, energy $E_t/E \approx 22\%$, and are responsible for the current and heat flow in the plasma which is at a small angle ($\Delta\theta \approx 15°$) with respect to the local magnetic field.

The distribution of figs. 9 and 10 are observed in the middle of the neutral sheet during reconnection. The tail of high energy (50–100 eV) electrons is thought to arise from the free acceleration of essentially collisionless tail electrons by the electric field E_y along B_{y0}. Outside the neutral sheet the transverse components (B_x, $B_z \neq 0$) prevent the free acceleration. This is clearly visible in fig. 11b which displays distribution functions, $\log f(v_y, v_z)$, at two different positions (x, z) pointed out by

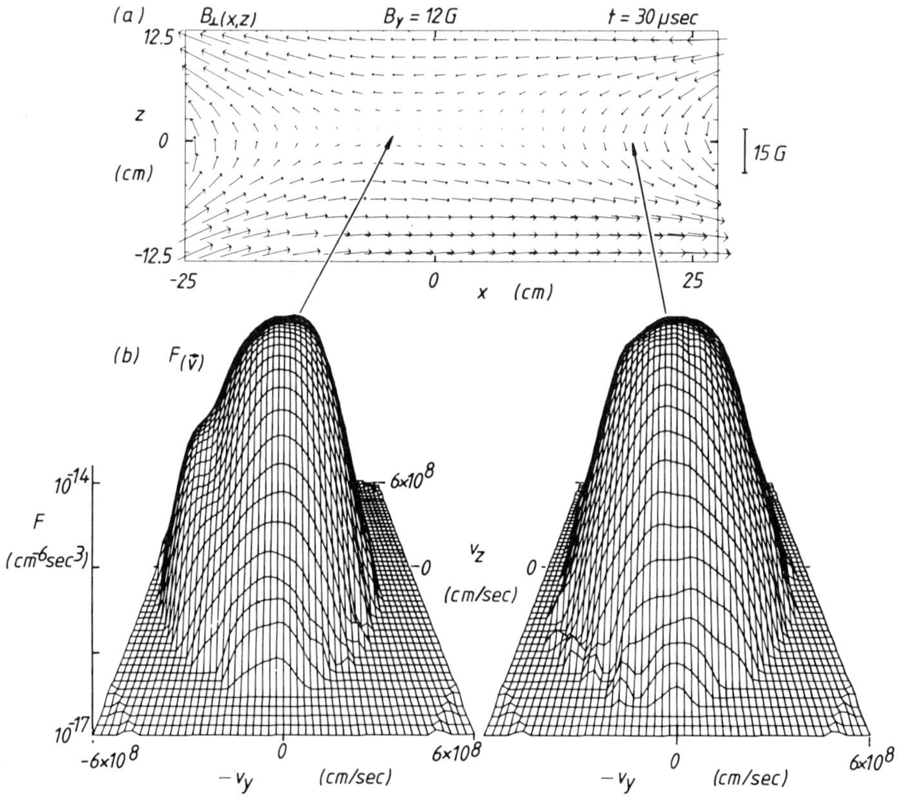

Fig. 11. Spatial variation of the distribution function in the neutral sheet. (a) Magnetic field topology with locations where $f(\boldsymbol{v})$ was measured. (b) Electron distribution $f(v_y, v_z)$ on a logarithmic scale with reference plane three orders of magnitude below the maximum. Note the presence of an energetic electron tail in the middle of the neutral sheet and absence of runaway electrons at the sheet edge.

arrows in the simultaneously measured magnetic field map $\boldsymbol{B}_\perp(x, z)$ of fig. 11a. It is interesting to note that the bulk electron temperature is higher at the side of the neutral sheet than in the middle where the energy is transferred to runaway electrons.

The presence of runaway electron tails in a current carrying plasma can give rise to various kinetic instabilities. For example, the emission of plasma waves and lower hybrid waves in low density tokamak discharges is well known [30]. Auroral kilometric radiation is associated with energetic electron precipitation [31]. Oblique whistler waves can be excited by parallel beams [32]. Here, we are particularly interested in low frequency electromagnetic instabilities ($0.1 < f < 1$ MHz) in order to establish the relation with the observed unstable whistler modes. Whistler instabilities arise from wave-particle interactions by cyclotron resonance $[(\omega - n\omega_c)/k_\parallel \approx v_\parallel]$ and Cerenkov resonance $[\omega/k_\parallel = v_\parallel]$ [33]. We have, therefore, investigated low frequency fluctuations in the distribution function in the velocity regime satisfying wave–particle resonance.

Fig. 12 (color plate) displays in the bottom plane $[(v_\parallel, v_\perp) = (v_y, v_z)]$ the rms fluctuations of the electron distribution, given by

$$f_{\rm rms}(\boldsymbol{v}) = \left\{ \frac{1}{N} \sum_{i=1}^{N} [f_i(\boldsymbol{v}) - \langle f(\boldsymbol{v}) \rangle]^2 \right\}^{1/2} \tag{7}$$

The average distribution, $\langle f(\boldsymbol{v}) \rangle = (1/N)\, \Sigma_{i=1}^{N} f_i(\boldsymbol{v})$, is shown for comparison as a contour plot in the middle plane and topographically in the upper plane. The distributions are displayed on a logarithmic scale so as to obtain a wide dynamic range in $f(\boldsymbol{v})$ (1000:1, 3 dB per contour, $f_{\rm rms\,max}/\langle f(0) \rangle \approx 0.2$). The observations show that near the runaway electron tail of $\langle f(\boldsymbol{v}) \rangle$ the fluctuation level is enhanced as expected from kinetic instability theories. The characteristic velocity of the tail particles is $v_\parallel \approx 3.5 \times 10^8$ cm/s close to the fundamental ($n = 1$) cyclotron resonance ($\omega_c/k_\parallel \approx 4 \times 10^8$ cm/s). Neither the second resonance ($n = 2$) nor the Cerenkov resonance ($n = 0$, $\omega/k_\parallel \approx 1.5 \times 10^7$ cm/s) involves a significant number of resonant particles. Thus, it appears that the tail electrons excite whistlers by cyclotron resonance. This plausible model awaits a final proof, i.e., the cross correlation between particle and wave fluctuations, $\langle \delta f \delta B \rangle$.

5. Discussion and conclusion

In the present experiment the classical reconnection problem has been investigated. Because of a well-controlled laboratory plasma, refined plasma diagnostics and the power of digital data processing we could extract information previously not available. We found that, to first order, the magnetic field topology and fluid flow are established in accord with theory. However, measurements of the resistivity and fluid acceleration showed orders of magnitude discrepancies with classical fluid parameters [12, 13]. A study of the microscopic processes such as fluctuations and distributions has been undertaken. The electron distribution in the current sheet is found to be anisotropic due to a tail of runaway electrons. This is a very important result since it shows that MHD theories break down in the neutral sheet. For example, the resistivity now is determined by the collisionality of the runaway electrons rather than the bulk electron temperature. The resistivity may become spatially very inhomogeneous due to changes in the distribution function.

Runaway electrons are a source of various microinstabilities one of which has been analyzed in great detail. The magnetic fluctuations have been identified from their dispersion and polarization to consist of a broad spectrum of whistler modes. The growth of the wave spectrum modifies, in turn, the electron distribution. The tail electrons lose some of their energy and momentum to the waves which results in an effective resistivity for the plasma current. The background electrons acquire energy through wave absorption which saturates the instabilities [34]. Ions are coupled to electrons by space charge electric fields.

Thus, the energy transfer during reconnection in collisionless plasmas involves various nonlinear wave–particle processes which cannot be analyzed in the framework of fluid theory. Refined experiments are necessary to clarify these "anomalous" transport processes.

Acknowledgments

The authors acknowledge helpful discussions with Dr. C. Kennel and technical support from Drs. N. Wild, D. Whelan, and Mr. J.M. Urrutia. This work was supported in part by the National Science Foundation (ATM 81-19717, PHY 82-09524), the National Aeronautics and Space Administration (NAGW-180), and the Air Force Office of Scientific Research (F19628-82-K0019).

Appendix A

In order to view figs. 4–8 and 10 stereoscopically, a simple viewing device is helpful. It consists of two identical lenses of focal length $f \approx 15$ cm and diameter $d \approx 2.5$ cm, mounted approx. 7 cm apart and held about 10–15 cm symmetrically above the figure. A divider between the two figures and lenses may be helpful in separating the two images.

A commerically made inexpensive viewer has been purchased by the authors from the Taylor Merchant Corporation, 212 W. 35th Street, New York, N.Y. 10001 (Stereopticon Model 707).

References

[1] E.R. Priest, Solar Phys. 47 (1976) 41.
[2] S.-I. Akasofu, Physics of Magnetospheric Substorms (Reidel, Dordrecht, Holland, 1977).
[3] D.S. Spicer, Space Science Revs. 31 (1982) 351.
[4] H. Alfvén, J. Geophys. Res. 81 (1976) 4019.
[5] J.W. Dungey, Cosmic Electrodynamics (Cambridge Univ. Press, London, 1958).
[6] B.U.Ö. Sonnerup, in: Space Plasma Physics: The Study of the Solar System, vol. 2 (National Academy of Sciences, Washington D.C., 1979), p. 879.
[7] H.P. Furth, J. Killeen and M.N. Rosenbluth, Phys. Fluids 6 (1963) 459.
[8] R.W. Fredricks, Space Sci. Revs. 17 (1975) 449.
[9] T.J. Dolan, Fusion Research (Pergamon Press, New York, 1982).
[10] R.L. Stenzel and W. Gekelman, J. Geophys. Res. 86 (1981) 649.
[11] W. Gekelman and R.L. Stenzel, J. Geophys. Res. 86 (1981) 659.
[12] W. Gekelman, R.L. Stenzel and N. Wild, J. Geophys. Res. 87 (1982) 101.
[13] R.L. Stenzel, W. Gekelman and N. Wild, J. Geophys. Res. 87 (1982) 111.
[14] R.L. Stenzel, W. Gekelman and N. Wild, J. Geophys. Res. 88 (1983) 4793.
[15] W. Gekelman and R.L. Stenzel, UCLA Plasma Physics Group Report PPG 737 June 1983 (to appear in J. Geophys. Res.).
[16] P.J. Baum and A. Bratenahl, in: Advances in Electronics and Electron Physics, vol. 54, C. Marton, ed. (Academic Press, New York, 1980), pp. 1–67.
[17] A.G. Frank, in: Neutral Current Sheets in Plasmas, vol. 74, Proceedings of the P.N. Lebedev Physics Institute, N.G. Basov, ed. (Consultants Burea, New York, 1976), p. 107.
[18] N. Ohyabu, S. Okamura and N. Kawashima, J. Geophys. Res. 79 (1974) 1977.
[19] R.L. Stenzel and W.F. Daley, United States Patent No. 4,216,405, August 1980.
[20] R.L. Stenzel, W. Gekelman, N. Wild, J.M. Urrutia and D. Whelan, Rev. Sci. Instrum. 54 (1983) 1302.
[21] V.M. Vasyliunas, Rev. Geophys. Space Phys. 13 (1975) 303.
[22] R.L. Stenzel, W. Gekelman and N. Wild, Geophys. Res. Lett. 9 (1982) 680.
[23] T. Sato and T. Hayashi, Phys. Fluids 22 (1977) 1189.
[24] T.W. Speiser, Planet. Space Sci. 18 (1970) 613.
[25] J.S. Bendat and A.G. Piersol, Random Data: Analysis and Measurement Procedures (Wiley Interscience, New York, 1981).
[26] R.A. Helliwell, Whistlers and Related Ionospheric Phenomena (Stanford Univ. Press, Stanford, 1965).
[27] J.C. Lee and F.W. Crawford, J. Geophys. Res. 75 (1970) 85.
[28] T.R. Sanderson and J. Henrion, Space Science Instrum. 1 (1975) 351.
[29] R.L. Stenzel, R. Williams, R. Aguero, K. Kitazaki, A. Ling, T. McDonald and J. Spitzer, Rev. Sci. Instrum. 53 (1982) 1027.
[30] Z.G. An, C.S. Liu, Y.C. Lee and D.A. Boyd, Phys. Fluids 25 (1982) 997.
[31] J.E. Maggs, J. Geophys. Res. 83 (1978) 3173.
[32] H. Matsumoto and Y. Yasuda, Phys. Fluids 19 (1976) 1513.
[33] A.B. Mikhailovskii, Theory of Plasma Instabilities, vol. 1 (Consultants Bureau, New York, 1974).
[34] D.A. Whelan and R.L. Stenzel, Phys. Rev. Lett. 50 (1983) 1133.

Physica 12D (1984) 145–153
North-Holland, Amsterdam

CONTOUR DYNAMICS – AN INTERFACE METHOD FOR STUDYING THE EVOLUTION OF LARGE DENSITY GRADIENT IONOSPHERIC PLASMA CLOUDS

Edward A. OVERMAN II and Norman J. ZABUSKY
Institute for Computational Mathematics and Applications, Department of Mathematics and Statistics, University of Pittsburgh, Pittsburgh, PA 15261, USA

A 2-D model is introduced of an ionospheric plasma cloud with a piecewise-constant density gradient and a diffusive-regularized boundary. Using contour dynamics (a boundary integral method) only the boundary between the regions of different densities need to be followed so that the 2-D model reduces to the 1-D motion of closed curves. Numerical results are shown including a comparison of the finite-difference and contour dynamics models.

1. Introduction

Plasma clouds in the ionosphere have been studied theoretically and experimentally for over a decade since barium was first injected into the ionosphere from rockets. The barium atoms are ionized by the sun and are visible from the ground at twilight and at dusk. The resulting plasma is low-β, collision dominated and magnetic-field aligned. It is driven in directions orthogonal to the earth's magnetic field by ambient electric fields. Pictures taken up the magnetic field lines are shown in fig. 1 for one release. The observed instability and structuring ("striations" or "bifurcations") have been attributed to the $\boldsymbol{E} \times \boldsymbol{B}$ "gradient drift" instability [1, 2, 3]. That is, the denser regions of the cloud are shielded from the ambient electric field and so move slower than the low density regions. This causes the density gradient to steepen on the backside of the cloud and then become unstable.

In studying these clouds analytically attention has focused mainly on the linear stability of one-dimensional (in density) clouds. For short times the linear calculations are in qualitative agreement with optical emissions of early-time field experiments [4, 5]. In computational studies finite-difference methods were used to examine the evolution of two-dimensional clouds seeded with small-amplitude perturbations [6, 7, 8]. These computations were performed in rectangular regions on which periodic or inflow–outflow boundary conditions were used. In addition, flux corrected transport [9] was needed in some of these runs [7, 8] to avoid the dispersive ripples that occur when steep gradients are not adequately resolved and which can even cause the density to go negative.

Recently, we have taken a different approach. We model the plasma cloud by one or more piecewise-constant distributions of density. In this way we can more accurately represent the steep density gradient which drives the striations. Also, since in this model we need only follow the motion of the one-dimensional curves separating the regions of different densities, we can more easily study the cloud's motion analytically and computationally. In the remainder of this paper we describe the equations of motion for the plasma cloud and for the contour dynamics model, discuss the linear stability analysis for a circular region, discuss "normal derivative" dissipation which models the steep density gradient on the backside of the cloud, and compare runs using finite-differences and contour dynamics.

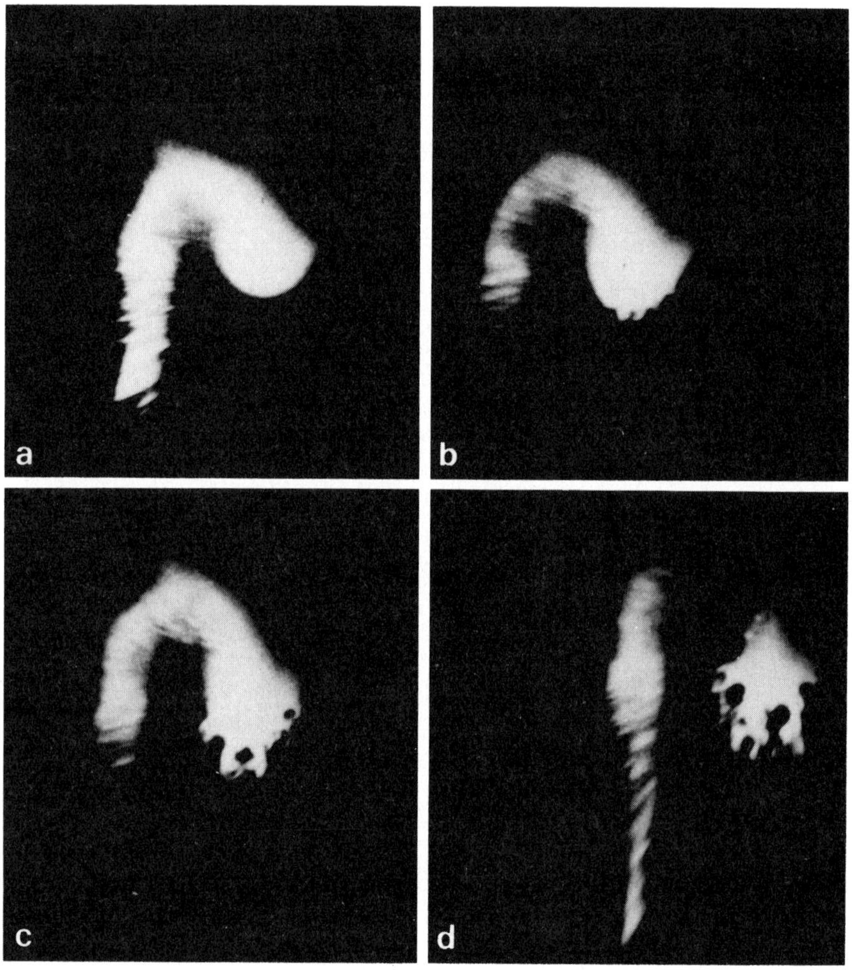

Fig. 1. The evolution of a plasma cloud in the ionosphere. (Figure courtesy of Morris B. Pongratz, Los Alamos National Laboratory.)

2. Equations of motion

The equations of motion of an idealized ionospheric plasma cloud may be represented by the two-space-dimensional, (x, y), coupled partial differential equations [2, 10]

$$\nabla \cdot (N \nabla \Phi) = 0, \tag{1a}$$

$$\partial_t N + \boldsymbol{v} \cdot \nabla N = \nu \nabla^2 N, \tag{1b}$$

$$\boldsymbol{v} = \boldsymbol{e}_z \times \nabla \Phi / B_0 = (-\boldsymbol{e}_x \partial_y + \boldsymbol{e}_y \partial_x) \Phi / B_0, \tag{1c}$$

where $N = N(x, y)$ is the density of the ions, $B_0 \boldsymbol{e}_z$ is the magnetic field which is large and unaffected by the cloud, and $\boldsymbol{E} = -\nabla \Phi$ is the electric field which has an ambient component, $E_0 \boldsymbol{e}_x$, which "drives" the plasma. The dissipation parameter ν arises from ion-neutral collisions.

In contour dynamics we assume that N is a piecewise-constant function and follow the evolution of smooth contours ∂D at which N is discontinuous. For simplicitly in the description of the equations we will only let N take on two values, i.e.,

$$N(x, y, t) = \begin{cases} N_- & \text{for } (x, y) \in D \\ N_+ & \text{for } (x, y) \notin D \end{cases} \tag{2}$$

where $D(t)$ is a bounded, simply connected region in $\mathbb{R}^2$.

The elliptic equation (1a) becomes

$$\begin{aligned} &\nabla^2\Phi_- = 0, \quad \text{for } (x, y) \in D, \\ &\nabla^2\Phi_+ = 0, \quad \text{for } (x, y) \notin D, \end{aligned} \tag{3}$$

where

$$\Phi(x, y, t) = \begin{cases} \Phi_-(x, y, t), & \text{for } (x, y) \in D(t), \\ \Phi_+(x, y, t), & \text{for } (x, y) \notin D(t), \end{cases} \tag{4}$$

and the boundary conditions become

$$\Phi_+(x, y, t) \approx -E_0 x, \quad \text{for } |(x, y)| \gg 1, \tag{5a}$$

$$\Phi_+(x, y, t) = \Phi_-(x, y, t), \quad \text{for } (x, y) \in \partial D, \tag{5b}$$

$$N_+ \partial_n \Phi_+(x, y, t) = N_- \partial_n \Phi_-(x, y, t), \quad \text{for } (x, y) \in \partial D, \tag{5c}$$

where ∂_n is the derivative in the direction of the outward normal to the boundary ∂D. Using the single- and double-layer Green's function of Laplace's equation, respectively, we find that

$$\begin{aligned} \Phi_-(p) &= c_1 E_0 x(p) + c_2 \int_{\partial D} \Phi_-(s)\, \partial_{n_s} \ln r(p, s)\, \mathrm{d}s, \\ \partial_n \Phi_-(p) &= c_1 E_0 \partial_n x(p) + c_2 \int_{\partial D} \partial_n \Phi_-(s)\, \partial_{n_p} \ln r(p, s)\, \mathrm{d}s, \end{aligned} \tag{6}$$

where

$$\begin{aligned} c_1 &= -2N_+/(N_- + N_+), \\ c_2 &= -(1/\pi)(N_- - N_+)/(N_- + N_+) \end{aligned}$$

and where $s, p \in \partial D$, $r(p, s)$ is the straight line distance between p and s, and ∂_{n_s} and ∂_{n_p} are the derivatives in the direction of the outward normal to the boundary at s and p, respectively. Numerically, (6) is discretized by the trapezoidal rule and solved by Gauss–Seidel iteration.

If $\nu = 0$, the evolution for N, eq. 1b, indicates that points on an isodensity contour remain on the contour. Thus ∂D is advected by

$$(\mathrm{d}/\mathrm{d}t)(x, y) \equiv (\dot{x}, \dot{y}) = (-\partial_y \Phi_-, \partial_x \Phi_-). \tag{7}$$

When computations were performed with (7) we found a great sensitivity to initial conditions and numerical processes. In a very short time (inversely proportional to the number of points on the contour) "kinks" developed on the contours. This indicated that the model was ill-posed. To overcome such phenomena we introduced a physically motivated small-scale regularization procedure [11] which mimics the component of $\nu \nabla^2 N$ in (1b) tangent to the contour [11, 12]. Hence we advect ∂D by

$$(\dot{x}, \dot{y}) = (-\partial_y \Phi_-, \partial_x \Phi_-) + \nu \partial_s^2 (x, y), \tag{8}$$

where ∂_s^2 is the second derivative with respect to arc length and corresponds to the tangential part of ∇^2 in (1b). Numerically, we solve (8) implicitly by predictor–corrector.

3. Linear stability analysis for a circular cloud

The electric field within a circular one-level cloud, eq. (2), is

$$\boldsymbol{E} = \boldsymbol{e}_x 2E_0/(1 + \lambda),$$

where $\lambda = N_-/N_+$. Thus the cloud drifts downward with a speed of $2v_0/(1 + \lambda)$ where $v_0 = E_0/B_0$. Thus we transform to a frame of reference moving downward with this speed, i.e., $(\xi, \eta) = (x, y - 2v_0 t/(1 + \lambda))$ and then introduce polar coordinates (r, φ) where the polar angle is measured counterclockwise from the positive η-axis. (This will decouple the sine and cosine coefficients in the Fourier decomposition of the boundary.) We assume that the boundary of the piecewise-constant region is

$$R(\varphi, t) = \rho^{(0)}(t) + \varepsilon\left[\alpha_0(t) + \sum_{m=1}^{\infty}\left(\alpha_m(t)\cos m\varphi + \beta_m(t)\sin m\varphi\right)\right] + \mathcal{O}(\varepsilon^2). \tag{9}$$

We find [12] that for times t such that

$$(\mathrm{d}R/\mathrm{d}\varphi)/R \ll 1$$

$$\rho^{(0)}(t) = \left(\left(\rho^{(0)}(0)\right)^2 - 2\nu t\right)^{1/2}, \tag{10a}$$

$$(\mathrm{d}/\mathrm{d}\tau)(\alpha_m, \beta_m) = m(\alpha_{m+1}, \beta_{m+1}) - Q(m^2-1)(\alpha_m, \beta_m) \tag{10b}$$

where $\gamma = 2v_0 r_0^{-1}(\lambda-1)/(\lambda+1)^2$, $\tau = \gamma t$, and $Q = \nu/(r_0^2\gamma)$. Qualitatively, eq. (10) shows the same behavior as runs using the finite-difference representation. That is [7, 12], the time for the onset of striations goes like λ^{-1}, the minimum dissipation required to prevent striations goes like λ^{-1}, and the wavenumber of maximum amplitude decreases as a function of time.

4. "Normal derivative" dissipation

To obtain better quantitative agreement we are attempting to model the steep density gradient on the backside of the real cloud. That is, the dissipative term in eq. (1b), $\nu\nabla^2 N$, causes (1) small-scale structures to decay and also (2) the cloud itself to spread out and the maximum density gradient to decrease. The first effect we have modeled by $\nu\partial_s^2(x, y)$ in eq. (8). The second effect can be modeled as shown in fig. 2 where two (or more) contours simulate the maximum density gradient on the cloud. For a steep (i.e., $\delta \ll r$) axisymmetric cloud we find [13] that for the two-dimensional heat equation, $N_t = \nu\nabla^2 N$, δ increases as a function of time to first order as

$$(\mathrm{d}/\mathrm{d}t)\delta = (\pi/2)\nu/\delta. \tag{11}$$

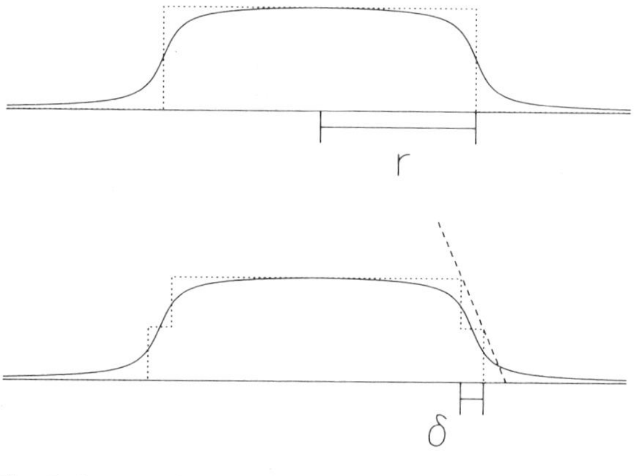

Fig. 2. Representing an axisymmetric cloud (the solid line) by one or two contours (the dotted lines). When two contours are used we can model the maximum density gradient of the cloud (the dashed line) by using "normal derivative" dissipation.

We include (11) in (8) by subtracting $(\pi/4)(\nu/\delta^2(x, y))\boldsymbol{\delta}(x, y)$ from the right hand side of (8) for both contours where $\boldsymbol{\delta}(x, y)$ is the vector from the point (x, y) on the contour to the closest point on the other contour. In this way we apportion the effect of (11) evenly between the two contours.

In fig. 3 we show the evolution of a piecewise-constant cloud where the density, N, is 1 outside the cloud, 2 between the two contours, and 3 inside the cloud. The initial perturbation is 0.1% at the 40th mode. Note how the growth of the perturbation can be seen very accurately in the curvature plots. In fig. 4 we use one contour with density 1 outside and 3 inside the contour so that we cannot model the "normal derivative" dissipation. Note that the perturbation grows at a much faster rate and gives rise to many more fingers.

In fig. 5 we show the evolution of a cloud using the continuum model, eqs. (1), written by S.T. Zalesak of the Naval Research Laboratory [14]. The parameters are different from those used in the contour dynamics runs but the same qualitative behavior can be seen even in the curvature plots. Work is continuing on understanding the differences and similarities in the two models and in using the contour dynamics model to obtain an analytical understanding of this structuring.

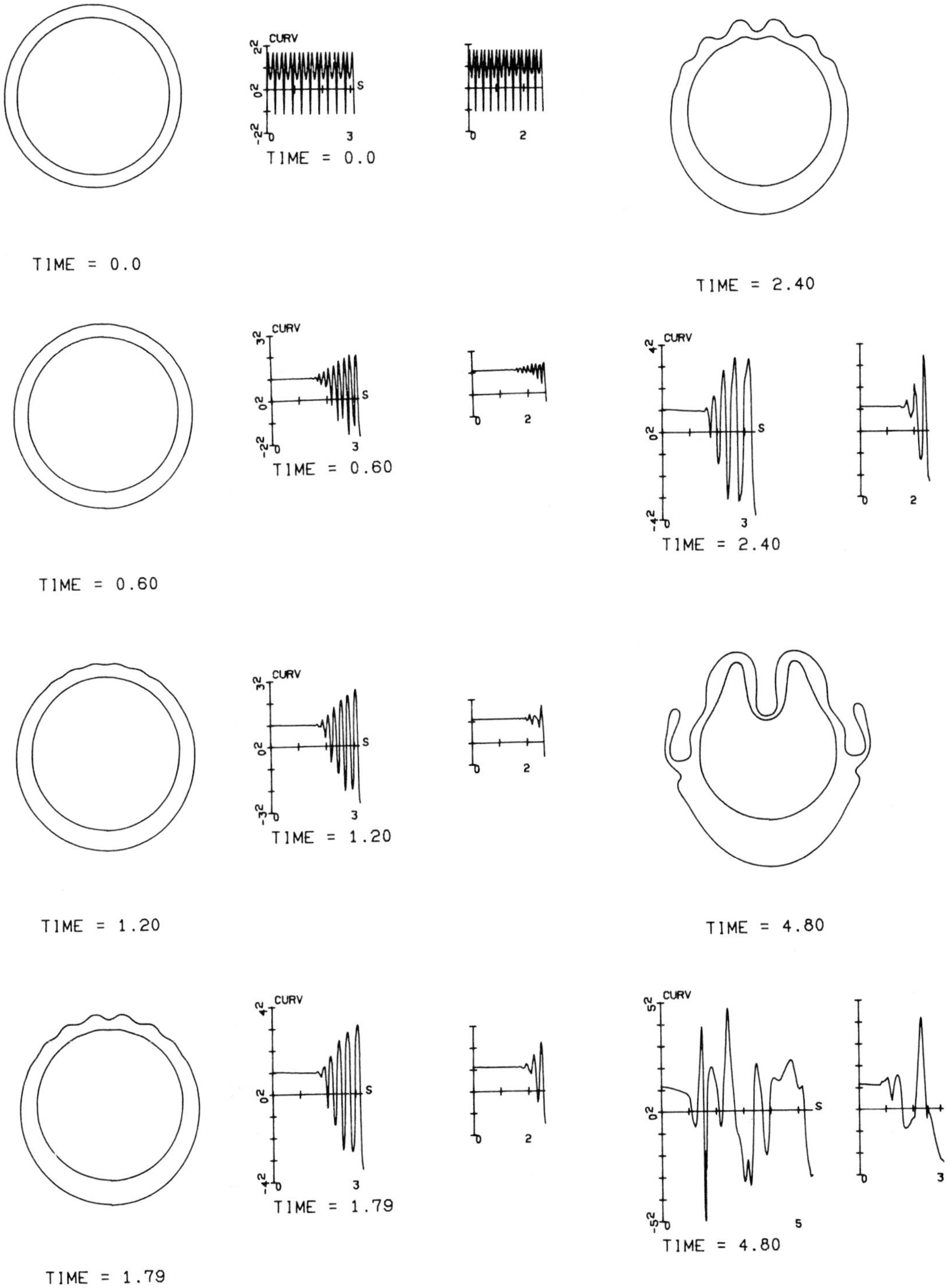

Fig. 3. The time evolution of a two contour model using "normal derivative" dissipation. We also plot the curvature (note that the vertical scale is $\ldots, 0^2, 1^2, 2^2, 3^2, \ldots$) where the bottom of the cloud is $s = 0$ and the top is the maximum value of s (the contour is symmetric about the vertical axis).

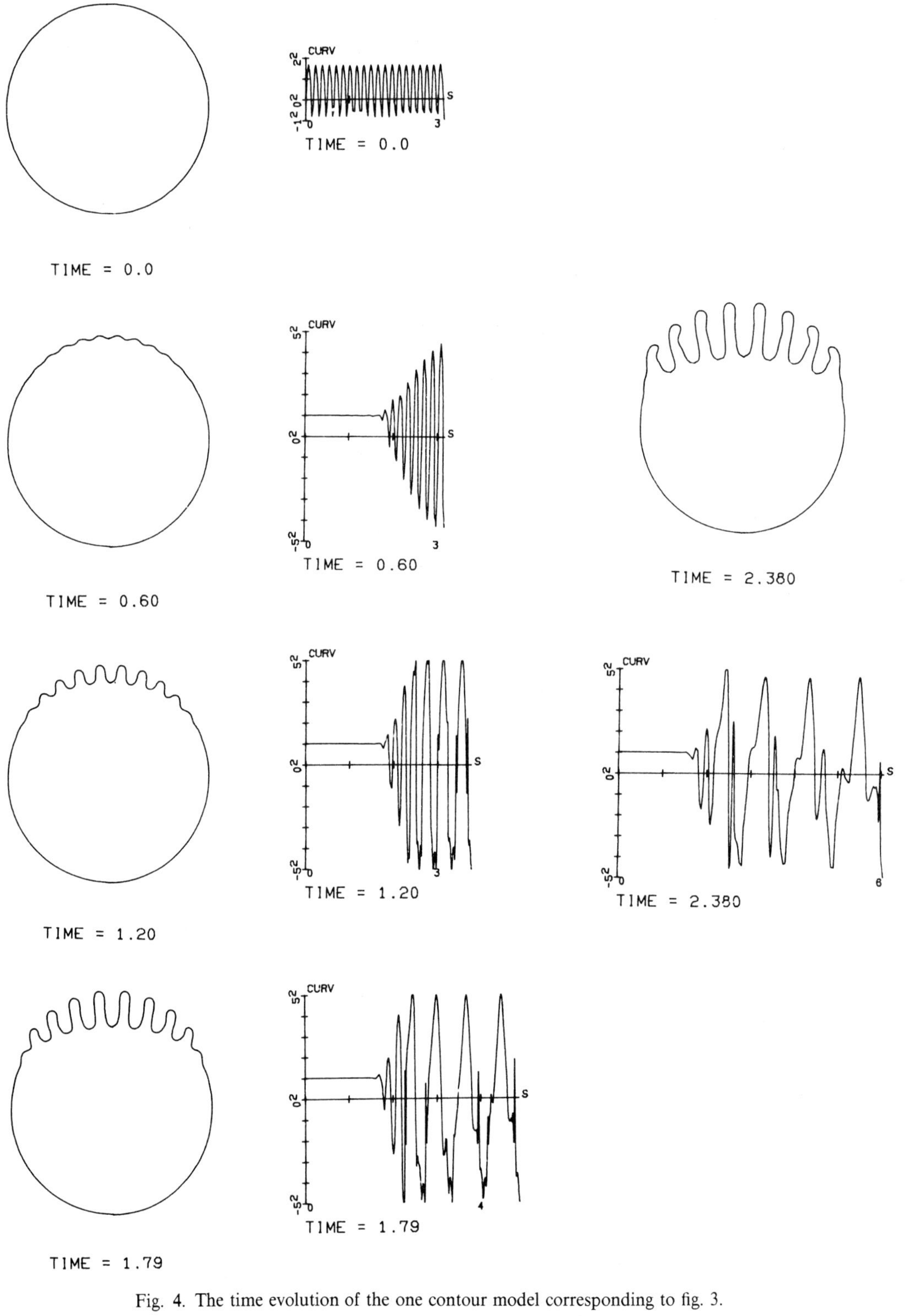

Fig. 4. The time evolution of the one contour model corresponding to fig. 3.

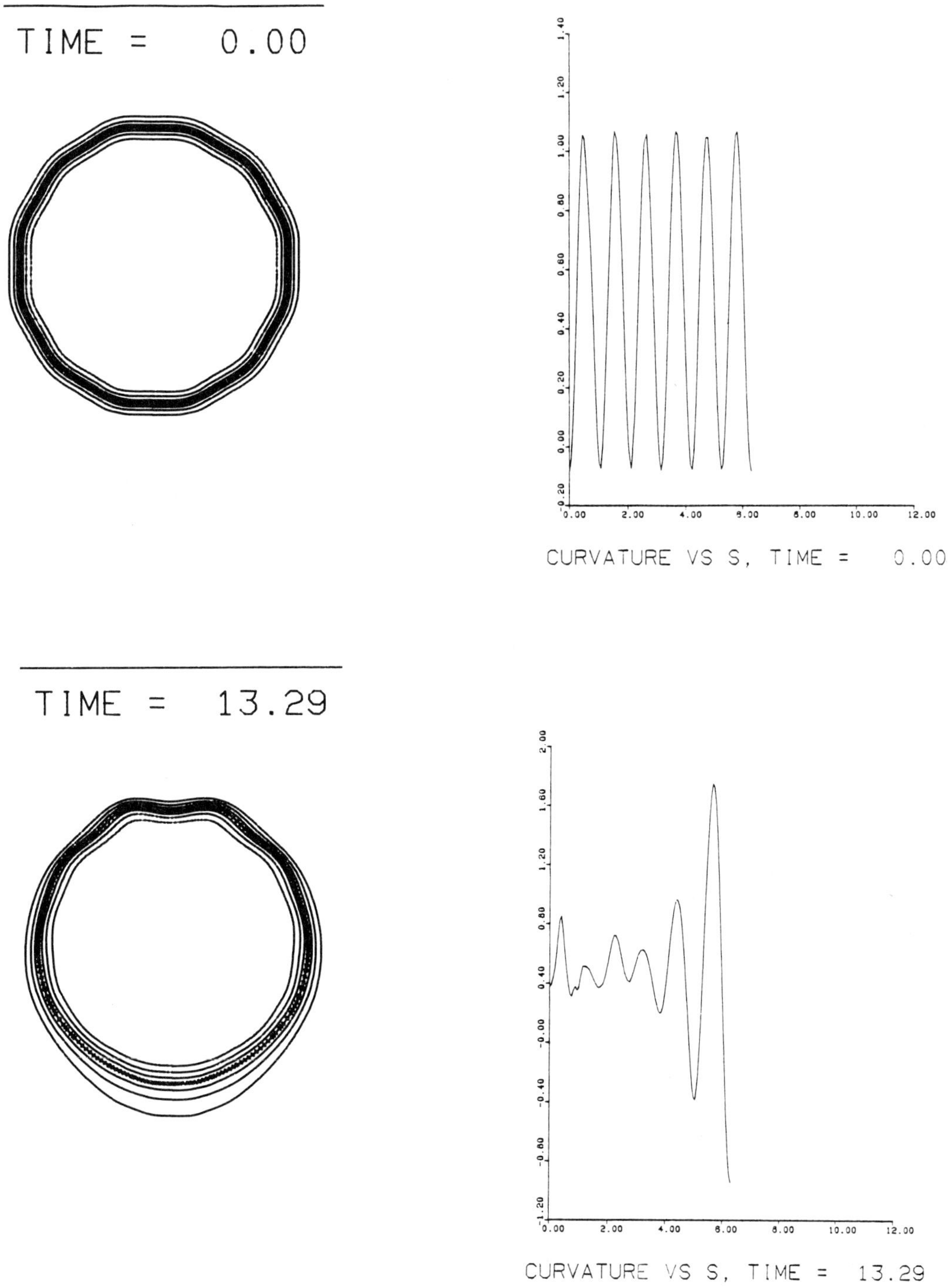
TIME = 0.00
CURVATURE VS S, TIME = 0.00
TIME = 13.29
CURVATURE VS S, TIME = 13.29

Fig. 5

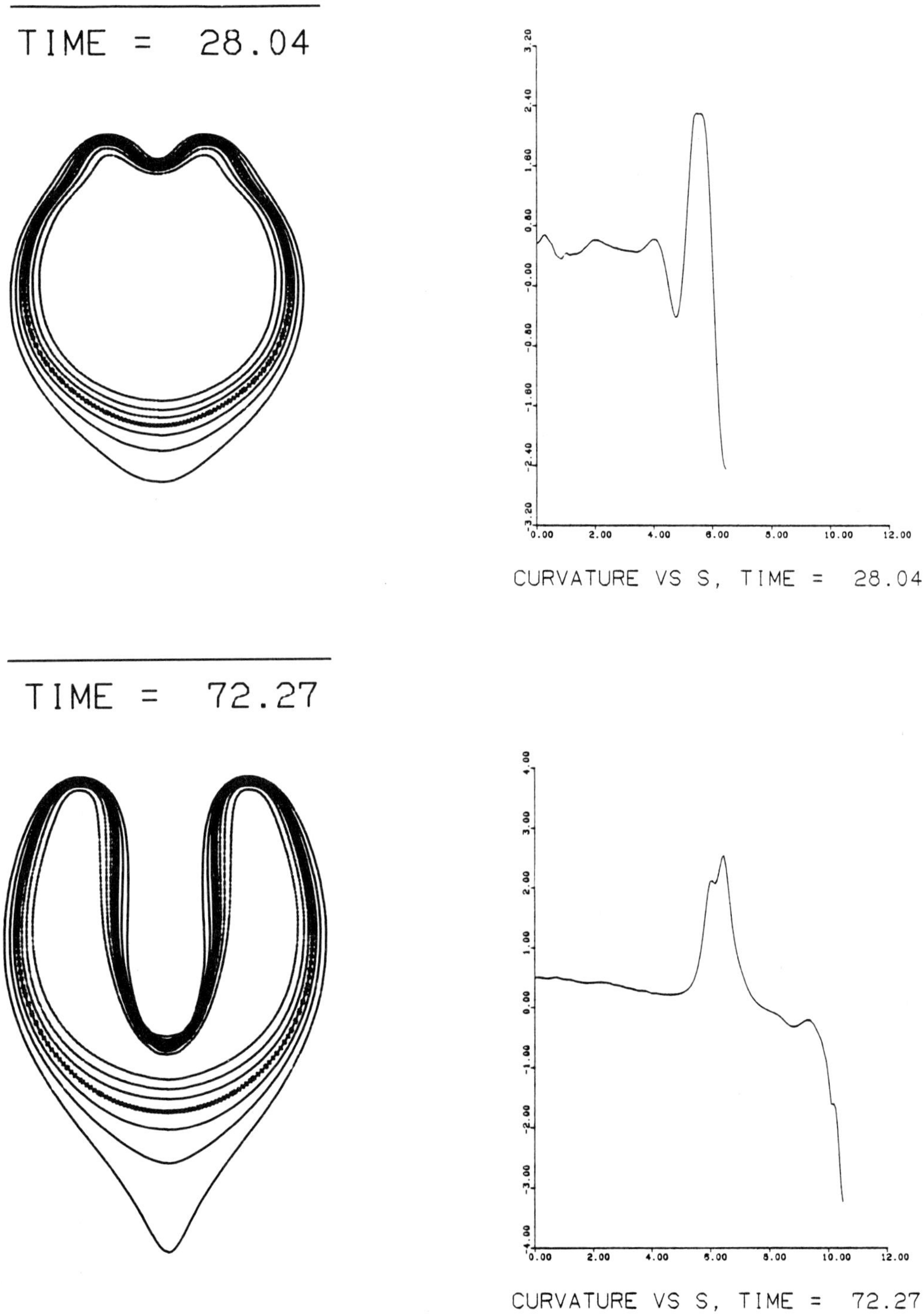

Fig. 5. The time evolution of a cloud using the continuum model, eqs. (1). The parameters are not the same as in figs. 3 and 4. The code was written by S.T. Zalesak [14]. Note that the scale on the curvature axis varies from one plot to another.

Acknowledgements

This work was supported by the Defense Nuclear Agency. The work of EAO was supported under ONR contract N00013-78-C-0074. The work of NJZ was supported under contract N00014-82-C-2030.

References

[1] L.M. Linson and G. Haerendel, J. Geophys. Res. 75 (1970) 3211.
[2] H.J. Völk and G. Haerendel, J. Geophys. Res. 76 (1971) 4541.
[3] A. Simon, Phys. Fluids 6 (1963) 382.
[4] N.W. Rosenberg, J. Geophys. Res. 76 (1971) 6856.
[5] T.N. Davis, G.J. Romick, E.M. Wescott, R.A. Jeffries, D.M. Kerr and H.M. Peek, Planet. Space Sci. 22 (1974) 67.
[6] N.J. Zabusky, J.H. Doles III and F.W. Perkins, J. Geophys. Res. 78 (1978) 711.
[7] B.E. McDonald, S.L. Ossakow, S.T. Zalesak and N.J. Zabusky, J. Geophys. Res. 86 (1981) 5775.
[8] B.E. McDonald, M.J. Keskinen, S.L. Ossakow and S.T. Zalesak, J. Geophys. Res. 85 (1980) 2143.
[9] S.T. Zalesak, J. Comput. Phys. 31 (1979) 335.
[10] F.W. Perkins, N.J. Zabusky and J.H. Doles III, J. Geophys. Res. 78 (1973) 697.
[11] N.J. Zabusky and E.A. Overman II, J. Comput. Phys. 52 (1984) 351.
[12] E.A. Overman II, N.J. Zabusky and S.L. Ossakow, Phys. Fluids 26 (1983) 1139.
[13] E.A. Overman II and N.J. Zabusky, in progress.
[14] S.T. Zalesak, "Small Scale Structures in the Earth's Ionosphere: Theory and Numerical Simulation", NRL Memorandum Report 5119 (1983).

Physica 12D (1984) 154–162
North-Holland, Amsterdam

1.4. SOLIDS

PATTERN FORMATION BY SHOCK PROCESSES*

J.W. SHANER
Los Alamos National Laboratory, Los Alamos, New Mexico 87545, USA

Shock waves in condensed media often produce and leave behind periodic patterns and textures. These patterns have been observed both in real time and in post-mortum examination. In many cases the patterns can be related to analogous pattern-forming mechanisms in classical fluid dynamics, such as the Rayleigh–Taylor and Helmholtz instabilities. In other cases, the textures arise from peculiarities in the dynamic stress state immediately behind the leading edge of the shock wave.

Periodic waves in the interface between two shock-welded metals have a close resemblance to the classical Helmholtz instability. From a practical point of view, these waves are crucial to the formation of a good bond.

Impulsive acceleration of an interface can result in the Meshkov instability, which forms patterns qualitatively similar to the Rayleigh–Taylor instability driven by continuous acceleration. However, the patterned stress state left behind after a shock crosses a perturbed interface can result in perturbation growth for shock propagation in either direction across the interface.

Even in homogeneous media, the non-hydrostatic component of the stress behind a shock can drive a pattern forming instability. Adiabatic shear banding has been proposed as a mechanism to explain both the patterns observed in shock-compressed and recovered metal samples and the apparent loss of macroscopic shear strength of shocked ceramics. New optical photographs of shocked quartz support this mechanism.

1. Introduction

Shock waves in condensed media are usually associated with very destructive and chaotic explosions. However, even these destructive processes exhibit a wide variety of patterns formed directly behind the shock front. In some cases one can even recover, upon release of the shock pressure, pieces in which the shock-induced patterns are preserved.

Experiments over the past few decades have shown that a surprising degree of order and coherence persists behind a shock wave, at least until waves reflected from irregular boundaries seriously complicate the flow. For example, with the assumption that the stress behind a shock is hydrostatic, measurements of shock propagation have provided a good first approximation to much of our high pressure equilibrium equation of state data [1]. Furthermore, direct flash X-ray diffraction measurements have shown that even crystalline order is preserved behind a shock wave in solids, and that this order is coherent with respect to the unshocked crystal structure as well [2].

In this paper I shall review and highlight three kinds of macroscopic patterns which occur in shock wave processes in condensed media. These patterns have in common that they evolve in a microsecond, or less, and that each is analogous to an instability in conventional fluid mechanics. The first is a wavy interface left behind when two metals are forced to collide at high velocity. These waves have features qualitatively like those found in the classical Kelvin–Helmholtz instability associated with shear flow in fluids. The second pattern I discuss is a periodic lamellar structure which has often been observed in metals and ceramics subjected to shock compression and recovered upon release. This phenomenon is related to a thermomechanical fluid instability usually referred to as adiabatic shear banding. The third pattern I discuss evolves after an initially planar shock wave

* Work supported by the United States Department of Energy.

passes a rippled interface separating two materials. This configuration is similar to the Rayleigh–Taylor instability problem. However, since for strong shocks even condensed media must be considered compressible, interesting non-linearities occur early in the growth of the initially perturbed interface.

2. Waves in explosive welding

In order to form good weld joints between metals, two essential conditions must be met. The metal surfaces must be uncontaminated (e.g. by oxide layers) and they must be brought into intimate contact. In fusion welding, the metals are actually melted and contaminations are allowed to float away from the joint region. In pressure welding, any surface layer of contaminants is broken up, allowing intimate contact, at elevated pressure, of the clean metals. Neither of these processes is very effective if the metals have widely different melting points or plastic flow strengths.

During World War I, ordinance specialists noticed that under certain conditions, bullets and shell fragments bonded to the metal target plates they hit. Although these may have been the first observations of explosive welding, the first published account did not appear until 1949 [3]. During the next two decades, a number of people began to exploit some of the unique features of explosive welding to make bonds between metals which would be impossible by other means [4]. At the same time, carefully controlled experiments, where impact velocity and obliquity of collision were controlled, showed that ripples formed at the interface under the same conditions that favored good weld joints [5, 6]. These ripples are the subject of this section.

The ripples formed at the interface of two metals undergoing high velocity oblique impact were first studied systematically by Abrahamson [6]. He used the configuration sketched in fig. 1a. Flat-nosed steel bullets were fired at thin metal targets inclined at an angle, θ, with respect to the front surface. In

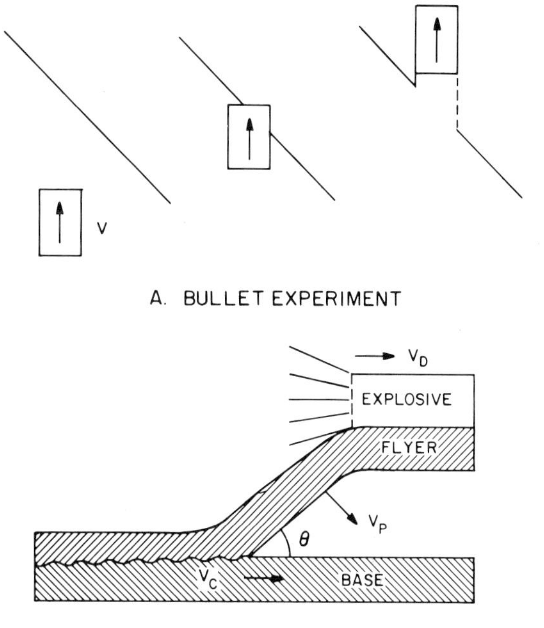

Fig. 1. a) A flat-nosed bullet, fired obliquely into a target plate can develop ripples in the bullet face; b) The most useful configuration for explosive welding results in ripples at the weld joint.

a coordinate system fixed at the point of contact, the target looks like an incident jet of velocity $V/\sin\theta$ and angle of incidence θ with respect to the plane of the bullet face. Abrahamson found that if $V/\sin\theta$ was less than the speed of sound in the target, or if it was supersonic but with θ greater than some critical value, waves were formed at the interface. If $V/\sin\theta$ was supersonic and θ was less than the critical value, no waves were formed.

For the subsonic case, simple conservation of momentum implies that the incident jet bifurcates into flows to the right and to the left along the bullet face in fig. 1a. For supersonic flow with θ above a critical value, we also have conditions for a jet forming and flowing to the left away form the point of contact. For supersonic flow with smaller than critical obliquities, no such jet is formed, and the incident flow does not bifurcate. This behavior of supersonic flow is well known from the development of explosively driven shaped-charge jets [7].

The correlation of a wavy interface with bifurcated flow and a jet emerging from the point of contact and flowing to the left in fig. 1a suggests the Kelvin–Helmholz instability as a mechanism [8, 9]. In fact there must be a large gradient in horizontal flow velocity at the planar bullet face. On the other hand, the large gradients along the bullet surface seriously complicate a quantitative analysis.

The configuration shown in fig. 1b is now that most commonly used for welding and cladding, especially for very large pieces [10]. In this configuration, V_p is the velocity given to the flyer plate by detonating explosive, while θ is the angle of impact. V_p can be varied by changing the type of explosive or the standoff between the flyer and the base plate. For steady flow, the point of contact moves with the detonation velocity ($V_c = V_D$), and V_p and θ are constants. This geometry can easily be transformed to that of fig. 1a, and therefore, the jetting conditions are similar. For conditions favoring good weld joints, jets have actually been observed [11].

In the welding process, the function of the jet emerging from the contact point is easy to understand. The jet scrubs material from both the flyer and base plates allowing clean metal surfaces to come into contact at high pressures. For optimum welding, one needs a jetting configuration, and impact velocities high enough to cause metals to flow plastically but not so high that material at the interface melts. Rapid melting and resolidification seriously weakens the bond [10].

The most detailed study of the waves on the welded interface was carried out by Bahrani et al. [12]. In this study they found that as the wave amplitude grew, significant non-linearities appeared. By plating the steel base and flyer plates with copper and nickel, they were able to observe the effects of the jet and where in the interface profile remaining surface material wound up. Nickel from the flyer plate was concentrated in a vortex pattern behind the peaks of the waves, while copper from the base plate was moved ahead of the peak of the wave. Both copper and nickel were mixed in a vortex on the leading edge of the wave. The leading and trailing vortices are also asymmetric.

At the present time, the only theoretical and numerical analyses of these waves in weld joints have been qualitative or badly oversimplified. The analysis by Bahrani et al. [12], although in qualitative agreement with their experiments is not quantitative. The quantitative analyses, based on Kelvin–Helmholtz instability theory are successful at predicting wavelengths for the ripples, but these calculations do not address the vorticity and nonlinear wave growth [8, 9]. Detailed calculations of the nonlinear instability of parallel shear flows perhaps come closest to the experimental situation, except that we do not have a strictly parallel shear flow [13]. Finally, the effects of interfacial energy, material strength, and compressibility have not been considered quantitatively. In conclusion we can say that we have enough empirical results to make good weld joints explosively. However, we do not yet have the analytical capability to predict the kinds of non-linearities which develop in the waves at the bonded interface.

3. Adiabatic shear banding

Another pattern often found in materials recovered after shock compression and release consists of roughly periodic lamellar structures along specific crystallographic planes. These features have been known for a long time in metals [14]. Recently, spatially periodic luminescence also been observed directly behind a shock front in quartz [15]. Some crystallites in a polycrystalline sample show many such features, while others do not. The crystallite orientation with respect to the shock or release wave fronts are obviously important. The connection between crystalline and stress tensor orientation is the key to the generation of these patterns.

In the simplest model for steady shock wave propagation in condensed media we consider the stress to be a simple scalar pressure. However, in even the earliest work on shocks in solids the authors realized that solids could support non-hydrostatic stresses [1]. Since the compression due to a shock wave is uniaxial in the direction of propagation, the natural choice of principle stresses are the longitudinal component, along the propagation direction, and the transverse components, assumed equal for isotropic materials. We expect that the longitudinal component will be greater than the transverse components, so long as the material can support shear stresses. With this stress configuration, the planes in which the shear stresses are maximum will lie at 45° to the longitudinal direction.

Under these conditions, as the shock strength is increased, the longitudinal component of the stress will increase, while the transverse component is determined solely by Poisson's ratio, until a point called the Hugoniot elastic limit is reached. At this point, the shear stresses in the planes at 45° to the propagation direction are the maximum the material can support, and stronger shocks will result in plastic deformation such that the resolved shear stresses remain bounded. This picture is the basis of the von Mises yield criteria often used in hydrodynamic flow calculations.

Within the framework of this simple model we would expect that as the shock strength increased, the longitudinal and transverse stresses would differ by an amount such that the maximum resolved shear stresses would be just enough to start plastic flow.

We show in fig. 2 the various stress strain paths we expect if the material behaves hydrostatically or with a constant difference between longitudinal and transverse stress above the Hugoniot elastic limit.

For a number of years, however, we have suspected that the stress state behind a strong shock approaches the hydrostat (i.e. no stress deviators). This has been especially evident for brittle materials, but more recent results suggest the same phenomenon for metals. In particular, by looking at the ability of shock-compressed metals to support elastic recompression and decompression waves,

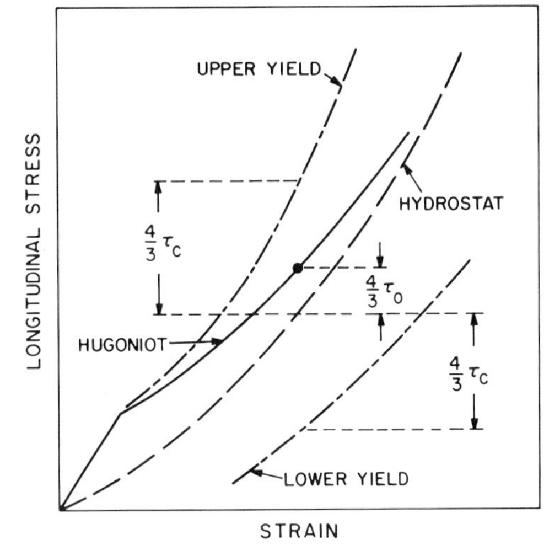

Fig. 2. Longitudinal stress vs strain for various compression processes. The Hugoniot represents a possible track for shock compression. τ_c represents the maximum stress deviator the material can support. If the stress behind a shock front had no tendency to approach a hydrostatic state, compression would be along the upper yield curve.

Asay has shown that the intermediate stress–strain curve in fig. 2 may be more appropriate for metals [16].

A model developed by Grady shows the qualitative features resulting in the periodic structures [17]. If an initially homogeneous material is allowed to evolve from initially uniform shear, it may develop a shear banding instability under appropriate conditions. If, for example, the effective viscosity decreases with temperature, a local high temperature will cause concentration of the dissipative energy due to shear flow. As a result further temperature rises will further concentrate the shear flow.

For a two dimensional problem with material velocity, U, in the y-direction and gradients only in the x-direction, the flow equations are

$$\frac{\delta U}{\delta t} - \frac{1}{\rho_0}\frac{\delta \tau}{\delta x} = 0 \tag{1}$$

and

$$\frac{\delta T}{\delta t} - D\frac{\delta^2 T}{\delta x^2} = \frac{\tau}{\rho_0 C}\frac{\delta U}{\delta x}\,. \tag{2}$$

In eq. (1), ρ_0 is the density, and τ is the shear stress. In eq. (2), D is the thermal diffusivity, which limits the local temperature rise, and C is the heat capacity. These equations must be completed by a third describing the relation of shear stress, velocity gradient and temperature. For example, Grady modeled the effective viscosity to give

$$\tau = [\eta_0 \, e^{-a(T-T_0)}] \frac{\delta U}{\delta x} . \qquad (3)$$

By linearizing eqs. (1)–(3) for small perturbations from a homogeneous state, and assuming solutions of the form $e^{\alpha t} e^{i\beta x}$, one obtains a curve for α vs. $\beta = 1/\lambda$ as shown in fig. 3. The growth rate is limited for small wavelengths by the assumed thermal conduction.

The largest difficulty in applying this model quantitatively is establishing a realistic constitutive relation (3). With reasonable assumptions about the viscosity model, Hayes and Grady have shown that agreement can be obtained between the solution to eqs. (1)–(3) and the observed lamellar structure in shocked aluminum with a wavelength of the order of 1–10 μm [18]. The predicted growth rate at this wavelength is around $10^8\ s^{-1}$, which is also appropriate for a shock process.

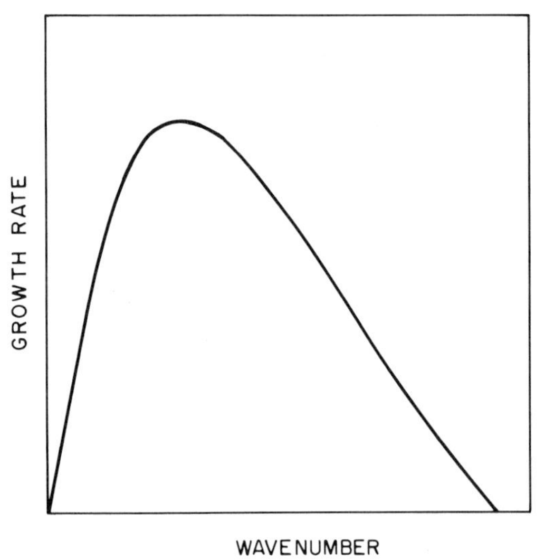

Fig. 3. Schematic of the solution to the linearized version of eqs. (1)–(3). The growth rate for adiabatic shear bands is limited at large wavenumber by thermal conduction.

A much more detailed analysis, especially of eq. (3) will be necessary before an understanding of the phenomenon of shear banding can be complete. For example, one would expect localized work hardening to play the same role of damping the instability as does thermal conduction. In the case of brittle solids one needs to include the surface energy generated by fractures, unless the confining pressure is so great that even brittle materials behave plastically. However, the adiabatic shear banding model does explain the apparent macroscopic hydrodynamic nature of solids under shock compression, and the varied patterns appearing during and upon recovery from shock waves.

4. Richtmyer–Meshkov (R–M) instability

When a shock wave passes an interface separating two fluids of differing density, an instability analogous to the classical Rayleigh–Taylor instability occurs. Since the shock accelerates the interface in the direction of propagation, any initial perturbations will grow if the acceleration is directed from the light fluid to the heavier fluid. However, there are several qualitative differences between the continuous acceleration and the impulsive acceleration cases.

In the original paper, Richtmyer solved numerically the hydrodynamic flow associated with the shock wave, or impulsive, acceleration of the interface between two fluids [19]. The configuration is shown in fig. 4. An initial perturbation of the interface of the form $a_0 \cos kx$ was assumed. In the case of continuous acceleration, the time dependent amplitude is given by

$$\ddot{a} = k g(t) a(t) \left(\frac{\rho_2 - \rho_1}{\rho_2 + \rho_1} \right) , \qquad (4)$$

where $g(t)$ is the acceleration of the system and ρ_1 and ρ_2 are the densities [20]. If the acceleration is impulsive, as with a shock passing the interface,

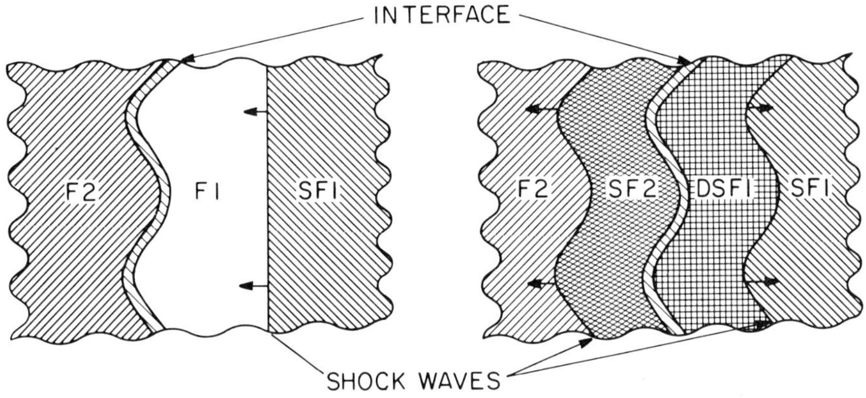

Fig. 4. Configuration for observing Richtmyer–Meshkov instabilities. On the left, a shock is incident from fluid 1 (F1). The shocked fluid is denoted SF1. On the right, after this shock has passed the fluid interface, the transmitted shock separates fluid 2 (F2) from the shocked fluid 2 (SF2). The reflected shock separates shocked fluid 1 (SF1) from the double shocked fluid (DSF1).

then

$$g(t) = U\delta(t)\,, \tag{5}$$

where U is the material velocity jump across the shock. Integrating eq. (4) then gives

$$\dot{a} = kUa_0\left(\frac{\rho_2 - \rho_1}{\rho_2 + \rho_1}\right). \tag{6}$$

Although this expression is an oversimplification from several points of view, which will be discussed later, results derived from it agreed with Richtmyer's numerical calculations for γ-law gases, so long as he chose densities and an initial amplitude appropriate for the time just after the shock-interface interaction. For low amplitude shocks in ideal gases, the density ratio between singly shocked dense fluid (SF2 in fig. 4) and doubly shocked light fluid (DSF1 in fig. 4) is within a few percent of the initial density ratio. Therefore, the most sensitive parameters in eq. (6), are a_0, which can be compressed by a factor of two even with 0.1–0.2 MPa shocks in ideal gases, and the material velocity at the interface which may be less than half of that following the first shock in fluid 1. Uncertainty in the choice of these parameters arises as a result of comparison of numerical results for compressible fluids with an analytical expression for incompressible fluids.

The simplified expression (6) does show a fundamental difference between the impulsive and constant acceleration cases. In the constant acceleration case, the perturbations grow only if the acceleration is directed from the light fluid to the heavy fluid. Otherwise the perturbation oscillates. On the other hand, eq. (6) indicates that for an impulsive acceleration the perturbation of an interface grows no matter which direction the acceleration, or the direction of shock propagation, is. If a shock passes from a low density to a high density fluid the perturbation amplitude grows. If the shock propagates from a high density to a low density fluid, the amplitude passes through zero and grows with opposite phase.

Nine years after the Richtmyer work, Meshkov published experiments measuring interface perturbation growth with weak shocks ($\approx$0.1 MPa) in ideal gases ($\approx$0.1 MPa air, He, Freon, and CO_2) [21]. In these experiments he observed the linear growth of an initial perturbation when the shock passed from a light gas to a heavy gas. He also observed the reversal in phase and subsequent amplitude growth when the shock passed from the heavy to the light gas. These results, although in qualitative agreement with eq. (6), showed growth rates only 0.5 times the predictions. The source of this discrepancy was presumed to be either an experimental problem associated with the film separating the gases, or the onset of non-linearity ($ka \sim 1$). No further experiments have been done to confirm these hypotheses.

An interesting feature found both in the numerical calculations and in a close examination of the experimental data is that even the initial growth of the perturbations is non-linear. The growth rate oscillates by 10–20% around the value predicted by eq. (6). Richtmyer already identified this feature as due to the overstable nature of a rippled shock front in compressible fluids. In reference to fig. 4, if the shock velocity in fluid 2 is slower than the incident shock in fluid 1, the transmitted shock will be rippled with an initial phase the same as that of the interface. When such a shock wave propagates in a compressible fluid (i.e. finite sound velocity), pressure is concentrated behind the lagging part of the shock front by convergence. Therefore, the stress field immediately behind a rippled shock will have a coherent texture with respect to the ripples. In the subsequent flow, signals can propagate forward to the shock front, making the shock front non-steady, and backward to the original interface, superimposing another texture on this interface.

The overstable, or oscillatory nature of a rippled shock front has been measured directly in metals by Sakharov and Mineev [22, 23]. In these experiments, a rippled shock front was driven into metals at a stress level high enough that the metals behaved macroscopically like fluids. The phase of the ripples reversed before the shock front became planar indicating that the perturbation oscillates as it decays. This feature is a manifestation of the textured stress field left behind by non-planar shocks in a compressible medium.

Since the textured stress field behind a rippled shock depends on the nature of the ripples in the shock front, and since this texture can effect the initial non-linear growth of a perturbed interface, we have to look more closely at what variety of configurations of interface and stress waves are possible. The calculations of Richtmyer and Meshkov's experiments utilized only gamma-law ideal gases. For this restrictive equation-of-state, the density, ρ, is proportional to molecular weight, m. Similarly, the sound velocity, C, is proportional to $(T/m)^{1/2}$, where T is the temperature. In the limit of weak shocks, the shock impedance, $I = \rho C$, is then simply proportional to $m^{1/2}$. The shock impedances, shock velocities, and densities of the two fluids all play a role in the hydrodynamic evolution of the system.

The ideal gas equation-of-state allows only two of six qualitatively different configurations for the R–M instability. For example, we have already seen that if $\rho_2 > \rho_1$, the initial interface perturbation grows without change in phase. On the other hand, if $\rho_1 > \rho_2$, the initial perturbation changes phase before growing. Similarly, if $C_2 < C_1$, the transmitted shock wave is in phase with the initial interface perturbation. Therefore, the initial regions of highest stress in the textured stress field behind the rippled transmitted shock will be near the part of the interface concave toward fluid 2. On the other hand if $C_2 > C_1$, the initial ripples in the transmitted shock are of opposite phase with respect to the interface perturbations, and the stress field texture induced by the rippled shock is the opposite from the $C_2 < C_1$, case. For ideal gases, if $\rho_2 > \rho_1$, then C_2 must be less than C_1, and vice versa. However, it is easy to find pairs of materials with more general equations-of-state for which this role does not hold.

One further complexity concerns the relative shock impedance of the fluids. For the ideal gas equation of state, the denser fluid necessarily has the greater shock impedance. Therefore, if $\rho_2 > \rho_1$, then the wave reflected back into fluid 1 is a second shock, as shown in fig. 4. Again it is easy to find pairs of materials for which $\rho_2 > \rho_1$, but $I_2 < I_1$. Then, although the interface perturbations still grow without change of phase, the reflected wave is a rarefaction. For a rippled rarefaction we should expect the stress field texture left behind to be roughly opposite in phase to that left by a shock. All of these cases are summarized in table I.

Table I
Possible R–M configurations

Case	Density	Sound velocity	Impedance	Shock	Interface	Reflected Wave	Example
1	$\rho_2 > \rho_1$	$C_2 > C_1$	$I_2 > I_1$	$-$[b]	$+$[a]	S[c]	Cu/PMMA
2	$\rho_2 > \rho_1$	$C_2 < C_1$	$I_2 > I_1$	$+$	$+$	S	IDEAL GASES
3	$\rho_2 > \rho_1$	$C_2 < C_1$	$I_2 < I_1$	$+$	$+$	R[d]	Teflon/Be
4	$\rho_2 < \rho_1$	$C_2 > C_1$	$I_2 > I_1$	$-$	$-$	S	Be/Teflon
5	$\rho_2 < \rho_1$	$C_2 > C_1$	$I_2 < I_1$	$-$	$-$	R	IDEAL GASES
6	$\rho_2 < \rho_1$	$C_2 < C_1$	$I_2 < I_1$	$+$	$-$	R	PMMA/Cu

[a] In phase with original perturbation.
[b] Opposite phase with respect to original perturbation.
[c] Shock wave.
[d] Rarefaction wave.

At this point, we should point out that even the complexities of table I are not sufficient to cover all of the potential non-linearities in this originally simple hydrodynamic flow. Careful measurements by Sturtevant [24] and Meshkov [25], have demonstrated that the convergent parts of shock waves, even in ideal gases, undergo a complex time history, including the formation of Mach reflections, or three-shock intersection points. In the classic Mach reflection problem, we know that a slip line develops behind the triple shock interaction point, with a large flow velocity gradient across the slip line. In some cases this slip line is clearly unstable (Helmholtz instability), and persistent vortices can be left behind [26]. Meshkov has also shown that similar complex wave interactions can develop in a rippled rarefaction wave [27].

Although we have listed the gross features which influence the non-linear growth of the R–M instabilities in table I, the complexity of the details of flow associated with rippled waves may result in a much richer variety of possibilities. Fluid viscosity and interfacial tension will undoubtedly complicate the R–M instability in fluids with weak shocks. We expect, for example, that if the interfacial energy density in the mixing region is comparable with the energy of compression by the shock wave, then surface tension will play a role in the subsequent flow. More careful measurements and reliable and detailed hydrocode calculations will be necessary to observe all of the non-linear phenomena associated with this instability.

We have outlined three qualitatively different kinds of non-linear instabilties resulting in pattern formation behind shock waves in condensed media. Each instability has an analog in more conventional fluid dynamics. A great deal of analytical work, particularly numerical calculations, will be necessary before we will be able to claim a detailed predictive capability for any of these pattern forming mechanisms. However, the strong analogies to better known fluid instabilities have given us significant insight into the processes at work even on the submicrosecond time scale.

References

[1] J.M. Walsh and R.H. Christian, Phys. Rev. 97 (1955) 1544.
[2] Q. Johnson and A.C. Mitchell, Phys. Rev. Lett. 29 (1972) 1369.
[3] L.R. Carl, Met. Prog. 46 (1944) 102.
[4] R. Crossland and A.S. Bahrani, Contemp. Phys. 9 (1968) 71.
[5] W.A. Allen, J.M. Mapes and W.G. Wilson, J. Appl. Phys. 25 (1954) 675.
[6] G.R. Abrahamson, J. Appl. Mech. 83 (1961) 519.
[7] J.M. Walsh, R.G. Shreffler and F.J. Willig, J. Appl. Phys. 29 (1953) 349.
[8] J.N. Hunt, Phil. Mag. 17 (1968) 669.
[9] Yu.A. Gordopolow, A.N. Dremin and A.N. Mikhailor, Fiz. Goreniya i. Vzryua 14 (1978) 77.
[10] S.H. Carpenter and R.H. Whittman, Ann. Rev. of Matls. Sci. 5 (1975) 177.

[11] O.R. Bergmann, G.R. Cowan and A.H. Holtzman, Trans. Metal. Sos AIME 236 (1964) 646.
[12] A.S. Bahrani, T.J. Black and B. Crossland, Proc. Reyal Soc. A296 (1967) 123.
[13] J.L. Robinson, J. Fluid Mech. 63 (1974) 723.
[14] C.S. Smith, Trans. Metall. Soc. AIME (1958).
[15] P.J. Brannon, C.H. Konrad, R.W. Morris, E.D. Jones and J. Asay, Sandia Report SAND82-2469 (1983).
[16] J. Asay, private communication.
[17] D. Grady, J. Geophys. Res. 85 (1980) 913.
[18] D.B. Hayes and D.E. Grady, in: Shock Waves in Condensed Matter – 1981, (P412 AIP, New York, 1982).
[19] R.D. Richtmyer, Comm. on Pure and Appl. Math XIII (1960) 297.
[20] G. Taylor, Proc. Royal Soc. (London) A201 (1950).
[21] E.E. Meshkov, Izv. ANSSSR. Mekh. Zhidkosti i Gaza 4 (1960) 151.
[22] A.D. Sakharov, R.M. Zaidel', V.N. Mineev and A.G. Oleinik, Soviet Physics-Dokl. 9 (1965) 1091.
[23] V.N. Mineev and E.V. Savinov, Soviet Physics JETP 25 (1967) 411.
[24] B. Sturtevant and V.A. Kulkarny, J. Fluid Mech. 73 (1976) 651.
[25] E.E. Meshkov and V.N. Mokhov, Fiz Goreniya i Vzryua 18 (1982) 93.
[26] R.E. Duff, Proc. Symposium in Appl. Math 13 (1962) 77.
[27] B.A. Klopov and E.E. Meshkov, Fiz. Goreniya i. Vzryua 18 (1982) 96.

Physica 12D (1984) 163–170
North-Holland, Amsterdam

WAVE PROPAGATION IN INELASTIC MEDIA

L.G. MARGOLIN
Los Alamos National Laboratory, Los Alamos, New Mexico 87545, USA

The calculation of wave propagation is often more difficult in solids than in fluids. The difficulty is associated with the need for a rate dependent constitutive law to describe inelastic material response. We begin by characterizing the essential features of inelastic behavior. We then contrast two types of constitutive models. Microphysical models are based on describing the microphysical processes responsible for inelasticity and have many advantages over macroscopic models based on plasticity theory. We describe a particular microphysical model for brittle fracture and from this description outline the general elements of the microphysical approach. It is shown that such models have the structure of a viscoelastic solid. The relaxation time depends on internal (i.e. microstructural) variables. It then follows that the inelasticity is due to the evolution of these internal variables.

In numerical simulations of wave propagation, the speed and attenuation of a front may depend sensitively on the mesh spacing. The origin of this effect and a general technique to overcome it are described. The technique is called gradient scaling and is a generalization of the familiar idea of shock capturing.

1. Introduction

This paper is principally concerned with modeling the dynamic response of solids. At low stress levels most solids behave elastically. That is, changes in the size or shape of an elastic body produce forces that act to restore the body to its original state. At higher stress levels, solids exhibit a variety of inelastic behaviors.

The inelastic response of solids can often be associated with changes in the microstructure of the solid material. For example, fracture can be described in terms of the growth, interaction, and coalescence of microcracks that are present in brittle solids. Thus, one might expect that a constitutive model for brittle materials would be based on crack mechanics. However, most constitutive models of inelastic behavior, including fracture, are macroscopic – i.e., are based on plasticity theory – and ignore the microstructure.

Constitutive models that contain a detailed description of microstructure are called microphysical models. The microphysical approach is a relatively recent and attractive alternative for representing inelasticity. Various papers [1–4] describe particular microphysical models for brittle, porous, and granular materials. In the first part of this paper, we will provide a more general viewpoint, discussing the common elements and features of microphysical models.

The key element of the microphysical approach is the use of effective elastic moduli to relate the microscopic detail to macroscopic material properties. The effective moduli allow us to write the constitutive law in the form of an equation of state. That is, stress is uniquely determined as a function of strain plus certain internal variables that rise naturally from the theory.

The rate equations derived from a constitutive law based on effective moduli have the form of a Maxwell (viscoelastic) solid with variable relaxation times. In this form, we can explicitly identify the inelasticity with the evolution of the microstructure.

There are numerical problems associated with simulating wave propagation in a rate-dependent solid. In particular, the finite size of a computation cell leads to a dependence of the results on zoning. In the second part of this paper, the origin of this effect, which is not seen in fluids or elastic solids, is identified. A new technique, gradient scaling, is described that eliminates the mesh dependence.

To summarize, the calculation of wave propagation is often more difficult in solids than in fluids. The difficulties, both theoretical and numerical, are associated with the use of a rate-dependent constitutive law to describe inelastic material response. We begin by defining some of these terms.

2. Definitions and concepts

In solid mechanics, one considers stress–strain curves. Both stress and strain are symmetric two-index tensors. Each of the six independent stress components may depend on any of the six components of strain. Thus, a stress–strain curve is really a projection of a higher dimensional surface. However, the trace of the stress tensor is related to the pressure.

$$P = -\tfrac{1}{3}(\sigma_{11} + \sigma_{22} + \sigma_{33})\,. \tag{1}$$

Also the trace of the strain tensor is the dilatation (the change in volume per unit volume)

$$\frac{\Delta V}{V} = (\epsilon_{11} + \epsilon_{22} + \epsilon_{33})\,. \tag{2}$$

Qualitatively then, the stress–strain curve is analogous to a pressure-volume curve.

Now consider loading a sample along some path in strain space, and then unloading along the same path. If the sample loads and unloads along the same curve in stress–strain space, then the material has behaved elastically. If the loading and unloading paths are different, then the material has behaved inelastically. The term elastic can be equated with reversibility whereas inelastic implies the dissipation of energy. Most solids exhibit both elastic and inelastic response; the distinction is not one of material properties, but of the level of loading. Thus, we should speak of elastic behavior rather than an elastic material.

A constitutive law is a generalization of the idea of an equation of state. By equation of state, we mean a functional relation between thermodynamic variables. An example is the polytropic gas,

$$P = (\gamma - 1)\rho I\,. \tag{3}$$

Given the current value of density (ρ) and of specific internal energy (I) the current value of pressure (P) is uniquely determined.

A constitutive law is a differential relation between thermodynamic variables. An example is the Maxwell solid,

$$\frac{\mathrm{d}\sigma_{ij}}{\mathrm{d}t} = M_{ijkl}\frac{\mathrm{d}\epsilon_{kl}}{\mathrm{d}t} - \frac{\sigma_{ij}}{\tau}\,. \tag{4}$$

The value of stress (σ_{ij}) at any time is not uniquely determined by the current value of strain (ϵ_{kl}), but depends on the history of loading – that is, on the path through strain space. An equation of state can, of course, be written in differential form. However, the equation of state is exactly integrable whereas the constitutive law is not in general.

Let us now consider a simple experiment in which a sample of some brittle material is put into a testing machine and the ends pulled until the sample fails. We arrange to measure both stress and strain throughout the experiment. Intuitively, one might expect that the maximum stress – sometimes called the failure stress or strength – would be a material property. In fact, the measured failure stress is not constant, but depends upon the details of the experiment. For example, the larger the rate of strain the stronger the material appears. Also, the bigger the sample the weaker it appears.

In spite of such simple and clearcut experiments, most constitutive models are based on the idea of the existence of the failure stress as a material property. These theories use the fracture stress to define a yield surface in a plasticity theory.

The reason for this is largely historical. The failure stress does indeed approach a constant value for very low values of strain rate, and for constant sample size. This regime, the quasistatic regime, is where constitutive models were first employed to investigate questions of design stability and integrity. At higher strain rates, these models do poorly. Most of the effort in the field is concerned with constructing ever more elaborate models, incorporating more parameters. However, these models still have limited validity. In particular, they do not scale properly with sample size and strain rate, and they work for only a limited number of simple loading paths.

These models do not extend because they ignore the underlying physical mechanisms of inelastic behavior. For example, fracture is really the growth and coalescence of microcracks. Plastic models ignore the cracks and attempt to describe fracture by relaxing the stress field directly. Another class of constitutive models, the microphysical models, do describe fracture in terms of the details of crack growth. In these models, the crack growth alters the material properties and this alteration leads to stress relaxation. In the next section, we give a brief description of a particular microphysical model for fracture. From this description, we will abstract the key elements and common features of microphysical models in general.

3. The bedded crack model

The bedded crack model (BCM) is a constitutive relation for brittle and quasibrittle materials [1, 5, 6]. The BCM describes fracture in terms of the growth and interaction of microcracks contained in brittle materials. There are two aspects to this approach. First, we must account for the effects of the macroscopic stress field on the cracks. Second, we must account for the effects of the cracks on the macroscopic response of the material.

The BCM begins by postulating that brittle materials contain tiny microcracks, and that a statistical characterization of these is a material property that can be determined by experiment (e.g., by section and counting). This characterization is a distribution function giving the probable number of cracks per unit volume as a function of crack size and orientation.

The effect of the stress on the cracks is a microscopic process. Based on the stress, some of the cracks may begin to grow. The BCM contains a generalization of the Griffith theory of cracks [7] which provides a criterion for deciding which cracks can grow. Further results from this theory determine the speed at which cracks grow. Thus, as loading progresses, the crack distribution function may become complex. The BCM employs analytic fits to the distribution function.

The effect of the cracks on the stress field is contained in the dependence of material properties on the cracks. Effective modulus theory [8, 9] describes the straining of a cracked body in terms of the strain of the matrix plus the extra strain due to the opening and displacement of crack faces. The effective moduli relate the details of the microstructure to the macroscopic material properties.

The effective moduli depend on the elastic properties of the matrix plus the crack distribution function. In particular, certain dimensionless integrals of the distribution function arise naturally from the theory. These integrals play the role of internal or structural variables, which change in time as the distribution function evolves.

The BCM is meant for use in a numerical solid mechanics code. The constitutive law is integrated in time and space along with the dynamical equations that represent conservation of mass, energy, and momentum. At the beginning of each cycle and for every cell, we have values for the stress tensor and the velocity field. Also, we have a fit for the crack distribution function in every cell. The cycle, in its simplest form, consists of the following steps:

1) The current values of velocity are used to calculate the strain rate tensor.

2) The crack distribution function is updated in time from current values of stress and based on the generalized Griffith criterion.

3) The new effective moduli are calculated from the new distribution function.

4) The new stresses are calculated from the strain rates, the effective moduli, and the old stresses.

5) The new stresses are used to update the velocity field based on the momentum equations.

These steps advance all the variables through a small increment in time. The mass equation is updated independently of the BCM. For some materials, there may be a dependence of the pressure on the energy.

Further details of the model can be found in the references. However, on the basis of this brief description, it is now possible to isolate the key elements. First, we must identify the inelastic behavior with some microstructural elements. In the BCM, these are the cracks. Second, we must characterize the microstructure in terms of its elements. In our model, we used a statistical distribution function. Third, we must describe the evolution of an element due to macroscopic processes. In the BCM, this is done with the Griffith criterion for crack growth. Finally, we must relate the details of the microstructure to macroscopic material properties. This was done by constructing effective elastic moduli.

The effective moduli are the heart of our microphysical model. Independent of the details of the microstructure and its evolution, the use of effective moduli gives a common form of the constitutive relation and to the representation of inelasticity. We explore these consequences in the next section.

4. Effective elastic moduli

The effective elastic moduli are found from static solutions of the displacement field for a body with a (statistically) prescribed microstructure and subject to a macroscopically uniform stress field. The calculation begins with a macroscopic definition of strain as the integral of displacement (u_i) over the outer surface (S_0) of a body,

$$\epsilon_{ij} = \frac{1}{V} \int S_0 \, [u_i n_j] \, \mathrm{d}s \, . \tag{5}$$

Here the brackets signify the symmetrized product and V is the volume of the body.

Next one converts the surface integral through the divergence theorem

$$\epsilon_{ij} = \frac{1}{V} \int \left[\frac{\partial u_i}{\partial x_j} \right] \mathrm{d}V - \frac{1}{V} \int_{S_i} [u_i n_j] \, \mathrm{d}S \, , \tag{6}$$

where S_i is the inner surface of the body – for example, the surfaces of the cracks or pores or other microstructure. The volume integral reflects the constitutive response of the matrix material. Usually the matrix is assumed linear elastic

$$\sigma_{ij} = M_{ijkl} \epsilon_{kl} \, . \tag{7}$$

To evaluate the surface integral, we need to find the response of the microstructural element to external stress – for example, the displacement of a crack or a spherical pore to external loading. This integral will depend on certain parameters like crack radius or orientation. Hence, the integral must be averaged over the appropriate distribution function.

For a material containing cracks bedded parallel to the x–y plane, the effective compliance C_{zzzz} is related to the compliance of the matrix material C^0_{zzzz} by [8]

$$C_{zzzz} = C^0_{zzzz} \left(1 + \frac{16}{3} (1 - \nu^2) \gamma \right) . \tag{8}$$

Here ν is Poisson's ratio and γ is the third moment of the crack density distribution function $N(c, t)$.

$$\gamma = \overline{Nc^3} = \int_0^\infty N(c, t) c^3 \, \mathrm{d}c \, . \tag{9}$$

In fact, γ is one of the internal variables of the BCM.

It is worth distinguishing between effective elastic moduli and tangent moduli. A tangent modulus is the local slope of the stress–strain curve at a given value of stress or strain. Since the stress is not uniquely determined by strain, the tangent modulus is also not uniquely determined, but depends on the history of loading.

By contrast, the effective elastic moduli are uniquely determined once the microstructure is specified. These moduli describe the response of a material with a given structure that is not changing – that is, an elastic response. The essence of the effective moduli is that they are an equation of state for stress as a function of strain plus internal variables. The additional variables make the constitutive relation exactly integrable.

One may make this analogy with an ideal gas. A pressure–density relation for an ideal gas is not unique, but appears to depend on the history. Introduction of another variable, the internal energy, leads to an integrable relation and a unique determination of pressure as a function of density and internal energy. One cannot push the analogy too far, however, for energy is a macroscopic variable and obeys a conservation law. The internal variables of our theory are structural variables and evolve according to microscopic laws.

Inelasticity results from the evolution of the microstructure. In rate form, the constitutive law has the form

$$\frac{\mathrm{d}\epsilon_{ij}}{\mathrm{d}t} = \frac{\mathrm{d}}{\mathrm{d}t}\,[C_{kjkl}\,\sigma_{kl}]\,. \tag{10}$$

Inverting for the stress rate

$$\frac{\mathrm{d}\sigma_{kl}}{\mathrm{d}t} = (C^{-1})_{klij}\left[\frac{\mathrm{d}\epsilon_{ij}}{\mathrm{d}t} - \sigma_{mn}\frac{\mathrm{d}C_{ijmn}}{\mathrm{d}t}\right]. \tag{11}$$

Thus, the constitutive relation has the form of a viscoelastic (Maxwell) solid with a variable relaxation time.

In this sense, the microphysical theory has reproduced a macroscopic law. In the process, however, it has given a meaning and a method of calculation to the parameters of the theory. Finally, it is worth emphasizing that microphysical models based on the effective moduli lead to viscoelastic models rather than the plastic models often employed for inelastic solids.

5. Gradient scaling

Calculations of wave propagation with the BCM have shown a disturbing numerical effect. Results of computer simulations, and, in particular, of wave speed and attenuation behind fronts, appear sensitive to the zoning – i.e., to the size of computational cells. This effect is numerical in the sense that the results would reach some asymptotic values as the zones get small enough. In many calculations, however, small enough was one to two orders of magnitude smaller than we could afford to run. These limitations are imposed by the size of memory and the increase in running time which more cells require.

In numerical fluid dynamics there are at least two techniques for dealing with mesh dependency. One is shock fitting where the shock is explicitly tracked. This is not suitable for rate-dependent solids where the front is not a true discontinuity (i.e., is not much smaller than a cell width) and where the state behind the front depends on the shape of the front.

A second technique is adaptive rezoning of the type described by Karl-Heinz Winkler in this conference. In this technique, cells move through the fluid so as to always follow and resolve the front. This is a reasonable choice. However, there are difficulties associated with advecting some of the fracture quantities for which there is no corresponding conservation law. Moreover, it does not seem necessary to resolve the front unless one is really interested in its structure (which we are not).

Thus, we developed a new method that allows us to eliminate mesh effects while still calculating on a coarse mesh. The method is introduced here in the context of rate-dependent solids. However, it is applicable anytime one has strongly rate-dependent processes coupled to the conservation laws of continuum mechanics.

First, we must identify the source of the problem. To do this, we introduce the model equations

$$\frac{d\sigma}{dt} = K\frac{d\epsilon}{dt} - \frac{\sigma}{\tau} \tag{12}$$

and

$$\rho\frac{du}{dt} = \frac{\partial\sigma}{\partial x} + \lambda\frac{\partial^2 u}{\partial x^2}. \tag{13}$$

Here K is a modulus, τ is a relaxation time, λ is the coefficient of (artificial) viscosity, σ is a stress, u is a material velocity.

It is clear that the rate dependence is important since our mesh dependence does not exist for fluids nor elastic solids. Also, the artificial viscosity is introduced since it is necessary to damp numerical oscillations in the solution.

(The viscosity is artificial because its magnitude is much larger than the physical viscosity of the solid and because λ is chosen to scale with the size of the computation cells. More details can be found in refs. 10 and 11.)

The model equations are considerable simpler than BCM. For example, they are one dimensional. Also, they represent the nearly incompressible limit where the density ρ is constant. Finally, in choosing the Maxwell solid, we have allowed the relaxation time to be constant.

Does this simple system exhibit the mesh dependence? We wrote a small program and solved a simple problem analogous to the piston problem in fluids. The problem consists of a slab of viscoelastic material of length L. The boundary at $x = L$ is held rigidly at rest while the boundary at $x = 0$ commences to move with fixed velocity U at time $t = 0$. A compressional stress wave propagates to the right.

In figs. 1a and b, there are two plots. The solid line represents an asymptotic problem calculated with one thousand mesh points. The triangles represent a coarser calculation done with only twenty-five mesh points. In fig. 1a we see that the material velocity of the coarse calculation is much higher than in the finely zoned calculation. The

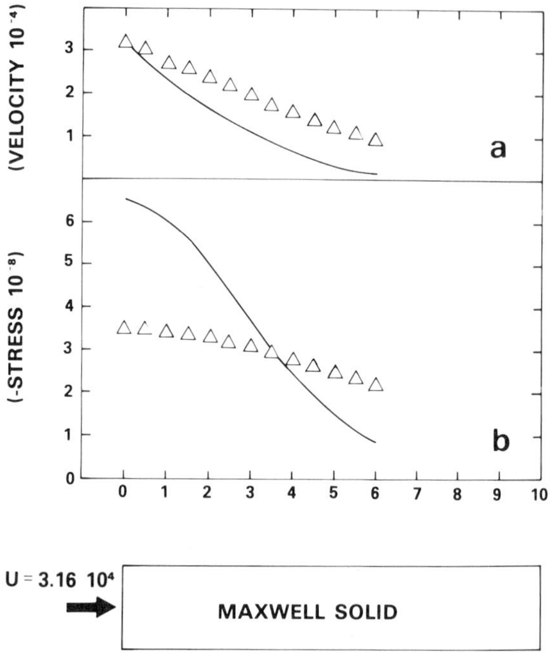

Fig. 1. A comparison of stress and velocity profiles for two calculations. The solid line is a finely zoned calculation and the triangles are from a coarsely zoned calculation. No gradient scaling is used in these calculations and great sensitivity to the zoning is seen.

comparison of stresses in fig. 1b shows even worse agreement.

The model eqs. (12) and (13) certainly exhibit mesh dependence. Furthermore, this dependence is mostly independent of the details of the numerical (discrete) approximations to the equations. This last fact begins to point the finger at the true culprit – the artificial viscosity. Indeed, the origin of the effect becomes clear from the following exercise. It is possible to construct a wave equation for either velocity or stress from the model equations. For example, after cross differentiating, we derive

$$\rho\frac{\partial^2 u}{\partial t^2} + \frac{\rho}{\tau}\frac{\partial u}{\partial t} = \left(K + \frac{\lambda}{\tau}\right)\frac{\partial^2 u}{\partial x^2} + \lambda\frac{\partial^3 u}{\partial x^2 \partial t}. \tag{14}$$

(We have not distinguished between the partial and the total derivatives since, in the incompressible limit, convection is ignored.)

The interesting thing about this equation is the appearance of the term

$$\frac{\lambda}{\tau}\frac{\partial^2 u}{\partial x^2}. \tag{15}$$

This term would not exist in fluids where $\tau \to \infty$. Nor would it appear in the physical equations where $\lambda = 0$. Furthermore, its effect is precisely what we are seeking, for it represents an increase in the wave speed, and has its origin in the artificial viscosity, which is a numerical effect, and whose magnitude is tied to the size of the computational cells.

Having identified the problem, how do we correct it? The most obvious answer would seem to be to redefine the material modulus.

$$K' = K - \frac{\lambda}{\tau}. \tag{16}$$

In fact, this does not work. The simulations with this correction lead to the correct velocities, but not the correct stresses. The reason is subtle and involves the boundary conditions. We must alter the wave equation to make it correspond to the physical equations *with the correct boundary conditions.*

In our problem, and in most numerical codes, we apply velocity boundary conditions. Thus, in the model equations, the constitutive law will yield the correct stress if the velocity is correct. So we should not alter the constitutive equation.

Instead, we introduce a scaling factor α into the momentum equation. Now our model set is

$$\frac{d\sigma}{dt} = K\frac{d\epsilon}{dt} - \frac{\sigma}{\tau} \tag{17}$$

and

$$\rho\frac{du}{dt} = \alpha\frac{\partial\sigma}{\partial x} + \lambda\frac{\partial^2 u}{\partial x^2}. \tag{18}$$

Constructing the wave equation for velocity, we

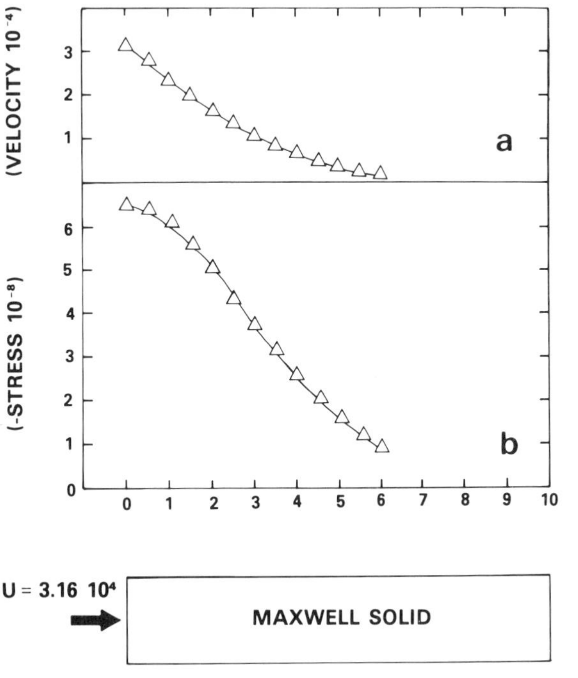

Fig. 2. The same comparison as in fig. 1. Gradient scaling is employed in the coarsely zoned calculation which effectively eliminates the sensitivity to the zone size.

find we must choose

$$\alpha = 1 - \frac{\lambda}{K\tau}. \tag{19}$$

Then, we get the correct dispersion relationships and, with the correct boundary conditions for velocity, get the proper velocity solution. With the proper velocity, eq. (17) gives the proper stress.

Figs. 2a and b show a comparison of a coarsely and finely zoned calculation, but with the gradient scaling parameter α. Now one can see excellent agreement. The method of gradient scaling allows us to use a coarse mesh and yet eliminate the mesh dependence.

6. Conclusions

A proper formulation of the constitutive relations for inelastic material behavior is essential to calculating wave propagation in inelastic media. The microphysical models represent an attractive alternative to the phenomenologic models based on plasticity. The microphysical models are based on a description of the physical processes leading to inelasticity.

The key elements of a microphysical model are

1) identification of inelasticity with some microstructure in the material;

2) a statistical description of the microstructure in terms of its elements;

3) a mechanical model for the evolution of an individual element as a function of the macroscopic stress; and

4) effective moduli which relate the details of the microstructure to the macroscopic material response.

By microstructural unit, we mean the individual cracks, pores, grains, etc. of which the microstructure is comprised.

The microphysical models lead to an equation of state for the material even during inelastic behavior. Stress is always a uniquely determined function of strain plus certain internal variables which arise naturally from the effective modulus theory. In rate form, the microphysical models lead to viscoelasticity where the relaxation times vary with loading history. In this form, inelasticity is seen to arise directly from the evolution of the internal variables.

The numerical simulation of waves in viscoelastic solids exhibits a sensitivity to the discrete zoning. This effect arises from the coupling of the stress relaxation to the artificial viscosity necessary to produce a smooth numerical solution. The technique of gradient scaling is based on forcing the dispersion relation of the numerical equations to correspond to that of the physical equations. Gradient scaling is a simple and inexpensive method to eliminate mesh dependence in numerical simulations.

References

[1] L.G. Margolin, "Numerical Simulation of Fracture," Proc. Int. Conf. Constitutive Laws, C.S. Desai, ed. Tucson, Arizona, 1983.

[2] L. Seaman, D.R. Curran and D.A. Shockey, "Computational Models for Ductile and Brittle Fracture", J. Appl. Phys. 47 (1976) 4814–4826.

[3] M.M. Carroll, "Compaction of Dry or Fluid-Filled Porous Materials", J. Eng. Mech. Div. (ASCE) 106 (1980) 969–990.

[4] P.A. Cundall and O.D.L. Strack, "Development of Constitutive Laws for Soil Using the Distinct Element Method", Third Int. Conf. Numerical Methods in Geomechanics, Aachen, 1979.

[5] L.G. Margolin, "Calculations of Cratering Experiments with the Bedded Crack Model", the American Physical Society Symposium on Shock Waves in Condensed Matter (AIP Proceedings No. 78), Menlo Park, California, 1981.

[6] L.G. Margolin and T.F. Adams, "Numerical Simulation of Fracture", Proc. 23rd U.S. Symp. Rock Mechanics, Berkeley, California, 1982.

[7] B.R. Lawn and T.R. Wilshaw, Fracture of Brittle Solids (Cambridge Univ. Press, New York, 1975).

[8] L.G. Margolin, "Elastic Moduli of a Cracked Body", Int. J. of Fracture 22 (1983) pp. 65–79.

[9] A. Hoenig, "Elastic Moduli of a Nonrandomly Cracked Body", Int. J. of Solids and Structures 15 (1979) 137–154.

[10] J. Von Neumann and R. D. Richtmeyer, "A Method for the Numerical Calculations of Hydrodynamic Shocks", J. Appl. Physics 21 (1950) 232–237.

[11] M.L. Wilkins, "Use of Artificial Viscosity in Multidimensional Fluid Dynamic Calculations", J. Comp. Phys. 36 (1980) 281–303.

SECTION 2

BOUNDARIES OF REACTION AND TRANSFORMATION

Physica 12D (1984) 173–181
North-Holland, Amsterdam

2.1. COMBUSTION

FRONTS AND PATTERNS IN LOW MACH NO. COMBUSTION—A REVIEW

J. BUCKMASTER
University of Illinois, Urbana Champaign Campus, Urbana, IL 61801, USA

Fronts or interfaces arise in a number of natural ways in combustion, and the mathematical limits that formally lead to these structures are discussed. Patterns arise if the fronts and the associated combustion field are unstable, and the most important mechanisms are described.

1. Introduction

The purpose of this essay is to describe, in an elementary fashion, the natural way in which combustion fits into the theme of the conference. Before describing how fronts can arise in flame theory, however, it is appropriate to make some general introductory remarks about the subject.

Flames can be classified in a number of ways. This discussion will be restricted to low Mach number flames, thereby excluding the rich phenomena associated with detonation, a marriage of combustion with shock fronts. In addition only laminar combustion will be considered, although the concepts that will be introduced certainly have meaning for weakly turbulent flames. Thus we shall be concerned with low Mach number flames, of both the premixed and diffusion variety.

The distinction between diffusion flames and premixed flames is not always clear; diffusion flames can have some of the characteristics of premixed flames. Nevertheless, there is a general distinction. In a premixed flame the chemical constituents whose reaction defines and characterizes the combustion are thoroughly mixed at some point or time remote from the burning region; in a diffusion flame the reactants are supplied separately and react at a diffusion-smeared interface. For the premixed flame, reaction occurs if the mixture is hot enough; for the diffusion flame the reactants must also come together.

The most common example of a premixed flame is the blue/green cone that is seen at the mouth of a Bunsen burner in which city gas is burnt. The fuel – essentially methane – is supplied through a nozzle at the base of the tube. Should the air hole be open, air containing oxygen is drawn into the tube and mixes with the fuel during passage up the tube. It is this mixture which burns in the conical flame at the mouth of the tube. Typically, not all of the fuel is burnt in this way – insufficient oxygen is drawn into the tube for complete combustion to occur. Thus if air is available beyond the tube mouth (the usual case) the excess fuel will burn in a diffusion flame downstream of the cone. A Bunsen flame typically has both premixed and diffusion components. If the air-hole is closed only the diffusion flame is seen; if there is no oxygen beyond the tube mouth only the premixed flame is seen.

A candle supports a pure diffusion flame. A small pool of liquid wax at the top of the stick wets the base of the exposed wick. Capillary action draws this liquid up the wick where it evaporates through proximity to the flame so that there is a diffusive flux of fuel vapor away from the wick. It is this, combined with a diffusive flux of oxygen towards the wick, that sustains the flame.

In order to understand the behavior of flames it is necessary to examine them in a mathematical framework as well as a physical one. Justification of the equations for a reacting mixture of gases is a

0167-2789/84/$03.00 © Elsevier Science Publishers B.V.
(North-Holland Physics Publishing Division)

complicated matter, even when reduced to its bare essentials [1], so that we shall be content with little more than the setting down of a system adequate for describing low Mach number flames in a dilute mixture. The restriction to low Mach numbers leads to profound simplifications which will be identified in due course; by confining our attention to a dilute mixture, one for which the major component is an inert such as nitrogen, the equations are essentially those of a single fluid and the diffusion laws are essentially those of a binary system.

Overall mass and momentum balances are described by

$$\frac{\partial \rho}{\partial t} + \nabla \cdot (\rho \boldsymbol{v}) = 0, \tag{1}$$

$$\rho \frac{\mathrm{d}\boldsymbol{v}}{\mathrm{d}t} = -\nabla p + P\left[\nabla^2 \boldsymbol{v} + \tfrac{1}{3}\nabla(\nabla \cdot \boldsymbol{v})\right]. \tag{2}$$

These equations have been non-dimensionalized in a manner which is not important for our purposes and P is a Prandtl number. A more realistic form could be adopted for the viscous stresses without affecting our discussion.

When we turn to energy conservation the small Mach number restriction is of importance for it leads to the conclusion that the energy is predominantly thermal in nature. Thus, with T the temperature, we have

$$\rho \frac{\mathrm{d}T}{\mathrm{d}t} - \nabla^2 T = \Omega. \tag{3}$$

The source term Ω represents the heat generated by the various reactions and is discussed later.

The equation of state is also much simplified. To leading order the pressure is uniform and we shall take it constant, consistent with (3)†, so that we have Charles's law

$$\rho T = \text{constant}. \tag{4}$$

†Otherwise a term proportional to $\mathrm{d}p/\mathrm{d}t(t)$ must be added to the energy equation.

p in eq. (2) represents a perturbation pressure which vanishes as the Mach number, i.e. v, goes to zero.

Finally we have a mass balance for each of the reactants and these have the form

$$\rho \frac{\mathrm{d}Y_i}{\mathrm{d}t} - \frac{1}{L_i}\nabla^2 Y_i = \Omega_i; \tag{5}$$

L_i is a Lewis number and Ω_i accounts for the various reactions.

Eqs. (1)–(5) are quite general and can be used to tackle a wide range of combustion problems. For many of these, quite complicated descriptions of the chemical kinetics, represented by Ω and Ω_i, are necessary. Should one want to describe the production of oxides of nitrogen in an internal combustion engine, then a large number of different species (the Y_i) must be considered, together with a large number of reactions. For the most part such problems can not lie within the scope of a mathematical discussion since early recourse to a powerful computer is necessary. But profound insight, essentially of a qualitative nature, can be obtained into a remarkably large number of important combustion problems by adopting a breath-takingly simple-minded approach to the kinetics, which may be expressed most simply as

Reactants → Products + Heat.

In other words the most important effect of the multitude of reactions that occur in reality is that certain components of the gas are consumed (the oxygen, the fuel) and heat is liberated in a proportional amount. For a premixed flame it can be enough to suppose that there is but one component – the 'mixture' – denoted by Y. Then $i = 1$ in eq. (5) and writing

$$\frac{\Omega_1}{\alpha} = \Omega = \mathscr{D} Y \mathrm{e}^{-\theta/T} \tag{6}$$

assumes an irreversible one-step elimination of Y.

Here $\mathscr{D}$ is a parameter known as the Damkohler number and θ, which defines the sensitivity of the reaction to temperature, is the activation energy. $\mathscr{D}$ and θ play a central role in our discussion.

For a diffusion flame the fuel and oxygen are supplied separately so that the simplest model requires two components ($i=1,2$). Analogous to (6) we then have

$$\frac{\Omega_1}{\alpha_1}=\frac{\Omega_2}{\alpha_2}=\Omega=\mathscr{D}Y_1Y_2^{-\theta/T}; \qquad (7)$$

if either reactant is absent there can be no reaction.

So far so good – we now have a reasonable set of equations on which to base our discussion. But even these equations, despite the far-reaching simplifications that have been adopted, present too formidable an obstacle in many cases and then more drastic simplification is necessary. The so-called constant density model† discards the fluid mechanics (eqs. (1), (2) and (4)) and sets $\rho=$ constant in the remainder. To the extent that a particular phenomenon arises from hydrodynamical interactions, the constant density model is of no value in its study. But to the extent that, fundamentally, the phenomenon arises because of the physics intrinsic to eqs. (3), (5), (6) and (7), the model is illuminating.

2. Combustion fronts

Flames are not always yellow flickering clouds – sometimes they consist of thin sheets or fronts. The conical premixed flame on a Bunsen burner is one example; and if one looks closely at the base of a candle flame one will see a thin blue sheet close to the wick, a front of a different sort, albeit the same color. There are in fact three ways in which fronts can arise in combustion, and in this section we shall describe their basic nature.

†Some workers refer to it as the constant density approximation, but since typical density variations within a flame are seven-fold, this is not an appealing description, despite the fact that the model can be justified in the limit of vanishingly small heat release, when overall temperature variations are small.

2.1. *Damkohler number asymptotics* ($\mathscr{D}\to\infty$)

The parameter $\mathscr{D}$ that appears in the formulas (6) and (7) depends on the pressure and on the size of the flame and is often very large. It is then natural to consider the asymptotic limit $\mathscr{D}\to\infty$. In the case of diffusion flames the limit is a non-trivial singular one, named the Burke–Schumann limit after a famous 1928 paper [2]. To see this it is only necessary to examine the equation governing the reactants, to wit

$$\frac{1}{\mathscr{D}}\left[\frac{\mathrm{d}Y_i}{\mathrm{d}t}-\frac{1}{L_i}\nabla^2Y_i\right]=\alpha_iY_1Y_2\mathrm{e}^{-\theta/T}, \qquad i=1,2. \qquad (8)$$

In the limit $\mathscr{D}\to\infty$ this yields

$$Y_1Y_2=0 \qquad (9)$$

except possibly in thin layers, since the small parameter $\mathscr{D}^{-1}$ multiplies the highest derivative. Thus the combustion field is naturally divided into two parts in one of which Y_1 identically vanishes, in the other Y_2. These regions are separated by a thin reaction zone in which there is a balance between diffusion and chemical reaction. This front, or interface, is called a Burke–Schumann flame sheet. The structure of the combustion field in the neighborhood of the flame sheet is sketched in fig. 1. T, Y_1 and Y_2 are continuous at the front but their derivatives jump so that the sheet acts as a source of heat and a sink of reactants.

Large well-defined Burke–Schumann flame sheets can easily be created by the kitchen experimenter. If the shallow depression at the bottom of a 2-lb coffee can is filled with denatured alcohol

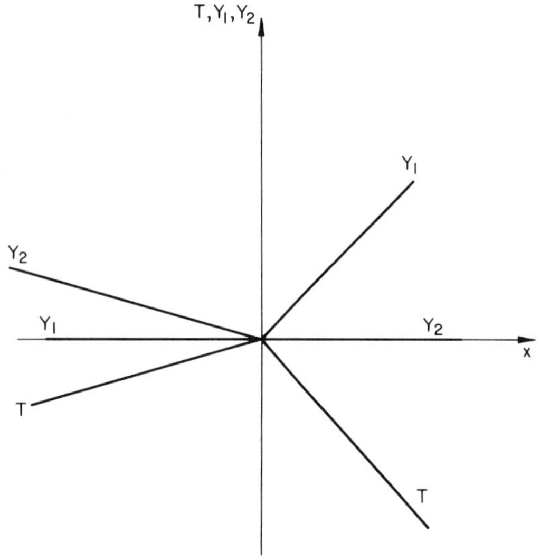

Fig. 1. The neighborhood of a Burke–Schumann flame sheet.

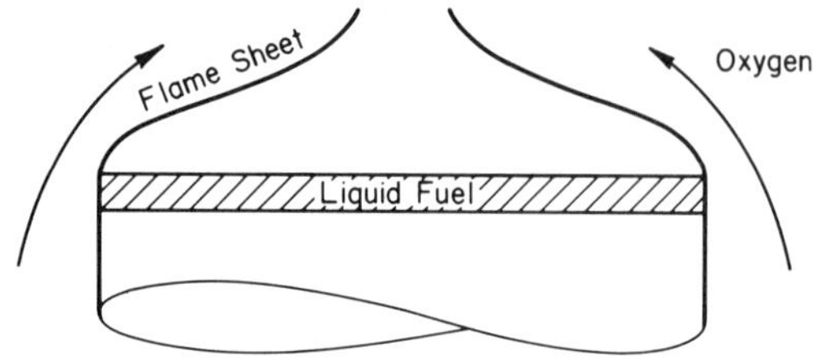

Fig. 2. Demonstration of a Burke–Schumann flame sheet.

which is ignited, a flame sheet originates at the lip of the can (fig. 2). Buoyancy induced convection sustains the flux of oxygen to the flame; for a detailed analysis the reader is referred to Clarke and Riley [3].

2.2. *Activation energy asymptotics* ($\theta \to \infty$)

We turn now to the second way in which fronts can arise, and this is relevant for both premixed and diffusion flames. The quantity θ/T is often quite large everywhere in the combustion field (> 10–20, say) so that the exponential which appears in the reaction rate Ω is very small. This gives rise to an unusual type of asymptotics which can be used to unravel a large variety of combustion phenomena, [1]. Consider, for example, the premixed flame for which $\mathscr{D}$ is defined by eq. (6). It is clear that if the exponential is everywhere very small there can be no reaction unless $\mathscr{D}$ is correspondingly large. This suggests that we examine a distinguished limit by writing

$$\mathscr{D} = e^{\theta/T_*}, \tag{10}$$

where T_* is a temperature that characterises the magnitude of $\mathscr{D}$. Sometimes (it depends on how the equations have been nondimensionalized) $\mathscr{D}$ is a function of θ in which case (10) identifies the nature of this dependence; in the more usual case where $\mathscr{D}$ is an independent parameter, (10) simply defines T_*.

Ω can now be written as

$$\Omega = Y e^{\theta(1/T_* - 1/T)} \tag{11}$$

and the consequence of taking the limit $\theta \to \infty$ depends on the relative magnitude of T and T_*. In regions where $T < T_*$ we have $\Omega \to 0$; reaction is frozen because the temperature is low. In regions where $T > T_*$ the exponential factor in (11) is unbounded. Since outside of thin layers (temporal or spatial) there is nothing to balance an unbounded reaction it follows that $Y \to 0$ and from (5) we have, once again, $\Omega \to 0$, but for quite a different reason. Finally, in regions where T is close to T_* (cf. $T = T_*$) Ω can be nonvanishing.

If the asymptotic development is pursued in detail it is typically found that Ω behaves, in the limit, just like a δ-function, defining a thin flame-sheet of temperature T_*. This flame sheet is a front for the burnt gas. The conical flame sitting on a Bunsen burner is an example.

2.3. *The hydrodynamic limit*

The third way in which fronts can arise is by examining premixed flames on a large scale. Formally we write

$$t = \frac{1}{\varepsilon} t', \qquad r = \frac{1}{\varepsilon} r', \tag{12}$$

in the governing equations and consider the limit

$\varepsilon \to 0$. This leads to a combustion field in which a sharp front separates fresh cold gas from hot burnt gas. Thus on one side

$$T = T_{\mathrm{f}}, \qquad Y = Y_{\mathrm{f}}, \tag{13}$$

whereas on the other

$$T = T_{\mathrm{b}}, \qquad Y = 0. \tag{14}$$

The flow-field on each side is described by Euler's equations

$$\frac{\partial \rho}{\partial t} + \nabla \cdot (\rho \boldsymbol{v}) = 0, \qquad \rho \frac{\mathrm{d}\boldsymbol{v}}{\mathrm{d}t} = -\nabla p \tag{15}$$

and will be irrotational on the cold side if the supply is irrotational; vorticity can be generated at the front. The solutions on each side of the front are connected by conserving the mass and momentum fluxes through the front, and by conserving energy by setting

$$T_{\mathrm{b}} = T_{\mathrm{f}} + Y_{\mathrm{f}}. \tag{16}$$

In addition the speed of the front relative to the cold gas (the flame speed) must be specified. An explicit formula for the flame speed can be deduced by a one-dimensional analysis of the front structure in the limit $\theta \to \infty$, or the speed can be determined by other means (numerical computations or experiment).

The conical Bunsen burner flame is actually an activation energy flame sheet imbedded within a hydrodynamic front (fig. 3). Examined on a scale defined by the tip curvature it is the former; on a scale defined by the tube diameter, the latter.

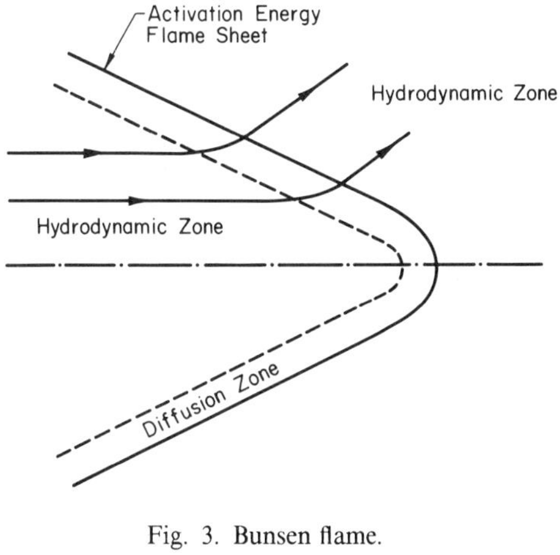

Fig. 3. Bunsen flame.

3. Patterns in combustion

The fronts that we have described can appear as thin smooth sheets. If the combustion field is unstable, however, two-dimensional structure can be generated in the plane of the sheet, giving rise to patterns. This is primarily a characteristic of premixed flames, to which we shall confine the discussion.

3.1. *Hydrodynamic instabilities*

Darrieus and Landau independently considered the linear stability of plane combustion fronts in the hydrodynamic limit. They discovered, surprisingly, that such fronts are unconditionally unstable. The manifestation of the instability can be understood by considering how a perturbed front would move in the absence of induced gas velocities (fig. 4). Since the front propagates normal to itself at the uniform flame speed it is clear that

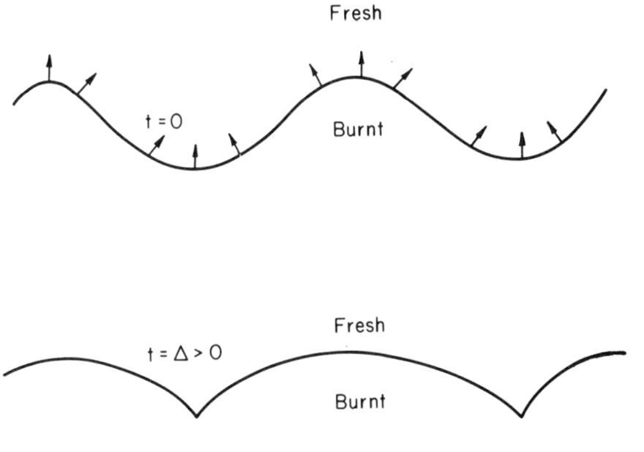

Fig. 4. Hydrodynamic corrugations.

sharp corners will emerge on those portions of the front that are initially convex to the burnt gas. Curvature is a stabilizing influence so that the corrugated front shown in fig. 4b can persist. One way of generating large scale premixed flames (albeit unsteady ones) in the laboratory is to ignite a large volume of mixture at a point thus creating an outwardly propagating spherical front. Sharp folds are often seen on such fronts, a consequence of the hydrodynamic instability. It is worth noting that a flame travelling in a horizontal cylindrical tube is strongly curved and is concave towards the burnt gas. This also is a manifestation of the hydrodynamic instability.

3.2. *Thermal-diffusive instabilities*

Evolution theory, in tracing the origin of species back to a starting point, has to deal with the question of how it is that complexity can arise from a primordial soup. Turing, in his well-known 1953 paper [4], showed that a two-component reactive–diffusive system can be unstable if the two components have different diffusion coefficients. In this way very complicated structure can arise. It is not surprising, then, that for combustion systems, Lewis numbers different from 1 can generate instabilities. They can be understood in the context of the constant density model.

Consider a plane combustion front. Its structure, in the limit $\theta \to \infty$, is sketched in fig. 5. The temperature rises from the value T_f in the fresh mixture to the value T_b $(= T_f + Y_f)$ at the flame sheet and is constant thereafter. Similarly Y falls from Y_f to zero. Consider now small perturbations of this structure, proportional to $e^{\alpha t + iky}$ where y is measured parallel to the undisturbed flame sheet. α is determined from the dispersion relation

$$2K^2(1-K) + \left[(1-K)^2 - 4k^2\right] \frac{Y_f \theta}{2T_b^2}(\mathscr{L}-1) = 0, \tag{17}$$

$$K \equiv \sqrt{1 + 4\alpha + 4k^2},$$

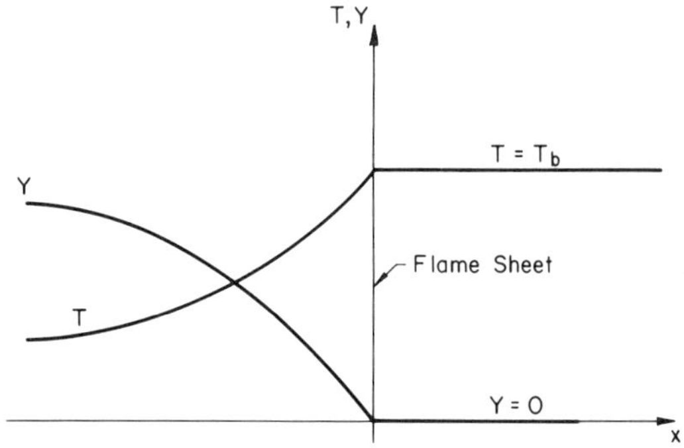

Fig. 5. Plane combustion wave.

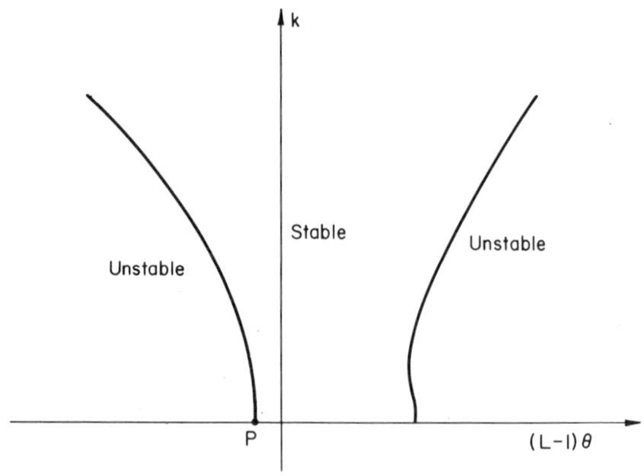

Fig. 6. Stability boundaries for the plane combustion wave.

a result due to Sivashinsky [1]. The corresponding stability boundaries are drawn in fig. 6 which reveals that if $|\mathscr{L}-1|$ is sufficiently large instability is assured. Indeed, only $\mathscr{O}(1/\theta)$ values are required so that if it wasn't that $\mathscr{L}$ is close to 1 for many gases, the thermal-diffusive or Turing–Sivashinsky instability would be ubiquitous. The two stability boundaries are associated with quite different behavior and we shall discuss them separately.

The most important feature of the right boundary is that it is rather inaccessible; for most gases $\mathscr{L}$ is not large enough. However, there are substances, called thermites, solids that burn to form solids, for which $\mathscr{L} = \infty$ suggesting that they are firmly in the unstable domain. It is only a suggestion since the analysis that leads to (17) is only valid for finite $\mathscr{L}$, but a modified analysis can be carried out which essentially duplicates the right half of fig. 6 provided $\mathscr{L}$ is omitted from the abscissa variable†.

A second important feature of the stability boundary is that α does not vanish on it. Neutrally stable modes are pulsations if k is zero, travelling waves otherwise. Consider now an infinite thermite cylinder of radius a whose surface is insulated. If a is small, the only allowable disturbances will be one-dimensional ($k = 0$) or those with small wavelengths (k large). Then the mode with $k = 0$ will be the most unstable one so that the manifestation of the instability will be plane pulsations. During the course of these pulsations the flame temperature varies so that the material is burnt unevenly. This gives rise to a banded structure (fig. 7a). If a is increased, modes of larger wavelength (i.e. smaller k) are admissable and it can be (see fig. 6) that these are more unstable than the one-dimensional mode. The instability will lead to travelling waves which give rise to helical bands (fig. 7b) rather than plane ones.

The right stability boundary can play an important role in conventional flames that are attached to burners. The crucial role of a burner is to fix or stabilize a flame, to prevent it propagating upstream (flashback) or being swept downstream (blowoff). It does this by extracting heat from the flame. Should the flame be displaced towards the burner this heat loss increases, diminishing the flame-speed, so that the flame gets pushed back downstream. If the flame is displaced away from the burner the heat loss decreases, the flame speed increases, and the flame propagates back upstream to its equilibrium position. At the same time, however, as Margolis [6] first showed, heat loss to the burner can displace the right stability boundary to the left, making it more accessible. Thus a burner can stabilize a flame in a global sense (i.e. prevent blow-off or flash-back) while, at the same time, inducing pulsations. Buckmaster [7] showed that as the mass flux through the burner is decreased, the stability boundary first moves to the left but then moves back to the right. Thus reduction in the mass flux can first trigger pulsations and then eliminate them. This dual phenomenon has recently been observed in the laboratory (M. Gorman, private communication).

†This analysis is not activation energy asymptotics but models the reactions by a δ-function whose strength is suggested by asymptotics. See Buckmaster and Ludford [5] and the original references cited there.

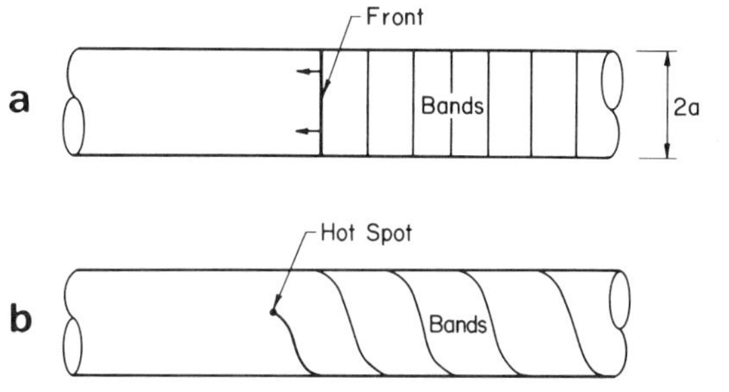

Fig. 7. Patterns in thermites.

3.3. *Cellular flames*

We turn now to the left stability boundary of fig. 6 which is relevant when the Lewis number is sufficiently smaller than 1. It is quite accessible and leads to cellular flames, so called because the flame sheet displays a cellular structure superficially similar to the cells seen in Benard or Marangoni convection. Flame cells, however, are highly unsteady, continuously growing and subdividing.

The connection between theory and experiment is most vividly demonstrated by means of Sivashinsky's equation,

$$\frac{\partial \phi}{\partial t} + 4\nabla^4\phi + \nabla^2\phi + \tfrac{1}{2}(\nabla\phi)^2 = 0. \tag{18}$$

This equation is deduced from a bifurcation analysis valid in the neighborhood of the point P in fig. 6, and ϕ is the displacement of the flame sheet from the equilibrium position. Numerical solutions of (18) are strikingly similar to the observed phenomena [8].

Cellular flames come in a variety of forms of which the nominally plane unbounded flame in a uniform flow, as described by (18), is but one

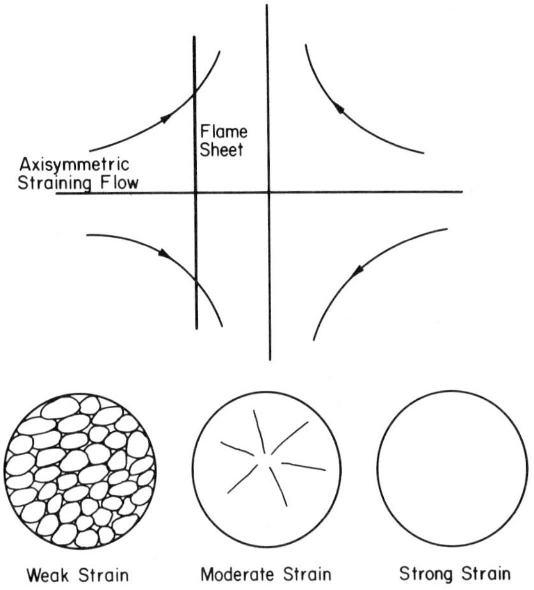

Fig. 8. Flame in a straining flow.

example. An axisymmetric straining flow (fig. 8) tends to suppress the instability, especially structure within the flame sheet that is perpendicular to the flow direction. Thus if the strain is weak unsteady cells are observed, but stationary radial ridges whose crests point towards the burnt gas occur at moderate strain rates (fig. 8). Large rates of strain can completely suppress the instability. This phenomenon is observed in flames travelling upwards in vertical tubes filled with lean hydrogen mixtures. Buoyancy induces a straining flow near the nose of the flame and radial ridges are observed (fig. 9). These dark ridges separate ribbons of flame. Towards the skirt of the flame the straining rate is reduced so that the tendency to suppress the instability is diminished. Indeed, the ribbons break up into small balls of flame which persist, rising behind the main body of the flame. A sequence of drawings, of which fig. 9 is a rough composite, are shown in [9].

It would be unthinkable to discuss cellular flames, no matter how briefly, without making mention of polyhedral flames. These are a striking manifestation of cellular instability seen on Bunsen burners. The cone divides into a number of petals which can be stationary and can sometimes spin, often at remarkable speed (several revolutions per second). The stationary manifestation can be understood by modifying equation (18) to account for heat loss to the burner rim [10] so that we have

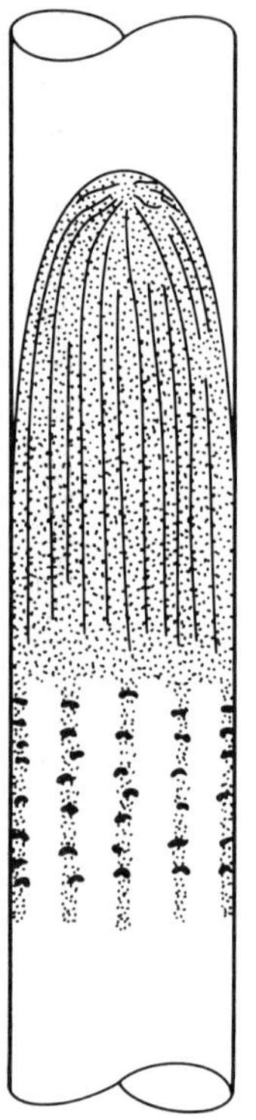

Fig. 9. Flame rising in a tube.

$$\frac{\partial \phi}{\partial t} + 4\nabla^4\phi + \nabla^2\phi + \tfrac{1}{2}(\nabla\phi)^2 + \beta\phi = 0. \qquad (19)$$

The neutral stability curve for this equation shows that the most unstable wave number is non-zero (cf. fig. 6); this wavenumber essentially determines the number of petals. Spinning flames can also be described by (18) but there are unresolved questions [10] which will not be discussed here.

In this brief discussion we have separated hydrodynamic effects from the thermal-diffusive effect. Elsewhere in these proceedings, P. Clavin discusses what happens when these phenomena are treated simultaneously.

Acknowledgements

This work was supported by the Army Research Office and by the National Science Foundation.

References

[1] J.D. Buckmaster and G.S.S. Ludford, Theory of Laminar Flames (Cambridge Univ. Press, New York 1982).
[2] S.P. Burke and T.E.W. Schumann, Industrial and Engineering Chemistry 20 (1928) 998–1004.
[3] J.F. Clarke and N. Riley, J. Fluid Mechanics 74 (1976) 415–431.
[4] A.M. Turing, Philosophical Truncations of the Royal Society of London B237 (1953) 37–72.
[5] J.D. Buckmaster and G.S.S. Ludford, Lectures on Mathematical Combustion SIAM (NSF/CBMS lecture series), (1983).
[6] S.B. Margolis, Combustion Science and Technology 22 (1980) 143–169.
[7] J.D. Buckmaster, Stability of the Porous Plug Burner Flame, SIAM Journal on Applied Mathematics 43 (1983) 1335–1349.
[8] G.I. Sivashinsky, in Annual Review of Fluid Mechanics (1983).
[9] B. Lewis and G. von Elbe, Combustion, Flames and Explosions of Gases (Academic Press, New York 1961).
[10] J.D. Buckmaster, Polyhedral Flames – An Exercise in Bimodal Bifurcation Analysis, SIAM Journal on Applied Mathematics 44 (1984) 40–55.

Physica 12D (1984) 182–197
North-Holland, Amsterdam

FLAME FRONTS WITH COMPLEX CHEMICAL NETWORKS

Paul C. FIFE
Department of Mathematics, University of Arizona, Tucson, AZ 98721, USA

and

Basil NICOLAENKO
Center for Nonlinear Studies and Theoretical Division, Los Alamos National Laboratory, Los Alamos, NM 87545, USA

For laminar flame fronts with general complex chemistry networks, we present a systematic method to reduce the detailed chemistry in the limit of high activation energy asymptotics. We detail the method on a global multistep hydrocarbon model. Generic flame profile structures are discussed, and extensions to networks with zero activation energies are outlined.

1. Introduction

Complex chemistry in flames involves a mix of reactions with both high and small activation energies. Since the pioneering works of Frank-Kamenetskii [11] and Zeldovich [16, 17], it has been recognized that *large* activation energies E_i determine a hierarchy of scales

$$\varepsilon_i = \frac{RT_+}{E_i}\frac{T_+}{T_+ - T_-}, \quad \varepsilon_i \ll 1 ,$$

where T_- is the cold mixture temperature, and T_+ a typical adiabatic flame final temperature. The ε_i are the inverse of the (large) Zeldovich numbers.

In this paper we develop the asymptotic flame theory with complex chemistry first outlined in [7–8]. We specifically consider the case of finite ratios $E_i/E_j \neq 1$ for the subclass of High Activation Energy (H.A.E.) reactions. Cases where $E_i \approx E_j$, but the pre-Arrhenius factors are not the same order of magnitude, are also covered by our theory.

We first explain concepts which are common to *all* reactions networks: admissible reactant allocations, which yield the possible final burned states of the flame; and admissible reactant sub-allocations to the network, which determine feasible chemical network pathways as one passes from the unburned state to the burned state through the front (section 1.2). The latter correspond to subgroups of reactions which go to completion at temperatures lower than the final one. The notion of set of admissible allocations (section 2.2) turns out to be a convenient tool.

Next, to determine the actual pathways within the network and along the flame profile, we construct normalized power gauge functions (section 1.3) for both H.A.E. and Z.A.E. (Zero Activation Energy) reactions. Unnormalized power rate functions have already been used implicitly by Frank-Kamenetskii [11] and Liñan [12]. We derive systematic comparison principles based on these gauge functions.

In section 2, we detail our method on a global hydrocarbon two-step model. We evidence some generic flame structures present in complex networks: merged flames, competition flames and tandem flames.

In section 3, we outline the methodology for general networks

0167-2789/84/$03.00 © Elsevier Science Publishers B.V.
(North-Holland Physics Publishing Division)

1.1. *The flame front equations for a chemical network*

Let n chemical species $A_1, A_2, \ldots, A_n$ be involved in r reactions $R_1, R_2, \ldots, R_r$; symbolically

$$\sum_{j=1}^{n} \nu(j,i)A_j \to \sum_{j=1}^{n} \lambda(j,i)A_j, \quad i = 1, \ldots, r. \tag{1.1}$$

where $\nu(j,i)$ (resp. $\lambda(j,i)$) is the stoichiometric coefficient of species j on the left side (resp. right side) of reaction i. Both ν and λ are nonnegative integers; $\nu(j,i)$ is zero if species j is not a reactant in reaction i, while $\lambda(j,i)$ is zero if it is not a product. Each reaction yields heat releases $Q_1, Q_2, \ldots, Q_r$.

The flow field is described by the temperature T and the vector of rescaled mass concentrations (fractions):

$$\boldsymbol{Y} = (Y_1, Y_2, \ldots, Y_n), \tag{1.2}$$

where the rescaled Y_j, $j = 1, \ldots, n$, are the actual mass fractions (concentrations) of species j divided by the corresponding molecular mass m_j (this rescaling somewhat simplifies the flame front eqs. (1.6). The full state vector $\boldsymbol{U}$ is

$$\boldsymbol{U} = (T, \boldsymbol{Y}) = (T, Y_1, \ldots, Y_n). \tag{1.3}$$

Each reaction can be characterized by an n-dimensional reaction vector $\boldsymbol{V}_i$, $i = 1, \ldots, r$, which represents the net change in the number of molecule grams (moles) of various species in reaction R_i:

$$\begin{aligned}\boldsymbol{V}_i &= (\lambda(1,i) - \nu(1,i), \lambda(2,i) - \nu(2,i), \ldots, \\ &\quad \lambda(n,i) - \nu(n,i)) \\ &= (V_i^{(1)}, V_i^{(2)}, \ldots, V_i^{(n)}).\end{aligned} \tag{1.4}$$

The extended $(n+1)$-dimensional reaction vector $\boldsymbol{K}_i$, $i = 1, \ldots, r$, is defined as

$$\boldsymbol{K}_i = (Q_i, \boldsymbol{V}_i); \tag{1.5}$$

Example 1.1. Consider the network of sequential reactions

$R_1 : A_1 \to A_2$

$R_2 : A_2 \to P$ (P inert product)

then $\boldsymbol{Y} = (Y_1, Y_2)$, $\boldsymbol{U} = (T, Y_1, Y_2)$, $\boldsymbol{V}_1 = (-1, +1)$, $\boldsymbol{V}_2 = (0, -1)$, $\boldsymbol{K}_1 = (Q_1, -1, +1)$, $\boldsymbol{K}_2 = (Q_2, 0, -1)$.

Example 1.2. Consider the network of competing reactions

$R_1 : A_1 \to P_1$
$R_2 : A_1 \to P_2$ (P_1, P_2 are both inert products)

then $\boldsymbol{U} = (T, Y)$, $\boldsymbol{K}_1 = (Q_1, -1)$, $\boldsymbol{K}_2 = (Q_2, -1)$.

With the above vector notations, the equations for a one-dimensional traveling flame front, in a system of coordinates moving with the front, and in the constant pressure approximation, reduce to the system of $(n+1)$ equations in $T, Y_1, \ldots, Y_n$:

$$-(D(\boldsymbol{U})\boldsymbol{U}_x)_x + M\boldsymbol{U}_x = \sum_{i=1}^{r} \omega_i(\boldsymbol{U})\boldsymbol{K}_i; \tag{1.6}$$

where $D(\boldsymbol{U})$ is a $(n+1) \times (n+1)$ diffusion matrix (positive definite;) M is the *unknown* mass flux, which equals wave velocity if the density is unity. The scalar functions ω_i are the rates of the reactions R_i. The system 1.6 is closed by the perfect gas law,

$$\bar{P}_c = (\rho T)\Big(\sum_j Y_j\Big), \tag{1.7}$$

where $\rho(x)$ is the mixture global density and $\bar{P}_c$ the constant pressure in the isobaric, small Mach number limit. Finally, the cold state

$$\boldsymbol{U}^- = \boldsymbol{U}(-\infty) = (T_-, Y_1^-, \ldots, Y_n^-) \tag{1.8}$$

is prescribed at $x = -\infty$.

The reaction rates ω_i are given by the usual law

of mass action and Arrhenius factors,

$$\omega_i = \prod_{j=1}^{n} Y_j^{\nu(j,i)} B_i \exp\left(-\frac{E_i}{T}\right), \tag{1.9}$$

where E_i is the activation energy of reaction i and B_i the corresponding pre-Arrhenius factor; strictly speaking, B_i is a slowly varying function of T and ρ, hence of T and Y_j through (1.7); we will in the sequel neglect its dependence w.r.t. Y_j, taking for instance the context of a reacting mixture diluted in some inert species.

In this presentation, we omit reversible reactions. The extension to reversible reactions has been completed and will be presented in [9].

We are concerned with exploiting the high activation energies E_i to obtain approximate wave front solutions. The E_i determine scales

$$\varepsilon_i = \frac{T_+}{E_i}\frac{T_+}{T_+ - T_-} \ll 1, \tag{1.10}$$

where T_+ is a typical final (burned) temperature and the ε_i are inverse Zeldovich numbers (T has been renormalized so that the gas constant is unity). The sharp sensitivity of Arrhenius terms $\exp(-E_i/T)$ w.r.t. variations of T enable a systematic High Activation Energy (H.A.E.) asymptotic analysis. The procedures are linked to properties of the underlying reaction network, which supplies the vectors $\boldsymbol{K}_j$. These procedures are well known in the case when the network is modeled by a single reaction A→P. (See [2] and the review article [3] in these proceedings.) An approach similar to the one in this paper was used in treating two-reaction networks of competing, sequential, and competing–fuel types in [7] and [8], and it was clear that the extension of this approach to more complex schemes could uncover a baffling array of possibilities. A systematic formalism was called for, and hopefully was provided in part by the authors in [8].

The methods are those of asymptotic analysis. Roughly speaking, to the lowest order this means developing rational procedures for replacing the original problem with a limit problem in which the ε_i do not appear explicitly. The limit problem is expected to yield an approximate solution, and is easy to solve. In the limit, flame sheets appear as free boundaries with $\boldsymbol{U}$ continuous across, but with jumps in the gradients. A rigorous justification of this expectation has been given recently for the case of a single reaction [1]. Reactions are concentrated on flame sheets (internal layers) with pure convective-diffusive flow in between, if there are no reactions with $E_i = 0$:

$$(D(\boldsymbol{U})\boldsymbol{U}_x)_x - M\boldsymbol{U}_x = 0$$

between the flame sheets. The flame sheets correspond to a Dirac-function in temperature limiting behavior of the ω_i as $\varepsilon_i \to 0$. If there is a mix of reactions with H.A.E. $E_i \gg 1$, $i = 1, \ldots, p$ and with zero activation energies (Z.A.E.) $E_i = 0$, $i = p+1, \ldots, r$, there are still flame sheets (free boundaries) associated to the R_i, $i = 1, \ldots, p$; but the flow in between is convective-diffusive-reactive due to the Z.A.E. reactions. The latter "thicken" the wave front profiles.

Many other parameters besides ε_i enter the problem, for example, the elements of the matrix D and of the unburned state vector of the gas, $\boldsymbol{U}^-$. Throughout the paper, we assume that their magnitudes do not interfere with the asymptotics. Particularly, we assume that any constants not depending explicitly on ε_i, B_i, or E_i (as defined in (1.9) and (1.10)) are $\mathcal{O}(1)$ quantities. This especially refers to elements of D and D^{-1}, reference dimensionless T_j, their differences $T_i - T_j$ and the heat releases Q_i. These restrictions are relaxed, and a more meticulous account of the other parameters is taken in [10].

For background and previous work on asymptotic methods in flame theory, the reader is referred to [2] and [3] in these proceedings. A number of papers have been written on the multiple reaction case; see the references in [2], [4], [12], and [17].

We remark that the consideration of realistic chemical networks avoids the so-called "cold boundary difficulty" [2], where $\boldsymbol{U}^-$ is not an equilibrium state of the chemical network. In realistic chemical networks, the initial zero concentrations of intermediate radicals species insures equilibrium at the cold side; a good example is the autocatalytic chain network investigated by Liñan [12]:

$$R_1: A_1 + A_2 \rightarrow 2A_2$$
$$R_2: 2A_2 + M \rightarrow 2P + M\,, \tag{1.11}$$

where P and M are inert; the initial concentration of the intermediate radical A_2 is null. Here, $\boldsymbol{K}_1 = (Q_1, -1, +1)$ and $\boldsymbol{K}_2 = (Q_2, 0, -2)$.

1.2. *The reactant allocations and suballocations to a chemical network*

The final state on the burned side

$$\boldsymbol{U}^+ = \boldsymbol{U}(+\infty) = \boldsymbol{U}(T_+, Y_1^+, \ldots, Y_n^+) \tag{1.12}$$

is actually coupled with the initial state $\boldsymbol{U}^-$ through conservation laws which are the equivalent of the Rankine–Hugoniot conditions for nonreactive flows. To see this, integrate (1.6) from $x = -\infty$ to $x = +\infty$, use $\boldsymbol{U}_x = 0$ at $x = +\infty$ to obtain

$$M(\boldsymbol{U}^+ - \boldsymbol{U}^-) = \sum_{i=1}^{r} \Omega_i \boldsymbol{K}_i\,, \tag{1.13a}$$

where Ω_i is the integrated rate function for R_i along the flame profile (proportional to the thermal power due to R_i):

$$\Omega_i = \int_{-\infty}^{+\infty} \omega(\boldsymbol{U}(x))\,\mathrm{d}x\,; \tag{1.13b}$$

now setting $\alpha_i = \Omega_i/M$,

$$\boldsymbol{U}^+ - \boldsymbol{U}^- = \sum_{i=1}^{r} \alpha_i \boldsymbol{K}_i\,, \tag{1.14}$$

hence the difference between the asymptotic state vectors lies in the *linear span of the extended reaction vectors* $\boldsymbol{K}_i$.

Moreover, the α_i, $i = 1, \ldots, r$, must satisfy constraints such that the *final concentration vector* $\boldsymbol{Y}^+$ *ensures that the network is in equilibrium*:

$$Y_k^+ = 0 = Y_k^- + \sum_{i=1}^{r} \alpha_i V_i^{(k)} \tag{1.15}$$

for some integer subset $\{k\}$ such that $\omega_i(\boldsymbol{U}^+) = 0$, $\forall i$. Because of (1.14) and (1.15), M must be considered as unknown nonlinear eigenvalue for the general system (1.6).

We will call the sequence of positive or null numbers $\alpha_1, \alpha_2, \ldots, \alpha_r$ an *allocation* of the initial concentration vector $\boldsymbol{Y}^-$ to the network. To make this point of view more systematic, we first need a basic assumption about the reaction vectors:

Assumption 1.
If

$$\sum_{i=1}^{r} \alpha_i V_i = 0 \text{ and } \alpha_i \geq 0, \text{ then } \alpha_i \equiv 0\,. \tag{1.16}$$

This assumption assures, as we shall see, that equilibrium can only be reached when each reaction rate is zero. It can easily be removed to cover reversible reactions [9].

Definition 1. $Y \in \mathscr{E}$ is an *equilibrium state* if, for all T,

$$\sum_{i=1}^{r} \omega_i(\boldsymbol{Y}, T)\boldsymbol{V}_i = 0\,. \tag{1.17}$$

It follows from (1.16) that this can only happen if each $\omega_i = 0$.

For each $i = 1, \ldots, r$, we denote by $J(i)$ the set of all indices j of species A_j entering reaction R_i (appearing on the left side of the arrow). At this point we assume $\omega_i = 0$ if and only if $Y_j = 0$ for some $j \in J(i)$. We symbolize the equilibrium cone for each reaction, and the total equilibrium cone, as follows:

$$\mathscr{E}_i = \{ \boldsymbol{Y} \geqslant 0\colon\ Y_j = 0 \text{ for some } j \in J(i) \} ,$$

$$\mathscr{E} = \bigcap_{i=1}^{r} \mathscr{E}_i .$$

(This is, of course, a cone imbedded in a set of coordinate hyperplanes.)

For simplicity only we require

Assumption 2. $\boldsymbol{V}_i \notin \mathscr{E}_i$, for all i.

Definition 2. Given an input state $\boldsymbol{Y}^- \geqslant 0$, an *allocation* of $\boldsymbol{Y}^-$ to the network is a set of numbers $\alpha_i \geqslant 0$, $i = 1, \ldots, r$, such that

$$\boldsymbol{Y}^- + \sum_{i=1}^{r} \alpha_i \boldsymbol{V}_i \in \mathscr{E} . \tag{1.18}$$

Example. Consider the network of competing reactions described in Example 1.1. Clearly $\mathscr{E} = \{0\}$. The T and Y_1 components at eq. (1.14) are

$$T_+ = T_- + \alpha_1 Q_1 + \alpha_2 Q_2; \quad 0 = Y_1^+ = Y_1^- - \alpha_1 - \alpha_2;$$

but each α_i can take values only between zero and Y_1^- and their sum must be Y_1^-. Setting $\alpha_1 = \theta Y_1^-$, $0 \le \theta \le 1$, we have $\alpha_2 = (1-\theta) Y_1^-$ and $T_+ = T_- + Y_1^-(\theta Q_1 + (1-\theta) Q_2)$.

The final temperature can be any temperature between $T_1 = T_- + Y_1^- Q_1$, adiabatic temperature of a pure reaction R_1, and $T_2 = T_- + Y_1^- Q_2$, adiabatic temperature of a pure reaction R_2.

Indeed this network of competing reactions is the canonical example of *nonuniqueness of allocations.* In more complex networks, there is usually a whole array of possible allocation vectors which are restricted to the boundary of a specific convex set (simplex) defined by

$$\begin{aligned} &\alpha_i \geqslant 0, \quad i = 1, \ldots, r \\ &Y_j^- + \sum_{i=1}^{r} \alpha_i V_i^{(j)} \geqslant 0, \quad j = 1, \ldots, n . \end{aligned} \tag{1.19}$$

(see section 2 for examples of such convex sets). In general, we must consider the final states $\boldsymbol{U}^+(\alpha_1, \ldots, \alpha_r)$ and $T_+(\alpha_1, \ldots, \alpha_r)$ as functions of the α_i. Nonunique allocations reflect the possibility of different chemical pathways in the network.

Another consideration is that experimental and numerical simulations show that within a flame profile, there is a sequence of temperatures where some subgroups of reactions are completed, while some other reactions are still frozen [13]. For instance, in heavy hydrocarbon flames [13] peak concentrations of a typical radical occur at a distinctly different temperature than the temperature band where most of the fuel is consumed [14]. This leads to the notion of suballocations (to be developed in sections 2 and 3) such that a partial subgroup of reactions go to completion at temperatures lower than T_+, within a partial subnetwork.

1.3. *Normalized power functions*

The actual chemical pathways and chemical subnetworks in a flame are determined by the competing rates ω_i; the latter can vary sharply not only in absolute but also in relative magnitude as temperature increases across the flame front. Hence the need for some normalized gauge functions to systematically exploit these relative variations.

Consider now the integrated rate functions Ω_i in eq. (1.13):

$$\Omega_i = \int_{-\infty}^{+\infty} \omega_i(\boldsymbol{U}(x))\, \mathrm{d}x ; \tag{1.20}$$

writing the first component in 1.13a yields the exact relation

$$M(T_+ - T_-) = \sum_{i=1}^{r} Q_i \Omega_i , \tag{1.21}$$

i.e., the relative contribution of each reaction to the temperature profile is its heat release Q_i weighted by its integrated rate function. But if the reaction R_i with *high* activation energy E_i is *isolated* then classical H.A.E. analysis for a single step reaction [2] going to completion at temperature T, gives

$$\Omega_i(T) = C_i \varepsilon_i (B_i \exp - E_i/T)^{1/2}, \tag{1.22}$$

where C_i is some proportionality constant. Also, the rate function ω_i is concentrated in the flame sheet. For networks with partial H.A.E., the ω_i for the H.A.E. reactions are still concentrated in a sequence of flame sheets (possibly merged) at various temperatures and it can be shown with a layer analysis that the Ω_i are still of the order of magnitude given by (1.22).

Rather than (1.22), we shall use the following choice for normalized power gauge functions $H_i(T)$ to compare the relative powers of reactions R_i:

Definition 3. *Power gauge functions*:

$$H_i(T) = \bar{\varepsilon}\left[\ln B_i - \frac{E_i}{T}\right], \tag{1.23a}$$

$$\bar{\varepsilon} = \min \varepsilon_i = \min \frac{T_+}{E_i} \frac{T_+}{T_+ - T_-}, \quad i = 1, \ldots, r. \tag{1.23b}$$

In (1.23a) we allow T to vary, since the flame sheet temperatures are a priori unknown (nonunique allocations). *We also allow the same definitions* (1.23*a*–*b*) *whether reaction* R_i *is H.A.E. or Z.A.E. Indeed*

1) if R_i is H.A.E., then

$$H_i(T) = \bar{\varepsilon}(2 \ln \Omega_i + \mathcal{O}(\ln \bar{\varepsilon})),$$

which ensures that the variation of $H_i(T)$ over the whole flame temperature profile is $\mathcal{O}(1)$;

2) if R_i is Z.A.E., then

$$H_i = \bar{\varepsilon} \ln B_i \approx \text{constant},$$

as $B_i(T)$ varies weakly with T; nevertheless we should keep in mind that physical B_i may be very large indeed, anywhere from 10^{11} to 10^{19}; so the normalized H_i are still $\mathcal{O}(1)$ or smaller.

Remark 1.1. The power gauge functions $H_i(T)$ are monotone increasing in T. Any linear combination of $H_i(T)$ is a monotone (decreasing or increasing) function of T.

Our basic comparison principle is based on the following inequalities: if

$$H_p(T) > H_q(T)$$

at a given T, and the difference is $\mathcal{O}(1)$, then

$$\Omega_p(T) \gg \Omega_q(T)$$

and R_q has negligible effect compared to R_p, at temperature T. An equilibrium configuration might still be obtained for some distinguished temperatures T_p and T_q such that $H_p(T_p) = H_q(T_q)$ and $\Omega_p(T_p)/\Omega_q(T_q) = \mathcal{O}(1)$. We will detail the systematic use of the comparison principle to construct flame fronts in sections 2 and 3.

As a simple example, we show how the power gauge functions resolve the nonuniqueness of allocations in the competing reactions network (example 1.1). Recall that T_+ can vary anywhere between T_1 and T_2 (single reaction adiabatic temperatures). Let $E_1 \gg 1$, $E_2 \gg 1$ and $E_1/E_2 = \mathcal{O}(1)$; three cases are possible:

1) $H_1(T) > H_2(T)$ for all $T \in [T_1, T_2]$; then R_1 globally quenches R_2, and there is a single flame sheet at T_1; the allocation is $(Y_1^-, 0)$;

2) $H_2(T) > H_1(T)$ for all $T \in [T_1, T_2]$; then there is a single flame sheet at T_2, with allocation $(0, Y_1^-)$;

3) $H_1(T_*) = H_2(T_*)$ for some T_*, $T_1 < T_* < T_2$; T_* is unique because of remark 1.1, and the two reactions are at equilibrium at the flame sheet temperature $T_+ \equiv T_*$. With $T_* = \theta T_1 + (1-\theta)T_2$, $0 < \theta < 1$, the allocations are $\alpha_1 = \theta Y_1^-$, $\alpha_2 = (1-\theta) Y_2^-$. The full layer analysis of this competition flame has been done by the authors in [7]. In section 2, we show that competition type flame regimes can in fact be generic for more complex networks, due to competing subnetworks.

Finally extending formula (1.23) to Z.A.E. reactions ensures that our comparison principle overlaps classical pseudo-steady state analysis: if $H_p(T) > H_q(T)$ for R_p and R_q both Z.A.E.. then for all temperature ranges $\omega_p \gg \omega_q$ (as H_p and H_q are nearly constant) and the usual pseudo-steady state reduction of R_p can be done.

2. A global hydrocarbon two-step model

2.1. *The two-step Dryer–Westbrook model*

Hydrocarbon flames are usually modelled by global one-step models:

$$\text{Fuel} + n_1\,O_2 \rightarrow n_2\,CO_2 + n_3\,H_2O\,, \tag{2.1a}$$

where n_1, n_2, n_3 are positive integers, and the global reaction rate ω is

$$\omega = B(T)\exp\left(-\frac{E}{RT}\right)[\text{Fuel}]^{\alpha}[O_2]^{\beta}; \tag{2.1b}$$

the pre-Arrhenius factor B and the exponents α and β are usually adjusted to fit calculated flame parameters to some experimental data; α and β turn out to be positive *fractional* numbers. As pointed out in [15], the single-step model predicts flame speeds reliably over a considerable range of reactions; but it systematically overpredicts the total heat of reaction and the final adiabatic temperature. Moreover, at temperatures characteristic of hydrocarbon fuels (≈ 2000 K), substantial amounts of CO and H_2 exist in equilibrium in the combustion products with CO_2 and H_2O. It is also well recognized that typical hydrocarbons burn in a sequential manner; i.e., the fuel is partially oxidized to CO and H_2, which are not appreciably consumed until all of the hydrocarbon species have disappeared [6]. From these observations, and following a similar model for methane by Dryer and Glassmann [5], Westbrook and Dryer [15] have proposed the following global two-step reaction mechanism:

$$R_1 : C_nH_m + \left(\frac{n}{2} + \frac{m}{4}\right)O_2 \rightarrow n\,CO + \frac{m}{2}\,H_2O\,,$$
$$R_2 : CO + \tfrac{1}{2}O_2 \rightleftarrows CO_2\,, \tag{2.2}$$

with empirically derived reaction rates for both reactions; for R_2:

$$\omega_2 = 10^{14.6}\exp\left(-\frac{40}{RT}\right)[CO]^1[H_2O]^{0.5}[O_2]^{0.25} \tag{2.3a}$$

and for R_1:

$$\omega_1 = B_1\exp\left(-\frac{E_1}{RT}\right)[\text{Fuel}]^{\alpha}[O_2]^{\beta}\,. \tag{2.3b}$$

For Octane C_8H_{18}, $B_1 = 5.7 \times 10^{11}$, $E_1 = 30.0$ kcal/mol, $\alpha = 0.25$ and $\beta = 1.5$. In general, for all hydrocarbons except methane CH_4, $E_1 \approx 30$ kcal, and $B_1 \ll B_2 = 10^{14.6}$. So

Remark 2.1. For the global two-step model (2.2), the power gauge functions $H_1(T)$ and $H_2(T)$ *always intersect* at some T_* such that

$$\bar{\varepsilon}\left(\ln\frac{B_2}{B_1}\right) = \bar{\varepsilon}\left(\frac{E_2 - E_1}{T_*}\right), \tag{2.4}$$

since $B_2 \gg B_1$ and $E_2 - E_1 \approx 10$ kcal. Competing flames and subnetworks are possible.

In the sequel, we systematically investigate the simplified network

$$R_1 : \text{Fuel} + \text{Oxygen} \rightarrow R + P_1\,,$$
$$R_2 : R + \text{Oxygen} \rightarrow P_2\,, \tag{2.5}$$

where P_1 and P_2 are inert products (e.g., H_2O and CO_2) and R is an intermediate "radical" species (e.g., CO). We take Y_f, Y_{ox} and Y_r to be the respective (rescaled) concentrations of fuel, oxygen, and radical. We take the simplified extended reaction vectors (stoichiometric coefficients set equal to one):

$$\boldsymbol{K}_1 = (Q_1, -1, -1, +1)\,,$$
$$\boldsymbol{K}_2 = (Q_2, 0, -1, -1)\,, \tag{2.6}$$

with both $Q_i > 0$. Notice the alternation ± 1 for the last component, as the Radical is produced by R_1, but consumed by R_2. The flame front equations for the state vector $\boldsymbol{U} = (T, Y_f, Y_{ox}, Y_r)$ are

$$-(D(U)U_x)_x + MU_x = \omega_1(U)\boldsymbol{K}_1 + \omega_2(U)K_2\,, \quad (2.7)$$

with the reaction rates

$$\omega_1 = B_1 Y_f Y_{ox} \exp(-E_1/T)\,,$$
$$\omega_2 = B_2 Y_{ox} Y_r \exp(-E_2/T)\,, \quad (2.8)$$

and

$$E_1 \gg 1, \quad E_2 \gg 1, \quad \frac{E_2}{E_1} = \mathcal{O}(1)\,.$$

The initial state is

$$\boldsymbol{U}^- = (T_-, a, b, 0)\,, \quad (2.9)$$

with $Y_f^- = a$, $Y_{ox}^- = b$ and $Y_r^- = 0$ (which is the physical situation). Notice that we do not assume reversibility of R_2; the reversible case is fully treated in [9].

2.2. *The convex set of admissible allocations*

Let α_1 and α_2 be the allocations to R_1 and R_2 with $\boldsymbol{\alpha} = (\alpha_1, \alpha_2)$; possible final states are

$$\boldsymbol{U}^+(\boldsymbol{\alpha}) = \boldsymbol{U}^- + \alpha_1 \boldsymbol{K}_1 + \alpha_2 \boldsymbol{K}_2,$$

i.e.

$$T_+(\boldsymbol{\alpha}) = T_- + \alpha_1 Q_1 + \alpha_2 Q_2\,, \quad (2.10)$$

$$Y_f^+ = a - \alpha_1 \geqslant 0\,,$$
$$Y_{ox}^+ = b - \alpha_1 - \alpha_2 \geqslant 0\,, \quad (2.11)$$
$$Y_r^+ = \alpha_1 - \alpha_2 \geqslant 0\,;$$

inequalities (2.11) together with the conditions $\alpha_1 \geqslant 0$, $\alpha_2 \geqslant 0$ define the convex set of eventual allocations in the positive quadrant α_1, α_2; but we must require that equilibrium be attained. Since the reactions are not reversible, chemical equilibrium corresponds to some points (α_1, α_2) on the boundary of this simplex. However, equilibrium for the chemical network can be reached in two different ways: first *both* $Y_f^+ = 0$ and $Y_r^+ = 0$, which corresponds to an oxygen rich initial mixture. Or $Y_{ox}^+ = 0$, which corresponds to a fuel rich mixture. More precisely:

1) In the oxygen rich case (fig. 1), $a < b/2$ and $Y_f^+ = a - \alpha_1 = 0$, $Y_r^+ = \alpha_1 - \alpha_2 = 0$, so

$$\alpha_1 = \alpha_2 = a = Y_f^-\,; \quad (2.12)$$

the final state is

$$Y_f^+ = 0\,, \quad \text{(fuel depletion)}\,,$$
$$Y_{ox}^+ = b - 2a > 0\,, \quad \text{(excess oxygen)}\,, \quad (2.13)$$
$$Y_r^+ = 0\,,$$

(the radical concentration pulses to a maximum and then decreases to zero);

$$T^+ = T^- + a(Q_1 + Q_2)\,. \quad (2.14)$$

For the oxygen rich case, notice that the admissible allocation (a, a) is *unique* (edge of the convex set). The case $a = b/2$ corresponds to an initially stoichiometric concentration $\boldsymbol{Y}^-$.

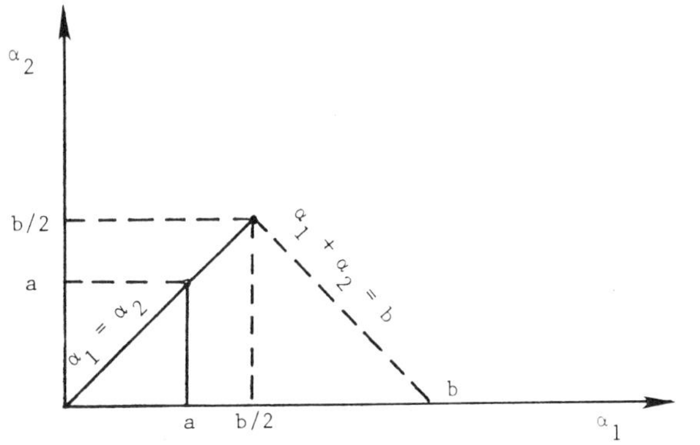

Fig. 1. The oxygen rich case, $a < b/2$.

2) In the fuel rich case (fig. 2), $a > b/2$, $Y^+_{ox} = b - \alpha_1 - \alpha_2 = 0$; since $\alpha_2 \geqslant 0$, we have $b/2 \leqslant \alpha_1 \leqslant b$, in fact:

$$b/2 \leqslant \alpha_1 \leqslant \min(a, b)\,; \tag{2.15}$$

the final allocations are

$$\begin{aligned} Y^+_f &= a - \alpha_1\,, \quad \text{(excess fuel)}\,, \\ Y^+_{ox} &= 0\,, \quad \text{(oxygen depletion)}\,, \\ Y^+_r &= 2\alpha_1 - b\,. \end{aligned} \tag{2.16}$$

There usually is some residual radical concentration on the burned side of the flame. The key feature of the fuel rich case is that *allocations are not unique*: (α_1, α_2) covers a whole closed interval on the line $\alpha_1 + \alpha_2 = b$, with (2.15) and

$$b - \min(a, b) \leqslant \alpha_2 \leqslant b/2\,.$$

The final temperature is now parameterized by α_1:

$$T_+(\alpha_1) = T_- + \alpha_1 Q_1 + (b - \alpha_1) Q_2\,; \tag{2.17}$$

to fix ideas, take $Q_2 \geq Q_1$ and $a \leqslant b$, so $\min(a, b) \equiv a$; then $T_+(\alpha_1)$ is a *decreasing* function of α_1, which reaches its minimum T_{min} at $\alpha_1 = a$ and its maximum T_{max} at $\alpha_1 = b/2$,

$$\begin{aligned} T_{min} &= T_- + aQ_1 + (b - a)Q_2\,, \\ T_{max} &= T_- + b/2(Q_1 + Q_2)\,. \end{aligned} \tag{2.18}$$

2.3. *Flame configurations in the oxygen rich case* $(a \leqslant b/2)$

This is characterized by a unique final temperature (2.14),

$$T_+ = T_- + a(Q_1 + Q_2)$$

and a unique final allocation $\boldsymbol{\alpha} = (a, a)$. Flame configurations are determined by the relative ordering of the power gauge functions (1.23) $H_1(T)$ and $H_2(T)$ at $T = T_+$. We consider the case $H_1(T_+) = H_2(T_+) + \mathcal{O}(\bar{\varepsilon})$ as nongeneric; and in fact, the inequalities introducing cases 1–6 are meant to imply that the two sides differ by a quantity $> \mathcal{O}(\bar{\varepsilon})$.

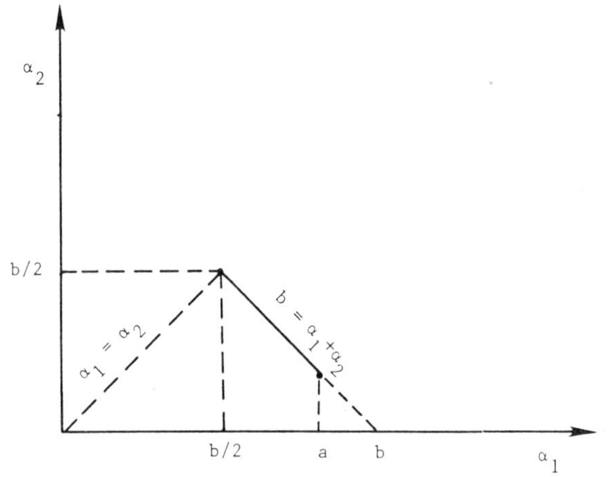

Fig. 2. The fuel rich case, $a > b/2$.

Case 1. $H_1(T_+) > H_2(T_+)$: *Sequential flames*

R_1 would tend to dominate R_2 at T_+; what we mean is that if R_1 went to completion at T_+, then R_2 would have negligible rate in comparison, and we would appear to have a one-step reaction. This would contradict the fact that both reactions must operate in order to achieve T_+. A stable flame configuration exists only if R_1 is completed in some flame sheet of temperature $T_1 < T_+$, such that the

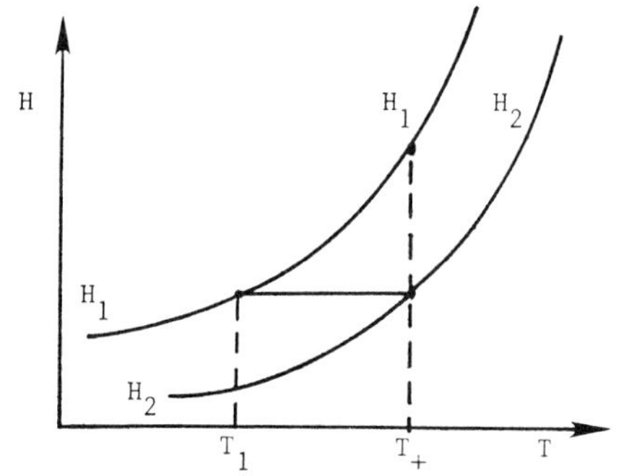

Fig. 3. Case 1.

power H_1 is *lowered to a power equilibrium configuration*,

$$H_1(T_1) = H_2(T_+) ;$$

such a temperature T_1 exists and is unique because of remark 1.1.

However, to be physically attainable, T_1 must be larger than the adiabatic temperature $T_- + aQ_1 = \hat{T}_1$ associated with R_1 in isolation [7, 8]. For simplicity, we assume that is so; if not, total equilibrium is only attained in stages (2 successive flames travelling at different velocities), rather than in a single steady flame.

The flame configuration is sequential with R_1 completed in a flame sheet (1) at T_1, where all the fuel is burned out; this is followed by a flame sheet at T_+, where R_2 consumed all the radical created by the sheet (1). The temperature sequence (T_1, T_+) equates the power rates and induces a stable chemical pathway along the flame profile: first R_1 with R_2 frozen, then R_2 with R_1 completed. We remark that there is a *suballocation* $\alpha^{(1)} = (a, 0)$ to the flame sheet (1), and a final allocation $\alpha^{(2)} = \alpha = (a, a)$ to the sheet (2).

Remark 2.1. In practice the flame profile will be thickened because of Z.A.E. involving other radicals.

Remark 2.2. The principle of determining a sequence of temperatures (possibly equal) to equate all the power gauge functions H_i, systematically induces stable chemical pathways and subnetworks in general complex networks.

Case 2. $H_2(T_+) > H_1(T_+)$: *Pseodoreaction* $\tilde{R}_1$

R_2 is now much faster than R_1 at T_+. We might think of R_2 first taking place at some $T_2 < T_+$, such that $H_2(T_2) = H_1(T_+)$, followed by R_1 at T_+. However, the corresponding suballocation would be (0, 0); that is, such a separated flame sheet cannot exist, as there is no initial radical concentration ($Y_r^- = 0$) to trigger it. The only possible structure is a single flame layer at T_+, with a pseudo-reaction $\tilde{R}_1$ characterized by the same rate function ω_1, but with an instant heat release $Q_1 + Q_2$. The radical produced by $\tilde{R}_1$ is instantly burned by R_2 within the same flame layer. Strictly speaking, this is a situation with merged layers, but to *lowest order*, R_2 does not affect the layer analysis at T_+.

2.4. *Flame configurations in the fuel rich case* ($a > b/2$)

Since allocations are not unique (fig. 2), we must compare $H_1(T)$ and $H_2(T)$ over the *whole range* $[T_{min}, T_{max}]$ as defined in (2.17)–(2.18). Hence, a larger variety of possible configurations than in the oxygen rich situation.

Case 3. $H_1(T) > H_2(T)$ *over* $[T_{min}, T_{max}]$: *sequential flames*

R_1 is much faster than R_2 over the whole permissible range of final temperatures. To determine allocations uniquely, we introduce the following physical principle:

Maximum Allocation Principle. If $H_j(T) > H_i(T)$ *over some permissible range of final temperatures, then from all possible allocations, a stable chemical pathway maximizes* α_j *for reaction* R_j *at the expense of* α_i.

In this case, maximizing α_1 for R_1 uniquely yields

$$\alpha_1 = a, \quad \alpha_2 = b - a ,$$

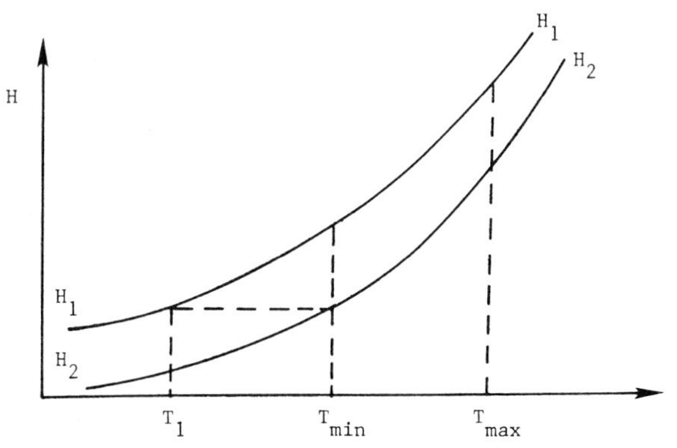

Fig. 4. Case 3.

i.e., the rightmost point on the interval of possible allocations (fig. 2). Moreover, $T_+ \equiv T_{\min} = T_- + aQ_1 + (b-a)Q_2$, and $Y^+ = (0, 0, 2a - b)$. Since at $T_{\min}$, R_1 is still much faster than R_2, the situation is similar to case 1. R_1 is completed in a flame sheet (1) at T_1 such that

$$H_1(T_1) = H_2(T_{\min}),$$

while R_2 is completed in a flame sheet (2) at $T_{\min}$. (Again, we assume that $T_1 > \hat{T}_1 = T_- + aQ_1$.)

Case 4. *$H_2(T) > H_1(T)$ over $[T_{\min}, T_{\max}]$: pseudo-reaction $\tilde{R}_1$*

R_2 is much faster than R_1 over the whole permissible range of final temperatures. Applying the Maximum Allocation Principle to R_2 maximizes α_2:

$$\alpha_2 = b/2, \quad \text{hence } \alpha_1 = b/2,$$

i.e., the leftmost point on the interval of possible allocations (fig. 2) and $T_+ \equiv T_{\max} = T_- + b/2 \times (Q_1 + Q_2)$, $Y^+ = (a - b/2, 0, 0)$. Since at $T_+ \equiv T_{\max}$, R_2 is still much faster than R_1, the flame configuration is similar to case 2: pseudo-reaction $\tilde{R}_1$ at $T_{\max}$, with instant heat release $Q_1 + Q_2$.

Case 5. *Genuine (stable) competition flame*

Here, there exists a unique T_*, $T_{\min} < T_* < T_{\max}$, where both power gauge functions are equal,

$$H_1(T_*) = H_2(T_*);$$

α_1 is uniquely determined by (2.17),

$$T_* \equiv T_+ = T_- + \alpha_1 Q_1 + (b - \alpha_1)Q_2.$$

In this fuel rich context, there is genuine competition between the subnetworks R_1 and R_2 for the initial oxygen concentration.

The layer analysis at T_* is entirely similar to the canonical example 1.2 of competing reactions [7], hence the genericity of the latter. To insure stability of this competition flame, we need

$$H_2(T_{\min}) > H_1(T_{\min})$$

and

$$H_1(T_{\max}) > H_2(T_{\max}).$$

The reasons will be made clear by the next case.

Case 6. *A non-uniqueness case*

This is again the case of intersecting power curves at T_*, but now

$$H_2(T_{\max}) > H_1(T_{\max})$$

and

$$H_2(T_{\min}) < H_1(T_{\min}).$$

We have *three flame configurations possible*: a competition flame at T_* (as in case 5); sequential

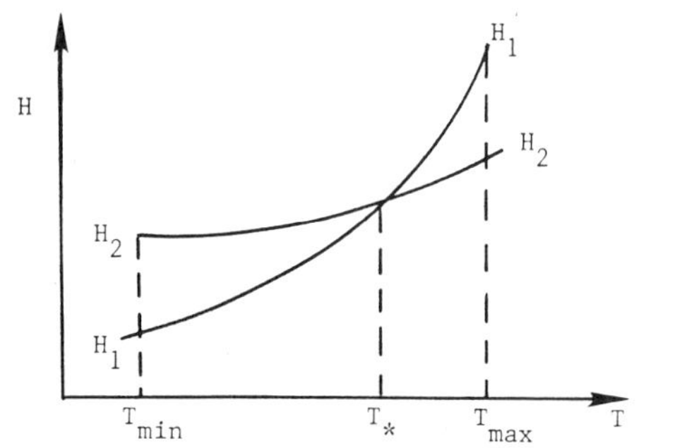

Fig. 5. Case 5.

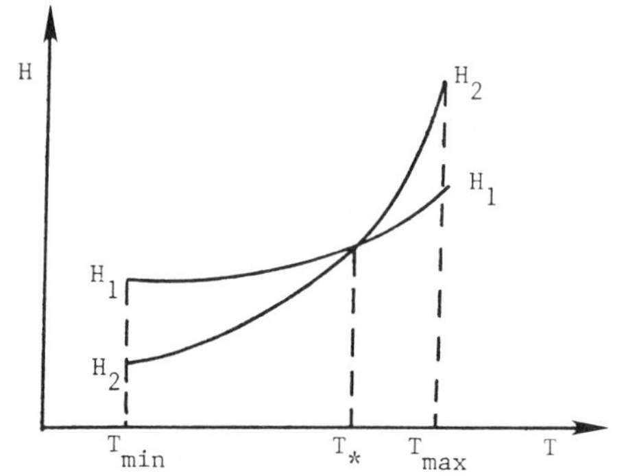

Fig. 6. Case 6.

flames with R_2 completed at T_{min} (as in case 3); or a pseudo-reaction $\tilde{R}_1$ with a merged single flame layer at T_{max} (as in case 4).

We conjecture that the competition flame at T_* is unstable for reasons given in [7]; and that both flames at T_{min} and T_{max} are possible, depending on initial transient regimes.

The previous analysis has been done with the assumption $b/2 < a < b$, to simplify the algebra. The same cases occur for $a > b$.

3. Extending the method to general networks

The detailed study of the hydrocarbon model evidences several features generic to complex networks:

- depending on whether the initial mixture is fuel rich or fuel lean, the same network can exhibit tandem (sequential) flame sheets, or genuine competition flame sheets; or merged flame sheets dominated by a single reaction (the other ones being quasi-instantaneous);
- for competition flames to occur, we need not only power gauge functions to intersect at some common temperature, but also a *whole range* of possible allocations; this is possible only if some deficient reactant is common to the two competing subnetworks;
- for some power gauge curve configurations, three flame regimes are a priori possible; such non-uniqueness can only be resolved by a dynamical stability analysis, and is reminiscent of non-uniqueness and hysteresis in isothermal chemical reactors.

With these points in mind, we will survey the extension of the method to general networks (without reversible reactions).

Remark 3.1. We consider the case of *three* power gauge functions intersecting at the same T_* as *non-generic*.

Remark 3.2. For networks with reversible reactions, another type of flame sheet can appear, where the sheet temperature locks onto the equilibrium temperature of some reversible reaction [9].

3.1. *The hierarchy of power rate functions*

Let an allocation be given:

$$\boldsymbol{\alpha} = (\alpha_1, \alpha_2, \ldots, \alpha_r) \tag{3.1}$$

and

$$T_+(\boldsymbol{\alpha}) = T_- + \sum_{i=1}^{r} \alpha_i Q_i \tag{3.2}$$

Our basic principle is to construct a temperature sequence

$$T_1, T_2, \ldots, T_r$$

such that power gauge functions are *equal* along this temperature sequence (fig. 7). Specifically,

1) set $T_r \equiv T_+(\boldsymbol{\alpha})$;

2) pick the reaction R_m which has the *minimal* power at T_+: $H_m(T_+) = \min H_i(T_+)$, $i = 1, \ldots, r$, $\alpha_i \neq 0$; (3.3)

3) define the T_i by

$$H_i(T_i) = H_m(T_+),\ 1 \leqslant i < r,\ i \neq m .$$

There exists a unique temperature sequence $\{T_i\}$ satisfying (3.3) for a given α; this is ensured by remark 1.1. We now *reorder* the reactions in order of increasing T_i, so that

$$T_1 \leqslant T_2 \leqslant \cdots \leqslant T_r = T_+ , \tag{3.4}$$

where the new T_r corresponds to the former reaction R_m. The new sequence $\{T_i\}$ defined in (3.4) equates the power rates of the reactions. It induces a hierarchy within the reactions and generates a stable chemical pathway in the flame profile. It is entirely possible that some of the T_i coincide, corresponding to competition flame sheets (cf.

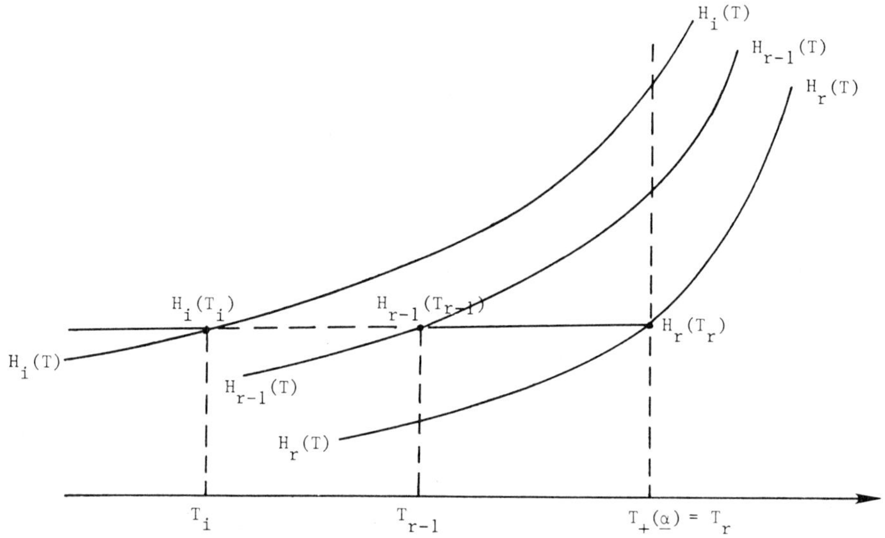

Fig. 7. Temperature sequence such that power gauge functions are equal along this sequence.

cases 5 and 6, section 2.4). Fig. 8 shows such a configuration. Here the possible allocations are a priori nonunique, with an allowable range for $\boldsymbol{\alpha}$ on the boundary of the convex set of allocations. T_+ and $\boldsymbol{\alpha}$ are actually determined by setting $T_+ = T_*$, intersection of the power gauge curves of R_r and R_{r-1}, the two reactions with the *smallest power gauge functions* over the interval $[T_{min}, T_{max}]$ of possible final temperatures.

To cover such a situation, let $\{\tilde{T}_i\}$, $i = 1, \ldots, p \leqslant r$ be the possibly shorter sequence of *distinct* temperatures among the T_i in (3.4):

$$\tilde{T}_1 < \tilde{T}_2, \ldots, < \tilde{T}_{p-1} < \tilde{T}_p = T_+ ; \tag{3.5}$$

we denote their number by p.

It also may happen that two of the T_j almost coincide, that is, $0 < |T_i - T_j| = \mathcal{O}(\varepsilon_i)$. This case is

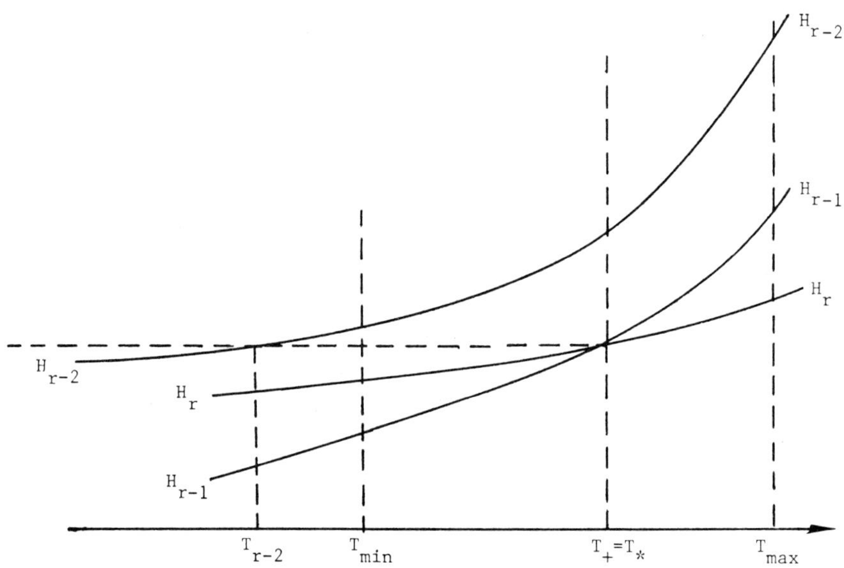

Fig. 8. Competition flame sheets.

considered non generic, and is hereby excluded from consideration. Thus, we assume that two temperatures T_i, T_j either coincide (this may be required by the feasibility criterion given later) or differ by a quantity of larger order than $\varepsilon_i + \varepsilon_j$. We also assume that at each $\tilde{T}_i$, all but perhaps two of the quantities $H_k(\tilde{T}_i)$ are distinct.

Our attempt will be to construct a combustion front in which, roughly speaking, p sets of reactions go to completion at the temperatures $\tilde{T}_i$ in that same order as one proceeds across the front's profile in the (upstream) direction of increasing temperatures. For that purpose, we use the notion of suballocation.

3.2. *Suballocations to subnetworks*

Definition 4. For each $l \leqslant p$, suballocations $\alpha^{(l)}$ are defined to be allocations for the subnetwork consisting only of the R_i for $\tilde{T}_i \leqslant \tilde{T}_l$. Let k_l be the number of such reactions.

Thus, $\alpha^{(l)} = (\alpha_1^{(l)}, \ldots, \alpha_{k_l}^{(l)})$, but for comparison purposes, we extend them to be sequences of length r by defining $\alpha_i^{(l)} = 0$ for $i > k_l$. We say $\boldsymbol{\alpha}^{(l)} \leqslant \boldsymbol{\alpha}^{(l')}$ if $\alpha_i^{(l)} \leqslant \alpha_i^{(l')}$ for all i.

The situation we envisage is when there exist suballocations $\boldsymbol{\alpha}^{(l)}$ for all $l \leqslant p$ satisfying

$$\boldsymbol{\alpha}^{(l)} \leqslant \boldsymbol{\alpha}^{(l+1)}, \quad l = 1, \ldots, p-1; \quad \boldsymbol{\alpha}^{(p)} = \boldsymbol{\alpha}. \tag{3.6}$$

There is a further condition which we shall require the $\boldsymbol{\alpha}^{(l)}$ to satisfy which is a *generalized maximum allocation principle*: Let us fix $l \leqslant p$ and $\alpha^{(l-1)}$. There may be many suballocations $\boldsymbol{\alpha}^{(l)} \geqslant \boldsymbol{\alpha}^{(l-1)}$). We say that $\boldsymbol{\alpha}^{(l)}$ is "maximal" if it allocates a maximal amount of reactants to those reactions with the largest values of H at the temperature $\tilde{T}_l$. This, simple put, is the principle that among competing reactions, the one with the largest rate (as defined by the functions $H_i(T)$) wins out. The precise definition is as follows. First, suppose all the quantities $H_k(\tilde{T}_l)$ ($\tilde{T}_l$ fixed) are distinct. Let k_1 be the index that yields the largest of them; then k_2 the next largest, etc. Out of all possible suballocations $\boldsymbol{\alpha}^{(l)} \geqslant \boldsymbol{\alpha}^{(l-1)}$, $\alpha_{k_1}^{(l)}$ is required to be maximal. With it known, $\alpha_{k_2}^{(l)}$ is then maximal, etc. In this case, $\boldsymbol{\alpha}^{(l)}$ is uniquely determined. Next, suppose two of the $H_k(\tilde{T}_l)$ coincide, say $H_{k_1}(\tilde{T}_l) = H_{k_2}(\tilde{T}_l)$. Then instead of maximizing $\alpha_{k_1}^{(l)}$, we maximize $\alpha_{k_1}^{(l)} + \alpha_{k_2}^{(l)}$. In this respect, we generalize the principle introduced in section 2, cases 3 and 4.

For such suballocations, k_l sets of reactions go to completion at $\tilde{T}_l$. We can now determine the temperature at which the suballocation $\boldsymbol{\alpha}^{(l)}$ would bring the corresponding subnetwork to equilibrium, if no reaction outside that subnetwork were present. We call it

$$\hat{T}_l = T_- + \sum_{i=1}^{k_l} \alpha_i^{(l)} Q_i\,. \tag{3.7}$$

We are now ready to give the exact mathematical criterion to select from all possible final allocations those which are *feasible*:

Definition 5. The allocation $\boldsymbol{\alpha}$ is *feasible* if

1) there exists for each $l \leqslant p$, suballocations $\boldsymbol{\alpha}^{(l)}$ which are maximal and satisfy $\alpha^{(p)} = \alpha$, and

2) $\tilde{T}_i > \hat{T}_i$ for $i < p$.

The last condition ensures that there is a flame front with tandem sheets traveling at the same speed. If condition (2) is not satisfied, then it may be possible to find two or more fronts traveling at *different* speeds, such that the combined effect of the faster followed by the slower is to attain the burned state $U^+(\boldsymbol{\alpha})$, the suballocations going to completion in the prescribed order.

Example. Consider again case 3 in the fuel rich situation of section 2. There, $\boldsymbol{\alpha} = (a, b-a)$ is determined so that $\alpha_1 = a$ maximizes allocations for the reaction R_1 with maximal power over $[T_{min}, T_{max}]$. This fixes $T_+ \equiv T_{min}$. This suballocation $\boldsymbol{\alpha}^{(1)} = (a, 0)$ is such that $H_1(T_1) = H_2(T_{min})$, and

$$\boldsymbol{\alpha}^{(1)} = (a, 0) < \boldsymbol{\alpha} = (a, b-a)\,;$$

the allocation is feasible if:

$$\hat{T}_1 = T_- + aQ_1 < T_1 ,$$

in which case a steady tandem profile exists.

Once feasible suballocations have been determined it remains to analyze the mathematical asymptotic limit problems, that is:

1) solve the various internal layer equations for each flame sheet;

2) determine in a self-consistent manner the transmission conditions between different sheets (in particular their geometric separation);

3) determine the mass flux eigenvalue M.

For details, we refer to our papers [9] and [10]. We simply remark that if only H.A.E. reactions are involved, then the *last* layer equation on the burned side (T_+) uniquely determines M. If this layer involves a single pseudoreaction, M is given by the usual explicit formulas [17, 2]. If it involves a competition flame layer, the analysis is more complex and only implicit formulas exist [7]. Nevertheless, successive flame sheet separations are calculated working backwards from the burned side. In any case, the qualitative structure of the profile can be immediately found without performing layer analysis; and if only H.A.E. reactions are present, then only one layer analysis is necessary to determine M.

3.3. *Some remarks on the methodology for Z.A.E. reactions*

We briefly review how our methodology extends to a mix of H.A.E. and Z.A.E. reactions on the simple chain reaction model (1.11) investigated at length by Liñan [12]:

$$\begin{aligned} &R_1 : A_1 + A_2 \rightarrow 2A_2 , \\ &R_2 : 2A_2 + M \rightarrow 2P + M , \end{aligned} \qquad (1.11)$$

with $\boldsymbol{K}_1 = (Q_1, -1, +1)$, $\boldsymbol{K}_2 = (Q_2, 0, -2)$, $Q_1 > 0$, $Q_2 > 0$. The branching reaction R_1 has H.A.E. $E_1 \gg 1$; whereas the recombination reaction R_2 has Z.A.E. The initial concentration of the radical A_2 is $Y_2^- = 0$. The *unique* final allocation is

$$\alpha_1 = Y_1^-, \quad \alpha_2 = Y_1^-/2 ,$$

with final temperature $T_+ = T_- + Y_1^-(Q_1 + Q_2/2)$. Notice that the adiabatic temperature of a pure flame (1) involving only R_1 is $T_1 = T_- + Y_1^- Q_1$.

By considering the normalized power gauge functions,

$$H_1(T) = \bar{\varepsilon}\left(\ln B_1 - \frac{E_1}{T}\right),$$

$$H_2(T) = \bar{\varepsilon} \ln B_2 ,$$

$$\bar{\varepsilon} = \varepsilon_1 = \frac{T_+}{E_1} \frac{T_+}{T_+ - T_-},$$

and comparing them over the temperature interval $[T_1, T_+]$, we immediately exhibit the three flame structures studied by Liñan:

1) if $H_2 > H_1$ over $[T_1, T_+]$, there is a merged flame structure at T_+; the radicals do not diffuse out of the flame sheet;

2) if $H_1 > H_2$ over the interval, we have a pure flame (1) at T_1, followed by very slow radical recombination; this is reflected by a *suballocation*:

$$\boldsymbol{\alpha}^{(1)} = (Y_1^-, 0) ,$$

3) if $H_1(T_*) = H_2(T_*)$ at some unique T_* in the interval, we have a competition flame situation: a thin layer at T_* for R_1, imbedded in a thick radical flame.

The general methodology for H.A.E. plus Z.A.E. reactions will be presented elsewhere. We simply remark that the generic flame cases evidenced in section 2 are still present, provided that one allows for thick radical flames.

Acknowledgements

This research partially supported by the Center for Nonlinear Studies, Los Alamos National Laboratory; work also performed under the auspices of the U.S. Department of Energy under contract W-7405-ENG-36 and contract KC-04-02-01, Division of Basic and Engineering Sciences; and partial support was provided by NSF Grant number MCS-8202056.

References

[1] H. Berestycki, B. Nicolaenko and B. Scheurer, Traveling Wave Solutions to Reaction Diffusion Systems Modeling Combustion, Contemporary Mathematics, vol. 17 (A.M.S., Providence, 1983), pp. 189–207.

[2] J. Buckmaster and G.S.S. Ludford, Theory of Laminar Flames (Cambridge Univ. Press, London, 1982).

[3] J. Buckmaster, Physica 12D (1984) 173 (these proceedings).

[4] P. Clavin, Dynamical Behavior of Premixed Flame Fronts in Laminar and Turbulent Flow, Progress in Energy and Combustion Science, to appear, 1984.

[5] F.L. Dryer and I. Glassman, High-temperature oxidation of CO and CH_4, Fourteenth Symposium (International) on Combustion (The Combustion Institute, Pittsburgh, 1972), p. 987.

[6] F.L. Dryer and C.K. Westbrook, "Chemical kinetic modelling for combustion applications," presented at the Propulsion and Energetics Panel 54th Specialists Meeting, Cologne, West Germany, Oct. 1979. NATO AGARD Conference Proceedings No. 275. University of California Lawrence Livermore National Laboratory report UCRL-8177, September 1979.

[7] P.C. Fife and B. Nicolaenko, "The singular perturbation approach to flame theory with chain and competing reactions," Ordinary and Partial Differential Equations, W.N. Everitt and B.D. Sleeman, eds., Lecture Notes in Mathematics, No. 964 (Springer, Berlin, 1982) pp. 232–250.

[8] P.C. Fife and B. Nicolaenko, Asymptotic Flame Theory with Complex Chemistry, Contemporary Mathematics, vol. 17 (A.M.S., Providence, 1983), pp. 235–255.

[9] P.C. Fife and B. Nicolaenko, "Asymptotic Flame Theory with Reversible Reaction Networks," Proc. I.N.R.I.A. Conf. on Combustion, Paris, 1983, to appear.

[10] P.C. Fife and B. Nicolaenko, "Asymptotic Flame Theory with Complex Chemical Networks," in preparation.

[11] D.A. Frank-Kamenetskii, Diffusion and Heat Transfer in Chemical Kinetics (Plenum, New York, 1965).

[12] A. Liñan, "A theoretical analysis of premixed flame propagation with an isothermal chain reaction," Instituto Nacional de Tecnica Aerospacial "Esteban Terradas" (Madrid), USAFOSR Contract No. E00AR 68-0031, Technical Report No. 1, 1971. Also P. Clavin and A. Liñan, to appear.

[13] C.K. Westbrook and F.L. Dryer, "Chemical Kinetics Modelling of Hydrocarbon Combustion," University of California Lawrence Livermore National Laboratory report UCRL-88651, 1983.

[14] C.K. Westbrook and W.J. Pitz, "Prediction of Laminar Flame Properties of Propane-Air Mixtures," Proc. 9th International Colloquium on Dynamics of Explosions and Reactive Systems, Poitiers, France, 1983, to appear.

[15] C.K. Westbrook and F.L. Dryer, Simplified Reaction Mechanisms for the Oxydation of Hydrocarbon Fuels in Flames, Combustion Sc. and Tech. 27 (1981) 31–44.

[16] Ia.B. Zeldovich, Theory of Flame Propagation, J. Phys. Chem. 22 (1948) 27–48.

[17] Ia.B. Zeldovich, G.I. Barenblatt, V.B. Librovich and G.M. Mahviladze, Mathematical Theory of Combustion and Detonation (Nauka, Moscow, 1980).

Physica 12D (1984) 198–211
North-Holland, Amsterdam

COMPUTATIONAL MODELING OF FLAME STRUCTURE*

Robert J. KEE
Applied Mathematics Division, Sandia National Laboratories, Livermore, CA 94550, USA

and

James A. MILLER
Combustion Chemistry Division, Sandia National Laboratories, Livermore, CA 94550, USA

In this paper we discuss the need to model the detailed structure of a flame. That is, we argue the value of tracing the elementary reaction steps that are responsible for the creation of pollutant species and the release of heat. Accomplishing this task requires the computational solution of equations describing the conservation of mass, momentum, energy, and chemical species. In the course of our development we compare the computational approach with that of large activation energy asymptotic analysis. In the second half of the paper we concentrate on the computational consequences of flame modeling. Typically the governing equations are large and stiff systems of partial differential equations. Computational solution requires strongly stable implicit numerical algorithms, and we discuss these methods. We also discuss the adaptive meshing strategies that are required to resolve accurately the structure of thin flames in relatively large domains.

1. Introduction

Our purpose in this paper is to discuss a computational approach to combustion modeling. We do this in the context of considering the structure of premixed laminar flames. Such flames are used widely for the investigation of chemical kinetic processes in combustion. From the viewpoint of the combustion chemist these flames provide a valuable laboratory test bed in which to study the interplay of complex chemical kinetic processes together with fluid mechanical transport processes. In general, the chemistry which occurs in a premixed laminar flame is more like that which occurs in a practical combustor than that in any other readily achievable laboratory condition.

Understanding laboratory measurements in combustion environments is nearly impossible without accompanying theoretical modeling interpretation. There is an extensive and growing literature on modeling of premixed flames. Both computational and asymptotic analytical approaches have found considerable success. However, each approach has its strengths and its weaknesses. Moreover, there is relatively little overlap in the utility of the two approaches.

How any given flame phenomenon should be analyzed depends strongly on the kind of questions the analyst wants to answer. In our opinion, the strength of the asymptotic approach is that it provides an ability to study the stability of flames. Variations of the basic approach have been particularly successful at predicting various pulsating, spinning, and cellular flame structures. The price it pays to do so, however, is that it gives up all but a rough qualitative description of the chemical processes. Therefore, from the standpoint of a combustion chemist, the asymptotic analysis misses the very thing that is important. For example, he may want to know what pollutants are formed, and what effect various chemical additives can

*Research sponsored by the Department of Energy, Office of Basic Energy Sciences.

have. It is these investigations that require computational approaches.

In this paper we first describe the chemical structure of premixed flames. We do this by drawing on examples from our modeling work on hydrogen, hydrocarbon, and ammonia flames. We hope to leave the reader with an understanding of the important chemical interactions that occur in a variety of flames. In doing so we hope to demonstrate the value of determining the chemical makeup of flames and the power of a computational approach. We then discuss some of the problems that the computational analyst encounters as he designs and implements computer programs to solve the governing equations of combustion.

2. Premixed flame equations

In this section we write the equations that govern one-dimensional premixed flame behavior. We use this system because it is representative of a broad class of combustion phenomena. Moreover, in the following section we discuss several conclusions that can be drawn about flame behavior by solving these equations.

2.1. *Transport equations*

The equations governing steady, isobaric, quasi-one-dimensional flame propagation may be written as follows:

$$\rho u A = \dot{M} = \text{constant}, \tag{1}$$

$$\dot{M}\frac{\mathrm{d}T}{\mathrm{d}x} = \frac{1}{c_p}\frac{\mathrm{d}}{\mathrm{d}x}\left(\lambda A \frac{\mathrm{d}T}{\mathrm{d}x}\right) - \frac{A}{c_p}\sum_{k=1}^{K}\rho Y_k V_k c_{pk}\frac{\mathrm{d}T}{\mathrm{d}x} - \frac{A}{c_p}\sum_{k=1}^{K}\dot{\omega}_k h_k W_k, \tag{2}$$

$$\dot{M}\frac{\mathrm{d}Y_k}{\mathrm{d}x} = -\frac{\mathrm{d}}{\mathrm{d}x}(\rho A Y_k V_k) + A\dot{\omega}_k W_k \quad (k=1,\ldots,K), \tag{3}$$

$$\rho = \frac{p\overline{W}}{RT}. \tag{4}$$

In these equations x denotes the spatial coordinate; T, the temperature; Y_k, the mass fraction of the kth species; p, the pressure; u, the velocity of the fluid mixture; ρ, the mass density; W_k, the molecular weight of the kth species; $\overline{W}$, the mean molecular weight of the mixture; R, the universal gas constant; λ, the thermal conductivity of the mixture; c_p, the constant pressure heat capacity of the mixture; c_{pk}, the constant pressure heat capacity of the kth species; $\dot{\omega}_k$, the molar rate of production by chemical reaction of the kth species per unit volume; h_k, the specific enthalpy of the kth species; V_k, the diffusion velocity of the kth species; and A, the cross-sectional area of the stream tube encompassing the flame (increasing due to thermal expansion).

In a typical problem we consider between 30 and 40 chemical species and between 100 and 200 reversible elementary chemical reactions. The net chemical production rate $\dot{\omega}_k$ of each species results from a competition between all the chemical reactions involving that species. We consider that each reaction proceeds according to the law of mass action and the forward rate coefficients are in the modified Arrhenius form

$$k_\mathrm{f} = A T^{\beta}\exp\left(\frac{-E_\mathrm{A}}{RT}\right). \tag{5}$$

We also consider reactions in which certain species have an enhanced third body collision efficiency in recombination or dissociation reactions, and reactions that have pressure-dependent rate constants. The details of the chemical reaction equations and the thermochemical properties are found in the users manual for CHEMKIN, the code package that we developed to evaluate these expressions (Kee, Miller, and Jefferson [1]).

In addition to chemical reaction rates, we must also be concerned with the transport properties of the species. We determine the thermal conductivity of the gas mixture from the individual component conductivities using Wilke's semiempirical formula [2]. An extended Eucken–Hirschfelder correction for polyatomic species is used in computing the single component conductivities. We consider the diffusion velocity V_k in three parts,

$$V_k = \mathscr{V}_k + \mathscr{W}_k + V_c. \tag{6}$$

$\mathscr{V}_k$ is the ordinary diffusion velocity and is given in the Curtiss–Hirschfelder [3] approximation by

$$\mathscr{V}_k = -D_k \frac{1}{X_k} \frac{dX_k}{dx}, \tag{7}$$

where X_k is the mole fraction, with the multi-component diffusion coefficient D_k given explicitly in terms of the binary diffusion coefficients $\mathscr{D}_{jk}$ by

$$D_k = \frac{1 - Y_k}{\sum_{j \neq k}^{K} X_j / \mathscr{D}_{kj}}. \tag{8}$$

Non-zero thermal diffusion velocity $\mathscr{W}_k$ is included only for the low molecular weight species H, H_2, and He; the trace, light component limit is employed in determining $\mathscr{W}_k$,

$$\mathscr{W}_k = \frac{D_k k_{T_k}}{X_k} \frac{1}{T} \frac{dT}{dx}. \tag{9}$$

The thermal diffusion ratio (Chapman and Cowling, [4]) is approximated as

$$k_{T_k} = \sum_{\substack{j=1 \\ j \neq k}}^{K} k_{kj}, \tag{10}$$

where the sign of k_{kj} is such that the lower molecular weight species diffuse from regions of lower to higher temperature. The velocity V_c (independent of species but a function of x) is included to insure that the mass fractions sum to unity (or equivalently $\sum_{k=1}^{K} Y_k V_k = 0$).

This transport model is essentially the same as that recommended by Coffee and Heimerl [5] in their extensive investigation of approximate transport models in hydrogen flames. Stockmayer potentials are used throughout in evaluating transport properties. A detailed description of the transport property expressions and the evaluation algorithms can be found in Kee, Warnatz, and Miller [6].

2.2. *Boundary conditions*

We consider two different types of flames: burner-stabilized flames and adiabatic, freely propagating flames. The equations governing both situations are the same, but the boundary conditions differ. In both cases the appropriate boundary conditions may be deduced from the early work of Hirschfelder and Curtiss [7]. For burner-stabilized flames, $\dot{M}$ is a known constant, the temperature and mass flux fraction ($\epsilon_k = Y_k + \rho Y_k V_k A / \dot{M}$) are specified at the cold boundary, and vanishing gradients are imposed at the hot boundary. For freely propagating flames $\dot{M}$ must be determined from the solution to the problem. An additional boundary condition is needed to determine $\dot{M}$ – e.g., the temperature gradient vanishes at the cold boundary.

3. Detailed chemical kinetics and flame structure

In this section we wish to point out a number of simple features of premixed flames that can be understood and predicted only by consideration of a detailed chemical kinetic mechanism. The examples we discuss here are drawn from our studies of hydrogen flames (Smooke, Miller and Kee [8]), acetylene flames (Miller, et al. [5]), and ammonia flames (Miller, et al. [10]). A kindred discussion has recently been presented by Lovachev [11]. His article contains many historical references and is recommended to the interested reader.

The fuel whose oxidation mechanism is best understood is molecular hydrogen (H_2). Hydrogen flames also exhibit most of the features of flames of other fuels, and for that reason we begin by considering their characteristics. Fig. 1 shows profiles of temperature and several species computed for a stoichiometric hydrogen–air flame at atmospheric pressure. Two points are worthy of note.

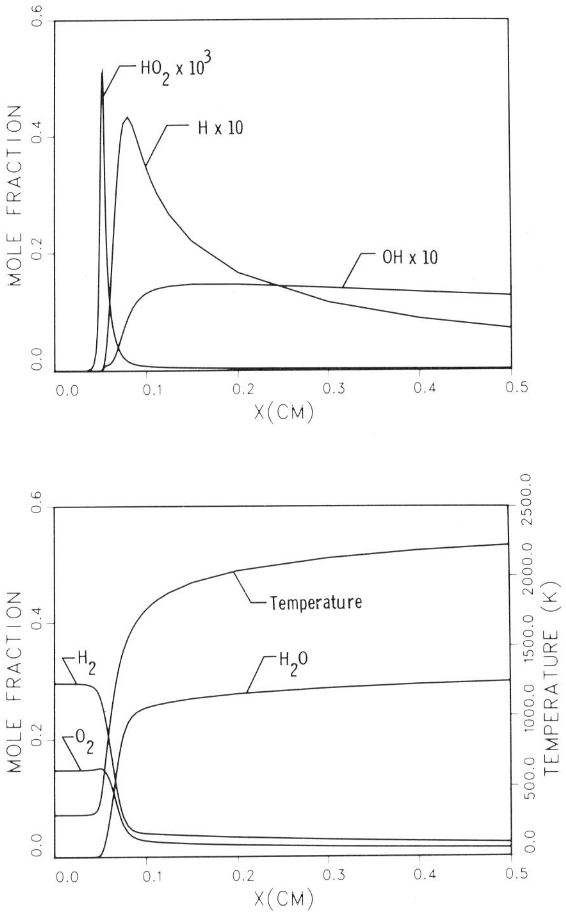

Fig. 1. Computed species and temperature profiles in a stoichiometric atmospheric pressure, hydrogen–air flame.

First, the temperature rises rapidly from its pre-flame value at the flame front (at about $x = 0.05$ cm on the figure) but approaches the adiabatic flame temperature of 2379 K only far downstream (the temperature is 2373 K at $x = 4$ cm and 2378 K at $x = 10$ cm). The second, and related, point is the appearance of relatively sharp peaks of free-radical species (H, O, HO_2, and to a lesser extent OH) in the reaction zone. These two points are connected by the fact that in the early part of the flame the fuel and oxidizer are primarily consumed by the following rapid chain branching sequence of reactions (i.e., net production of active free radical species):

$$OH + H_2 \leftrightarrows H_2O + H, \tag{R1}$$

$$H + O_2 \leftrightarrows OH + O, \tag{R2}$$

$$O + H_2 \leftrightarrows OH + H. \tag{R3}$$

At about $x = 0.08$ cm to 0.09 cm reactions (R1) through (R3) come to a state of "partial equilibrium," i.e., forward and reverse reactions proceed at approximately equal rates. At this point the following slower three-body recombination reactions (i.e., reactions that reduce the number of free radicals and produce stable species as products) begin to dominate the flame chemistry, since the faster processes now have almost no net effect:

$$H + O_2 + M, O_2, N_2 \leftrightarrows HO_2 + M, O_2, N_2 \tag{R4}$$

$$H + H + M \leftrightarrows H_2 + M, \tag{R5}$$

$$H + OH + M \leftrightarrows H_2O + M, \tag{R6}$$

The free radical concentrations (H atoms are particularly affected) peak because reactions (R1) through (R3) result in a growth of free radicals, reactions (R4) through (R6) result in a net loss.

Fig. 2 shows plots of $(\mathrm{d}T/\mathrm{d}x)_i$, where

$$\left(\frac{\mathrm{d}T}{\mathrm{d}x}\right)_i = \frac{\dot{h}_i}{\rho u c_p}, \tag{11}$$

and $\dot{h}_i$ is the local volumetric enthalpy deposition rate for the ith reaction. The ith reaction refers to the net of the forward and reverse steps. The quantity $(\mathrm{d}T/\mathrm{d}x)_i$ may be interpreted as the contribution of the ith reaction to the local temperature gradient. If summed over all reactions and integrated over all space, it yields the net temperature rise in the flame. However, if the integration is stopped in the middle of the flame, heat conduction keeps the "instantaneous" temperature from being obtained. An examination of fig. 2 shows the dominance of recombination reactions throughout the flame in releasing heat. The roughly 800 K temperature rise beyond $x = 0.08$ cm is due almost exclusively to three-body recombinations, which

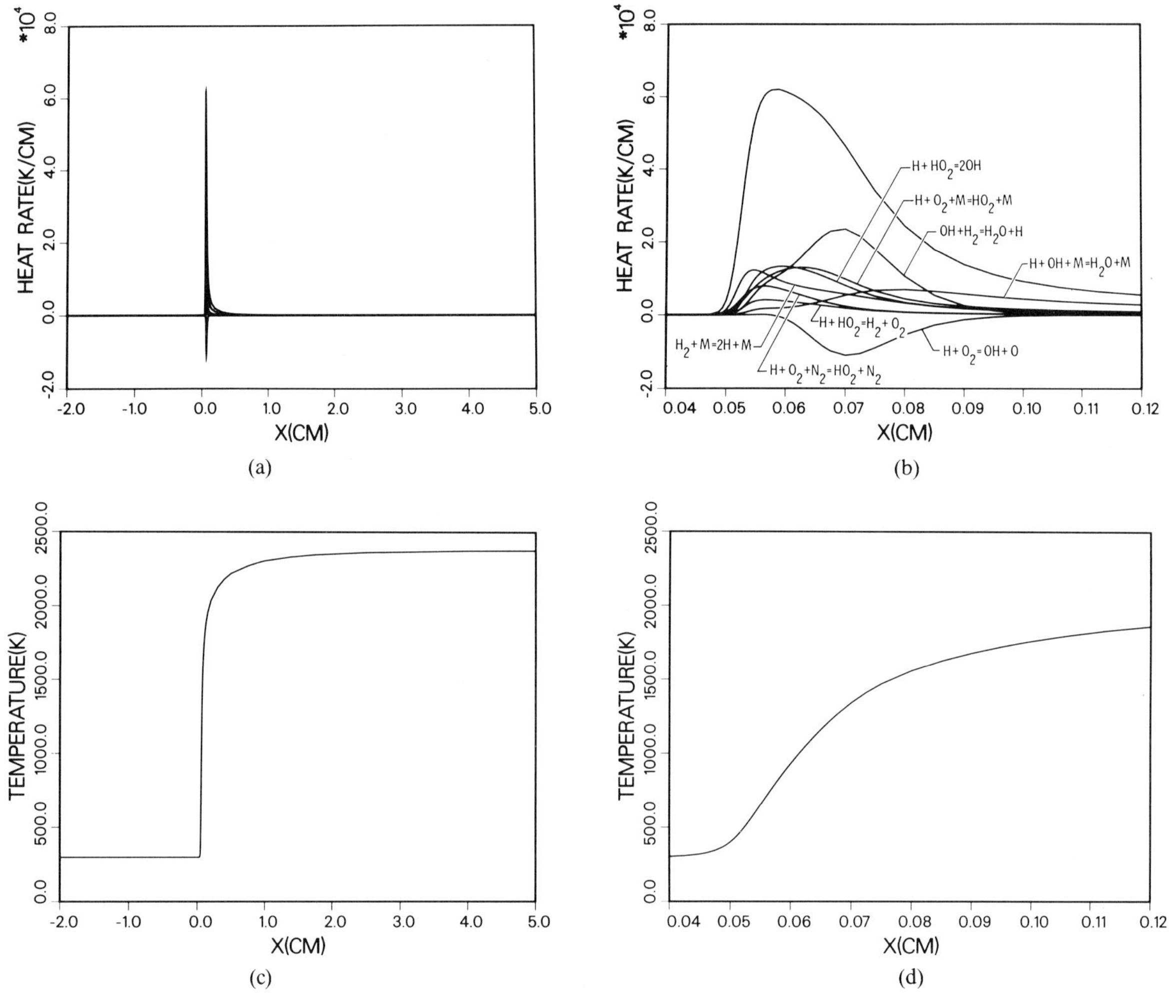

Fig. 2. Computed temperature profiles and heat release rates in a stoichiometric atmospheric pressure, hydrogen–air flame. The profiles are plotted on two different scales. On the scale of 5 cm the flame looks very "thin", but on a scale of 0.1 cm the profile is seen to have considerable structure. The unlabeled curve in (b) is the net chemical heating rate.

are inherently slow. This is the reason behind the very slow approach of the temperature to its equilibrium value.

The behavior described above is to be contrasted with that implied by the normal "activation energy asymptotic" approach (see, for example, the review by Margolis and Matkowsky [12]). In the latter approach, reactant consumption, and thus the energy deposition from chemical reaction, are confined to a very narrow (normally a delta function) region of space imbedded in a larger convection-diffusion layer. That is, the characteristic length scale for transport processes is assumed to be very large compared to the characteristic length scale for chemical reaction. Most often, however, this is not the case when a realistic, detailed kinetic description is employed. This is seen clearly in fig. 1, where the region of high temperature gradient (and hence diffusion of heat) is essentially the same size as the region of heat release by chemical reaction. In fact, in flames where recombination-dissociation reactions are important in releasing heat, characteristic length scales for chemical reaction may even be larger than those for transport processes. Moreover, most reactions important in flames in fact have relatively small activation energies. Many even have rate coefficients which decrease with increased temperature (e.g. recombination reactions). As a rule of thumb one can normally assume that the important reaction with the largest activation energy is reaction (R2). For this reaction $E_A = 16$ kcal/mole, which leads to a non-dimensional activation energy (for the flame temperature of fig. 1) of only $\tilde{E}_A \approx 3.4$. The above considerations can be a source of error in the "large activation energy assumption" for many situations of practical interest.

Referring again to fig. 1, we see that the HO_2 profile is very sharply peaked in the early, cooler regions of the flame. This is due to the fact that above $T \approx 750$ K (at atmospheric pressure) reaction (R4), which forms HO_2, is dominated by reaction (R2), the chain branching step. At higher temperatures in the flame, when (R2) equilibrates, (R4) again is able to compete, but by this time there are enough hydrogen atoms available that the HO_2 is short-lived, being converted to OH by the reaction

$$H + HO_2 \leftrightarrows OH + OH. \tag{R7}$$

The competition between (R2) and (R4) is responsible for a wide variety of combustion phenomena. Most notably, the second explosion limit in hydrogen oxidation is due to this competition. Also, the reduction in flame speed at pressures between one and ten atmospheres in hydrocarbon–air flames is due to this competition. As pressure increases, the cross-over temperature at which $k_4[M]$ becomes larger than k_2 increases (the k's are the forward rate coefficients for the subscripted reactions, and $[M]$ is the total concentration of the gas). Consequently in a greater portion of the flame the chain branching sequence initiated by reaction (R2) is replaced by a less effective sequence (at most chain propagating) initiated by (R4). Thus the flame speed drops, as is shown for acetylene–air flames in fig. 3. This effect was first pointed out in the literature by Westbrook and Dryer [13]. However, if the oxidizer is pure oxygen, higher pressures are required to reduce the flame speed significantly (Miller, et al. [10]). This effect is just the opposite of that expected from the simple reaction order concept, from which one deduces

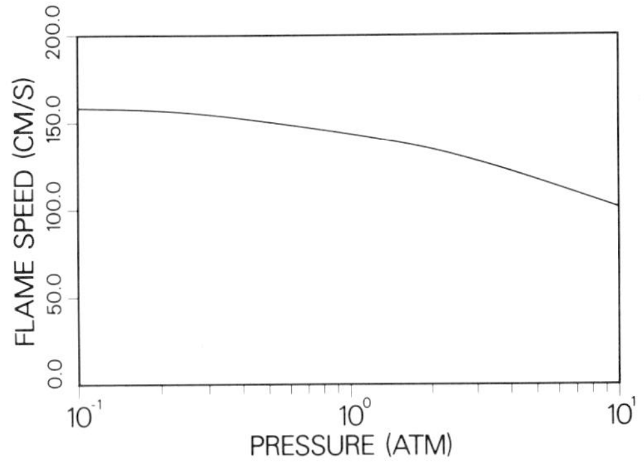

Fig. 3. Computed acetylene–air flame speeds plotted as a function of pressure.

$$V_f \propto p^{(n/2)-1}, \tag{12}$$

where V_f is the flame speed and n is the reaction order. From eq. (12) one would expect an increased contribution from third order reactions (i.e. three body recombinations) to be accompanied by an increase, not a decrease, of flame speed with pressure.

Many other important flame phenomena can trace their origins to a competition for H atoms between reactions (R2) and other interfering reactions. The simplest and best known of these is chemical flame inhibition by halogenated compounds. For example, it is well known that the introduction of hydrogen bromide (HBr) into reactant mixtures inhibits flame propagation. This is due directly to the fact that the reaction

$$HBr + H \leftrightarrows H_2 + Br$$

competes with (R2) for H atoms and thus inhibits chain branching in the flame.

To some extent one can discern the significance of recombination (or dissociation) reactions from very simple considerations. For example, an approximate over-all reaction for hydrogen oxidation can be written as

$$2H_2 + O_2 \leftrightarrows 2H_2O,$$

from which it can be seen that three molecules must be reduced to two in the flame. That is, recombination must occur to reach final products. As we have already noted, this is why the approach to the adiabatic flame temperature is so slow in fig. 1. The simple balancing of an over-all reaction, as done here, is often not sufficient to determine whether recombination or dissociation is required in the oxidation process. In general one must examine the fixed pressure, fixed enthalpy equilibrium composition. If the mean molecular weight of the products ($\overline{W} = \Sigma_k W_k X_k$) is larger than that of the reactants, recombination must take place. Conversely, if $\overline{W}$ is smaller for products, dissociation must take place. Note that $\overline{W}$ is the reciprocal of the mole number, the number of moles per gram of gas, which in turn is proportional to the number of molecules per gram of gas.

Under most conditions the oxidation of hydrocarbons requires a small increase in the mole number (decrease in $\overline{W}$); however, it is well known experimentally that hydrocarbon flames have recombination zones similar to those of hydrogen flames. This state of affairs then requires that rapid dissociative (mole number increasing) reactions take place in the main reaction zone of the flame. Identification of these reaction steps is crucial to the development of adequate reaction mechanisms. The reactions usually responsible for the "mole number overshoot" are

$$HCO + M \leftrightarrows H + CO + M \tag{R8}$$

or

$$CH_2 + O_2 \leftrightarrows CO_2 + H + H. \tag{R9}$$

Note that reaction (R8) is very fast for a direct bond breaking reaction. This is due to the fact that the H—CO bond in formyl (HCO) is very weak. In fact, both the activation energy ($E_A \approx 15$ kcal/mole) and the pre-exponential factor ($A \approx 10^{14}$ cm^3/mole · s) are comparable to those of the chain branching step (R2). A rate coefficient which is comparable to that of (R2) is required, of course, to produce the overshoot. Under certain conditions, in reactions such as the following, the pressure dependent dissociation of small hydrocarbon free radicals can also be responsible for increasing the mole number,

$$C_2H_3(+M) \leftrightarrows C_2H_2 + H(+M),$$
$$C_2H_5(+M) \leftrightarrows C_2H_4 + H(+M).$$

Other flames can behave differently from what we have described above. Flames of H_2—N_2O—Ar, for example, typically require virtually no change in mole number. Concomitantly, free radical concentrations (OH to be specific) in these flames at atmospheric pressure reach total chemical equilibrium within millimeters of the flame front. That is, there is no recombination zone. Even more interesting behavior is shown by our calculations for NH_3—O_2 flames. These flames typically require a significant increase in the mole number. However, the reaction most responsible for the mole number increase,

$$HNO + M \leftrightarrows H + NO + M,$$

has a relatively large activation energy (~ 50 kcal/mole). Consequently, in this case there is a long dissociation (not recombination) zone and we find that the temperature overshoots the adiabatic flame temperature. The adiabatic flame temperature is approached from above only at large distances from the flame front. All the heat is released by fast bimolecular reactions, with the slow endothermic dissociation process bringing the system to equilibrium.

The construction of reaction mechanisms for combustion processes is presently a very active field of research. The ultimate goal of such research is to predict quantitatively, as well as to understand qualitatively, parameters such as flame speed, flammability limits, ignition limits, detonability limits, etc. Equally important is to discover how species that are important in environmental chemistry are produced in combustion processes. An example of such a fundamental understanding comes from our own work on ammonia–oxygen flames (Miller et al. [10]). It is reasonably well established that the "global" mechanism by which nitrogen bound chemically in a fuel molecule is converted to nitric oxide in flames can be described by the following diagram:

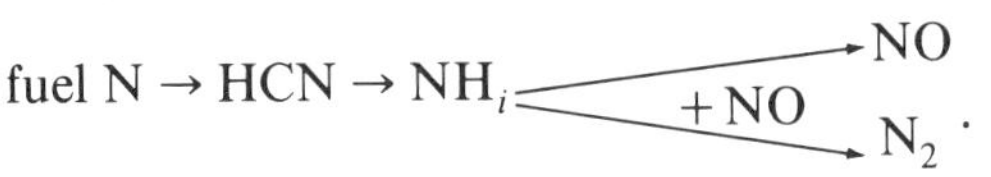

Fuel nitrogen is rapidly converted to hydrogen cyanide, which in turn reacts to produce amine species. The amine species may be oxidized to form nitric oxide or react with nitric oxide to produce molecular nitrogen. However, the chemical identity of the NH_i species in the diagram has been a source of controversy. By constructing a reaction mechanism from experimental data on elementary reaction steps, supplementing this data with theoretical information, and finally by comparing the computed results with a wide range of flame data, we have been able to make some definite statements about the identity of the "critical amine intermediate." Its identity depends on the equivalence ratio of the flame. For very lean flames NH_2 is most important, but for richer flames the emphasis shifts to NH and then to N atoms.

4. Implicit computational methods

The equations that describe combustion processes can be characterized as large, strongly nonlinear, stiff systems of partial differential equations with inherently disparate time and length scales. Each of the above characteristics has an influence on the applicability and choice of computational algorithms. We consider here both steady state and transient problems, although the examples in the previous section dealt only with steady flame problems. We will see, however, that many of the computational difficulties are common to both steady and time dependent problems.

4.1. *Stiffness and implicit methods*

One of the first concepts that often comes to mind when explaining the computational solution of chemical kinetics or combustion problems is stiffness. Curtiss and Hirschfelder [14] first recognized the stiffness problem in chemical kinetics, and they showed the value of using implicit solution methods for producing computationally efficient solutions. Their method consisted of an implicit trapeziodal rule with essentially no error control. Since that time considerable progress has been made in the development of stiff solution methods. Among these are implicit Runge-Kutta, Extrapolation, and the Backward Differentiation Formula (BDF) methods.

To date the BDF methods have made the most impact on the chemistry community. The methods were first implemented in general purpose software by Gear [15], and later into very robust, user friendly, widely available computer codes by Hindmarsh [16]. These codes are variable-order, variable-step methods that provide the user with assurance that solutions are accurate to within a specified error tolerance and are computed in a "computationally optimal" manner. The BDF methods have been developed and used principally to solve systems of ordinary differential equations (ODE), and we take advantage of these methods (and software) by pursuing a method-of-lines approach to the solution of partial differential equations (PDE).

Stiffness is often misinterpreted, yet it is a relatively easily understood concept. Stated simply, a problem is stiff if the time step required to maintain numerical stability with an explicit integration formula is small compared to the time step that would be required if local accuracy were the only consideration. In other words, a stiff problem is one in which the solution is slowly varying, even though the characteristic time scales in the equations are relatively very much shorter. It is often the case that a problem begins as non-stiff in an initial transient region where the solution is changing on the same scale as the characteristic scales, but the problem becomes stiff as the solution asymptotically approaches an equilibrium condition.

Explicit methods are those in which the solution at an advanced time level can be determined directly in terms of solutions at previous time levels. Implicit methods, on the other hand, are those in which the solution at the advanced time level cannot be written directly in terms of previous solution values; typically a system of nonlinear equations must be solved to obtain the solution at the advanced time level. The stability properties of the two basic approaches differ significantly. For linear problems, many implicit methods are stable for arbitrarily large time steps, whereas the explicit methods are restricted in their time steps by the shortest characteristic time scale in the problem (even if the solution is not changing). For higher order methods and for nonlinear problems, which are the ones of interest, the implicit methods also have a stability restriction, but it is much less stringent than that for explicit methods. The advantage of explicit methods is that the computational cost per step is small, and so long as they are able to take reasonably sized steps, they are more efficient than implicit methods. Implicit methods require the solution of a system of nonlinear equations at each step (a computationally expensive procedure), but for stiff problems they compensate for that inefficiency by allowing much larger time steps than are possible with explicit methods.

Stiffness has two major origins in combustion problems. The first stems from the inherently disparate time scales associated with the chemical kinetics. That is, some species are created or destroyed by chemical reaction at much higher rates than others. For example, free radical species (like H atoms) are usually significantly more reactive than stable species. The second stems from the very fine mesh intervals that are required in the "flame zone." This disparity in mesh spacing introduces disparate characteristic time scales into the solution procedure, and hence restricts the time steps that can be used in explicit methods. The well known time step restriction $2\Delta t/\Delta x^2 < 1$ for the explicit Euler solution of the heat equation is a consequence of mesh stiffness.

In order to take advantage of the theory, methods, and software that have been developed for the solution of stiff ODEs (see Petzold [17]), we approach the solution of time dependent combustion problems in a method-of-lines formulation using the implicit BDF methods. In the method-of-lines procedure parabolic partial differential equations are converted to ODEs by discretizing the spatial derivatives; the resulting ordinary differential equations are then solved using ODE software. However, we find considerably more flexibility in the method by posing the discretized problems as systems of differential-algebraic equations (DAE), rather than standard form ordinary differential equation systems, $\phi' = F(t, \phi)$. Situations often occur in which the governing equations either have more than one time derivative (e.g., the momentum equation for compressible fluid flow), or no time derivatives at all (e.g., boundary conditions, or various algebraic constraint conditions). In these cases the DAE formulation is much more natural for the method-of-lines approach than is the standard form ODE formulation.

Stated generally, the DAE system is written as

$$G(t, \phi, \phi') = 0. \tag{13}$$

In this equation G is the spatial difference operator, and the dependent variable vector ϕ is typically composed of the temperature, velocity, and species mass fractions at each mesh point. We note at this point that if the entire ϕ' vector is exactly zero then we have a steady state problem. Our solution of the steady state problem is analogous to one time step of a transient problem.

For illustrative purposes here we demonstrate how to solve the DAE system with a first order backward Euler method, which is representative of the more complex BDF methods. After time discretization, eq. (13) becomes

$$G\left(t, \phi_{n+1}, \frac{\phi_{n+1}-\phi_n}{\Delta t}\right)=0. \tag{14}$$

To advance from time level n to time level $n+1$, eq. (14) represents a system of nonlinear algebraic equations for ϕ_{n+1}. Solution is accomplished via a modified Newton iteration. That is, we first evaluate the Jacobian matrix of the system,

$$J=\frac{\partial G}{\partial \phi_{n+1}}, \tag{15}$$

and then repetitively solve for ϕ_{n+1}^{k+1} in

$$J \times\left(\phi_{n+1}^{k+1}-\phi_{n+1}^{k}\right)=-G\left(t, \phi_{n+1}^{k}, \frac{\phi_{n+1}^{k}-\phi_n}{\Delta t}\right), \tag{16}$$

where k is the iteration index. The initial iterate ($k=0$) comes from a predictor step, which is usually computed as an extrapolation from the solution at previous time steps. The iteration continues until it converges, i.e., until the correction computed for the next iterate is sufficiently small. For the sake of computational efficiency, the Jacobian is only computed periodically rather than at each step and at each iteration, as would be done in a full Newton's method. The rate of convergence is monitored, and when it is sufficiently slow, or not converging, a new Jacobian is computed. If the iteration is still too slow, then the time step is reduced. Once the iteration has converged for a given step, then an error estimate is made. Based on the error estimate, it is determined whether or not the time step Δt was too large. If it was, the step is rejected, the time step reduced, and the whole procedure repeated. If the errors were much smaller than required, then the time step for the next step would be increased.

In practice, we do not use a backward Euler method. Instead, we use a variable order, variable step method based on the Backward Differentiation Formulas. The method has been implemented by Petzold [18] in a general purpose differential-algebraic equation solution code called DASSL. The order ranges from one to five, and the time steps are chosen so that the truncation errors in the time integration are maintained to within a user specified tolerance.

The straightforward implementation of the method-of-lines approach has served us well in one dimensional problems, but some modifications are required for multi-dimensional problems. At a minimum the linear equations in the Newton iteration eq. (16) must be solved differently. A two-dimensional problem with a simple five point differencing scheme leads to a Jacobian that is block penta-diagonal, with the outer blocks being far from the diagonal. A diagram of such a matrix structure is shown in fig. 4a. Even for problems with only a few species, say 10, and a modest 50 by 50 grid network, there are 1.25 million non-zero entries in the Jacobian. This alone is larger than the memory on many scientific supercomputers, and prohibits the direct use of Gaussian elimination to solve the linear equations. Alternatives are defect correction methods and out-of-core sparse linear equation solvers. Defect correction methods are iterative methods that repeatedly solve a related, but in some way simpler, problem until the solution to the original problem is obtained.

For two-dimensional problems, moreover, single step methods, rather than multistep methods (such as BDF), may be the method of choice. One reason has to do with adaptive gridding. It may prove desirable to rezone the mesh at the end of every few steps, rather than to compute a continuous rezoning (this is discussed more later). In this case, the discrete system of equations changes at each step since they are discretized on a different mesh. Therefore, a multistep method would have to be restarted after each mesh change, and hence

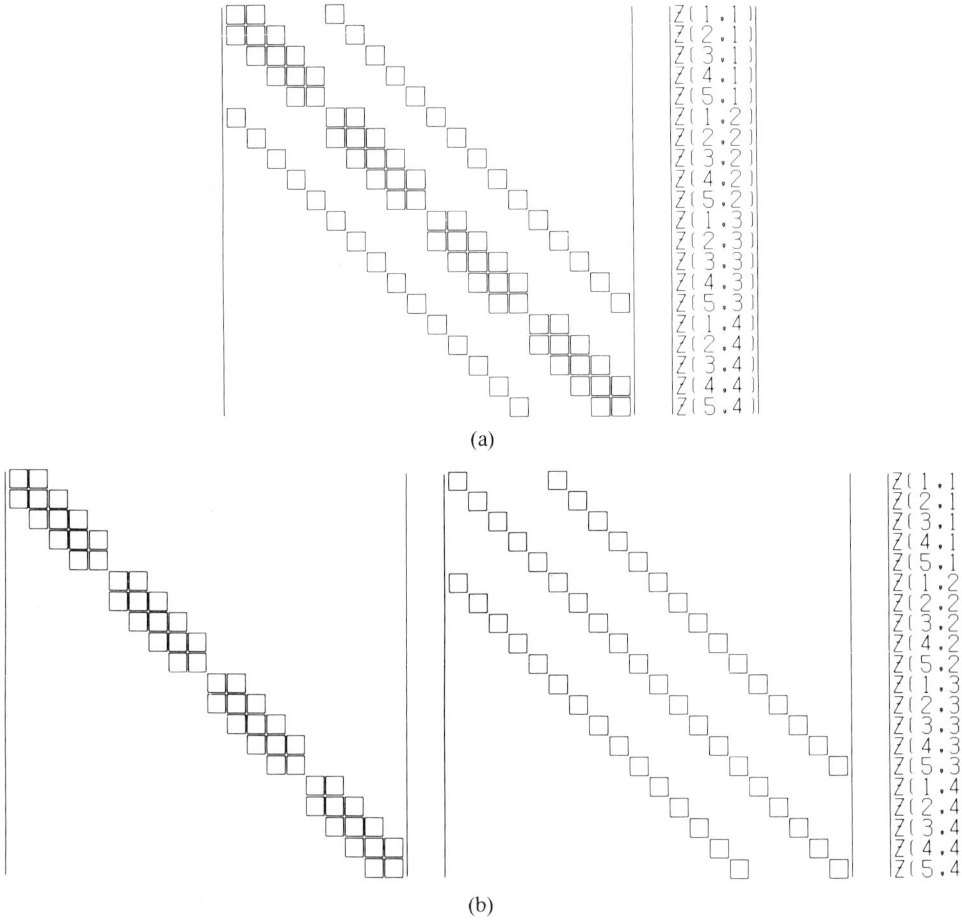

Fig. 4. Diagrams illustrating the structure of Jacobian matrices for a two-dimensional 4 by 5 rectangular mesh. The vector Z represents the dependent variable vector, and the indices represent the mesh location: a) The full block penta-diagonal Jacobian; b) ADI factorization of the Jacobian into systems of block tri-diagonal matrices.

it would be restricted to a very low order. It would probably be more efficient to use a higher order one-step method, such as a fourth order implicit Runge-Kutta method. The higher order should allow the code to take relatively long time steps, but there would be no penalty for changing the equations often. This topic is the subject of current research.

4.2. *Mesh adaptation*

Disparate time and length scales lead to localized behavior, which requires mesh adaptation. In other words, flames are relatively thin compared to the size of combustion chambers, and to compute flame structure accurately, the mesh must be finer in the region surrounding the flame. Since it is not known beforehand where the flame will be, the mesh must dynamically adjust itself to the flame during the course of the solution. Without adaptation, either so many points are needed that the problem does not fit into the computer, or the mesh spacing is so coarse that the solution is inaccurate in the vicinity of the flame. In premixed flames the inaccuracy usually manifests itself in flame speeds that are too high, flames that are too thick (i.e., the species profiles are too broad), and peak heights of the radical intermediate species that are too low. Examples of these effects in hydrogen flames are found in Smooke [15],

Smooke, Miller, and Kee [20], and Dwyer, Smooke and Kee [21].

We consider the mesh adaptation problem as one of equidistributing a positive weight function among the available mesh intervals (see Dwyer, Smooke, and Kee [21]). In equation form, this concept is written as

$$\int_{j}^{j+1} f\,\mathrm{d}x \leq W, \quad j = 1, \ldots, J-1, \tag{17}$$

where j is the mesh index, f is some function of the solution, and W is some equidistribution criterion. The most desirable quantity to equidistribute is the truncation error in the discrete solution. Unfortunately, this is relatively difficult to accomplish, so typically some related quantity is chosen, for example, the arc-length of the solution, or the change in slope or curvature of the solution within each mesh interval. While these adaptation strategies do not measure the truncation error directly, they do accomplish the required objective; they place more mesh points in regions where the solution is changing rapidly, and this serves to control the discretization error.

Adaptation strategies fall into two broad categories according to whether they fix or vary the number of mesh intervals. With a fixed number of mesh intervals the mesh moves in space so as to equidistribute the weight function. The spatial location of the mesh points may be determined via a coordinate transformation. The new independent variables become the mesh indices, and auxiliary equations are appended to the system to describe the adaptation criteria. Examples of these methods may be found in Dwyer, Kee, and Sanders [22], Gelinas and Doss [23], or Brackbill and Saltzmann [24].

Using a variable number of points, mesh intervals are added and deleted as needed both to equidistribute the weight function and to bring the weight function to within a specified tolerance (see Smooke and Koszykowski [25]). In two or three dimensions the points still may be arranged to maintain a checkerboard-like mesh. This approach has the benefit that its simple data structure can be optimized for various computer architectures, especially for vector machines like the CRAY. However, the approach wastes some points by placing them in regions where they are required only to maintain the logically rectangular mesh structure. Storage-efficient approaches that discard superfluous points must explicitly identify the neighboring mesh points. The complex data structures and the indirect referencing that result are inefficient, especially on vector computers.

In principle the adaptation algorithm may be applied either implicitly or explicitly. That is, the mesh point locations can either be computed at each time step simultaneously with the solution, or they can be computed at the beginning of the step and held fixed during the step. The implicit approach would seem desirable from an accuracy viewpoint, and since we are already considering implicit methods, there should not be too much additional overhead incurred. However, there are stability problems. We have found that using an implicitly coupled transformation approach directly leads to an unstable system of equations. For example, take a stable equation, say Burgers equation, and make an independent variable transformation from the physical spatial coordinates to the arc length of the solution, and then solve simultaneously for the physical coordinate and the velocity. What we find is that the solution to this coupled system is apparently unstable to any perturbations; as the solution proceeds it drifts slowly away from the true solution.

Our remedy for the stability problem has been to apply the adaptation procedure explicitly but still compute the solution implicitly (Dwyer et al. [22] or Smooke and Koszykowski [25]). Thus on any step the solution is computed on a fixed mesh, and the mesh is changed for the next step. Others have developed different strategies. Winkler (reported elsewhere in this volume) introduces what he calls a "stiffness factor," which requires a certain relative spacing between adjacent mesh points. Also, the "penalty functions" introduced by Gelinas and Doss [23], or the "smoothness mapping" in the Brackbill and Saltzman [24] approach apparently serve a similar function in stabilizing the method.

4.3. *Linear algebra considerations*

Both the steady and the transient problems require solution of many systems of nonlinear algebraic equations. Assuming that a method like Newton's method is used for solution, then the local linear system can be characterized by a large sparse system of algebraic equations. Further, assuming a logically rectangular data structure for a two dimensional mesh, then this Jacobian matrix has a block penta-diagonal structure (fig. 4a). The block size is equal to the number of species plus whatever other dependent variables are included, such as velocity and temperature. There are as many block rows and columns as there are mesh points, but at most only five nonzero blocks in each block row or column. In combustion problems the system is strongly nonlinear because of the exponential dependence of reaction rate on temperature. Moreover, the Jacobian will be computationally expensive to evaluate, because the reaction rates involve complicated relationships.

The size of the systems effectively prohibits the direct solution by in-memory Gaussian elimination-type algorithms. Instead, an iterative method needs to be used. The best approach is probably some sort of a defect correction method [27]. In such an approach, a related, but simpler, problem is solved repeatedly until a solution to the original problem emerges. Factored methods such as Alternating Direction Implicit (ADI) or Block Line Successive Over Relaxation (BLSOR) are examples of these methods.

The well-known ADI splitting is illustrated in fig. 4. The system represented by the two factored matrices is only approximately equal to that represented by the original matrix, but it is much easier to solve. The factored system corresponds to one dimensional implicit systems along each mesh row and column. The method retains enough of an implicit nature to provide the needed stability. The ADI method is often considered a one-step procedure for solving parabolic PDEs; after one sweep through the mesh rows and one through the mesh columns, the step is accurate to order Δt, which is the accuracy of the matrix factorization (see Warming and Beam [26]). Therefore, further iteration does not improve the order of the result. However, compared to other splittings, ADI is unusual in this regard, and thinking of the method as a defect correction is more general. Often the approximate problem needed for the defect correction does not have such a physical interpretation as the ADI method.

Another reason to consider iterative methods has to do with the expense of computing Jacobian matrices. We hope to be able to use the same Jacobian for several iterations or time steps. The Jacobian will not be accurate after the first iteration, so the methods must be iterative and they must be reliable when using approximate Jacobians. We note here that most production stiff ODE software (e.g. Hindmarsh [16]) is designed to use methods that are stable and accurate with approximate Jacobians.

Because combustion problems are highly nonlinear, convergence of iterative algorithms is often a problem. An initial iterate or starting estimate of the solution is required. In the case of the transient problem the previous solution or a predictor step provides a good estimate. However, in the case of a steady problem, making this estimate is often quite difficult since full species and temperature fields are required. Usually, one does not know either where the flame is located or the concentrations of most of the minor species to within several orders of magnitude. The defect correction iteration continues to improve the iterates if the method is converging. However, it is often the case that the starting guess is not sufficiently accurate and the methods do not converge. We impose various damping strategies to help make the iteration more robust, but often there is no recourse but to try a new starting guess. Iterative strategies for such problems are related to a large literature on optimization and minimization algorithms, which are the subject of much current research.

Acknowledgements

We wish to acknowledge our close interactions with Drs. Linda Petzold, Mitchell Smooke, and Joe Grcar on the numerical and computational aspects of combustion modeling.

References

[1] R.J. Kee, J.A. Miller and T.H. Jefferson, CHEMKIN: A General Purpose Problem-Independent, Transportable, Fortran, Chemical Kinetic Code Package, Sandia National Laboratories Report SAND80-8003 (1980).

[2] C.R. Wilke, A Viscosity Equation for Gas Mixtures, J. Chem. Phys. 18 (1980) 517.

[3] C.F. Curtiss and J.O. Hirschfelder, Transport Properties of Multicomponent Gas Mixtures, J. Chem. Phys. 17 (1949) 550.

[4] S. Chapman and T.G. Cowling, The Mathematical Theory of Non-Uniform Gases, third ed. (Cambridge Univ. Press, Cambridge, 1970).

[5] T.P. Coffee and J.M. Heimerl, A Transport Algorithm for Premixed, Laminar, Steady-state Flames, Comb. and Flame 43 (1981) 273.

[6] R.J. Kee, J. Warnatz and J.A. Miller, A Fortran Computer Code Package for the Evaluation of Gas-Phase Viscosities, Conductivities, and Diffusion Coefficients, Sandia National Laboratories Report SAND83-8209 (1983).

[7] J.O. Hirschfelder and C.F. Curtiss, The Theory of Flame Propagation, J. Chem. Phys. 17 (1949) 1076.

[8] M.D. Smooke, J.A. Miller and R.J. Kee, Determination of Adiabatic Flame Speeds by Boundary Value Methods, Comb. Sci. and Tech. 34 (1983) 79.

[9] J.A. Miller, R. E. Mitchell, M.D. Smooke and R. J. Kee, Toward A Comprehensive Chemical Kinetic Mechanism for the Oxidation of Acetylene: Comparison of Model Predictions with Results from Flame and Shock Tube Experiments, Nineteenth Symposium (International) on Combustion (The Combustion Institute, Pittsburgh, PA, 1982) p. 181.

[10] J.A. Miller, M.D. Smooke, R.M. Green and R.J. Kee, Kinetic Modeling of the Oxidation of Ammonia in Flames, Comb. Sci. Tech. 34 (1983) 149.

[11] L.A. Lovachev, Chemical Kinetics in Flames – A Review, Comb. Sci. Tech. 25 (1981) 49–69.

[12] S.B. Margolis and B.J. Matkowsky, Nonlinear Stability and Bifurcation in the Transition from Laminar to Turbulent Flame Propagation, Comb. Sci. and Tech. 34 (1983) 45.

[13] C.K. Westbrook and F.L. Dryer, Prediction of Laminar Flame Properties of Methanol–Air Mixtures, Combustion and Flame 37 (1980) 171.

[14] C.F. Curtiss and J.O. Hirschfelder, Integration of Stiff Equations, Proc. Natl. Acad. Sci. (U.S.A.) 38 (1952) 235–243.

[15] C.W. Gear, Numerical Initial Value Problems in Ordinary Differential Equations (Prentice-Hall, Englewood Cliffs, NJ, 1971).

[16] A.C. Hindmarsh, Linear Multistep Methods for Ordinary Differential Equations, Lawerence Livermore Laboratory Report, UCRL-51186 (1972).

[17] L.R. Petzold, Multistep Methods: An Overview of Methods, Codes, and New Developments, Proc. Summer Computer Simulation Conference, Vancouver, B.C., Canada, and Sandia National Laboratories Report, SAND83-8673 (1983).

[18] L.R. Petzold, A Description of DASSL: A Differential-Algebraic System Solver, Proc. IMACS World Conference, Montreal, and Sandia National Laboratories Report, SAND82-8637 (1982).

[19] M.D. Smooke, Solution of Burner-Stabilized Premixed Laminar Flames by Boundary Value Methods, J. Comp. Phys. 48 (1982) 72.

[20] M.D. Smooke, J.A. Miller and R.J. Kee, Numerical Solution of Burner Stabilized Pre-mixed Laminar Flames by an Efficient Boundary Value Method, in: Numerical Methods in Laminar Flame Propagation, N. Peters and J. Warnatz, eds. (Vieweg, Wiesbaden, 1982).

[21] H.A. Dwyer, M.D. Smooke and R.J. Kee, Adaptive Gridding for Finite Difference Solutions to Heat and Mass Transfer Problems, Numerical Grid Generation J.F. Thompson, ed. (North-Holland, New York, 1982).

[22] H.A. Dwyer, R.J. Kee and B.R. Sanders, Adaptive Grid Methods for Problems in Fluid Mechanics and Heat Transfer, AIAA J. 18 (1980) 1205–1212.

[23] R.J. Gelinas and S.K. Doss, The Moving Finite Element Method: Applications to General Partial Differential Equations with Multiple Large Gradients, J. Comp. Phys. 40 (1981) 202–249.

[24] J.U. Brackbill and J.S. Saltzman, Adaptive Zoning for Singular Problems in Two Dimensions, J. Comp. Phys. 46 (1982) 342–368.

[25] M.D. Smooke and M.L. Koszykowski, Fully Adaptive One-Dimensional Mixed Initial-Boundary Value Problems with Applications to Unstable Problems in Combustion, Sandia National Laboratories Report, SAND83-8219 (1983).

[26] R.F. Warming and R.M. Beam, On the Construction of Implicit Factored Schemes for Conservation Laws, SIAM-AMS Proceedings – Symposium on Computational Fluid Dynamics, II (1977).

[27] J.M. Hyman, Numerical Methods for Nonlinear Differential Equations, Nonlinear Problems: Present and Future, A.R. Bishop, D.K. Campbell and B. Nicolaenko, eds. (North-Holland, New York, 1982) pp. 91–107.

Physica 12D (1984) 212–214
North-Holland, Amsterdam

2.2. SOLIDIFICATION

MORPHOLOGICAL AND HYDRODYNAMIC INSTABILITIES DURING PHASE TRANSFORMATIONS

Robert F. SEKERKA
Carnegie-Mellon University, Pittsburgh, PA 15213, USA

Extended abstract

Fundamental aspects of the theory of morphological stability may be ascertained from a number of review articles [1–8]. The basic theory pertains to the shape of the interface that separates a growing phase from a nutrient phase under conditions for which the growth is controlled by diffusion of some dissolved species and/or heat conduction. Most analyses begin by considering the growth of a phase of some simple shape (the unperturbed problem), perturbing this shape, ascertaining whether the perturbations will grow (unstable) or decay (stable), and then trying to find the patterns and shapes, including their size scale, to which the system will evolve. Instability results essentially from the "point effect of diffusion," according to which "composition gradients" are steeper near protrusions and more shallow near depressions. Surface tension tends to play a stabilizing role.

The first problem to be analyzed in detail [9] was the stability of a sphere growing by diffusion *or* heat flow. The sphere is perturbed by spherical harmonics and the amplitudes of these harmonics are determined as a function of time in the quasi-steady-state approximation, according to which growth is assumed to be negligible on a time scale in which the concentration or temperature fields relax to their steady-state values. Instability takes place at a critical radius of the sphere; this radius is inversely proportional to the supersaturation and increases roughly as the square of the order of the spherical harmonic. The wavelength of the fastest growing perturbation is proportional to the geometric mean of a capillarity length and a diffusion length, the latter characterized in this case by the radius of the sphere.

Instabilities also take place during the solidification of binary alloys and their analysis involves simultaneous diffusion and heat flow. For unidirectional solidification [10], instability takes place whenever, at the solid–liquid interface, a conductivity weighted average of the temperature gradients in the solid and the liquid exceeds the gradient of the liquidus temperature by an amount that depends on a stability function [11] that relates to capillarity. The wavelength of the fastest growing sinusoidal perturbation is proportional to the geometric mean of a capillarity length and a diffusion length, the latter equal to the quotient of the solute diffusivity and the unidirectional growth velocity.

Similar results obtain whether or not one uses the quasi-steady-state approximation or a fully time dependent analysis [12, 13]. Moreover, interfacial defects such as grain boundary grooves [14, 15] can serve as potent initial perturbations. The spreading of instabilities from initial perturbations can be described by a Green's function [16].

Subsequent to instability, the shape of the growing interface can sometimes become restabilized in a cellular mode. More severe degrees of instability usually result in tree-like growth forms known as dendrites. Analysis of the shapes of cells and dendrites is a non-linear problem, but the growth velocities of primary dendrites can be calculated from a linear stability analysis [17–19].

When the nutrient phase is a fluid, as in the case for solidification, there can be convection in addi-

0167-2789/84/$03.00

tion to diffusion and heat flow. Forced convection [20, 21] results in enhanced stability for perturbations with wavevectors parallel to the flow. Natural convection [22] results in cellular hydrodynamic modes of long wavelength, morphological modes of short wavelength, and coupled hydrodynamic-morphological modes of intermediate wavelength. The interaction of convection and morphological instability has been reviewed recently [23].

During solidification of alloys at very rapid rates, local equilibrium can cease to exist at the solid–liquid interface. This can give rise to non-equilibrium incorporation of solute, which, in turn, can lead to morphological instabilities that are oscillatory in time [24]. These instabilities occur for values of growth parameters for which the interface would ordinarily (if local equilibrium prevailed) be stabilized by capillarity but would be unstable if capillarity were neglected (modified constitutional supercooling).

Non-linear aspects of morphological stability theory were first formulated [25] in terms of a Landau equation for a single Fourier mode but later in terms of series expansions that involve coupled Fourier modes [26–28]. Criteria for the formation of cellular steady states can be formulated for conditions close to stability, but the existence and stability of these cells seems to depend sensitively on details of the model, e.g., on the ratio of thermal conductivities in solid and liquid. Various approaches to pattern selection have been hypothesized on the basis of marginal stability [29, 30] and subjection of the system to a source of noise [31, 32].

A non-linear morphological stability problem that can be treated in detail [33, 34] is the instability and eventual segmentation of initially circular cylindrical rods in directionally solidified eutectics by means of interfacial diffusion. For this transport mechanism, the partial differential equation that determines the interface dynamics can be written entirely in terms of local (interfacial) properties. Numerical integration allows one to follow the process all the way to rod segmentation (which occurs much faster than would be predicted by a linear analysis) and to deduce the behavior of various modes via Fourier analysis of the resulting shapes.

Acknowledgement

The author is grateful to the National Science Foundation, Division of Materials Research, for financial support under grant DMR78-22462.

References

[1] R.F. Sekerka, J. Crystal Growth 3, 4 (1968) 71–81.

[2] A.A. Chernov, Soviet Physics Crystallography 16 (1972) 734–753.

[3] R.F. Sekerka, in: Crystal Growth: An Introduction, P. Hartman, ed. (North-Holland, Amsterdam, 1973), pp. 403, 443.

[4] R.T. Delves, in: Crystal Growth, B.R. Pamplin, ed. (Pergamon, Oxford, 1974), pp. 40–103.

[5] A.A. Chernov, J. Crystal Growth 24/25 (1974) 11–31.

[6] D.J. Wollkind, in: Preparation of Properties of Solid State Materials, W.R. Wilcox, ed. (Dekker, New York, 1979), pp. 111–191.

[7] J.S. Langer, Rev. Mod. Phys. 52 (1980) 1–28.

[8] R.F. Sekerka, in: Encyclopedia of Materials Science and Engineering (Pergamon, Oxford), to be published.

[9] W.W. Mullins and R.F. Sekerka, J. Appl. Phys. 34 (1963) 323–329.

[10] W.W. Mullins and R.F. Sekerka, J. Appl. Phys. 35 (1964) 444–451.

[11] R.F. Sekerka, J. Appl. Phys. 36 (1965) 264–268.

[12] R.F. Sekerka, in: Crystal Growth, H.S. Peiser, ed. (North-Holland, Amsterdam, 1967), pp. 691–702.

[13] R.F. Sekerka, J. Phys. Chem. Solids 28 (1967) 983–994.

[14] S.R. Coriell and R.F. Sekerka, J. Crystal Growth 19 (1972) 90–104.

[15] S.R. Coriell and R.F. Sekerka, J. Crystal Growth 19 (1973) 285–293.

[16] R.F. Sekerka, J. Crystal Growth 10 (1971) 239–250.

[17] J.S. Langer, R.F. Sekerka and T. Fujioka, J. Crystal Growth 44 (1978) 414–418.

[18] J.S. Langer and H. Müller-Krumbhaar, J. Crystal Growth 42 (1977) 11–14.

[19] J.S. Langer and H. Müller-Krumbhaar, Acta. Met. 26 (1978) 1681–1687, 1689–1695, 1697–1708.

[20] R.T. Delves, J. Crystal Growth 3, 4 (1968) 562–568.

[21] R.T. Delves, J. Crystal Growth 8 (1971) 13–25.

[22] S.R. Coriell, M.R. Cordes, W.J. Boettinger and R.F. Sekerka, J. Crystal Growth 49 (1980) 13–28.

[23] S.R. Coriell and R.F. Sekerka, J. Physico-Chemical Hydrodynamics 2 (1981) 281–293.

[24] S.R. Coriell and R.F. Sekerka, J. Crystal Growth 61 (1983) 499–508.

[25] D.J. Wollkind and L.A. Segel, Phil. Trans. Roy. Soc. London 268 (1970) 351–380.
[26] J.S. Langer and L.A. Turski, Acta. Met. 25 (1977) 1113–1119.
[27] J.S. Langer, Acta. Met. 25 (1977) 1121–1137.
[28] Rajiv, Mathur, Doctoral Thesis, Carnegie-Mellon University, (1981), "A Non-Linear Analysis and Numerical Simulation of Directional Solidification in Binary Alloys".
[29] J.S. Langer and H. Müller-Krumbhaar, Phys. Rev. A 27 (1983) 499–519.
[30] G. Dee and J.S. Langer, Phys. Rev. Letters 50 (1983) 383–386.
[31] Michel Kerszberg, Phys. Rev. B 27 (1983) 3909–3912.
[32] Michel Kerszberg, Phys. Rev. B 27 (1983) 6796–6810.
[33] T.F. Marinis and R.F. Sekerka, in: Conference on In Situ Composites – III, J.L. Walter, M.F. Gigliotti, B.F. Oliver and H. Bibring, eds. (Ginn and Company, 1979), pp. 86–96.
[34] R.F. Sekerka and T.F. Marinis, in: Proc. Int. Conf. Solid–Solid Phase Transformations, H.I. Aaronson, D.E. Laughlin, R.F. Sekerka and C.M. Wayman, eds. (AIME, Warrendale, PA, 1982), pp. 67–84.

Physica 12D (1984) 215–240
North-Holland, Amsterdam

INTERFACIAL PATTERNS DURING PLANE FRONT ALLOY SOLIDIFICATION

David J. WOLLKIND
Department of Applied Mathematics, Washington State University, Pullman, Washington 99164-2930, USA

Rukmini SRIRANGANATHAN*
Department of Mathematics, University of Idaho, Moscow, Idaho 83843, USA

and

David B. OULTON
Department of Mathematics, Virginia Wesleyan College, Norfolk, Virginia 23502, USA

An investigation is made of the development of hexagonal cells on an initially planar interface of solid growing uniformly into the liquid phase during the unidirectional solidification of a dilute binary alloy. The model employed postulates three-dimensional diffusion of solute and heat under the simplifying assumptions that there is no convection in the liquid phase, no solute diffusion in the solid phase, both these phases are infinite in extent, latent heat is negligible, and the thermal fields satisfy Laplace's equation in each phase. A nonlinear stability analysis is performed analogous to that originally developed for the study of Bénard convection cells. The main results of this analysis are that, for a fixed value of the liquid temperature gradient, the structure of the planar interface evolves from nodes (circular depressions) to bands (elongated cells) to dome-shaped hexagonal cells as the solidification speed is increased. These results, which can only be obtained provided the variation of interfacial surface free energy with solute concentration is taken into account, are in excellent agreement with the experimental observations of Morris and Winegard [8, 9] in regard to interface morphologies and solute segregation under their solidification conditions.

1. Introduction and formulation of the solidification problem

The classical subject of linear stability theory analyzes disturbances to an equilibrium state by neglecting all terms except those which are linear in the perturbation quantities. The resulting class of eigenvalue problems has been under intensive study for many years. Physically, the main fruits of linear theory are the determination of the critical conditions for the onset of instability to initially infinitesimal disturbances and the determination of the critical wavelength of those disturbances which are most likely to grow first.

If the spatial pattern of the growing disturbance is to be predicted, if the effect of large disturbances is to be estimated, or if the long-time behaviour of growing perturbations is to be ascertained, then the nonlinear terms must be taken into account. The required nonlinear stability theory of phenomena governed by systems of partial differential equations has been the subject of increasingly vigorous investigation in the decades since the end of the Second World War. Perhaps the bulk of research has been directed toward understanding certain prototype problems in hydrodynamics. One of these is the Bénard problem of natural convection in a thin liquid layer heated uniformly from below and cooled from above by means of a gas layer (see fig. 1). The nonlinear work on this problem was surveyed in 1966 by Segel [1] and more recently by Drazin and Reid [2].

*Current address: Department of Mathematics, University of Wyoming, Laramie, Wyoming 82071, USA.

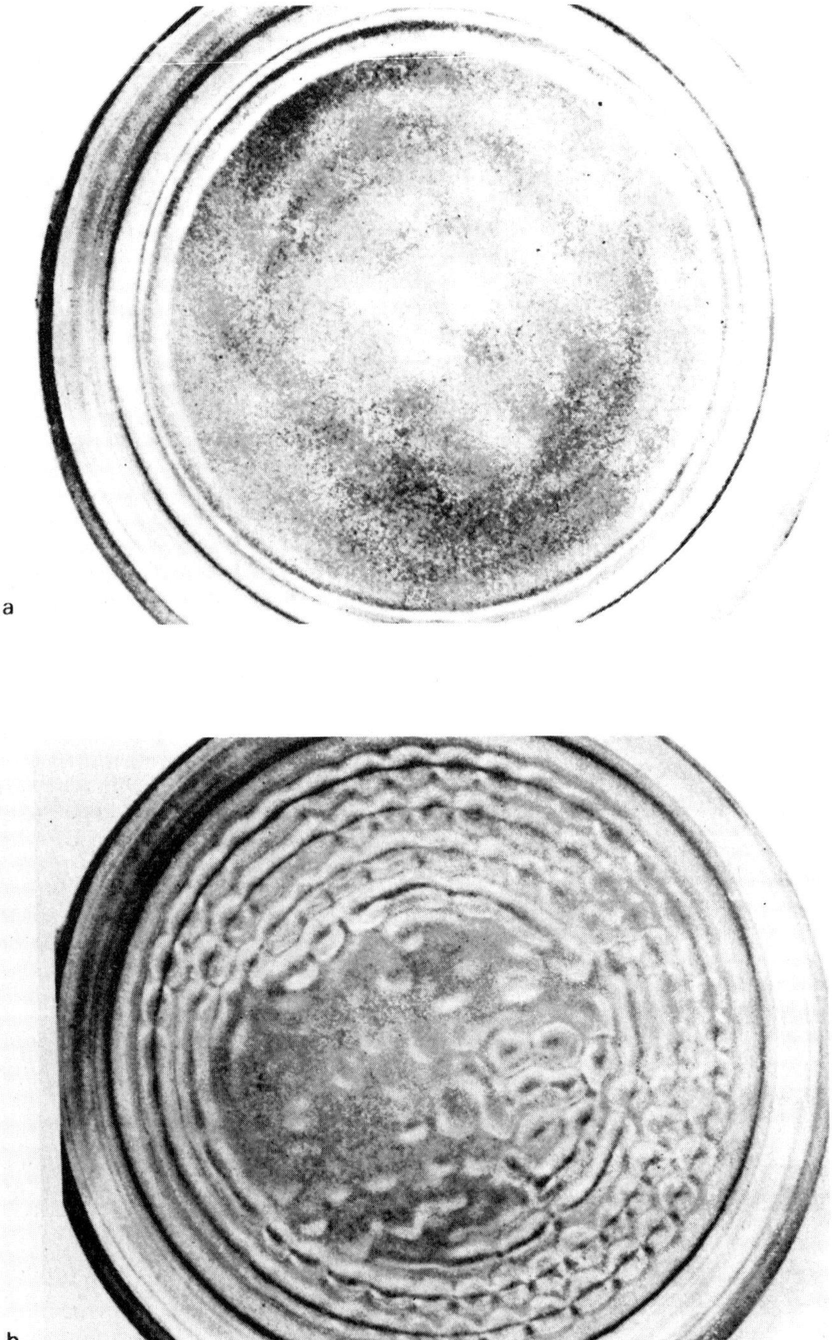

Figs. 1a, b.

c

Fig. 1. The development of hexagonal convection cells characteristic of free-surface flow in thin viscous layers of oil heated from below and cooled from above by the ambient air. Here photograph a) is of no motion; b) of toroidal rolls at the onset of instability; and c) of the fully formed steady-state cellular convection pattern (courtesy of H.L. Koschmeider from ref. 1).

Investigators of problems in nonlinear hydrodynamic stability are aware of the fact that their results can be adapted so that they apply to a broad class of physical problems. A member of this class was identified at the conference at which Segel's survey was presented when Kirkaldy [3] pointed out the similarity between patterns in Bénard convection and cellular patterns observed when an alloy is solidified in a temperature gradient (see fig. 2). This led to the first nonlinear stability analysis of the plane front solidification of a dilute binary alloy performed by Wollkind and Segel [4]. Since that analysis was restricted to a two-dimensional version of the solidification model, they were only able to predict a band-like structure of the solid–liquid interface (see fig. 2c). Inspired by the photograph of interfacial cells during solidification (see fig. 2d) from Kirkaldy's contribution to the above-mentioned conference proceedings [3], we made a theoretical investigation of the conditions under which a planar solid–liquid interface (see fig. 2a) breaks down and forms cells during the controlled unidirectional solidification of a dilute binary alloy [5]. We employed a three-dimensional version of the two-dimensional model of ref. 4 appropriate for performing nonlinear analyses on solidification problems [6]. Our primary purpose was to use a two-disturbance nonlinear stability analysis originally developed by Segel and Stuart [7] for studying Bénard convection cells in order to explain more fully the morphology and solute segregation of the solid–liquid interface observed experimentally during this transition from a planar to a cellular structure [8, 9]. In this paper, we refine our previous results by applying Segel's extended six-disturbance stability analysis for the Bénard problem [10] to our solidification problem.

In particular our three-dimensional model describes the following experimental procedure. A rod of homogeneous binary alloy with composition C_0 is solidified in such a way as to cause a temperature distribution in the solid and liquid phases as depicted in fig. 3 and so that the mean position of the

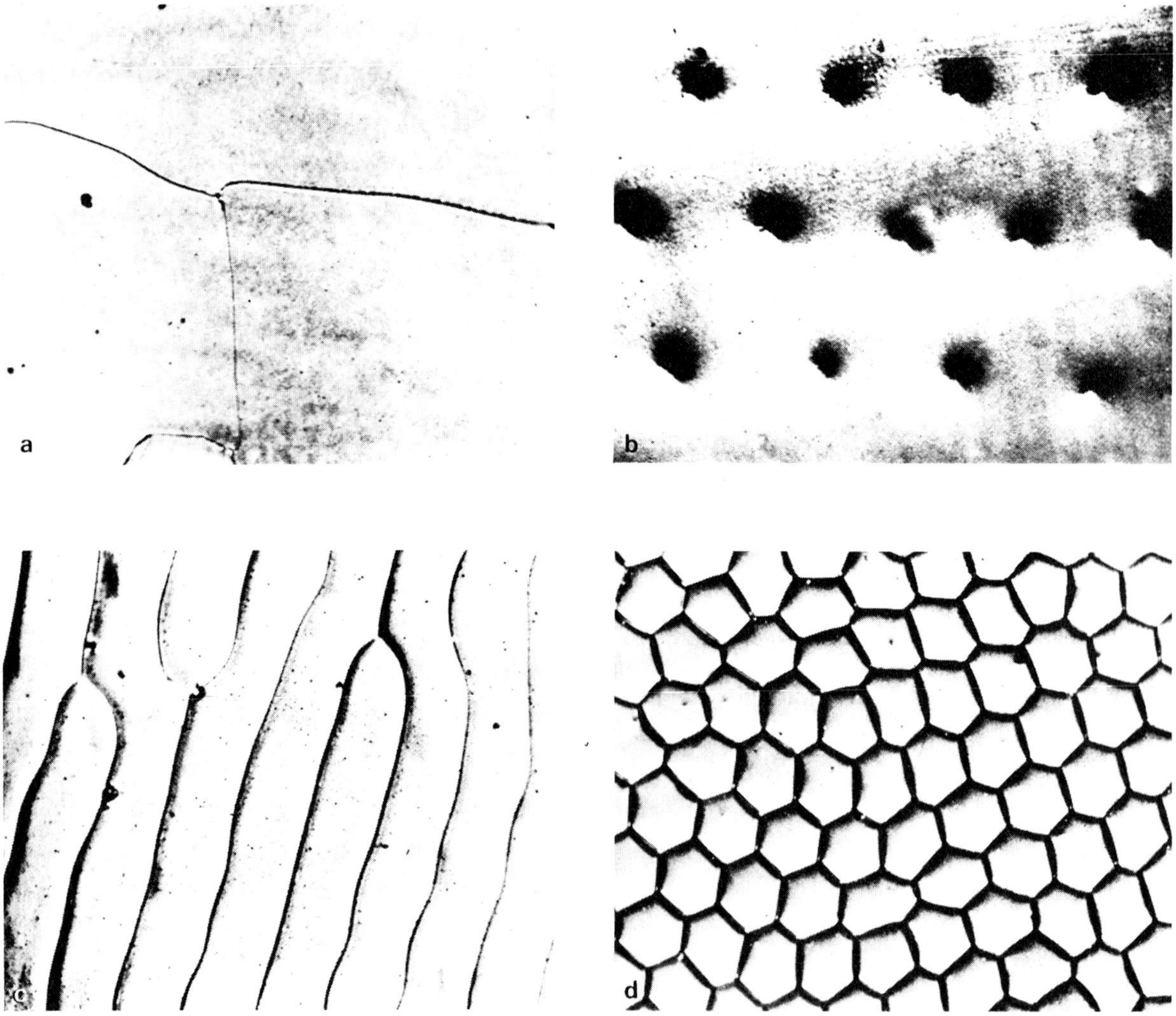

Fig. 2. Interfacial patterns occurring during Pb–Sn alloy solidification. Here photograph a) is of the planar interface; b) of nodes; c) of bands; and d) of cells, which develop sequentially as the solidification conditions intensify. Observe that the cusp-like projections appearing at the centers of the hexagonal close-packed array of nodes in b) are an artifice of the decanting procedure and influenced Tiller and Rutter [27] to use the term "pox" to describe such structures initially. We note that these photographs also inspired M. Kerszberg [32] to produce his recent work on pattern formation in directional solidification (courtesy of G. Cole and W.C. Winegard from ref. 3).

solid–liquid interface is constrained to advance at a constant specified speed. During this process some pains are taken to minimize convection. Our main goal is to predict under what conditions on the imposed liquid temperature gradient, G_L, and rate of solidification, V, the interfacial morphologies observed experimentally will exist. The latter consisted of a sequence which included nodes (circular depressions of liquid into the solid), bands (elongated cells), and dome-shaped regular hexagonal cells [8, 9].

We now turn to a detailed formulation of that model. Consider a three-dimensional situation in which a moving coordinate system, denoted by (x, y, z), is such that the x–y plane ($z = 0$) coincides with the mean position of an interface which in laboratory coordinates is assumed to be moving with constant speed V normal to this plane while the z coordinate measures the distance from that mean position and time is

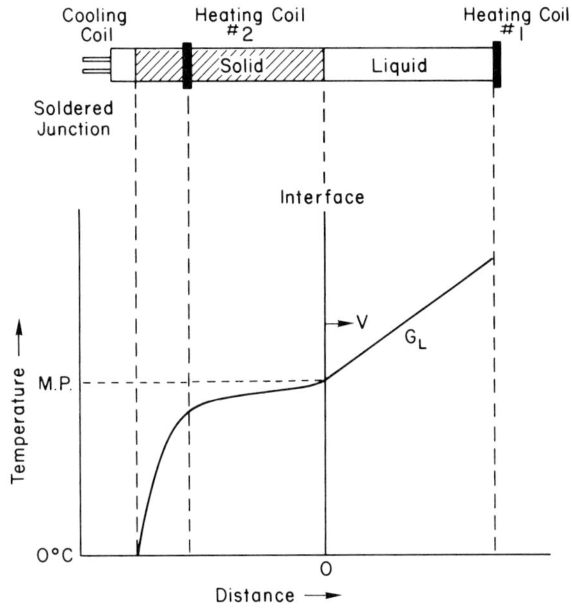

Fig. 3. A schematic representation of the experimental solidification situation and the relevant temperature distribution under consideration [27].

denoted by t. For the development that follows we shall use the additional nomenclature given below:

$C \equiv$ concentration per unit volume of solute in the liquid,

$T(T') \equiv$ temperature in the liquid (solid),

$\mathscr{D} \equiv$ solute diffusion coefficient in the liquid,

$\kappa_L(\kappa_S) \equiv$ thermal diffusivity of the liquid (solid),

$T_M \equiv$ melting temperature of the pure solvent in K,

$m \equiv$ slope of the liquidus line on the phase diagram,

$\gamma \equiv$ interfacial surface free energy per unit area,

$\gamma_0 \equiv$ interfacial surface energy per unit area,

$\gamma_P = \gamma_0 - sT_0 - \mu C_0 \equiv$ interfacial surface free energy per unit area associated with a planar interface which is at temperature $T_0 = T_M + mC_0$ and of concentration C_0 where s and μ are interfacial surface entropy and the coefficient of solutal capillarity respectively,

$L \equiv$ latent heat of fusion per unit volume,

and

$k \equiv$ solute segregation coefficient.

Then the interface satisfies $z = \zeta(x, y, t)$ and we have the following governing equations in three-dimensions:

For $z > \zeta(x, y, t)$ (in the liquid):

$$\mathscr{D}\nabla^2 C + V\frac{\partial C}{\partial z} - \frac{\partial C}{\partial t} = 0, \quad \nabla^2 T = 0; \tag{1a,b}$$

for $z < \zeta(x, y, t)$ (in the solid):

$$\nabla^2 T' = 0, \quad \text{where } \nabla^2 = \frac{\partial^2}{\partial x^2} + \frac{\partial^2}{\partial y^2} + \frac{\partial^2}{\partial z^2}; \tag{2}$$

with boundary conditions:
For $z=\zeta(x, y, t)$ (at the interface):

$$T=T', \quad T=T_M+mC+\frac{T_M}{L}H(\gamma,\zeta), \tag{3a, b}$$

$$\kappa_S\frac{\partial T'}{\partial n}=\kappa_L\frac{\partial T}{\partial n}, \quad \mathscr{D}\frac{\partial C}{\partial n}=(k-1)C(V+\zeta_t), \tag{3c, d}$$

where

$$H(\gamma,\zeta)=\gamma\kappa \quad \text{and} \quad \frac{\partial}{\partial n}=\frac{\partial}{\partial z}-\zeta_x\frac{\partial}{\partial x}-\zeta_y\frac{\partial}{\partial y},$$

with

$$\gamma=\gamma_P-s(T-T_0)-\mu(C-C_0)$$

and

$$\kappa=\left[\zeta_{xx}\left(1+\zeta_y^2\right)-2\zeta_x\zeta_y\zeta_{xy}+\zeta_{yy}\left(1+\zeta_x^2\right)\right]\left(1+\zeta_x^2+\zeta_y^2\right)^{-3/2};$$

for $|z|\to\infty$ (far field conditions):

$$C\to C_0(z)=C_0\left[1+(k-1)(1-\mathrm{e}^{-Vz/\mathscr{D}})\right], \quad T\to T_0(z)=T_0+G_Lz, \quad \text{as } z\to\infty; \tag{4a, b}$$

$$T'\to T_0'(z)=T_0+\frac{\kappa_L}{\kappa_S}G_Lz, \quad \text{as } z\to-\infty. \tag{4c}$$

In the above system time derivatives have been eliminated from (1b), (2), and (3c) by replacing the usual diffusion of heat equation with Laplace's equation in each phase [11] and neglecting the effect of latent heat at the interface [12]. This approximation is valid for many alloy solidification situations in which $\mathscr{D}\ll\kappa_L$ and the basic temperature gradients in both phases are positive [13]. In particular, it is necessary, as indicated implicitly by the form of (4c), that the solid temperature gradient G_S satisfy $G_S\simeq\kappa_LG_L/\kappa_S$ which was true for the experiments of [9] since Morris and Winegard found $G_S/G_L\simeq\kappa_L/\kappa_S=0.56$ at the low growth velocities V under investigation. Although we have taken γ to be isotropic [11–13] in (3b), unlike many models, which, in addition, consider $\gamma\equiv\gamma_P$ [13–15], this one admits the possibility of the variation of γ with interfacial temperature and concentration [16–18].

In section 2 we summarize the theoretical results of our two-disturbance nonlinear stability analysis of ref. 5. Section 3 deals with the application of the six-disturbance stability analysis of ref. 10 to our solidification problem. In section 4 we interpret our results with respect to the experimental observations of Morris and Winegard [8, 9] and suggest some further refinements of both our model and the nonlinear analysis in order to explain the development of cells as reported by Takahashi et al. [19] and the cell–dendrite transition [20].

2. The two-disturbance method of nonlinear stability analysis

In the far field conditions of (4), $C_0(z)$, $T_0(z)$, and $T_0'(z)$ are the components of the planar interface solution to the basic system of equations (1)–(3) which represents the uniform growth of a planar interface ($\zeta\equiv0$) of solid into the molten phase. It is the stability of this solution to a particular class of perturbations with which we were concerned in ref. 5. After Segel and Stuart [7], we assumed that, to third order terms in the amplitude functions $A(\tau)$ and $B(\tau)$, the interface satisfied the equation [5]

$$\begin{aligned} z = \zeta(x, y, t) &= A(\tau)\cos(\omega x) + B(\tau)\cos(\omega x/2)\cos(\sqrt{3}\,\omega y/2) \\ &+ A^2(\tau)\left[\zeta_{2000} + \zeta_{2040}\cos(2\omega x)\right] + A(\tau)B(\tau)\left[\zeta_{1111}\cos(\omega x/2)\cos(\sqrt{3}\,\omega y/2)\right. \\ &\left.+ \zeta_{1131}\cos(3\omega x/2)\cos(\sqrt{3}\,\omega y/2)\right] + B^2(\tau)\left[\zeta_{0200} + \zeta_{0220}\cos(\omega x) + \zeta_{0202}\cos(\sqrt{3}\,\omega y)\right. \\ &\left.+ \zeta_{0222}\cos(\omega x)\cos(\sqrt{3}\,\omega y)\right] + A^3(\tau)\left[\zeta_{3020}\cos(\omega x) + \zeta_{3060}\cos(3\omega x)\right] \\ &+ A^2(\tau)B(\tau)\left[\zeta_{2111}\cos(\omega x/2)\cos(\sqrt{3}\,\omega y/2) + \zeta_{2131}\cos(3\omega x/2)\cos(\sqrt{3}\,\omega y/2)\right. \\ &\left.+ \zeta_{2151}\cos(5\omega x/2)\cos(\sqrt{3}\,\omega y/2)\right] + A(\tau)B^2(\tau)\left[\zeta_{1200} + \zeta_{1220}\cos(\omega x)\right. \\ &+ \zeta_{1240}\cos(2\omega x) + \zeta_{1202}\cos(\sqrt{3}\,\omega y) + \zeta_{1222}\cos(\omega x/2)\cos(\sqrt{3}\,\omega y) \\ &\left.+ \zeta_{1242}\cos(2\omega x)\cos(\sqrt{3}\,\omega y)\right] \\ &+ B^3(\tau)\left[\zeta_{0311}\cos(\omega x/2)\cos(\sqrt{3}\,\omega y/2) + \zeta_{0331}\cos(3\omega x/2)\cos(\sqrt{3}\,\omega y/2)\right. \\ &\left.+ \zeta_{0313}\cos(\omega x/2)\cos(3\sqrt{3}\,\omega y/2) + \zeta_{0333}\cos(3\omega x/2)\cos(3\sqrt{3}\,\omega y/2)\right], \end{aligned} \tag{5}$$

where

$$\frac{\mathrm{d}A}{\mathrm{d}\tau} = \sigma A - aB^2 - A\left(a_1 A^2 + a_2 B^2\right) \tag{6a}$$

and

$$\frac{\mathrm{d}B}{\mathrm{d}\tau} = \sigma B - 4aAB - B\left[2a_2 A + \tfrac{1}{4}(a_1 + 2a_2)B^2\right]. \tag{6b}$$

Here we introduced an interface z scale such that $\mathscr{D}/V \equiv 1$ unit of deviation; $\tau = V^2 t/\mathscr{D} \equiv$ nondimensional time; and $\omega = 2\pi/\lambda$, λ being the wavelength of the disturbance. Further, the ζ_{ijpq} were constants such that each was the coefficient of a term in (5) of the form $A^i(\tau)B^j(\tau)\cos(p\omega x/2)\cos(q\sqrt{3}\,\omega y/2)$ and $\zeta_{1020} = \zeta_{0111} = 1$ while $\zeta_{0000} = 0$. Finally, σ was the growth rate of linear stability theory [13] and a, a_1 and a_2 were solvability constants to be determined. We also assumed analogous expansions for $C(x, y, z, t)$, $T(x, y, z, t)$, and $T'(x, y, z, t)$ consistent with that of (5) where now the coefficient of each term corresponding to a ζ_{ijpq} of the latter was a function z and $i = j = p = q = 0$ corresponded to the planar interface solution of (4).

Then after substituting these solutions into our basic system of equations (1)–(4) and expanding the boundary conditions of (3) in a Taylor series about $A = B = 0$, we obtained a series of ordinary differential equation systems in z with boundary conditions evaluated at $z = 0$ and as $|z| \to \infty$, one for each term of (5) proportional to a nonzero $A^iB^j\cos(p\omega x/2)\cos(q\sqrt{3}\,\omega y/2)$. The explicit forms of these systems can be found in ref. 21. Each of these systems was reduced to a single differential equation for $C_{ijpq}(z)$ with boundary conditions containing ζ_{ijpq}. The rest of our analysis in this section is concerned with determining the coefficients in the amplitude equations of (6) and analyzing the long-time behavior of the solutions of those equations. In what follows we shall present a brief sketch of the procedure used with an emphasis on results and refer the reader to refs. 5 and 21 for specific details.

2.1. *The determination of* σ

The systems proportional to $A(\tau)$ and $B(\tau)$ were identical to the problem of linear stability theory. Hence from [13], we concluded that σ satisfied the secular equation

$$\sigma + \tfrac{1}{2} + (k - 1/2)(Q + \beta\alpha^2) = \left[1 - (Q + \beta\alpha^2)\right]\sqrt{\sigma + \alpha^2 + \tfrac{1}{4}}\,, \tag{7a}$$

where

$$\alpha = \omega\frac{\mathscr{D}}{V} \equiv \begin{matrix}\text{nondimensional}\\ \text{wavenumber}\end{matrix}, \quad \beta = \frac{VT_{\mathrm{M}}\Gamma_{\mathrm{P}}}{\mathscr{D}C_0 m(k-1)}, \quad Q = \frac{2G_{\mathrm{L}}\mathscr{D}}{VC_0(1 + \kappa_{\mathrm{S}}/\kappa_{\mathrm{L}})m(k-1)}, \tag{7b}$$

with $\Gamma_{\mathrm{P}} = \gamma_{\mathrm{P}}/L \equiv$ capillarity constant, while

$$C_{1020}(z) = C_{0111}(z) = C_0(1-k)\left[1 - (Q + \beta\alpha^2)\right]\mathrm{e}^{-m_1 Vz/\mathscr{D}}, \tag{7c}$$

with

$$m_1 = \tfrac{1}{2} + \sqrt{\alpha + \alpha^2 + \tfrac{1}{4}}\,.$$

Since we shall ultimately be interested in plotting various $\sigma \geq 0$ loci in V–G_{L} space for a particular alloy of fixed concentration, we proceeded as follows. Setting $\sigma = c \geq 0$, we solved (7a) for Q, obtaining

$$Q_c(\alpha^2) = -\beta\alpha^2 + \frac{\sqrt{c + \alpha^2 + \tfrac{1}{4}} - \tfrac{1}{2} - c}{k - \tfrac{1}{2} + \sqrt{c + \alpha^2 + \tfrac{1}{4}}}\,. \tag{8}$$

This curve has a maximum at $\alpha = \alpha_c$ where α_c is such that

$$\frac{\mathrm{d}Q_c}{\mathrm{d}\alpha^2}(\alpha_c^2) = 0 \Rightarrow \beta = \frac{k + c}{2\sqrt{c + \alpha_c^2 + \tfrac{1}{4}}\left[k - \tfrac{1}{2} + \sqrt{c + \alpha_c^2 + \tfrac{1}{4}}\right]^2} \tag{9a}$$

and that maximum value is given by

$$Q_c = Q_c(\alpha_c^2). \tag{9b}$$

We note that for $Q > Q_c$ there exists no α^2 such that $\sigma > c$ while for $Q < Q_c$ there exists a band of such α^2. Now since $\alpha_c = \alpha_c(V)$ from (9a) and hence $Q_c = Q_c(V)$ from (9b), we can define a function

$$G_c(V) = \frac{C_0 m(k-1)(1 + \kappa_{\mathrm{S}}/\kappa_{\mathrm{L}})}{2\mathscr{D}} VQ_c(V), \tag{10a}$$

such that the curve in V–G_{L} space given by

$$G_{\mathrm{L}} = G_c(V) \tag{10b}$$

corresponds to $\sigma = c$. In particular for $\sigma = 0$, the curve corresponding to $\sigma = 0$ in V–G_{L} space

$$G_{\mathrm{L}} = G_0(V) \tag{11}$$

is marginal in the sense that $\sigma > 0$ if and only if $G_{\mathrm{L}} < G_0(V)$. In what follows after Wollkind and Notestine

[6], we shall identify the ω contained in (5) with the so-called critical wavenumber of linear theory. That is we shall take $\omega = \alpha_0 V/\mathscr{D}$, where

$$\beta = \frac{k}{2\sqrt{\alpha_0^2 + \frac{1}{4}}\left[k - \frac{1}{2} + \sqrt{\alpha_0^2 + \frac{1}{4}}\right]^2}. \tag{12}$$

We further note that consistent with the development just completed those σ satisfying (7a) which are complex can be characterized by $\operatorname{Re}\sigma < 0$ and will occur only for those α such that $\alpha < k$ [13]. Hence if $\alpha_0 > k$ or, equivalently from (12), if

$$\beta < 4k\Big/\left[\sqrt{1+4k^2}\,(2k - 1 + \sqrt{1+4k^2}\,)^2\right],$$

σ may be considered real. Since this condition is satisfied for the low growth velocities V under investigation in the experiments of ref. 9, we are justified in assuming σ is real for the purposes of our analysis.

2.2. *The determination of a, a_1 and a_2*

There were eight second order systems. Six of these could be solved in a straightforward manner. The other two, which were related, contained the solvability constant a. The determination of this constant and ultimately the solutions of these systems was reduced to an examination of an equation of the form

$$a - \sigma\zeta_{0220}(\sigma) = r(\sigma), \tag{13}$$

where $r(\sigma)$ is a known function which depends on the solutions of the first order systems, as well as on the planar interface solution, evaluated at $z = 0$, and is given explicitly in [21]. Then from (13) we can deduce that

$$a = r(0) \tag{14a}$$

and, upon substitution of this value for a back into (13), that

$$\zeta_{0220}(\sigma) = \frac{r(0) - r(\sigma)}{\sigma}, \tag{14b}$$

or, in particular,

$$\zeta_{0220}(0) = -\frac{\mathrm{d}r(0)}{\mathrm{d}\sigma}, \tag{14c}$$

since our procedure for calculating a_1 and a_2 to be outlined below only required the solutions of our second order systems evaluated at $\sigma = 0$.

Although there were fifteen third order systems it was only necessary for us to consider explicitly two specific ones which contained the solvability constants a_1 and a_2, respectively. As in the calculation of a, the determination of these constants can be reduced to an examination of equations of the form

$$a_1 - 2\sigma\zeta_{3020}(\sigma) = r_1(\sigma) \tag{15a}$$

and

$$a_2 - 2\sigma\zeta_{1220}(\sigma) = r_2(\sigma). \tag{15b}$$

Hence

$$a_i = r_i(0), \quad \text{for } i = 1, 2, \tag{16a, b}$$

where the $r_i(0)$ are known functions depending on the planar interface solution and the solutions of the first and second order systems evaluated at $z = 0$ and with $\sigma = 0$ (see [21] for the explicit forms of these functions).

We now summarize the conditions under which the relevant critical points of the amplitude equations of (6) will exist and be stable. In particular the critical points (A_0, B_0) of (6), which are those points such that $\mathrm{d}A/\mathrm{d}\tau = \mathrm{d}B/\mathrm{d}\tau = 0$, can be catalogued as follows [7]:

I:
$$A_0 = B_0 = 0,$$

II:
$$A_0 = \pm\sqrt{\sigma/a_1}, \quad B_0 = 0,$$

III:
$$A_0 = \frac{-2a \pm \sqrt{4a^2 + \sigma(a_1 + 4a_2)}}{a_1 + 4a_2}, \quad B_0 = 2A_0,$$

IV:
$$A_0 = \frac{-2a \pm \sqrt{4a^2 + \sigma(a_1 + 4a_2)}}{a_1 + 4a_2}, \quad B_0 = -2A_0,$$

V:
$$A_0 = -\frac{4a}{2a_2 - a_1}, \quad B_0 = \pm 2\sqrt{\left[\sigma - 16a^2 a_1 (2a_2 - a_1)^{-2}\right]/(a_1 + 2a_2)}\,.$$

Then, if $A(\tau)$ and $B(\tau)$ equilibrate,

$$\lim_{\tau\to\infty} A(\tau) = A_0 \quad \text{and} \quad \lim_{\tau\to\infty} B(\tau) = B_0; \tag{17a}$$

hence the interface to lowest order satisfies

$$z = A_0 \cos(\omega x) + B_0 \cos(\omega x/2) \cos(\sqrt{3}\,\omega y/2). \tag{17b}$$

Thus I corresponds to a planar morphology of the interface while each of the critical points of II represents an equivalent observable two-dimensional *band-like* structure of the interface aligned parallel to the y-axis with a band-width of $\lambda = 2\pi/\omega$ [5]. In what follows we use a superscript (1) to denote the "+" solution and (2), the "−" solution (we shall also employ this designation to distinguish between the critical points of III as well).

The critical points of III and IV can be shown to yield equivalent hexagonal morphologies of the interface of the form [5]

$$z = A_0 f(x, y) \quad \text{where} \quad f(x, y) = \cos(\omega x) + 2\cos(\omega x/2)\cos(\sqrt{3}\,\omega y/2)$$
$$= \cos(\omega x) + \cos\left(\omega\left[x - \sqrt{3}\,y\right]/2\right) + \cos\left(\omega\left[x + \sqrt{3}\,y\right]/2\right), \tag{18}$$

and since the stability analysis to be performed on each critical point depends only on B_0^2 (see below), we lose no generality in considering only the critical points of III, which we can see from figs. 4 and 5 give rise to an interface possessing an hexagonal structure with circular elevations at the centers of the hexagons should $A_0 > 0$ and depressions if $A_0 < 0$ with both the diameter of these regular hexagons and the distance between adjacent centers equal to $2\lambda/\sqrt{3}$. Further, defining

$$C(x, y, z) = \lim_{t \to \infty} C(x, y, z, t),$$

we note, that for an interface of the form of (18), the interfacial solute concentration to first order is given by [5]

$$C(x, y) = C(x, y, A_0 f) = C_0\left[1 + \frac{(k-1)g(\alpha_0^2)}{k + g(\alpha_0^2)} A_0 f(x, y)\right], \quad \text{with} \quad g(\alpha_0^2) = \sqrt{\alpha_0^2 + \tfrac{1}{4}} - \tfrac{1}{2}. \tag{19}$$

In the stability analysis that follows we shall concentrate on the critical points of II and III and defer until section 3 a discussion of the critical points of V, which represent cells bounded by some closed nonhexagonal contour. We investigate the local stability of these critical points by considering solutions of (6) of the form

$$A(\tau) = A_0 + \varepsilon X e^{\rho\tau} + \mathcal{O}(\varepsilon^2) \tag{20a}$$

and

$$B(\tau) = B_0 + \varepsilon Y e^{\rho\tau} + \mathcal{O}(\varepsilon^2), \quad |\varepsilon| \ll 1. \tag{20b}$$

Upon substituting these solutions into (6), neglecting terms of $\mathcal{O}(\varepsilon^2)$, and cancelling the common factor $\varepsilon e^{\rho\tau}$, we obtain a system of two linear homogeneous equations in the constants X and Y. In order to ensure

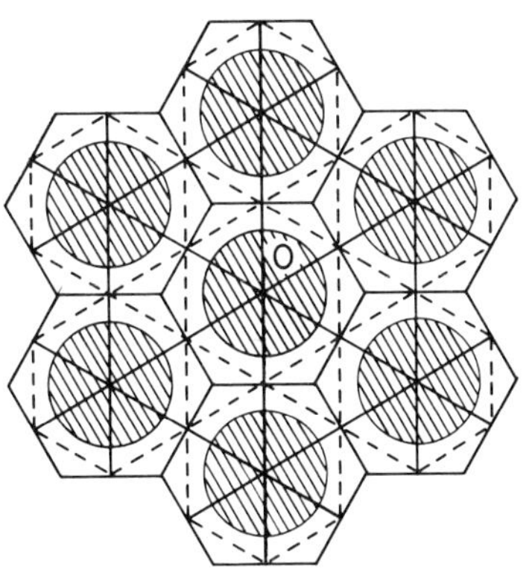

Fig. 4. A plot of $z = f(x, y)$ relevant to hexagonal symmetry. The shaded regions lie above the plane $z = 0$ and the unshaded ones, below it. Here point O corresponds to the origin [33].

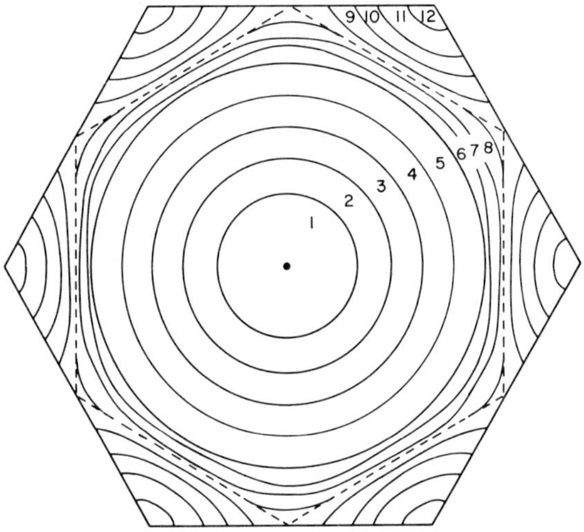

Fig. 5. An enlargement of the hexagonal cell containing the origin (the point at the center) with representative level curves. The origin corresponds to a z value of 3 while the curves labeled 1,2,…,12 are for values of $z = 2.25$, 1.50, 0.75, 0, −0.60, −0.75, −0.90, −0.975, −1.08, −1.20, −1.32, −1.44, respectively. The value of z on the inscribed hexagon (shown by broken lines as in fig. 4) is −1 while at the vertices of the principal hexagon $z = -1.50$ [34].

the existence of nontrivial solutions for this system we equate the determinant of its coefficients to zero which yields the following equation that ρ must satisfy:

$$\rho^2 + \mathscr{A}\rho + \mathscr{B} = 0, \tag{21a}$$

where

$$\mathscr{A} = -2\sigma + 4aA_0 + (3a_1 + 2a_2)A_0^2 + \tfrac{1}{4}(3a_1 + 10a_2)B_0^2$$

and

$$\mathscr{B} = \left(\sigma - 3a_1A_0^2 - a_2B_0^2\right)\left[\sigma - 4aA_0 - 2a_2A_0^2 - \tfrac{3}{4}(a_1 + 2a_2)B_0^2\right] - 8B_0^2(a + 2a_2A_0)^2. \tag{21b}$$

We say that a critical point (A_0, B_0) is locally stable provided (17a) holds which is guaranteed for solutions of the form of (20) if $\mathrm{Re}\,\rho < 0$ for all ρ satisfying (21). This in turn is true when [22]

$$\mathscr{A} > 0 \quad \text{and} \quad \mathscr{B} > 0, \tag{22}$$

which serves as our stability criteria in what follows.

We first note that for critical point I

$$\mathscr{A} = -2\sigma \quad \text{and} \quad \mathscr{B} = \sigma^2, \quad \text{or } \rho = \sigma, \sigma.$$

Hence the planar interface is locally stable provided $\sigma < 0$. Since we are ultimately interested in examining nonplanar morphologies where Wollkind and Segel [4] predicted bands for their two-dimensional nonlinear stability analysis, we now restrict ourselves to the case where σ, $a_1 > 0$, which are necessary conditions for the stability of the critical points of II. Further, to guarantee the existence of the critical points of III, we

Table I
Two-disturbance stability results for critical points II

a	$2a_2 - a_1$	Stability behavior
+	−	$A_0^{(1)}$ is stable for $0 < \sigma < \sigma_1$ $A_0^{(2)}$ is unstable
+	+	$A_0^{(1)}$ is stable for $\sigma > 0$ $A_0^{(2)}$ is stable for $\sigma > \sigma_1$
−	+	$A_0^{(1)}$ is stable for $\sigma > \sigma_1$ $A_0^{(2)}$ is stable for $\sigma > 0$
−	−	$A_0^{(1)}$ is unstable $A_0^{(2)}$ is stable for $0 < \sigma < \sigma_1$

also require that $a_1 + 4a_2 > 0$. Thus, in what follows, we shall ignore the possibility of subcritical ($\sigma < 0$) hexagonal structures [10], as well as that of the stability of the critical points of III when $a_1 + 4a_2 < 0$.

We next investigate the stability behavior of the critical points of II and III. The results of this investigation [21], which depend only on the signs of a and $2a_2 - a_1$, are catalogued in tables I and II where

$$\sigma_1 = \frac{16a_1a^2}{(2a_2 - a_1)^2} \quad \text{and} \quad \sigma_2 = \frac{32(a_1 + a_2)a^2}{(2a_2 - a_1)^2}. \tag{23}$$

3. The six-disturbance method of nonlinear stability analysis

Motivated by a generalization of an interface of the form of (5) and after Segel [10], we now assume that to lowest order the interface satisfies the equation

$$\begin{aligned} z = \zeta(x, y, t) = {} & A(\tau)\cos(\omega x) + B(\tau)\cos(\omega x/2)\cos(\sqrt{3}\,\omega y/2) + D(\tau)\cos(\omega x/2)\sin(\sqrt{3}\,\omega y/2) \\ & + E(\tau)\sin(\omega x) + F(\tau)\sin(\omega x/2)\cos(\sqrt{3}\,\omega y/2) + G(\tau)\sin(\omega x/2)\sin(\sqrt{3}\,\omega y/2), \end{aligned} \tag{24}$$

Table II
Two-disturbance stability results for critical points III

a	$2a_2 - a_1$	Stability behavior
+	−	$A_0^{(1)}$ is stable for $\sigma > \sigma_2$ $A_0^{(2)}$ is stable for $\sigma > 0$
+	+	$A_0^{(1)}$ is unstable $A_0^{(2)}$ is stable for $0 < \sigma < \sigma_2$
−	+	$A_0^{(1)}$ is stable for $0 < \sigma < \sigma_2$ $A_0^{(2)}$ is unstable
−	−	$A_0^{(1)}$ is stable for $\sigma > 0$ $A_0^{(2)}$ is stable for $\sigma > \sigma_2$

for which

$$\frac{\mathrm{d}A}{\mathrm{d}\tau} = \sigma A - a(B^2 + D^2 - F^2 - G^2) - A\left[a_1(A^2 + E^2) + a_2(B^2 + D^2 + F^2 + G^2)\right], \tag{25a}$$

$$\frac{\mathrm{d}B}{\mathrm{d}\tau} = \sigma B - 4a(AB + EF) - \tfrac{1}{2}(2a_2 - a_1)DFG$$
$$- B\left[2a_2(A^2 + E^2) + \tfrac{1}{4}(a_1 + 2a_2)(B^2 + D^2 + F^2) + \tfrac{1}{4}(3a_1 - 2a_2)G^2\right], \tag{25b}$$

$$\frac{\mathrm{d}D}{\mathrm{d}\tau} = \sigma D - 4a(AD + EG) - \tfrac{1}{2}(2a_2 - a_1)BFG$$
$$- D\left[2a_2(A^2 + E^2) + \tfrac{1}{4}(a_1 + 2a_2)(B^2 + D^2 + G^2) + \tfrac{1}{4}(3a_1 - 2a_2)F^2\right], \tag{25c}$$

$$\frac{\mathrm{d}E}{\mathrm{d}\tau} = \sigma E - 2a(BF + DG) - D\left[a_1(A^2 + E^2) + a_2(B^2 + D^2 + F^2 + G^2)\right], \tag{25d}$$

$$\frac{\mathrm{d}F}{\mathrm{d}\tau} = \sigma F + 4a(AF - BE) - \tfrac{1}{2}(2a_2 - a_1)BDG$$
$$- F\left[2a_2(A^2 + E^2) + \tfrac{1}{4}(a_1 + 2a_2)(B^2 + F^2 + G^2) + \tfrac{1}{4}(3a_1 - 2a_2)D^2\right], \tag{25e}$$

and

$$\frac{\mathrm{d}G}{\mathrm{d}\tau} = \sigma G + 4a(AG - DE) - \tfrac{1}{2}(2a_2 - a_1)BDF$$
$$- G\left[2a_2(A^2 + E^2) + \tfrac{1}{4}(a_1 + 2a_2)(D^2 + F^2 + G^2) + \tfrac{1}{4}(3a_1 - 2a_2)B^2\right], \tag{25f}$$

where the forms of the amplitude equations can be deduced by following the procedure of ref. 23.
Alternatively, motivated by the form of the interface of (18), we can rewrite (24) as [10]

$$z = \zeta(x, y, t) = A_1(\tau)\cos\left[\omega x + \theta_1(\tau)\right] + A_2(\tau)\cos\left[\omega(x - \sqrt{3}\,y)/2 - \theta_2(\tau)\right]$$
$$+ A_3(\tau)\cos\left[\omega(x + \sqrt{3}\,y)/2 - \theta_3(\tau)\right], \tag{26}$$

and replace (25) by

$$\frac{\mathrm{d}A_1}{\mathrm{d}\tau} = \sigma A_1 - 4aA_2A_3\cos(\theta_1 + \theta_2 + \theta_3) - A_1\left[a_1A_1^2 + 2a_2(A_2^2 + A_3^2)\right], \tag{27a}$$

$$\frac{\mathrm{d}A_2}{\mathrm{d}\tau} = \sigma A_2 - 4aA_1A_3\cos(\theta_1 + \theta_2 + \theta_3) - A_2\left[2a_2(A_1^2 + A_3^2) + a_1A_2^2\right], \tag{27b}$$

$$\frac{\mathrm{d}A_3}{\mathrm{d}\tau} = \sigma A_3 - 4aA_1A_2\cos(\theta_1 + \theta_2 + \theta_3) - A_3\left[2a_2(A_1^2 + A_2^2) + a_1A_3^2\right], \tag{27c}$$

$$A_1\frac{\mathrm{d}\theta_1}{\mathrm{d}\tau} = 4aA_2A_3\sin(\theta_1 + \theta_2 + \theta_3), \tag{27d}$$

$$A_2\frac{\mathrm{d}\theta_2}{\mathrm{d}\tau} = 4aA_1A_3\sin(\theta_1 + \theta_2 + \theta_3), \tag{27e}$$

$$A_3\frac{\mathrm{d}\theta_3}{\mathrm{d}\tau} = 4aA_1A_2\sin(\theta_1 + \theta_2 + \theta_3). \tag{27f}$$

We next wish to examine the stability of the critical points of these equivalent systems. It is much easier to find equilibrium solutions to (27) than to the equivalent set (25) while it is best to use (25) for stability calculations because the nonlinear terms contained therein are polynomials and simpler results can be utilized [10]. Proceeding in this manner we find that the critical points consist of (see appendix 3 of ref. 10)

(i) the planar interface solution which is equivalent to critical point I of the previous section;

(ii) a one-parameter family of bands aligned parallel to the y-axis which is equivalent to the critical points of II and translations thereof, plus two similar families of bands making angles of $\pm 60°$ to these for which superposition with a member of the original family at equilibrium is not possible;

(iii) hexagonal structures equivalent to the critical points of III and translations thereof;

(iv) closed cells equivalent to the critical point of V (they can be shown to be equivalent in the same manner as the critical points of III and IV were [5]) and translations plus rotations by $\pm 60°$ thereof.

Since axis translations and rotations cannot affect stability we need only examine the stability of the critical points of the previous section to deduce the morphology of the interface. Performing a local stability analysis of these critical points for system (25) analogous to that described in section 2 we find four new conditions in addition to those of our two-disturbance analysis (see appendix 4 of ref. 10). For critical point I the four new roots ρ are

$$\sigma, \sigma, \sigma, \sigma;$$

hence as before the planar interface solution is stable if and only if $\sigma < 0$.

For the critical points of II, the four new roots are

$$\sigma - 4aA_0 - 2a_2A_0^2, \quad 0, \quad \sigma + 4aA_0 - 2a_2A_0^2, \quad \sigma + 4aA_0 - 2a_2A_0^2.$$

The zero root is a consequence of our critical points occurring on a two-dimensional equilibrium surface in the six-dimensional phase space and can be disregarded for our purposes as can the other zero roots which will arise in this section [10]. The nonzero roots must all be negative for stability. Since our previous two-disturbance conditions, from which table I was generated, were [21]

$$\sigma > 0, \quad \sigma - 4aA_0 - 2a_2A_0^2 < 0,$$

while $A_0^{(2)} = -A_0^{(1)}$, we see that stable bands will only occur where *both* $A_0^{(1)}$ and $A_0^{(2)}$ are stable.

For the critical points of III, the four new roots are

$$0, \quad 0, \quad 12aA_0, \quad 2\left[4aA_0 + (2a_2 - a_1)A_0^2\right].$$

Since our previous two-disturbance stability criterion, from which we generated table II, was [21]

$$4aA_0 + (2a_2 - a_1)A_0^2 < 0,$$

we obtain the additional condition that $aA_0 < 0$ for stability of hexagonal structures. Upon examination of figs. 4 and 5 and eq. (19) we conclude that, for alloys having $k < 1$, hexagonal structures with $A_0 > 0$ correspond to *cells* with dome-shaped centers projecting into the liquid and higher solute concentrations at the boundaries [24] while those with $A_0 < 0$ correspond to a close-packed arrangement of *nodes* or circular depressions of solute rich liquid into the solid [25] (see fig. 6). We also note that the $A_0^{(1)}$ and $A_0^{(2)}$ of table II

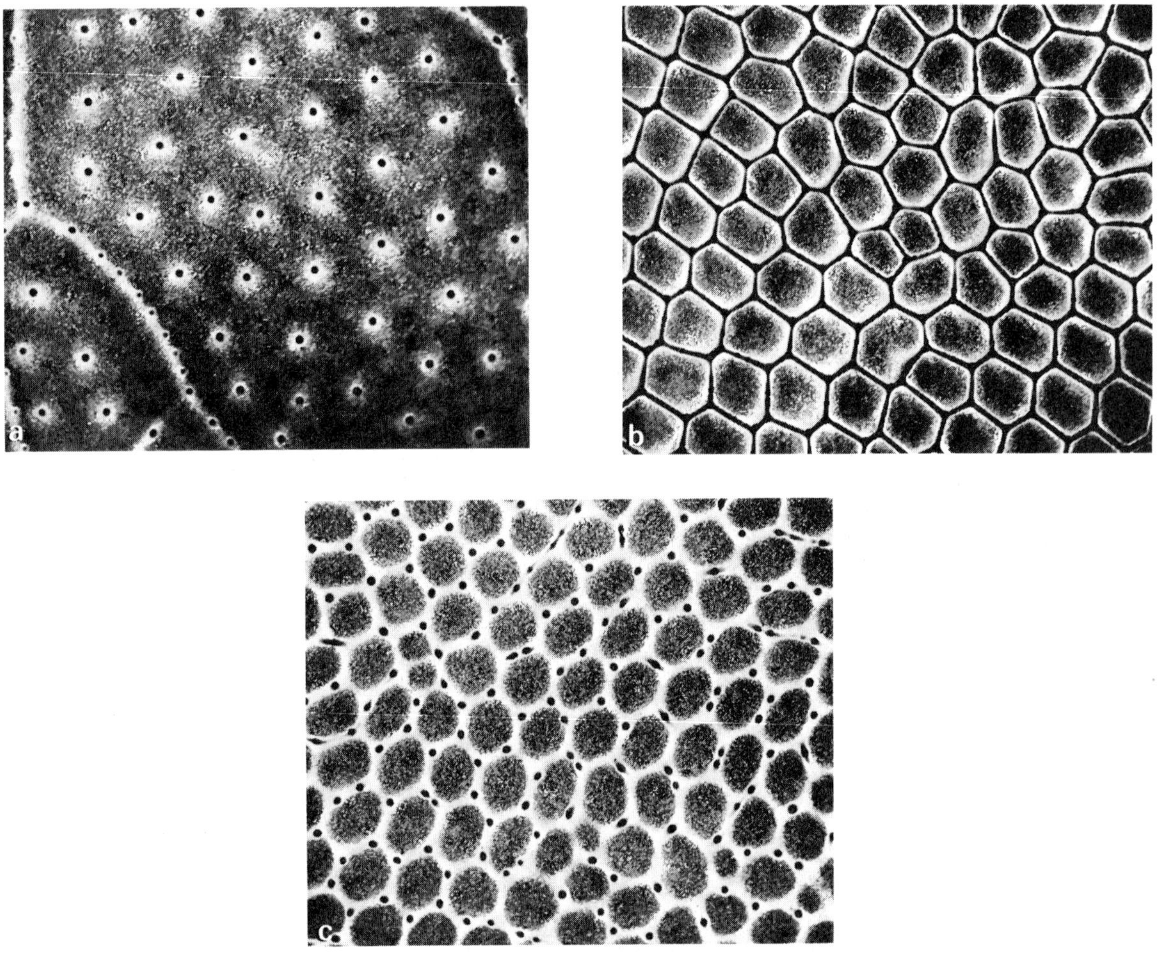

Fig. 6. Photomicrographs of a) nodes on a cross-section taken near a quenched interface; b) cells on a cross-section taken at the quenched interface; and c) cells on a cross-section taken 1 mm before the quench. Here the dark regions are those of highest solute concentration, the white regions those of next highest concentration, and the gray regions those of lowest concentration (courtesy of Morris and Winegard [9]).

correspond to cells and nodes respectively since $A_0^{(1)} > 0$ and $A_0^{(2)} < 0$ for the parameter range under consideration. Hence from our new stability criterion deduced above, we see that stable *nodes* can only occur when $a > 0$ and stable *cells* only when $a < 0$. Interpreting the combined results of tables I and II in conjunction with the observations contained in the previous two paragraphs, we arrive at the conclusions about morphological stability summarized in table III.

We have deferred until now a discussion of the stability of the critical points of V (as mentioned earlier we lose no generality by considering only the "+" solution). We first observe that critical point V is only defined for $\sigma > \sigma_1$. The two-disturbance analysis of the previous section yields the following stability condition for the parameter range under examination in this paper:

$$\frac{2a_2 - a_1}{a_1 + 2a_2}(\sigma - \sigma_1)(\sigma - \sigma_2) > 0,$$

Table III
Six-disturbance stability results for bands, nodes, and cells

a	$2a_2 - a_1$	Stability behavior
+	−	nodes are stable for $\sigma > 0$
+	+	nodes are stable for $0 < \sigma < \sigma_2$ bands are stable for $\sigma > \sigma_1$
−	+	cells are stable for $0 < \sigma < \sigma_2$ bands are stable for $\sigma > \sigma_1$
−	−	cells are stable for $\sigma > 0$

Table IV
Values of the solvability constants versus V

V (10^{-3} cm/s)	a	a_1	a_2	$a_1 + 4a_2$	$2a_2 - a_1$
3.3	5.1×10^{-2}	6.1×10^{-2}	-2.5×10^{-2}	-4.0×10^{-2}	-1.1×10^{-1}
3.5	4.8×10^{-2}	7.0×10^{-2}	-1.2×10^{-2}	2.1×10^{-2}	-9.4×10^{-2}
3.8	4.5×10^{-2}	7.7×10^{-2}	1.7×10^{-4}	7.8×10^{-2}	-7.7×10^{-2}
4.1	4.2×10^{-2}	8.3×10^{-2}	1.1×10^{-2}	9.5×10^{-2}	-6.0×10^{-2}
4.4	3.9×10^{-2}	8.8×10^{-2}	2.1×10^{-2}	1.7×10^{-1}	-4.5×10^{-2}
4.7	3.6×10^{-2}	9.2×10^{-2}	3.0×10^{-2}	2.1×10^{-1}	-3.1×10^{-2}
5.1	3.3×10^{-2}	9.5×10^{-2}	3.8×10^{-2}	2.5×10^{-1}	-1.8×10^{-2}
5.5	3.0×10^{-2}	9.7×10^{-2}	4.5×10^{-2}	2.8×10^{-1}	-6.6×10^{-3}
5.9	2.8×10^{-2}	9.8×10^{-2}	5.1×10^{-2}	3.0×10^{-1}	3.6×10^{-3}
6.4	2.5×10^{-2}	9.8×10^{-2}	5.5×10^{-2}	3.2×10^{-1}	1.3×10^{-2}
6.9	2.2×10^{-2}	9.7×10^{-2}	5.9×10^{-2}	3.3×10^{-1}	2.0×10^{-2}
7.5	2.0×10^{-2}	9.6×10^{-2}	6.1×10^{-2}	3.4×10^{-1}	2.6×10^{-2}
8.2	1.7×10^{-2}	9.3×10^{-2}	6.2×10^{-2}	3.4×10^{-1}	3.1×10^{-2}
9.0	1.5×10^{-2}	9.1×10^{-2}	6.3×10^{-2}	3.4×10^{-1}	3.4×10^{-2}
9.8	1.3×10^{-2}	8.7×10^{-2}	6.2×10^{-2}	3.3×10^{-1}	3.7×10^{-2}
10.7	1.0×10^{-2}	8.3×10^{-2}	6.0×10^{-2}	3.2×10^{-1}	3.7×10^{-2}
11.7	8.2×10^{-3}	7.9×10^{-2}	5.8×10^{-2}	3.1×10^{-1}	3.7×10^{-2}
12.9	6.1×10^{-3}	7.4×10^{-2}	5.5×10^{-2}	2.9×10^{-1}	3.6×10^{-2}
14.2	4.0×10^{-3}	6.9×10^{-2}	5.1×10^{-2}	2.7×10^{-1}	3.3×10^{-2}
15.7	2.1×10^{-3}	6.4×10^{-2}	4.7×10^{-2}	2.5×10^{-1}	3.0×10^{-2}
17.4	2.4×10^{-4}	5.8×10^{-2}	4.2×10^{-2}	2.2×10^{-1}	2.5×10^{-2}
19.3	-1.5×10^{-3}	5.2×10^{-2}	3.6×10^{-2}	2.0×10^{-1}	2.0×10^{-2}
21.5	-3.2×10^{-3}	4.7×10^{-2}	3.1×10^{-2}	1.7×10^{-1}	1.4×10^{-2}
23.9	-4.7×10^{-3}	4.1×10^{-2}	2.5×10^{-2}	1.4×10^{-1}	8.3×10^{-3}
26.7	-6.2×10^{-3}	3.5×10^{-2}	1.9×10^{-2}	1.1×10^{-1}	1.8×10^{-3}
29.9	-7.5×10^{-3}	3.0×10^{-2}	1.2×10^{-2}	8.0×10^{-2}	-4.8×10^{-3}
33.6	-8.7×10^{-3}	2.5×10^{-2}	6.5×10^{-3}	5.0×10^{-2}	-1.2×10^{-2}
37.8	-9.8×10^{-3}	1.9×10^{-2}	6.5×10^{-4}	2.2×10^{-2}	-1.8×10^{-2}
42.6	-1.1×10^{-2}	1.5×10^{-2}	-4.9×10^{-3}	-4.7×10^{-3}	-2.5×10^{-2}

while the six-disturbance analysis yields the four new roots

$$0, \quad 0, \quad \frac{a_1 - 2a_2}{a_1 + 2a_2}(\sigma + \sigma_3), \quad 2\frac{(2a_2 - a_1)}{a_1 + 2a_2}(\sigma - \sigma_2),$$

where

$$\sigma_3 = 16a^2 \frac{(a_1 + 4a_2)}{(2a_2 - a_1)^2}.$$

Since the fourth root being negative is incompatible with our two-disturbance stability condition, critical point V is unstable for the parameter range in question.

Finally, upon examining critical points II, III, and V as defined at the beginning of section 2, we observe that by taking $\sigma = \mathcal{O}(a^2)$, consistent with (23), we can guarantee that the nonzero components of these critical points will be of $\mathcal{O}(a)$ provided $a_1, a_1 + 4a_2 = \mathcal{O}(1)$ as $a \to 0$. Thus, if $|a|$ is small in comparison with both a_1 and $a_1 + 4a_2$, it is permissible to have truncated the amplitude equations by retaining only those terms in (6) through third order [7]. An examination of table IV introduced in the next section shows that these conditions on a, a_1, and $a_1 + 4a_2$ are fulfilled for the solidification of the dilute binary alloy under investigation.

4. Comparison between theory and experiment

Once a particular alloy of a given concentration is selected, the solvability constants a, a_1, and a_2 defined by (14a), (16a), and (16b), respectively, at $\sigma = 0$ can be calculated as functions of V alone since they now depend only on the quantity α_0 which is determined implicitly by V through relation (12). Then we can use the stability criteria given in table III to deduce interfacial morphology and, in conjunction with our knowledge of the behavior of σ, represent these morphologies graphically in the V–G_L plane.

Toward that end, we assigned the following typical values to the relevant parameters [13]:

$$T_0 = 500 \text{ K}, \quad C_0 = 0.01\% \text{ by wt. of solute}, \quad m = -2°\text{C}/\% \text{ by wt. of solute},$$
$$\kappa_S/\kappa_L = 2, \quad \Gamma_P = 10^{-8}\,\text{cm}, \quad \mathscr{D} = 10^{-5}\,\text{cm}^2/\text{s}, \tag{28}$$

and evaluated a, a_1, and a_2 versus V for various values of k, $\partial\gamma/\partial T$, and $\partial\gamma/\partial C$ [21]. We observe that for $\gamma \equiv \gamma_P$, i.e., $\partial\gamma/\partial T = \partial\gamma/\partial C = 0$, conditions for stable hexagonal cells were not met for any value of k. In particular, we found that, for $\gamma \equiv \gamma_P$, $a > 0$ for all V and hence, given the necessary condition for the stability of cells mentioned in the previous section, this guarantees the instability of such structures. We also observe that the problem appeared to be virtually independent, in a quantitative sense, of our choice for $\partial\gamma/\partial T$ but depended crucially on the parameters $\partial\gamma/\partial C$ and k. In particular for $\partial\gamma/\partial C < 0$ stable hexagonal structures only existed for $k > 1$ while for $\partial\gamma/\partial C > 0$ they only existed for $k < 1$ [21]. In the analysis that follows we shall consider an alloy for which $\partial\gamma/\partial C = 2.7\gamma_P/C_0$, $k = 0.1$, and, for the sake of definiteness, $\partial\gamma/\partial T = -10\gamma_P/T_0$ (note that $-\gamma_0 < T_0\partial\gamma/\partial T + C_0\partial\gamma/\partial C < \gamma_P$). The values of the quantities needed to identify the various interfacial morphologies in the V–G_L plane are given in tables IV and V

Table V
The behavior of $\sigma = 0$, σ_1, and σ_2 in the V–G_L plane

V (10^{-3} cm/s)	G_0 (°C/cm)	σ_1	G_1 (°C/cm)	σ_2	G_2 (°C/cm)
3.5	7.2	2.9×10^{-1}	4.2	4.7×10^{-1}	3.1
3.8	7.7	4.2×10^{-1}	3.5	8.3×10^{-1}	1.1
4.1	8.2	6.3×10^{-1}	2.3	1.4	− 2.6
4.4	8.7	1.0	− 0.9	2.6	− 6.7
4.7	9.3	2.0	− 5.8	5.2	−17.6
5.1	9.9	5.0	−20.6	1.4×10^{1}	−42.1
5.5	10.5	3.2×10^{1}	—	9.4×10^{1}	—
5.9	11.2	9.0×10^{1}	—	2.7×10^{2}	—
6.4	12.0	6.2	− 3.2	1.9×10^{1}	—
6.9	12.9	1.9	− 1.2	6.2	−38.7
7.5	13.8	8.7×10^{-1}	0.3	2.9	−24.0
8.2	14.7	4.7×10^{-1}	3.0	1.5	−15.7
9.0	15.8	2.7×10^{-1}	7.3	9.2×10^{-1}	− 5.6
9.8	17.0	1.7×10^{-1}	10.5	5.6×10^{-1}	0.6
10.7	18.2	1.0×10^{-1}	13.8	3.5×10^{-1}	5.2
11.7	19.6	6.1×10^{-2}	15.9	2.1×10^{-1}	9.8
12.9	21.1	3.4×10^{-2}	19.1	1.2×10^{-1}	13.9
14.2	22.6	1.6×10^{-2}	21.0	5.7×10^{-2}	18.3
15.7	24.3	5.1×10^{-3}	23.5	1.8×10^{-2}	22.0
17.4	26.2	8.3×10^{-5}	26.0	2.9×10^{-4}	26.0
19.3	28.1	4.8×10^{-3}	28.0	1.6×10^{-2}	25.9
21.5	30.2	3.6×10^{-2}	25.6	1.2×10^{-1}	16.8
23.9	32.4	2.1×10^{-1}	9.1	6.8×10^{-1}	−18.6
26.7	34.8	6.3	—	1.9×10^{1}	—
29.9	37.3	1.1	−47.5	3.3	—
33.6	39.8	2.2×10^{-1}	3.9	5.6×10^{-1}	−26.1
37.8	42.3	8.9×10^{-2}	22.3	1.9×10^{-1}	7.4

and these morphologies are represented graphically in fig. 7. Here we have adopted the notation

$$G_i(V) = G_{\sigma_i}(V), \quad i = 1 \text{ and } 2, \tag{29}$$

where σ_1, σ_2 are as defined in (23), and in fig. 7 have characterized the loci $G_L = G_0(V)$, $G_1(V)$, and $G_2(V)$ by $\sigma = 0$, σ_1, and σ_2, respectively.

We now wish to make an interpretation of these results with respect to the experimental observations of Morris and Winegard [8, 9]. In particular the latter made a very careful study of the sequence of morphologies from which regular cell structure develops as a function of V for a dilute binary Pb–Sb alloy with $k < 1$ [9] as well as a comparison of microsegregation and morphology of cells formed during the solidification of alloys having solutes with $k < 1$ and $k > 1$ [8]. As a prelude to this discussion we reexamine our morphological predictions in some detail. The level curves of $z = f(x, y)$ in figs. 4 and 5 for z in the range $0 \le z < 3$ are circular in shape (the shaded regions of fig. 4); for $-1 \le z < 0$, oval to hexagonal; and for $-1.50 < z < -1$, they are centered about the vertices. The principal hexagon which forms the boundary of the cells is not itself a level curve but is at a variable depth ranging from $z = -1.50$ at the vertices to $z = -1$ at the mid-points of each edge. Hence, from figs. 4 and 5 and (19), we see that for cells with $k < 1$ the highest solute concentration should occur at the vertices with the lowest concentrations occurring at the

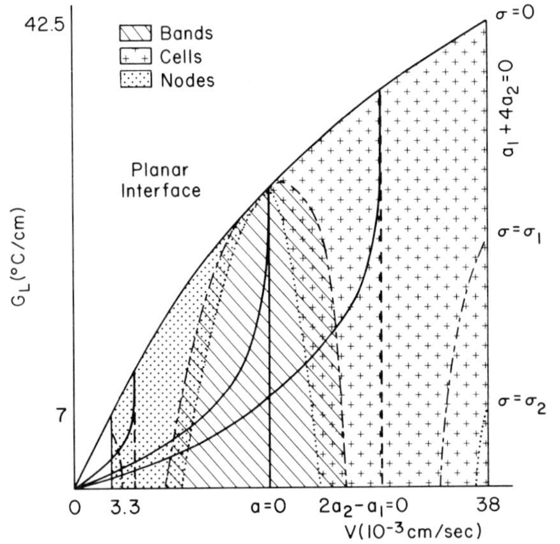

Fig. 7. A plot of G_L versus V denoting the predicted interfacial morphologies. Here the curves starting at the origin and asymptotically approaching the vertical lines $2a_2 - a_1 = 0$, $a = 0$, and $2a_2 - a_1 = 0$, respectively, are schematic representations of what those lines might be transformed into should the Landau coefficients be considered functions of σ. This figure has been drawn for $\partial\gamma/\partial C = 2.7\gamma_P/C_0$. Observe that as $\partial\gamma/\partial C > 0$ decreases the vertical lines would be shifted to the right [21].

circular or sometimes oval caps. This is consistent with the findings of Morris and Winegard [9] during the growth of cells for such alloys (see fig. 6c) and also would seem to verify one of their conjectures regarding cellular morphology. Morris and Winegard [9] photographed regular cells in two cross-sections taken near the position where the growth front was quenched which we have reproduced in fig. 6b, c. At the quenched interface, fig. 6b, the cells appear to have straight boundaries forming regular hexagons. In fig. 6c, typical of sections taken before the quench, the cell caps appear to have an oval or circular cross-section while the vertices are marked by dark circles and the edges by white regions. Due to the lighting conditions the most solute rich regions appear dark, the next richest appear white, and the lowest, gray [9]. Morris and Winegard [9] conjectured that the hexagonal shape of the regular cells (quenched interface of fig. 6b and, by inference, the decanted interface shape of fig. 2d) was produced by the lateral growth of the cell cap during quenching and that the cell caps are more circular in cross-section during growth. Our theoretical predictions would seem to bear out this conjecture. In the case of nodes, which Morris and Winegard [9] describe as circular wells of liquid penetrating the solid, the shaded regions of fig. 4 now represent depressions. Since from (19) these regions are solute rich, they each appear as a dark circle surrounded by a white annulus in fig. 6a while the unshaded regions of fig. 4 which are now depleted of solute appear as an uniformly gray background. Further, these nodes are of uniform circular cross-section and arranged in a close-packed formation (see fig. 6a) consistent with the prediction of fig. 4, an arrangement first remarked upon by Sharp and Hellawell [25]. The fact that only regions of high solute concentration are delineated in such photographs is dramatically demonstrated in [8] where a cellular structure for an Al–Ti dilute alloy with $k = 13$ appears as dark circles surrounded by uniformly gray boundaries, which lack any demarcations at the vertices and are quite broad when compared to cells in alloys having $k < 1$. This is consistent with our predictions from (19) and figs. 4 and 5 of solute segregation for alloys with $k > 1$.

Having examined the morphology and solute segregation of our predicted hexagonal structures, we interpret our results of fig. 7 within the framework of the experiments of [9]. Morris and Winegard [9] found that for a *fixed* value of G_L in the 10 to 12°C/cm range the sequence of morphologies observed during the transition from a planar to a cellular interfacial structure, as V was increased, depended strongly upon the crystallographic orientation of the planar interface. In particular for their face centered cubic crystalline lattice they found that when growth was in a ⟨100⟩ direction this sequence consisted of a planar interface, a regularly-spaced array of nodes, curved elongated cells the boundaries of which were random in orientation, and finally hexagonal cells while for growth in a ⟨110⟩ direction the sequence was planar interface, elongated or "two-dimensional" cells, and then regular cells. Taking $G_L = 11$°C/cm in fig. 7, we would predict a sequence of planar interface, nodes, bands, and then cells. We note that in the overlap regions where bands and nodes or bands and cells, respectively, are both stable the initial conditions determine which of these possible morphologies are actually observed [7, 10]. We further observe from our results of section 3 that not only are two-dimensional cells aligned parallel to the y-axis stable structures in a region where we predict bands but also similar families of two-dimensional cells, making angles of $\pm 60°$ to these, are stable as well (see fig. 4 where the three sets of parallel lines intersecting at angles of 60° are representative of these families of bands). Then, as the initial conditions vary from point-to-point on the interfacial surface (see fig. 8), such families of two-dimensional cells could give rise to polygonal arcs which would appear quite random in orientation. Hence, we now identify our band region with those structures described by Morris and Winegard as "curved" elongated cells and conclude that our sequence of morphologies for $G_L = 11$°C/cm corresponds to growth in the ⟨100⟩ direction.

More precisely, our choice for γ_P, and hence Γ_P, corresponds to growth in the ⟨100⟩ direction for $G_L = 11$°C/cm and these qualitative results would hold for any constant G_L up to approximately that value where the line $a = 0$ intersects the curve $\sigma = 0$ in fig. 7. In order to determine how a different choice for γ_P would affect our results in a quantitative manner, it is necessary to understand how fig. 7 was deduced in [21]. Sriranganathan's [21] final results depended only on the values of two nondimensional quantities H, a dimensionless temperature gradient, and U, a dimensionless solidification rate, which were related to G_L and V, respectively, by

$$G_L = \frac{T_0}{\Gamma_P} H \quad \text{and} \quad V = \frac{\mathscr{D}}{\Gamma_P} U. \tag{29}$$

If Γ_P were increased then the plot of G_L versus V in fig. 7 would correspondingly be reduced in scale. Thus $G_L = 11$°C/cm could now intersect the curve $\sigma = 0$ in that region where $\sigma = \sigma_1$ is virtually tangent to that curve. In that event, the sequence of morphologies, given the proper initial conditions (see fig. 8), would be

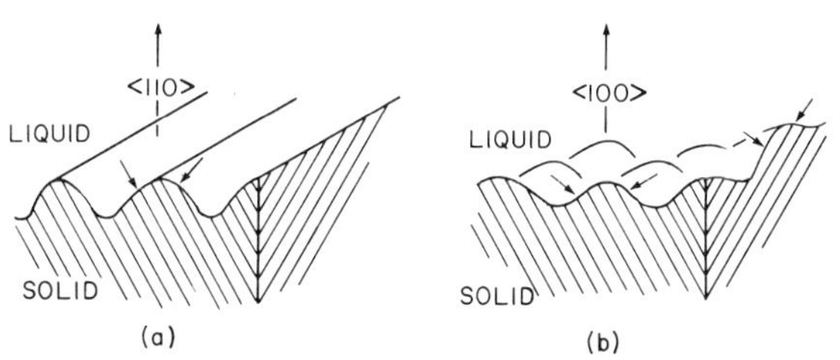

Fig. 8. Schematic representation of developing perturbations from a planar interface for a) the ⟨110⟩ growth direction (elongated cells develop); and b) the ⟨100⟩ growth direction (nodes develop). It is plausible that bands developing from a uniform planar interface would all be aligned parallel as is the case for the ⟨110⟩ growth direction while those developing from nodes would give rise to all three possible band types due to the nonuniformity of initial conditions as one moves from point to point on the interface (courtesy of Morris and Winegard [9]).

planar interface, parallel two-dimensional cells, and regular cells. This, of course, is the sequence Morris and Winegard found to be associated with growth in the $\langle 110 \rangle$ direction. Such an interpretation would be consistent with the conjecture of Sharp and Hellawell [26] that $\gamma_{110} > \gamma_{100}$ and is the most simplistic way to introduce the effect of anisotropy implicitly into an isotropic model.

We also note that if we hold V constant in the range from 10×10^{-3} to 22×10^{-3} cm/s and decrease G_L along such a vertical line in fig. 7 for $\sigma > 0$, the morphological behavior is very reminiscent of the sequence of convection patterns predicted in the Bénard problem as R, the Rayleigh number, is increased [7, 10]. In particular for V in this range such that $a > 0$ ($V < 18 \times 10^{-3}$ cm/s), our sequence is nodes, nodes or bands, and bands while for $a < 0$ ($V > 18 \times 10^{-3}$ cm/s) it consists of cells, cells or bands, and bands. For the Bénard problem the comparable transition states as R increases are hexagons, hexagons or rolls, and rolls with the flow direction in the center of the hexagons opposite in sign to a [7, 10].

As a final conclusion we wish to demonstrate explicitly the range of spacing characteristic of the various interfacial morphologies predicted by our model. Recall that the width of a band is given by λ while both the distance between adjacent nodes and the diameter of a cell is given by $d = 2\lambda/\sqrt{3}$, where

$$\lambda = \frac{2\pi}{\omega} = \frac{2\pi}{\alpha_0} \frac{\mathscr{D}}{V}. \tag{30}$$

Since α_0 is defined implicitly as a function of V by (12), it is possible, using fig. 7, tables IV and V, and the definition of d given above, to determine those spacings for nodes, bands, and cells versus V from (30). These results are summarized in tables VI, VII and VIII. From those tables it is apparent that the distance between two adjacent nodes and the width of bands decrease much more rapidly with increasing V then does the diameter of cells. This is in agreement with the experimental observations of Morris and Winegard

Table VI
Distance between adjacent nodes versus V

V (10^{-3} cm/s)	d (μm)
3.5	213
3.8	204
4.1	194
4.4	185
4.7	175
5.1	167
5.5	158
5.9	150
6.4	142
6.9	134
7.5	127
8.2	119
9.0	112
9.8	106
10.7	100
11.7	94
12.9	88
14.2	82
15.7	77
17.4	72

Table VII
Width of bands versus V

V (10^{-3} cm/s)	λ (μm)
5.5	137
5.9	130
6.4	123
6.9	116
7.5	110
8.2	103
9.0	98
9.8	92
10.7	86
11.7	81
12.9	76
14.2	71
15.7	67
17.4	62
19.3	58
21.5	54
23.9	51
26.7	47

[9] who found the spacing characteristic of all three morphologies to decrease with increasing V but noted that cells seemed to be a much more stable structure in this regard than either nodes or elongated cells. Also our results for nodes and cells compare favorably with Morris and Winegard's [9] observation that the nodes appearing at the onset of instability were typically spaced about 200 μm apart and Tiller and Rutter's [27] measurements of the range of diameters of cells. Further, that the transition from elongated to regular cells proceeds by the folding of the interface down the center of the original cells thus producing cells of *one-half* the original spacing which fold laterally as V increases to form regular cells [9] is consistent with our results. In fact, we selected the value of $\partial\gamma/\partial C$ which has been employed for the exposition in this section precisely because that selection yielded results in good quantitative agreement with experimentally observed spacing of interfacial morphologies. Hence this model is predictive in the sense that it can be used to determine $\partial\gamma/\partial C$, a heretofore difficult quantity to measure experimentally [28].

Table VIII
Diameter of cells versus V

V (10^{-3} cm/s)	d (μm)
19.3	67
21.5	63
23.9	58
26.7	54
29.9	51
33.6	47
37.8	44

In summary our six-disturbance analysis has basically shown the three-dimensional model under investigation to be compatible with the formation of nodes, bands, and cells during plane front solidification of dilute binary alloys. Our results, which are presented graphically in fig. 7, can only be expected to be asymptotically valid in a limited $\sigma > 0$ neighborhood of the marginal curve $\sigma = 0$ in that figure by virtue of the way a, a_1 and a_2 were calculated. Moreover our $\sigma \approx 0$ analysis was only for the case of $a_1 + 4a_2 > 0$. We close by suggesting two extensions of the analysis presented in this paper. Although our predictions are in good agreement with the experiments of Morris and Winegard for a fixed value of G_L, they are not in such good agreement with the experiments of Takahashi et al. [19] in which both G_L and V were varied (see fig. 9). In order to provide agreement with such data it is attendant upon us to extend our results so that they apply far from the marginal curve in V–G_L space (note from table V that σ remains reasonably small for a considerable distance from the marginal curve in fig. 7). Recently, we have devised a method for doing so and it requires a departure from those procedures used by all researchers in nonlinear stability theory since Landau's initial contribution in 1944 [3]. When solving for Landau-type coefficients such as a in the amplitude equations, one always obtains equations such as (13), (15a), and (15b) of the generic form

$$a(\sigma) = \sigma\zeta(\sigma) + r(\sigma), \tag{31}$$

where $\zeta(\sigma)$ is the interfacial perturbation at that order and $r(\sigma)$ is a known function. We, as has traditionally been done for the last forty years, in this paper have taken

$$a(\sigma) \equiv a(0) = r(0) \tag{32a}$$

– that is, assumed a to be *constant* – and then solved for $\zeta(\sigma)$ obtaining

$$\zeta(\sigma) = \begin{cases} \dfrac{r(0) - r(\sigma)}{\sigma}, & \sigma \neq 0 \\ -\dfrac{dr(0)}{d\sigma}, & \sigma = 0. \end{cases} \tag{32b}$$

In order to extend our results to those $\sigma > 0$ reasonably far from the marginal curve in the V–G_L plane, it will be necessary for us to consider

$$\zeta(\sigma) \equiv \zeta(0) = -\frac{dr(0)}{d\sigma} \tag{33a}$$

and then calculate

$$a(\sigma) = r(\sigma) - \frac{dr(0)}{d\sigma}\sigma; \tag{33b}$$

that is, to assume the Landau-type coefficients are *not* constant. Then, for those σ such that $\sigma^{1/2}$ and $|a|$ are small in comparison with both a_1 and $a_1 + 4a_2$, it will still have been permissible to have truncated the amplitude equations. With the Landau coefficients determined by (33b) rather than (32a), the vertical lines in fig. 7 should then be transformed into curves passing through the origin (see fig. 7) in agreement with the observations of Takahashi et al. [19]. The curves $\sigma = \sigma_1$ and $\sigma = \sigma_2$ were calculated by determining how the linear problem behaved as σ varied through relevant positive values. Now we shall be able to obtain a similar behavior for the nonlinear problem. In addition by subsequent curve fitting we can deduce for

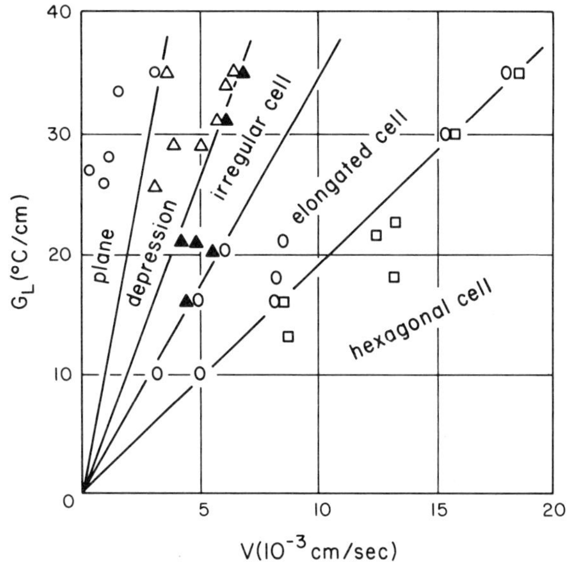

Fig. 9. The morphological transition of the solid–liquid interface from a plane to hexagonal cells in 99.997% Al for the experiments of Takahashi et al. [19]. The slopes of the lines measured in units of 10^3 °C s/cm^2 are 10, 5.1, 3.4, and 1.9 in order of transition.

which values of $b > 0$ a line of the form

$$G_L = bV \tag{34}$$

best fits the data separating the planar interface from nodes, nodes from bands, and bands from cells. Further we can determine precise elevations and solute segregation ratios for cell caps, edges, and vertices and, upon comparison of these and the slopes (see fig. 9) with the experimental findings of Takahashi et al. [19], select that value of $\partial\gamma/\partial C$ which provides the closest agreement with those results. Such predictions are of considerable technological importance. Controlled solidification of alloys and other binary substances is used to grow single crystals, refine materials, and obtain uniform (or non-uniform) composition within a material [29]. The most important industrial applications are growth of crystals for metal oxide semiconductors (MOS's) and growth of oxides for optical devices in laser systems [29]. The technology can be used to make crystals for jewels, and similar principles arise in casting processes in the steel and glass industries [30]. Quantitative predictions of solid–liquid interface morphologies, including information on cell size and intracellular solute distribution, are therefore extremely valuable to industrial researchers.

Our second extension involves redoing the analysis of the amplitude equations for $a_1 + 4a_2 < 0$ and at the same time considering the possibility of the occurrence of subcritical ($\sigma < 0$) structures. This most likely includes the region of dendritic growth [31] and critical point V almost certainly plays a fundamental role in the cell-dendrite transition [20]. Flemings [29] gives as good a rationale as any for the benefits of investigating this particular topic:

"Quantitative description of effects of solidification variables on the cell-dendrite transition is one that has eluded investigators in spite of the many studies conducted over the last 15 years.... This entire topic remains a good one for theoretical and experimental study."

Such an analysis may require incorporating the full interfacial balance of heat condition [6]

$$\kappa_S \frac{\partial T'}{\partial n} - \kappa_L \frac{\partial T}{\partial n} = \mathscr{L}[1 + (\Delta\Gamma)\kappa](V + \zeta_t), \tag{35}$$

where $\Delta\Gamma = (\gamma - \gamma_0)/L$ and $\mathscr{L} = L/(\rho_0 c_0)$ with $\rho_0 \equiv$ density and $c_0 \equiv$ specific heat at constant pressure, into our model by replacing (3c) by (35) as well as introducing the appropriate anisotropic form [11] for $H(\gamma, \zeta)$ in (3b).

References

[1] L.A. Segel, Nonlinear hydrodynamic stability theory and its applications to thermal convection and curved flows, in: Nonequilibrium Thermodynamics, Variational Techniques, and Stability, R.J. Donnelly, R. Herman, and I. Prigogine, eds. (Univ. of Chicago Press, Chicago, 1966), pp. 165–197.

[2] P.G. Drazin and W.H. Reid, Hydrodynamic Stability (Cambridge Univ. Press, Cambridge, 1981).

[3] J.S. Kirkaldy, Interface stability during steady-state phase transformations, in: Nonequilibrium Thermodynamics, Variational Techniques, and Stability, R.J. Donnelly, R. Herman, and I. Prigogine, eds. (Univ. of Chicago Press, Chicago, 1966), pp. 281–283.

[4] D.J. Wollkind and L.A. Segel, A nonlinear stability analysis of the freezing of a dilute binary alloy, Phil. Trans. Roy. Soc. London Ser. A 268 (1970) 351–380.

[5] R. Sriranganathan, D.J. Wollkind and D.B. Oulton, A theoretical investigation of the development of interfacial cells during the solidification of a dilute binary alloy: Comparison with the experiments of Morris and Winegard, J. Crystal Growth 62 (1983) 265–283.

[6] D.J. Wollkind and R.D. Notestine, A nonlinear stability analysis of the solidification of a pure substance, IMA J. Appl. Math. 27 (1981) 85–104.

[7] L.A. Segel and J.T. Stuart, On the question of the preferred mode in cellular thermal convection, J. Fluid Mech. 13 (1962) 289–306.

[8] L.R. Morris and W.C. Winegard, Segregation and morphology of cells during freezing, J. Inst. Metals 97 (1969) 220–222.

[9] L.R. Morris and W.C. Winegard, The development of cells during the solidification of a dilute Pb–Sb alloy, J. Crystal Growth 5 (1969) 361–375.

[10] L.A. Segel, The nonlinear interaction of a finite number of disturbances to a layer of fluid heated from below, J. Fluid Mech. 21 (1965) 359–384.

[11] S.R. Coriell and R.F. Sekerka, The effect of the anisotropy of surface tension and interface kinetics on morphological stability, J. Crystal Growth 34 (1976) 157–163.

[12] J.S. Langer, Instabilities and pattern formations in crystal growth, Rev. Mod. Phys. 52 (1980) 1–28.

[13] D.J. Wollkind, A deterministic continuum mechanical approach to morphological stability of the solid–liquid interface, in: Preparation and Properties of Solid State Materials, vol. 4, W.R. Wilcox, ed. (Dekker, New York, 1979), pp. 111–191.

[14] R.T. Delves, Theory of interface stability, in: Crystal Growth, B.R. Pamplin, ed. (Pergamon, Oxford, 1975), pp. 40–103.

[15] R.F. Sekerka, Morphological stability, in: Crystal Growth: An Introduction, P. Hartman, ed. (North-Holland, Amsterdam, 1973), pp. 403–443.

[16] J.W. Gibbs, Collected Works, vol. 1 (Longmans, Green and Co., London, 1928).

[17] J. Frenkel, Kinetic Theory of Liquids (Dover, New York, 1955).

[18] H.P. Greenspan, On the deformation of a viscous droplet caused by variable surface tension, Stud. Appl. Math. 57 (1977) 45–58.

[19] T. Takahashi, A. Kamio and Nguyen An Trung, Morphology of the solid–liquid interface and redistribution of solutes in unidirectional solidified aluminum alloys, J. Crystal Growth 24/25 (1974) 477–483.

[20] L.R. Morris and W.C. Winegard, The cell to dendrite transition, J. Crystal Growth 6 (1969) 61–66.

[21] R. Sriranganathan, A Theoretical Investigation of Interfacial Cellular Structure during the Solidification of a Dilute Binary Alloy, Thesis, Washington State University (1982).

[22] J.V. Uspensky, Theory of Equations (McGraw-Hill, New York, 1948).

[23] L.A. Segel, The structure of nonlinear cellular solutions to the Boussinesq equations, J. Fluid Mech. 21 (1965) 345–358.

[24] B. Chalmers, Physical Metallurgy (Wiley, New York, 1959).

[25] R.M. Sharp and A. Hellawell, Solute distributions at non-planar, solid–liquid growth fronts, I, Steady-state conditions, J. Crystal Growth 6 (1970) 253–260.

[26] R.M. Sharp and A. Hellawell, Solute distributions at non-planar, solid–liquid growth fronts, IV, Ripening of cells and dendrites behind the growth front, J. Crystal Growth 11 (1970) 77–91.

[27] W.A. Tiller and J.W. Rutter, The effect of growth conditions upon the solidification of a binary alloy, Can. J. Phys. 34 (1956) 96–121.

[28] J.W. Cahn, private communication.

[29] M.C. Flemings, Solidification Processing (McGraw-Hill, New York, 1974).

[30] J.R. Ockendon and W.R. Hodgkins, Moving Boundary Problems in Heat Flow and Diffusion (Clarendon, Oxford, 1975).

[31] S. O'Hara, The metastable solid–liquid phase boundary, J. Appl. Phys. 37 (1966) 3783–3786.

[32] M. Kerszberg, Pattern formation in directional solidification, Phys. Rev. B 27, in press.

[33] E.F. Keller and L.A. Segel, Initiation of slime mold aggregation viewed as an instability, J. Theor. Biol. 26 (1970) 399–415.

[34] S. Chandrasekhar, Hydrodynamic and Hydromagnetic Stability (Clarendon, Oxford, 1961).

Physica 12D (1984) 241–244
North-Holland, Amsterdam

SIMPLE MODELS OF INTERFACE GROWTH

Richard C. BROWER
Lyman Lab of Physics, Harvard University, Cambridge, MA 02138, USA

David A. KESSLER
Los Alamos National Laboratory, Los Alamos, NM 87544, USA

and

Joel KOPLIK and Herbert LEVINE
Schlumberger-Doll Research, Ridgefield, CT 06877, USA

We propose a simple class of models for interface growth. These models are characterized by the idea that the growth velocity be locally determined by the interface geometry. Study of these systems should clarify the mechanisms involved in pattern formation.

There are a host of patterns formed in nature by the time evolution of an interface between two domains. One need only refer to the many exciting papers presented at this conference for confirmation of this fact. However, the science of the study of these patterns is still in a primitive stage. This is at least partially because the equations obeyed by systems with moving domain boundaries are often exceedingly complex. Such systems are analytically almost intractable and computationally very costly, largely due to their intrinsic non-locality.

In this talk, we wish to introduce the idea of using simple, geometrically motivated models for interfacial growth. We will postulate that the velocity be determined locally by the interfacial geometry. Clearly, this class of models is not meant as a fully realistic alternative to actual evolution equations which will correctly reproduce all the intricacies of the long-time development of the system. Nevertheless, we propose that these models will prove to be useful for understanding the relevance of various analytic structures (steady-state solutions, linear stability, etc.) as well as for building intuition about the mechanisms whereby real patterns emerge.

For definiteness, let us focus on the solidification process in two dimensions [1]. If we demand locality and also demand invariance under rotations and translations, we are led to the form [2]

$$\hat{n}\cdot\frac{\mathrm{d}\boldsymbol{x}}{\mathrm{d}t} = V(\kappa) + \gamma\frac{\partial^2\kappa}{\partial s^2}, \tag{1}$$

where $\boldsymbol{x}$ denotes the interface position, κ is the curvature and s is the arclength along the interface. First, we discuss how one might choose the "potential" V and then return to the importance of γ. V must embody some of the macroscopic features of the particular system under investigation. A standard calculation [3] shows that the growth rate of a spherical crystal (circular in 2-d) behaves as $1/R$, for large R. For eq. (1) to incorporate this, $V \sim \kappa$ for small curvature. Similar arguments [2] suggest that for very large κ, V must become negative and that positive curvatures grow faster than negative ones. This suggests

$V=\kappa+\alpha k^2-\beta\kappa^3$. In this talk, we will mostly consider $\alpha=\beta=0$.

One solution of (1) is a circle with $r(t)=\sqrt{2t+r^2(0)}$. To see the meaning of γ, we perform a stability analysis around this solution. One only need consider perturbations normal to the original surface. In this case, this amounts to

$$\boldsymbol{x}=\hat{r}(r(t)+\epsilon(t,\theta_0)), \tag{2}$$

where θ_0 is the polar angle of the original circle. We derive the equation

$$\sigma_m=(m^2-1)\left(\frac{1}{r^2}V'\left(\frac{1}{r}\right)-\frac{\gamma m^2}{r^4}\right)$$

for the eigenfunctions $\epsilon^m=\epsilon_0\cos m\theta_0 e^{\sigma_m t}$. The first term is destabilizing for all m, an effect directly analogous to the Mullins–Sekerka instability [3]. The second term sets a limit to the range of m values which will grow. In this regard, γ plays the role of the surface tension in the full growth equations.

The solution of the solidification equations which appears to have the most relevance for the actual time development of the system is the dendrite crystal [4]. This is a steady state solution of the zero surface tension equations corresponding to a parabola which uniformly translates along the axis of symmetry. These solutions are thought to be responsible for the branched structure of the familiar hexagonal "snowflake" crystal [5]. As discussed in great detail by Langer and Muller-Krumbhaar [4], dendritic growth poses an interesting pattern selection problem. There exists a complete family of dendritic solutions, with the translation velocity determined by the radius of the parabolic tip. Experimentally one observes a unique velocity [6]; this velocity may be related to the point of marginal stability of the linear fluctuations around the dendrite [7]. However, this remains only a hypothesis consistent with the data and has not yet been convincingly demonstrated.

For our local model with $\gamma=0$, the relevant steady state solution is best considered by introducing θ, the angle between the normal and the $\hat{y}$-axis. In terms of this angle, $\kappa=\partial\theta/\partial s$. Then, the solution is

$$v\cos\theta=\kappa, \tag{3}$$

which can be integrated to yield

$$x_0=\rho\theta, \quad y_0=\rho\ln\cos\theta, \tag{4}$$

with $\rho=1/v$. Near the tip, this is approximately parabolic with radius ρ. In this two-dimensional case, y_0 goes to infinity at a finite value of x_0. It is amusing to note that the analogous system in three dimensions $V=\kappa_m$, (the mean curvature), gives rise to a dendrite which asymptotically approaches a paraboloid of revolution.

We now turn to the stability analysis. Assuming $\boldsymbol{x}=\boldsymbol{x}_0+\hat{n}_0\delta(\theta_0,t)$, we can derive

$$\dot{\delta}=-v^2\cos^2\theta_0(\delta''+\delta). \tag{5}$$

The solutions of (5) are

$$\delta_k(\theta_0,t)=\sqrt{\cos\theta_0}\,P^{ik}_{1/2}(\sin\theta_0)\,e^{Wt},$$
$$W=v^2(1/4+k^2), \tag{6}$$

where P is an associated Legendre polynomial. All modes grow in time, representing the instability of the dendrite to tip splitting.

We now redo the stability analysis including the γ term. Asymptotically far from the tip, a perturbation of the form $\delta\sim(\pi/2-\theta_0)^{ip}$ will have a growth rate

$$W=v^2(p^2-\tilde{\gamma}p^4-ip), \tag{7a}$$
$$\tilde{\gamma}=\gamma v^2. \tag{7b}$$

Spectral equations of this type give rise to critical value of $\tilde{\gamma}$ below which the first mode with a positive value $\mathrm{Re}\,W$ appears. This is the marginal stability point γ^*, which for fixed γ defines a marginal velocity $v^*=\sqrt{\gamma^*/\gamma}$.

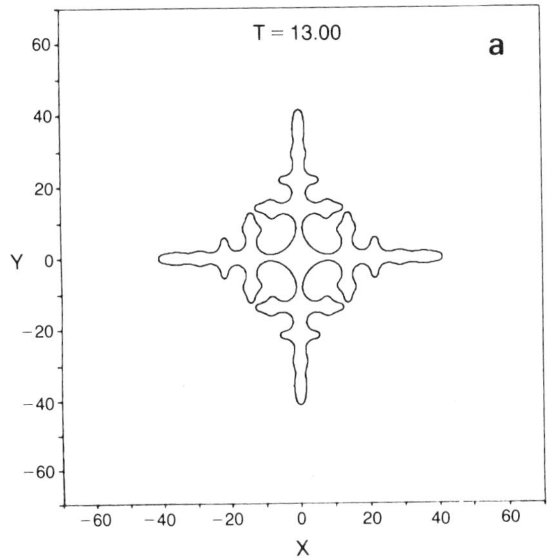

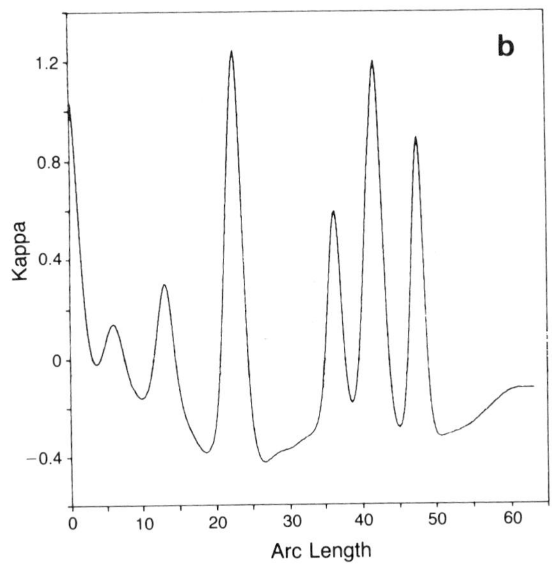

Fig. 1. Growth with parameters $\alpha = 4$, $\beta = 1.5$, $\gamma = 1$, and 15% anisotropy. a) real space; b) curvature (one octant).

We have just shown how our models contain a family of steady-state solutions with a linear fluctuation operator which gives rise to a point of marginal stability. If these features are enough to give rise to stable dendritic growth at a selected velocity, our model should also exhibit this growth pattern when simulated on the computer. We will describe the details of our computer simulations elsewhere [8], but let us sketch some of the crucial conclusions. We find that steady-states dendritic growth is only possible in the presence of a symmetry breaking perturbation such as crystalline anisotropy. (A similar conclusion appears to apply to the boundary-layer model of crystal growth [9].) In the presence of moderate anisotropy in the growth rate, the dendrite undergoes a series of stable oscillations, resulting in the emission of a sequence of equally spaced sidebranches along the original stem. A typical resulting shape is shown in figs. 1a and 1b, in real and curvature spaces respectively. The tip velocity as a function of time is plotted in fig. 2, where the oscillatory nature of the sidebranching process is clearly visible.

Let us return to the general considerations discussed at the outset of this talk. Using the same reasoning, we can use local models to represent other classes of interface growth dynamics. These would incorporate such features as area conservation (vortex contour-dynamics [10]), substrate-induced breaking of either translation or rotation symmetry and whatever other effects one might deem important. The major goal of the study of these models is to pinpoint which mechanisms

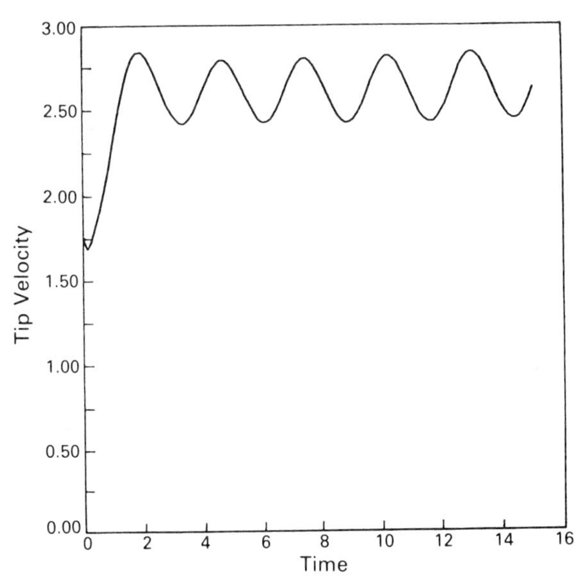

Fig. 2. Tip velocity as a function of time for the growth depicted in fig. 1.

control which features of the long-time development. If there exists any concept of universality in pattern formation, study of these models may prove indispensable for its elucidation. One will then be able to return to the actual evolution equations, formulate the proper questions and have a much better handle on possible forms the answers may take.

Acknowledgements

We would like to thank the Santa Barbara group (E. Ben-Jacob, N. Goldenfeld and J.S. Langer) for sharing with us some of their results prior to publication and also acknowledge useful discussions with R. Sekerka and H. Aref.

References

[1] For a review, see J.S. Langer, Rev. Mod. Phys. 52 (1980) 1.
[2] For a more detailed description of our results to date, see R. Brower, D. Kessler, J. Koplik and H. Levine, Geometrical Models of Interface Evolution, I, Phys. Rev. A 29 (1984) 1335; Geometrical Approach to Moving Interface Dynamics, Phys. Rev. Lett. 51 (1983) 1111.
[3] W.W. Mullins and R.F. Sekerka, J. Appl. Phys. 35 (1964) 444.
[4] J.S. Langer and H. Muller-Krumbhaar, Acta Metallurgica 26 (1978) 1681, 1689, 1697.
[5] U. Nakaya, Snow Crystals (Harvard Univ. Press, Harvard, 1954).
[6] S.C. Huang and M.E. Glicksman, Acta Metallurgica 29 (1981) 701, 717.
[7] For similar behavior in a toy problem see G. Dee and J.S. Langer, Phy. Rev. Lett. 50 (1983) 383.
[8] R. Brower, D. Kessler, J. Koplik and H. Levine, in preparation.
[9] E. Ben-Jacob, talk presented at this conference, and E. Ben-Jacob, N. Goldenfeld, J.S. Langer and G. Schoen, Phys. Rev. Lett. 51 (1983) 1930.
[10] E.A. Overman and N.J. Zabusky, Physica 12D (1984) 145 (these proceedings).

Physica 12D (1984) 245–252
North-Holland, Amsterdam

STRING MODELS OF INTERFACIAL PATTERN FORMATION

Eshel BEN-JACOB, Nigel GOLDENFELD, J.S. LANGER and Gerd SCHÖN
Institute for Theoretical Physics, University of California, Santa Barbara, California 93106, USA

We introduce a new approach to describe the dynamics of interfacial pattern formation. The central idea is that local properties may dominate the dynamics of the interface in many realistic situations. Here we have chosen to present a model for dendritic solidification where the basic mechanism which allows pattern formation – the Mullins–Sekerka instability – is well understood. In this and many other cases where the dynamics is driven by an additional field such as thermal or chemical diffusion, the coupling of this field to the interface may be phenomenologically described by a boundary layer with its own dynamics.

With the inclusion of the non-equilibrium effect of crystalline anisotropy, our preliminary results appear to reproduce the observed dynamics of dendritic growth.

1. Introduction

There is a great deal of interest in dynamical pattern-forming systems far from equilibrium. While the long-term goal of such research is to discover what new principles, if any, govern non-equilibrium systems, currently most effort is directed towards the development of numerical and analytical techniques for dealing with the equations commonly encountered. In this paper we introduce a new, physically motivated approach for describing interfacial pattern formation. The central idea is to treat the interface as a dynamical entity (called a "string" in two dimensions) and to use physical arguments to deduce equations of motion. It seems that pattern-forming systems of this kind rely on a delicate imbalance between two or more competing fields, so that the string interacts with one or more additional fields. In many cases at least one of the fields is diffusive and may be visualized as a boundary layer around the string.

In dendritic solidification [1], we are confronted with a dynamical system in which the mechanism of pattern-formation is intimately connected with the stability of the interface, a situation which may prove to have some degree of generality, not only in physical systems, but also in other fields such as chemistry and biology. The essential ingredient for dendritic solidification is the coupling between the interface and the diffusion field, which may be either thermal or chemical or both. Traditionally, attention has been focused on the diffusion field whose dynamics is nonlocal in both space and time due to the presence of the interface as a free boundary. The present investigation is motivated by the idea that many of the most interesting features of dendritic pattern formation may best be studied by ignoring the fully nonlocal properties of the diffusion field and focusing on the dynamics of the interface itself.

The purpose of this contribution is to show how to construct a boundary-layer model for interfacial pattern formation by discussing the explicit example of dendritic solidification. We also discuss a computational scheme for string models, which is capable of following the dynamics well into the nonlinear large-amplitude regime. Finally, we present our preliminary results. With the presence of crystalline anisotropy, arising through the inclusion of molecular attachment kinetics, they appear to reproduce qualitatively the observed behavior of dendrites.

2. Basic physics of dendritic solidification

The underlying physics of the solidification process is simply stated. To ease the presentation, we will treat only the case of thermal diffusion. We consider, in fig. 1, a crystal growing into its supercooled melt, so that the solid is actually at a higher temperature than the melt. In terms of the melting temperature T_M, the temperature of the melt far from the interface T_∞, the latent heat L and the heat capacity C, the dimensionless undercooling is defined by

$$\Delta = (T_M - T_\infty)/(L/C) \tag{1}$$

and is generally less than 0.1 for easily accessible experimental conditions. Neglecting thermal diffusion in the solid, the dimensionless temperature field

$$u(x, t) = (T(x, t) - T_\infty)/(L/C) \tag{2}$$

obeys the usual diffusion equation with the boundary condition at the interface that the rate of latent heat generation is the rate at which heat enters the

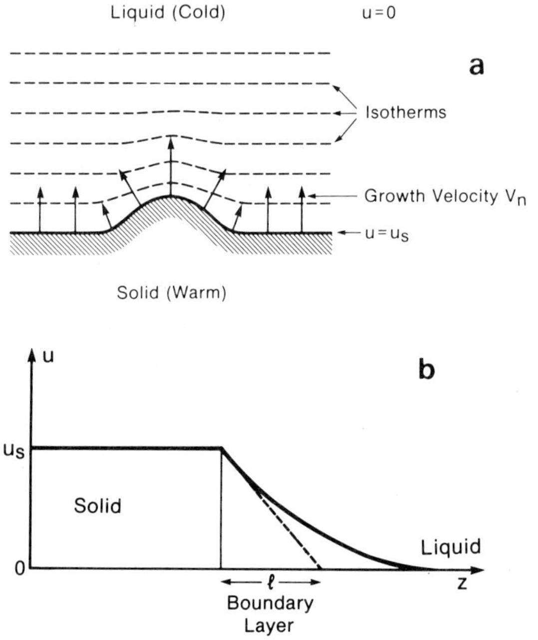

Fig. 1. (a) Schematic illustration of solidification, showing the Mullins–Sekerka instability, and various quantities defined in the text; (b) temperature vs. Z, the distance from the interface along the local normal. The solid line is the solution of the full moving boundary diffusion problem; the dashed line is the approximation used in the boundary layer model.

melt:

$$V_n = -D\frac{\partial u_s}{\partial n}, \tag{3}$$

where D is the thermal diffusion coefficient and V_n is the velocity of the interface along its normal. For present purposes, it is convenient to neglect heat fluxes in the solid. The value of u at the interface is given by the Gibbs–Thomson expression

$$u_s = \Delta - d_0 K. \tag{4}$$

The term $-d_0K$ is best thought of as arising from the work required to create a bump of radius $1/K$ on an otherwise flat interface at temperature Δ; and d_0 is a capillary length proportional to the surface tension of the interface and is typically about 10 Å. It is important to note that eq. (3) is simply a statement of equilibrium thermodynamics, indicating that we are implicitly coarse-graining our length and time scales in the usual way such that local equilibrium may be assumed. Later we will mention a crude correction to (3) due to non-equilibrium effects.

The conventional formulation given above can easily be shown to exhibit the Mullins–Sekerka shape instability [2]. As illustrated in fig. 1, the tip of a small protruding bulge on an initially flat solidification front experiences a larger temperature gradient than a point on the flat front; consequently heat diffuses rapidly away from it, causing it to grow out even further. Surface tension acts in such a way as to stabilize the small wavelength instabilities. In summary, the pattern forming mechanism derives from a competition between the tendency to maximize the arc length of the interface in order to facilitate diffusion and the tendency to minimize the arc length due to the surface tension.

A further aspect of the pattern forming mechanism is that the velocity of the dendrite tip is experimentally a unique, reproducible function of the undercooling Δ. In contrast, a linear stability analysis around a parabolic needle-crystal solution of the equations allows a continuous family of stable propagating states. Thus a selection criterion must be operative. A principle of marginal stability [3] discussed recently in the literature, appears to be consistent with experiment but has yet to be given a firm basis in theory.

3. The boundary layer model

In the limiting case of Δ just smaller than unity, the diffusion field away from the interface decays to its asymptotic value within a distance which is small compared to the local radius of curvature of the interface [4]. In this limit, it is reasonable to consider the diffusion as occurring only in a thin boundary layer at the interface, the outer edge of the boundary layer being an isotherm at $u = 0$ (see fig. 1).

It is convenient to describe the string not by its real-space Cartesian coordinates, but by its curvature

$$K = \frac{\mathrm{d}\Theta}{\mathrm{d}s}, \tag{5}$$

where Θ is the angle between the normal to the string and a fixed arbitrary direction in space, and s is the arc length displacement.

The equations of motion, following a particular point moving along the outward normal are

$$\frac{\mathrm{d}K}{\mathrm{d}t} = -\left(K^2 + \frac{\partial^2}{\partial s^2}\right)V_\mathrm{n}, \tag{6a}$$

$$\frac{\mathrm{d}s}{\mathrm{d}t} = \int_0^s \mathrm{d}s' K V_\mathrm{n}. \tag{6b}$$

We approximate the boundary condition (3) by

$$V_\mathrm{n} = u_\mathrm{s}/l, \tag{7}$$

where $l(s)$ is the thickness of the boundary layer, measured along the normal to the interface. It is convenient to introduce the quantity

$$h(s, t) = u_\mathrm{s} l \tag{8}$$

which is the heat content density in the boundary layer. This satisfies a heat balance equation

$$\frac{\mathrm{d}h}{\mathrm{d}t} = V_\mathrm{n}(1 - u_\mathrm{s}) + D\frac{\partial}{\partial s} l \frac{\partial u_\mathrm{s}}{\partial s} - K V_\mathrm{n} h, \tag{9}$$

which states that the latent heat generated at the interface either heats up the solidifying melt, diffuses away along the boundary layer, or is redistributed along the boundary layer as the interface grows. The factor $1 - u_\mathrm{s}$ in the first term accounts for the fraction of heat which does not enter the boundary layer because it is retained by the new layer of solid formed from the melt. The quantity $l(\partial u_\mathrm{s}/\partial s)$ is just the heat current in the boundary layer and is a stabilizing term. The final term arises because h is a heat content per unit length of the interface, so that there is a contribution due to the change in the differential arc length as given by eq. (6b).

Eqs. (4), (6), (7), (8), and (9) completely specify the boundary layer model in its simplest form. Elsewhere [4], we have shown that these equations accurately reproduce the known behavior of eqs. (1) and (4) in simple geometries such as the growing circle, and exhibit the Mullins–Sekerka instability. Furthermore, in the limit of vanishing capillary length, the boundary layer model possesses a family of unstable parabolic needle-crystal solutions corresponding to the Ivantsov [5] solution of the Stefan problem. We can clearly see that in the boundary layer model the capillary length enters eq. (9) as a singular perturbation, suggesting that the steady state needle-crystal solutions are destroyed for nonzero capillary length. A detailed examination supports this conclusion, implying that the marginal stability criterion may require re-interpretation.

4. Numerical method

We turn now to a few comments about numerical techniques for dealing with string dynamics. These remarks have to be considered in the context of the fact that, to the best of our knowledge, it has not been possible to numerically integrate the full diffusion problem with eqs. (1)–(4) far enough to see dendritic behavior. The difficulties of dealing with moving boundary problems using conventional grid or finite element techniques stem in part from the representation and tracking of the interface, and have been discussed elsewhere in these proceedings [6]. The underlying philosophy of our approach, however, obviates these difficulties at the expense of introducing new problems due to the parameterization of the interface. At first sight, eqs. (6)–(9) are a set of coupled partial differential equations in one space and one time dimension. Unfortunately, this is misleading because the spatial variable itself obeys an equation of motion coupled to the remaining equations.

Suppose that we have somehow discretized (6)–(9) on a spatial grid of points $S_i(t) = S_{i-1}(t) + \Delta S_i(t)(i = 1, 2, \ldots, N)$ taking into account the fact that ΔS_i is truly a function of the label i. The time evolution of the discretized equations drives the grid points away from the dendrite tip and other regions of large positive curvature. This tendency is undesirable not only because the computational error is unevenly distributed along the string, but also because of "numerical surface tension". The grid size artificially restricts the length scales probed by the dynamics, and even if there exists a natural short-wavelength cut-off in the system, the modification of the dispersion relation induced by the grid can couple to the long-wavelength behavior via the nonlinearity. In the present example this can lead to an over-stabilization of the dendrite tip.

We have dealt with these difficulties by using interpolation to redistribute the grid points evenly along the string after each time step. Furthermore, we continuously add new grid points as the string expands so that the grid spacing remains constant. The most stable way that we have found is to interpolate the string in the space of (s, K) rather than real space. Our time step algorithm is a fully implicit scheme with linearization, so that the only limitation on the time step is the rate of change of the quantities we compute.

5. Results and discussion

We have numerically integrated the boundary layer model for values of Δ in the range 0–1 starting from a variety of initial shapes. These include circles, ellipses, and closed curves of the parametric form

$$K(s) = a + b \cos ms \,. \tag{10}$$

In order to minimize the effects of transients we chose the initial h field such that

$$h(s, t = 0) = h_0[K(s)] \,, \tag{11}$$

where $K(s)$ is given by (10) and $h_0(K)$ is the steady state solution of (9) for a growing circle of radius K^{-1}. We scaled the variables such that lengths are measured in units of $d_0 \Delta^{-3}$ and time is measured in units of $d_0^2 D^{-1} \Delta^{-8}$.

The limit of small Δ is relatively simple and will be discussed first*. We have verified that the

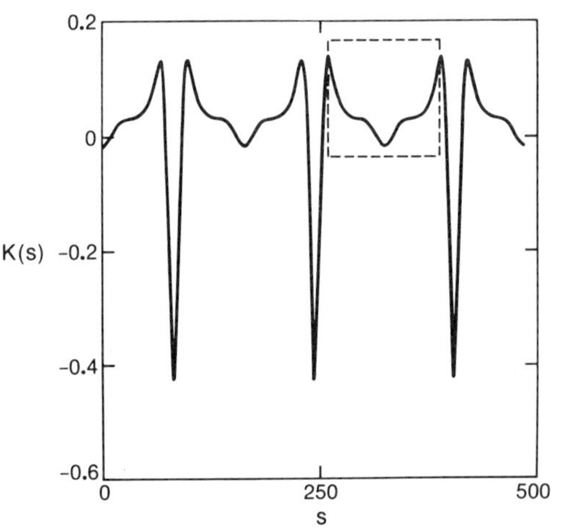

Fig. 2. κ vs. s at $t = 435$ in the small-Δ limit, showing the Mullins–Sekerka instability. The structure corresponds to the formation of grooves in real space.

* This limit is taken after the equations are expressed in terms of the scaled variables.

interface undergoes the Mullins–Sekerka instability, the effect being most conspicuous in (s, K) space, as shown in fig. 2. In real space, fig. 3, the interface develops grooves which resemble those seen in very mildly undercooled crystals. Ultimately the opposite sides of the grooves intersect but this is not important from our point of view. Correlations are absent between points close in two-dimensional space, but distant along the string, so that the model is only valid as long as the groove width is much larger than the width of the boundary layer at the groove walls. The small-Δ limit is not a realistic model for dendritic growth, since the derivation of the model implicitly assumed that $\Delta \approx 1$. Nevertheless, its mathematical features are related to those of the boundary layer model for $\Delta \approx 1$, and it may actually be a better description of some pattern forming sysem other than a dendrite. For example, a single cell organism, Micrasterias, evolves by grooving in quite a similar way to the small-Δ model. The grooves create fingers which themselves become grooved at the tip, thus splitting into two fingers. If continued indefinitely, a fractal or Cayley-tree-like structure would form, but in practice the growth ceases as the food supply becomes exhausted. The small-Δ model continues to grow, of course, and the structure in (s, K) space becomes increasingly complex and suggestively fractal-like. In order to follow the evolution far enough for this to become apparent, it was necessary to successively evolve smaller and smaller portions of the string with more and more grid points, and then reconstitute the final curve, as shown in fig. 4.

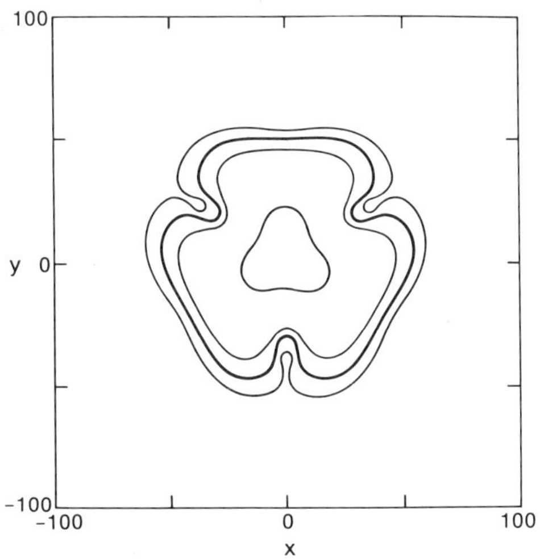

Fig. 3. Time-order sequence of the pattern of fig. 2 plotted in real space, showing grooving. The thick line corresponds to the (s, κ) graph in fig. 2.

For $\Delta \neq 0$, the string does not appear to evolve towards a more dendritic structure (see fig. 5, where the initial shape was an ellipse). Instead, it is clear that the growth is isotropic, and that the tip flattens, presumably because the diffusion term is too effective in stabilizing the string. We were careful to check that this result is not an artifact of our numerical scheme. What seems to be lacking in the model is a mechanism allowing instabilities to grow only in certain preferred directions as is observed experimentally. The most plausible origin of such an effect is the crystalline anisotropy. In order to verify that this could be important, we modified the boundary layer model in a simple way by multiplying the velocity V_n by the factor

$$f(\Theta) = \frac{1 + \alpha \cos m\Theta}{1 + \alpha}, \tag{12}$$

where m is the crystalline symmetry. This factor inhibits growth in directions other than that of the

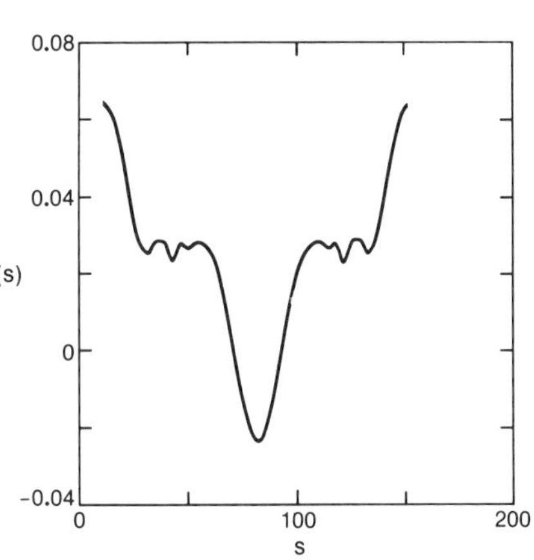

Fig. 4. The region in fig. 2 enclosed by the dashed square at $t = 520$, showing the emergence of a complex fractal-like pattern in (s, κ) space.

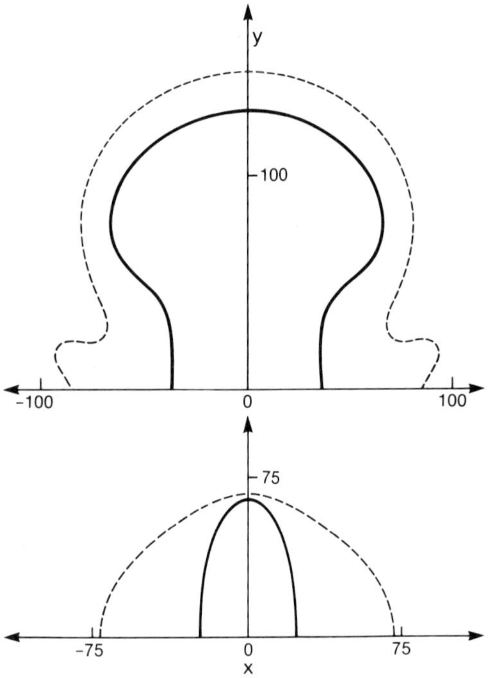

Fig. 5. Evolution of an ellipse at $t = 0$ (lower diagram) until $t = 100$ (upper diagram), with $\Delta = 0.5$ and no anisotropy. The solid line is the interface, the dashed line is the edge of the boundary layer.

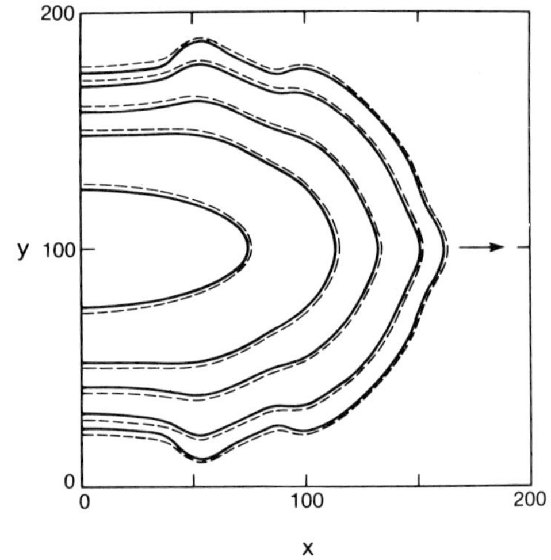

Fig. 6. Sequence of dendrite-like structures with $\Delta = 0.99$, $\alpha = 0.25$, and $m = 4$, at times 0, 50, 75, 100, 115. The solid line is the interface: the dashed line is the edge of the boundary layer.

dendrite propagation and symmetry-related directions. For $\Delta \approx 1$, this model shows dendritic behavior, with a smooth tip starting to leave a trail of side-branches behind it (fig. 6). By following the tip velocity as a function of time, we are able to check that the system is out of the initial transient stage. The behavior is particularly interesting in (s, K) space, as shown in fig. 7. The evolution of the dendrite appears as a periodic state modulated by an envelope function propagating away from the tip back down the dendrite. This is reminiscent of the behavior of certain one-dimensional partial differential equations which have been studied in connection with other pattern forming systems [7]. We have not yet succeeded in calculating the tip velocity for the boundary layer model, so that a comparison with the marginal stability hypothesis is not presently available.

It is also interesting to compare the structures generated by a six-fold symmetric starting configuration with a real snowflake. In fig. 8 we present the time development of such a crystal, showing that many of the qualitative morphological features of a real snowflake are indeed reproduced by the anisotropic boundary layer model. Although real snowflakes are not grown at $\Delta \approx 1$, it is quite likely that the parameters in our phenomenological model, particularly Δ itself, require interpretation and are not necessarily the actual experimental values.

The boundary layer model we have described takes no account of non-equilibrium effects. At large undercoolings, the growth rate increases, and the assumption of local equilibrium at the interface may break down as molecular attachment kinetics become the rate determining step. It is precisely in connection with such effects that the crystalline anisotropy is expected to play a crucial role. We can account for it in the first approximation, by assuming that the effective temperature at the interface is reduced in an anisotropic manner. Thus we replace eq. (4) by

$$u_s = \Delta - d_0 K - V_n \tilde{f}(\Theta) , \tag{13a}$$

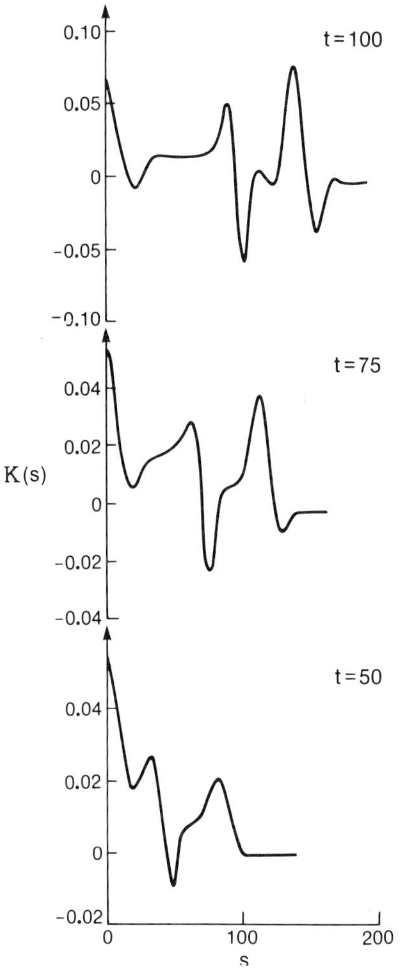

Fig. 7. Evolution of the dendrite-like structure shown in fig. 6, plotted in (s, κ) space. The tip of the dendrite is at $s = 0$.

Fig. 8. Evolution of a snowflake-like structure (top) compared to sketches of real snowflakes. $\Delta = 0.99$, $\alpha = 0.25$, and $m = 6$. The penultimate stage did not evolve directly from the earlier stage (due to lack of computer time), but was inserted by hand.

where

$$\tilde{f}(\Theta) = \beta(1 - \cos m\Theta) . \tag{13b}$$

This result can be obtained from a detailed treatment. Eq. (13) is more difficult to handle numerically, but our preliminary results show similar behavior to that found with the cruder inclusion of anisotropy discussed above.

It is still not completely clear what the role of the anisotropy actually is. One possibility is that an infinitesimal anisotropy is all that is required to trigger instabilities in the preferred directions, while suppressing growth in other directions, and that the actual amplitude is not crucial. Alternatively, the amplitude of the anisotropy may need to be above some threshold in order to ensure the suppression of instabilities in the unfavored directions. Work currently in progress should be able to clarify this point.

Finally, we mention that a modest extension of

the above model enables us to describe directional solidification and the growth of single cell organisms such as the Micrasterias. In directional solidification it is necessary to include the chemical diffusion, while in the Micrasterias an additional complication is the nonlinear behavior of V_n as a function of the difference in chemical potential of (e.g.) calcium across the cell wall. We hope to report on these systems in the near future.

Acknowledgments

This research was supported in part by the National Science Foundation, grant number PHY77-27084 and by DOE contract DE-AM03-76SF00034.

References

[1] J.S. Langer, Rev. Mod. Phys. 52 (1980) 1.
[2] W.W. Mullins and R.F. Sekerka, J. Appl. Phys. 34 (1963) 323; 35 (1964) 444.
[3] J.S. Langer and H. Müller-Krumbhaar, Acta Metall. 26 (1978) 1681.
[4] E. Ben-Jacob, N.D. Goldenfeld, J.S. Langer and G. Schön, Phys. Rev. A 29 (1984) 330.
[5] G.P. Ivantsov, Dokl. Akad. Nauk SSSR 58 (1947) 567.
[6] J.M. Hyman, Physica 12D (1984) 396 (these proceedings).
[7] E. Ben-Jacob, H. Brand, G. Dee, L. Kramer and J.S. Langer, to be published.

Physica 12D (1984) 253–261
North-Holland, Amsterdam

NONPLANAR INTERFACE MORPHOLOGIES DURING UNIDIRECTIONAL SOLIDIFICATION OF A BINARY ALLOY

G.B. McFADDEN and S.R. CORIELL
National Bureau of Standards, Washington, DC 20234, USA

During directional solidification of a binary alloy, a planar solid–liquid interface may become unstable and develop into a cellular non-planar interface, exhibiting periodic structure transverse to the growth direction. Steady state two-dimensional temperature, solute concentration, and interface shapes are calculated numerically and the solute inhomogeneity (microsegregation) in the solidified material obtained. Specific results are presented for an aluminum alloy containing silver for solidification velocities of 0.01 and 1.0 cm/s, which correspond to the constitutional supercooling and absolute stability regimes, respectively.

1. Introduction

Directional solidification is an established method for the production of alloys of technological importance [1–6]. In this process, there is uniform relative motion of the alloy sample and its thermal environment, so that at steady state the solid–liquid interface propagates at constant velocity. Growth velocities range from 10^{-4} to 1 cm/s with temperature gradients up to 500 K/cm. At equilibrium, the solute concentration in the liquid is different from the solute concentration in the solid. Thus, there is either rejection or preferential incorporation of solute by the solid at a moving solid–liquid interface; this leads to a concentration gradient in the liquid in the vicinity of the solid–liquid interface.

For a planar-crystal melt interface moving at constant velocity, there is a simple solution for which the temperature and concentration fields depend only on the distance from the interface, and the resulting solid is of uniform composition. When the planar interface is unstable, the composition of the solid is non-uniform with spatial inhomogeneities of the order of 10 μm (microsegregation). In this paper, we investigate the nature of the planar to cellular transition and examine its dependence on parameters that can be controlled during solidification.

A treatment of the linear stability of a planar interface was given by Mullins and Sekerka in 1964 [7], see also [8], which extended the constitutional supercooling criterion originally given in [2]. A further analysis by Coriell et al. includes the effects of convection in the liquid [9]. Nonlinear stability analyses have been performed for the two-dimensional case by Wollkind and Segel [10], and Dee and Mathur [11], and for the three-dimensional case by Sriranganathan et al. [12]. Kerszberg [13] computes several terms in a modal expansion of the interface and considers the question of wavelength selection. Studies of the dependence of the concentration field on a given nonplanar interface shape have been performed by Carlson et al. [14], Coriell and Sekerka [15], and Coriell et al. [16]. A nonlinear free boundary problem in crystal growth has been studied using finite elements by Ettouney and Brown [17]. In that work the interface shape is determined by heat transfer effects, and the concentration field is computed once the interface location is known.

The method to be presented here is a finite difference formulation of the equations describing diffusion of solute in the liquid and diffusion of heat in the liquid and solid. A self-consistent interface location is computed by an iterative procedure that effectively introduces an artificial time dependence in the problem. The procedure reproduces the correct linear instabilities for the planar interface, and computes nonplanar steady state solutions in the limit of large artificial time. The results described here correspond to the two-dimensional case, although the method is equally applicable to three dimensions. Finite element methods are being applied to this problem by Ungar and Brown [18].

2. Theory

We consider solidification at constant velocity V in the $\bar{Z}$ direction, and choose a coordinate system $(\bar{X}, \bar{Z})$ moving with the interface. We assume the steady-state interface shape can be expressed in the form $\bar{Z}=\bar{W}(\bar{X})$. The temperature and solute concentration in the liquid region $\bar{Z}>\bar{W}(\bar{X})$ are denoted by $\bar{T}$ and $\bar{c}$, and the temperature in the solid region $\bar{Z}<\bar{W}(\bar{X})$ is denoted by $\bar{T}_s$. We assume that no diffusion of solute occurs in the solid, since typical values of the solute diffusion coefficient are generally orders of magnitude smaller in the solid than in the liquid. The concentration of solute in the solid is given by its values at the interface $\bar{Z}=\bar{W}(\bar{X})$. The equations governing conservation of solute and heat are then

$$-V\,\partial\bar{c}/\partial\bar{Z}=D\left(\left(\partial^2\bar{c}/\partial\bar{X}^2\right)+\left(\partial^2\bar{c}/\partial\bar{Z}^2\right)\right),\quad \bar{Z}>\bar{W}(\bar{X}), \tag{1}$$

$$-V\,\partial\bar{T}/\partial\bar{Z}=K\left(\left(\partial^2\bar{T}/\partial\bar{X}^2\right)+\left(\partial^2\bar{T}/\partial\bar{Z}^2\right)\right),\quad \bar{Z}>\bar{W}(\bar{X}), \tag{2}$$

$$-V\,\partial\bar{T}_s/\partial\bar{Z}=K_s\left(\left(\partial^2\bar{T}_s/\partial\bar{X}^2\right)+\left(\partial^2\bar{T}_s/\partial\bar{Z}^2\right)\right),\quad \bar{Z}<\bar{W}(\bar{X}). \tag{3}$$

The constants D, K, and K_s are solute diffusivity in the liquid and thermal diffusivity in the liquid and solid, respectively. We will assume that the thermal properties of the liquid and solid are the same, so that $K=K_s$. We assume that during the phase change local equilibrium is maintained at the solid liquid interface. The Gibbs–Thomson equation,

$$\bar{T}=T_m+M\bar{c}-T_m\Gamma\mathscr{K},\quad \text{at } \bar{Z}=\bar{W}(\bar{X}), \tag{4}$$

then relates the temperature, concentration of solute in the liquid, and interface mean curvature $\mathscr{K}$. Here T_m is the melting point of the pure material $(c=0)$, and Γ is a capillary length. M is the slope of the liquidus curve and gives the alteration of melting point due to the presence of solute. The temperature is continuous at the interface,

$$\bar{T}=\bar{T}_s,\quad \text{at } \bar{Z}=\bar{W}(\bar{X}), \tag{5}$$

while there is a jump in concentration across the interface. The phase diagram implies the relation $\bar{c}_s=\tilde{k}\bar{c}$, where the distribution coefficient $\tilde{k}$ is the ratio of solute concentration in the solid to that in the liquid. Conservation of energy is expressed by the condition

$$-LV_n=k\left(\left(\partial\bar{T}/\partial\bar{\nu}\right)-\left(\partial\bar{T}_s/\partial\bar{\nu}\right)\right),\quad \text{at } \bar{Z}=\bar{W}(\bar{X}), \tag{6}$$

where L is the latent heat per unit volume released upon solidification, k is the thermal conductivity in the liquid and solid, and the normal component of the velocity $V_n=V\nu_z$. The normal derivative $\partial/\partial\bar{\nu}$ is taken with $\nu=(\nu_x,\nu_z)$ pointing into the liquid. Conservation of solute at the interface takes the form

$$D\,\partial\bar{c}/\partial\bar{\nu}=-V_n(1-\tilde{k})\bar{c},\quad \text{at } \bar{Z}=\bar{W}(\bar{X}). \tag{7}$$

The solution to eqs. (1)–(7) corresponding to a planar interface $\bar{Z}=\bar{W}(\bar{X})=0$ are

$$\bar{c}(\bar{Z})=c_\infty\left[1+((1-\tilde{k})/\tilde{k})\exp\left(-V\bar{Z}/D\right)\right], \tag{8}$$

$$\bar{T}(\bar{Z})=T_m+Mc_\infty/\tilde{k}+\left(KG_L/V\right)\left[1-\exp\left(-V\bar{Z}/K\right)\right], \tag{9}$$

$$\bar{T}_s(\bar{Z})=T_m+Mc_\infty/\tilde{k}+(K/V)(G_L+LV/k)\times\left[1-\exp\left(-V\bar{Z}/K\right)\right], \tag{10}$$

where c_∞ is the solute concentration far from the interface and G_L is the temperature gradient in the liquid at the interface. If the solution (8)–(10) is unstable, there may be additional solutions to the system (1)–(7) that satisfy the same far-field boundary conditions and are periodic in the $\overline{X}$-direction.

3. Numerical method

Although the equations (1)–(3) are linear, the boundary conditions (4)–(7) are nonlinear in the concentration and temperature fields and the unknown interface location $\overline{Z} = \overline{W}(\overline{X})$. We employ an iterative approach that is similar to methods used in other free boundary problems [19–20]. The procedure is started with a guess for the interface location; for example, a slightly perturbed planar interface or the interface obtained in a previous calculation using different growth conditions. For this location of the interface, the solute and temperature fields are computed using equations (1)–(3) with boundary conditions (5)–(7) and appropriate far-field boundary conditions. For the assumed interface location, the Gibbs–Thomson equation (4) will generally not be satisfied. The Gibbs–Thomson equation is used to generate an improved estimate for the interface shape, and the procedure is then repeated until a given convergence criterion is met. The successive approximations generate a sequence of fields and interface shapes that can be viewed as an evolution in artificial time to a steady state that satisfies eqs. (1)–(7).

For a given periodic interface shape $\overline{Z} = \overline{W}(\overline{X})$ with fixed wavelength λ, the concentration field is computed as follows. If the length scale is nondimensionalized by λ and the concentration field by c_∞, the equations are

$$\beta\, \partial c/\partial z + \partial^2 c/\partial x^2 + \partial^2 c/\partial z^2 = 0, \qquad H_c > z > w(x), \quad (11)$$

$$\partial c/\partial \nu + \nu_z \beta(1 - \tilde{k}) c = 0, \quad z = w(x), \quad (12)$$

$$\partial c/\partial z + \beta c = \beta, \quad z = H_c, \quad (13)$$

$$\partial c/\partial x = 0, \quad x = 0, \tfrac{1}{2}. \quad (14)$$

Here the parameter $\beta = V\lambda/D$, and the height H_c is chosen large enough that the concentration is independent of H_c and is uniform in x near $z = H_c$. The boundary condition (13) gives the proper far field behavior seen in (8) and also insures overall conservation of solute [16]. For large β, a boundary layer of width $\mathcal{O}(1/\beta)$ is formed near $z = w(x)$. The domain $w(x) \le z \le H_c$ is mapped to a rectangular computational domain by the sequence of transformations

$$\tilde{x} = x,$$
$$\tilde{z} = H_c(z - w(x))/(H_c - w(x)), \quad (15)$$
$$\xi = \tilde{x},$$
$$\eta = 1 - \exp(-\beta \tilde{z}). \quad (16)$$

The transformed equation (11) for $C(\xi, \eta)$ takes the form

$$(eC_\xi + fC_\eta)_\xi + (fC_\xi + gC_\eta + \beta C)_\eta = 0, \quad (17)$$

where

$$e = (H_c - w(\xi))/(\beta H_c(1 - \eta)),$$
$$f = -(1 - (\tilde{z}/H_c))w'(\xi),$$
$$g = \beta H_c(1 - \eta) \times \left(1 + \left(1 - (\tilde{z}/H_c)^2\right) w'(\xi)^2\right) \Big/ (H_c - w(\xi)).$$

Eq. (17) is approximated by second-order accurate central differences for $C_{ij} \approx C(i\Delta\xi, j\Delta\eta)$, using discrete values $w_i \approx w(i\Delta\xi)$ in the coefficients e, f, and g. An approximate solution to the finite difference equations is obtained at each step of the overall iteration scheme by using 10–20 sweeps of an alternating-direction-implicit method [21]. The results were checked independently using the ELLPACK software system [22].

The temperature is made dimensionless by the factor λG_L. The dimensionless temperature equations are

$$\alpha\, \partial T/\partial z + \partial^2 T/\partial x^2 + \partial^2 T/\partial z^2 = 0, \qquad H_T > z > w(x), \quad (18)$$

$$\alpha\,\partial T_s/\partial z + \partial^2 T_s/\partial x^2 + \partial^2 T_s/\partial z^2 = 0,$$

$$-H_T < z < w(x), \quad (19)$$

$$T = T_H, \quad z = H_T, \quad (20)$$

$$T_s = T_c, \quad z = -H_T, \quad (21)$$

$$T - T_s = 0, \quad z = w(x),$$

$$\partial T/\partial \nu - \partial T_s/\partial \nu = -LV_n/(kG_L), \quad z = w(x), \quad (22)$$

$$\partial T/\partial x = \partial T_s/\partial x = 0, \quad x = 0, \tfrac{1}{2}, \quad (23)$$

where $\alpha = V\lambda/K$, and the constants T_H and T_c correspond to the temperatures of the planar solution (9) and (10) far from the interface at $z = H_T$ and $z = -H_T$, respectively. We remove the discontinuous behavior of the temperature gradient at the interface by introducing the auxiliary temperature field

$$T_p(x,z) = \begin{cases} 0, \quad z > w(x), \\ (LV/(kG_L))(z - w(x))/ \\ \left(1 + w'(x)^2\right) \\ \quad + (1/2)A(x)(z - w(x))^2, \quad z < w(x), \end{cases}$$

where $A(x) = -(LV/(kG_L))[\alpha + w'(x)^2(\alpha + 3w''(x)) - w''(x)]/(1 + w'(x)^2)^3$. The function

$$\tilde{T}(x,z) = \begin{cases} T(x,z) - T_p(x,z), \quad z > w(x), \\ T_s(x,z) - T_p(x,z), \quad z < w(x), \end{cases}$$

is then twice continuously differentiable for $-H_T < z < H_T$ and satisfies an equation

$$\alpha\,\partial\tilde{T}/\partial z + \partial^2\tilde{T}/\partial x^2 + \partial^2\tilde{T}/\partial z^2 = \tilde{F}(x,z),$$

$$-H_T < z < H_T, \quad (24)$$

where $\tilde{F}$ is a continuous function of x and z which follows from the preceding definitions. We use the fast elliptic solver GENBUN from the FISHPACK library [23] to solve this separable equation. Values of $\tilde{T}$ along the interface $z = w(x)$ are obtained by interpolation in z with cubic splines [24] along each vertical line $x = i\,\Delta x$.

Given approximations $w^{(n)}$, $C^{(n)}$, and $T^{(n)}$ to the solution, an $(n+1)$st approximation $w^{(n+1)}$ is obtained from the Gibbs–Thomson equation

$$0 = T_m/(\lambda G_L) + (Mc_\infty/(\lambda G_L))C + (T_m\Gamma/(\lambda^2 G_L))w''(x)\left(1 + w'(x)^2\right)^{-3/2} - T = F[w, w', w''], \quad (25)$$

where the functional F depends both explicitly and implicitly on $w(x)$ through the fields C and T. The updated value is computed in the form

$$a w_{tt} + b w_t = F, \quad (26)$$

where

$$w_{tt} = \left(w_i^{(n+1)} - 2w_i^{(n)} + w_i^{(n-1)}\right)/\Delta t^2$$

$$w_t = \left(w_i^{(n+1)} - w_i^{(n)}\right)/\Delta t,$$

and $F = F_i^{(n)}$ is obtained from $C^{(n)}$, $T^{(n)}$, and $w^{(n)}$.

Second-order accurate central differences are used for w' and w'' in the curvature term. We take $\Delta t = \Delta x$, $a = T_m\Gamma/(\lambda^2 G_L)$, and choose b empirically so that the iterates converge rapidly [19] to a steady state corresponding to $w_t = w_{tt} = 0$ in (26).

A linear stability analysis of the planar solution (8)–(10) to the artificially time dependent system (11)–(14), (18)–(23), and (26) shows that the conditions for marginal stability are identical to those corresponding to the time dependence in the true physical problem [8]. A perturbation of the unstable planar solution can then lead to a computed nonplanar solution with finite amplitude distortions in the interface shape.

Most of the results shown here are obtained on a mesh of 32 × 32 points in each of the liquid and solid regions. The solution is computed over half a wavelength in the horizontal direction. Running time on the NBS Univac 1100/82 using single precision arithmetic is approximately thirty minutes for the full development of a nonplanar solution from an unstable planar interface. Convergence typically proceeds as shown in fig. 1, where the solid curve shows the logarithm of the

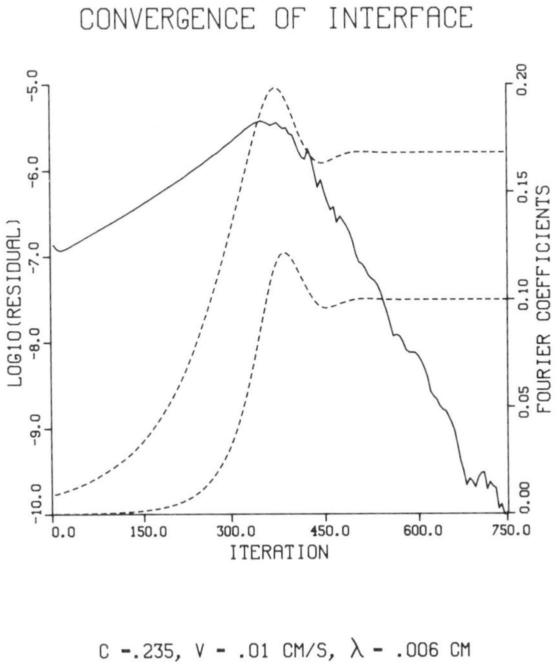

Fig. 1. Iterative convergence of the interface shape for c_∞ = 0.235 wt. %, V = 0.01 cm/s, and λ = 0.006 cm. The solid curve represents the base 10 logarithm of the maximum residual in the Gibbs–Thomson equation. The two-dotted curves show the amplitude of the main and first harmonic Fourier components of the interface shape.

maximum residual of the Gibbs–Thomson equation as a function of iteration number, and the two dotted curves show the development of the first two Fourier components of the interface shape.

4. Numerical results and discussion

We consider an aluminum–silver system [25] with the thermal properties of the solid taken to be identical to those of the actual liquid, see table I. The temperature gradient in the liquid is fixed at 100 K/cm. Fig. 2 shows linear stability results for the planar interface in terms of the critical bulk concentration C_∞^* as a function of the solidification velocity V. The planar interface is stable for bulk concentrations below the curve, and unstable for values above the curve. The left side of the curve corresponds to the constitutional supercooling criterion, while the curve rises on the right side due to the stabilizing effects of surface tension at higher

Table I
Aluminum–silver properties used in calculations

Property	Symbol	Value
Solute diffusion coefficient	D	4.5×10^{-5} cm^2/s
Distribution coefficient	$\tilde{k}$	0.4
Liquidus slope	M	-1.7 K/wt. %
Surface energy parameter	$T_m\Gamma$	1.0×10^{-5} cm · K
Latent heat of fusion	L	1.08×10^3 J/cm^3
Thermal conductivity of liquid	k	9.04×10^{-1} J/(s · cm · K)
Thermal diffusivity of liquid	K	3.52×10^{-1} cm^2/s

velocities (absolute stability [7]). We present calculations of cellular interfaces for V = 0.01 cm/s and V = 1.0 cm/s.

For a given solidification velocity and imposed temperature gradient, the critical bulk solute concentration at the onset of instability depends on the wavelength λ of the sinusoidal perturbation of the interface. We refer to this curve as $C_\infty(\lambda)$ and denote the minimum value $C_\infty(\lambda^*)$ as C_∞^*; C_∞^* was shown in fig. 2 as a function of solidification velocity V. For V = 0.01 cm/s, C_∞^* = 0.2235 wt. % and $\lambda^* = 7.54 \times 10^{-3}$ cm. For bulk concentrations higher than C_∞^* there is a range of linearly unstable wavelengths that is roughly centered about the marginal value λ^*. In fig. 3 we show the steady state interface shape for c_∞ = 0.235 wt. % and $\lambda = 6.0 \times 10^{-3}$ cm, which is slightly smaller than the wavelength of the most unstable linear mode ($\lambda = 6.6 \times 10^{-3}$ cm) at this concentration. The

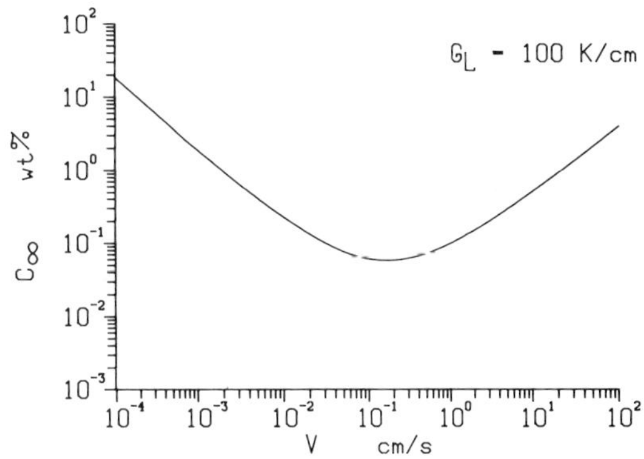

Fig. 2. The bulk silver concentration as a function of the interface velocity at the onset of linear instability for a planar interface in an aluminum–silver alloy. The planar interface is stable below the curve, unstable above it.

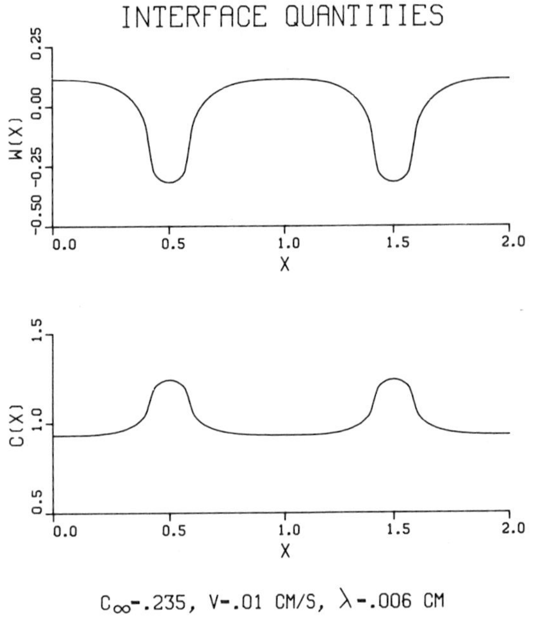

Fig. 3. The interface shape and distribution of solute in the solid. The liquid is above and the solid is below the indicated interface. Two wavelengths are shown.

solute accumulation along the interface is highest in the grooves of the cell, and the interface shape is distinctly non-sinusoidal, as is indicated by the Fourier coefficients shown in fig. 4. If we fix the wavelength of the cell at $\lambda = 6.0 \times 10^{-3}$ cm and increase the bulk concentration from the marginal value for this wavelength, $C_\infty(\lambda) = 0.2258$ wt. %,

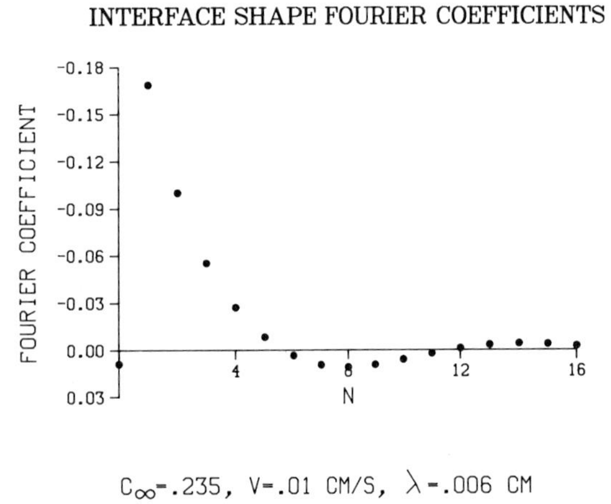

Fig. 4. Fourier coefficients of the interface shape shown in fig. 3.

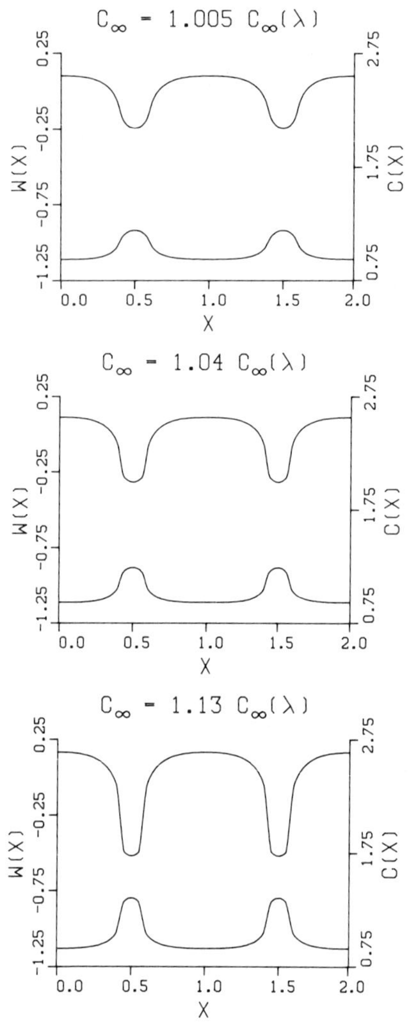

Fig. 5. Plots of interface shape and distribution of solute in the solid for a given wavelength as the bulk concentration is increased. The top curve in each plot is the interface shape, the bottom curve is the concentration of solute. Two wavelengths are shown.

the cell depth and solute distribution in the solid change as shown in fig. 5. The three plots represent solutions computed for $c_\infty = 1.005 C_\infty(\lambda)$, $c_\infty = 1.04 C_\infty(\lambda)$, and $c_\infty = 1.13 C_\infty(\lambda)$, respectively. The cell groove deepens and the accompanying solute segregation increases rapidly as the bulk concentration is raised.

If the bulk concentration is fixed at $c_\infty = 0.235$ wt. % and the wavelength of the cell is changed, the results shown in fig. 6 are obtained. The vertical axis gives the amplitude of the interface deformation in terms of the depth $(\max \overline{W}(\overline{X}) - \min \overline{W}(\overline{X}))/\lambda^*$. The range of linearly unstable wavelengths is 4.65×10^{-3} cm $\leq \lambda \leq 1.32 \times 10^{-2}$ cm. Nonplanar solutions exist at wavelengths shorter than those predicted by linear theory. At $\lambda = 4.25 \times 10^{-3}$ cm, for example, an initial sinusoidal perturbation of sufficiently large amplitude will develop into a nonplanar steady state, while the planar solution is stable to small perturbations. The marginal amplitude is indicated with a "+" in fig. 6. Different nonplanar solutions are obtained for $\lambda = 4.0 \times 10^{-3}$ cm and $\lambda = 8.0 \times 10^{-3}$ cm, so that there may be more than one solution with a given periodicity. Note that the maximum cell depth occurs for wavelengths smaller than λ^*. The fastest growing linear mode also shows a shift to shorter wavelengths.

An initial disturbance of wavelength 9.0×10^{-3} cm initially grows in accord with linear theory, and the first harmonic with half this wavelength is also excited. Eventually, however, the main harmonic stops growing and then decays to zero amplitude, while the first harmonic asymptotes at a finite amplitude, resulting in a periodic solution with wavelength $\lambda = 4.5 \times 10^{-3}$ cm. This behavior is observed for all wavelengths in the range 9.0×10^{-3}cm $\leq \lambda \leq 13.1 \times 10^{-3}$ cm. The triangular data points shown in fig. 6 correspond to solutions

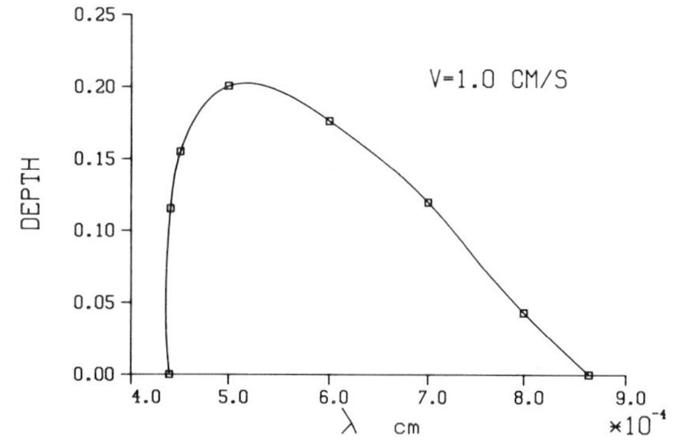

Fig. 7. Maximum depth of the interface as a function of wavelength for $V = 1.0$ cm/s and $c_\infty = 1.11$ C_∞^*.

whose initial perturbation of wavelength λ give rise to steady solutions with wavelength $\lambda/2$. These solutions are computed using 64 mesh points in the x-direction so that the number of mesh points per wavelength ($\lambda/2$) remains the same.

Fig. 7 shows results similar to fig. 6, but corresponding to a solidification velocity of 1.0 cm/s and bulk concentration $c_\infty = 0.110$ wt. %, where $C_\infty^* = 0.099$ wt. % and $\lambda^* = 6.15 \times 10^{-4}$ cm. The range of linearly unstable wavelengths is 4.33×10^{-4}cm $\leq \lambda \leq 8.67 \times 10^{-4}$ cm, and the numerical procedure reproduces this range to within 1.6%. In contrast to the lower velocity, we have not found

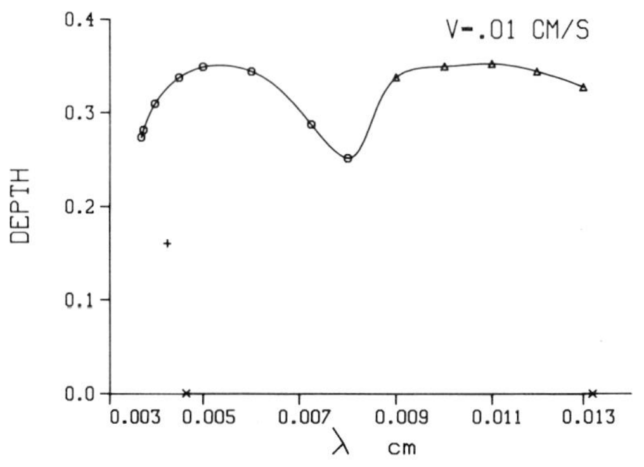

Fig. 6. Maximum depth of the interface as a function of wavelength for $V = 0.01$ cm/s and $c_\infty = 1.05$ C_∞^*. The range of linearly unstable wavelengths is indicated by crosses on the x-axis.

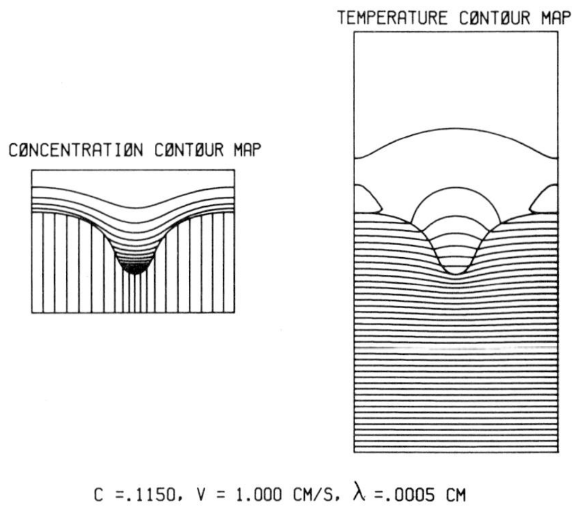

Fig. 8. Concentration and temperature contours in the liquid and solid for $V = 1.0$ cm/s and $c_\infty = 1.16$ $C_\infty^* = 0.115$ wt. %. The solid–liquid interface is shown also.

solutions outside the wavelength range indicated by linear stability theory. This difference between the two velocities is not unexpected based on the non-linear stability analysis of Wollkind and Segel [10].

Fig. 8 shows the concentration field, temperature field, and interface shape for a sample with $c_\infty = 0.115$ wt. % and $\lambda = 5.0 \times 10^{-4}$ cm. The distortion of the isotherms near the interface is due to the heat released by the liquid upon solidification. The concentration contours emphasize the large buildup of solute in the interface depression. The vertical lines in the solid indicate the extent of solute segregation; each band contains the same amount of solute. Fig. 9 shows the interface solute distribution in more detail. Note that the minimum concentration does not occur at the maximum of the interface. Solute distributions with local maxima of concentration at positions corresponding to the peak of the solid–liquid interface have been observed experimentally at high growth velocities, and this has been attributed to non-equilibrium conditions prevailing at the interface [26]. It is interesting that a similar feature can be

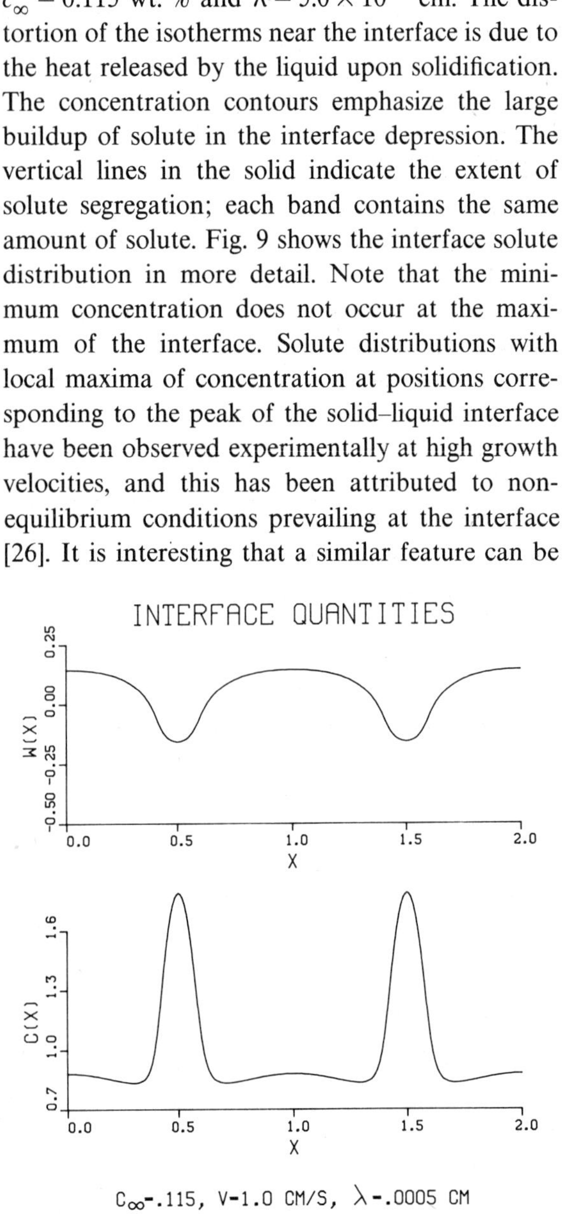

Fig. 9. The interface shape and distribution of solute in the solid for the case shown in fig. 8. Two wavelengths are shown.

obtained under the equilibrium conditions assumed here.

These results indicate the extreme non-linearity of the cellular solidification problem. For example, in fig. 9 there is more than a factor of two difference between the maximum and minimum solute concentration in the solidified material for a bulk concentration 16% above the critical concentration. Even with the assumption of local equilibrium at the interface and equal thermal properties for solid and liquid, cellular solidification depends on six dimensionless parameters. A relatively small range of parameters has been investigated so far. We believe the steady state solutions we compute would also be obtained by integrating the true time-dependent equations. The numerical method used by Ungar and Brown [18] allows the computation of both stable and unstable solutions, and provides information about their linear stability as well. The stability of cell shapes of a given wavelength to perturbations of longer wavelength has not been examined here but is an area of active research [13, 27]. Time-dependent and three-dimensional calculations and methods to treat even greater degrees of non-linearity are areas for further research.

Acknowledgements

It is a distinct pleasure to acknowledge numerous contributions of R.F. Sekerka and R.F. Boisvert to various aspects of this problem. The authors have also benefited from conversations with W.J. Boettinger, R.A. Brown, G. Dee, M. Kerszberg, J.S. Langer, R.G. Rehm, R.J. Schaefer, and D.J. Wollkind. This work was supported in part by the Defense Advanced Research Projects Agency.

References

[1] M.C. Flemings, Solidification Processing (McGraw-Hill, New York, 1974).

[2] W.A. Tiller, J.W. Rutter, K.A. Jackson and B. Chalmers, Acta Met. 1 (1953) 428.

[3] R.F. Sekerka, Crystal Growth: An Introduction, P. Hartman, ed. (North-Holland, Amsterdam, 1973), p. 403.
[4] R.T. Delves, Crystal Growth, vol. 1, B.R. Pamplin, ed. (Pergamon, Oxford, 1974), p. 40.
[5] D.J. Wollkind, Preparation and Properties of Solid State Materials, vol. 4, W.R. Wilcox, ed. (Marcel Dekker, New York, 1979), p. 111.
[6] J.S. Langer, Rev. Mod. Phys. 52 (1980) 1.
[7] W.W. Mullins and R.F. Sekerka, J. Appl. Phys. 35 (1964) 444.
[8] R.F. Sekerka, Crystal Growth, H.S. Peiser, ed. (Pergamon, Oxford, 1967), p. 691.
[9] S.R. Coriell, M.R. Cordes, W.J. Boettinger and R.F. Sekerka, J. Crystal Growth 49 (1980) 13.
[10] D.J. Wollkind and L.A. Segel, Phil. Trans. R. Soc. 268 (1970) 351.
[11] G. Dee and R. Mathur, Phys. Rev. B27 (1983) 7073.
[12] R. Sriranganathan, D.J. Wollkind and D.B. Oulton, J. Crystal Growth 62 (1983) 265.
[13] M. Kerszberg, Phys. Rev. B 27 (1983) 3909.
[14] F.M. Carlson, L-Y. Chin, A.L. Fripp and R.K. Crouch, Materials Processing in the Reduced Gravity Environment of Space, G.E. Rindone, ed. (Elsevier, New York, 1982), p. 629.
[15] S.R. Coriell and R.F. Sekerka, J. Crystal Growth 46 (1979) 479.
[16] S.R. Coriell, R.F. Boisvert, R.G. Rehm and R.F. Sekerka, J. Crystal Growth 54 (1981) 167.
[17] H.M. Ettouney and R.A. Brown, J. Crystal Growth 58 (1982) 313.
[18] L.H. Ungar and R.A. Brown, Phys. Rev. B29 (1984) 1367.
[19] F. Bauer, O. Betancourt and P. Garabedian, A Computational Method in Plasma Physics (Springer, New York, 1978).
[20] P.R. Garabedian and G.B. McFadden, AIAA Journal 20 (1982) 289.
[21] E. Isaacson and H.B. Keller, Analysis of Numerical Methods (Wiley, New York, 1966).
[22] J.R. Rice, Mathematical Software III, J.R. Rice, ed. (Academic Press, New York, 1977), p. 319.
[23] P.N. Swarztrauber, SIAM J. Numer. Anal. 11 (1974) 1136.
[24] C. de Boor, A Practical Guide to Splines (Springer, New York, 1978).
[25] R.J. Schaefer, W.J. Boettinger, F.S. Biancaniello and S.R. Coriell, Lasers in Metallurgy, K. Mukherjee and J. Mazumder, ed. (Metallurgical Society of AIME, 1981), p. 43.
[26] T.Z. Kattamis, Z. Metallkde 61 (1970) 856.
[27] G. Dee and J.S. Langer, Phys. Rev. Lett. 50 (1983) 383.

Physica 12D (1984) 262–269
North-Holland, Amsterdam

DIRECTIONAL SOLIDIFICATION FRONTS: STRUCTURE AND DYNAMICS

Michel KERSZBERG
Physics Department, Harvard University, Cambridge, MA 02138

and

Xerox Palo Alto Research Center, Palo Alto, CA 94304, USA

The equations of motion for a two-dimensional solidification front are established. Beyond a threshold velocity the front develops spatial structure. The resulting patterns are displayed and compared with experiment. When noise is added to the equations, pattern selection effects are observed which can be explained phenomenologically.

1. Introduction

Although crystal growth morphology, as illustrated by the snowflake, has fascinated man for a long time [1], it is only recently that definite progress has taken place in our qualitative and quantitative understanding of the related physics [2]. In this contribution, we present an analysis of directional solidification [3], a problem which is of great technological importance, and affords some insights into the more general questions raised by crystal growth.

In a typical directional solidification experiment, a metallic ingot is driven at a constant velocity past an oven; the metal melts upon entering the furnace and resolidifies thereafter. This last phenomenon occurs in an interesting way. When the ingot velocity is low, the solidification front remains planar; however, at increased speeds, corrugated structures of various types are observed, which will be described below in greater detail [3, 4]. Our aim in what follows will be to report on some recent advances in the comprehension of solidification patterns.

In section 2, we introduce the fundamental relevant equations. A first step in their analysis then consists in trying to explain the stationary features of the solidification front, i.e., the stable shape of the interface, as seen in a comoving referential. A major stumbling block here has been the complex nature of the observed patterns. In particular, even slightly above threshold velocity, the front appears to be sharply cusped: this signals trouble for the customary single-mode perturbation theory, which leads naturally to sinusoidally modulated structures. In section 3, we introduce a general, multimode approach, which leads us to markedly improved results: these are described in section 4, where various methods are used in order to truncate the multimode expansion. In section 5, another, in certain ways more drastic approximation will be described which gives rise to a completely integrable equation of motion for the interface; the equation has solutions in the form of isolated time-independent cusp-like solitons, periodic arrays of infinitely deep cusps, oscillating solutions, etc. Finally, section 6 is devoted to the long-time dynamics of the interface; the dynamical problem arises because the behavior of the front seems to depend on the initial conditions, leading to stationary corrugated states with a *non-unique* spatial periodicity. This is the famous "pattern selection" question. Here, we investigate, by numerical simulation, the behavior of the system when external noise is applied. It is found that under these circumstances "selection" takes place, in the sense that *one* wavenumber only appears to survive in the long-time limit.

0167-2789/84/$03.00 © Elsevier Science Publishers B.V.
(North-Holland Physics Publishing Division)

2. Physics of directional solidification

To be definite, consider an interface between the solid and liquid phases of a binary alloy, forced to move at a constant average velocity V_z toward the liquid side. Assume that relative diffusion of the constituents is the limiting kinetic factor, i.e., growth is hampered only by the impurity having to diffuse away from the solid. In practice, an excellent approximation [5] is that diffusion occurs *only* in the liquid, while convective effects may be neglected altogether. In the referential where the interface is immobile, diffusion is described by

$$\frac{\partial \mu}{\partial t} = D\nabla^2\mu + V_z\frac{\partial \mu}{\partial z}, \tag{1}$$

where μ is the impurity's relative chemical potential. If the front moves slowly enough [3], this becomes

$$D\nabla^2\mu + V_z\frac{\partial \mu}{\partial z} \approx 0. \tag{2}$$

Eq. (2) is of course linear; the nonlinearity is introduced by the boundary conditions. These specify

(a) the value of μ on the interface, μ_s: if $z_s(x, y)$ is the interface position, this is [6]

$$\mu_s(x, y) = \mu_0 - \Gamma_s(x, y), \tag{3}$$

where μ_0 is the chemical potential at liquid–solid coexistence when the interface is flat, and $\Gamma_s(x, y)$ the mean curvature of $z_s(x, y)$, i.e.,

$$\Gamma_s(x, y) = \nabla \cdot \left(\frac{\nabla z_s}{\sqrt{1 + |\nabla z_s|^2}} \right); \tag{4}$$

(b) μ at "infinity" on the liquid side, which may be taken as constant;

(c) the way in which z_s changes as a result of diffusion; since the diffusion flux is proportional to $\nabla\mu$, the velocity of the interface will be given by

$$\dot{z}_s \approx (\nabla\mu)_z + (\nabla\mu)_x\frac{\partial z_s}{\partial x} + (\nabla\mu)_y\frac{\partial z_s}{\partial y}. \tag{5}$$

Now, in spite of the linearity of (2), the problem defined by (2), (3) and (5) is such that a planar interface (z_s = constant) is *always* unstable to deformations of sufficiently long wavelengths! This is due physically to the "point effect," whereby any protuberance on the front causes the flux around it to be enhanced and growth there to proceed faster. In order to stabilize the planar interface, at least for low velocities, one must impose on the interface a thermal gradient $G = \partial T/\partial z$, which is assumed given [3]. Because coexistence at different temperatures implies different μ, we now have, instead of (3),

$$\mu_s(x, y) = \mu_0 - \Gamma_s(x, y) - \text{const} \times z_s. \tag{6}$$

In the next section, we show how the system (2), (5), (6) can be used in deriving an autonomous *equation of motion* for the front, in the form

$$\dot{z}_s = F[\{z_s(x, y)\}], \tag{7}$$

where F is a *functional* of z_s, which depends parametrically on V_z as well as on thermodynamic quantities. The *non-local* nature of (7) is a direct consequence of the elliptic character of eq. (2), which in turn leads to the "point effect;" the fact that F is a functional thus reflects an essential aspect of the system's physics.

3. Equations of motion for a solidification front

It will be clear from (5) that, in order to determine $\dot{z}_s$, we must compute the chemical potential. Although the problem of solving a partial differential equation with boundary conditions imposed on a slightly deformed contour

$$z_s = \sum_{\boldsymbol{k}} \varepsilon_{\boldsymbol{k}} e^{i\boldsymbol{k}\cdot\boldsymbol{x}} \tag{8}$$

would appear to be a generic one in mathematical physics, a simple and systematic method of solution has been published [7] only recently. The idea is to write successive differential problems for the partial derivatives $\partial^n\mu/\partial\varepsilon_{k_1}\cdots\partial\varepsilon_{k_n}$, and solve these to zeroth order in the ε_k's. The partial derivatives then generate the Taylor expansion for μ. To illustrate the principle of the solution, let us consider for simplicity only one mode $\varepsilon_k=\varepsilon$. Let us concentrate on the variation of μ with ε; denoting $\partial\mu/\partial\varepsilon\equiv\mu_\varepsilon$, we have, from (2),

$$D\nabla^2\mu_\varepsilon+V_z\frac{\partial\mu_\varepsilon}{\partial z}=0. \tag{9}$$

In finding the boundary condition associated with (9), remember that varying ε by $\delta\varepsilon$ leads to a motion of the interface

$$\delta z_{\mathrm{s}}=\delta\varepsilon\,\mathrm{e}^{\mathrm{i}\boldsymbol{k}\cdot\boldsymbol{x}}, \tag{10}$$

so that

$$\frac{\partial\mu_{\mathrm{s}}}{\partial\varepsilon}\equiv(\mu_{\mathrm{s}})_\varepsilon=(\mu_\varepsilon)_{\mathrm{s}}+(\nabla_z\mu)_{\mathrm{s}}\mathrm{e}^{\mathrm{i}\boldsymbol{k}\cdot\boldsymbol{x}}, \tag{11}$$

where $(\mu_\varepsilon)_{\mathrm{s}}$ denotes the partial derivative μ_ε evaluated on the surface. Thus the boundary condition for (9) reads

$$(\mu_\varepsilon)_{\mathrm{s}}=(\mu_{\mathrm{s}})_\varepsilon-(\nabla_z\mu)_{\mathrm{s}}\mathrm{e}^{\mathrm{i}\boldsymbol{k}\cdot\boldsymbol{x}}, \tag{12}$$

in which $(\mu_{\mathrm{s}})_\varepsilon$ can be computed explicitly from (6). There remains to solve (9) with (12) on the unperturbed ($\varepsilon=0$) contour, thus determining μ_ε everywhere. It is easy to see that we can establish in like manner all the derivatives $(\partial\mu/\partial\varepsilon)_{\varepsilon=0}$, $(\partial^2\mu/\partial\varepsilon^2)_{\varepsilon=0}$, etc, and thus obtain

$$\mu=\mu|_{\varepsilon=0}+\left.\frac{\partial\mu}{\partial\varepsilon}\right|_{\varepsilon=0}\varepsilon+\tfrac{1}{2}\left.\frac{\partial^2\mu}{\partial\varepsilon^2}\right|_{\varepsilon=0}\varepsilon^2+\cdots. \tag{13}$$

It is now a "simple" matter to write (7) explicitly [3]. The result can be expressed as an autonomous system of first-order differential equations for the amplitudes ε_k:

$$\dot{\varepsilon}_k=f_k(\{\varepsilon\}). \tag{14}$$

The detailed form of (14) has been given elsewhere [3]. The functions f are given as polynomials of infinite order in ε; the first-order term has the following approximate structure

$$\dot{\varepsilon}_k\approx k\left[-\frac{V_z^2}{k}+V_z-k^2-\text{constant}\right]. \tag{15}$$

It is apparent from (15) that, while below a certain velocity V_{zc}, all amplitudes have a negative linear growth rate, above V_{zc} a range of perturbations may grow, thus destabilizing the flat front ($\varepsilon=0$). The nonlinear terms in (14) play a crucial role in restabilizing the interface in some new stationary state. It turns out that the nonlinearity leading to the point effect is very strong, and that high-order terms become rapidly important. Not unrelated to this, the lack of stability is such that not only *one* mode ε_k, but also its harmonics $\varepsilon_{2k},\varepsilon_{3k},\ldots$ etc. are necessary in order for the system to reach stationarity. Usually, in physics, one considers an expansion scheme as useful if and only if the expansion can be truncated after a small number of terms; such is not the case here. However, a perturbative expansion appears at the moment, because of the nonlocality of the problem, as the only way to proceed [3]. We now describe briefly two truncation schemes which we have used in computing the solutions displayed on figs. 1 and 2.

Fig. 1. The stationary one-dimensional cellular interface structure, as calculated from truncated eq. (14). The liquid phase is on the top of the picture. The figure illustrates the solution at fourth order precision in ε (see text), which compares rather well with some experimental results [8].

Fig. 2. Two-dimensional interface structures, computed as stationary solutions of eq. (14) truncated to third order. The solutions are represented as contour plots; the solidification front advances out of the page, with the letter "H," indicating a high or warm point, while "L" denotes a low or cold point on the surface. a) This solution can be described as a triangular array of "pits." b) In this configuration, the interface is covered by a hexagonal array of "grooves," which meet at the lowest points. There is a third possible solution with the same parameter values, i.e., one where the one-dimensional solution (fig. 1) is extended in an array of infinite straight "grooves" running perpendicular to the page.

4. Stationary structure of the solidification interface

We first study one dimensional (1D) interfaces, i.e., those described by $z_s(x)$. Such fronts have been the object of careful experiments [8]; in general, at large velocity, they restabilize in periodic arrays of sharp cusps, known as cellular structures. It turns out that in order to reproduce such patterns with good accuracy, it is indispensable to keep up to sixteen modes, i.e., $\varepsilon_k, \ldots, \varepsilon_{16k}$ – that is the most our computer would take anyway – and retain terms in (14) which are formally of order four or five in ε. The stationary structures are computed by setting $\dot{\varepsilon}_k = 0$. A typical result is displayed in fig. 1. It compares quite well with the experimental findings. In the figure, we have scaled all lengths in such a way that the interface appears exactly as it would be seen under a microscope; the cellular spacing is about 20 μm. Note that the spacing could be chosen over a *range* of wavelengths: the fact that the fundamental period is not unique is a manifestation of the pattern selection problem, about which we shall have more to say in section 6. As far as we know, this is the first time that a theoretical approach is able to reproduce the sharp cellular patterns observed in solidification experiments. Elsewhere in these proceedings, McFadden reports some interesting results obtained by direct numerical solution of the one-dimensional interface problem [9].

For obvious reasons, computations in the 2D case cannot possibly include as many harmonics or extend to such high orders as was feasible in one dimension. Thus, we have limited ourselves here to a three-mode, third order ε_k expansion. It is gratifying, however, to note that even so, we are able to reproduce many of the qualitative features observed experimentally [10]. Three types of stationary patterns emerge from the calculation. The first one is a simple generalization of the 1D structure found previously, where the cusps of fig. 1 may now be seen to extend perpendicular to the page as linear equidistant grooves. Such cellular fronts, albeit defected, are seen in experiments. The second and third species of 2D cells are depicted in fig. 2 as contour plots. The interface should be imagined as parallel to the paper, and moving out of it; constant altitude lines go from low (L) to high (H). We see that these two structures both possess sixfold symmetry; one (a) may be thought of as a triangular array of "dimples," while the other (b) resembles a hexagonal network of grooves, with the grooves meeting at the L points. We should remark here that, while in the experiments, only one type of pattern is seen under any given set of external conditions, the three 2D structures described above are equally valid stationary solutions of eqs. (14), and may in principle coexist. The question of the relative stability of these three patterns is an open and difficult one. Furthermore, each 2D pattern may be constructed (just as the 1D ones) with a range of fundamental periodicities; in contrast with the 1D case, we are unable at the present time to say anything about wavelength selection in 2D structures: our pattern selection studies (see section 6) have, out of computational necessity, been confined to one-dimensional systems.

5. A completely integrable model

We shall now outline cursorily how one can simplify our evolution equations (14) in such a way as to obtain an integrable nonlinear equation. We restrict ourselves to 1D interfaces and consider only the first- and second-order terms in the equations, but we keep all modes k. It turns out that the second-order contributions take a simple form in the limit where the constant in eq. (6), which depends on the applied thermal gradient as well as on various thermodynamic parameters, is much greater than one. This is not very realistic physically, but we shall forsake plausibility in order to explore the resulting model, which is interesting in itself. Eq. (14) then becomes, after some rescaling

$$\dot{\varepsilon}_k = \left(\nu + |k| - |k|^3\right)\varepsilon_k - c\sum_{k_1}|k_1|\varepsilon_{k_1}\varepsilon_{k-k_1}, \tag{16}$$

where the linear terms correspond to those displayed in eq. (15), while ν and c are constants. We now transform (16) back to real space; some manipulations lead to

$$\partial_t z_s = \left[\nu - \mathrm{i}\partial_x\left(1 + \partial_{xx}^2\right)\right] z_s - \mathrm{i}\beta z_s \partial_x z_s + \text{non-local terms}, \tag{17}$$

with β a constant, and where the appearance of the nonlocal corrections, (i.e., integrals over x) stems from the elliptic nature of the basic equations. One can show that the nonlocal terms vanish provided that, for some x_0,

$$\int \mathrm{d}x\, z_s(x - x_0)\sin kx = 0 \tag{18}$$

for all k. If we limit ourselves to functions z_s which satisfy (18), we may thus ignore the nonlocal corrections altogether; if we accept functions for which the integral in (18) does not vanish, but remains small, there is at least a way to treat the nonlocal terms by perturbation theory, the integral in (18) representing a natural small parameter. Now, the local part of (17) has a number of interesting characteristics, which we have only begun to explore; in what follows we simply present some particular solutions of (17); let us stress at the outset that investigation of these solutions' properties, such as stability, is a complicated exercise in singular perturbation theory and is only in its preliminary stages.

We solve (17) by noting that *upon setting* $\nu = 0$, it can be transformed into a real equation through either

$$x \to \mathrm{i}x \tag{19a}$$

or

$$t \to \mathrm{i}t. \tag{19b}$$

Using (19), we find equations which can be solved through application of the techniques developed by Hirota [11] and Nakamura [12]. Whether the solutions are stable upon reinstating $\nu \neq 0$ *has not yet been determined*. Of particular interest is the stationary solution

$$z_s = \frac{k^2 - 1}{6} - \frac{k^2}{1 + \cos kx}, \tag{20}$$

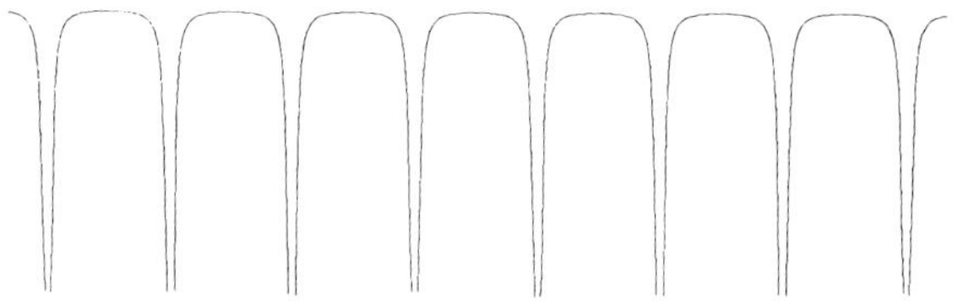

Fig. 3. A stationary interface structure obtained from our simple, completely integrable model [eq. (17)]. The solution exhibits infinitely deep cusps, a result which we have experimental reason to assume is not altogether unphysical [8].

which describes an array of infinitely deep cusps (see fig. 3) very reminiscent of some experimental observations [8]. Note that (20) does *not* have a Fourier transform, and is thus the result of a highly nontrivial resummation of the perturbation expansion we started with; z_s in (20) satisfies the restriction (18), if the integral in the latter is interpreted as a *principal part*: physically, this amounts to denying – quite reasonably – any meaning to the cusps beyond a certain depth.

Another remarkable stationary solution is the isolated, or soliton-like cusp

$$z_s = z_0 + \partial_{xx}^2 [\log \theta(\eta, \tau)],$$

where θ is the theta-function

$$\theta(\eta, \tau) = \sum_{n=-\infty}^{\infty} \exp(2\pi i n \eta + \pi i n^2 \tau), \tag{21}$$

and $\eta = ikx$, τ = integration constant. Isolated grooves have been observed [10] on 2D interfaces. In a similar fashion, it is possible to find time-dependent solutions of (17), in particular, oscillating ones; their physical significance is not, however, clear at the present time. The same may be said of several "exploding" solutions which grow without bound as $t \to \infty$. We believe exploring the features of the model equation (17) is a fascinating and potentially very rewarding enterprise; in particular, (17) is the first non-linear equation that this author knows of, which has a nontrivial pattern selection problem and for which some fully non-linear solutions are known in a well-defined limit.

6. Wavenumber selection under external noise

We now discuss the problem which has been surfacing time and time again in our study of directional solidification, that is, pattern selection. More specifically, we limit ourselves to the question of wavenumber selection in 1D systems. We shall proceed by analyzing in detail the effect of Gaussian white noise on the dynamics of the solidification front. As we shall see, noise seems to generate a selection effect: there appears to exist a preferred wavenumber which the system likes to occupy when "shaken" ever so slightly. This is interesting [13] in part because the selection problem's scope goes much beyond the physical situation considered here: selection is probably a relatively general phenomenon in nonlinear dynamics.

We study the equation (14), with additive Gaussian white noise $\zeta_k(t)$, $\langle \zeta_k(t) \zeta_{k'}(t') \rangle = A\delta(k+k')\delta(t-t')$, thus

$$\dot{\varepsilon}_k = f_k(\{\varepsilon\}) + \zeta_k(t). \tag{22}$$

We proceed [3] by numerical simulation of eq. (22), taking between 30 to 40 modes k, $k = n\delta$, with δ a small wavenumber which sets the effective size of the system, $L \sim \delta^{-1}$. We truncate eq. (22) formally at third order in ε; the procedure used to generate the noise $\zeta(t)$ and integrating (22) has been described in ref. 3.

The outstanding feature of the simulations is that the solidification interface unequivocally displays a unique final wavenumber, which we denote by k_s (this wavenumber depends of course on the externally imposed average velocity V_z of the front). In what follows we illustrate the selection phenomenon with examples. Let us first note, however, that the noise level required is very small; for instance, the noise in the simulation summarized in fig. 4 has a variance which is about 6.5 percent of the largest amplitude present at each moment in the spectrum. The effect of noise is to help induce the strongly nonlinear events which are ultimately responsible for wavelength change. Some of these

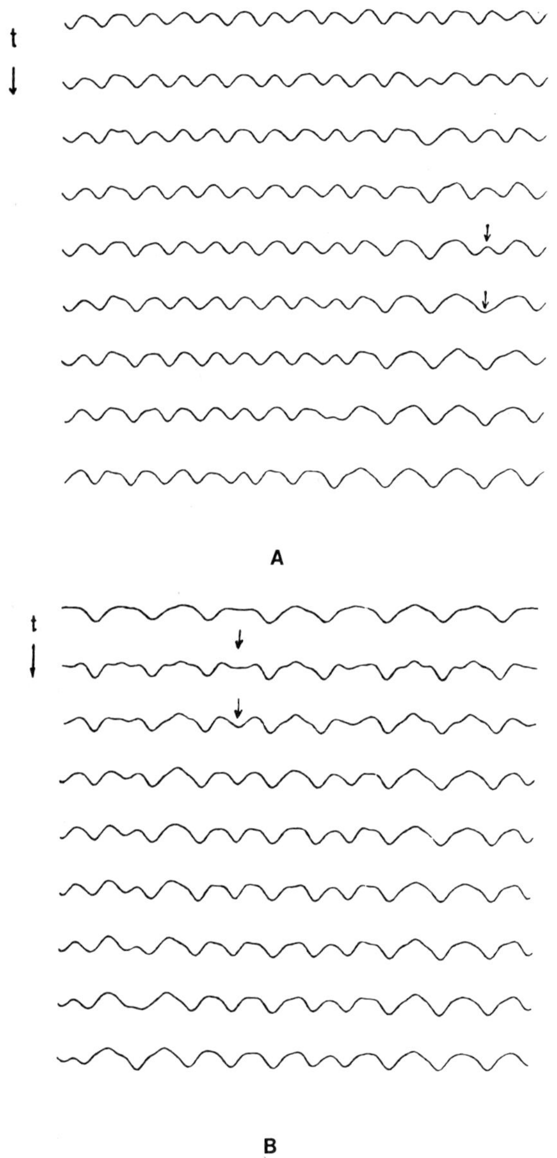

Fig. 4. Simulated time evolution under noise of a one-dimensional interface, as described by eq. (22). The noise standard deviation is 6.5%; the liquid side is on top of each profile, while time increases from top to bottom. Evolution of the wavenumber occurs through cell "deaths" (fig. 4a, arrows) or "divisions" (fig. 4b), leading to a unique final state, which begins to fill the axis in both pictures.

events closely resemble the phase "slips" believed to be involved in supercurrent decay [14], and we shall term them cell "deaths": they correspond to the *disappearance* of a cell from the front (see arrow on fig. 4a). Alternatively, a cell may divide under the influence of the noise, and leave behind two "daughters" (see arrow on fig. 4b). The striking phenomenon is that ultimately, one wavenumber seems to "nucleate" through this process and invades the whole interface. We shall not go into further details regarding the simulations; the interested reader may consult [3]. Here we should merely like to point out that the selection phenomenon may well be sensitive in its fine features to the noise spectrum, and it would be a fascinating problem to study the effect of a noise source with correlations other than those we have used.

The theoretical analysis of the "death" and "division" process is extremely difficult; it is possible that these events will ultimately be analyzable in a simplified model. This was partly the motivation behind the construction of eq. (17) (see previous section). In the remainder of this review, we present a simple phenomenological analysis which seems to explain with fair quantitative success the location of the selected wavenumber.

The system of 30 to 40 equations we are considering [see eq. (22)] is, of course, extremely complex. However, a major simplification occurs because, as can be observed on the various "snapshots" of fig. 4, the interface tends to divide into regions which are characterized each by a relatively well-defined local wavenumber $k(x, t)$. We exploit this observation by first rewriting the equations of motion (22) for a subset of modes, namely some fundamental k_0 and its harmonics $\varepsilon_i \equiv \varepsilon_{ik_0}$, $i = 1, 2, \ldots$,

$$\dot{\varepsilon}_i = f_{i;\,k_0}(\{\varepsilon_k\}) + \zeta_i(t), \tag{23}$$

and then introducing a slow (x, t) dependence of k_0:

$$\dot{\varepsilon}_i = f_{i;\,k_0(x,t)}(\{\varepsilon_i\}) + \zeta_i'(t). \tag{24}$$

It should be understood that (24) summarizes a series of complicated effects. In principle, to deduce it from (23) would require the formal elimination of the ε_k, $k \neq ik_0$ modes. This will result in new contributions to the fast Gaussian noise–which is why we write ζ' instead of ζ–but also in memory, or non-Markovian effects. The main point here is the assumption that, on a slow time scale, the only way the system remembers a "cell death" is through the fact that the local density of cells $\sim k_0(x, t)$ has changed, thence the fluctuations in k_0. We have thus circumvented the difficulty of having to analyze the "cell deaths" themselves: but can we learn something useful from this reduced description?

The response elicited in each region as k_0 fluctuates depends on the sensitivity of the (nonlinear) growth rates $\dot{\varepsilon}_i$ to k_0. To first order, and neglecting phase factors, the response of $\dot{\varepsilon}_i$ to perturbations will be

$$\delta\dot{\varepsilon}_i = \left(\frac{\partial f_{i;k_0}}{\partial k_0}\right)_{\varepsilon_j} \delta k_0 + \sum_j \left(\frac{\partial f_{i;k_0}}{\partial \varepsilon_j}\right)_{k_0} \delta\varepsilon_j, \qquad (25)$$

where the ε_i assume their stationary values $\varepsilon_i(k_0)$. Consistent with our previous hypotheses, we assume that the fluctuations $\delta\varepsilon_j$ average out to zero rather rapidly, leaving $\delta\dot{\varepsilon}_i(\text{slow}) = (\partial f_i/\partial k_0)_\varepsilon \delta k_0$. It now appears reasonable to surmise that a region where some of the $\partial f_i/\partial k_0$ vanish would exhibit a diminished sensitivity to the noise δk_0, that is, to the broad spectrum noise which generates cell-death type events. It turns out that $(\partial f_1/\partial k_0)_{\varepsilon_i}$ does indeed vanish at a wavenumber $k_0 = k^*$, and for all values of the various parameters which we have checked, k^* was found to coincide with k_s, the wavenumber selected under noise in the numerical experiments. It thus seems plausible to assume that the interface ultimately freezes around $k_0 = k^*$ for lack of an efficient "escape" mechanism linking scale fluctuations to the other slow motions of the system [15].

7. Conclusions

We have reported several recent results in the theory of directional solidification. Some progress has been made both in determining the stationary structure and the dynamics of the solidification interface. Specifically, the "cusp" problem, relating to the complicated shape of cellular fronts, has been analyzed with a measure of success. The dynamical problem of wavenumber selection under noise has received a clear "experimental" (i.e., numerical) answer, and may have begun to yield to theoretical study. Finally, a simplified, completely integrable model presented in section 5 may, in the near future, lead to interesting clues regarding the intriguing problems posed by the variety of solidification patterns.

Acknowledgements

The author's work at Harvard was partially supported by NSF under grant No: DMR 82-07431, and by a Ch. Weizmann Postdoctoral Fellowship.

References

[1] J. Kepler, De Nive Sexangula, Opera, Frisch ed., VII (1611) 715.
[2] R.F. Sekerka, Physica 12D (1984) 212 (these proceedings).
[3] M. Kerszberg, Phys. Rev. B 27 (1983) 3909 and in press.
[4] D.J. Wollkind, Physica 12D (1984) 215 (these proceedings), and J. Cryst. Growth, to appear.
[5] D.J. Wollkind and L.A. Segel, Phil. Trans. R. Soc. Ser. A 268 (1970) 351.
[6] See, e.g., L.D. Landau and E.M. Lifshitz, Statistical Physics, 2nd ed., (1958) p. 460 ff.
[7] D.D. Joseph and R.L. Fosdick, Arch. Ration. Mech. Anal. 49 (1973) 321.
[8] K.A. Jackson, in: Solidification (Am. Soc. Metals, Metals Park, Ohio, 1971).
[9] J. McFadden, Physica 12D (1984) 253 (these proceedings) and to be published.
[10] See, e.g., J.S. Kirkaldy, in: Non-Equilibrium Thermodynamics: Variational Techniques and Stability, R.J. Donnelly et al., eds. (Univ. of Chicago Press, Chicago, 1966).
[11] R. Hirota, in: Lecture Notes in Mathematics, vol. 515 (Springer, Berlin, 1976).
[12] A. Nakamura, J. Phys. Soc. Japan 47 (1979) 1701.
[13] V. Datye and J.S. Langer, Phys. Rev. B 24 (1981) 4155.
[14] D.E. McCumber and B.I. Halperin, Phys. Rev. B 1 (1970) 1054.
[15] M. Kerszberg, Phys. Rev. B28 (1983) 247.

Physica 12D (1984) 270–278
North-Holland, Amsterdam

A NUMERICAL APPROACH TO DIRECTIONAL SOLIDIFICATION

G. DEE
Institute for Theoretical Physics, University of California, Santa Barbara, California 93106, USA

We describe a numerical technique for solving the integrodifferential equations which describe the motion of the solidification interface. The technique enables us to study both the stationary states and dynamics of the interface for the symmetric model. Comparison is made between existing analytical results and the numerical calculations.

1. Introduction

Many systems in nature exhibit pattern formation. In metallurgy, the process of directional solidification [1] (D.S.) is such a system. Any realistic model of this system must involve many parameters and must include a realistic description of the dynamics of the system. For the purpose of studying realistic problems like this, we decided to develop numerical procedures to solve the integrodifferential equations which can be formulated to describe the motion of the solidification interface [2]. We apply these numerical methods to the so-called symmetric model [2,3]. This model has similar equations of motion and hence the same qualitative behavior as more realistic models [4]. Another simplifying feature is that the model has only two parameters.

In this paper we will develop numerical techniques to apply to the problem of directional solidification which are

(a) not limited to small amplitude deformations of the solidification interface;

(b) capable of scanning the whole parameter space of the system.

We will apply these techniques to the symmetric model and investigate the stationary states of this model. We will investigate the stability of these states. We will study the dynamics of the system with particular emphasis on the problem of pattern selection. The problem of pattern selection exists here because of the fact that in a region of parameter space where stable pattern states are permitted, the system has a band of allowed states. Hence the question, what set of initial conditions leads to a particular final state or set of final states? We will not answer these questions here but we will illustrate the problem of pattern selection to demonstrate possible uses of the numerical methods to study these questions.

The integro-differential equation for the interface motion is described in section 2 as well as the stability analysis of the planar solidification interface. In section 3 we describe three numerical schemes which can be applied to models of this type. In section 4 we apply these techniques to the symmetric model.

2. The model

2.1. *The equations of motion*

The situation we wish to discuss is shown in fig. 1. A dilute binary alloy sample is being pulled through a stationary thermal gradient G at a constant velocity v. As in Rayleigh–Bénard convection there exists a

0167-2789/84/$03.00 © Elsevier Science Publishers B.V.
(North-Holland Physics Publishing Division)

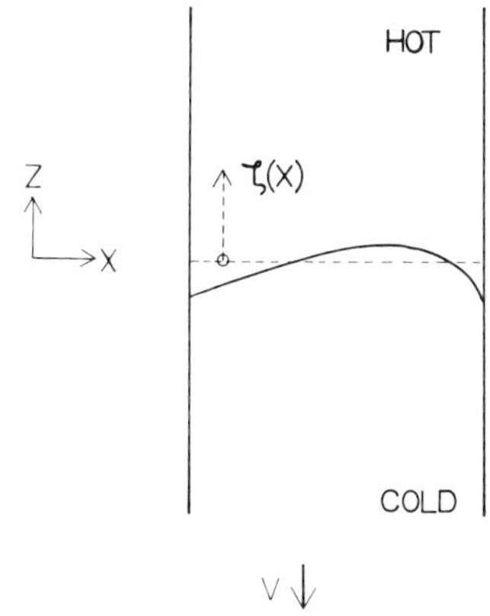

Fig. 1. A binary alloy sample is being pulled through a fixed temperature gradient in the negative z direction at a constant velocity v. The displacement of the solidification interface away from the planar configuration is denoted by $\zeta(x, t)$.

control parameter ν, analogous to the Rayleigh number R, which in the directional solidification problem is proportional to v and inversely proportional to G. On increasing ν through ν_c, the planar non-faceted solidification interface undergoes a transition to a periodic pattern state. In this discussion we will deal with a simple model of this type of system which has the same qualitative behavior as the more realistic models. The model is called the symmetric model. The name derives from the fact that the physical properties of the high and low temperature phases are assumed to be the same. The important physical feature of the D.S. problem is the competition between the surface tension of the interface, a stabilizing effect, and the destabilization caused by the rejected solute in front of the moving interface. The symmetric model has this feature. The model has two parameters, ν and d_0, where d_0 fixes the length scale of the instability that occurs at ν_c.

A complete solution of this problem would involve solving for both the solute and thermal diffusion fields in front of and behind the advancing solidification interface [6]. If one is only interested in the dynamics of the interface itself, we can project the full problem on to one involving the interface only, using Green's function methods [3, 4]. The equation one obtains from this procedure is an integro-differential equation for the interface displacement away from the planar state. For the symmetric model this equation is of the form

$$\int_{-\infty}^{\infty} \mathrm{d}x_1 \int_{-\infty}^{t} \mathrm{d}t_i G(x, t, \zeta, x_1, t_1, \zeta_1) \frac{\partial \zeta_1(x_1, t_1)}{\partial t_1}$$
$$= d_0 \Theta(\zeta) - \frac{\zeta}{\nu} - \int_{-\infty}^{\infty} \mathrm{d}x_1 \int_{-\infty}^{t} [G(x, t, \zeta, x_1, t_1, \zeta_1) - G(x, t, 0, x_1, t_1, 0)], \tag{2.1}$$

where we have considered a one-dimensional interface for simplicity, $\zeta(x, t)$ is the interface displacement (see fig. 1), d_0 and ν are the parameters mentioned above, $\Theta(\zeta)$ is the local curvature and is of the form

$$\Theta(\zeta) = \frac{\nabla^2 \zeta}{\left(1 + (\nabla \zeta)^2\right)^{3/2}} \tag{2.2}$$

and $G(x, t, \zeta, x_1, t_1, \zeta_1)$ is the relevant Green's function for this problem and is of the form

$$G(x, t, \zeta, x, t_1, \zeta_1) = f_1(x, t, x_1, t_1) f_2(t, \zeta, t_1, \zeta_1), \tag{2.3}$$

where

$$f_1 = \frac{1}{(4\pi\tau)} \exp\left[-\frac{(x-x_1)^2}{4\tau} - \tau\right], \tag{2.4}$$

$$f_2 = \exp\left[-\frac{(\zeta-\zeta_1)^2}{4\tau} - (\zeta-\zeta_1)\right] \tag{2.5}$$

and

$$\tau = \frac{t-t_1}{2}, \tag{2.6}$$

where all lengths and times are in units of the solute diffusion length l and the solute diffusion time $l^2/2D$, respectively, where $l = 2D/v$ and D is the solute diffusion coefficient.

2.2. *Stability of the planar interface*

The planar interface $\zeta_s = 0$ is a stationary solution of eq. (2.1). If we investigate the stability [3, 5] of this solution against perturbations of the form

$$\zeta \propto e^{\omega t + ikx}, \tag{2.7}$$

we can solve for the linear growth rate ω as a function of k, d_0, and ν. The growth rate ω has a maximum at a finite k as a function of k for fixed d_0 and ν. For ν less than ν_c, $\omega(k, d_0, \nu)$ is less than zero and hence the planar interface is stable. For $\nu = \nu_c$ the planar interface is unstable at $k = k_c$ and for $\nu > \nu_c$, a finite band of wave vectors exists for which the planar interface is unstable. The upper and lower bounds of this band of unstable wave vectors is specified by the zeros of $\omega(k, \nu, d_0)$ as a function of k for fixed ν and d_0. The locus of these zeros of the function ω in the ν–k plane for fixed d_0 is known as the neutral stability curve and such a curve is shown in fig. 3 for $d_0 = 5 \times 10^{-3}$.

3. The numerical scheme

3.1. *Finite size effects*

Clearly, any solution of eq. (2.1) will involve performing integration on the variables x_1 and t_1 over infinite and semi-infinite intervals. In developing a numerical scheme we will consider a system of finite size. Therefore we must investigate the effects of the system size, L say, on the system. To do this we will repeat the linear stability calculation presented in the last section in a finite system. Thus we will write

$$\int_{-\infty}^{\infty} \mathrm{d}x_1 \quad \text{as} \quad \int_{x-L/2}^{x+L/2} \mathrm{d}x_1. \tag{3.1}$$

Hereafter $\int \mathrm{d}x_1$ will imply integration over the finite system.

We can then compute $\nu_c(L, d_0)$ as a function of L. A plot of $\nu_c(L)$ is shown in fig. 2. We see that for a value of L greater than 10 diffusion lengths $\nu_c(L)$ is close to $\nu_c(\infty)$. This can be understood from the fact that the integral kernel G given by eq. (2.3) falls off rapidly as a function of $|x - x_1|$ and $(t - t_1)$ where we have assumed that ζ is a bounded function of x and t. Therefore in evaluating these integrals we will not

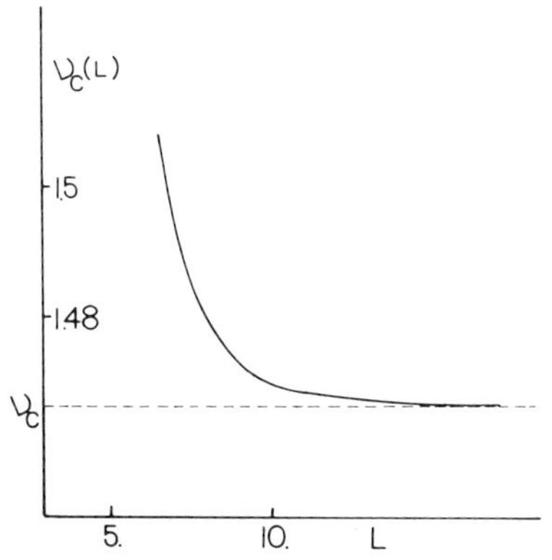

Fig. 2. A plot of the critical value of the parameter ν as a function of the size of the system L.

have to consider the whole range of the integration variables. We will consider a system of length L and we will impose periodic boundary conditions at each end.

3.2. *An implicit scheme*

Collecting all integrals involving $\zeta_1(x_1, t_1)$ together we rewrite them in the following form:

$$\int \mathrm{d}x_1 \int_{-\infty}^{t} \mathrm{d}t_1 f_1 f_2 \left(1 - \frac{\partial \zeta_1}{\partial t_1}\right) = \int \mathrm{d}x_1 \int_{-\infty}^{t} \mathrm{d}t_1 f_1 \left[f_2 \left(\frac{\partial \zeta_1}{\partial t_1} - 1 \right) \right]_{t_1 = t}$$
$$+ \int \mathrm{d}x_1 \int \mathrm{d}t_1 f_1 \left\{ f_2 \left(\frac{\partial \zeta_1}{\partial t_1} - 1 \right) - \left[f_2 \left(\frac{\partial \zeta_1}{\partial t_1} - 1 \right) \right]_{t_1 = t} \right\}. \tag{3.2}$$

We now perform the integration over t_1 in the first term on the right-hand side of (3.2) and then substitute the resulting expression for (3.2) back into (2.1) to obtain

$$\int \mathrm{d}x_1 A(x, x_1, \zeta, \zeta_1(x_1, t)) \frac{\partial \zeta_1}{\partial t}(x_1, t) = F[\zeta(x, t)] - \int \mathrm{d}x_1 E(x, x_1, \zeta, \zeta_1(x_1, t))$$
$$- \int \mathrm{d}x_1 M(x, x_1, t, \zeta, \zeta_1(x_1, t)), \tag{3.3}$$

where

$$A(x, x_1, \zeta, \zeta_1(x_1, t) = \frac{\mathrm{e}^{-(\zeta - \zeta_1(x_1, t))}}{\pi} K_0(z), \tag{3.4}$$

$$F(\zeta(x, t)) = d_0 \Theta(\zeta) - \frac{\zeta}{\nu}, \tag{3.5}$$

$$E(x, x_1, \zeta, \zeta_1(x_1, t)) = \frac{1}{\pi} \left[K_0(z) \mathrm{e}^{-(\zeta - \zeta_0(x_1, t))} - K_0(|x - x_1|) \right], \tag{3.6}$$

$$M(x, x_1, t, \zeta, \zeta_1(x_1, t)) = \int \mathrm{d}t_1 f_1 \left\{ f_2 \left(1 + \frac{\partial \zeta_1}{\partial t_1}\right) - \left[f_2 \left(1 + \frac{\partial \zeta_1}{\partial t_1}\right) \right]_{t_1 = t} \right\}, \tag{3.7}$$

$$z = \sqrt{(x - x_1)^2 + (\zeta - \zeta_1(x_1, t))^2}, \tag{3.8}$$

and K_0 is a Bessel function. The function M will be termed the memory term for obvious reasons. If we now consider a system with N equally spaced points eq. (3.3) can be approximated by the following expression

$$\sum_j w_j A_{ij}\left(\frac{\partial \zeta}{\partial t}\right)_j = F_i[\zeta_i] - \sum_j w_j E_{ij} - \sum_j w_j M_{ij}, \tag{3.9}$$

where the w_j are weights of the integration rule used and A_{ij}, F_i, E_{ij}, and M_{ij} are the discretized versions of the functions described above. The sum on j is over all points x_j for which $|x_i - x_j| < L/2$.

The integral over t_1 in the memory term is also performed numerically and over times such that $(t - t_1) < \tau^*$ where τ^* will be on the order of a few diffusion times. The integral kernels E and M have been written in a form so that they are finite for $x = x_1$ and $t = t_1$. We will use a forward difference representation of the time derivative of ζ on the left-hand side of eq. (3.9) and a backward difference representation of the time derivative of ζ in the memory term. Eq. (3.9) is now in the form of a matrix equation and it can be solved for the time derivative of ζ at each point and hence for ζ at the new time step.

The addition of the memory term makes life difficult due to the amount of computation necessary to evaluate this term. Therefore we next consider cases where we can reduce the calculational complexity of eq. (3.9).

3.3. *Slow dynamical regime*

There are two dynamical situations which arise for which ζ varies slowly as a function of t. The first case arises when $\nu \gtrsim \nu_c$. This results from the fact that the dynamics is driven by the band of unstable wave vectors and for ν close to ν_c, $\omega(k, \nu \approx \nu_c)$ is small for these values of k. The second case arises when the system approaches a stable fixed point. In both the above cases we can expand $\zeta_1(x_1, t_1)$ about $t_1 = t$ and keep lowest order terms in the time derivatives of ζ. We then obtain an equation of the form

$$\sum_j w_j \tilde{A}_{ij}\left(\frac{\partial \zeta}{\partial t}\right)_j = F_i - \sum_j w_j E_j, \tag{3.10}$$

where $\tilde{A}_{ij}$ is the discretized version of the function

$$\tilde{A}(x, x_1, t, \zeta, \zeta_1(x_1, t)) = \frac{1}{\pi} e^{-(\zeta - \zeta_1)}\left[K_0(z)\left(1 - \frac{(\zeta - \zeta_1)}{2}\right) - \frac{z}{2} K_1(z)\right], \tag{3.11}$$

where again K_1 is a Bessel function.

3.4. *Stationary states*

Eqs. (3.9) and (3.10) can be used to determine the dynamical behavior of the model. If we are only interested in the stationary states of the system, then an explicit scheme of the following form can be used:

$$\zeta_i^{(M+1)} = \zeta_i^{(M)} + \Delta t\left(F_i - \sum_j w_j E_j\right), \tag{3.12}$$

where the M indicates the Mth time step and the quantity Δt is the size of the time step. We will use an initial condition of the form

$$\zeta^{(0)} = \alpha_0 \cos k_0 x. \tag{3.13}$$

If we choose α_0 and Δt appropriately then (3.12) will converge rapidly to a stationary solution of Eq. (3.9). The numerical schemes described by (3.12) and (3.10) will be used in the next section and they provide us with a means to search the parameter space of the model for stable states.

4. The results

4.1. *Stationary states*

Periodic pattern solutions of eq. (2.1) are of the form

$$\zeta_s = A_0 + \sum_{n=1}^{\infty} A_n \cos nk_0 x. \tag{4.1}$$

Close to ν_c analytical techniques [3, 4] can be used to study these states. This is possible due to the fact that the solution (4.1) is small and is dominated by the amplitude A_1 of the fundamental. Close to ν_c, states exist for all values of k for which the planar interface is unstable. However, only a subset of these states are stable against the so-called Eckhaus instability [7]. This subset of stable solutions can be determined theoretically for $\nu > \nu_c$ and is shown in fig. 3.

In order to test the numerical techniques, we compared the stationary solutions generated by (3.10) and (3.12) with the theoretical solutions for ν close to ν_c and $d_0 = 5 \times 10^{-3}$. No significant differences were observed. We then increased ν to 1.5 and repeated the comparison. In fig. 4 we show a plot of the amplitude A_1 as a function of k. The solid and dotted lines indicate stable and unstable states as predicted by the theory. The filled and open circles indicate the stable and unstable numerically generated solutions. Small deviations between the theoretical and numerical solutions occur at points where A_1 has its largest values. These deviations signal the breakdown of the theory. At larger values of ν the deviations increase. At these large values of ν we can now use our numerical techniques to study the stationary states. For example, in fig. 5 we show a stationary state at $\nu = 2$.

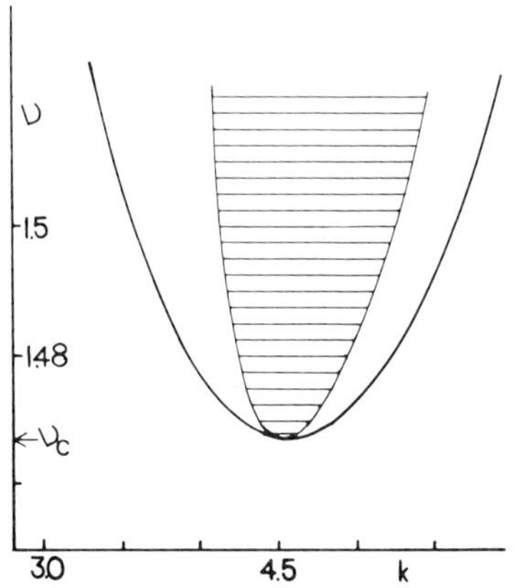

Fig. 3. The heavy line denotes the neutral stability curve for a value of $d_0 = 5 \times 10^{-3}$. The shaded area is the domain of stable periodic solutions as calculated using amplitude expansion techniques.

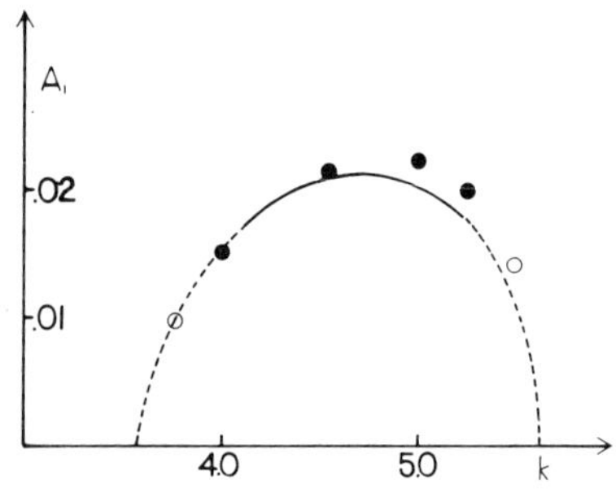

Fig. 4. A plot of the amplitude of the fundamental A_1 as a function of wave number k. The solid and broken curves indicate stable and unstable theoretical values of A_1, respectively. The filled and open circles indicate the values of A_1 for stable and unstable states respectively as computed numerically for a value of $d_0 = 5 \times 10^{-3}$.

4.2. *Dynamics*

We next turn our attention to the problem of pattern selection. In other pattern forming systems it has been observed [8] that large sets of initial conditions evolve to a specific state under a particular dynamical selection mechanism. We would like to study these types of selection mechanisms in this model. In the following examples we will illustrate these dynamical processes but due to the small system size we will not be able to make any definitive conclusions. However, more extensive studies are under way. Both of the examples described below are relevant to real D.S. experiments.

In the first example we will consider a system which is in a planar state at $\nu_i < \nu_c$. We then change ν to $\nu_f = 1.5$, where $\nu_f > \nu_c$. The state $\zeta = 0$ is now unstable. We will perturb this state with a small localized asymmetric perturbation at the origin and study the subsequent evolution. Clearly, the planar state will decay into one of the allowed states. The question of interest is which state does it select and is the selection mechanism sensitive to the initial perturbation? The evolution of the system is shown in figs. 6a and 6b. We see that the initial perturbation grows and propagates into the region occupied by the stationary unstable state. The final state chosen is in the central region of the band of allowed stable states and close to the fastest growing mode. Clearly, in a system this size, we cannot ascertain whether or not the mechanism of propagation into an unstable state selects a specific state and we will leave this question for future work.

As a second example we will consider the initial conditions where $\nu_i > \nu_c$ and the system is one of the allowed stable periodic states. We then increase the pulling velocity v. This causes a change in the parameters ν and d_0. The parameter ν increases to a larger supercritical value and d_0 also increases. As a consequence of the change of d_0, the system may now find itself in a domain of the ν–k plane where it is an allowed stationary state but unstable. Therefore, this state will decay as a consequence of the Eckhaus instability. We can therefore ask the same questions as posed in the last example. The evolution of such an initial state is shown in figs. 7a, b, and c. Again the final state chosen is in the central domain of allowed

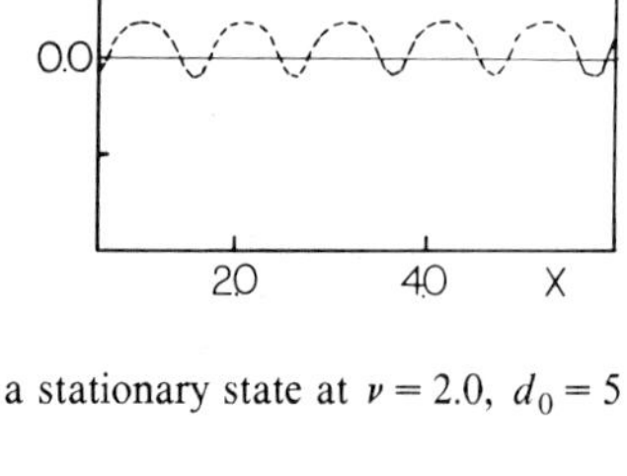

Fig. 5. A plot of the interface profile $\zeta(x)$ for a stationary state at $\nu = 2.0$, $d_0 = 5 \times 10^{-3}$, and fundamental wave number $k = 6.0$.

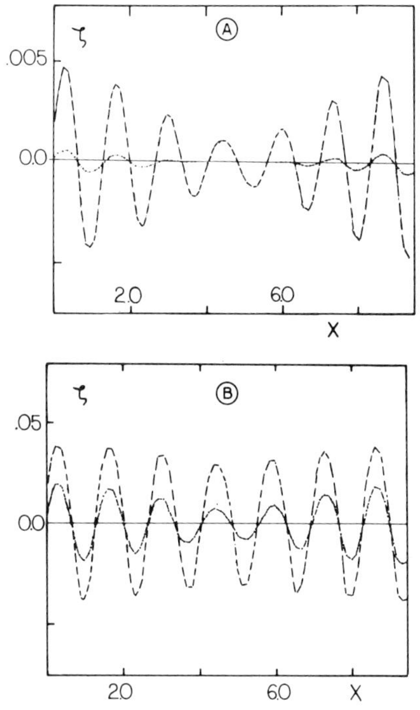

Fig. 6. We show the evolution of the system from the state $\zeta_s = 0$ to a periodic state with fundamental wave number $k \approx 4.54$ for $\nu = 1.5$ and $d_0 = 5 \times 10^{-3}$. In fig. 6a we show the initial perturbation growing and propagating into the unstable state at times $t = 11$ and 33 where the dotted line indicates the earlier time. Fig. 6b shows the continued evolution of the system towards the stable state at times $t = 77$ and 99.

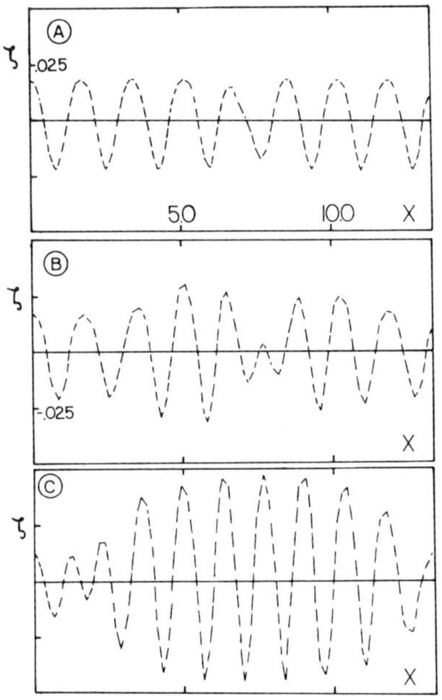

Fig. 7. We show the evolution of an unstable stationary state at $k = 3.75$ after it has been subjected to a small perturbation which disturbed the local wave number in the vicinity of $x = 7$. Figs. 7a, b, and c show the system at times $t = 3.2$, 33.2, and 66.2, respectively. At longer times the system evolves to a stable stationary state at $k \approx 4.68$. The parameters ν and d_0 have the values 1.5 and 5×10^{-3}, respectively.

states. As in the previous example we will have to repeat this experiment in a larger system to make any definitive statements about pattern selection under these conditions.

We are now applying the methods described in section 3 to more realistic models of D.S. We are also studying the problem of pattern selection in such systems. We are also interested in the dynamics of such systems for $v \gg v_c$.

Acknowledgements

I would like to thank Dr. J.S. Langer for his helpful comments and encouragement in the completion of this work.

This research was supported by the U.S. Department of Energy, under contract no. DE-AM03-765F00034, and by the National Science Foundation, grant no. PHY77-27084, supplemented by funds from the National Aeronautics and Space Administration.

References

[1] J. Langer, Rev. Mod. Phys. 52 (1980) No. 1.
[2] J.S. Langer and L.A. Turski, Acta Metall. 25 (1977) 1113.
[3] J.S. Langer, Acta Metall. 25 (1977) 1121.
[4] G. Dee and R. Mathur, to be published in Phys. Rev. B, June, 1983.
[5] W.W. Mullins and R.F. Sekerka, J. Appl. Phys. 34 (1963) 323.
[6] See the paper by G. McFadden and S. Coriell, Physica 12D (1984) 258 (these proceedings).
[7] See A.C. Newell and J.A. Whitehead, J. Fluid Mech. 38 (1969) 279.
[8] G. Dee and J.S. Langer, Phys. Rev. Lett. 50 (1983) 383; E. Ben-Jacob, H. Brand, and L. Kramer, to be published; E. Ben-Jacob, H. Brand, G. Dee, L. Kramer, and J.S. Langer, to be published.

Physica 12D (1984) 279–294
North-Holland, Amsterdam

SURFACE UNDULATIONS IN EXPLOSIVE CRYSTALLIZATION: A NONLINEAR ANALYSIS OF A THERMAL INSTABILITY

Wim VAN SAARLOOS and John D. WEEKS
AT&T Bell Laboratories, Murray Hill, New Jersey 07974, USA

We argue that surface undulations observed after self-sustained rapid crystallization of amorphous films result from a thermal instability that induces a periodically varying crystallization rate. Its physical origin is discussed for a simple nonlinear heat conduction model which yields good agreement with experimental observations. The model displays a rich variety of interesting physical phenomena. It contains a co-dimension two bifurcation point in whose vicinity detailed analytic results can be obtained. A numerical analysis of the nonlinear oscillations shows that these, in turn, bifurcate via a series of period-doubling bifurcations.

1. Introduction

Recently, there has been much interest in instabilities occurring during crystal growth, particularly those leading to dendritic growth and the morphological instabilities of the crystal–melt interface [1]. In this paper† we analyze a much simpler thermal instability which can explain phenomena observed during the rapid crystallization of metastable amorphous films. Instead of being a morphological instability along the amorphous to crystalline (a–c) interface of the type discussed at this conference by Sekerka [2], this is a thermal instability in the direction of propagation of the a–c front, and occurs even for straight fronts. As a result, its wavelength is given in terms of a thermal length only.

Under favorable experimental conditions, layers of amorphous Sb [3–8], Ge [9–24], Si [24–28] and other materials [29, 30] can crystallize very rapidly when crystallization is initiated by a laser pulse or by impact with a stylus. The name explosive crystallization stems from the fact that the speed of the crystallization front can be of the order of meters per second. Since crystallization rates are strongly dependent on temperature, it was realized long ago [4] that this phenomenon is associated with the temperature rise induced by the latent heat liberated in the a–c transition. Whether a self-sustained crystallization wave can occur in a particular case therefore depends on an energy balance. If the latent heat liberated is too small or the heat loss to the substrate too great, the crystallization wave cannot sustain itself and dies out. One then has to initiate the process repeatedly by scanning the layer with a laser beam [15–20, 26–28] so that the crystallization front proceeds in bursts: the front first outruns the laser beam, then slows down and stops, after which the process repeats itself when the heat from the moving laser beam catches up [31].

In materials like Sb [3–8], Bi [30] or Yb [30], on the other hand, *self-sustained* crystallization waves have been observed that propagate through the whole layer (often a few micron thick) after being initiated at a single spot. In several of the latter experiments the crystalline phase exhibits periodic variations in the height of the layer [4, 8]. Far from the initiation point, the structure is often essentially one-dimensional, resembling a frozen-in pattern of parallel water waves near a beach. Wickersham *et al.* [8] noted that the undulations could have a thermal origin, since the wavelength λ in their experiments was roughly equal to the thermal length κ/V_{av} (here κ is the thermal diffusivity which we assume for simplicity is the same

†A brief account of this work was given in Phys. Rev. Lett. 51 (1983) 1046.

0167-2789/84/$03.00 © Elsevier Science Publishers B.V.
(North-Holland Physics Publishing Division)

in the a and c phases, and V_{av} the average crystallization speed). In this paper, we argue that they indeed result from a thermal instability of a steadily advancing a–c front to one with an oscillating growth rate.

In the next section a simple model for explosive crystallization, already studied by Gilmer and Leamy [32], is introduced. The steady state solutions, corresponding to a–c fronts propagating with constant speed, are discussed. In section 3 we analyze the stability of the steady state solutions in qualitative terms, deferring the explicit linear stability analysis to an appendix. It is shown that under certain conditions an oscillatory (Hopf) bifurcation can occur, which leads to solutions with a periodically varying growth rate. The physical mechanism leading to this behavior is discussed. A detailed analytic treatment of the nonlinear behavior of our model is possible in the neighborhood of a particular point (a co-dimension two bifurcation point) where time dependences are slow. This is carried out in section 4. In section 5 we compare our predictions with the experimental observations, and comment on the validity of the model. Section 6 contains a more detailed discussion of successive period doubling bifurcations that were found in a numerical analysis of the nonlinear oscillating solutions.

2. Steady state propagation

In view of the experimental observations that the surface undulations far from the initiation point are often parallel, we analyze the propagation of a straight front, neglecting temperature differences in the vertical and lateral direction and any possible coupling to density and height differences. (We will come back to this in section 4.) In the laboratory frame, the propagation is taken to be in the positive x' direction; it is more convenient, however, to use the coordinate frame $x = x' - \int^t dt' V(t')$ co-moving with the front so that the a phase is at $x > 0$ and the c phase at $x < 0$. Here V is the interface velocity or growth rate. Using this frame, we follow Gilmer and Leamy [32] and assume a balance equation of the form

$$\frac{\partial T}{\partial t} = \kappa \frac{\partial^2 T}{\partial x^2} + V \frac{\partial T}{\partial x} - \Gamma(T - T^0) + qV\delta(x). \quad (1)$$

Eq. (1) expresses the change in the temperature T in terms of four contributions: The first term on the right-hand side gives the heat conduction through the layer; the second term results from the transformation to the moving frame; the third one crudely describes the heat loss to the substrate at temperature T^0, and replaces the contribution from radiation and heat conduction to the substrate which should be taken into account in a more realistic treatment. The fourth contribution is the source term due to the latent heat LV liberated per unit of time and area at the a–c boundary (here L is the latent heat per unit volume [33]; if C is the specific heat per unit volume, $q = L/C$). The delta function ensures that the latent heat is released only at the a–c boundary at $x = 0$, and is equivalent to the boundary condition $\kappa\, \partial T/\partial x|_{0-} - \kappa\, \partial T/\partial x|_{0+} = qV$. Eq. (1) is supplemented with the boundary condition $T(x, t) \to T^0$ for $x \to \pm\infty$.

Eq. (1) is incomplete until the growth rate V is specified. As usual, we assume that V is an explicit function only of the boundary temperature [34] $T^b(t) \equiv T(x = 0, t)$,

$$V = V(T^b(t)). \quad (2)$$

This "constitutive relation" introduces a nonlinear dependence on T in eq. (1) which causes the instability we discuss below. The appearance of a term $V\,\partial T/\partial x$ with V independent of the position x is typical of moving boundary problems.

For concreteness, we consider a dependence of V on T^b as sketched in fig. 1. This is the type of growth rate observed for crystal growth from a melt [34]. The a–c growth rate is thought to be of a similar form, although accurate measurements are only available on the low temperature side where V increases with T^b as $V \propto \exp(-Q/T^b)$. It is precisely on this side where the thermal instability discussed below occurs.

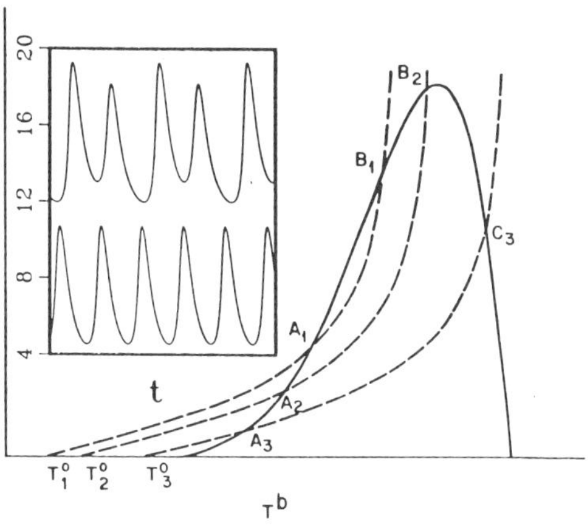

Fig. 1. Solid line: a typical curve for the a–c growth velocity as a function of the a–c boundary temperature. The dashed lines give the relationship between V_{ss} and T_{ss}^{b} for which eq. (1) has steady state solutions. Inset: V as a function of time for two different values of T^0 which produce oscillatory growth rates. The upper curve (obtained by adding 8 units to V) is a "period 2" solution.

In the steady state, when the a–c boundary is propagating with a constant velocity V_{ss}, it is most convenient first to solve eq. (1) without recourse to eq. (2). In this case, eq. (1) is linear and is solved by an exponential profile of the form $T = T_{ss}^{b}\exp(-ax)$ for $x>0$ and similarly for $x<0$, provided that V_{ss} and T_{ss}^{b} are related by [32]

$$V_{ss} = 2(\kappa\Gamma)^{1/2}\left((q/\Delta T_{ss}^{b})^{2} - 1\right)^{-1/2}. \tag{3}$$

Here $\Delta T_{ss}^{b} \equiv T_{ss}^{b} - T^{0}$ is the temperature rise at boundary. This function is drawn in fig. 1 with a dashed line for several values of T^0, keeping κ and Γ fixed. For very large steady state velocities, ΔT_{ss}^{b} approaches the value $q = L/C$, since the heat loss to the substrate is much smaller than the latent heat liberated per unit of time; energy conservation then dictates that $C\Delta T_{ss}^{b} \simeq L$.

Not all hypothetical V_{ss} satisfying eq. (3) are consistent with the physically possible growth rates given in eq. (2). Thus the actual steady state velocity and boundary temperature are determined by the intersection of these curves, as shown in fig. 1, though the *stability* of these solutions must be investigated. As in the experiments [8], different steady state values are obtained by changing T^0. (In principle, one could alternatively change the value of $\kappa\Gamma$.) If T^0 is too low, the two curves do not intersect, and self-sustained explosive crystallization is not possible.

If one considers V as a given function of time, eq. (1) is a linear partial differential equation for T, whose Green's function can be derived in the standard way [35]. Using this result, the formal solution to eq. (1) can then be written as

$$\begin{aligned}\Delta T^{b}(t) = {} & q\int_{-\infty}^{0} \mathrm{d}x\,(4\pi\kappa\tau)^{-1/2} \\ & \times \exp\left[-\Gamma\tau - x^{2}/(4\kappa\tau)\right] \\ & + \int_{-\infty}^{+\infty} \mathrm{d}x\,(4\pi\kappa t)^{-1/2}\exp\left\{-\Gamma t\right. \\ & \left. - \left(x - \int_{0}^{t} V(t')\,\mathrm{d}t'\right)^{2} \Big/ 4\kappa t\right\} \Delta T(x,0),\end{aligned} \tag{4}$$

where $\tau \equiv \tau(x, t)$ is implicitly given by

$$x = -\int_{t-\tau}^{t} \mathrm{d}t'\, V(T^{b}(t')). \tag{5}$$

Eqs. (4) and (5) have the following interpretation in the lab frame. The heat released at previous positions of the moving a–c boundary spreads out due to conduction and heat loss as a damped Gaussian, and so can still contribute to the present boundary temperature. Eq. (5) determines the time τ elapsed since the boundary passed a position a distance $|x|$ to the left of the present position of the boundary, and the first term in the right-hand side of eq. (4) gives the sum of all the contributions to T^b from the heat previously released. The second term in the right-hand side of eq. (4) results from the initial conditions at $t=0$. This interpretation of eqs. (4) and (5) will be helpful in the discussion of the linear stability analysis below.

3. Stability analysis of the steady state solutions

We now investigate the stability of the possible steady state solutions obtained above. We argue qualitatively that an oscillatory (Hopf) bifurcation of one of the steady state solutions can occur, and discuss the physical origin of the resulting oscillatory solutions. The detailed linear stability analysis supporting the qualitative discussion of this section is given in appendix A.

In the following, we will distinguish between the three different types of intersections, labeled A, B and C in fig. 1. Consider C first. As already argued by Gilmer and Leamy [32] the steady state solutions at this point should be *stable*, since a small increase in T^b results in a smaller velocity from eq. (2), and the resulting decrease in liberation of latent heat will bring the boundary temperature back to its steady state value at C. Using the same argument, we expect the points A to be *unstable*.

However, the most interesting behavior occurs for points B, where changes in the heat flow and heat loss as well as the heat generated by a perturbation must be taken into account in a stability analysis. Consider the steady state at B_2 where the slope of the growth rate curve is small but positive. Here a large increase in T^b, and so a relatively large increase in heat loss and heat flow away from the boundary, is associated with only a small increase in velocity (and so heat generation). We therefore expect this point to be *stable*. For intersections with a large slope, like B_1, the situation is reversed and these points should therefore be *unstable*, provided that Γ is not too large.

These elementary considerations are supported by the explicit linear stability analysis of appendix A, the main results of which are summarized in fig. 2. In this stability diagram, $(\Delta T_{ss}^b/V_{ss})(dV/dT^b) \equiv \alpha$, the (dimensionless) slope of the growth rate curve in the steady state point, is plotted along the horizontal axis and $(\Delta T_{ss}^b/q)^2 = (4\kappa\Gamma/V_{ss}^2 + 1)^{-1} \equiv \beta$ along the vertical axis. Here β is a measure of the ratio of the thermal diffusion time $4\kappa/V_{ss}^2$ and the relaxation time Γ^{-1} and varies between 0 and 1. In agreement with the physical arguments given above, steady state points C are stable, and those of type A unstable. The ones of type B become unstable in the case $\beta > 2/3$ (or $\Gamma < V_{ss}^2/8\kappa$) when

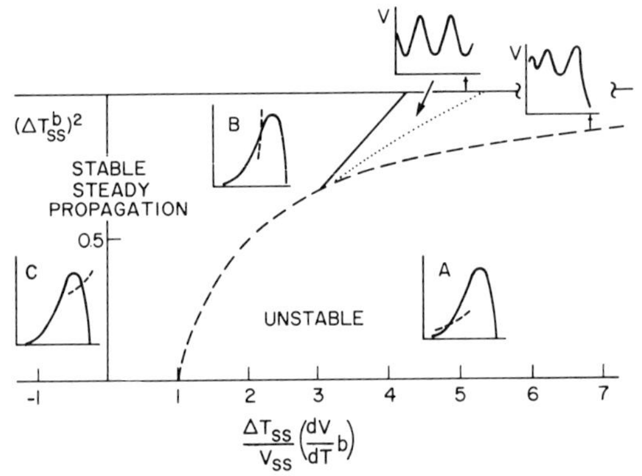

Fig. 2. Stability diagram for the various steady state solutions whose locations on the growth rate curve are schematically indicated. The (dimensionless) slope of the function (2) in the steady state point is plotted along the horizontal axis and $(\Delta T_{ss}^b)^2$ (in units of q^2) along the vertical axis. The dashed line separates the points denoted by A and B in fig. 1. The steady state points B to the right of the solid line are unstable. However, in this region the full nonlinear equations can admit either stable oscillatory solutions (to the left of the dotted line) or solutions that die out (for points B to the right of the dotted line). The behavior of the velocity in both cases is schematically indicated.

the dimensionless slope (plotted along the horizontal axis) reaches some value between 3 and $2+\sqrt{5} \approx 4.236$. The instability occurring when the point B moves to the right of the solid line is of oscillatory (Hopf) type: for values of the slope larger than the critical value, the boundary temperature and speed of the front oscillate periodically in time (see the inset of fig. 1). (These solutions will be discussed in more detail later.)

It is easy to understand why the instability at steady state points B on the solid line is of Hopf type so that the linearized equations have a purely imaginary eigenvalue. At points B, the slope of the growth rate curve dV/dT^b is, by definition, smaller than the slope of the curve of steady state points: $dV/dT^b < dV_{ss}/dT_{ss}^b$†. Suppose on the contrary that the eigenvalue of the equations, linearized around the steady state solutions B, could be real. Then there would exist solutions of the form $T^b \approx$

†Points of type A are defined by $dV/dT^b > dV_{ss}/dT_{ss}^b$, and the dashed line in fig. 2 is given by the equation $dV/dT^b = dV_{ss}/dT_{ss}^b$.

$T_{ss}^b + \epsilon \exp(\omega t)$ and $V \simeq V_{ss} + (dV/dT^b)\epsilon \exp(\omega t)$ with ω real and, close to the solid line, arbitrarily small. In the limit $\omega \to 0$, however, the perturbed temperature field must obey the same equation as the steady state solutions, and hence one would also have up to linear order in ϵ, $V \approx V_{ss} + (dV_{ss}/dT_{ss}^b)\epsilon \exp(\omega t)$ for ω small. Since at points B, $dV/dT^b < dV_{ss}/dT_{ss}^b$, this is obviously impossible. We therefore conclude that near the solid line in the diagram, ω must be complex.

The above analysis implies the following picture. At high enough substrate temperatures, T_3^0 in fig. 1 say, the intersection at A is unstable and the one at C stable. Hence, after initiation of the process, the a–c front will propagate with a speed determined by the intersection C. If T^0 is lowered to T_2^0, say, the intersection B_2 is still stable since the slope of the growth rate curve is relatively small, and the front continues to advance with constant speed. However, if T^0 is lowered to T_1^0, say, the slope of the growth rate curves is large enough that steady state propagation is no longer possible. In this case we find that self-sustained explosive crystallization fronts propagate with a periodically varying growth rate.

In a numerical analysis of these nonlinear oscillatory solutions, we have found that two things can happen in cases where T^0 is lowered even more so that the slope becomes even larger. One effect arises from the fact that the amplitude of the oscillations grows larger and larger as T^0 is lowered. Indeed, it finally becomes so large that the front stops propagating because T^b drops too much at the minima. This is schematically indicated in fig. 2 by the dotted line. Solutions to the right of this line eventually die out. Thus, self-sustained explosive crystallization does not occur in every case where the two curves in fig. 1 intersect. This nonlinear effect is analyzed in the next section in the neighborhood of the bifurcation point where the three lines of fig. 2 intersect. There we will determine the slope of the dotted line near this point. Away from this point, the conditions under which the oscillations die out depend on the detailed shape of the growth rate curve. Secondly, it turns out that the periodic solutions undergo period doubling bifurcations in the region of fig. 2 between the solid and dotted line. These bifurcations will be discussed in section 6.

Having understood the stability, it remains to understand the physical origin of the oscillations and to get an estimate for their period. In the steady state, when eq. (5) reduces to $|x| = V_{ss}\tau$, the main contribution in the first integral in (4) comes from those positions x for which $\Gamma\tau + x^2/4\kappa\tau \lesssim 2$, or

$$\tau \lesssim \frac{2}{\Gamma + V_{ss}^2/4\kappa} \equiv \tau_m. \tag{6}$$

Contributions to T^b from positions further back than $d_m \equiv V_{ss}\tau_m$ are essentially negligible. In the regime $\Gamma \ll V_{ss}^2/4\kappa$, τ_m sets the timescale over which changes in T^b take place. Hence we expect the period of the oscillations to be of the order of $\tau_m \approx 8\kappa/V_{ss}^2$. Note that the "memory time" τ_m becomes shorter the larger the velocity is, since the boundary moves more rapidly out of the region that can be reached by heat conduction. To understand the oscillatory solutions, consider a small perturbation with an increased velocity around a steady solution in the case of small damping ($\Gamma \ll V_{ss}^2/4\kappa$). A small decrease in τ makes a Gaussian $\tau^{-1/2}\exp(-y^2/4\tau)$ narrower: the center increases whereas the wings at $|y| > (2\tau)^{1/2}$ decrease. For the (slightly damped) Gaussians in the integrand of eq. (4) with $|x| < 2\kappa/V_{ss} = d_m/4$ the perturbation therefore results in an increase of T^b, while it decreases the contributions of those at distances further away†. Thus, when the growth rate speeds up, the boundary can move ahead so rapidly that the heat diffusion from positions not immediately behind the boundary can not keep up. Consequently, T^b and the front velocity drop, after which more heat diffuses towards the boundary, and the velocity can rise again.

†In the opposite case ($\Gamma > V_{ss}^2/4\kappa$) all damped Gaussians in eq. (4) respond similarly to a perturbation and all disturbances decay monotonically.

4. Behavior close to the bifurcation point $\alpha = 3$, $\beta = 2/3$

In the last section, we found two lines in the linear stability diagram where the stability changes. Along the solid line drawn in fig. 2 where $\beta = (\alpha - \alpha^{-1})/4$, we found a Hopf bifurcation. Hence, for perturbations in T^{b} of the form $T^{\mathrm{b}} = T^{\mathrm{b}}_{\mathrm{ss}} + \varepsilon e^{2\omega t/\tau_{\mathrm{m}}}$, the two eigenvalues $\omega_{\pm}$ along the solid line satisfy $\mathrm{Re}\,\omega_{\pm} = 0$, $\mathrm{Im}\,\omega_{\pm} \neq 0$. Moreover, one eigenvalue always vanishes along the dashed line, where $\beta = 1 - \alpha^{-1}$ and the A and B points coincide. The two lines intersect at $\alpha = 3$, $\beta = 2/3$. As a result, both eigenvalues vanish at this point, which means that it is a co-dimension two bifurcation point [36–40]. Using the fact that the time-dependences are slow, we can carry out a detailed analytic investigation of the behavior of the system in the neighborhood of such a point. Note that the growth rate curve (2) and the curve of steady state points (3) have equal slopes at points on the dashed line in fig. 2. Thus we analyze how the change of stability at B takes place in the special case in which the curves (2) and (3) are nearly tangent, so that the intersections A and B lie close together. Readers primarily interested in physical implications of our analysis can skip to section 4.3.

4.1. *Reduction to the normal form of a co-dimension two bifurcation*

For small $\delta\alpha \equiv \alpha - 3$ and $\delta\beta \equiv \beta - 2/3$ we explicitly find from the result for $\omega_{\pm}$, given in eq. (A.7)

$$\frac{2}{\tau_{\mathrm{m}}}\omega_{\pm} \approx -(18\delta\beta - 5\delta\alpha) \pm \sqrt{24(\delta\alpha - 9\delta\beta)}\,. \quad (7)$$

The two slow modes associated with these small eigenvalues give the dominant behavior for long times. The time dependence of the system in the neighborhood of the bifurcation point is therefore described by a second order equation of the form

$$\ddot{v} - \lambda_2\dot{v} - \lambda_1 v = F(v, \dot{v}), \quad (8)$$

with λ_1 and λ_2 small and F vanishing at the bifurcation point. Here $v \equiv V/V_{\mathrm{ss}} - 1$ is the deviation from the steady state velocity and the nonlinear function F is yet to be determined. From the requirement that the eigenvalues of the $\dot{v} = v = 0$ fixed point in (8) coincide with (7), we get up to lowest order the identification

$$\lambda_1 = 24(\delta\alpha - 9\delta\beta), \quad \lambda_2 = -2(18\delta\beta - 5\delta\alpha). \quad (9)$$

In the absence of symmetries, the function F will be of the form $F = a_1 v^2 + a_2 v\dot{v} + a_3\dot{v}^2$, keeping only terms of second order in smallness. The coefficients a_1, a_2 and a_3 are determined in appendix B by expanding the integral solution about the bifurcation point. We find

$$a_1 = -72\left(1 - \frac{1}{18}V''\right), \quad a_2 = 144, \quad a_3 = -75/2, \quad (10)$$

where

$$V'' \equiv \frac{\left(\Delta T^{\mathrm{b}}_{\mathrm{ss}}\right)^2}{V_{\mathrm{ss}}}\frac{\mathrm{d}^2 V}{(\mathrm{d}T^{\mathrm{b}})^2} \quad (11)$$

is the (dimensionless) second derivative of the growth rate curve in the steady state.

By performing a transformation to the variable $y = v + \frac{1}{2}a_3 v^2$ in eq. (8), we can eliminate the term proportional to $\dot{v}^2$, while to second order all the other terms are unchanged. Introducing the variables $y_1 = y$, $y_2 = \dot{y}$, eq. (8) can thus be rewritten as the set of first order equations

$$\dot{y}_1 = y_2, \quad (12)$$

$$\dot{y}_2 = \lambda_1 y_1 + \lambda_2 y_2 + a_1 y_1^2 + a_2 y_1 y_2, \quad (13)$$

and it is convenient to study the trajectories in the y_1, y_2 plane. The above set of equations is the normal form of one of the two possible co-dimension two bifurcations with quadratic nonlinearities [37].

4.2. *Analysis of the normal form*

Eqs. (12) and (13) have two fixed points, $y_1 = 0$, $y_2 = 0$ and $y_1 = -\lambda_1/a_1$, $y_2 = 0$, whose stability is related by a symmetry: if we perform the transformation $w_1 = y_1 + \lambda_1/a_1$, $y_2 = w_2$, then w_1 and w_2 obey, up to the relevant order, the same equations as y_1 and y_2 but with λ_1 replaced by $-\lambda_1$ and λ_2 replaced by $\lambda_2 - \lambda_1 a_2/a_1$. Hence the global behavior of eqs. (12) and (13) can be obtained using only the behavior for $\lambda_1 < 0$ by interchanging the role of the fixed points and rescaling λ_2.

We summarize the results of a linear stability analysis of eqs. (12) and (13) and the connection with the stability analysis of section 3 as follows: (i) If $\lambda_1 > 0$, the $y_1 = y_2 = 0$ fixed point is unstable (for $\lambda_1 > 0$, $\delta\alpha > 9\delta\beta$ implies that we are below the dashed line in fig. 2, and the point $y_1 = y_2 = 0$ corresponds to an A point). Because of the symmetry, the $y_1 = -\lambda_1/a_1$, $y_2 = 0$ fixed point is always unstable for $\lambda_1 < 0$. (ii) If $\lambda_1 < 0$ and $\lambda_2 < 0$, the fixed point $y_1 = y_2 = 0$ is stable (a B point left of the solid line). Along the line $\lambda_1 < 0$, $\lambda_2 = 0$ there is a Hopf bifurcation, and for $\lambda_1 < 0$, $\lambda_2 > 0$, the fixed point is unstable (a B point right of the solid line).

The limit cycle that bifurcates from the $y_1 = y_2 = 0$ fixed point along the line $\lambda_2 = 0$, $\lambda_1 < 0$ does not exist for all values of λ_1 and λ_2 in the second quadrant: if the limit cycle grows so large that it crosses the other unstable fixed point at $y_2 = 0$, $y_1 = -\lambda_1/a_1$, it becomes unstable. Following Guckenheimer [36, 37], we now determine the values of the parameters for which this happens.

We take advantage of the smallness of λ_1 and λ_2 by introducing the scaling

$$\lambda_1 = \mu^2 \Lambda_1, \quad \lambda_2 = \mu^2 \Lambda_2, \quad \mu t = T,$$
$$y_1 = \mu^2 Y_1, \quad y_2 = \mu^3 Y_2, \tag{14}$$

so that eqs. (12) and (13) become

$$\dot{Y}_1 = Y_2, \tag{15}$$
$$\dot{Y}_2 = \Lambda_1 Y_1 + a_1 Y_2^2 + \mu(\Lambda_2 Y_2 + a_2 Y_1 Y_2). \tag{16}$$

We will investigate the limit cycle around the $Y_1 = Y_2 = 0$ fixed point in the left half plane $\Lambda_1 < 0$. For $\mu = 0$ the above equations reduce to the Hamilton equations corresponding to the Hamiltonian

$$H = \tfrac{1}{2} Y_2^2 - \tfrac{1}{2}\Lambda_1 Y_1^2 - \tfrac{1}{3} a_1 Y_1^3. \tag{17}$$

To zeroth order in μ, the integral curves of eqs. (15) and (16) are therefore the lines of constant energy $H(Y_1, Y_2) = \text{constant}$.

Since areas are conserved by a Hamiltonian flow, the contraction or expansion of an area A under the flow given by eqs. (15) and (16), depends only on the terms proportional to μ,

$$\frac{\mathrm{d}A}{\mathrm{d}t} = \mu \int_A \mathrm{d}Y_1 \mathrm{d}Y_2 \, (\Lambda_2 + a_2 Y_1). \tag{18}$$

Obviously, the area enclosed by a limit cycle does not change in time,

$$\mu \int_{\text{l.c.}} \mathrm{d}Y_1 \mathrm{d}Y_2 \, (\Lambda_2 + a_2 Y_1) = 0. \tag{19}$$

When evaluating this expression to lowest order in μ, the limit cycle may be approximated by its $\mu = 0$ limit, i.e. by the integral curve of the Hamiltonian system. Thus, for a given set of parameters, eq. (19) enables us to select the limit cycle from the set of curves $H(Y_1, Y_2) = \text{constant}$ [41].

As stated before, the limit cycle disappears if it intersects the second unstable fixed point at $Y_1 = -\Lambda_1/a_1$ (its period then goes to infinity). To determine the set of parameters for which this happens up to lowest order in μ, we therefore have to evaluate eq. (19) in the approximation that the limit cycle is the integral curve $H(Y_1, Y_2) = H(-\Lambda_1/a_1, 0)$, or

$$Y_2 = \pm \left[\frac{2a_1}{3} \left(Y_1 + \frac{\Lambda_1}{a_1} \right)^2 \left(Y_1 - \frac{\Lambda_1}{2a_1} \right) \right]^{1/2} \quad (\Lambda_1 < 0). \tag{20}$$

Using this result in eq. (19) we find that the limit cycle in the left half plane $\Lambda_1 < 0$ disappears for parameter values such that

$$\int_{-\Lambda_1/a_1}^{\Lambda_1/2a_1} \mathrm{d}Y_1\, 2\left[\frac{2a_1}{3}\left(Y_1+\frac{\Lambda_1}{a_1}\right)^2\left(Y_1-\frac{\Lambda_1}{2a_1}\right)\right]^{1/2} \times(\Lambda_2+a_2Y_1)=0, \quad (\Lambda_1<0), \tag{21}$$

or, after an elementary integration,

$$\frac{\Lambda_2}{\Lambda_1}=\frac{1}{7}\frac{a_2}{a_1} \quad (\Lambda_1<0). \tag{22}$$

If a_2/a_1 is negative, this result shows that the limit cycle disappears along a straight line in the second quadrant of the λ_1, λ_2 plane, indicated by L_3 in fig. 3. Taking also into account the behavior in the right half plane ($\lambda_1>0$) which follows from the symmetry of the equations discussed after eq. (13), we now arrive at the following global picture in the neighborhood of the bifurcation point in the case $a_1/a_2<0$. In region IV of fig. 3, the system exhibits one stable and one unstable fixed point. The stable fixed point bifurcates along the line L_3 ($\lambda_2=0$, $\lambda_1<0$) to a stable limit cycle as depicted in region III, and along the line L_4 ($\lambda_2/\lambda_1=a_2/a_1$) to the limit cycle of region II. When λ_2 becomes larger for fixed λ_1, the size of the limit cycle grows until it finally encounters the second unstable fixed point. This happens along the lines L_1 ($\lambda_2/\lambda_1=a_2/7a_1$) and L_2 ($\lambda_2/\lambda_1=6a_2/7a_1$). For even larger values of λ_2 we are in region I, and since there are then no stable fixed points and limit cycles, higher order terms have to be taken into account for a complete description of the system.

Eqs. (12) and (13) are also invariant under a change of sign in y_2, λ_2, a_2 and t. Hence, the behavior of the system for $a_1/a_2>0$ can be obtained from the one for $a_1/a_2<0$ by inverting the stability of all fixed points and limit cycles because of the change in sign of the time t. In particular, there are no stable limit cycles in the case $a_1/a_2>0$.

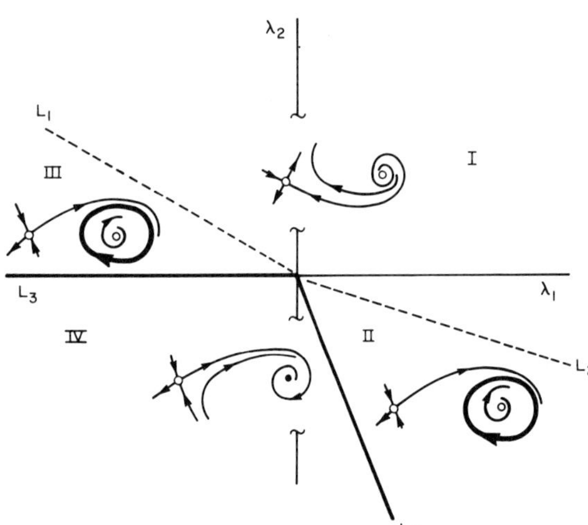

Fig. 3. Schematic phase portrait of the behavior of the normal form of the codimension two bifurcation with quadratic nonlinear terms, showing the different characteristic behavior for various values of λ_1 and λ_2 in the case $a_1/a_2<0$. Following ref. 39, the typical phase space orbits and the attractors in the y_1, y_2 plane are drawn with the convention that solid circles denote stable fixed points, open ones unstable fixed points, and a thick solid line a stable limit cycle. Orbits are drawn as thin solid lines. In the case $a_1/a_2>0$ there are no stable limit cycles.

4.3. *Implications for our model*

In our model, we have from eq. (10) $a_1/a_2=-\frac{1}{2}(1-V''/18)$. In cases of practical interest, V'' is small and this ratio is negative. (If $V''>18$ the curvature of the growth rate curve is so large that (2) and (3) have three rather than two intersections.) Then the global behavior near the bifurcation point is as depicted in fig. 3, and comparison with fig. 2 shows the following. According to eq. (9), the line $\lambda_2=0$, $\lambda_1<0$ where the Hopf bifurcation occurs coincides with the slope of the solid line ($\mathrm{d}\beta/\mathrm{d}\alpha=5/18$) near the point $\alpha=3$, $\beta=2/3$ in fig. 2. In fig. 3, the stable limit cycle in region III disappears along the dotted line L_1, and consequently in fig. 2 along a line with slope $\mathrm{d}\beta/\mathrm{d}\alpha=59/342$ for $V''=0$, and the region to the right of the dotted line in fig. 2 is related to region I of fig. 3. Since there are in general only two steady state solutions in our model, the absence of a limit cycle and stable fixed points in region I is in agreement with the numerical observation described in section 3 that the a–c front can cease propagating even though the two curves in fig. 1 intersect.

We emphasize that figs. 2 and 3 display different results. Fig. 3 is only valid in the neighborhood of the bifurcation point and shows the *global* nonlinear behavior of the solutions, which depends on the interplay of the two nearby fixed points. Fig. 2, on the other hand, gives the *linear* stability of *each* steady state point regardless of the stability and location of the other intersection. As a result, the analogues of the lines L_2 and L_4 are absent in the region below the dashed curve in fig. 2, since they describe the behavior of a nearby point of type B, not of the one of type A.

5. Comparison with experimental observations

Since there are considerable density differences between the a and c phase, it is likely that the predicted oscillations in T^b and V are the cause of the surface undulations and compositional variations, through differences in grain size or the extent of completeness of the a–c transformation [42]. Note that the slope of the growth rate is large when the oscillatory instability occurs. As a result there are large variations in the growth rate (of order 10 to 50 percent) which would be expected to affect the morphology of the growing crystal. Assuming this to be the case we can compare our results with the following experimental observations of Wickersham et al. [8].

i) In agreement with the arguments given in section 3, the linear stability analysis shows that the period of the oscillations is of the order τ_m. Consequently, we get for the wavelength λ of the oscillations in the regime where $\Gamma \ll V_{ss}^2/4\kappa$ [see eq. (A.9)] $\lambda \approx 6\kappa/V_{av}$. Here we have replaced V_{ss} by the average velocity. This estimate is in reasonable agreement with the experimental values measured [8].

ii) Let T^* be the substrate temperature above which self-sustained explosive crystallization is possible. For T^0 slightly above T^*, say $T^0 = T_1^0$, as shown in fig. 1, the slope at B_1 is steep and the growth rate will oscillate. If T^0 increases to T_2^0, say, the point B_2 becomes stable. Thus, in our model, the oscillations occur for T^0 slightly above T^* and cease at higher temperatures, much like the experimental observation that the "surface roughness was found to decrease as the triggering temperature was increased above T^*" [8].

In principle, our predictions may be tested experimentally by monitoring the growth rate using the differences in optical properties in the a and c phase [20].

As discussed in section 3 the oscillatory behavior can only occur in the regime $\Gamma < V_{ss}^2/8\kappa$, where the thermal diffusion time κ/V_{ss}^2 is much shorter than the relaxation time Γ^{-1} associated with heat losses. This means that for the effect to be seen in an experiment, the heat loss to the substrate within a time of the order of the thermal diffusion time ought to be relatively small. Layers that are too thin can not satisfy this condition. In the case that radiation is negligible, one can make a rough (conservative) estimate of how thick a layer should be. Let us assume that the substrate is a good heat conductor so that it remains always at temperature T^0. After the a–c boundary has passed, the temperature on the bottom part of the layer will decrease as a result of the heat conduction to the substrate. The thickness of the region affected after at time τ_m is of the order of $\sqrt{\kappa\tau_m}$, and the effect of heat loss is small if this length is smaller than the thickness d of the layer. Since $\kappa\tau_m \approx 8\kappa^2/V_{ss}^2 \approx \lambda^2/8$, the instability should thus occur in layers for which $d \gtrsim \lambda$.

In explosive crystallization in Ge, the a phase probably melts first before crystallizing [32, 22, 15]. Whether the same happens in Sb is not clear, but experiments on Yb and Bi [30] at liquid He temperatures seem to rule out the presence of a liquid zone in these materials. In our analysis, we have not taken the possible existence of such a liquid zone into account. However, since these liquid regions are probably much smaller than the wavelength of the undulations [32, 22, 15], we believe their presence would not modify the thermal instability qualitatively.

Up to now, we have based the analysis on the assumption that the a–c boundary remains straight. To investigate qualitatively the possibility of having a Mullins–Sekerka [43, 1, 2] type instability along the front, consider a stable steady state on the left hand side of the growth rate curve (e.g. point B_2 in fig. 1). The temperature profile in front of the a–c boundary drops off rapidly, roughly as $T^{b}_{ss}\exp(-V_{ss}x/\kappa)$. Hence, if there is a small perturbation along the front, the boundary at points bulging forward will be at a lower temperature and will therefore grow less rapidly (we neglect the effect of curvature on the growth rate). This will act to counter the perturbation. In steady state points of type C, however, the situation is different: here the growth will be more rapid at a spot bulging forward. The straight front will now be unstable and break up into protrusions with a typical length determined by a capillary length and the thermal length κ/V_{ss} (the slope of the temperature in front of the boundary). The observation by Leamy et al. [22] that the surface of explosively crystallized Ge layers showed undulations that "are composed of froth-like bubbles that are aligned in irregular rows" [22] may be related to such effects.

In conclusion, when the growth rate increases with increasing temperature we expect the thermal instability discussed before without major changes resulting from a morphological instability along a front. However, on the right-hand side of the growth rate curve, there is no thermal instability, but there will exist significant morphological instabilities that invalidate the use of a simple one-dimensional model.

6. Period-doubling bifurcations

In a numerical study of the nonlinear periodic solutions given by eqs. (1) and (2), we have found that these, in turn, bifurcate via a sequence of period-doubling bifurcations. These Feigenbaum sequences [45] have been analyzed mostly in models with a finite-dimensional phase space. The model studied here for the "infinite-dimensional" temperature field is simple enough to permit a detailed numerical analysis. Here, we discuss an example for one set of parameters obtained by an algorithm described in appendix C. In a future publication we hope to come back to the derivation of an approximate set of ordinary differential equations in a three-dimensional phase space for this system.

In this section, all quantities will be given in dimensionless units: temperatures are measured in units of q, time in units of Γ^{-1}, length in units of $\sqrt{\kappa/\Gamma}$, and hence velocities in units of $\sqrt{\kappa\Gamma}$. The function $V(T^{b})$ used in our specific calculations is

$$V = D\mathrm{e}^{-Q/T^{b}}\left(1-\mathrm{e}^{-L(1/T^{b}-1/T^{M})}\right). \tag{23}$$

Such a functional dependence of V on T^{b} is typical for the growth of a crystal from a melt [34]. In the example discussed here, $T^{M}=1.8$ and $L=3$, and D and Q are adjusted such that the maximum growth velocity of 12 is reached for $T^{b}=1.6$.

In our numerical calculations, we vary the substrate temperature T^{0}. If we decrease T^{0} starting from a relatively large value, the steady state solution becomes unstable at $T^{0}=0.431626$. At this point, $(\Delta T^{b}_{ss})^{2}\approx 0.934$. For lower values of T^{0}, we first observe stable periodic solutions of the type shown in the inset of fig. 1. When lowering T^{0} even more, period-doubling bifurcations are observed. The bifurcation to a period 2 solution occurs at $T^{0}=0.426884$. We have been able to follow the successive bifurcations to period 4, 8 and 16 at T^{0} = 0.4266355, 0.4266035 and 0.4265965, respectively†. These values seem to converge rapidly to the Feigenbaum ratio δ. At slightly lower temperatures we enter an apparently chaotic regime where no periodicity is evident. (Lowering T^{0} still further finally causes us to cross the dotted line in fig. 2 and the crystallization wave eventually stops.)

†These values of T^{0} depend slightly on the size of the time step used in the calculation, but the qualitative picture is the same for any small step size.

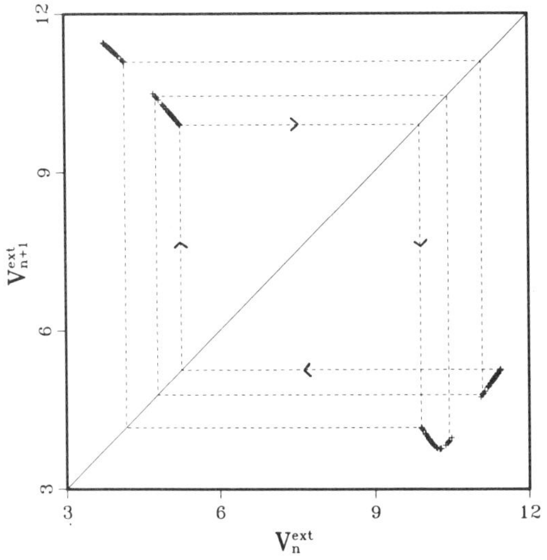

Fig. 4. Plot of V_{n+1}^{ext} as a function of V_n^{ext}. All data points obtained after the initial value term in eq. (4) becomes negligible are included. Here $T^0 = 0.426586$, slightly below the value of T^0 where chaotic behavior sets in. A few iterations of the map are indicated by the dashed lines.

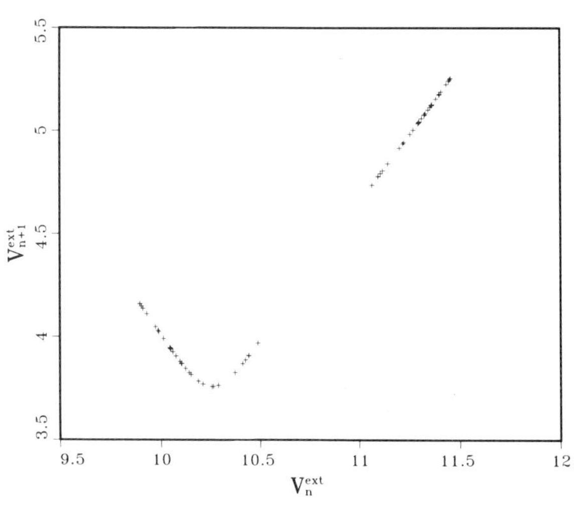

Fig. 5. Enlargement of the lower right hand corner of fig. 4.

Let V_n^{ext} be the value of the velocity V in the nth extremum (maximum or minimum) of a solution of the equations (1) and (2). Empirically, we found that the above observations can be understood within the context of the map obtained by plotting V_{n+1}^{ext} as a function of V_n^{ext}. As shown in figs. 4 and 5, these points turn out to lie on a single curve. Hence in this regime the model is well described by a single valued function f such that

$$V_{n+1}^{\mathrm{ext}} = f\left(V_n^{\mathrm{ext}}\right). \tag{24}$$

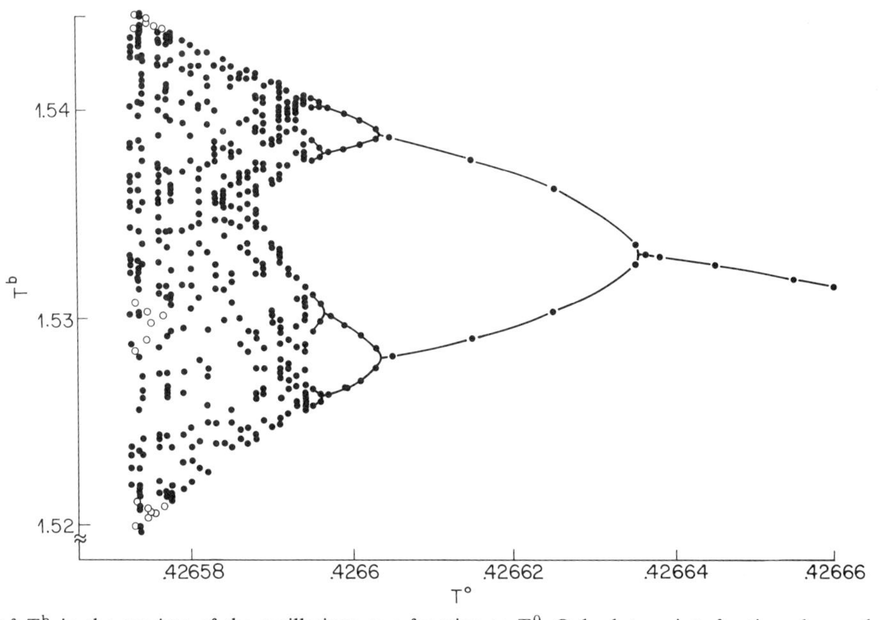

Fig. 6. Values of T^{b} in the maxima of the oscillations as a function to T^0. Only data points for times larger than about 20 basic periods of the system are included, so that the solutions have relaxed towards the stable periodic solutions, if these exist. Since only points on the upper branch where $11 \lesssim V_n^{\mathrm{ext}} \lesssim 11.5$ are shown, the points on the right are actually period 2 solutions. Open circles denote period 6 and 12 solutions. The solid line is merely drawn to guide the eye.

Moreover, this one-dimensional map has indeed a parabolic minimum, characteristic of the Feigenbaum map [45] that explains the occurrence of period doublings. A plot of $T^{\rm b}$ in the successive maxima of the solutions also shows the characteristic behavior of period doubling bifurcations (fig. 6). Note the following features:

i) There are no indications that a higher dimensional map is needed to describe the system better, because the points defined by the above map appear to lie on a single line to a high degree of accuracy. The fact that the memory time $\tau_{\rm m}$, defined in eq. (6), is of the same order as the period of the oscillations, is probably the reason this system is described so accurately by a one-dimensional map.

ii) The function f appears to be independent of the initial conditions; moreover, as soon as the initial value term in eq. (4) is negligible, the points of the map lie on the single curve. Thus the system converges very rapidly to the behavior described by (24).

iii) The existence of a minimum in the function f may be related to curvature of the growth rate curve. Even though the maximum velocity is never reached in our numerical solutions (in the calculations described here, the maximum is 12), we did not observe a minimum in f and period doubling bifurcations in calculations when we had modified the function $V(T^{\rm b})$ so that it increased linearly with $T^{\rm b}$ for large boundary temperatures.

Acknowledgements

We are grateful to P.C. Hohenberg, G.H. Gilmer, K.A. Jackson and D.S. Fisher for helpful discussions and to J. Guckenheimer for useful correspondence. We particularly benefitted from a conversation with P.C. Hohenberg on co-dimension two bifurcations.

Appendix A

Linear stability analysis

In this appendix we consider a small perturbation $\Delta T(x,0)$ around the steady state solution $T_{\rm ss}(x)$ at $t=0$, and investigate the behavior of $T^{\rm b}$ for large t. We assume that $\Delta T(x,0)$ is negligible for x large enough. In that case, the dominant contribution from the initial value term in eq. (4) decays for large t as

$$t^{-1/2}{\rm e}^{-2t/\tau_{\rm m}}. \tag{A.1}$$

The asymptotic decay of $T^{\rm b}$ is therefore at least of this order. Our strategy is to try to find a more slowly decaying mode of the form

$$T^{\rm b}(t)=T^{\rm b}_{\rm ss}+\epsilon{\rm e}^{2\omega t/\tau_{\rm m}},\quad V(t)=V_{\rm ss}+\epsilon V'{\rm e}^{2\omega t/\tau_{\rm m}}, \tag{A.2}$$

with ${\rm Re}\,\omega>-1$. Here $V'\equiv {\rm d}V/{\rm d}T^{\rm b}$ is the derivative of (2) in the steady state point and ϵ is assumed to be small. We have verified that the steady state solutions are indeed stable with perturbations decaying asymptotically as (A.1) if there are no solutions of the form (A.2).

By substituting eq. (A.2) into eq. (5) and linearizing, one finds for the resulting change $\delta\tau(x,t)$ in τ

$$\delta\tau=\frac{-\epsilon V'\tau_{\rm m}}{2\omega V_{\rm ss}}\exp(2\omega t/\tau_{\rm m})(1-\exp(2\omega x/V_{\rm ss}\tau_{\rm m})), \tag{A.3}$$

where we used the fact that $x=-V_{\rm ss}\tau$ in the steady state. Upon linearization of eq. (4), one gets by neglecting the initial value term

$$\epsilon e^{2\omega t/\tau_m} = \frac{q\epsilon V'}{4\kappa^{1/2} V_{ss}^{1/2}\omega} e^{2\omega t/\tau_m}$$
$$\times \int_{-\infty}^{0} dx\,(1-\exp(2\omega x/V_{ss}\tau_m))\frac{1}{|x|^{3/2}}\left(\frac{1}{2}V_{ss} - \left(\Gamma - \frac{V_{ss}^2}{4\kappa}\right)x\right)\exp(2x/V_{ss}\tau_m)$$
$$= \frac{qV'\epsilon}{2(2\pi\kappa\tau_m)^{1/2}\omega} e^{2\omega t/\tau_m}\int_0^{\infty} dy\,(1-e^{-y\omega})\left(-\frac{d}{dy}(y^{-1/2}e^{-y}) - \frac{V_{ss}^2}{4\kappa}\tau_m y^{-1/2}e^{-y(1+\omega)}\right). \quad \text{(A.4)}$$

After a partial integration, we are left with elementary integrations. Evaluating these, we arrive at the basic relation

$$1 = \alpha\left[-\frac{2\beta}{\omega} + \left(1+\frac{2\beta}{\omega}\right)(1+\omega)^{-1/2}\right]. \quad \text{(A.5)}$$

The parameters α and β used here have the same meaning as in section 3; $\alpha \equiv qV'/(8\kappa\tau_m)^{1/2} = V'\Delta T_{ss}^b/V_{ss}$ is the dimensionless slope of the growth rate curve and $\beta \equiv V_{ss}^2\tau_m/4\kappa = (4\kappa\Gamma/V_{ss}^2+1)^{-1} = (\Delta T_{ss}^b/q)^2$. Eq. (A.5) can be written in the "semiquadratic" [44] form

$$(1+\omega)^{1/2} = (\omega+2\beta)/(\omega/\alpha+2\beta). \quad \text{(A.6)}$$

After taking the square of this equation, and rearranging, one arrives at a quadratic equation for ω with roots

$$\omega_{\pm} = -\tfrac{1}{2}(1+4\beta\alpha-\alpha^2) \pm \tfrac{1}{2}\sqrt{(1-\alpha)^2\left[(1+\alpha)^2 - 8\alpha\beta\right]}. \quad \text{(A.7)}$$

As a result of taking the square of (A.6), $\omega_{\pm}$ either solve eq. (A.6) or the equation $(1+\omega)^{1/2} = -(\omega + 2\beta)/(\omega/\alpha+2\beta)$, and we must verify that the solutions (A.7) indeed satisfy (A.6).

The following results follow in a straightforward way from eqs. (A.6) and (A.7):

i) For $\alpha > (1-\beta)^{-1}$, ω_+ and ω_- both solve eq. (A.6) and are real. Since $\omega_+ > 0$, there is a mode of the form (A.2) that is unstable. One can show from eq. (3) and the definitions of α and β that

$$\frac{dV}{dT^b} \gtrless \frac{dV_{ss}}{dT_{ss}^b} \leftrightarrow \alpha \gtrless (1-\beta)^{-1}, \quad \text{(A.8)}$$

where the $>$ and $<$ sign correspond to points of type A and B respectively. Hence the above result implies that steady state solutions of type A are always *unstable*. The line $\alpha = (1-\beta)^{-1}$ separating type A and type B solutions is drawn in figs. 2 and 7 with a dashed line.

ii) For points of type B ($\alpha < (1-\beta)^{-1}$), both ω_+ and ω_- are solutions of eq. (A.6) with $\mathrm{Re}\,\omega_{\pm} > 0$ if $\beta < (\alpha - \alpha^{-1})/4$. At the line $\beta = (\alpha-\alpha^{-1})/4$, drawn in figs. 2 and 7 with a solid line, $\mathrm{Im}\,\omega_+$ varies continuously from 0 at the point $\alpha = 3$, $\beta = 2/3$ to $(1+\sqrt{5})^{3/2}/\sqrt{2} \approx 4.12$ if $\beta = 1$. In the limit $\Gamma \ll V_{ss}^2/4\kappa$, when $\beta \approx 1$, comparison of the above result with (A.2) shows that in this regime, the period of the oscillations is about

$$\frac{\pi}{4.12}\tau_m \approx 0.76\tau_m \approx \frac{6.1\kappa}{V_{ss}^2} \quad \text{(A.9)}$$

justifying our earlier contention that the period is of the order of τ_m.

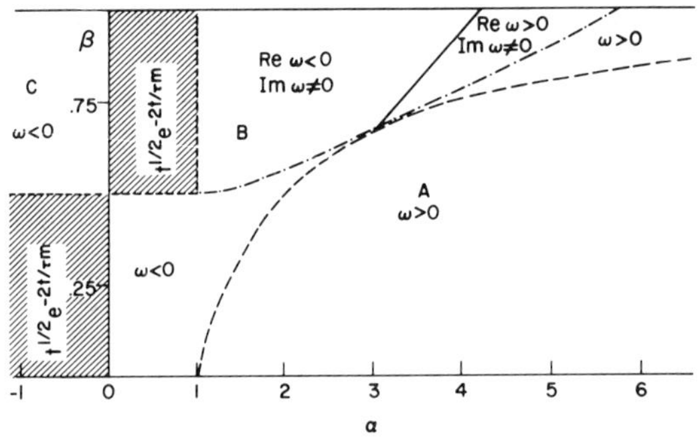

Fig. 7. The largest eigenvalue ω of the equations linearized around a steady state solution. At the solid line ($\beta=(\alpha-\alpha^{-1})/4$), $\operatorname{Re}\omega=0$ while $\operatorname{Im}\omega\neq0$. The chain-dotted line ($\beta=(\alpha+1)^2/8\alpha$) separates the points where ω is complex from those where ω is real, and the dashed line ($\alpha=(1-\beta)^{-1}$) separates the points of type A and B. For $0<\alpha<1$, $\beta>1/2$ and $\alpha<0$, $\beta<1/2$, the perturbations decay asymptotically as (A.1), as indicated in the figure.

All point of type B to the left of the solid line are stable. However, for $\alpha<1$, $\beta>1/2$, $\omega_\pm$ do not solve eq. (A.6) and all perturbations decay as (A.1).

iii) For the points of type B with $\beta>(\alpha+1)^2/8\alpha$, $\operatorname{Im}\omega_\pm\neq0$. The line $\beta=(\alpha+1)^2/8\alpha$ is drawn in fig. 7 with a dotted-dashed line. The conditions on the real part of $\omega_\pm$, which controls the stability, have been discussed in ii).

iv) Points of type C ($\alpha<0$) are always stable. For $\beta>\frac{1}{2}$, only ω_- solves eq. (A.6) and perturbations decay asymptotically as in (A.2), while for $\beta<\frac{1}{2}$, the asymptotic decay is as in (A.1).

Appendix B

Derivation of the coefficients (10)

To derive the coefficients a_1, a_2 and a_3 in eq. (10), we expand (4) near the point $\alpha=3$, $\beta=2/3$. Using the variable $v\equiv V(t)/V_{ss}-1$, we write $V(t-\tau)=V_{ss}(1+v-\dot v\tau+\frac{1}{2}\ddot v\tau^2+\cdots)$. Using τ as integration variable in (4) and neglecting the initial value term, we get with the aid of this expansion

$$\Delta T^{\rm b}(t)=\frac{q}{(4\pi\kappa\Gamma)^{1/2}}\int_0^\infty \mathrm{d}\tau\,\tau^{-1/2}\exp\left[-\tau-\frac{V_{ss}^2\tau}{4\kappa\Gamma}\left(1+v-\frac{1}{2}\tau\dot v+\frac{1}{6}\tau^2\ddot v\right)^2\right]$$
$$\times V_{ss}\left(1+v-\tau\dot v+\tfrac{1}{2}\tau^2\ddot v+\cdots\right)=\Delta T^{\rm b}_{ss}\int_0^\infty \mathrm{d}\tau\,\tau^{-1/2}\mathrm{e}^{-\tau(1+\gamma)}\Big\{1+v(1-2\tau\gamma)+(\gamma\tau-1)\dot v\tau$$
$$-\tau^2\ddot v\left(\frac{2\tau}{3}-1/2\right)-v^2(6\tau-8\tau^2)+8\dot v v\tau-\frac{5}{2}\tau^3\dot v^2+\cdots\Big\}. \tag{B.1}$$

Here $\gamma\equiv V_{ss}^2/4\kappa\Gamma=\beta/(1-\beta)$ and only terms of order v, $\dot v$, $\ddot v$, v^2, $\dot v v$ and $\dot v^2$, have been retained. In the terms proportional to $\ddot v$, v^2, $\dot v v$, and $\dot v^2$, we take $\gamma=2$, the values in the bifurcation point. After evaluation of the integrals we then arrive at

$$\Delta T^{\rm b}(t)=\Delta T^{\rm b}_{ss}\left\{1+(1-\beta)v+\frac{1}{4}\delta\beta\dot v-\frac{1}{6^3}\ddot v-\frac{1}{3}v^2+\frac{2}{3}v\dot v-\left(\frac{5}{12}\right)^2\dot v^2\right\}. \tag{B.2}$$

Expansion of $\Delta T^{\mathrm{b}}(t)$ using the variable α and V'' (defined in eq. (11)) yields

$$\Delta T^{\mathrm{b}}(t) = \Delta T^{\mathrm{b}}_{\mathrm{ss}}\left(1 + \alpha^{-1}v - \tfrac{1}{2}\alpha^{-3}V''v^2 + \cdots\right). \tag{B.3}$$

Combination of eqs. (B.2) and (B.3) finally yields

$$\ddot{v} - 54\delta\beta\dot{v} - 24(\delta\alpha - 9\delta\beta)v + 72\left(1 - \frac{1}{18}V''\right)v^2 - 144v\dot{v} + \frac{75}{2}\dot{v}^2 = 0. \tag{B.4}$$

From this follow the expressions for a_1, a_2 and a_3, given in eq. (10).

Notice that the coefficient of the term proportional to $\dot{v}$ does not agree with eq. (9). This is due to the fact that this factor gets renormalized by terms involving the third order derivative that have been left out in the above analysis. To see this, consider the equation $c\dddot{v} + \ddot{v} - \mu\dot{v} - \nu v = 0$, with c of order unity and μ and ν small. The characteristic polynomial then reads $c\omega^3 + \omega^2 - \mu\omega - \nu = 0$. Retaining only the dominant terms, we may write $c\omega^3 + \omega^2 - \mu\omega - \nu \approx (c\omega + 1)[\omega^2 - (\mu - \nu/c)\omega - \nu] = 0$. Thus, while in the neighborhood of the $\mu = \nu = 0$ critical point the system is well described by a second order equation, the coefficient of the $\dot{v}$ term gets renormalized by the coefficient of the $\dddot{v}$ term [46].

Appendix C

Description of the numerical solution

Our numerical analysis was based on solving eq. (4) with τ rather than x as the integration variable in the first integral. We used a fixed time step Δt; given the values of T^{b} and V at time $t, t - \Delta t, t - 2\Delta t, \ldots$, an initial guess for $V(t + \Delta t)$ was made. Using this estimate in the integrals in the right-hand side of eq. (5), $T^{\mathrm{b}}(t + \Delta t)$ was calculated using (4) and the resulting $V(t + \Delta t)$ using eq. (2). If the updated value of $V(t + \Delta t)$ differed more than 10^{-5} from the previous value, we recalculated $T^{\mathrm{b}}(t + \Delta t)$ using the updated values in the integral. The time integral in eq. (4) was cut off at $\tau = 1.4$ (the contribution from the rest of the integral is typically of the order of 10^{-8}), and the integral from $\tau = 0$ to $\tau = 1$ was done with equally spaced grid in the variable $\tau^{1/2}$ rather than τ. Integrals were calculated using Simpson's rule and the time step used in the calculation was $0.00032 \approx \tau_{\mathrm{m}}/500$.

References

[1] For general reviews see e.g. R.F. Sekerka, in: Crystal Growth, an Introduction, P. Hartman, ed. (North-Holland, Amsterdam, 1973); J.S. Langer, Rev. Mod. Phys. 52 (1980) 1.

[2] R.F. Sekerka, Physica 12D (1984) 212 (these proceedings).

[3] G. Gore, Phil. Mag. 9 (1855) 73.

[4] C.C. Coffin and S. Johnston, Proc. Roy. Soc. A146 (1934) 564.

[5] W. Lotmar, Helv. Phys. Acta 18 (1945) 369.

[6] A. Götzberger, Z. Physik 142 (1955) 182.

[7] F.M. Aymerich and A. Delunas, Phys. Stat. Sol. A31 (1975) 165.

[8] C.E. Wickersham, G. Bajor and J.E. Greene, Solid State Commun. 27 (1978) 17; C.E. Wickersham, Ph.D. Thesis, University of Illinois at Urbana-Champaign (1978).

[9] T. Takamori, R. Messier and R. Roy, Appl. Phys. Lett. 20 (1972) 201; J. Mat. Sci. 9 (1974) 159.

[10] R. Messier, T. Takamori and R. Roy, Solid State Commun. 16 (1975) 311.

[11] A. Mineo, A. Matsuda, T. Kurosu and M. Kikuchi, Solid State Commun. 13 (1973) 329.

[12] A. Matsuda, A. Mineo, T. Kurosu and M. Kikuchi, Solid State Commun. 13 (1973) 1165.

[13] M. Kikuchi, A. Matsuda, T. Kurosu, A. Mineo and K.J. Callanan, Solid State Commun. 14 (1974) 713.
[14] K.J. Callanan, A. Matsuda, A. Mineo, T. Kurosu and M. Kikuchi, Solid State Commun. 15 (1974) 119.
[15] R.B. Gold, J.F. Gibbons, T.J. Magee, J. Peng, R. Ormond, V.R. Deline and C.A. Evans, in: Laser and Electron Beam Processing of Materials, C.W. White and P.S. Peercy, eds. (Academic Press, New York, 1980), p. 221.
[16] J.C.C. Fan, H.J. Zeiger, R.P. Gale and R.L. Chapman, Appl. Phys. Lett. 36 (1980) 158.
[17] R.L. Chapman, J.C.C. Fan, H.J. Zeiger and R.P. Gale, Appl. Phys. Lett. 37 (1980) 292.
[18] H.J. Zeiger, J.C.C. Fan, B.J. Palm, R.P. Gale and R.L. Chapman, in: Laser and Electron Beam Processing of Materials, C.W. White and P.S. Peercy, eds. (Academic Press, New York, 1980), p. 234.
[19] R.L. Chapman, J.C.C. Fan, H.J. Zeiger and R.P. Gale, in: Laser and Electron-Beam Solid Interactions, J.F. Gibbons, L.D. Hess and T.W. Sigmon, eds. (North-Holland, Amsterdam, 1981), p. 81.
[20] H.J. Zeiger, J.C.C. Fan, B.J. Palm, R.L. Chapman and R.P. Gale, Phys. Rev. B25 (1982) 4002.
[21] C.E. Wickersham, Solid State Commun. 34 (1980) 907.
[22] H.J. Leamy, W.L. Brown, G.K. Celler, G. Foti, G.H. Gilmer and J.C.C. Fan, Appl. Phys. Lett. 38 (1981) 137; ref. 19, p. 89.
[23] R. Koba and C.E. Wickersham, Appl. Phys. Lett. 40 (1982) 672.
[24] O. Bostanjoglo, Phys. Stat. Sol. A70 (1982) 473.
[25] L.N. Aleksandrov and F.L. Edelman, Surf. Sci. 86 (1979) 222.
[26] G. Auvert, D. Bensahel, A. Georges, V.T. Nguyen, P. Henoc, F. Morin and P. Coissard, Appl. Phys. Lett. 38 (1981) 613.
[27] G. Auvert, D. Bensahel, A. Perio, V.T. Nguyen and G.A. Rozgonyi, Appl. Phys. Lett. 39 (1981) 724.
[28] R.A. Lemons and M.A. Bösch, Appl. Phys. Lett. 39 (1981) 343.
[29] S.V. Krishnaswamy, R. Messier, P. Swab, L.L. Tongson and K. Vedam, J. Electr. Mat. 10 (1981) 433.
[30] V.M. Kuz'menko and V.I. Mel'nikov, Sov. Phys. JETP 55 (1982) 474.
[31] Zeiger and coworkers (ref. 20) have analyzed this case in some detail. However, they make the physically questionable approximation that the temperature at the boundary remains constant despite possible variations in the interface velocity. This approximation is inconsistent with the mechanism leading to propagating oscillatory crystallization waves discussed here. Furthermore, as noted by the authors, it also causes numerical convergence problems and predicts unphysical negative interface velocities at low substrate temperatures.
[32] G.H. Gilmer and H.J. Leamy, ref. 15, p. 227.
[33] Measurements by Koba and Wickersham (ref. 23) suggest that L is rather the latent heat per unit volume less a correction term (probably due to heat radiation) inversely proportional to the height of the layer.
[34] See e.g. R.F. Sekerka, in: "Proc. Int. School of Crystallography," Erice, 1981, (Reidel, Dordrecht, 1982).
[35] See e.g., H.S. Carslaw and J.C. Jaeger, Conduction of Heat in Solids, (Oxford Univ. Press, London, 1959); P.M. Morse and H. Feshbach, Methods of Theoretical Physics (McGraw-Hill, New York, 1953).
[36] J. Guckenheimer, IEEE Trans. Circuits Syst. 27 (1980) 982.
[37] J. Guckenheimer, in: New Approaches to Nonlinear Problems in Dynamics, P.J. Holmes, ed. (SIAM, Philadelphia, 1980); "Multiple Bifurcation Problems of Codimension 2", preprint.
[38] M. Shaerer, SIAM J. Math. Anal. 11 (1980) 365.
[39] H.R. Brand, P.C. Hohenberg and V. Steinberg, Phys. Rev. A27 (1983) 591.
[40] P.H. Coullet and E.A. Spiegel, "Amplitude Equations for Systems with Competing Instabilities", preprint.
[41] A. Andronov, E. Leontovich, I. Gordon and A. Maier, The Theory of Bifurcations of Plane Dynamical Systems (Wiley, New York, 1973).
[42] For experimental evidence for these effects see Wickersham's thesis, ref. 8, p. 269.
[43] W.W. Mullins and R.F. Sekerka, J. Appl. Phys. 34 (1963) 323; 35 (1964) 444.
[44] D.J. Wollkind, in: Preparation and Properties of Solid State Materials, vol. 4, W.R. Wilcox, ed. (Marcel Dekker, New York, 1979).
[45] M.J. Feigenbaum, J. Stat. Phys. 19 (1978) 25; 21 (1979) 669.
[46] Therefore, α in eq. (5) of ref. 39 should be replaced by $\alpha + \beta\gamma$.

Physica 12D (1984) 295–302
North-Holland, Amsterdam

THE AGING OF NUCLEI IN A BINARY MIXTURE

E.A. COUTSIAS*
Department of Mathematics and Statistics, University of New Mexico, Albuquerque, NM 87131, USA

and

J.C. NEU
Mathematical Sciences Research Institute, University of California-Berkeley, Berkeley, CA 94720, USA

A theory of nucleation for a metastable binary mixture is developed, based on the Cahn–Hilliard theory of phase separation. It is shown that at least in its intermediate stages the transition from a supersaturated state to an energetically favored separated state follows the typical mechanism of surface tension governed nucleation processes and long-time similarity solutions describing the distribution of nuclei are found, analagous to the results of Lifshitz et al. It is shown that spatially uniform nucleation is destabilized by diffusion and the modifications to the nucleation process due to spatial inhomogeneity are found.

1. Introduction

In the evolution of first order phase transitions, there is a stage called Ostwald Ripening, or aging [1]. At the onset, homogeneously distributed nuclei of one phase are surrounded by supersaturated medium consisting of the other phase. The larger nuclei grow and deplete the supersaturation of the surrounding phase, and this leads to the decay of the smaller nuclei. The demise of the smaller nuclei replenishes the supersaturation of the surrounding phase and supports the continued growth of the large nuclei. This phenomenology is described quantitatively in an analysis of I. Lifshitz and V. Slyozov [2]. They find that in a medium with a spatially uniform distribution of nuclei which occupy only a small fraction of the total volume, the supersaturation and density of nuclei decrease with time like $t^{-1/3}$ and t^{-1} respectively, while the mean nuclear radius increases like $t^{1/3}$.

In practice, this description must be modified after sufficiently large times, due to processes such as coagulation or gravitational separation that might become important for large nuclei.

The subject of this analysis is an examination of Ostwald ripening in a binary mixture, where the transition from a homogeneous mixture to a spatially inhomogeneous, partially separated mixture is modeled by the Cahn–Hilliard theory [3]. We present new exact solutions of this theory which clearly correspond to the asymptotic results of [2].

Next, we consider the effects of spatial nonuniformity in the distribution of nuclei. This problem has been studied in the context of chemical precipitation [4] by use of linearized analysis. We present exact solutions of the full nonlinear theory which describe the development of localized spatial inhomogeneities. In the region of the inhomogeneity, the volume fraction occupied by nuclei grows like $t^{2/3}$, instead of asymptoting to a constant value as in the spatially homogeneous case.

2. Phase separation in binary mixture

We review the basics of Cahn–Hilliard theory [3] which describes phase separation in a binary

*This research partially supported by AFOSR Grant No. AFOSR-82-0277.

0167-2789/84/$03.00 © Elsevier Science Publishers B.V.
(North-Holland Physics Publishing Division)

mixture. The free energy φ per unit volume is a functional of c, the fractional concentration of one component. For isotropic material with small variations and gradients in c,

$$\varphi[c] \approx f(c) + \frac{\kappa}{2}|\nabla c|^2, \quad \kappa > 0, \tag{2.1}$$

where $f(c)$ is the free energy per unit volume in a medium with uniform concentration c and $(\kappa/2)|\nabla c|^2$ is the leading order effect of nonuniformities with small variations and gradients.

The dynamical evolution of small nonuniformities in c may be described as follows: The chemical potential μ per unit volume is

$$\mu = \frac{\delta\varphi}{\delta c} = f'(c) - \kappa\Delta c. \tag{2.2}$$

The concentration c is a locally conserved quantity whose flux is $-\Delta\mu$. Hence, c satisfies the transport equation

$$c_t - \Delta\left(f'(c) - \kappa\Delta c\right) = 0. \tag{2.3}$$

We discuss the stability of uniform states $c \equiv c_0$. From a linearized analysis of (2.3) we find that a harmonic perturbation with wavenumber k has growth rate

$$\sigma = -f''(c_0)k^2 - \kappa k^4. \tag{2.4}$$

If $f''(c_0) > 0$, then the medium is stable. If $f''(c_0) < 0$, then there is instability for $|k|^2 < k_c^2 \equiv -f''(c_0)/\kappa$ (fig. 1a).

Typically, we find that a homogeneous binary mixture is stable at high temperatures, but unstable to phase separation at low temperatures. There is a critical temperature such that $T < T_c \to f''(c) < 0$ for some range of c. Fig. 1b shows the typical form of the free energy $f(c)$ in this case. The darkened segment on the c-axis represents the range of unstable c values.

We review some basic results about steady equilibria. For an equilibrium state, the chemical potential μ assumes a steady, uniform value throughout the entire medium. From (2.2), we see

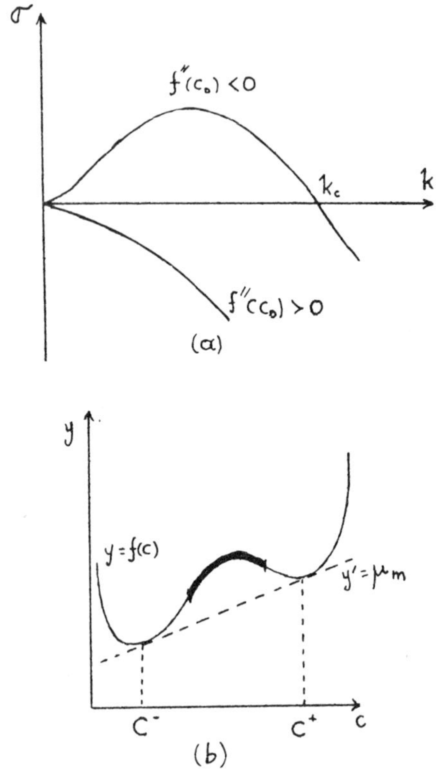

Fig. 1

that the steady equilibria satisfy

$$f'(c) - \kappa\Delta c = \mu. \tag{2.5}$$

States with uniform concentration c satisfy $f'(c) = \mu$. To determine the concentrations corresponding to a given value of chemical potential μ, we simply determine the points on the graph of $y = f(c)$ in fig. 1b where the slope is μ. Given μ with $\min f'(c) < \mu < \max f'(c)$, we see from fig. 1b that there are in general two concentrations $c = c^{\pm}(\mu)$, $c^- < c^+$, which correspond to uniform stable equilibria satisfying $f'(c) = \mu$, $f''(c) > 0$.

3. The critical nucleus

A uniform state $c \equiv c_0$ with c_0 below the unstable band may be metastable in the sense that finite amplitude perturbation applied in a localized region of space for a limited time may be sufficient to create a growing nucleus whose interior has a uniform value $c = c_0'$ above the unstable band. If the initial perturbation is too small, the medium will relax back to the uniform value c_0.

There is a steady, nonuniform solution of the equilibrium equation (2.5) which represents the *critical nucleus*. Physically, the critical nucleus represents a perturbation which is just large enough to escape dissolution, but not large enough to grow. The size of the critical nucleus is a function of the chemical potential μ in the medium surrounding the nucleus.

We may think of the planar profile $C(x)$† corresponding to the chemical potential μ_m as a critical nucleus of infinite radius. Accordingly, we expect that the critical nucleus corresponding to a chemical potential μ with $0 < \mu - \mu_m \ll \mu_m$ will have a radius $r_0 \gg 1$. In this limit, we expect that there is a transition zone about the spherical surface of radius r_0 where the concentration c adjusts from values close to $c^+(\mu)$ in the interior of the nucleus to the value $c = c^-(\mu)$ achieved at spatial infinity (fig. 2).

The radius r_0 is estimated in [3] in the limit $0 < \mu - \mu_m \ll \mu_m$ by a variational argument. The solutions of the equilibrium equation (2.5) which satisfy $c \to c^-(\mu)$ as $|x| \to \infty$ are stationary points of the free energy

$$G \equiv \int_{-\infty}^{\infty} \left(g(c) + \frac{\kappa}{2} |\nabla c|^2 \right) \mathrm{d}x^3. \tag{3.1}$$

Here,

$$g(c) \equiv f(c) - f(c^-(\mu)) - \mu(c - c^-(\mu)). \tag{3.2}$$

In the interior of the nucleus, we expect that the concentration $c(x)$ is near the uniform value $c^+(\mu)$, so that $g(c) \approx g(c^+(\mu))$ in $r < r_0$. For $0 < \mu - \mu_m \ll \mu_m$, we have $g(c^+(\mu)) \approx (C^+ - C^-)(\mu - \mu_m)$. Hence the contribution to G from the nucleus interior is approximately $(4\pi/3) r_0^3 (C^+ - C^-)(\mu - \mu_m)$. In (3.2), the term $(\kappa/2)|\nabla c|^2$ is significant only in the transition zone about $|x| = r_0$. Hence, we regard the integral $(\kappa/2)\int_{-\infty}^{\infty} |\nabla c|^2 \mathrm{d}x^3$ as a surface energy. Using the fact that the concentration profile is well approximated by the heteroclinic solution of the corresponding 1-D problem, $C(x)$, centered at $r - r_0$ [5], in the vicinity of the transition zone where $r - r_0 = \mathcal{O}(1)$, we find that the surface energy term is approximately $2\pi\kappa r_0^2 \int_{-\infty}^{\infty} C_x^2 \mathrm{d}x$. Summing the interior and surface contributions, we obtain for G the approximate value

†$C(x)$ is the solution of $f'(C) - \kappa C_{xx} = \mu_m$, $C(\pm\infty) = C^\pm$, with max C_x occurring at $x = 0$, that is it is the phase-plane trajectory joining the two saddle points of the system. For usual $f(C)$ it can be found in terms of elliptic functions.

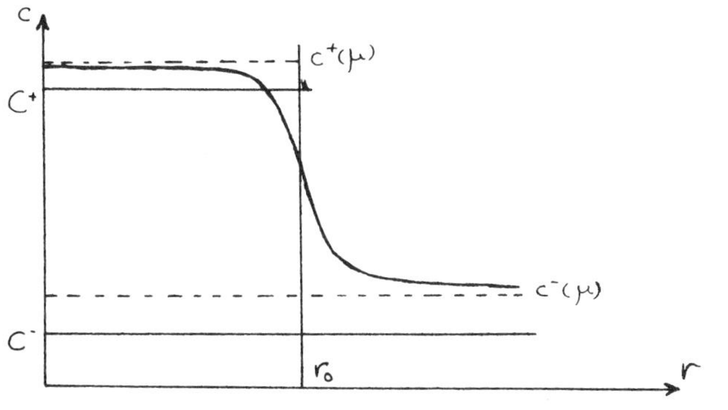

Fig. 2

$$G \approx -\tfrac{4}{3}\pi r_0^3 (C^+ - C^-)(\mu - \mu_m) + 2\pi\kappa r_0^2 \int_{-\infty}^{\infty} C_x^2 \mathrm{d}x. \tag{3.3}$$

G is stationary with respect to variations in r_0 if

$$\mu - \mu_m = \frac{H}{r_0}, \quad H \equiv \frac{\kappa}{C^+ - C^-} \int_{-\infty}^{\infty} C_x^2 \mathrm{d}x. \tag{3.4}$$

(3.4) is the relation between the size r_0 of the critical nucleus and the chemical potential μ in the limit $\mu - \mu_m \to 0^+$. This result is analogous to the Gibbs–Thomson formula for the critical radius of a liquid nucleus in vapor. H is analogous to surface tension, and $\mu - \mu_m$ is analogous to the supersaturation.

4. The time evolution of nuclei

We analyze the growth and decay of nuclei in a medium where the chemical potential μ is just above the critical value μ_m. We view the physical process as follows: Outside the nucleus, where c is close to $c^-(\mu)$ and the gradients of c are much smaller than in the transition layer, the transport process of c, represented by the Cahn–Hilliard equation (2.3), may be approximated by simple diffusion,

$$c'_t = D\left(c'_{rr} + \frac{2}{r}c'_r\right), \quad \text{in} \quad r > r(t), \tag{4.1}$$

where $c' \equiv c - c^-(\mu)$, $D \equiv f''(C^-)$, and $r(t)$ is the radius of the nucleus at time t. If the nucleus is growing, it feeds on the surplus c' in the exterior. If it is decaying, it augments the surplus. In other words, the surface of the nucleus acts as sink or source. The flux balance through the surface leads to the boundary condition

$$Dc'_r = \delta C \dot{r}(t), \quad \text{on} \quad r = r(t), \quad \delta C \equiv C^+ - C^-. \tag{4.2}$$

As a result of the diffusion process represented by (4.1)–(4.2), the chemical potential that the nucleus sees at its surface is not μ, but rather the value consistent with the local value of c' at the surface, which is approximately $\mu' \equiv Dc' + \mu^- - \mu_m$. If we assume that the nucleus is in a quasi-static equilibrium with this value, then (3.4) applies with $\mu = \mu'$, and we obtain a second boundary condition

$$Dc' = \frac{H}{r(t)} - \delta\mu, \quad \text{on} \quad r = r(t), \quad \delta\mu \equiv \mu^- - \mu_m. \tag{4.3}$$

Eqs. (4.1)–(4.3), together with the boundary condition $c' \to 0$ as $r \to \infty$, constitute a Stefan problem for determining $r(t)$ and $c'(r,t)$ in $r > r(t)$. The solution of (4.1) that goes to 0 as $|x| \to \infty$ is

$$c' = \frac{A}{r}. \tag{4.4}$$

Substituting into the boundary conditions (4.2)–(4.3) and eliminating A, we obtain an equation for the size of the nucleus, $r = r(t)$:

$$\dot{r} = \frac{1}{\delta C}\frac{1}{r}\left(\delta\mu - \frac{H}{r}\right). \tag{4.5}$$

This result for the time evolution of the nuclear radius in a binary medium has appeared in other contexts, most notably in describing the precipitation of crystals from ionic solutions [2, 4]. From (4.5), we read off the following results: If $r(t)$ is initially less than the critical size $r_0 = H/\delta\mu$, then $r(t)$ collapses to zero in finite time. If $r(t)$ is initially greater than r_0, then $r(t)$ grows like $t^{1/2}$ as $t \to \infty$.

5. Evolving distributions of nuclei

We consider a medium with a distribution of nuclei in which the volume fraction of nuclei is assumed to be small. The consequence of this assumption is that the diffusion zones around each nucleus have small radius compared with the internuclear distance. Any given nucleus sees a uniform value of supersaturation $\delta\mu$ around itself and grows in size according to the rate (4.5). The background supersaturation, on the other hand, feels the growth and decay of nuclei as sinks and sources. The macroscopic physics of this process may be described quantitatively as follows: Let $f(r, x, t)\,\mathrm{d}r\,\mathrm{d}x^3$ be the number of nuclei in volume $\mathrm{d}x^3$ whose radii are between r and $r + \mathrm{d}r$. $f(r, x, t)$ and the background concentration surplus $C(x,t) \equiv c(x,t) - C^-$ satisfy equations

$$f_t + \frac{D}{\delta C}\left(\frac{1}{r}\left(C - \frac{H}{Dr}\right)f\right)_r = 0, \tag{5.1}$$

$$C_t - D\Delta C = -\frac{4\pi}{3}\delta C \int_0^\infty r^3 f_t \,\mathrm{d}r. \tag{5.2}$$

(5.1) expresses the transport of nuclei in r space due to this growth or decay. (5.2) is a mass balance equation: Whatever material is gained by the nuclei must be taken from the concentration surplus field C. This is the meaning of the sink term on the r.h.s. The constants $\delta C, D, H$ in eqs. (5.1)–(5.2) may be absorbed by a rescaling of the variables. The units of the variables we adopt are given in the table below:

Variable	r	x	t	c	t
Unit	R	$\left(\frac{DH}{\delta CR^3}\right)^{1/2}$	$\frac{\delta CR^3}{H}$	$\frac{H}{DR}$	$\frac{H}{D\delta CR^5}$

(5.3)

Here, R may be thought of as a typical nuclear size in the initial distribution. In the units adopted above, (5.1)–(5.2) read

$$f_t + \left(\frac{1}{r}\left(c - \frac{1}{r}\right)f\right)_r = 0, \tag{5.4}$$

$$\left(c + \frac{4\pi}{3}\int_0^\infty r^3 f \, \mathrm{d}r\right)_t = \Delta c. \tag{5.5}$$

The systems (5.4)–(5.5) has been studied in various contexts. I. Lifshitz and V. Slyozov [2] considered the spatially uniform case. They show that a distribution f evolving according to (5.4)–(5.5) takes on a limiting form as $t \to \infty$, in which the mean radius of the nuclei grows like $t^{1/3}$ and the number of particles decays like t^{-1}. The concentration surplus C decays like $t^{-1/3}$. Later work has shown that spatially uniform solutions are unstable to spatial fluctuations. The instability appears to be the origin of the spatial inhomogeneities observed in precipitate patterns, because the initial distribution of nuclei is experimentally observed to be uniform [4].

We begin our analysis of (5.4)–(5.5) with a re-examination of spatially homogeneous distributions. We obtain exact solutions of the spatially homogeneous problem which clearly correspond to asymptotic solutions obtained by I. Lifshitz and V. Slyozov for large time.

In the spatially homogeneous case, $\Delta c = 0$ and (5.4) can be integrated once with respect to time, and we obtain an integral form of mass balance equation,

$$c + \frac{4\pi}{3}\int_0^\infty r^3 f \, \mathrm{d}r = C. \tag{5.6}$$

Here, c_0 is a constant independent of time. We may regard c_0 as the total amount of material per unit volume in excess of C^- contained either in the nuclei or the surrounding medium. (5.4) and (5.6) are the governing equations of $f(r,t)$ and $c(t)$.

We seek solutions of (5.4) and (5.6) as follows: (5.4) has similarity solutions of the form

$$f = t^{-n/3} f_n(s, \gamma), \quad s = rt^{-1/3}, \tag{5.7}$$

$$c = \gamma t^{-1/3}, \tag{5.8}$$

where γ and n are constants. Substitution of (5.7) and (5.8) into (5.4) gives

$$sPF_n' = Q_n F_n, \tag{5.9}$$

where

$$P \equiv 1 - \gamma s + \frac{s^3}{3}, \quad Q_n = 2 - \gamma s - \frac{n}{3}s^3. \tag{5.10}$$

It is necessary to determine which of the similarity solutions (5.7) are relevant to construction of physical distribution functions. Certain required properties may be anticipated from the characteristics of the transport equation (5.4). With c given by (5.8), the ODE of the characteristics reads

$$\frac{\mathrm{d}r}{\mathrm{d}t} = \frac{1}{r}\left(\gamma t^{-1/3} - \frac{1}{r}\right)$$

or

$$\frac{\mathrm{d}}{\mathrm{d}t}\left(\frac{1}{3}(r^3)\right) = \gamma r t^{-1/3} - 1. \tag{5.11}$$

Introducing $s = rt^{-1/3}$ as the dependent variable, (5.11) becomes

$$t\frac{\mathrm{d}s}{\mathrm{d}t} = -\frac{1}{5}\left(1 - \gamma s + \frac{s^3}{3}\right) = -\frac{1}{s^2}P, \tag{5.12}$$

where P is the cubic in s given by (5.10). If

$$\gamma > \left(\tfrac{3}{2}\right)^{2/3}, \tag{5.13}$$

then $P \equiv 1 - \gamma s + s^3/3$ has two positive zeros s^-, s^+ with $s^- < \gamma_{1/2} < s^+$. In this case, the characteristics in (s,t) space have the aspect depicted in fig. 3a. The corresponding structure of the char-

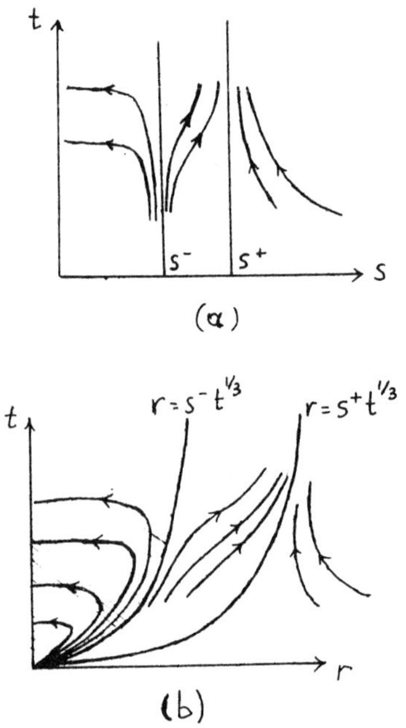

Fig. 3

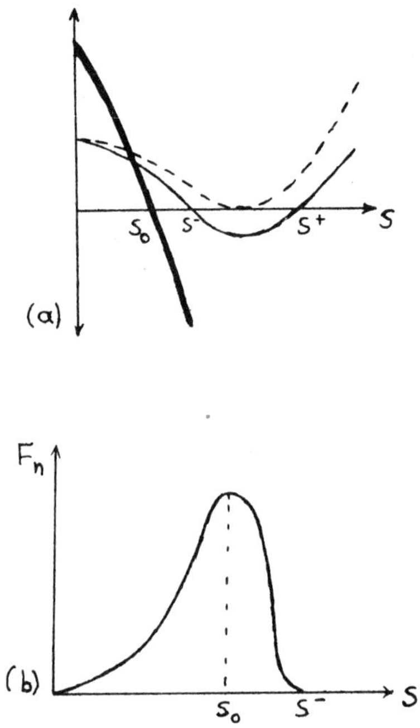

Fig. 4

acteristics in (r, t) space is depicted in fig. 3b. The characteristics in the shaded region of fig. 3b where $0 < r < s^- t^{1/3}$ correspond well with the known phenomenology of aging nuclei. Each characteristic in the shaded region eventually collides with the $r = 0$ axis. On the other hand, to the right of the curve $r = s^- t^{1/3}$ in fig. 3b, no characteristic terminates on the axis $r = 0$, so nuclei corresponding to these characteristics are never destroyed. Along these characteristics, $r \approx t^{1/3}$ as $t \to \infty$. Hence, the total amount of material in nuclei to the right of $r = s^- t^{1/3}$ would be proportional to t if such nuclei existed. Mass conservation would eventually be violated, hence there can't be nuclei in $r > s^- t^{1/3}$, and we must therefore seek solutions of (5.9) for F_n with $F_n \equiv 0$ in $s > s^-$.

We turn to the solution of F_n in $0 < s < s^-$. The cubic $Q_n \equiv 2 - \gamma s - (n/3)s^3$ in (5.10) is positive at $s = 0$ and has only one positive root s_0. In order for the solution for F_n to be free of a nonintegrable singularity at $s = s^-$, we must have $s_0 < s^-$. We show that there is a range of γ with $\gamma > (3/2)^{2/3}$ for which $s_0 < s^-$ if $n > 1$. If $\gamma = (3/2)^{2/3}$, then $s^- = (3/2)^{1/3}$ is a double root of $P(s)$. The dashed curve in fig. 4a depicts $P(s)$ for $\gamma = (3/2)^{2/3}$. The value of Q_n at $s = (3/2)^{1/3}$ is $2 - (3/2)(1 + n/3)$, which is negative if $n > 1$. In this case $s_0 < (3/2)^{1/3}$ and Q_n has the aspect depicted by the blackened curve. Hence, we have shown that $s^- - s_0 > 0$ will remain true for γ in some range of values above $(3/2)^{2/3}$. The solid curve just below the dashed curve depicts $P(s)$ for a γ in this range. In the case $s^- > s_0$, eq. (5.9) may be integrated to produce a solution for $F_n(s)$ which has the qualitative behavior shown in fig. 4b.

Since the transport equation (5.4) for $f(r, t)$ is linear, we may construct solutions for f by superposition of the similarity solutions (5.7) whose F_n functions meet the criteria discussed above. The superposition must satisfy the mass balance equation (5.6). In the similarity variables, (5.6) reads

$$\gamma t^{-1/3} + \frac{4\pi}{3} t^{4/3} \int_0^{s^-} s^3 f \, \mathrm{d}s = C. \tag{5.14}$$

Assuming that the F_n are normalized according to

$$\frac{4\pi}{3} \int_0^{s^-} s^3 F_n(s, \gamma) \, \mathrm{d}s = 1, \tag{5.15}$$

we see immediately from (5.14) that the superposition must be

$$f = c_0\left(t^{-4/3}F_4(s,\gamma) - \gamma t^{-5/3}F_5(s,\gamma)\right). \qquad (5.16)$$

Since $4.5 > 1$, we know from the previous discussion that there is a range of γ values with $\gamma > (3/2)^{2/3}$ for which acceptable F_4, F_5 functions are defined.

We note a basic limitation of this solution (5.16): Since $F_4, F_5 > 0$, we see that (5.16) gives negative f values for sufficiently small times. Hence, we expect that (5.16) really represents a limiting form of physically realizable solutions in the limit $t \to \infty$. For $t \to \infty$, (5.6) reduces to

$$f \approx c_0 t^{-4/3} F_4(rt^{-1/3}).$$

From this form of the asymptotic solution, the results of I. Lifshitz and V. Slyozov for the density of nuclei and mean nucleus radius follow immediately.

6. Spatially inhomogeneous aging processes

The method employed in section 5 to construct spatially independent solutions for f and c can be extended to the spatially dependent case. In equations (5.4), (5.5) we introduce the variable

$$s = rt^{-1/3} \qquad (6.1)$$

and write f and c in the forms

$$f = f(s,x,t), \quad c = t^{-1/3}\gamma(x,t). \qquad (6.2)$$

The equations for f and γ as functions of s, x, t are

$$\left(\partial_t - \frac{1}{3t}s\partial_s\right)f + \frac{1}{t}\left(\frac{1}{\delta}\left(\gamma - \frac{1}{\delta}\right)f\right)_s = 0, \qquad (6.3)$$

$$\partial_t\left(t^{-1/3}\gamma + \frac{4\pi}{3}\int_0^\infty s^3 f\,\mathrm{d}s\right) = t^{-1/3}\Delta\gamma. \qquad (6.4)$$

We obtain solutions of (6.3), (6.4) in the special case when γ is independent of t. That is,

$$\gamma = \gamma(x). \qquad (6.5)$$

In this case, for any $n > 1$, there is a range of γ values with $\gamma > (3/2)^{2/3}$ for which

$$f_n \equiv t^{-n/3}F_n(s,\gamma) \qquad (6.6)$$

is a solution of (6.3), where the F_n functions are identical to those employed in the spatially homogeneous analysis of section 5. More general solutions for f may be formed by superpositions of the f_n. These superpositions must satisfy the mass balance equation (6.4). From the form of (6.4), we see that it is sufficient to take a superposition of f_2, f_4 and f_5. The most general superposition of f_2, f_4, f_5 which satisfies the mass balance equation (6.4) is

$$f = \tfrac{3}{2}\Delta\gamma t^{-2/3}F_2(s,\gamma) + v(x)t^{-4/3}F_4(s,\gamma) - \gamma t^{-5/3}F_5(s,\gamma). \qquad (6.7)$$

Here, $v(x)$ is an arbitrary function of x. As in the case of the spatially independent solution (5.16), (6.7) yields solutions of the physical problem only during intervals of time for which f is positive.

The volume fraction occupied by nuclei is

$$V = \frac{4\pi}{3}\int_0^1 r^3 f\,\mathrm{d}r = \frac{4\pi}{3}t^{4/3}\int_0^\infty s^3 f\,\mathrm{d}s. \qquad (6.8)$$

Substituting into (6.8) the solution (6.7) for f and recalling the normalization condition $(4\pi/3)\int_0^\infty s^3 F_n \mathrm{d}s = 1$, we obtain for V the result

$$V = \tfrac{3}{2}t^{2/3}\Delta\gamma + v(x) - \gamma t^{-1/3}. \qquad (6.9)$$

We consider the instructive special case of an isolated inhomogeneity. We take

$$v(x) = v_0 > 0, \qquad (6.10)$$

$$\Delta\gamma = h(r), \quad r \equiv |x|, \qquad (6.11)$$

where $h(r)$ is nonnegative and $\int_0^\infty r^2 h(r)\,\mathrm{d}r < \infty$.

The solution of (6.11) is

$$\gamma=\gamma_0+\int_0^r \frac{1}{r'}\int_0^{r'} r''^2 h(r'')\,\mathrm{d}r''\mathrm{d}r', \tag{6.12}$$

where $\gamma_0=\gamma(0)$. For γ given by (6.12), we deduce that $\gamma_r>0$ and $\lim_{r\to\infty}\gamma=\gamma_\infty$ is finite and greater than γ_0. If we assume that γ_0, γ_∞ are both in the range of γ values for which the functions $F_2(s,\gamma)$, $F_4(s,\gamma)$, $F_5(s,\gamma)$ exist, then (6.7) provide a positive solution for f when t is sufficiently large. The values of c and V corresponding to this solution are given by

$$c=t^{-1/3}\gamma(r), \tag{6.13}$$

$$V=\tfrac{3}{2}t^{2/3}h(r)+v_0-\gamma(r)t^{-1/3}, \tag{6.14}$$

where $\gamma(r)$ is given by (6.12). Fig. 5 depicts snapshots of c and V at successive instants in time in a typical case.

We briefly discuss the physics contained in the results (6.13), (6.14). In the case of no spatial inhomogeneity with $h\equiv 0$, the volume fraction V occupied by nuclei tends to the constant v_0. If there is spatial inhomogeneity and $h\not\equiv 0$, then the volume fraction occupied by nuclei grows like $t^{2/3}$ in those regions where $h(r)\neq 0$. The reason for this growth in the spatially inhomogeneous case is the following: The voracious nuclei in the region of the inhomogeneity can feed upon the more abundant supersaturation in the distant surroundings, which are populated by smaller and less voracious nuclei.

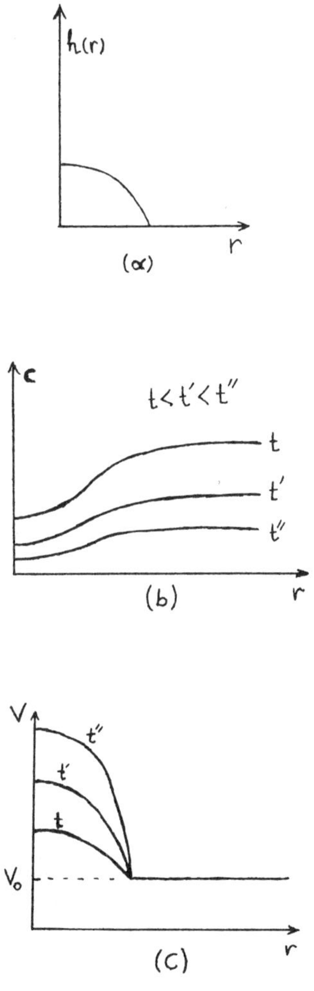

Fig. 5

References

[1] W. Dunning, Particle Growth in Suspensions, L. Smith, ed. (Academic Press, New York, 1973).

[2] I. Lifshitz and V. Slyozov, The kinetics of precipitation from supersaturated solid solutions, JPCS, V. 19, no. 1 (1961).

[3] J. Cahn and H. Hilliard, Free energy of a non-uniform system, I. Interfacial free energy, JCP, V. 28, no. 2 (1958); III. Nucleation, JCP, V. 31, no. 3 (1959).

[4] R. Lovett, P. Ortoleva and J. Ross, Kinetic instabilities in first order phase transitions, JCP, V. 69, no. 3 (1978); G. Venzl and J. Ross, Pattern formations in precipitation, JCP, V. 77, no. 3 (1982).

Physica 12D (1984) 303–304
North-Holland, Amsterdam

2.3. DIFFUSION, WETTING, BIOLOGICAL DIFFERENTIATION AND EXCITATION

FORMATION OF SPATIAL STRUCTURES IN CHEMICAL REACTIONS

John ROSS
Department of Chemistry, Stanford University, Stanford, California 94305, USA

Extended abstract

We have carried out an extensive series of experiments on macroscopic structure formation in chemical systems initially with spatial concentration gradients (Liesegang bands [1–4]) and without [5]. The purpose was the determination of the origin of pattern formation in both cases. Experiments [6–10] on systems with initial concentration gradients consisted of determining the temporal sequence of events up to completion of the final band pattern by visual observations and measurements of transmitted light, of scattered light, of deflection of a transmitted light beam and of gravity effects. Both nucleation and colloid formation are determined to occur in space continuously contrary to previous hypotheses (Ostwald theory). Band formation occurs within a zone of continuously formed colloid as observed by a sharp increase in localized gradients of the index of a refraction at the position of the prospective band and a decrease in turbidity in the regions on either side of the band. We conclude that ring formation is a post-nucleation process which occurs by a focussing mechanism and propose that that mechanism is due to the autocatalytic growth of colloidal particles coupled with diffusion. Experiments on the influence of gravity and the ring location confirm the presence of colloidal particles of several hundred Å in size which contribute to the formation of the visible structure.

We have made a systematic investigation of the dependence of precipitation patterns on the initial concentrations of reactants. In the presence of large gradients the locations of the bands are deterministic whereas in the presence of small concentration gradients the location of the bands becomes random. The experiments show the continuous progression of structure formation from high to zero initial concentration gradients and thus leads us to the suggestion that the origin of pattern formation is the same with and without initial gradients. The mesoscopic structure formed in the absence of initial gradient is shown to consist of independent colloidal particles and empty regions. These observations substantiate an instability theory of periodic precipitation processes.

We have analyzed the theory [11, 12] of colloidal growth in the presence of concentration gradients by means of postulated reaction diffusion equations. These equations are more realistic than previous ones used in formulations of the Ostwald hypothesis of Liesegang band formation but in limits contain the previous formulations. The numerical solutions of our equations predict nucleation to occur as a spatially continuous process in agreement with experiments. Spatial discontinuities in nucleation can only be obtained by unwarranted approximations or ad hoc assumptions on the boundary conditions.

We have made absorption experiments on chemical waves in the Belosouv–Zhabotinsky reaction by means of a Vidicon system which records absorption to a resolution of about 0.1 mm. Our preliminary results show oscillatory behavior in the amplitude of the wave as well as an oscillatory component in the velocity of the wave.

References

[1] R.E. Liesegang, Naturwiss. Wochenschr. 11 (1896) 353.
[2] E.S. Hedges and J.E. Myers, The Problem of Physico-chemical Periodicity (Longmans Green, New York, 1926).

0167-2789/84/$03.00

[3] E.S. Hedges, Liesegang Rings and Other Periodic Structures (Chapman and Hall, London, 1932).
[4] K.H. Stern, Chem. Rev. 54 (1954) 79.
[5] M. Flicker and J. Ross, J. Chem. Phys. 60 (1974) 3458.
[6] S. Kai, S.C. Müller and J. Ross, J. Chem. Phys. 76 (1982) 1392.
[7] S.C. Müller, S. Kai and J. Ross, SCI 216 (1982) 635.
[8] S.C. Müller, S. Kai and J. Ross, J. Phys. Chem. 86 (1982) 4294.
[9] S.C. Müller, S. Kai and J. Ross, J. Phys. Chem. 86 (1982) 4078.
[10] S. Kai, S.C. Müller and J. Ross, J. Phys. Chem. 87 (1983) 806.
[11] G. Venzl and J. Ross, J. Chem. Phys. 77 (1982) 1302.
[12] R. Lovett, P. Ortoleva and J. Ross, J. Chem. Phys. 69 (1978) 947.

Physica 12D (1984) 305–320
North-Holland, Amsterdam

FROM NONLINEAR WAVES TO SPIRAL AND SPECKLE PATTERNS:

NONEQUILIBRIUM PHENOMENA IN GEOLOGICAL AND BIOLOGICAL SYSTEMS*

P. ORTOLEVA**
Departments of Chemistry and Geology, Indiana University, Bloomington, Indiana 47405, USA

and

Geo-Chem Research Assoc., Inc., 400 East Third Street, Bloomington, Indiana 47401, USA

Patterns and waves in nonlinear biological and geological systems have certain universal properties but also some mathematically interesting differences. Models of a variety of these phenomena are reviewed and compared. Applications to developmental biology (early *Fucus* egg self-organization and Ca^{2+} fertilization waves in *Medaka* eggs) and geological phase transition kinetic patterns (Liesegang bands, metamorphic layering and stylolites) are presented.

1. Traveling waves

1.1. *A brief review of chemical wave phenomena*

Waves of chemical composition may propagate through reaction-transport systems in a variety of contexts [1–6]. These waves have a great variety of structures. In multiple time scale systems they can arise as propagating discontinuities with characteristic wave profiles related to the elementary catastrophes or singularities associated with the geometry of quasi-equilibrium surfaces (slow manifolds) in concentration space [5–8]. In other types of multiple time scale systems waves may propagate as discontinuities in concentration gradients. Both of these cases lead to interesting generalizations of the moving boundary problem of Stefan [6–10]. If there are stable limit cycles in the homogeneous kinetics, then allowing for transport leads to the propagation of wave trains [1, 11].

When waves interact with electrical fields, a number of interesting nonlinear effects can be observed [5, 10, 12]. Multiple types of waves may propagate when no such multiplicity exists in the field free system, wave annihilation, wave direction reversal and the formation of crescent shaped objects with free wave front ends that expand as time evolves may be observed as a result of wave–field interactions.

The geometry of wave forms in higher spatial dimensions can be quite interesting. Spiral waves, like the electro-crescent wave, exhibit the free wave front end phenomenon. Signals were found to potentially occur in single and multiple arm versions by studying their core structure [13]. Autonomous center waves of circular geometry (i.e. concentric circular fronts emanating from an autonomous core) were found to exist in certain frequency ranges separated by gaps of forbidden frequencies [14].

This review, too brief and subjective as it is, serves to show that nonlinear chemical wave mechanics is indeed a very rich field. We now compare three new wave phenomena as examples of rather distinct situations in which propagation can occur.

*Research supported in part by grants from the U.S. Department of Energy, Basic Energy Sciences, Engineering Research Program, Contract #DE-AC02-82ER12074, the U.S. National Science Foundation and the Petroleum Research Fund.

**A.P. Sloan Fellow, 1980–1982, NATO Fellow, 1982.

0167-2789/84/$03.00

1.2. *Unstable redox fronts and mineral deposition*

When a flow of reactive waters is forced through a rock, a zone of transition advances through the system. This may cause a variety of interesting wave phenomena. A most common example is when oxygenated meteoric waters from above invade a reduced rock containing pyrite or carbonaceous material. Under such a circumstance a front of redox potential advances into the rock. This seems simple enough. What is interesting is that morphological instabilities, one-parameter wave families, "decelerons" and banding may develop as we now show [15–17].

1.2.1. *A simple redox front*

Consider the simple case of waters containing an oxidant (O_2) at concentration X_m that invade a rock, of original pyrite content P_0, driven by a flow of velocity v. Then for the schematic process $X + P \rightarrow$ products (at rate W) we assume the equations

$$\frac{\partial X}{\partial t} = D\nabla^2 X - v\frac{\partial X}{\partial z} - W, \qquad (1)$$

$$\frac{\partial P}{\partial t} = -W. \qquad (2)$$

We have assumed the flow velocity is independent of the z-coordinate along which the flow occurs. The velocity of the front may be calculated exactly, independent of the details of W – only the X and P profiles depend on W as long as P_0 is independent of z and the front is far from the inlet. This result holds for arbitrary $v(x, y)$, $X_m(x, y)$ and $P_0(x, y)$, x and y being the coordinate perpendicular to the direction of the imposed water flow. Furthermore, this constant velocity front may be explicitly shown for a wide class of rate laws W to be unique and stable [15]. This uniqueness and stability is, in fact, what we might have expected. So, why the fuss?

1.2.2. *Decelerons and deposition*

When the system involves more reactions, the situation gets much more interesting. Consider the two mineral (A, B) scheme

$$\begin{aligned} &X + A \rightleftarrows Y, \\ &X + B \rightleftarrows Z, \end{aligned} \qquad (3)$$

involving mobile solute species X, Y, Z. We examine the analogue of the $X + P \rightarrow$ products scheme where the imposed waters are X-rich and poor in Y and Z. Hence the reactions in (3) are driven to the right by the invading waters. At first sight this is just a more complex version of the X–P system. Then we might expect that there exists a constant velocity front (of velocity u say) that advances as A and B are dissolved. If this is true then for a one dimensional system, along the x-axis, we have in analogy to (1) and (2),

$$\begin{aligned} &\left[D_X X' + (u-v)X + uA + uB\right]' = 0, \\ &\left[D_Y Y' + (u-v)Y - uA\right]' = 0, \\ &\left[D_Z Z' + (u-v)Z - uB\right]' = 0, \end{aligned} \qquad (4)$$

where "$'$" represents a derivative with respect to the spatial coordinate $x - ut$ moving with the wave. From (4) we see that there are three constants of the wave motion. Using the choices $X(-\infty) = X_m$, $Y(-\infty) = Y_m$, $Z(-\infty) = Z_m$, $Z(+\infty) = Z^{eq}(X(+\infty))$, $Y(+\infty) = Y^{eq}(X(+\infty))$ (where Z^{eq} and Y^{eq} are the X-dependent values of Z and Y at equilibrium with respect to B and A) we may evaluate the three constants of motion at $+\infty$ and $-\infty$ and equate the results. This yields three algebraic equations in two variables $X(+\infty)$ and u, for given $A(+\infty)$, $B(+\infty)$, X_m, Y_m and Z_m $A(-\infty) = B(-\infty) = 0$. Thus the system of three equations for two variables is overdetermined unless A_0, B_0 and the A and B equilibrium constants are related in a very special way. This proves that there is no constant velocity wave (except for the very special parameter choice).

The nonexistence of constant velocity fronts signals the onset of a new phenomenon – the deceleron deposition front. Indeed if there is no constant velocity front, then the reaction zone must be accelerating or decelerating. Certainly the front cannot advance faster than v, the water flow velocity. Thus the wave is decelerating as one of the minerals is deposited at the reaction zone. In fig. 1 we see a numerical simulation of this system that displays this phenomenon. The simulation involves solving equations like those in (1) and (2) (adapted for the present model) using rate laws of the type $kA^{2/3}[KX - Y]$ (for constants k and K) and the usual spatial discretization scheme [17]. (The 2/3 power law dependence on the mineral content per rock volume (A) accounts for a dependence of the rate of reaction on grain surface area.)

1.2.3. *One parameter front families*

From a system with nonexistence of constant velocity fronts we turn to one that apparently can support a family of fronts. Consider the oxidative dissolution of pyrite (P) and subsequent precipitation of geothite (G) via the schematic process

$$\begin{aligned} &X + P \rightleftarrows F + T, \\ &X + F \rightleftarrows G, \end{aligned} \tag{5}$$

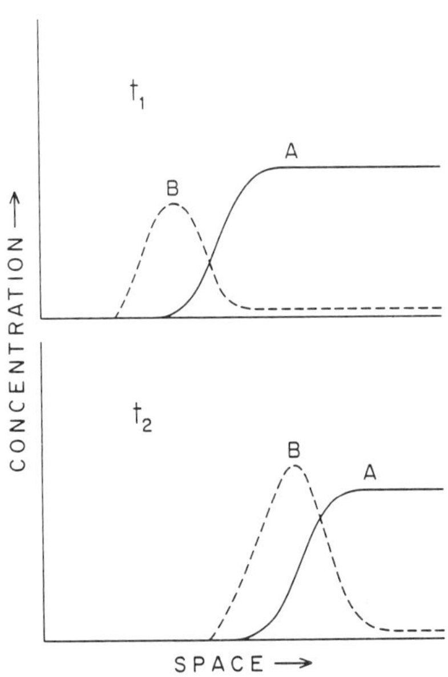

Fig. 1. Concentration profiles generated by the model of eq. (4). Note that A builds up at the B redox front and as a result the front slows down – i.e. no constant velocity front exists. This assemblage has been termed a decceleron [16, 17].

where F represents Fe^{2+} and T is thiosulfate (although the qualitative results are insensitive to such details). The initial data we wish to investigate is $X(-\infty) = X_m$, $T(-\infty) = 0$, $F(-\infty) = Q/X_m$, Q being the equilibrium constant for the goethite reaction so that the inlet water is in equilibrium with G, $X(+\infty)$ to be determined, $T(+\infty)F(+\infty) = KX(+\infty)$ (K being the pyrite equilibrium constant) and $P(-\infty) = 0$, $G(-\infty)$ to be determined.

An analysis of this system for the existence of constant velocity waves proceeds much like that for the previous example except that now we get the opposite extreme – we find that a one parameter family of fronts is indicated – the velocity is any value in a specified interval. (No existence proof for the corresponding family of profiles exists to date, however.) Numerical simulations indicate that from among this one parameter family a specific wave is selected for a wide range of initial data – apparently this is the only stable front.

The physical source of the one parameter continuum of solutions is easy to understand. The inlet waters are assumed to be in equilibrium with G. Hence any amount of G left behind the front (i.e. $G(-\infty)$) will satisfy the conditions at $-\infty$. For a situation wherein the inlet waters are undersaturated for G (i.e. $X_m F_m < Q$) there is only one front with a specific (and calculable) velocity and with $G(-\infty) = 0$.

1.2.4. *Unstable fronts and pattern formation*

The fronts of the P–G model of the above paragraph are found to be susceptible to a banding instability as follows. Since G is absent in the uninvaded rock, it must be nucleated. Thus consider the nucleation model reaction law

$$\frac{\partial G}{\partial t} = q(G + g)^{2/3}[FX - Q]. \tag{6}$$

The g factor models the nucleation of G; it vanishes for $FX < Q_n$. Q_n is the supersaturation product at which nucleation takes place, $Q_n \geq Q$.

With this we have the following possible scenario. Assume G is nucleated in a given region.

Then back (i.e. upstream) diffusion of F draws F to add to the growing G there. This then tends to repress nucleation of additional G downstream until a point sufficiently downstream where the consumption of the liberated F by upstream diffusion to the G growth is ineffective. Then a new band is nucleated. The band is expected to have a well defined width; its downstream termination due to the ever increasing $(G+g)^{2/3}$ factor in the upstream end of that band. All these conjectures are borne out in the simulation of fig. 2 from refs. 16 and 17. This banding model is essentially that of Ostwald in his explanation of Liesegang bands (see section 2.1 below). Bands of goethite in sandstone similar to those found here are commonly observed – see refs. 16 and 17. Two-dimensional patchwork patterns are also found in our simulations of this phenomenon. Such patterns are also found in nature – see ref. 18.

1.2.5. *Morphological instability of the planar front*

The planar redox interface is unstable in another way – via a coupling of the front with the driving percolation flow itself. Clearly, the passage of the redox front can lead to an increase in porosity since mineral grains (such as pyrite) are dissolved. If there is a fluctuation which increases the porosity, denoted φ (volume fraction of rock that is pore space) at some point on the redox interface, then flow through will be faster there. This increased flow-through will increase the rate of grain dissolution and hence the local φ disturbance amplifies to form a bump in the redox front. This is clearly a runaway effect that destabilizes the planar front, leading to the formation of front scalloping and fingering [19]. Scalloping is observed in some redox front mineral deposits (see ref. 19 for further details of the mathematical properties of this instability).

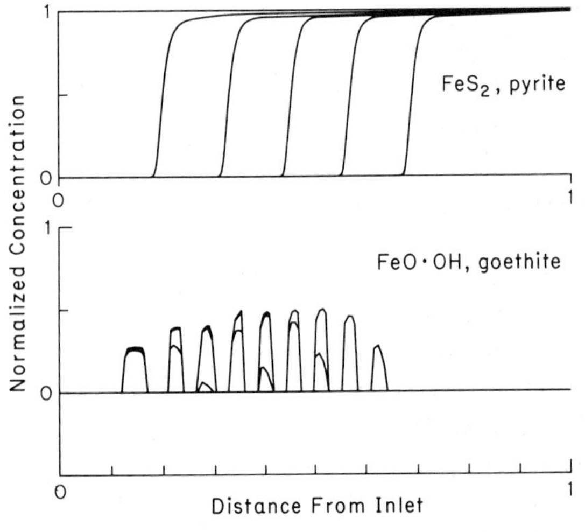

Fig. 2. Oscillatory mineral deposition pattern generated by Ostwald-type banding instability for the model of eq. (5). See refs. 16 and 17 for details.

1.3. *Autonomous waves – The* Ca^{2+} *stimulated* Ca^{2+} *release wave*

The propagating fronts of the previous sections had a velocity independent of the diffusion and reaction rate constants. Those waves were driven by the imposed fluid flow; their velocity is determined by the initial data (the original rock composition) and stoicheometry. Here we discuss an example of autonomous, reaction–diffusion waves that exist without a flow driving force. The speed of these autonomous waves depend directly on the presence of diffusion and the form of the rate law.

1.3.1. *The* Ca^{2+} *fertilization wave in Medaka egg*

A pulse of Ca^{2+} is observed to traverse the egg of the fish *Medaka* upon fertilization [20]. The wave emanates from the point of sperm entry and traverses the cell. The wave is not strongly effected by external Ca^{2+}. It involves an initial rise of calcium, released from a sequestered state in vesicles, and is then reabsorbed on a longer time scale. Here we review a model of the Ca^{2+} stimulated Ca^{2+} release that apparently drives the wave, emphasizing a simplified case of the more complete theory of ref. 20.

Through the mediation of the enzyme calmodulin a site S_m containing m sequestered

Ca^{2+} is induced to release the latter via ($C = Ca^{2+}$)

$$nC + S_m \to S_0 + (m+n)C. \tag{7}$$

The coefficient n arises because it is calmodulin in a state bound to n Ca^{2+} that is the activated enzyme for Ca^{2+} release. The process (7) is clearly a positive feedback for C release and appears to be the essence of the Ca^{2+} release wave. Assuming process (7) takes place with rate qC^nS_m we have for a C release front of velocity v in the wave fixed frame

$$DC'' + vC' + mqC^nS_m = 0, \tag{8}$$

$$vS_m' - qC^nS_m = 0. \tag{9}$$

Assuming that C is negligible ahead of the wave and S_m ahead has the value $S_m(+\infty)$ we have the boundary conditions $C(+\infty) = 0$, $C(-\infty) = mS_m(+\infty)$ and $S_m(-\infty) = 0$.

1.3.2. *Solving the* Ca^{2+} *wave equations*

These equations can be combined to obtain a constant of the steady wave motion,

$$[DC' + v(C + mS_m)]' = 0. \tag{10}$$

Using the boundary conditions far in advance of the arrival of the wave to evaluate the constant one obtains

$$DC' + v[C + mS_m - mS_m(+\infty)] = 0. \tag{11}$$

Thus (9) and (11) are a pair of coupled first order equations and are amenable to analysis in the C–S_m phase plane; With this, simple shooting techniques can be used to calculate the velocity [20].

This Ca^{2+} front is the leading edge of the Ca^{2+} fertilization pulse. To model the tail of the pulse, we must also include Ca^{2+} sequestering reactions [20].

1.3.3. *Comparison to flow driven and fully non-conserving waves*

The Ca^{2+} wave is dramatically different in mathematical structure from the constant velocity flow driven waves. In the Ca^{2+} front the velocity must be determined as a nonlinear eigenvalue problem (via (9) and (11) plus appropriate boundary conditions). The wave speed clearly depends on the diffusion coefficient and on the kinetics. In contrast the speed of the flow driven fronts does not depend on either diffusion or reaction rate laws.

The occurrence of a constant of the Ca^{2+} front motion is obvious physically (all the sequestered Ca^{2+} released turns up as free Ca^{2+}). It should be mentioned that this is not a typical occurrence in reaction–diffusion waves. For example, Ca^{2+} conservation breaks down by including the wake processes in the above Ca^{2+} model. For many examples of chemical wave motion without constants of the motion, see refs. 1–11.

2. Pattern formation

2.1. *Four classes of pattern self-organization*

Systems maintained out of equilibrium by contact with sources and sinks of matter and energy may spontaneously undergo transitions to and sustain patterned distributions of composition and temperature [6]. In contrast to this, systems undergoing a phase transition can form a pattern during the transient that becomes locked in at the late stage of the process. Here we briefly review some of these phenomena and contrast the general properties of these two routes to the self-organization of sustained patterns.

Leaving nonlinear pattern selection effects aside for the moment, we have been able to distinguish four types of pattern forming instabilities from the behavior of the linear stability analysis as follows [21]. We consider the stability of a uniform system to periodic patterned distributions of wave vector k. Four cases of the rate z of growth or decay of perturbations of wave vector k are shown in fig. 3. For intrinsic systems (case a) the perturbations of a given wave vector grow fastest and very short

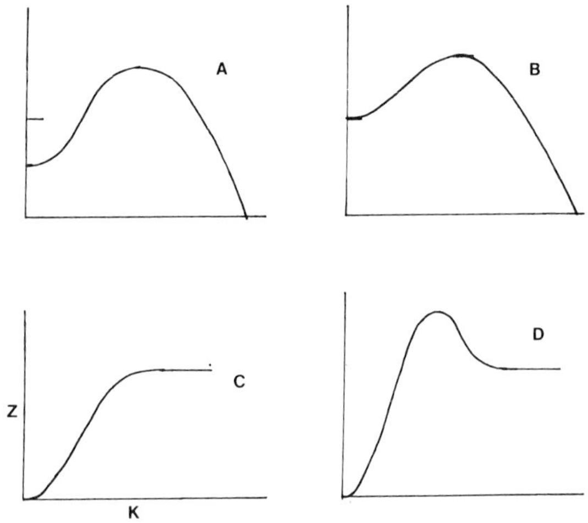

Fig. 3. Growth rates z of perturbations of wave vector k for various cases of pattern self-organization as discussed in section 2.1.

wavelength (large k) perturbations decay. The expected characteristic pattern length is determined by the reaction and transport coefficients – i.e. is intrinsic to the reaction-transport kinetics. A very different situation arises in case b. As such an extrinsic system becomes unstable, the first mode of instability is that of infinite wavelength and hence patterning sets in as the system parameter is changed further until the first mode of instability has a characteristic wavelength equal to the system size – i.e. pattern lengths are not inherent in the reaction transport network but are related to system size. As in the intrinsic case, large k perturbations decay ($z < 0$ for large k).

Another pathway to pattern self-organization is shown in the cases c and d. In the svegliabile system the most rapidly growing modes are those of vanishingly small wave length (i.e. large k). In this case patterns on a large length scale can be excited (and locked in by nonlinear interactions) by interaction of the system with boundaries, sources or overall gradients. The name *svegliabile* – Italian for "capable of being awoken" – was chosen to reflect this intimate dependence of the capacity to pattern on a macroscopic scale with some external influence. The final case is seen in frame d of fig. 3 and is termed cognito since the growth of microscopic ($k \to 0$) patterns is dominated by a fastest growing mode of well known ("cognito") length scale determined by the reaction-transport network. In the paragraphs to follow we give examples of all but the most well studied cases – the intrinsic systems.

2.2. *Liesegang and other precipitate patterns: CPG theory*

2.2.1. *Observations*

When coprecipitates interdiffuse, precipitation can occur in a variety of interesting banded and other structures. As we have seen in section 1.3. above, this can be understood in terms of the Ostwald supersaturation–nucleation–depletion cycle. However, a number of observations indicate that the Ostwald cycle cannot explain all such phenomena. Indeed banding can be induced in a pre-existing sol by infusing one of the coprecipitates into one end. Clearly nucleation is not related to this type of self-organization.

In an experiment suggested by the author and carried out in several laboratories, it was found that a homogeneous sol can break up into a mottled pattern of precipitate concentration [22]. In an apparently closely related phenomenon of "secondary banding" it is found that a given Liesegang precipitate band can itself evolve to a number of finer scale bands.

These precipitate phenomena were studied quantitatively via two types of experiments. In one experiment a tube containing auric ions (and in some cases a gold sol) was infused with oxalic acid. Oxalic acid reduces auric ions to elemental (atomic) gold that then leads to the growth of the gold sol. Light transmission through the tube along its length at various times shows that gold banding proceeds by the instability of a gold sol front (or coarsening wave for the case with the pre-existing sol) to the formation of bands. An example is shown in fig. 4. This indicates that banding is a post nucleation phenomenon in this auric ion–oxalic acid system.

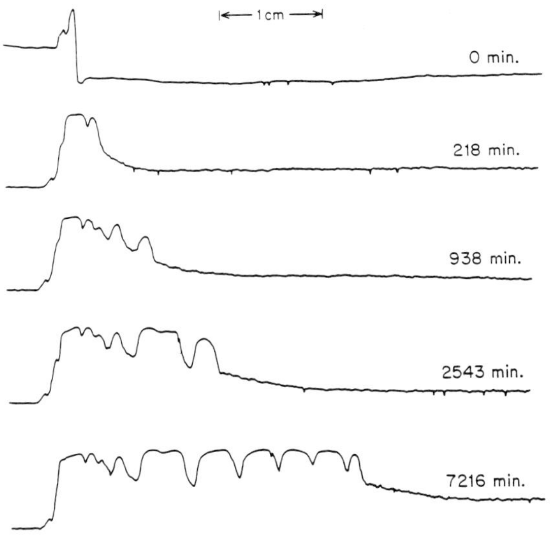

Fig. 4. Temporal evolution of gold bands as measured by light absorption – from ref. 27.

Two-dimensional experiments on PbI_2 sol aging showed that the uniform sol evolved to a mottled pattern with characteristic length scale that increased with decreasing initial PbI_2 content [22]. The temporal evolution of the correlation of transmitted light intensity in fig. 5 shows the temporal evolution of the precipitation patterns [23].

2.2.2. *CPG theory*

To explain these self organization phenomena a new mechanism was proposed. Since the precipitate content increases in some regions and decreases in others as the pattern develops, it is clear that the theory must be based on the kinetics of precipitate particle growth and dissolution.

The CPG (competitive particle growth) model of precipitate patterning is based on the observation that smaller precipitate particles require a higher solute concentration to survive than do larger ones because of surface tension. For the case of particle

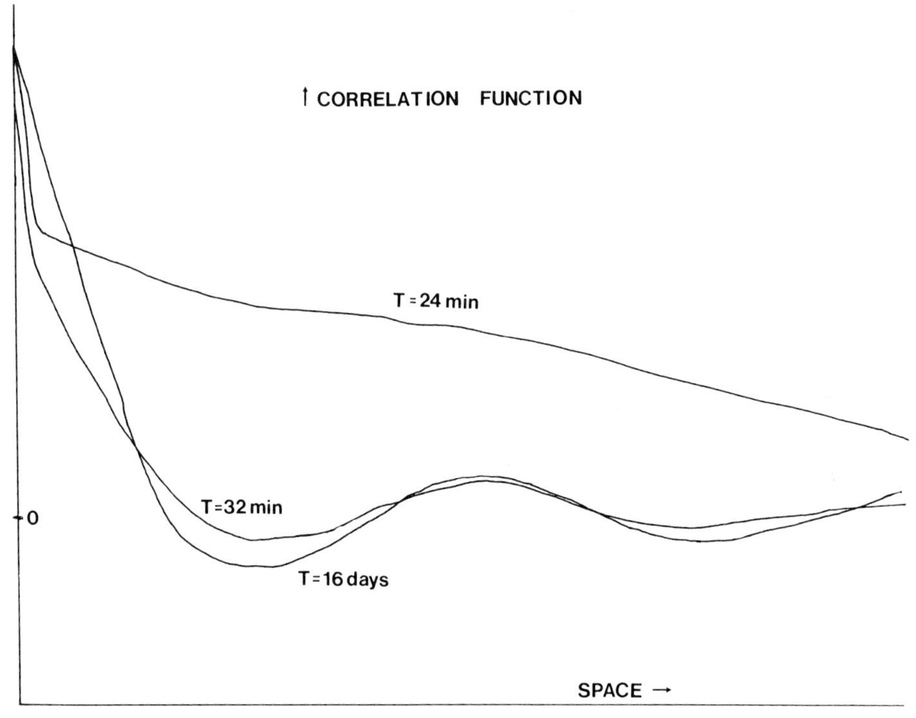

Fig. 5. Evolution of the correlation of transmitted light intensity from a two-dimensional precipitating system showing growth of correlation as discussed in section 2.2 – from ref. 28.

growth from a monoer of concentration c, the equilibrium concentration $c^{eq}(R)$ for spherical particles of radius R takes a form qualitatively like

$$c^{eq}(R) = c^{eq}(\infty)\left[1 + \frac{\Gamma R^2}{R^3 + R_c^3/2}\right], \tag{12}$$

where $c^{eq}(\infty)$ is the usual thermodynamic equilibrium concentration, Γ is a length related to the surface tension and R_c is the radius of the critical nucleus. For particles of radius R greater than R_c, $c^{eq}(R)$ is a monotonically decreasing function of R. Hence a super critical particle near a smaller one will grow at the expense of the latter as they both attempt to drive their environment to their respective equilibrium values of c. This dynamic leads to the coarsening of a sol as pointed out by Ostwald and modeled by Lifshitz and Slyozov.

However, it is found that the uniformly aging sol is unstable to large length scale inhomogenities – the CPG mechanism of precipitate self-organization – as follows.

If the average particle radius in one region is greater than elsewhere, then we expect that these larger particles grow at the expense of the particles in their immediate neighborhood. In two spatial dimensions this yields a maximum in average particle radius R surrounded by a ring of dissolution – i.e. smaller R. But in the ring further out, R is larger than the region of dissolution and hence a ring of particle growth is expected to arise next. This process should continue, creating a concentric ring pattern. Indeed these halo patterns are observed in the PbI_2 experiments of ref. 22.

This collective CPG dynamics was formulated into a simple mathematical model for the local average particle radius $R(\boldsymbol{r}, t)$ and the monomer concentration $c(\boldsymbol{r}, t)$ from which the particles are taken to grow. The particle growth law

$$\frac{\partial R}{\partial t} = k[c - c^{eq}(R)] \tag{13}$$

and continuity equation

$$\frac{\partial c}{\partial t} = D\nabla^2 c - 4\pi n \rho R^2 \frac{\partial R}{\partial t}, \tag{14}$$

were assumed to hold. Here k (dependent on R for diffusion limited growth situations) is a rate coefficient, D is the monomer diffusion coefficient while n and ρ are the number and solid molar density of the particles.

2.2.3. *Simulations of the CPG equations*

The deceivingly simple CPG model yield quite interesting behavior. Numerical simulations were carried out in one and two spatial dimensions using finite difference techniques. Fig. 6 shows a case in one spatial dimension wherein a single initial maximum in $R(x, 0)$ (of small amplitude, below the resolution of the figure) leads to the sequential development of satellite maxima and minima in R – i.e. to the induction of bands. In two spatial dimensions this leads to concentric rings, a quarter section of which is shown in the iso-R contour maps of the simulations of fig. 7.

The two-dimensional phenomena get very interesting indeed. A speckled pattern is shown in fig. 8. It was initiated by the interaction of a linear array of alternating maxima and minima in $R(x, y, 0)$. It was somewhat surprising that this simulation didn't simply lead to banding via a

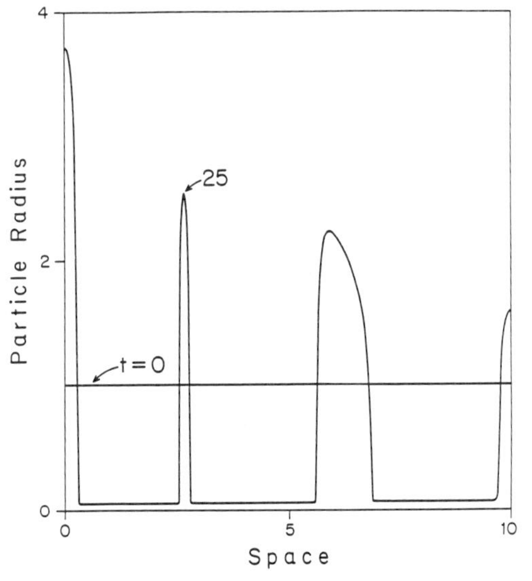

Fig. 6. Development of bands according to the CPG model of eqs. (13) and (14) showing dimensionless particle radius as a function of dimensionless space at indicated times.

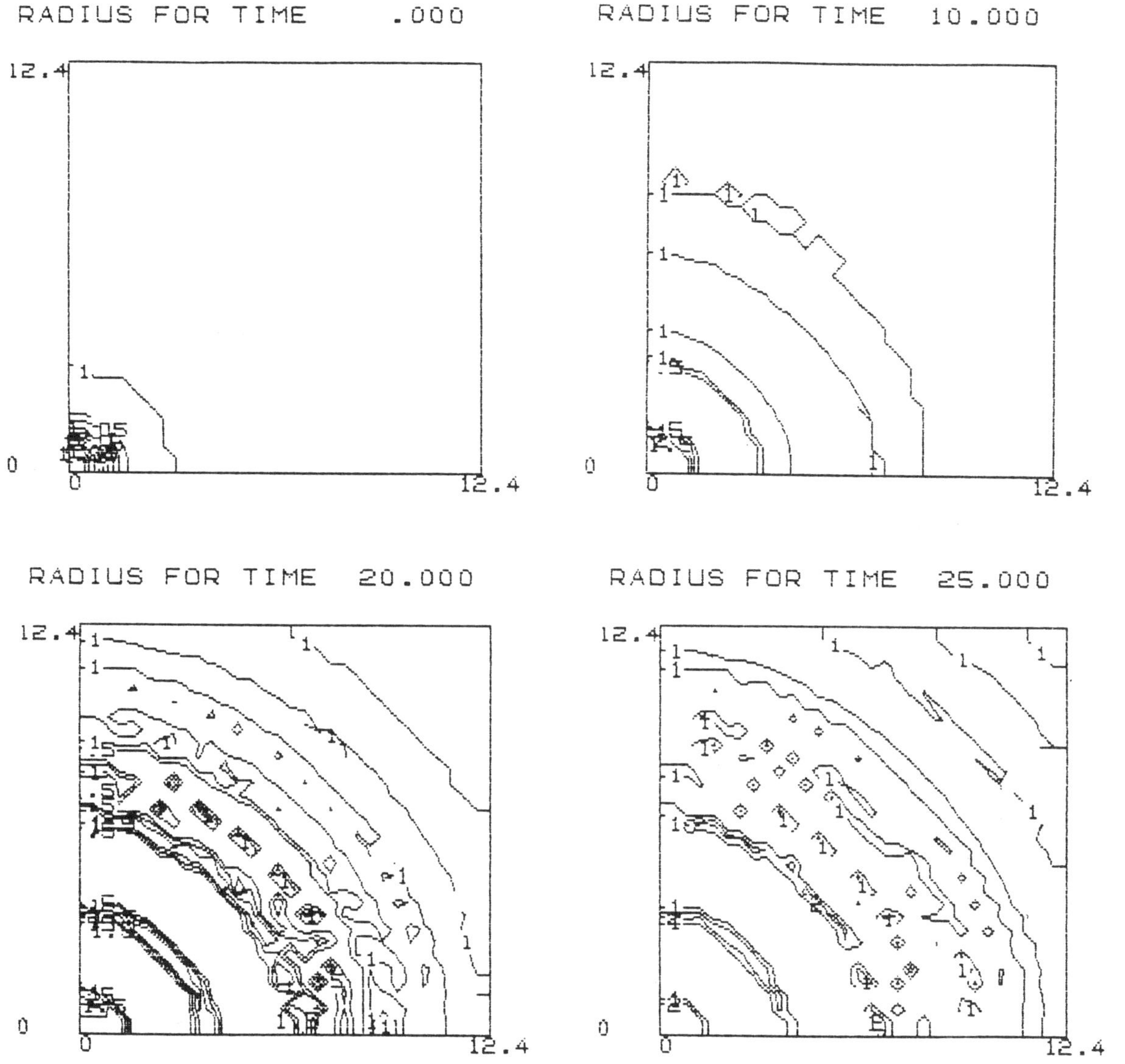

Fig. 7. CPG in two spatial dimensions showing the onset of a halo pattern as obtained in ref. 29 where eqs. (13) and (14) were solved numerically. Shown is the temporal sequence of the contours of dimensionless particle radius. Note a second instability leads to the breakup of bands laterally.

positive reinforcement of the initial R maxima (spots). This in fact does happen for different wavelengths of the initial line of induced R maxima and minima. Even more peculiar is the fact that a small spiral-like segment (less than one turn long) can lead to the unwinding as shown in fig. 9.

2.2.4. *Particle size distribution effects*

All particles in a homogeneous sol are not usually the same size. Thus a more complete CPG theory must account for a distribution of particle radii. Simulations of this type were carried out using equations like those in (13) and (14) except for a set of particle classes of radii $R_1, R_2 \ldots$.

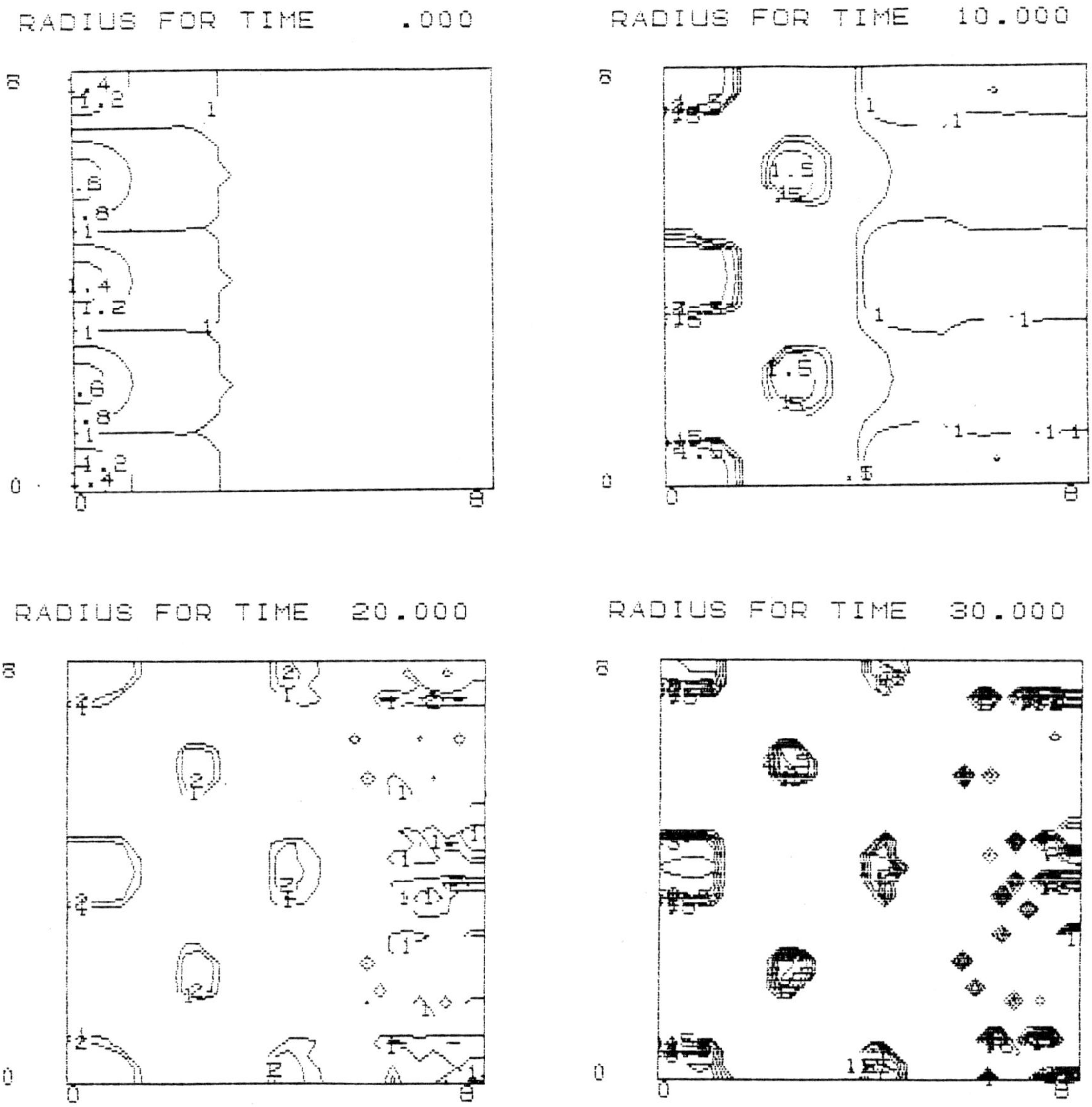

Fig. 8. As in fig. 7 except that an initial adjacent maximum and minimum induce a speckled and not a banded pattern as one might expect.

Statistics were accumulated and the local particle size distribution of Lifshitz and Slyozov was followed numerically (see ref. 24 for further details).

These simulations—see fig. 10—show that the satellite induction phenomenon could still occur when the sol was not monodispersed but that the number and amplitude of satellite bands decreased with increasing width of the initial particle size distribution. Since the satellite induction length depends on the particle size of the sol we expect this "wash-out" effect because the spectrum of expected wavelengths would tend to average out. However, for narrow but not monodispersed initial size distribution, a very interesting phenomenon can take place. In the region between a central R maximum imposed on the left and the first satellite we see in fig. 9 that particles in the tail of the initial size distribution survive and are left as "greedy giants" that "fed" off their immediate

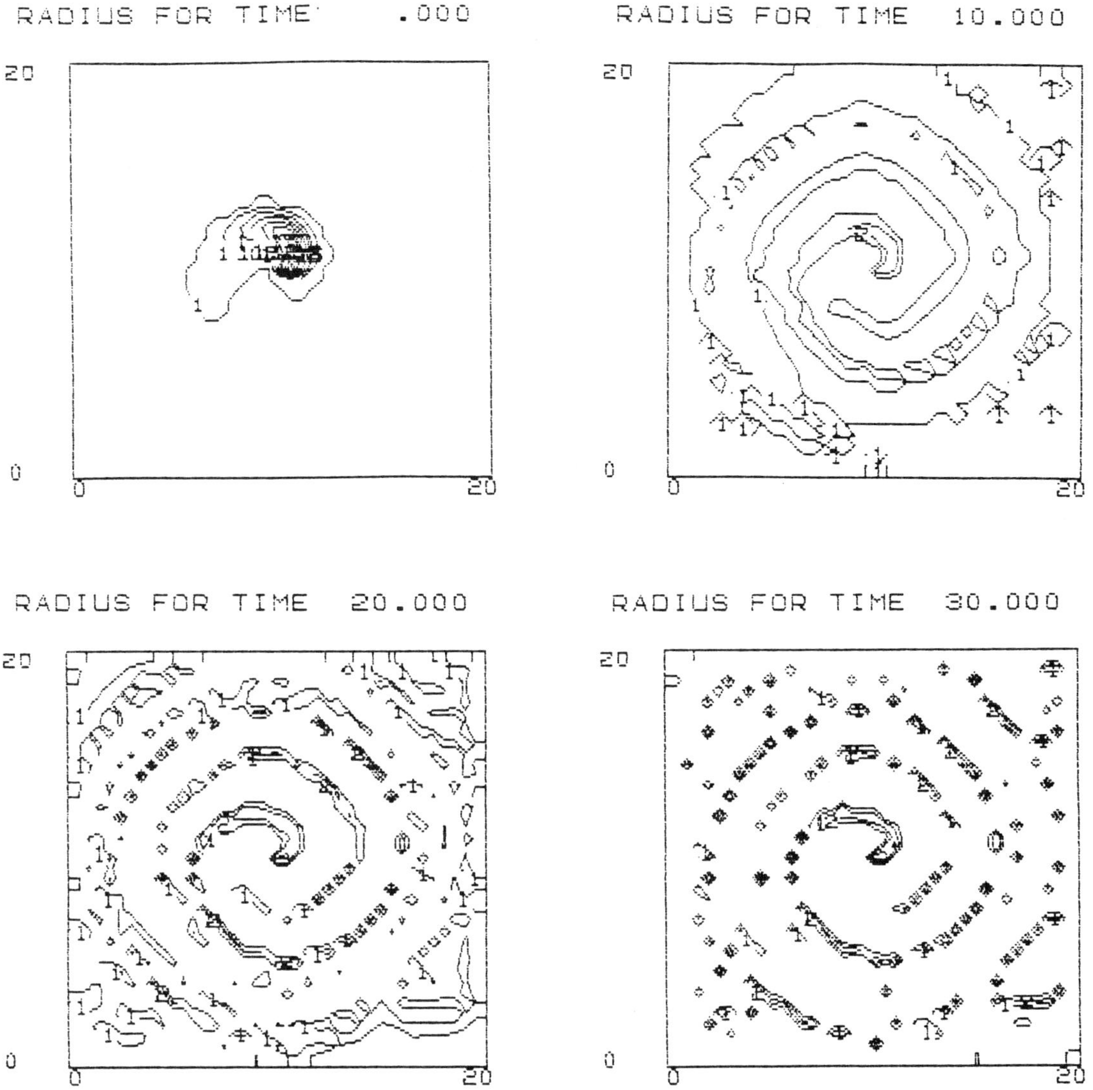

Fig. 9. Same as in fig. 7 except that a very small hook-like initial perturbation yields a spiral wave of precipitate concentration.

neighbors that were dissolving because of the initial R maximum to the left. These greedy giants were indeed observed in our PbI_2 experiments.

2.3. *Polycrystalline geo-patterns*

A rock is a polycrystalline medium. The distribution of mineral types or of pore space in these media can take on a number of fascinating spatial patterns. The evidence is growing in recent years that many of these repetitive patterns result not from imposed periodicity (such as the seasonal variations of sedimentary layering) but are the result of self-organization processes. Since these phenomena have been reviewed recently by Merino (1983) [25] and the author [21], we only briefly mention two of them here.

2.3.1. *Metamorphic layering*

It is found that under metamorphic conditions (pressures and temperatures intermediate between those at the surface and the molten state) rocks

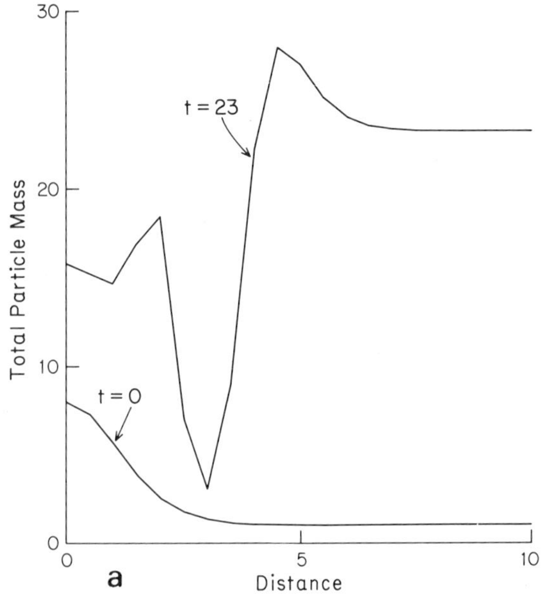

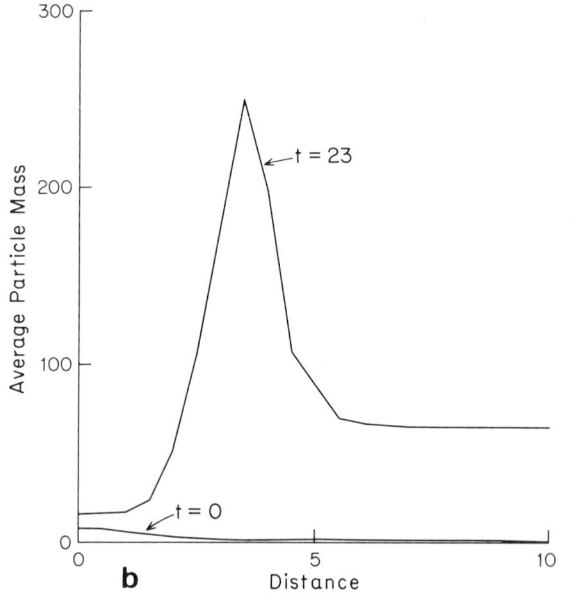

Fig. 10. Greedy giant phenomenon arising in band induction when the initial sol is not monodispersed as predicted by numerical solutions of the temporal evolution of the local particle size distribution carried out in ref. 24. Here an inversion effect produces larger particles on the right boundary than on the left, but it also produces another curious phenomenon. In the minimum, "wipe-out" eliminates all but a very few survivors (fig. 10a) and it is these survivors which grow to be the largest (fig. 10b) in the entire system. A particle-size profile shows the particles to be larger but much fewer in the minimum than in the maximum immediately to the right. This we refer to as the "greedy giants" phenomena and has been observed in our PbI_2 precipitation patterning experiments [8].

can become segregated compositionally into bands under the combined effects of stress, dissolution/crystal growth and transport of mineral dissolution products along grain boundaries.

A model of this self-organization phenomenon has been formulated mathematically as follows. A crystal in a rock under overall stress experiences a stress that depends on the number, size and elastic or plastic properties of the neighboring crystals. Since the solubility of a crystal depends on stress, it has been conjectured that the solubility (and hence rate of crystal growth) depends on the local texture (number density, size and other properties of crystals in the neighborhood). Thus the crystal growth dissolution laws depend on the concentrations of intergranular solutes and the local texture. The theory is mathematically similiar to the CPG theory with texture variables playing roles similiar to particle radius and dissolution product concentrations being analogous to monomer concentrations [26].

The theory predicts that the uniformally textured rock can become unstable to banding or other patterns in two distinct ways. In one case the system bands regularly only if it is subject to an overall gradient or interactive (but not oscillatory) boundary condition and another case where there is a well defined pattern length embedded in the reaction transport network as in cases c and d of fig. 3.

2.3.2. *Stylolites*

If a porous rock is subject to an overall stress, then grains in a region of higher porosity will be subjected to higher stress and hence be more soluble. This pressure solution effect leads to an instability since grains in regions of higher porosity will tend to dissolve, increasing the porosity still further, while grains in regions of lower porosity will be under lower stress and hence will grow (on the dissolution products of the grains in the neighboring higher porosity region) decreasing the porosity.

This feedback mechanism has been shown to lead to the instability of the state of uniform porosity and the formation of arrays of dissolution seams (stylolites) separated by regions of cementation [27]. This phenomenon has been modeled mathematically via equations of crystal growth and dissolution product transport. Again the mathematical structure of the theory is similiar to the CPG theory with porosity and dissolution product concentration playing the roles of particle radius and monomer concentration, respectively.

2.4. *Self-organization in the Fucus egg*

2.4.1. *The symmetry-breaking developmental phenomenon*

Fucus is a brown seaweed that lives in the intertidal zone. Its early development seems to be an excellent example of self-organization. We focus here on the first symmetry breaking event in the development of the egg – a spontaneous transition from a spherically symmetric to a dipolar configuration. A model of this event and its implications for normal and defect development are analyzed via bifurcation theory and presented here (see ref. 28 for details).

The one cell *Fucus* egg is shown schematically in fig. 11. There we see the normal developmental scenerio from the spherically symmetric state of the electrophysiology to the dipolar state and finally to the shape change wherein the rhizoid (proto-root-like) protrusion is formed. The subject of our interest here is the first symmetry breaking that precedes the shape change.

2.4.2. *The model*

The whole cell length scale character of the electrical and chemical symmetry breaking in early *Fucus* development suggests that the development can be understood as a pattern self organization involving the reaction transport network. The

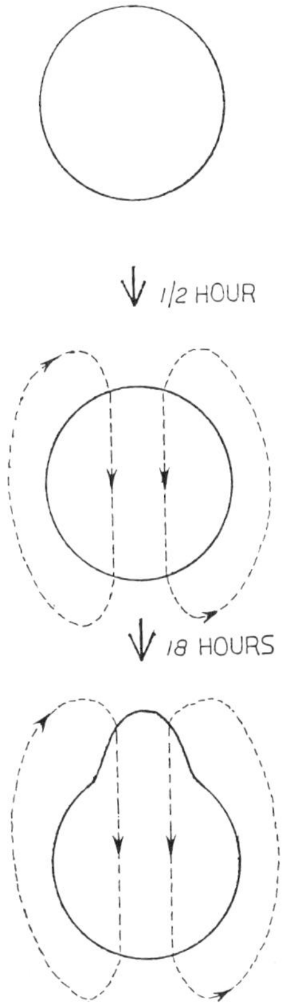

Fig. 11. Schematic development scenario for the onset of dipolar patterns in the single cell stage of the egg of *Fucus*. The egg retains its spherical shape while ionic current loops spontaneously form. At a later stage a bulge appears and then evolves into the rhizoid. The first transition is focused on in the theory presented here, modeling it as an electrophysiological symmetry breaking instability.

nongenetic nature of the transition and the possibility of orienting the developmental axis by external gradients (chemical, electrical or illumination) also suggest such a reaction-transport mechanism.

In a series of papers a set of reaction-transport equations were set forth and analyzed for pattern forming instability. The following processes: (1) electro-diffusion of mobile ions in the cell interior and environment; (2) permeation and pumping across the cell outer membrane; (3) the motion of the sites of cell membrane transport along the membrane; and (4) binding sites for ions (notably Ca^{2+}) in the cell interior.

Putting these factors together and using multiple time scale analysis it was shown that *Fucus* could be described by an equation of the form

$$\frac{\partial \Gamma}{\partial t} = \nabla \cdot \Gamma L \nabla \mu(\Gamma) \tag{15}$$

(see ref. 28 for limitations and the detailed arguments). Here Γ is the surface density (moles/area) of the Ca^{2+} leaks, L is a transport coefficient, ∇ is the gradient operator along the sphere and μ is the electro-chemical potential of Γ; $\mu(\Gamma)$ is found to take the form

$$\mu(\Gamma) = \mu^*(\Gamma) + \kappa\Gamma, \tag{16}$$

where κ is a linear operator that transforms the Γ-distribution into a voltage distribution (times the charge of the Γ species), and μ^* is the chemical part of μ (i.e. independent of the voltage). The explicit form of the operator κ is derived in ref. 27. There the values of constants appearing in κ are obtained from ion-transport measurements. Since Γ is not created or destroyed, we have

$$\int d^2\Omega\Gamma = A, \tag{17}$$

where $\int d^2\Omega$ indicates an integral over the spherical cell surface and A is a constant (the total number of moles of Γ on the cell surface).

2.4.3. *Cell development as a bifurcation phenomenon*

The development of the *Fucus* cell into asymmetric states was studied via a bifurcation analysis of the steady state solution of (15). Because there are no sources or sinks of Γ this yields

$$\mu^*(\Gamma) + \kappa\Gamma = \mu_0. \tag{18}$$

The constant μ_0 is determined by solving (18) consistent with the conservation condition (17).

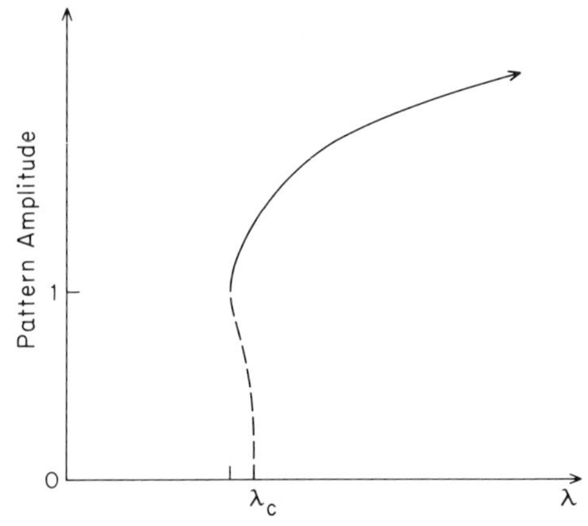

Fig. 12. Steady state solutions satisfying (17), (18) are found to be an example of an inverted bifurcation. Shown is the amplitude of the dipolar deviation from the spherically symmetric state as a system parameter λ passes through its critical value λ_{c1}. The s and u notation indicates stable and unstable branches.

The bifurcation analysis was carried out by focusing on a system parameter λ contained in κ. The theory was based on the assumption that the order of magnitude of the key nonlinear terms in the deviations from the spherically symmetric solution for Γ could be just balanced by the linear destabilizing terms. Results on inverted, imperfect

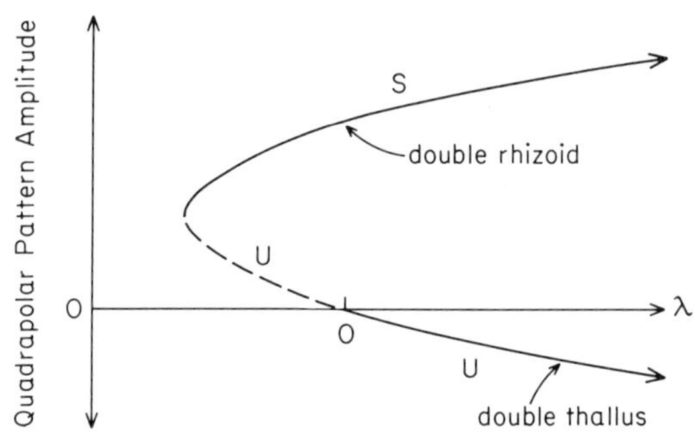

Fig. 13. Same as in fig. 12 except for quadrapolar states. The maxima and minima in Γ coordinate the location of rhizoid and thallus (proto-leaf-stem system). The double rhizoid defect state is observed in nature and is shown to be stable in the present theory. It corresponds to an accumulation of Γ at the "north" and "south" poles of the sphere. Another quadrapolar state is found and corresponds to the accumulation of Γ at the "equator." The theory shows that this state is unstable as indicated in fig. 14 and, indeed, this state is not observed in nature.

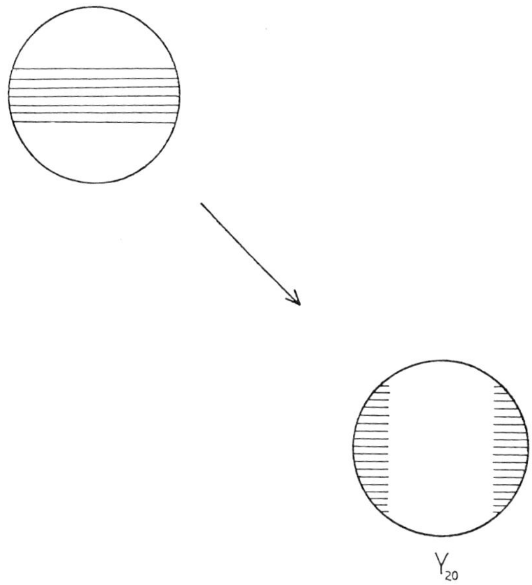

Fig. 14. The instability of the equatorial quadrapolar defect state is via a tendency for the equitorial accumulation to break up in an "east–west" concentration - bringing the cell to a rotated opposite pole-type accumulation that is the stable (and observed) state.

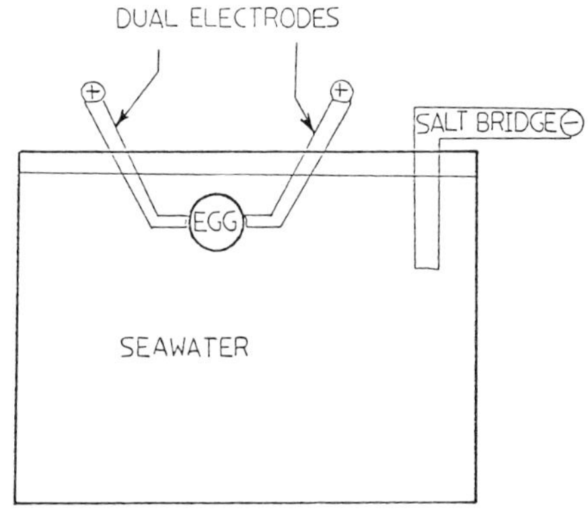

Fig. 16. The imposition of a quadrapolar field on the egg can be carried out as shown. This suggests the interesting possibility of inducing the double thallus defect state not observed in nature by giving that branch a range of stability as in fig. 17.

and secondary bifurcations were obtained and interpreted biologically. The findings are summarized in figs. 12–18 and their captions. There it is shown how the model can explain normal and

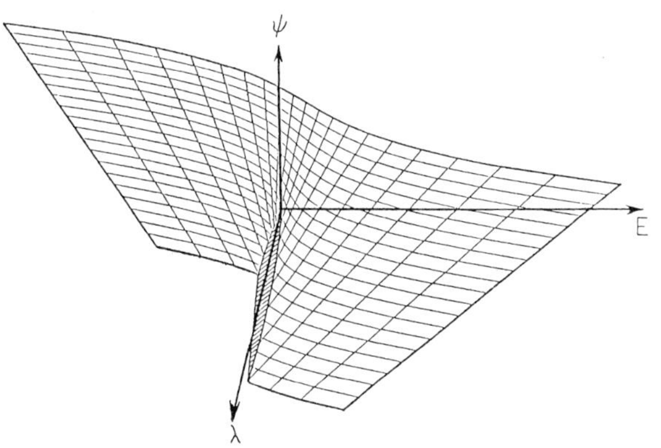

Fig. 15. Phase diagram showing the allowed (i.e. stable) branches of the dipolar state in the presence of an electric field of strength E for values of a system parameter λ. The theory shows that as observed experimentally, an applied electric field of dipolar geometry can orient the development of the egg. Here E is the field strength and ψ is the dipolar amplitude.

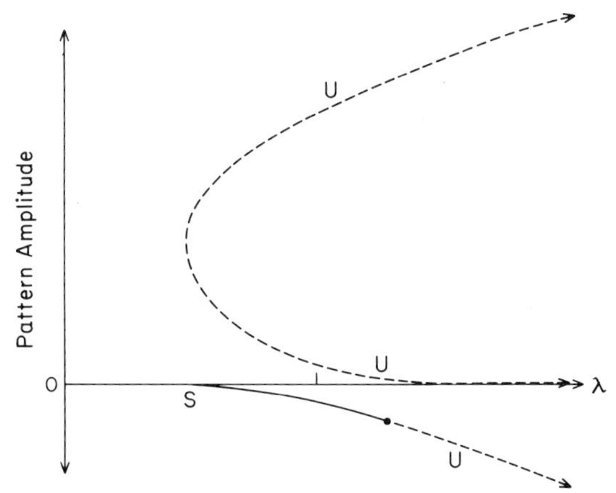

Fig. 17. The stabilization of the double thallus state as predicted by bifurcation theory on eqs. (17), (18) augmented by an addition term in (18) to account for the applied field induced by an experiment as in fig. 16. (See also fig. 18.)

defect observed modes of development and rule out certain defect modalities that could have been expected from symmetry considerations alone. The ability of externally applied electric fields to align the axis of differentiation and stabilize defect development modalities not observed in nature are also discussed in these figures.

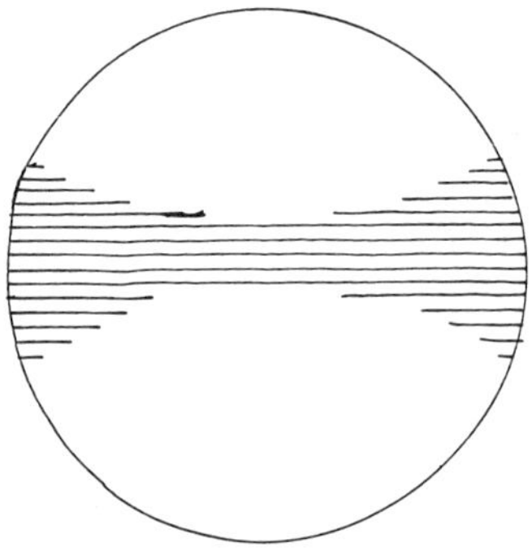

Fig. 18. The nature of the equatorial accumulation double thallus state stabilized by the applied quadrapolar field. This is a mixed mode state – i.e. it has a quadrapolar-type equitorial concentration of Γ but with an azymuthal asymmetry reminescent of a state of opposite pole accumulation of Γ.

References

[1] P. Ortoleva and J. Ross, Phase Waves in Oscillatory Chemical Reactions, J. Chem Phys. 58 (1973) 5673.

[2] D. Aronson and H.F. Weinberger, Lecture Notes in Math, 446 (Springer, Berlin, 1975).

[3] J. Tyson, The Zhabotinsky Reaction, in: Lecture Notes in Biomathematics (Springer, Berlin, 1976).

[4] P. Fife, J. Chem. Phys. 64 (1976) 554.

[5] P. Ortoleva and S. Schmidt, Variety and Structure of Chemical Waves, R. Field and M. Burger, eds. (Wiley, New York, 1983).

[6] G. Nicolis and I. Prigogine, Self Organization in Nonequilibrium Systems (Wiley, New York, 1977).

[7] P. Ortoleva and J. Ross, Theory of Propagation of Discontinuities in Kinetic Systems with Multiple Time Scales: Fronts, Front Multiplicity, and Pulses, J. Chem. Phys. 63 (1975) 3398.

[8] D. Feinn and P. Ortoleva, Catastrophe and Propagation, J. Chem. Phys. 67 (1977) 2119.

[9] S. Schmidt and P. Ortoleva, Asymptotic Solutions of

[10] R. Feeney and P. Ortoleva, Annihilation and Crescent Waves in BZ Wave Media in an Electric Field, in preparation.

[11] P. Ortoleva, Selected Topics from the Theory of Nonlinear Physico-Chemical Phenomena, vol. 4, Eyring, ed. (Academic Press, New York, 1978).

[12] R. Feeney and P. Ortoleva, Experimental Verification of the Competitive Particle Growth Theory of Liesegang Phenomena, in preparation.

[13] P. Ortoleva and J. Ross, On a Variety of Wave Phenomena in Chemical Reactions, J. Chem. Phys. 60 (1974) 5090.

[14] P. Ortoleva, Sh. Bose and Su. Bose, Dynamic Padé Approximants for Chemical Center Waves, J. Chem. Phys. 72 (1980) 4258.

[15] G. Auchmuty, J. Chadam, E. Merino, P. Ortoleva and E. Ripley, The Stability and Structure of Moving Redox Fronts, SIAM (submitted for publication).

[16] E. Merino, C. Moore, E. Ripley, G. Auchmuty, J. Chadam, J. Hettmer and P. Ortoleva, Kinetic Modeling of Redox Roll Fronts and their Instabilities, Geochimica et Cosmochimica Acta, submitted for publication.

[17] P. Ortoleva, G. Auchmuty, J. Chadam, J. Hettmer, E. Merino, C. Moore and E. Ripley, Redox Front Propagation and Banding Modalities, Physica D, submitted for publication.

[18] P. Strickholm, C. Moore and P. Ortoleva, Self Organization of Two Dimensional Redox Front Deposition Patterns, in preparation.

[19] J. Chadam, J. Goncalves, S. Hagstrom, J. Hettmer, D. Hoff, J. Johnson, R. Larter, E. Merino, P. Ortoleva and A. Sen, Porosity Instability at Reactive Percolation and Reactive Percolation Instabilities in Roll Front Propagation, in preparation.

[20] S. Schmidt, P. Ortoleva and H. Mahler, The Ca^{++} Fertilization Wave in Medaka: A Colmodular Model, in preparation.

[21] P. Ortoleva, Modeling Nonlinear Wave Propagation and Pattern Formation at Geological First Order Phase Transitions, (an invited review for the Proceedings of the NATO Conference on "Instabilities and Fluctuations in Chemical Engineering and Geological Systems," Austin, Texas, March, 1983).

[22] D. Feinn, P. Ortoleva, W. Scalf, S. Schmidt and M. Wolff, Spontaneous Pattern Formation in Precipitating Systems, J. Chem. Phys. 69 (1978) 27.

[23] J. Lechuna, J. Johnson, R. Spalader and P. Ortoleva, Development of Macroscopic Correlation in Precipitating Systems, in preparation.

[24] P. Strickholm and P. Ortoleva, Particle Size Distribution Effects in the CPG Model of Liesegang Banding, J. Chem. Phys., submitted for publication.

[25] E. Merino, Survey of Geochemical Self-patterning Phenomena, G. Nicolis and J.S. Turner, eds., Proceedings of the NATO Conference on Instabilities in Chemistry, Geology, and Materials Science; Austin, Texas, 1983.

[26] P. Ortoleva, P. Strickholm and E. Merino, The Self Organization of Metamorphic Layering, Geophysical Research, submitted for publication.

[27] P. Ortoleva, E. Merino and P. Strickholm, Stylolites in Sedimentary Rocks: A Pressure Solution Kinetic Theory and their Generation and Regular Spacing, published abstract.

[28] P. Ortoleva, El-Badewi and A. Strickholm, Chemical Modification of Early Fucus Egg Self-Organization, in preparation.

[29] P. Strickholm and P. Ortoleva, "Speckle and Spiral Precipitate Patterns: CPG Theory in Two Dimensions", Physica D (submitted for publication).

Physica 12D (1984) 321–332
North-Holland, Amsterdam

WAVEFRONT GEOMETRY IN EXCITABLE MEDIA

A.T. WINFREE
Department of Biological Sciences, Purdue University, West Lafayette, Indiana 47907, USA

In a certain kind of continuous medium, an isolated fragment of wavefront behaves peculiarly: its edge (a closed ring) remains stationary while the fragment's interior continues its forward propagation, curling into a scroll-like shape along the edge: a scroll ring. The edge (the axis of the scroll ring) is a 1-dimensional singularity in the resulting periodic wave field. From numerical simulations in 2-dimensional continua, and from laboratory experiments in 3-dimensional continua, this seems to be typical behavior for so-called "excitable" media. The implications include a bestiary of "organizing centers", each composed of a topologically distinct arrangement of linked and knotted scroll rings. Two examples are given. Exact description of these organizing centers awaits formulation of a tractable equation that embodies the essential features of an excitable medium.

1. Solitary waves in excitable media

In a so-called "excitable" medium, a disturbance of the local kinetic steady-state can propagate as a solitary pulse. The pulse typically approaches an asymptotically-standard shape and speed as it radiates away from the triggering perturbation (for examples see ref. 1, p. 234). Each volume element, as it is triggered by the encroaching front, contributes energy to the passing wave. It is usual for the wave's onset to be quite abrupt, while the local excitation dies away more gradually after the front passes. Because triggering of a volume element causes complete saturation of the local response, colliding pulses fail to pass through one another: both are annihilated.

So long as the wave has standard amplitude, its movement into a uniform isotropic medium can be reckoned using Huyghens' construction: each area element advances along the local normal at a standard rate. This simple picture is spoiled by the existence of points where there *is no* local normal, e.g. where the wave's amplitude (suitably defined for non-standard shapes) vanishes. Such singularities occur in a variety of intriguing linear wave fields as points (in two dimensions) or filaments (in three dimensions) [2–4]. In excitable (non-linear) media such loci play a distinctive role as organizing centers for continuing periodic wave patterns [5–9].

The purpose of this symposium contribution is simply to pose a riddle that begs for solution in the general case of pulse-propagating media, and to show some computer graphics purporting to sketch two distinct solutions for the special case of an excitable medium. These pictures are snapshots from a computer-animation videotape shown at the symposium.

2. Fragmented waves: what becomes of the pieces?

What becomes of the *edges* of a wave? Can a wave *have* edges? This riddle is posed in two parts as follows. Imagine any 3-dimensional medium capable of supporting solitary waves and periodic wave trains: the electro-magnetic continuum, for example, or any elastic material, or an excitable medium such as heart muscle or the Belousov–Zhabotinsky chemical reagent [1, 5–9]. Imagine one volume containing a solitary plane wave and another which remains for the moment uniformly quiescent. From the one, cut out a finite disk of wavefront and a surrounding cylindrical volume as suggested in fig. 1. Graft it into a cylindrical hole in the quiescent volume. If this graft is implanted

0167-2789/84/$03.00© Elsevier Science Publishers B.V.
(North-Holland Physics Publishing Division)

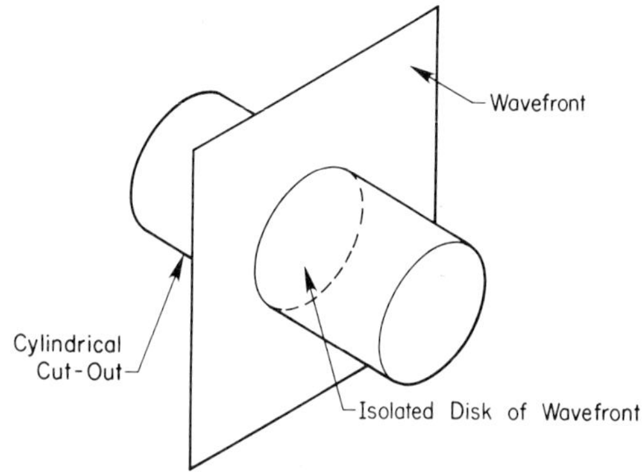

Fig. 1. A disk of wavefront is isolated for transfer into quiescent medium along with a cylindrical volume fore and aft.

instantaneously with geometrical precision, we create discontinuous initial conditions in the otherwise quiescent volume (fig. 2).

The first challenge is to envision a continuous approximation to this procedure. This is not too difficult: all aspects of the disturbance constituting the pulse are simply attenuated toward the quiescent state within a short distance from the graft's boundary [6].

The second challenge is to describe the subsequent evolution of wavefronts. Presumably, the initial disk of wavefront must exceed some minimum thickness if it is to retain its polarity: a mere film of wave *front* might only trigger excitation which would propagate away as a spherical shell in all directions. The fragment must also exceed some minimum area lest it promptly vanish, absorbed by the proportionally huge amount of quiescent medium along its perimeter. But is there a *critical* size above which some new wave pattern persists indefinitely? The answer is not known, but there are several hints, from computation of the two-dimensional analogue and from laboratory experiments.

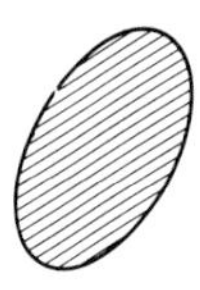

Fig. 2. The disk of wavefront and surrounding cylindrical volume of perturbed medium is grafted into quiescent medium. The middle of the disk will at first continue normal propagation, but what happens at the edges?

3.1. *Two-dimensional computations*

The computational experiments create a line-segment of wavefront (corresponding to a diameter of the disk above) in the planar concentration-field of a piece-wise linear caricature of a chemical excitable medium [10]. Concentration discontinuities are immediately smoothed by molecular diffusion. The middle of the wave segment continues to propagate forward, unaffected by peculiar conditions near its remote endpoints. Near either endpoint, excitation can spread not only forward, but also laterally into quiescent medium to the side. This lateral component at the wave's endpoint confers a persistent rotary motion. The center of rotation does not move significantly, so it becomes the center and source of an involute spiral wave (fig. 3). It can be shown that this shape satisfies the Huyghens' construction outside a tiny disk near the center [11]. Within that disk, concentration patterns are smooth. Diverse equivalent procedures with somewhat different initial conditions also result in spiral sources [1, 10].

3.2. *Two-dimensional experiments*

Manipulations that create wave fragments in Belousov–Zhabotinsky reagent do produce isolated stable sources of spiral waves near the origi-

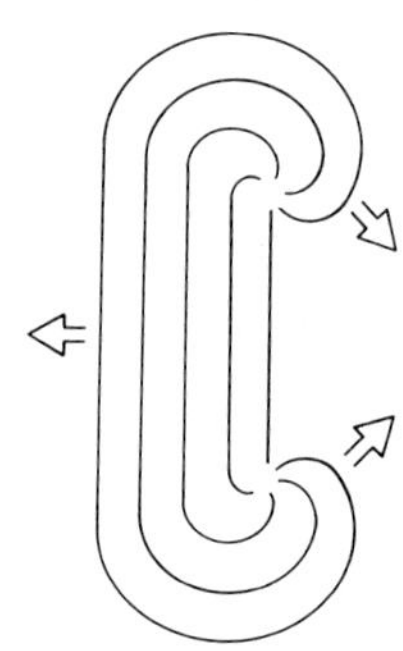

Fig. 3. A line-segment of wavefront grafted into quiescent medium continues normal propagation in its middle, but pivots around fixed points near the two edges.

nal endpoints (fig. 4). They prove unstable only if too close together or too close to an impermeable boundary of the medium. In other media it is conceivable that the wavefront would additionally grow or retract at its endpoint, so the "center of rotation" might not always be a fixed pivot. Can pulse-conducting media be classified by the behavior of these endpoints (their retraction, extension, or pivotal fixity)? Is there a finite range of parameters which result in rigid rotation about a fixed pivot? There may be: ostensibly rigid numerical or analytic rotors have been obtained in diverse systems [12–23]. Others have obtained "wobbly" rotation [10, 11, 18, 24–34]. In the following, we ignore the still-mysterious irregularity of some rotors, and describe only the idealized case of rigid rotation.

Fig. 4. In a chemical excitable reagent, a segment of wavefront pivots around its endpoints, forming mirror-image spiral waves. The chemical structure formed near each pivot is called a "rotor". The rotors are the sources of the two spiral waves. The frames are successive snapshots one minute apart in the order of reading English text. Photographed March 1974 in the author's laboratory by Fritz Goro for Scientific American magazine.

3.3. *Heuristic analysis of the 3-dimensional case*

The disk is a 3-dimensional analogue of the line-segment in two dimensions: rotate the line segment about its perpendicular bisector in the plane. It thus seems physically plausible that a surface of revolution constructed from the planar solution would satisfy the 3-dimensional problem. Such a solution might be assembled by moving fig. 3 through 3-space in any way that avoids self-intersections: 3-space is now seen as a stack of adjacent planes, each containing the 2-dimensional stable spiral solution. For example spin fig. 3 about its symmetry axis to make a scroll-shaped wave whose axis closes in a circle: it resembles a vortex ring. This is what might be expected to evolve from the initial conditions of fig. 2. Such "scroll rings" have been observed experimentally in Belousov–Zhabotinsky reagent [5, 35, 36] and computationally [37]. They seem stable on the time-scale of several rotations of the spiral cross-section, if not too small or too close to a bounding interface, just as counter-rotating spirals are stable in "2-dimensional" thin films if not too close together or too close to a bounding interface. (However, on a longer time scale (tens or hundreds of rotations), the ring contracts until it vanishes [36, 37].)

4. Computer graphics: two kinds of scroll ring wavefront

4.1. *The un-twisted scroll ring*

Fig. 5a shows a scroll ring constructed by rotating the planar spiral as suggested above. It is drawn in perspective with foreground parts painted brighter. Fig. 5b is a cut-away version from a different perspective. Fig. 5c is a view from the top (without letting background parts fade away).

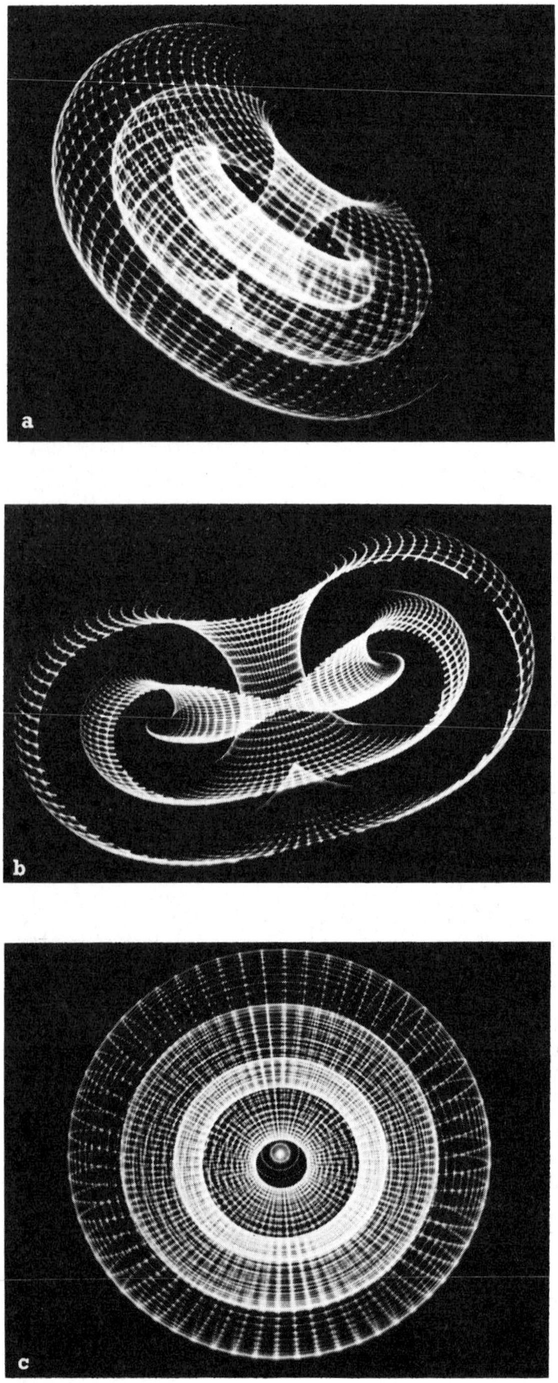

Fig. 5. a) Computer-constructed three-dimensional scroll wave, its axis closed in a ring to make a "scroll ring"; b) the same, cut away to reveal inner coils of the spirals; c) the same, viewed along the axle in projection.

This caricature is idealized in the following respects. Every cross-section in a half-plane bounded by the symmetry axis is an involute spiral. This curve represents only the *front* of a wave, part of a contour-line of uniform chemical concentration (or other state variable.) It is only a *part* of the contour line because the line continues through the central disk that surrounds an involute spiral's pivot point, and bends around behind the front to run parallel to the front, bounding the region of high concentration [1, 10, 11].

It is also idealized in that the construction terminates where the contour line simultaneously encounters both the symmetry axis and the corresponding segment of counter-rotating spiral on the other side. A succession of disjoint surrounding segments of spiral wave envelop this structure (fig. 6a). The outermost of these would contain the original disk (line-segment in this cross-section) of isolated wavefront. They are all concentric closed surfaces of revolution; the object shown in fig. 5 (interior to all the outer concentric bags) is only the *source* of those outer shells of wavefront. Fig. 6b gives an impression of the scroll ring slightly later, surrounded by one disjoint shell of wavefront. Fig. 6c shows it much later, enclosed within so many concentric shells that the outermost lies outside the field of view.

Fig. 7 shows a series of thin sections (a–h) bounded by planes parallel to the symmetry axis (the "axle") of fig. 5. Note that planes remote from the axle intercept the wavefront in closed rings, while those close to the axle reveal mirror-image spirals.

Fig. 8 shows a series of thin sections (a–g) bounded by planes not-quite-normal to the axle. Were they perfectly perpendicular to the axle, each plane would intersect the wavefront only in concentric perfect circles. The slight tilt was introduced deliberately to break some delicate symmetries that would seldom be observed in any real laboratory experiment; for example, some of the former "circles" now pair up to bound a single crescent-shaped region.

Fig. 9.

Fig. 17.

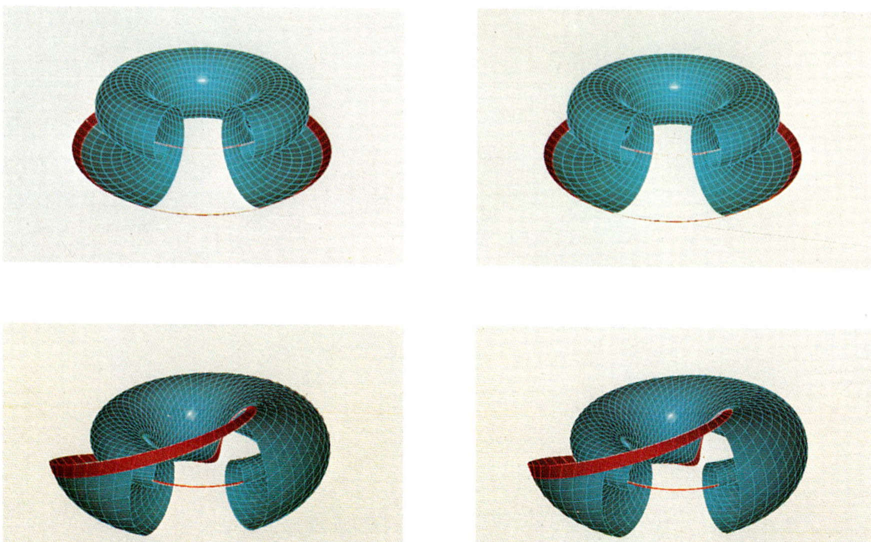

a

b

Fig. 15 (a, b).

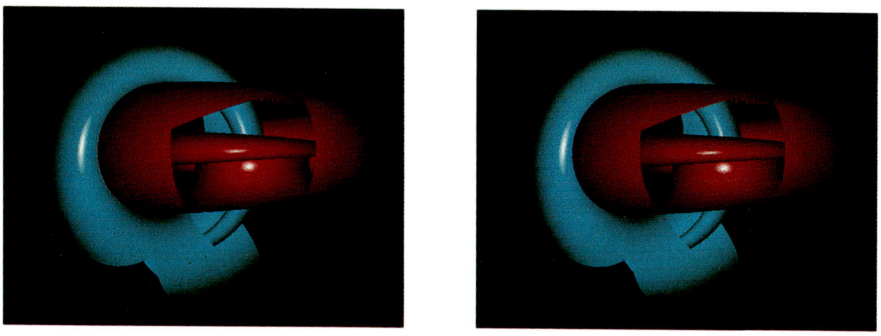

Fig. 18.

See captions overleaf.

Captions to color plate

Fig. 9. Slices of a wax scroll ring, rolled from a rectangular sheet red on one slide, blue on the other. Analogous to fig. 7.

Fig. 15. A scroll can be closed in a ring without (a) or with (b) a twist. The fine red line in each stereo pair is the ring singularity; the broad red band is the 45th cell along each spiral. The left image is for the left eye: just stare straight ahead, as though focusing on an object somewhat behind the page. If you can't, use a stereo viewer or cut out the right side picture, put it on the left, and read caption of fig. 13.

Fig. 17. Two tori of wax similar in appearance to fig. 16 are perpendicularly linked to suggest the manner of meshing of twisted scroll waves: they must both twist in the same sense.

Fig. 18. Wavefronts of linked twisted scroll rings joining along a collision interface. The windows are, of course, artificial. The two singular rings are perpendicular circles, each penetrating the other's center. The red and blue waves are identical, just rotated into linking positions so that their wavefront collide neatly. This entire structure surrounds itself in concentric sphere-like wavefronts composed of hemispheres (shaped like the two pieces of a tennis ball cover) from each source (not shown here).

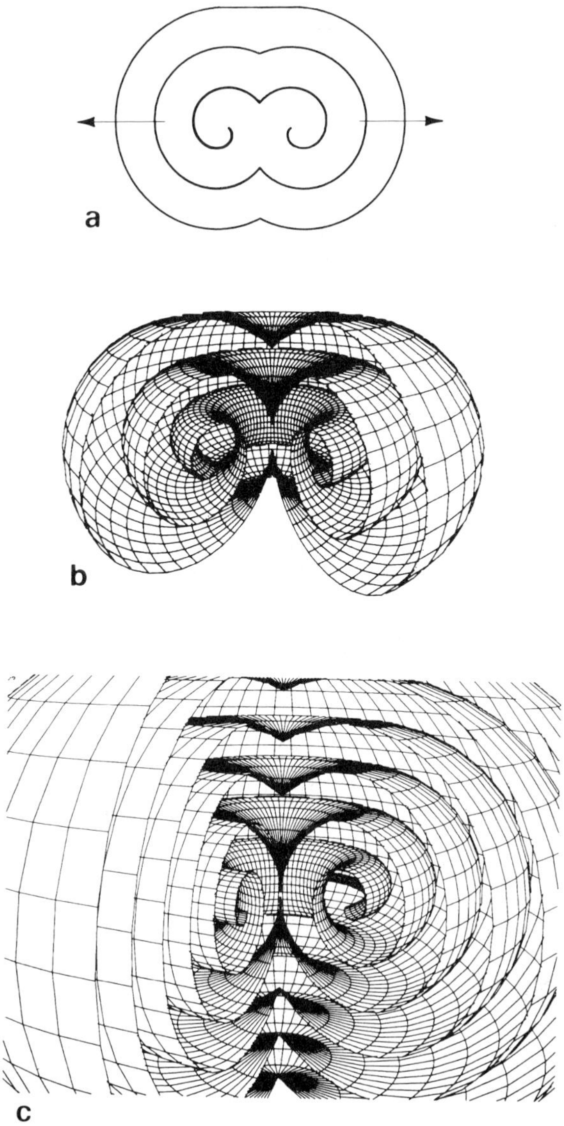

Fig. 6. a) A pair of adjacent spiral waves collide along a mirror axis. A single curve joins the two pivot points, but outside that curve, segments of wavefront join to make concentric closed curves (cross-sections of 3-dimensional "bags"). The outermost of them contains the original line-segment of wavefront. Rotate this picture about the mirror axis to obtain Figure 5 and the outer bags not shown there; b) the untwisted scroll ring somewhat later than in fig. 5: one spherical envelope of wavefront has separated itself from the interior surface. The wavefront was calculated by the same Fortran as for fig. 5; the shading, coloring, reflections, and hidden-surface processing were done by M. Prueitt's GRAFIC on the CRAY-1; c) much later, space is filled with concentric shells of wavefront surrounding the scroll ring and its attached piece of wavefront (fig. 5).

In the first slice of fig. 8a we only graze the bottom, torus-like bulge of the wavefront. In fig. 8b the window cut into that surface has opened enough to reveal the next turn of the scroll inside. In fig. 8c the section plane is cutting into that interior surface much as in fig. 8a, revealing a crescent of the inner edge of the scroll beneath. The outermost wave (the outermost turn of every spiral) is seen at this level only as two cylindrical segments: one interior and one exterior to the turn of the scroll that is just beginning to be windowed. In fig. 8d that window has enlarged enough to reveal most of the inside edge of the scroll ring, and we begin to see how the inside cylindrical segment joins the rest of the wave in the background. In fig. 8e only about 180° of the inside edge is still with us, and the segment of wave seen in the background of the previous slice is now conspicuous. Another disk-like piece of wavefront lies behind it, protruding into the central hole as a tiny white peak. In fig. 8f that hindmost disk is in plain view, obstructed only by a crescent of a foreground piece of wavefront. In the last slice we retain only a crescent of the hindmost piece of wavefront.

These pictures were made using an Evans and Sutherland Picture System 300, operating on a list of 20,000 (x, y, z) points computed by spinning a planar involute spiral around the symmetry axis. Steven Strogatz prepared the Apple computer algorithm for the (x, y, z) coordinates of constituent points. A less technology-intensive procedure entails rolling together a sheet of red wax and a sheet of blue wax to make a red/blue scroll. The scroll is then bent into a ring and closed, making sure the spirals join perfectly along the half-plane where ends meet to close the ring. The resulting scroll ring can then be sectioned with a hot knife [38] (fig. 9 analogous to fig. 7 in color).

4.2. *The linked pair of twisted scroll rings*

A related waveform that has *not* yet been sought experimentally is pre-figured by making a mirror-

Fig. 7. The wavefront shown in fig. 5 is shown in consecutive thin slices (a–h) parallel to the axle (the rotation axis referred to in fig. 6).

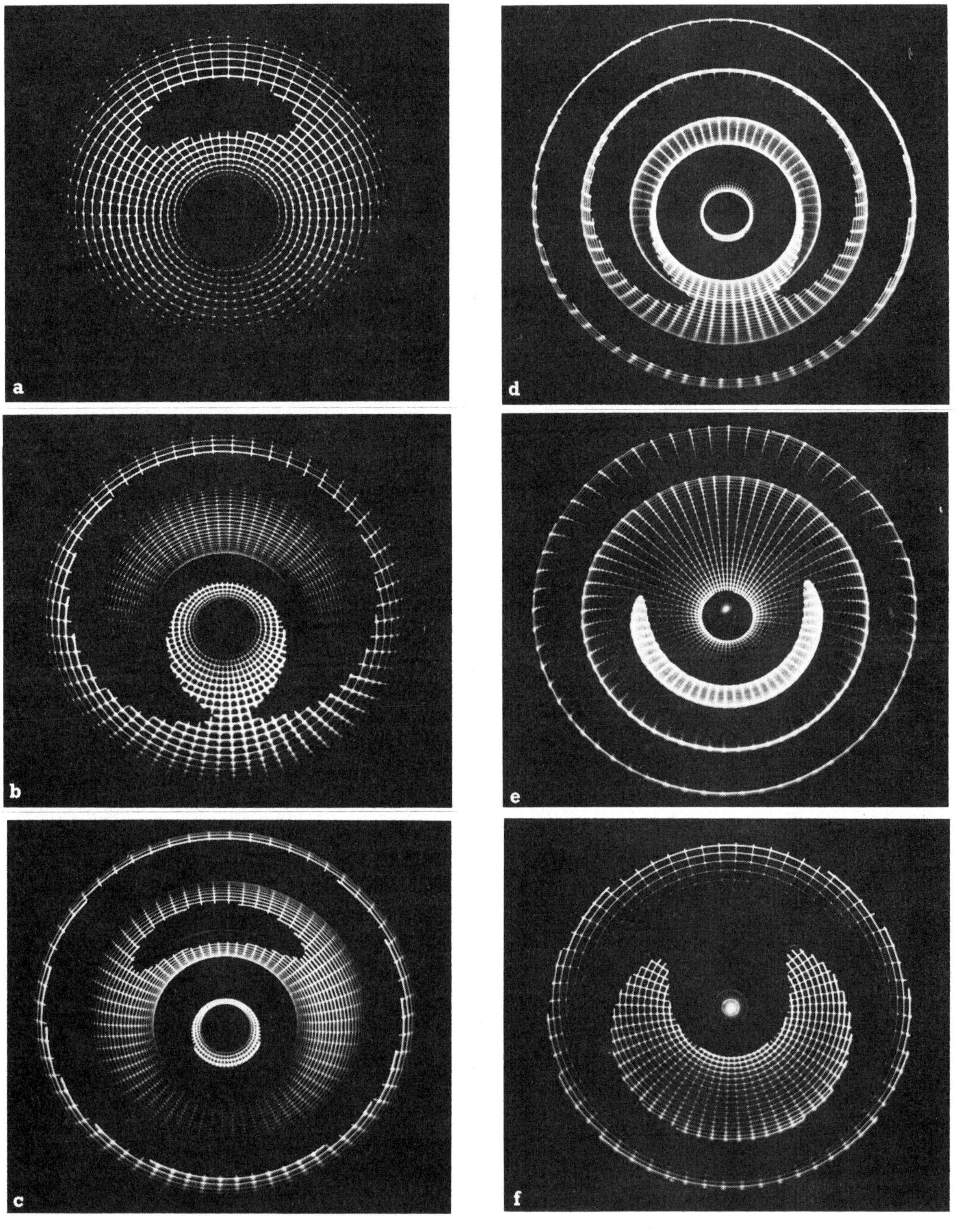

Fig. 8. The wavefront shown in fig. 5 is shown in consecutive thin slices (a–g) almost perpendicular to the axle.

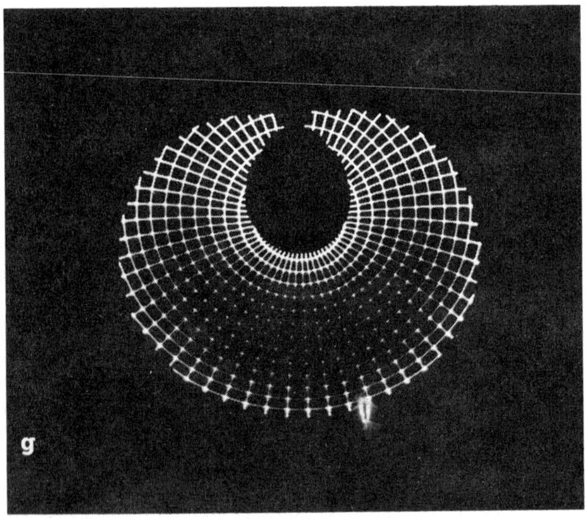

Fig. 8. continued

pair of parallel wax scrolls (fig. 10) and giving it a 360° twist before bending it into a ring (fig. 11). The topologically-different wave so constructed would develop only from initial conditions topologically different from the isolated disk of wavefront considered above: it turns out to require an initial segment of wavefront topologically equivalent to a once-twisted band (fig. 12) [7]. It thus has *two*, not one edge: the two edges are linked rings.

The various topologically-distinct scroll-shaped waves are trivially constructed in patterns selected by initial conditions. The real question is whether such initial conditions are physically realizable. The answer often seems to be "yes" for media whose local state is a pair of real numbers (for example, concentrations). There seem to be a lot of topologically distinct initial conditions that satisfy the geometrical requirements of physical chemistry [6–9].

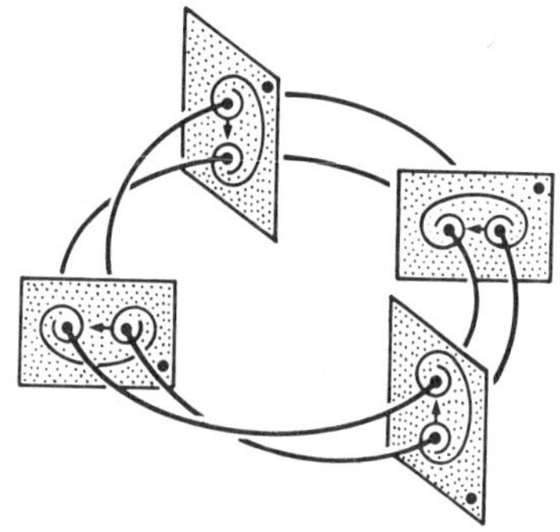

Fig. 11. A mirror-image pair of spirals is transported around a circular path to sweep out a pair of adjacent scrolls. Four local cross-section planes are shown. The pair is rotated as it is transported, so that each scroll is once-twisted in the same sense, and the two are mutually linked. Waves emerging from the pair in each cross-section plane eventually join in rings as in fig. 6a. If the radius of rotation is large enough the transported rings sweep out surfaces topologically equivalent to concentric tori, or otherwise to concentric spheres (not shown). From [7].

In the first, and as yet, only, mathematical solution for the once-twisted scroll ring in a reaction/diffusion medium [39], for convenience one ring is made immensely bigger than the other: only a segment of it is seen, vertically threading the smaller horizontal ring. This limiting case of the linked pair of twisted scroll rings is depicted in stereo perspective in fig. 13a, in monocular cutaway in fig. 13b, and in a view along the axle in fig. 13c. This construction is also peculiar in that the domain occupied by waves emitted from this vertical segment is made very, very narrow: waves from the two source rings collide along an intermediate surface which is nearly degenerate and is not visually distinguishable from the axle. Fig. 14 gives a clearer view, in stereo, of the singularity

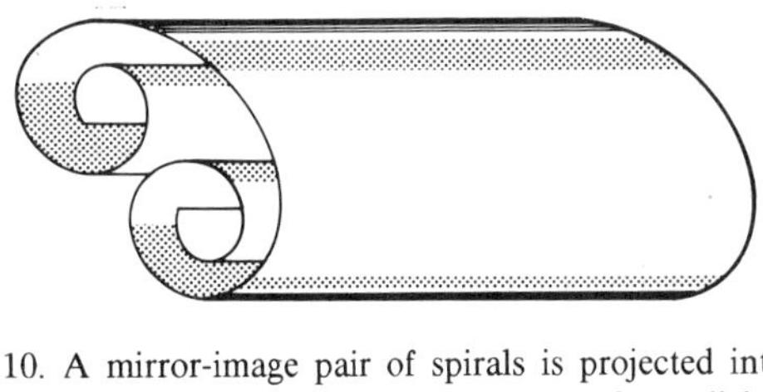

Fig. 10. A mirror-image pair of spirals is projected into the third dimension to become a mirror-pair of parallel scroll waves.

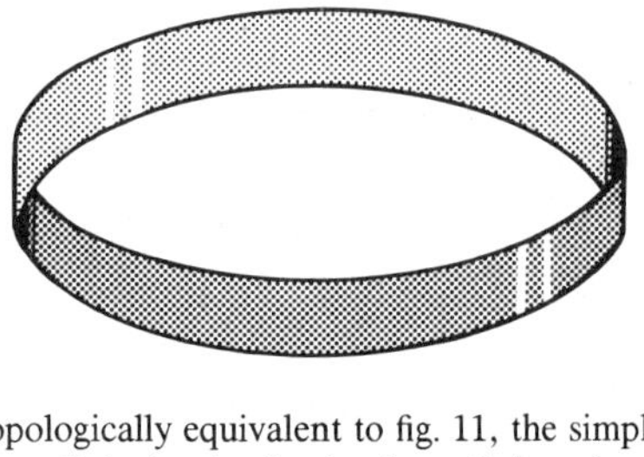

Fig. 12. Topologically equivalent to fig. 11, the simplest initial condition for a linked pair of twisted scroll rings (analogous to fig. 2 for the untwisted scroll ring) is a twisted cylindrical band.

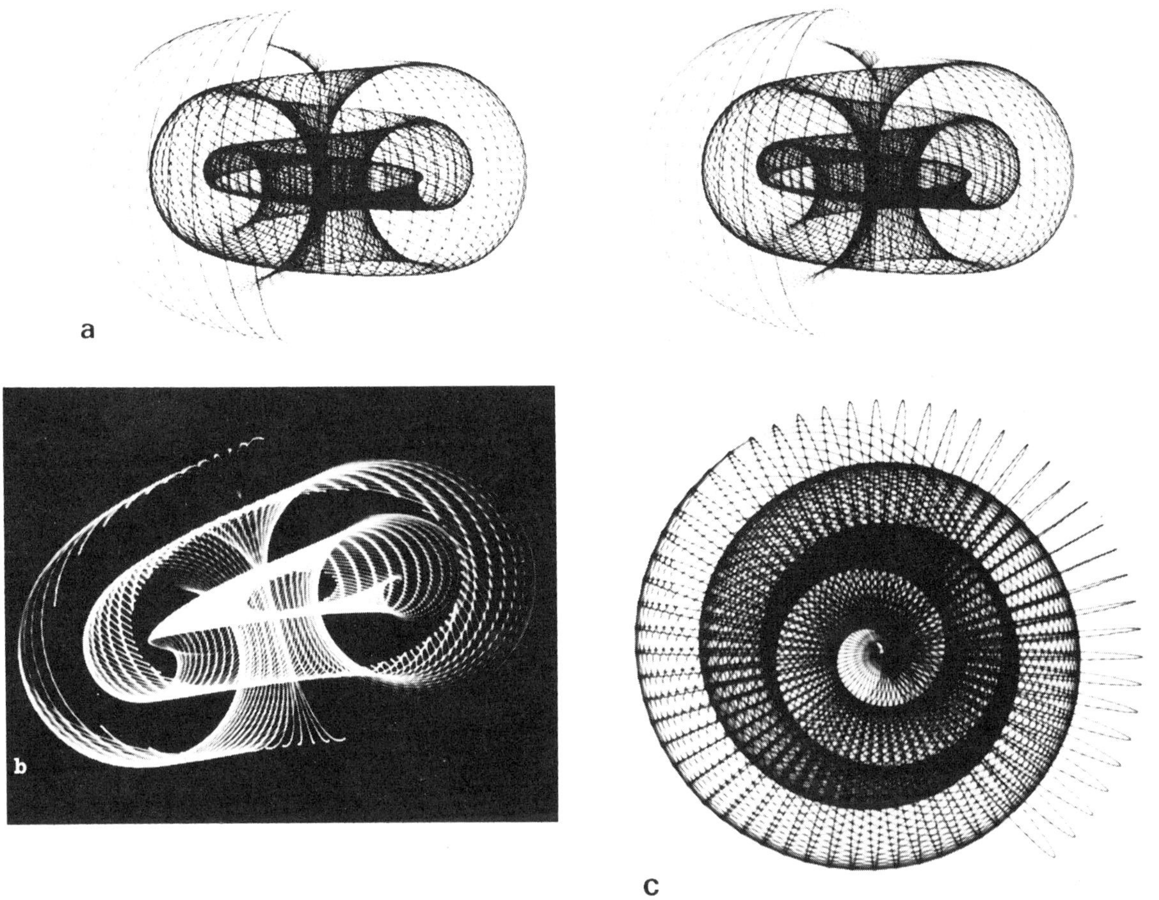

Fig. 13. a) A twisted scroll ring, in stereo. The linking ring is immense: only a segment of it is seen here, coinciding with the vertical axle. The left-side picture is for the right eye: cross eyes to fuse images in stereo; b) X-rayed in monocular cut-away for a different view of the interior; c) an X-ray view in projection along the axle. A spiral wave moves outward from a circular locus, the smaller scroll ring's axis. Another spiral wave moves inward, converging on the larger scroll axis, here seen in projection as a point. The figure was constructed by transporting a planar involute spiral to 60 positions around a circle; in this projection the spirals are seen as ribs, especially at the top where they are not cross-linked.

implicit in closing a scroll ring with one 360° twist of its spiral cross-section. As noted, a physically realistic singularity would be a wave source, but in this case the waves are not allowed to radiate any discernible distance before colliding with the incoming wave from the surrounding horizontal singular ring.

Despite its patent speciousness this limiting case has the advantages that:

1) the subtle wave structure surrounding one ring (the horizontal ring) occupies almost the entire space, so it is very plainly visible, unobscured by waves from the other source; the structure around the other ring is the same but it is squeezed out of view so far as possible; and

2) by rotating this limiting case about its axle, one advances every area element by the same distance, thus animating (rather than computing) the movement of the wave according to Huyghens' construction (see proof in [7]). Thus a videotape of the twisted scroll ring in motion can be (was) made using only the 3-D rotation hardware of the Evans and Sutherland Picture System 300.

The pair of linked twisted scroll rings is enveloped in closed bag-shaped wavefronts. We do not see them here because we do not even see the

a

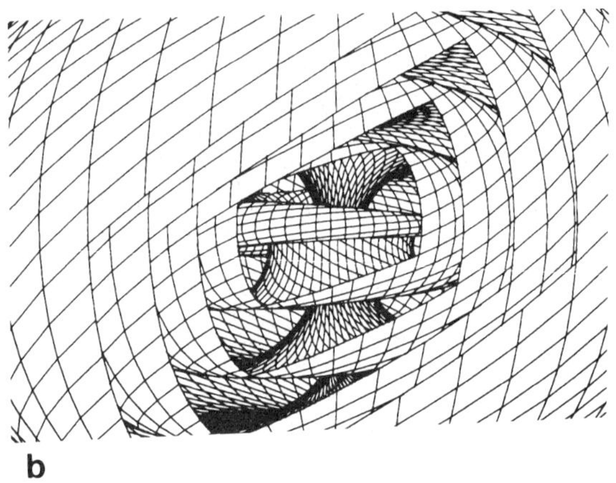
b

Fig. 14. The object shown in fig. 13 is depicted a) with a wedge cut out to reveal a helicoidal wave approaching the axial singularity; and b) with windows cut out, leaving structure near the singular axle ambiguous, so it could be replaced by an axial *source* singularity (cf. fig. 18).

complete wave structure internal to the concentric bags: it is incomplete because we do not see the entirety of the larger "ring" of wave-edge (coinciding with the axle in this limiting case.) The depicted wavefront terminates abruptly along an arbitrarily-chosen locus that resembles the opening of a snailshell in fig. 13. This is not intended to represent a real edge of the wavefront.

In fig. 13c especially you may notice another artifact: the cross-links joining adjacent spiral ribs are incomplete along a diagonal locus near the opening of the snailshell: this was done to indicate a locus of intersection with a symmetrically-disposed torus. The ribs and their cross-links, so conspicuous in fig. 13c, are in any case all artifacts of discretizing a continuous surface.

In the more symmetric case, a pair of identical scroll rings link one another perpendicularly. Each scroll ring is once twisted to the left (as in fig. 15b) or to the right (mirror-image of fig. 15b). Two such wave-fields can join along the common interface (a surface equidistant between the two linked circular axes) if and only if they have the same handedness [7]. Fig. 16 shows a bird's-eye view of a torus containing the twisted scroll of fig. 15b; the circles engraved on its surface are successive positions of the wavefront. Fig. 17 shows a pair of such tori perpendicularly linked: the wavefronts fit together if both are left-twisted or if both are right-twisted. Either of these mirror-image composite wavefronts seems to represent a viable mode of periodic time/space organization of an excitable medium, in that suitable planar cross-sections are involute spirals (up to their mutual collisions along the common interface) emerging from the 2-dimensional rotor source known to be stable in the laboratory and in computations.

In fig. 18, the wavefronts of two linked twisted scroll rings are shown joining where waves collide as they emerge from the two ring singularities. The waves from the horizontal ring source are shown in red; those from the vertical ring source are

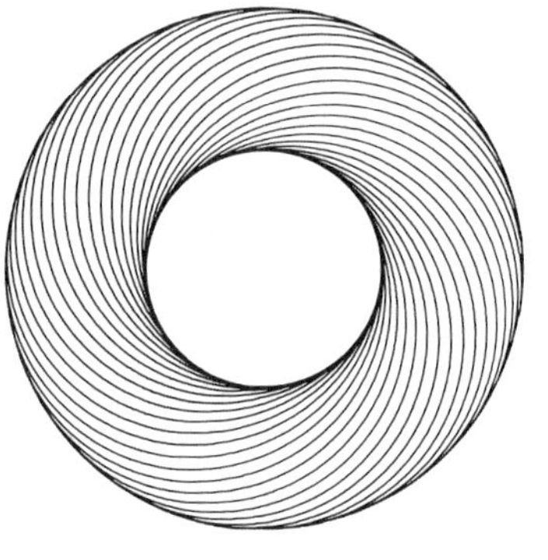

Fig. 16. A twisted scroll ring meets an enclosing toroidal surface along a closed curve which links the torus (the broad red band in fig. 15b). In this view from the top, that locus is seen in many consecutive positions as the scroll rotates (the wave propagates). From [1].

shown in blue. Windows are cut in both sets of waves for better visibility.

4.3. *Other kinds of scroll ring*

We have seen the simplest scroll ring (section 4.1) and a pair of linked scroll rings (each with one twist in the same direction: section 4.2). Any number of rings, even knotted rings, can be linked in arbitrary ways without violating the basic constraints of physico-chemical realizability, so long as each ring is given the appropriate number of positive or negative twists [7–9]. They thus suggest particle-like singularity-free solutions to the reaction-diffusion equation. These structures are distinguished by integer quantum numbers that characterize knotting and mutual linking. They *locally* (in suitable cross-sections) resemble the 2-dimensional spiral wave that seems characteristic of excitable media, so they seem likely to prove comparably stable. Evidence of slow instabilities due to *global* structure should be sought numerically. Analytical investigations are still impeded by the lack of a tractable dynamical equation for an excitable medium.

5. An urgent need: a mathematical prototype of excitable media

Of the known examples of excitable media (see [1]) not one lends itself to mathematical study as gracefully as do the sine-Gordon equation or the Schrödinger equation, for example. One pioneering effort [40] exploited a simple analytic form for the phase of the periodic wave outside its source in the center of each spiral, assuming that the pulse wave has standard amplitude everywhere; but it cannot have in the center of rotation. Such solutions must therefore be supplemented by demonstration that *within* the center of each spiral, chemically-realistic solutions can be arranged to smoothly join the outer standard-amplitude approximation. This demonstration is given in [6] as well as in earlier two-dimensional computations [10, 24, 25, 32, 41].

Nonetheless, mathematical investigation of these 3-dimensional organizing centers would be greatly facilitated by a model kinetic scheme for an excitable medium that would permit solution simultaneously for the standard-amplitude wave and its attenuation to the center of rotation. The highest priority should be given to the selection of such an equation. The requirements on the local kinetic equation, apart from analytical tractability, are:

1) It must involve two real-valued dynamical quantities analogous to chemical concentrations or deviations from steady-state concentration. A single variable is unable to support a rotating solution in simply-connected media (1); more than two would be unable to coordinate in the maintenance of scroll axis compositions unless their dynamics quickly reduces to a 2-dimensional "slow manifold" (6).

2) It must have a single, globally-attracting steady-state.

3) A tiny but finite departure from that steady-state must put the system on a trajectory that rapidly leads farther from steady-state before returning from this large excursion.

Contest entries should be mailed to Purdue University. The winner will be informed by mail. The first prize is immortalization. The equation will be named for the winner.

Acknowledgements

This paper was prepared while enjoying the hospitality of the Group in Theoretical Biology and Biophysics, Theoretical Division, Los Alamos National Laboratory. The computer graphics programs were prepared at Evans and Sutherland Corporation, Salt Lake City, courtesy of Bruce

Fox, Neil Harrington, and Michael Bailey (figs. 5, 7, 8, and 13), and at Los Alamos National Laboratory supported by the U.S. Department of Energy in collaboration with Melvin Prueitt and Steven Strogatz (figs. 6, 14, and 18). I thank the GC Dental and Industrial Corporation of Osaka, Japan, for a kilogram of dental wax, the U.S. National Science Foundation for grant CHE-8103322, and the John Simon Guggenheim Memorial Foundation for a year's leave to work.

References

[1] A.T. Winfree, The Geometry of Biological Time (Springer, New York, 1980).
[2] M.V. Berry in: Les Houches Lectures XXXV, R. Bahian, M. Kleman, and J.P. Poirier, eds. (North-Holland, Amsterdam, 1981), pp. 453–543.
[3] M.V. Berry in: Geometry of the Laplace Operator, Proc. Symp. Pure Maths. 36, R. Osserman and A. Weinstein, eds. (1980), pp. 13–28.
[4] J.F. Nye and M.V. Berry, Proc. Roy. Soc. Lond. A336 (1974) 165.
[5] A.T. Winfree, Science 181 (1973) 937.
[6] A.T. Winfree and S.H. Strogatz, Physica 8D (1983) 35.
[7] A.T. Winfree and S.H. Strogatz, Physica 9D (1983) 65.
[8] A.T. Winfree and S.H. Strogatz, Physica 9D (1983) 333.
[9] A.T. Winfree and S.H. Strogatz, Physica D, to be published.
[10] A.T. Winfree, Prog. Theor. Chem. 4 (1978) 1.
[11] A.T. Winfree, Science 175 (1972) 634.
[12] A.T. Winfree, SIAM AMS Proc. 8 (1974) 13.
[13] A.T. Winfree, Scientific American 230 (1974) 82.
[14] T. Erneux and M. Herschkowitz-Kaufman, J. Chem. Phys. 66 (1977) 248.
[15] D.S. Cohen, J.C. Neu and R.R. Rosales, SIAM J. Appl. Math. 35 (1978) 536.
[16] A.S. Michaelov and J.S. Uporov, Dokl. Akad. Nauk. CCCP 249 (1979) 733.
[17] Y.B. Zeldovich and B.A. Malomed, Dokl. Akad. Nauk. CCCP 254 (1980) 92.
[18] V.S. Zykov, Biofizika 25 (1980) 319.
[19] T. Erneux, J. Math. Biol. 12 (1981) 199.
[20] M.R. Duffy, N.F. Britton and J.D. Murray, SIAM J. Appl. Math. 39 (1980) 8.
[21] P.S. Hagan, SIAM. J. Appl. Math. 42 (1982) 762–786.
[22] A.S. Michaelov and V.I. Krinsky, Biofizika 27 (1982) 875.
[23] V.I. Krinsky and B.A. Malomed, Physica 9D (1983) 81.
[24] F.B. Gul'ko and A.A. Petrov, Biofizika 17 (1972) 261.
[25] A.I. Shcherbunov, N.I. Kukushkin and M.Y. Sakson, Biofizika 18 (1973) 519.
[26] K.I. Agladze, Physica D (1984).
[27] M.A. Allessie, F.I. Bonke and J.G. Schopman, Circ. Res. 33 (1973) 54.
[28] O. Rossler and C. Kahlert, Zeit. Naturf. 34 (1979) 565.
[29] B.Y. Kogan, V.S. Zykov and A.A. Petrov, in: Simulation of Systems, L. Dekker, G. Savastano, and G. Vansteenkiste, eds. (North-Holland, Amsterdam, 1980), p. 693.
[30] V.S. Zykov, Biofizika 25 (1980) 888.
[31] V.S. Zykov, in: Autowave Processes in Systems with Diffusion, M.T. Grechova, ed. (USSR Acad. Sci., Gorky, 1981), p. 85.
[32] F.J. van Capelle and D. Durrer, Circ. Res. 47 (1980) 454.
[33] Y. Kuramoto and S. Koga, Research Inst. Funda. Physics RIFP-440 (1981).
[34] R. Miura and R. Plant, in: Differential Equations and Applications in Ecology, Epidemics, and Population Problems, S.N. Busenberg and K.L. Cooke, eds. (Academic Press, New York, 1981).
[35] A.T. Winfree, Farad. Symp. Chem. Soc. 9 (1974) 38.
[36] B. Welsh et al., Nature 304 (1983) 611–614.
[37] A.V. Panfilov and A.M. Pertsov, Dokl. Akad. Nauk. CCCP (1984) in press.
[38] A.T. Winfree, Adv. Med. Biol. Phys. 16 (1978) 115.
[39] J. Gomatam, J. Phys. A 15 (1982) 1463.
[40] Y. Kuramoto and T. Yamada, Prog. Theor. Phys. 56 (1976) 724.
[41] T. Yamada and Y. Kuramoto, Prog. Theor. Phys. 55 (1976) 2035.

Physica 12D (1984) 333–350
North-Holland, Amsterdam

THE MECHANOCHEMISTRY OF CYTOGELS

G.F. OSTER
Department of Biophysics and Entomology, University of California, Berkeley, CA 94720, USA

and

G.M. ODELL
Department of Mathematics, Rensselaer Polytechnic Institute, Troy, NY 12181, USA

We present a mechanochemical model for force generation by actomyosin cytogel. The model is built around the observation that calcium ions both solate actomyosin gels and stimulate active contraction of the gel. The model cytogel can mimic a variety of morphogenetic pattern-producing phenomena.

1. Introduction

The shaping of an embryo is a consequence of how its constituent cells move and change shape. The basis for these cellular shape changes lies in the cells' cytoskeleton, and in particular in the actomyosin gel which generates the forces which deform the cell.

While the molecular constitution of cytoplasm is complex, it is clear that its contractile activity derives from the interaction between actin and myosin in a manner analogous to that in muscle fibers. A critical difference between the contractile ability of cytoplasm and that of striated muscle is that muscle cells have their actomyosin fibrils oriented in parallel arrays, while the contractile machinery in cortical cytoplasm is more randomly oriented. This raises the question of how directed mechanical action can arise from a more or less randomly isotropic medium.

It now appears that the mechanochemical behavior of cytogel is governed by its ability to coordinate its contractile activity with a gel–sol phase change [1–3]. In this paper we present a mathematical model for actomyosin gels. We adopt the viewpoint of contiuum mechanics; that is, we shall endeavor to capture the relevant macroscopic consequences of the molecular mechanism governing cytogel mechanics by constructing a viscoelastic medium with the appropriate constitutive behavior.

Before presenting our mathematical model for cytoplasmic viscoelasticity it is helpful to have a qualitative intuitive picture of the molecular events which underlie the model. Therefore, in section 2 we shall briefly describe a modification of the 'solation–contraction' hypothesis of Taylor et al. [1] which illustrates the essential features of our macroscopic model for cytoplasm. In section 3 we derive the model equations, and in section 4 we describe several applications of the model to cellular and embryonic morphogenesis.

2. Description of the model

Consider a small volume element of cytoplasm which is in mechanical equilibrium with its surroundings, as shown in fig. 1. Even with no myosin present, this volume element can sustain tensile

0167-2789/84/$03.00

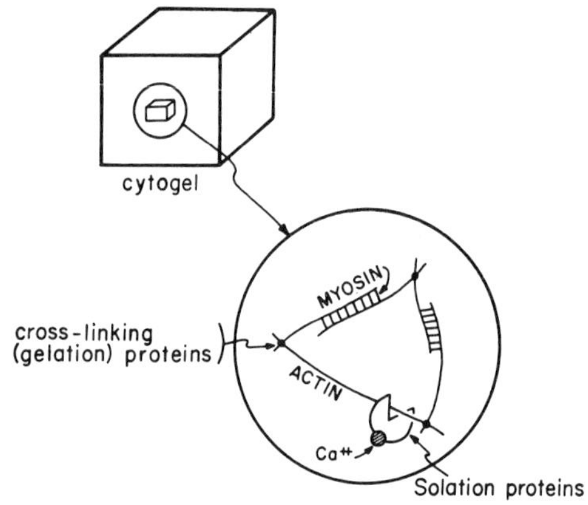

Fig. 1. A schematic microscopic view of a block of cytogel: actomyosin fibers are crosslinked by gelation factors, and severed by solation enzymes. The activity of both proteins can be affected by Ca^{++}, which also activates the actomyosin contractile machinery.

stresses passively since it consists of actin which is maintained in a gel state by cross-linking proteins (actin binding proteins and other gelation factors, cf. Taylor et al. [1] and Weeds [4]).

When myosin oligomers are present, partial solation of the actin gel facilitates active force generation: if the actin network is too tightly gelled there will be too few free ends to create many "contractile units", i.e. regions of actin fiber overlap cross-linked by myosin oligomers. Furthermore, a fully cross-linked actin gel can resist the deforming forces these contractile units generate. Consequently, the gel will be unable to generate contractile stresses unless it is at least partially solated first.

It is known, however, that calcium ions activate solation enzymes which break the actin network at points between the gel network nodes (e.g. gelsolin in amoebae, fragmin in *Physarum*: [1, 4, 5]). Thus, exposed to moderate (i.e. micromolar) Ca^{++}, the volume element partially solates. This creates free ends which can bind myosin to form contractile units, and allows deformation of the gel when the actomyosin fibers commence to contract.

In addition to activating the solation factors, Ca^{++} also triggers the actomyosin contractile units to commence generating tractions. It does this by several routes: (i) promoting the assembly of myosin monomers into oligomers, (ii) promoting the binding of the myosin oligomers to actin filaments, and (iii) by binding to calmodulin which, in turn, activates the kinase which phosphorylates the myosin light chain. Thus the same calcium signal which stimulates formation of contractile units simultaneously activates the myosin ATPase and initiates active contraction. The net effect, therefore, of increasing local calcium concentrations to the ten-micromolar range is to cause the volume element to contract isotonically against the tensions of the surrounding volume elements.

As the partially solated volume element contracts it transmits stress to the surrounding volume elements which are forced to expand until the system reaches a new mechanical equilibrium. In this new strained state the actin network in the surrounding elements is polarized toward the contraction site, and mechanical equilibrium is reached when the active contraction of the original volume element is balanced by the active and passive forces of its strained neighbors, as well as the passive compressive stress at the contracting site.

If the contraction process of the original volume element stimulated the surrounding volume elements to contract as well, then a contraction wave-front would spread out from the site of Ca^{++} stimulation. This can occur as follows. As the original volume element contracts, its internal hydrostatic pressure increases, and the solated cytoplasm within it is pumped into the surroundings. (The gel can be pictured as a sponge-like medium which, when it contracts, squeezes fluid out into the neighboring volume elements, a process called 'syneresis'.) This solated efflux carries with it some of the Ca^{++} which initiated the contraction, and so raises the Ca^{++} concentration of the surroundings. In addition, Fickian diffusion carries Ca^{++} to the surrounding gel. This, in turn, initiates the solation–contraction cycle in these volume elements. Thus the contraction tends to spread beyond the original site, but is still rather limited in extent, since the original Ca^{++} impulse is soon diluted below the level required to initiate the solation–contraction.

Each volume element can have its own internal store of sequestered Ca^{++} – which is usually assumed to be associated with membranous vesicles which permeate the cytoplasm. This sequestered Ca^{++} can be released by two mechanisms. First, if the Ca^{++} concentration pumped and diffusing from the original contracting volume element exceeds a threshold value, it can trigger a local autocatalytic cascade release of the sequestered Ca^{++} in the surrounding cytogel (so-called 'calcium stimulated calcium release'). Second, the act of stretching the surrounding volume elements can mechanically trigger the release of sequestered Ca^{++} by initiating a membrane depolarization of the sequestering vesicles. Either mechanism, or combinations of them, can sustain propagation of the contraction wavefront outward from its initiation site.

In every actomyosin-based contractile system studied so far Ca^{++}-initiated contraction seems to be an intrinsic feature, although the biochemical details differ in different systems. Thus the signal for local contraction of the cytogel is the raising of the local concentration of Ca^{++} above a threshold value. This can be accomplished by either an influx of high Ca^{++} bearing sol from neighboring regions, or by the mechanical deformation of the cytogel by neighboring contractions. Both of these mechanisms are known to operate in several contractile systems.

The stimulation by Ca^{++} of a Ca^{++} cascade release is familiar in many muscle systems [6, 7, 8] and is implicated in neural synaptic transmission [9]. Stretch activation is a characteristic of many smooth muscle systems and insect flight muscle. However, the most striking demonstration of stretch-induced mechanical action is found in the plasmodial strands of the slime mold *Physarum physarum* [10, 11]. We shall present a detailed analysis of the plasmodial oscillations in *Physarum* in a separate paper [12]; for our purposes here, however, we shall use the observations on this cytogel system to justify our basic assumptions about cytoplasmic contractility:

(1) Ca^{++} levels regulate the sol–gel equilibrium of the cytoplasm; that is, the number of actin–actin cross-links as well as the average lengths of the actin filaments.

(2) Ca^{++} initiates active contraction of the actomyosin gel.

(3) Calcium is sequestered locally in membranous vesicles.

(4) An autocatalytic release of Ca^{++} from these vesicles can be stimulated by raising local Ca^{++} levels above a threshold value (calcium stimulated calcium release).

(5) Release can also be stimulated by mechanically straining the cytogel ('stretch activation').

We emphasize that calcium is but one link in a complex biochemical scheme regulating contraction, and other substances can also trigger contraction by stimulating the Ca^{++} cascade (e.g. cAMP). Therefore, in the following discussion we shall use the symbol c to denote the concentration of "trigger chemical" – although we are thinking of calcium as the central, or rate-limiting substance. This amounts to assuming that the chemical kinetic cascade initiating contraction can be summarized in the dynamics of a single 'composite' chemical: c.

2.1. *The mechanochemical properties of cytogel*

Let us now examine what the above properties imply for the stress-strain behavior of the cytogel. Consider a cylindrical strip of cytogel of length L_0 and cross-section A_0. When the cylinder is subjected to an axial stress, σ, it stretches to length L and contracts laterally to area A. We define the Lagrangian strain, ϵ, in the z direction by

$$\epsilon \equiv \epsilon_{zz} = (L - L_0)/L_0. \tag{1}$$

We want to describe the mechanochemical response of the cylinder as the stress and calcium concentration is varied.

Fig. 2a shows the mechanical response of the cytogel cylinder. As the cylinder is stretched the stress–strain curve in region I initially rises faster than linear; this is because the strain aligns the fibers of the gel thus increasing the effective modulus of the cylinder. Eventually, a yield point is reached, and in region III the curve falls off as the fibers tear. The cylinder initially offers little resistance to compression, but eventually the stress–strain curve falls off to $-\infty$ (region III) since the cylinder cannot be compressed to zero length.

The response curve in fig. 2a supposes that the actomyosin contractile machinery of the cytogel has not been activated by calcium. Next we examine the mechanical response of the cylinder when the calcium concentration, c, is raised from zero. To do this we fix the cylinder at a length, L, and examine the isometric stress, σ, when the calcium concentration is varied, as shown in fig. 2b. In region I′ the active stress generated by the cylinder increases as the contractile machinery turns on (i.e. the myosin light chains are progressively phosphorylated). However, the same calcium which activates the contractile machinery also activates solation enzymes which clip the gel fibers, partially solating it. Initially, this solation is necessary to generate any active stress since it creates free-ended strands which can bind myosin and create "contractile units": segments of counter-oriented actin fibers cross-linked by myosin oligomers. Eventually, however, the action of these clipping enzymes so solates the gel that it can no longer sustain much stress, and the curve falls off in region II′. Thus fig. 2b is the isometric mechanochemical response curve of the cytogel cylinder.

The active stress has one other important property, which can be seen by clamping the calcium concentration, c, and examining the 'isochemical' stress-strain behavior, as shown in fig. 2c. The active stress generation of the cytogel cylinder is based on the sliding filament mechanism characteristic of actomyosin interactions. This means

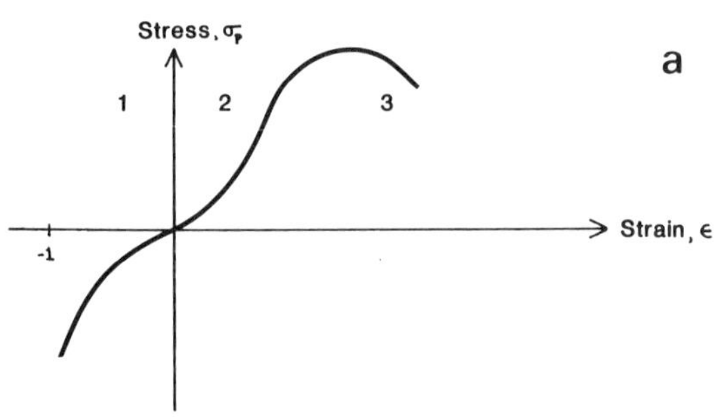

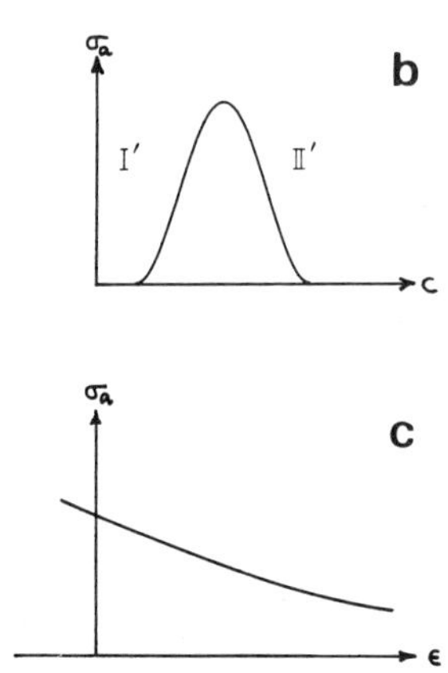

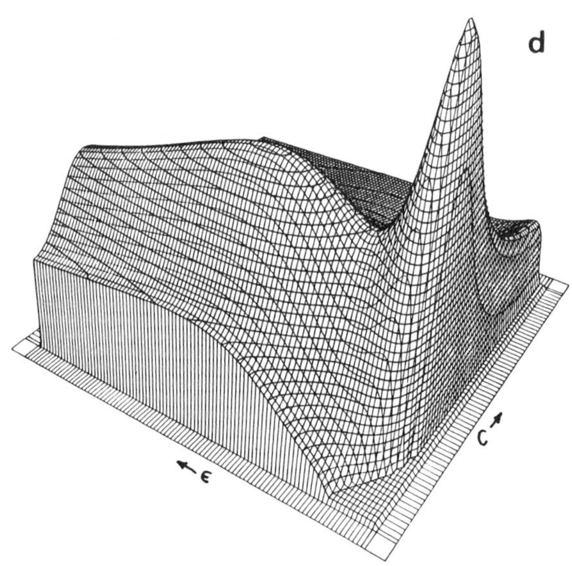

Fig. 2. a) The passive stress–strain behavior of a cytogel strip when $Ca^{++} = 0$. In region I the cytogel is in compression. In region 2 the fibers are in tension and are commencing to align, causing the apparent elastic modulus (i.e. the slope of the curve) to increase. In region 3 the fibers commence to yield under the imposed stress, causing the curve to level off, and eventually to fall as the fibers tear and separate. b) The active stress as a function of calcium concentration at constant strain (isometric). The curve rises as the contractile machinery is activated, then falls as the gel becomes so solated that it can no longer sustain tensile stresses. c) The active stress as a function of strain. The curve falls off at increasing strain because of the overlap effect: a sliding filament mechanism gets stronger as it shortens. d) The constitutive relation $\sigma(\epsilon, c)$ that characterizes the mechanochemical response of the cytogel strip.

that, as a contractile unit contracts the degree of overlap between adjacent actin fibers increases. This brings more myosin cross-bridges into play, and increases the sliding force between the fibers. Thus the contractile machinery of the cytogel cylinder has the property that it grows stronger as it contracts. This is shown in fig. 2c, where the active stress developed by the cylinder falls off as the strain, ϵ, increases.

We can assemble the mechanochemical properties of the cytogel cylinder by adding the active and passive stress contributions to obtain the response curve, $\sigma(\epsilon, c)$ shown in fig. 2d. Figs. 2a–c represent various cross-sections of this surface. We regard fig. 2d as an empirically determined surface, since deriving $\sigma(\epsilon, c)$ from basic principles is a formidable task. Since only the qualitative features of this response surface will be important in our model we shall parametrize it in a way that is convenient for mathematical and numerical purposes.

In our discussion so far we have spoken of the actin gel as a fixed entity. However, it too has its own dynamics since it can be assembled and disassembled by polymerization/depolymerization reactions. To the extent that the gel is solated and/or depolymerized, its constituents can also diffuse and be carried by the convective flow of cytosol. Thus we must include in our picture of cytoplasm its capacity to reorganize itself by disassembly, transport and reassembly.

With the above picture as a guide we now examine how the macroscopic viscoelastic properties of cytoplasm can be described mathematically.

3. The mathematical model for contractile cytogel

For simplicity, we shall describe the 1-dimensional case, and merely state the 3-dimensional equations; a more complete discussion can be found in [13].

3.1. *The stress tensor for a cytogel cylinder*

Consider a 1-dimensional cylinder of cytogel which is stretched from a length L_0 to L. If the cylinder were a passive elastic body, then we could characterize its mechanical behavior as a simple viscoelastic constitutive relation of the form

$$\sigma = E\epsilon + \mu\,\partial\epsilon/\partial t, \tag{2}$$

where σ is the stress (force/unit area) applied to the cylinder, $\partial\epsilon/\partial t$ is the rate of elongation of the cylinder, μ is the shear viscosity of the cytogel, and E its elastic modulus. Note that if A_0 is the original cross-sectional area of the cylinder, and A its strained cross-sectional area, then if the volume of the cylinder is conserved we have $A = A_0/(1+\epsilon)$. Therefore, the total force exerted by the cylinder is

$$\sigma A = \{A_0/(1+\epsilon)\}\{E\epsilon + \mu\,\partial\epsilon/\partial t\}.$$

In our subsequent discussion of the stress generated by the gel cylinder we shall understand that the net force generated is the stress multiplied by $A_0/(1+\epsilon)$.

We shall arrive at our constitutive relation for active cytogel by modifying this simple relation; a more complete derivation based on microscopic considerations will be presented elsewhere.

In eq. (2) the elastic modulus, E, is constant only for the simplest materials and for very small strains. There are several effects which invalidate the linear response law, of which we shall discuss two of the most important.

First, since the cytogel is a fibrous material as the gel is strained the fibers align, producing a nonlinear stress–strain response (fig. 2a). That is, as the cylinder of cytogel is stretched its effective elastic modulus appears to increase. We write the passive component of cytogel elasticity

$$\sigma_P = E_P(\epsilon, c)\epsilon. \tag{3}$$

However, at high strains, the gel eventually yields (i.e. the actin filaments tear and the actin–actin crosslinks break), so that

$$\lim_{\epsilon \to \infty} \sigma_P = 0.$$

Second, it is impossible to squeeze the cylinder to zero length, since the gel contains vesicles and other virtually incompressible components. This can be accounted for by including the entropic contribution to the passive cytogel elasticity; this is given by the equation [14]

$$\sigma_E = \chi RT(\lambda - 1/\lambda^2), \quad (4a)$$

where $\lambda = L/L_0 = \epsilon + 1$ is the stretch ratio, χ is the concentration of cross-links in the gel, and T is the absolute temperature. This can be written as

$$\sigma_E = E_E(\epsilon, c)\epsilon, \quad (4b)$$

where

$$E_E(\epsilon, c) = \left[\frac{\epsilon^2 + 3\epsilon + 3}{(\epsilon+1)^2}\right] RT\chi(c), \quad (4c)$$

where $\chi(c)$ will decrease as c increases and actin–actin crosslinks are broken. Eq. (4) ensures that

$$\lim_{\epsilon \to -1} \sigma_E = -\infty.$$

As we discussed above, the gel elasticity is a function of Ca^{++} concentration as well, for at very high Ca^{++} levels there is so much solation of the gel (i.e. so much filament cutting by the solation enzymes) that the active sliding filaments can no longer transmit tension very far, and the apparent elastic modulus finally decreases. Thus using the above expressions we can write the total elastic stress, σ, as

$$\sigma = E(\epsilon, c)\epsilon, \quad (5)$$

where $E(\cdot)$, the elastic modulus, incorporates the nonlinearities due to fiber alignment, entropic elasticity and solation as described in fig. 2.

Next we must model the active traction forces generated by the cytogel. We do this in the simplest possible way as follows. Let $n(x, t)$ be the density of contraction units. Then the contribution of the active stress is

$$\sigma_{ACTIVE} = \tau(n, \epsilon, c)n, \quad (6)$$

where $\tau(n, \epsilon, c)$, the traction per fiber. τ is an increasing function of fiber density, n, because the probability of a myosin oligomer finding itself at the overlap of actin filaments with oppositely directed polarity increases as the actin filament density increases. n itself can change due to polymerization and depolymerization of filaments, which we shall discuss below.

τ decreases with increasing strain because of the 'overlap effect' discussed above: actomyosin fibers generate contractile forces by a sliding filament mechanism, as a fiber contracts more myosin crossbridges are brought into play, and the contractile force increases. We assume that, for fixed n and c, $\tau(\epsilon) \propto 1/(1+\epsilon)$, i.e. τ is proportional to the cross-sectional area of the cylinder.

τ is also a sigmoidally increasing function of the free calcium, c, for it is calcium which triggers the contractile machinery. This is the coupling to the chemical kinetics governing calcium release, which we will discuss below.

Adding together the active and passive contributions to the stress we obtain the total elastic stress in the cytogel strip

$$\sigma_{ELASTIC}(\epsilon, c) = E(\epsilon, c)\epsilon + \tau(n, \epsilon, c)n. \quad (7)$$

The surface $\sigma_{ELASTIC}(\epsilon, c)$ is shown in fig. 2d; its key features are (i) the calcium 'window' wherein maximum contractile stress is developed, and (ii) the tension the strip can develop increases as it shortens; that is, it grows stronger as it contracts.

Fig. 2 suggests the kind of experiments that must be done to quantify the phenomenological constitution of cytogel. For example, an isolated piece of cytogel can be suspended, say, between needles in a Ca^{++} buffered medium. For each of many experimentally imposed distances between the needles, the force exerted on the needles by the cytogel is measured. The resulting data fix a constant-c cross section of the surface in fig. 2d. If a sequence of stress vs. strain measurements are made at each of a sequence of 'clamped' calcium concentrations in the bathing medium, then the entire surface can be determined. Mathematical recipes for $E(\epsilon, c)$ and $\tau(n, c)$ can then be found using nonlinear parameter optimization algorithms so as to fit eq. (7) to the data. Of course, we must bear in mind that we are using c, the free calcium concentration, as a surrogate for a complicated regulatory kinetic scheme.

We shall model the viscous stress associated with cytogel motion as simple linear friction:

$$\sigma_{\mathrm{VISC}} = \mu(c)\,\partial\epsilon/\partial t,$$

where μ is the viscosity. The viscosity of cytogel is dependent on local calcium levels, which activate solation and gelation enzymes. This sol–gel transformation is similar to a phase transition, and is characterized by a sigmoidal curve, as shown in fig. 3.

As we let the cylinder length, L, shrink to zero, we obtain the stress tensor for a volume element. Here, the infinitesimal strain is given by $\epsilon = \partial u/\partial x \equiv u_x$, where u is the displacement, $(L - L_0)$,

$$\sigma = \mu\,\partial\epsilon/\partial t + E(\epsilon, c)u_x + \tau(n, \epsilon, c)n. \tag{8}$$

3.2. *Long range contributions to the elastic stress*

We obtained the stress tensor (8) by a few straightforward modifications of a Hookian constitutive relation. However, there is one contribution to the stress tensor that goes beyond mere variations on conventional elasticity theory: in many situations we must include the effect of non-local interactions on the local stress. That is, conventional elasticity assumes that the stress at a point is a function only of the strain at that point. However, for a fibrous material such as cytogel, the stress at any point derives not only from the fiber tensions at that point, but from nearby points as well because fibers thread through the gel over relatively long distances. For example, in a passive linearly elastic medium, the local stress, σ, at a point p is a function of the average strain in the vicinity of p:

$$\sigma = E\langle\epsilon\rangle = E\langle u_x\rangle, \tag{9}$$

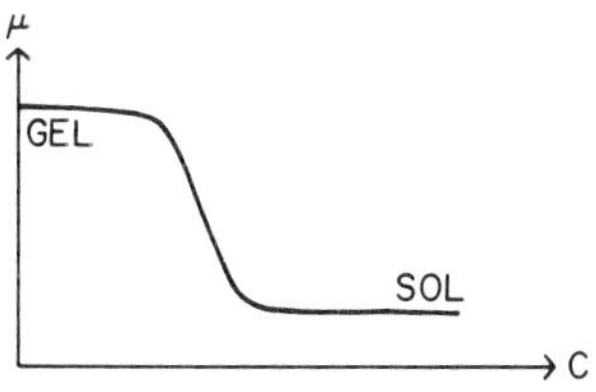

Fig. 3. The gel–sol phase transition is characterized by a sigmoidal drop in viscosity as calcium levels increase to the micromolar level.

where $\langle\cdot\rangle$ denotes a weighted average over a small neighborhood of p:

$$\langle\epsilon\rangle = \frac{1}{2r}\int_{x-r}^{x+r} W(x+r)\epsilon(r)\,\mathrm{d}r, \tag{10}$$

where W is the weighting function. This expression can be made more intuitively appealing as follows.

Consider a node in the gel network, as shown in fig. 1. If there are k fibers incident on the node, each of these fibers will in general have somewhat different elastic properties, and each will be stretched by different amounts and will generate different stresses:

$$\sigma_k = E_k\epsilon_k.$$

The average stress at that node is then

$$\langle\sigma_k\rangle = \frac{1}{k}\sum_k \sigma_k = \frac{1}{k}\sum_k E_k\epsilon_k. \tag{11}$$

When we pass to the continuum limit, we can obtain eq. (11) by defining the local average stress as $\sigma\langle x\rangle = \langle\sigma\rangle = \int\sigma(x+r)\,\mathrm{d}r/2r$, and a local average elastic modulus, $\langle E\rangle \equiv \int E(x+r)\,\mathrm{d}r/2r$. However, in order to define a local average strain we must take into account the network connectivity which enters into the geometric attenuation of the stress density about each point. Therefore, we must define a local average strain $\langle\epsilon\rangle$ as in eq. (11), with the weighting, $W(\cdot)$, accounting for the network connectivity and the distribution of internodal lengths.

Alternatively, by regarding σ as a flux, analogous to a diffusive flux of matter, then (10) says that the flux is directed not up the local displacement gradient, but rather up the local average gradient.

By expanding the integrand in a Taylor series about x one can show that σ is approximated by [15]

$$\sigma = Eu_x + E'u_{xxx} \equiv E(u_x - \beta u_{xxx}), \tag{12}$$

where $\beta \equiv -E'/E > 0$ to ensure that (12) represents passive elasticity. Othmer [15] gives recipes for computing the coefficients E and β from microscopic assumptions about the material. Here we need only be concerned with the qualitative features that spring from the constitutive relation.

With these modifications, we can finally write the stress tensor for the infinitesimal cylinder of cytogel as

$$\sigma = \mu(c)\,\partial\epsilon/\partial t + E(\epsilon, c)[u_x - \beta u_{xxx}] + n\tau(n, c, \epsilon). \tag{13}$$

3.3. *The equation of mechanical equilibrium for a cytogel cylinder*

In order that each volume element of the cylinder be in mechanical equilibrium with its surrounding volume elements the divergence of the stress field must exactly balance the body forces imposed on the volume element:

$$\operatorname{div}\sigma = \partial\sigma/\partial x = \boldsymbol{F}, \tag{14}$$

where $\boldsymbol{F}$ is the body force per unit volume of cytogel.

In the situation under study here the only body force stems from the attachment of the cortical cytogel to the cell membrane and to the subcortical cytoskeleton – microtubules, intermediate fibers and the trabecular lattice. Strictly speaking, these are surface forces, but in our 1-dimensional model they may be counted as body forces. We shall approximate them in the simplest possible way: as linear elastic restoring forces:

$$F = su. \tag{15}$$

Thus the equation of mechanical equilibrium governing the cytogel cylinder for small strains is

$$\partial/\partial x\{\mu(c)u_{xt} + E(u_x, c)[u_x - \beta u_{xxx}] + n\tau(n, \epsilon, c)\} - su = 0. \tag{16}$$

3.4. *The role of actin polymerization on spatial patterning*

The density function, $n(t, \boldsymbol{x})$ in the model equations accounts for the local concentration of contractile fibers. If we relax the constant volume constraint, but ignore the possibility of polymerization/depolymerization reactions, we can write a balance law for n simply as

$$\partial n/\partial t = -\partial/\partial x(nu_t). \tag{17}$$

However, these fibers, consisting of polymerized actin, myosin and associated proteins, can be assembled and disassembled by the cell in response to local chemical and mechanical conditions. The mechanisms the cell employs to regulate the state of its actomyosin cytogel has not been completely elucidated; however, we can investigate the role depolymerization can have on the behavior of the cytogel by modifying our model as follows.

Let $m(t, x)$ be the concentration of actin monomers, which we consider to be in local equilibrium with the polymerized fibers:

$$m \rightleftharpoons n. \tag{18}$$

To allow for depolymerization of the fibers we add to the fiber conservation equation (17) a reaction term:

$$\partial n/\partial t = -\partial/\partial x(n\,\partial u/\partial t) + P(m), \tag{19}$$

where $P(m)$ is the polymerization reaction which creates the fiber phase. Recalling that an actin polymer has different polymerization and dissociation rates at the 'barbed' (+) end and the 'pointed' (−) end, we model $P(m, n)$ by the simple polymerization kinetics [16]

$$\partial m/\partial t = -\partial n/\partial t \equiv P = k_+(m - m_+) - k_-(m - m_-), \tag{20}$$

where k_+ and k_- are the rate constants for addition to the (+) end of the polymer, and removal from the (−) end, respectively.

Next, we must write a balance equation for the monomer concentration, $m(t, x)$,

$$\partial m/\partial t = -\partial/\partial x(J_m) - P(m), \tag{21}$$

where J_m is the flux of monomers (#/cm^2 s). The flux term consists of two parts: (i) passive diffusion, and (ii) convection:

$$J_m = -D\partial m/\partial x + m v_c, \tag{22}$$

where D (cm^2/s) is the diffusion constant for actin monomers and v_c is the velocity with which monomers are convected along with the fluid phase of the cytoplasm (i.e. the cytosol). Strictly speaking, in order to model v_c we would need to construct a 2-phase flow model and solve for the pressure and velocity distributions. To avoid this complication we use the following approximate model.

We assume that the cytogel is a porous medium, and that the fluid velocity relative to the gel fibers is given by Darcy's law

$$v_c - \partial u/\partial t = \kappa \partial p/\partial x, \tag{23}$$

where κ is the permeability of the fibrous gel and p is the local hydrostatic pressure. Further, in order to avoid solving for the fluid (sol) motion we assume that the pressure is given as a unique function of the strain (i.e. equivalent to assuming a static equation of state),

$$p = p(\epsilon), \tag{24}$$

where $p(\epsilon)$ is a monotonically decreasing function of ϵ and $p(0) \equiv p_0$. Then (23) can be written as

$$v_c = \partial u/\partial t + k(\partial p/\partial \epsilon)(\partial \epsilon/\partial x). \tag{25}$$

Thus, since $\partial p/\partial e < 0$, when the gel goes into compression ($\epsilon < 0$) the increased hydrostatic cytosol motion can be directed opposite to the strain rate.

With (25) we can write the balance law for m entirely in terms of the strain rate:

$$\partial m/\partial t = D\partial^2 m/\partial x^2 - \kappa \partial/\partial x[\partial u/\partial t - m(\partial p/\partial \epsilon)(\partial \epsilon/\partial x)] - P(m). \tag{26}$$

3.4. 3-*Dimensional equations*

The higher dimensional generalization of the stress tensor (13) is not so straightforward, for we must account for the constraints of material frame indifference. However, for small strains, we can proceed by analogy with ordinary linear elasticity (e.g. Landau and Lifshitz, 1959 [17]) and write an approximate stress tensor as

$$\sigma = \left[\frac{E(\theta)}{2(1+\nu)}\left[\epsilon - \beta\nabla^3\epsilon + \frac{\nu}{1-2\nu}\theta \boldsymbol{I}\right] + \tau n \boldsymbol{I}\right] + \left[\mu_1\frac{\partial \epsilon}{\partial t} + \mu_2\frac{\partial \theta}{\partial t}\boldsymbol{I}\right], \tag{27}$$

where ν = Poisson's ratio, $\theta = \nabla \cdot \boldsymbol{u}$, the dilitation, μ_1 = the shear viscosity, μ_2 = the bulk viscosity, $\boldsymbol{I}$ = the identity tensor, and the strain, ϵ, is given by $\epsilon = \nabla \boldsymbol{u} + \nabla \boldsymbol{u}^T$. Here we have tacitly assumed that the fibrous gel is primarily a tension structure, i.e. shear forces are negligible, and so we can use the dilitation, θ, (a scalar quantity) to capture the nonlinearities due to fiber alignment.

The model equations in the higher dimensional case are formidable, even for linear analysis, and we shall present numerical studies in a subsequent paper.

3.5. *Chemical triggering of mechanical contractions*

The proximal event initiating fiber contraction is the release of a "trigger chemical", c, which is stored endogenously in the vicinity of the filaments, or as an influx from the outside. We have referred to c as free calcium, but we understand that calcium is but the surrogate – or rate limiting step – in the regulatory kinetics controlling contraction. Moreover, since we are only concerned with the morphogenetic phenomena associated with cell shape change we have not dealt with the (unknown) details of the mechanochemical transduction. Rather we have modeled the effect of the trigger chemical phenomenologically by specifying that the traction, τ, increases as the concentration of trigger chemical increases. Thus the trigger for cytogel contraction revolves around the control of the trigger chemical, c. The concentration of trigger chemical within each volume element of cytoplasm is governed by the conservation equation

$$\partial c/\partial t = -\nabla \cdot \boldsymbol{J} + A(c, \epsilon) - r(c), \tag{28}$$

where $\boldsymbol{J}$ is the flux of c, $A(c, \epsilon)$ is the autocatalytic release of c from its vesicular storage, and $r(c)$ is the rate c is resequestered.

The flux of c can be via diffusion and/or convection

$$\boldsymbol{J} = -D\nabla c + \boldsymbol{v}_c c,$$

where D is the diffusion coefficient for c and $\boldsymbol{v}_c$ is the convective fluid velocity discussed above (eq. (25)).

As we discussed above, the release of sequestered calcium can be triggered by calcium itself, or by mechanically straining the cytogel, which probably depolarizes the vesicle membranes and leads to the release cascade. The release curve, $A(c, \cdot)$ is thus a sigmoidally increasing function, similar to that of $\tau(c)$; the particular form we have employed is

$$A(c, \epsilon) = \alpha c^2/(1 + \beta' c^2) + \gamma\epsilon,$$

where γ is the release rate per unit strain and (α, β') characterize the autocatalytic release kinetics. The resequestration kinetics we assume to be first order:

$$r(c) = rc.$$

The kinetic equation governing c is thus

$$\partial c/\partial t = \alpha c^2/(1 + \beta' c^2) + \gamma\epsilon - rc - \nabla \cdot \boldsymbol{J},$$

where $\boldsymbol{J}$ is the flux of c from outside the volume element.

With these kinetics the conservation equation for c takes the form

$$\partial c/\partial t = D\nabla^2 c - \nabla \cdot (c\boldsymbol{v}_c) + \alpha c^2/(1 + \beta' c^2) + \gamma\epsilon - r(c). \tag{29}$$

The nature of the kinetics can be seen from the graph of this equation shown in fig. 4a. There can be one or two stable equilibria, depending on the parameter value. There are two mechanical equilibria corresponding to the two chemical equilibria: relaxed and contracted. It is enlightening to examine the behavior of a small strip of cytogel of length L on the (L, c) phase plane, as shown in fig. 4b. The nullclines are the loci

$$\alpha c^2/(1 + \beta' c^2) + \gamma\epsilon - rc = 0,$$
$$E(\epsilon, c)\epsilon + \tau(n, \epsilon, c)n = 0.$$

We can regard α as a 'developmental parameter' that can change slowly over time, for example, as the cell's trigger chemical machinery 'matures'. This change takes place on a time scale of the order of a division time – much longer than the dynamics of contraction. Therefore, we can represent the effect of this change by a family of curves in the (L, c) phase plane, as shown in fig. 4c. For each value of α we can compute the equilibrium states $(\bar{L}, \bar{c})$ from the intersection of the two nullcline curves. In fig. 4d we have plotted the locus of these steady states in the $(\bar{L}, \alpha)$-plane. The curve that emerges is divided into several domains. At low values of α there is but a single (relaxed) steady state. As α is increased the dynamics enters a region of 'excitability': a small perturbation is quickly damped out, but a perturbation beyond a threshold value initiates a large excursion which finally returns to the original equilibrium (see figs. 4e and 8c). Next, at a critical value of $\alpha = \alpha^*$, the system bifurcates: suddenly two new equilibria appear. In this region the system is 'bistable', and can exist in either the relaxed or contracted state. (Actually, as α increases toward α^*, a subcritical Hopf bifurcation occurs and the relaxed equilibrium becomes unstable before α reaches α^*.)

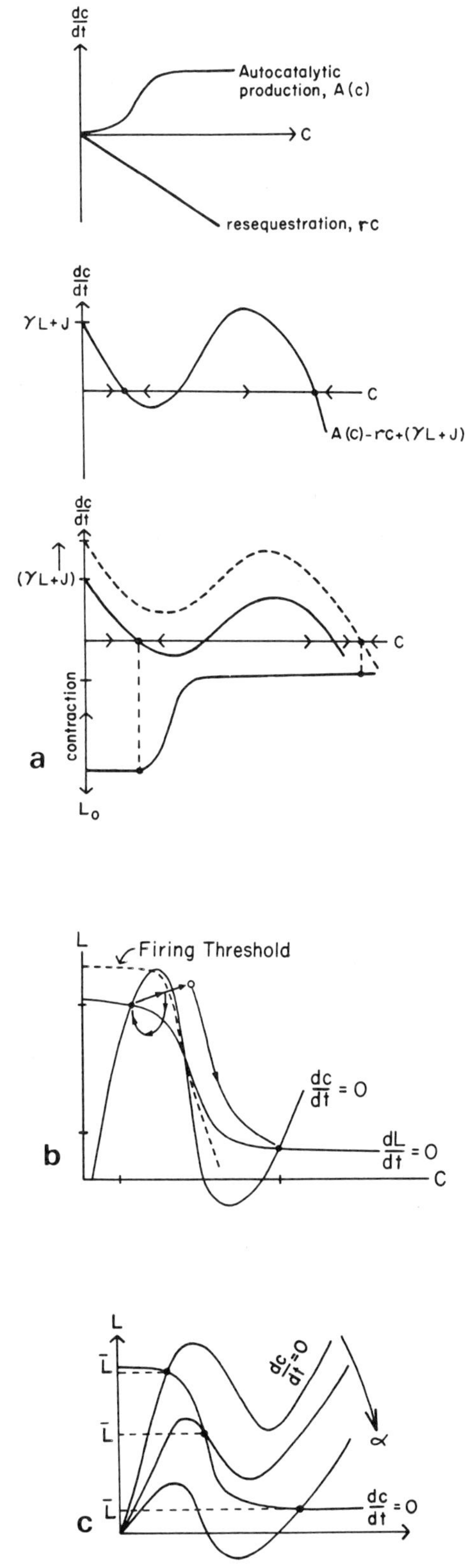

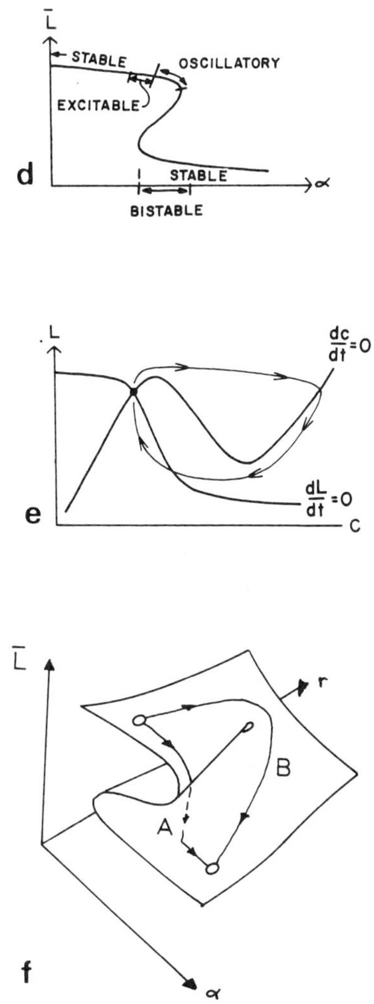

Fig. 4. a) The kinetics of c is governed by the autocatalytic release, $A(c)$, and the resequestration rate, rc. The kinetics has 1 or 2 stable equilibria, depending on the degree of stretch activation, γL and/or chemical flux, J. Corresponding to the two chemical equilibria there are two mechanical equilibria for the cytogel strip. b) The mechanochemical kinetics on the $(L\text{–}c)$ phase plane for the case where there are two stable attractors. c) Varying the kinetic parameter, α, varies the equilibrium length, $\bar{L}$. d) The locus of equilibrium states as a function of α passes through regimes of stability, excitability, bistability and oscillatory dynamics. e) The phase plane for excitable kinetics: small perturbations are damped, but larger perturbations traverse a large excursion before settling to the equilibrium state. f) The locus of equilibrium lengths for a cytogel element as a function of the autocatalytic amplification parameter, α, and the resequestration rate, r. Along path A the morphogenetic transition shown dashed (-----) occurs much faster than the slow changes in α. Along the developmental path B the "genetic" and "morphogenetic" rates are commensurable.

If we include a second slowly varying developmental parameter, say the resequestering rate, r, then the surface $\bar{L}(\alpha, r)$ can take on the appearance shown in fig. 4f (this is not a cusp catastrophe since the kinetic equations are not gradient flows).

Thus the model cytogel consists of the force equation (16), and the conservation equations for the actomyosin fibers, n (19), the monomers, m (26) and the trigger chemical, c (29). We emphasize that the various constitutive assumptions we have made on the form of $\tau(c)$, $A(c)$ etc. are generic: only the qualitative shape of the curves is relevant for the morphogenetic process we discuss; variations in the details of these curves will make no appreciable difference in the predictions of the model.

The above equations are far too complex to yield to analysis; however, one can obtain interesting specializations of the model corresponding to particular biological situations. In the following sections we shall use various approximations to the model equations to discuss a number of applications.

4. Applications of the model

In this section we describe how the model can be applied in a number of specific contexts. On the basis of the model's bifurcation behavior, we divide these applications into three general areas: spatial patterns, wave phenomena, and oscillations. Within each area we briefly describe a few biological settings wherein we feel that the model is relevant, and report our (sometimes meager) progress toward analyzing and/or simulating the situation.

4.1. *Spatial patterns*

4.1.1. *Linear analysis of spatial instabilities*

Consider first the simpler case wherein there is no polymerization/depolymerization, $P = 0$. The mechanical equilibrium of the cytogel cylinder is specified by the steady state solutions $u(x)$, $n(x)$ to eqs. (16) and (17). It is easy to see that the solutions $\{u = 0,\ n = n_0 = \text{constant}\}$ satisfy these equations. However, we are interested in the nonuniform solutions, and so we shall examine the behavior of the model as the cytogel commences to contract. That is, we examine the bifurcation behavior of the model by considering the calcium concentration as a parameter, and gradually increasing τ by increasing c.

First we linearize about the equilibrium solutions $\{u = 0,\ n = n_0\}$ to obtain the equations

$$-\beta E_0 u_{xxxx} + E_0 u_{xx} + \mu u_{xxt} - su + \tau n_x + n_0 \tau'(n_0) n_x = 0, \tag{30}$$

$$n_t + n_0 u_{xt} = 0, \tag{31}$$

where u and n now stand for the perturbation values, and the subscript 0 indicates evaluation at the equilibrium values.

It is sufficient to examine the behavior in the infinite domain, or with periodic boundary conditions. Therefore, we employ the substitutions

$$\{u(x,t), n(x,t)\} \sim e^{\lambda t + ikx}, \tag{32}$$

obtaining the dispersion relation

$$\lambda(k^2) = (1/\mu)\{n_0[\tau(n_0) + n_0\tau'(n_0)] - E_0 - E_0\beta k^2 - s/k^2\}, \tag{33}$$

whose graph is shown in fig. 5.

From fig. 5 we see that as the traction parameter, τ, is increased, the system becomes spatially unstable when

$$n_0[\tau(n_0) + n_0\tau'(n_0)] > E_0 + 2(E_0\beta s)^{1/2}, \tag{34}$$

that is, when the aggregate traction at a point (τn_0) exceeds the elastic restraining forces of the gel (E_0, β) and its anchors.

When τn_0 just exceeds the elastic restraining forces it can be shown that the system then evolves to a periodic array of fiber aggregations whose spacing can be estimated from the calculated

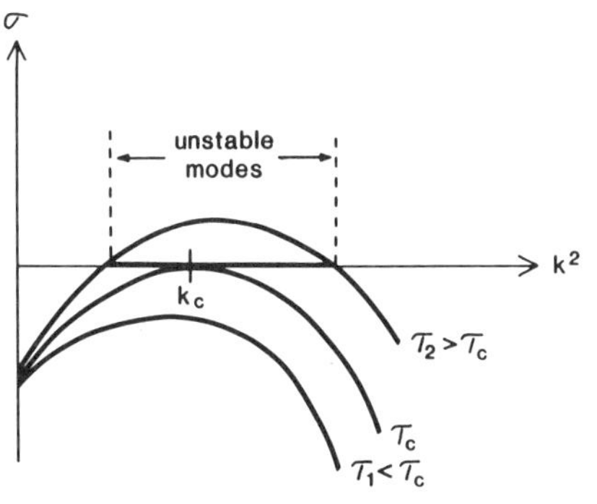

Fig. 5. The dispersion relation $\sigma(k^2)$ (eq. (33)). The system becomes linearly unstable when the traction, τ, increases to the point where the curve is tangent to the k^2 axis at k_c. The spacing of the foci are estimated from eq. (35).

critical wavenumber, k_c as

$$\text{spacing} \approx 2\pi/k_c = 2\pi(\beta E_0/s)^{1/4}. \tag{35}$$

From (34) we see that the long-range interaction effect contained in the biharmonic term is central to the formation of spatial patterns, as is the anchor elasticity, s.

The necessary condition for spatial patterns to form can be expressed as a condition on the dimensionless groupings of the parameters:

$$(\tau n_0/E_0) > 1 + 2(\beta s/E_0)^{1/2}. \tag{36}$$

Thus for a fixed τ, a reduction in β or s, for example, could result in patterning.

In the next section we shall describe some numerical simulations we have performed on a finite element version of the above model equations. These simulations verify that the model equations do indeed support spatial patterns as the linear analysis predicts – although the linear estimate of the wavelength is only accurate in certain cases. A substantial amount of additional analysis and simulations will be necessary to completely delineate the behavior of the model.

For the case when $P(m) \neq 0$, linearization of eqs. (16), (19) and (26) leads to a dispersion relation with the same qualitative features as fig. 5: a finite range of unstable wavelengths. Now, however, the critical wavelength and the spacing will depend on the parameters governing the mobility and equilibrium of the monomer phase as well as on the elastic properties of the gel. Indeed, one can now obtain spatial patterns even with $\beta = 0$, i.e. without long-range fiber tractions.

4.1.2. *Epithelial cell patterns*

In an earlier publication we investigated how cells in an epithelial layer coordinated their shape changes so as to bring about well defined global changes in the geometry of the layer [13].

In that investigation we represented a cell by a viscoelastic finite element model wherein the apical (top) surface was endowed with contractile properties. The finite element model we employed in that study can be derived independently, or as an approximation to the continuum model we have described here. Of course, there is no unique correspondence between finite element models and continuum equations, so physical reasoning has played a large part in constructing the simulations. One point here is worth noting: the long-range effect we have included in the continuum equations is automatically taken into account in the finite simulation equations; alternatively, by taking the limit of a finite difference scheme leads to the higher order terms (M. Slemrod, pers. comm.).

Epidermal placodes. The early morphogenesis of feather and scale primordia involves the formation of periodic thickenings of the epidermis, called placodes, which are underlain by aggregations of dermal cells, called papillae. In a separate publication we describe a model for dermal placodes [18]. Here we are concerned only with placode formation. Fig. 6 shows a model of an epidermal layer constructed from the finite element model. This consists of an array of cells attached to the substratum by elastic tethers, which represent both the fibrous basal on the cells and the collagen 'anchor filaments' which attach epidermal cells to the dermal layer [19, 20]. When the cellular array is triggered at one end two waves sweep over it: a fast mechanical stress wave followed by a slower

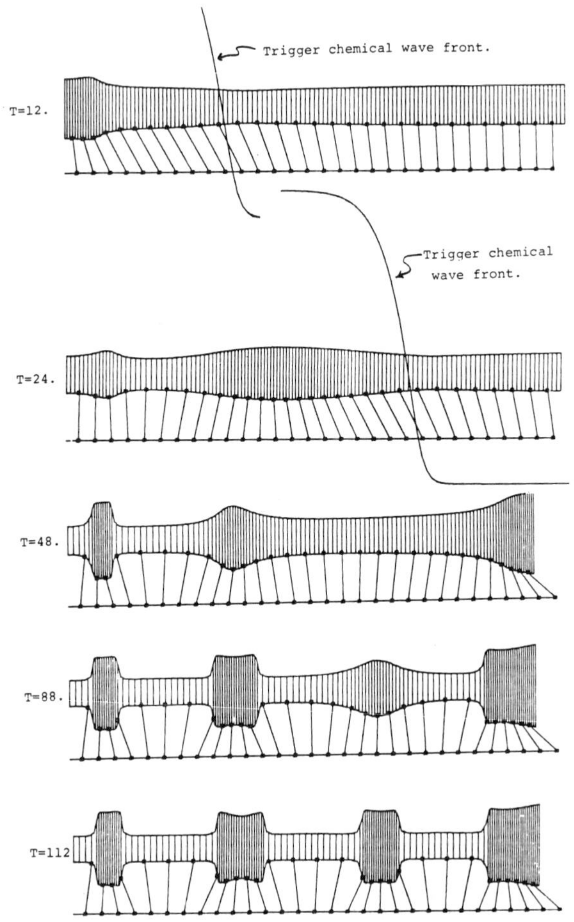

Fig. 6. Epidermal placode formation. An array of model epithelia is tethered to the substrate by a basal lamina and collagen anchor filaments – here represented by linear elastic restoring forces. The frames are taken from a computer movie of the solution to the model equations. The sequence shows the following phenomena: when the system is triggered at the left end a contraction wave sweeps over the layer from left to right, leaving behind it an activated region which commences to contract. Gradually a periodic array forms whose spacing depends on the viscoelastic parameters.

trigger chemical wave. Following the passage of the waves the epidermal cells are fired, and arrays of columnar cells (placodes) commence to form spontaneously. The triggering of the cells may be produced by either a chemical activation wave, which sweeps cell-by-cell across the sheet, or by a mechanical (i.e. stretch activation) wave which triggers the whole sheet almost simultaneously.

Folding of epithelial layers. In Odell et al. [13] we investigated the folding of sheets of epithelial cells in various settings using the finite element model described in [13]. In fig. 7 we reprint a simulation from that study showing the process of gastrulation in the sea urchin wherein an initially spherically symmetrical monolayer of cells invaginates to commence formation of the animal's internal structure. The finite difference model here consists of an array of passive viscoelastic elements and one active (i.e. triggerable) element at the apical end. The simulation represents a cross-section of the spherical blastula. When the topmost cell is chemically triggered a wave of deformation spreads out laterally and, because of the bending moment produced by the contraction of the apical surface, the sheet invaginates.

In our initial studies of epithelial folding we assumed that cells trigger one anothers contractions by mechanical activation (the stretch release term, $\gamma\epsilon$, in eq. (29)). In many situations, however, epithelial cells are chemically coupled by gap junctions which allow transport of small molecules and ions. If the chemical signal for contraction is passed

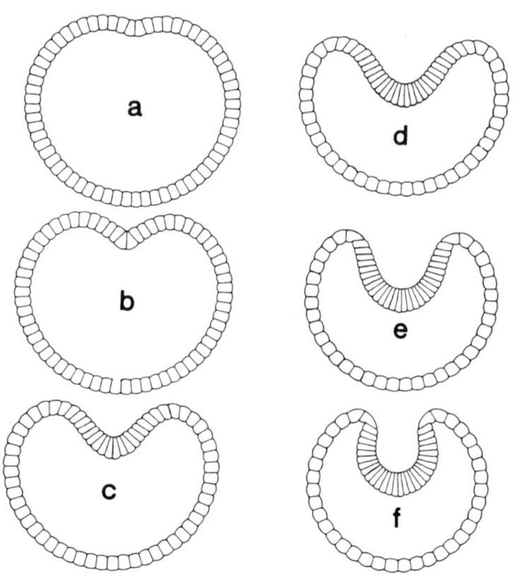

Fig. 7. Simulation of gastrulation in the sea urchin (see [13]). A hollow monolayer of epithelial cells is shown in cross-section. The frames are exerpts from a computer-generated movie showing the solution to the finite element model equations.

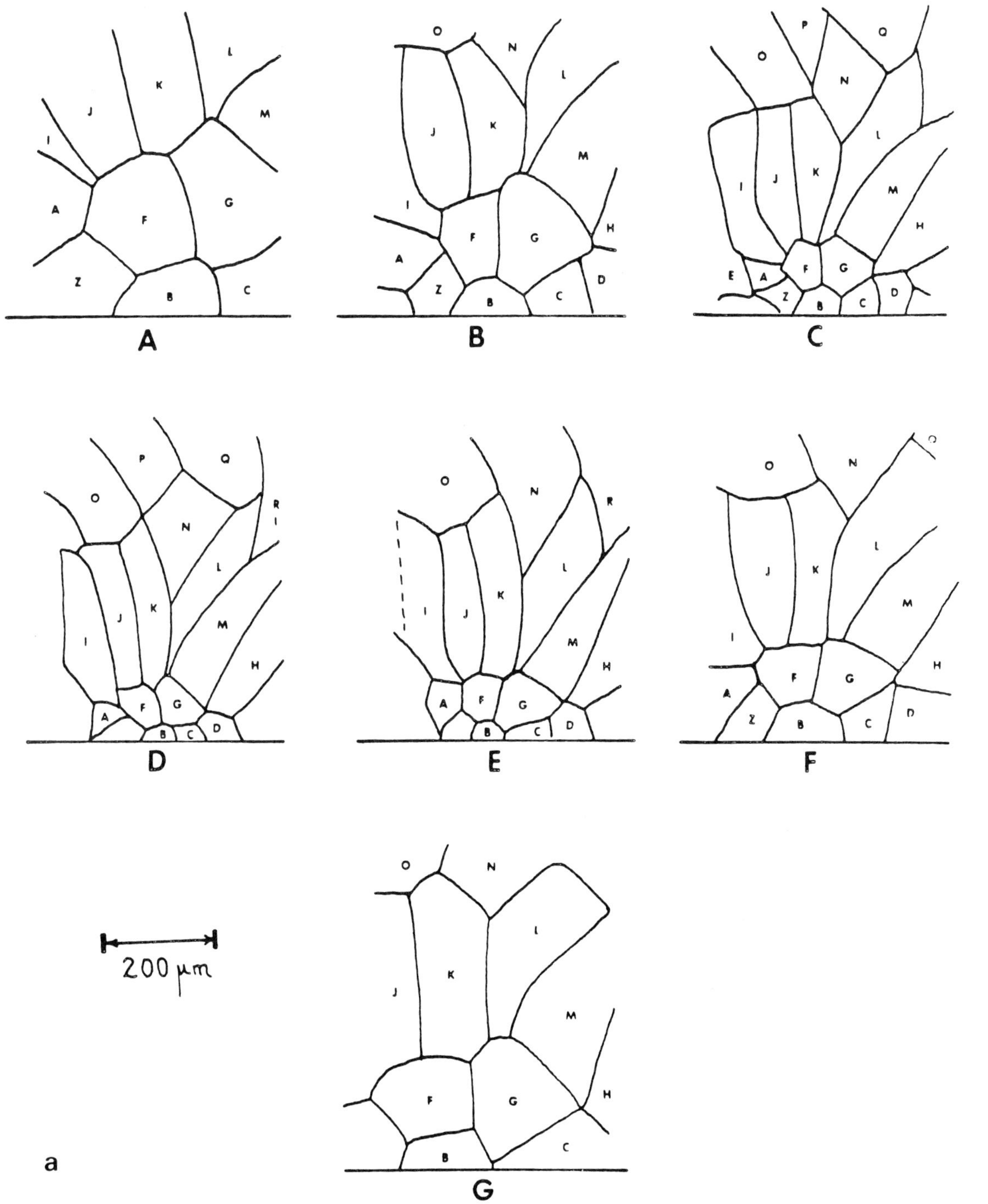

Fig. 8. a) Tracings of the sequence of cell shapes of a focal contraction on the surface of a salamander embryo (T. tarosa: late stage 8 blastula). Tracings were taken from a film provided to us by A. Jacobson, University of Texas, Austin. b) A plot of the circumference of the central cell (cell G) as a function of time. c) Simulation of (b) using an apical circumferential microfilament bundle modeled by the cytogel equations.

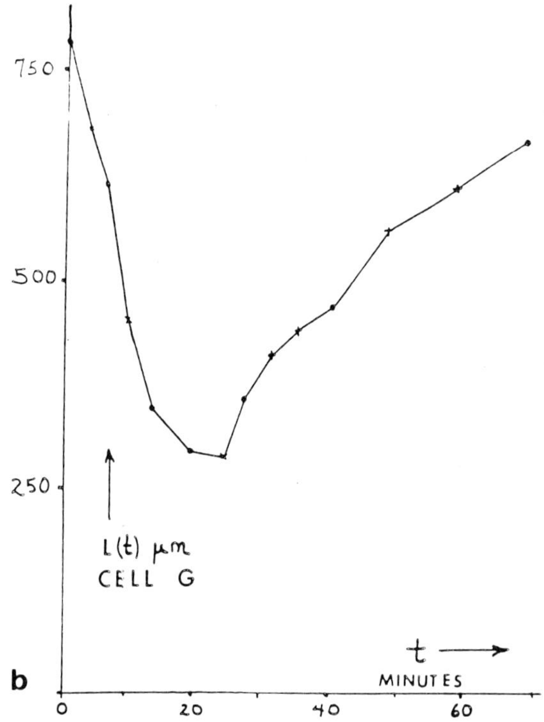

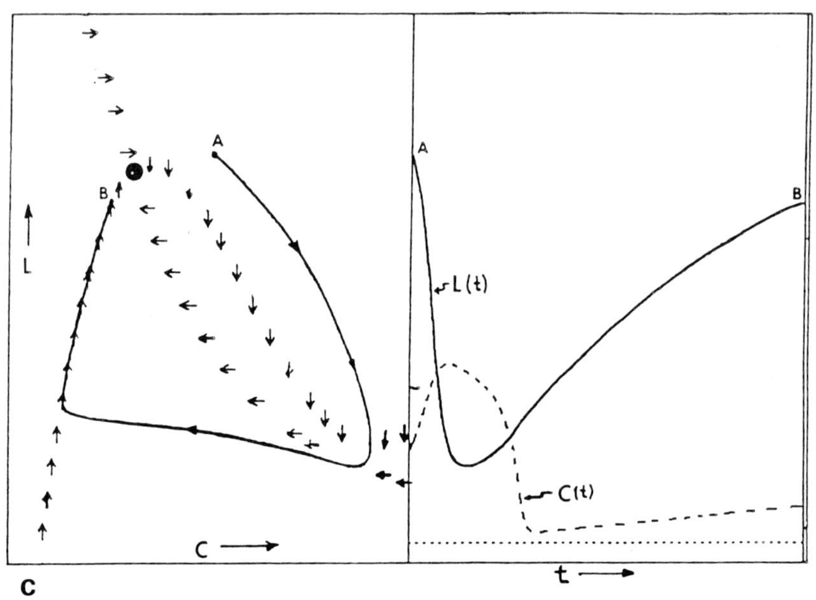

Fig. 8. continued

from cell to cell through these junctions, then one must include in the model a term for cell-to-cell diffusion. (This is not quite as straightforward as it seems, for one must account for the dilution effect of the trigger chemical.) In the following discussions both chemical and mechanical coupling may play a role, and either, or both, can be included in the model equations. In general, we find that direct chemical triggering of cytogel contraction produces much more localized cell shape changes than does stretch activation. That is, mechanical forces are equilibrated over many cell diameters and provide a mechanism for long-range coordination of cell movements, whereas gap-junction communication greatly localizes contraction triggering.

Focal contractions. Prior to and during neurulation in amphibia and other organisms the epithelial surface is punctuated by episodes of local and transient cell shape change, so-called focal contractions. Whatever the functional significance of this contractile activity, if any, they indicate clearly that the embryonic epithelium is a mechanically active surface. The dynamics of these focal contractions can tell us something about the mechanical properties of the epithelial sheet.

In fig. 8 we have plotted the time course of a typical focal contraction observed on a salamander embryo (stage 8 *Taricha tarosa*), as observed in time-lapse cinematography. (The films were supplied to us by A. Jacobson, pers. comm.). Fig. 8 also shows a graph of the time course of the model's simulation of such a contraction, produced by triggering a focal cell whose dynamics is in the 'excitable' regime. In order for the model to match the time course of the contraction, the mechanical activation parameter, γ, must be quite small so that most of the intercellular communication is via junction transport.

4.1.3. *Spatial patterns in cytogel*

Microvilli and stereocilia. On the surface of the 'hair cells' in the inner ear there are rod-like projections called stereocilia [20]. These projections, and the microvilli on the surrounding supporting cells are arranged in a quite regular hexagonal array. In [18] we show how the contractions of the actin cortex can lead to regular spatial arrays such as this. That is, a uniform density distribution of actin is spatially unstable, and will break up into periodic condensations. The linear analysis in the last section demonstrates this type of spatial instability; only a nonlinear analysis will show that this packing is hexagonal, although it is easy to show that the linearized solutions support hexagonal symmetry. In reality, it may not be necessary to demonstrate hexagonal packing numerically, since it appears that the aggregation foci may appear sequentially from an initiation site (the kinocilium). If this is true, the problem of establishing the stable spatial configuration is simpler, since we need analyze only the formation of one row at a time.

Other cytogel patterns. There are a number of other systems wherein actin gels may be involved in generating spatial patterns. For example, the pattern of tentacles on *Hydra* and anterior hairs on *Tetrahymena* commence as local protuberances arranged periodically around the anterior tip. From the linear analysis above it is easy to see that a cytogel annulus will give rise to periodic thickenings which could provide the nucleation sites for subsequent growth by other mechanisms. In general, we can expect that cytogel, when stimulated to contract, will not remain spatially homogeneous, but will develop spatial patterns according to the geometry and boundary conditions imposed on it.

4.1.4. *Contraction waves and oscillations*

If the model parameters are adjusted so that the mechanochemistry is in the 'excitable' regime, then the cytogel continuum can exhibit contraction waves and oscillations. These have counterparts in a number of biological situations which we briefly describe here.

Plasmodial oscillations in physarum. The plasmodial strands of the slime mold *Physarum* have received much attention because of their periodic contraction rhythms and oscillatory 'shuttle streaming'. In Oster and Odell [12] we present a model for plasmodial oscillations which consists of a hollow cytogel cylinder filled with solated cytoplasm. We show that if the kinetics is in the excitable regime, the model plasmodium will exhibit rhythmic contractions which drives a periodic fluid flow in the core of the cylinder. This model unites a variety of experimental observations on plasmodial dynamics and the phasing of contraction rhythms with chemical and morphological variables.

We are presently using the model cytogel to investigate a variety of other contractile phenomena, including cytokinesis in *Xenopus* and birds, epiboly in teleost fish, somite segmentation in amphibia, and lamellapodia in motile cells. We shall report on these in subsequent publications.

References

[1] D.H. Taylor, S. Hellewill, H. Virgin and J. Heiple, in: Cell Motility, S. Hatano, ed. (Univ. Park Press, Baltimore, 1979).

[2] J. Condeelis and D.L. Taylor, J. Cell Biol. 74 (1977) 901–27.

[3] D.H. Taylor, J. Condeelis and J. Rhodes, in: Proc. Conf. on Cell Shape (Alan Liss, New York, 1977), p. 581.

[4] A. Weeds, Nature 296 (1982) 811–815.

[5] T. Stossel, H. Hartwig, H. Yin and O. Stendahl, Biochem. Soc. Symp. 45 (1981) 51–63.

[6] L. Jaffe, Ann. N.Y. Acad. Sci. 339 (1980) 86–101.

[7] M. Endo, Physiol. Rev. 57 (1977) 71.

[8] E. Bittar, Membrane Structure and Function, vol. 4 (Wiley, New York, 1981).

[9] K. Kuba and S. Takeshita, J. Theor. Biol. 93 (1981) 1009–1031.

[10] N. Kamiya, in: Cell Motility: Molecules and Organization, S. Hatano, H. Ishikawa and H. Sato, eds. (Tokyo Press, Tokyo, 1979), pp. 399–414.

[11] K. Wohlfarth-Bottermann, J. Exp. Biol. 67 (1979) 49–59.

[12] G. Oster and G. Odell, Cell Motility, submitted.

[13] G. Odell, G. Oster, B. Burnside and P. Alberch, Dev. Biol. 85 (1981) 446–462.

[14] P. Flory, Principles of Polymer Chemistry (Cornell Univ. Press, Ithaca, N.Y., 1953)

[15] H. Othmer, J. Math. Biol., to appear.

[16] M. Kirschner, J. Gerhart, K. Hara and G. Ubbels, in: The Cell Surface: Mediator of Developmental Processes, S. Subtelny and N. Wessells, eds. (Academic Press, New York, 1980), pp. 187–215.

[17] L. Landau and E. Lifshitz, The Theory of Elasticity (Pergamon, London, 1959).

[18] G. Oster, J. Murray and G. Odell, J. Math. Biol., (in preparation).

[19] P. Sengel, Morphogenesis of Skin (Cambridge Univ. Press, Cambridge, 1976).

[20] A.J. Hudspeth, Scientific American 248 (1983) 86–93.

[21] N. Wessells, Tissue Interactions in Development, (Benjamin, Palo Alto, 1977).

Physica 12D (1984) 351–359
North-Holland, Amsterdam

WETTING, MULTILAYER ADSORPTION, AND INTERFACE PHASE TRANSITIONS

M.R. MOLDOVER and J.W. SCHMIDT
Thermophysics Division, National Bureau of Standards, Washington, D.C. 20234, USA

When two fluid phases coexist near their mutual critical point one fluid phase forms a layer which intrudes between the other fluid phase and any third phase that happens to be present. As the two fluid phases are taken away from their critical point a phase transition often occurs such that the intruding layer vanishes. We present recent evidence that the transition from the complete wetting (intruding layer) configuration to the incomplete wetting (three-phase contact) configuration is a first-order phase transition. The thickness of the intruding layers has been measured for diverse systems. These data are not fully understood. The intruding layers are extreme examples of multilayer adsorption. Certain theories predict that a "prewetting" transition from high adsorption to low adsorption is associated with the transition from complete wetting to incomplete wetting. Our search for this transition is inconclusive.

1. Introduction

In this short review, we shall describe the generic phase diagram predicted by Cahn's theory [1] of "Critical Point Wetting." We shall then describe the experimental efforts to test the relevance of his theory to a variety of real phase diagrams with emphasis on our own work. We will also note that the presence or absence of wetting layers does play a significant role in heterogeneous nucleation and in the solidification of ordered composite solids from melts under some conditions. Desirable solidification patterns have been obtained by varying the solidification conditions in a manner suggested by the generic phase diagram. We expect that other applications of the generic phase diagram to technology will appear.

2. The generic phase diagram

In 1977, Cahn used the van der Waals theory of interfaces to consider two-phase mixtures of fluids near their critical point in contact with an arbitrary third phase [1]. Cahn obtained the generic phase diagram which is shown in fig. 1 projected on to a temperature–density plane. The region of coexistence of two fluid phases near their critical (or consolute) point is shaded. The horizontal dashed line and the dashed curve connected to it represent the locations of the wetting phase transition and its associated prewetting phase transition for this system of fluids in the presence of some

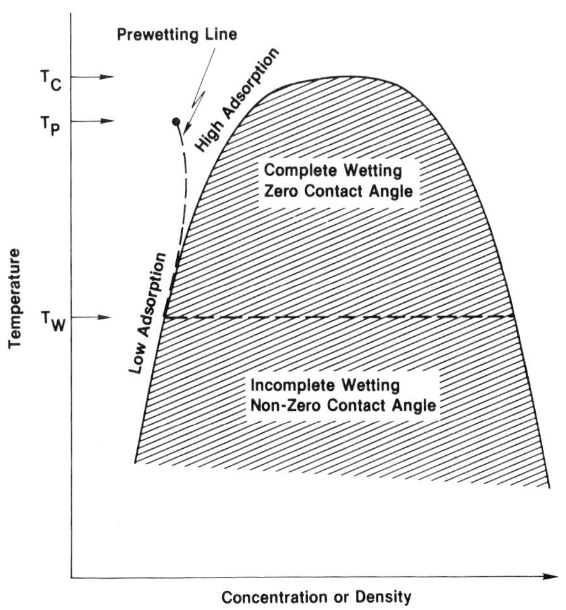

Fig. 1. Generic phase diagram. The first-order transition from complete wetting to incomplete wetting is indicated by the horizontal dashed line at T_w. The locus of (prewetting) transitions from high adsorption to low adsorption is indicated by the dashed line between T_w and T_p.

0167-2789/84/$03.00

particular third phase. (The third phase could be a solid, liquid, or gaseous phase.) If a different third phase were present, the locations of the horizontal dashed line and associated prewetting line would be changed. The prewetting line might well appear on the right side of the two-phase region; however, the topological features would remain unchanged unless a special symmetry were to occur.

In fig. 1 the horizontal dashed line is the transition from incomplete wetting to complete wetting which occurs at the temperature denoted by T_w. Below T_w, three phase contact occurs between the two critical fluids and the non-critical phase (see fig. 2). As T_w is approached from below, one of the dihedral angles (θ) characterizing the three phase contact gradually decreases and then vanishes at T_w. Above T_w, one of the two critical phases completely wets the non-critical phase and will form a film which intrudes between the other critical phase and the non-critical phase (fig. 2, top). This intrusion will be complete unless some additional field (e.g., gravity) is present to compete with the surface forces. The wetting transition is expected to be first order: a discontinuity occurs in

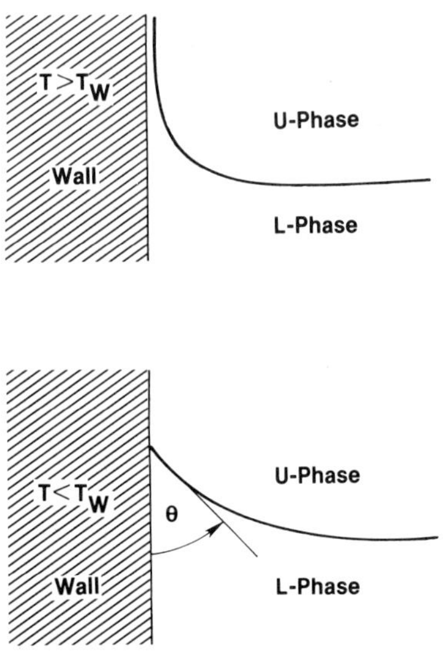

Fig. 2. The wetting transition when the non-critical phase is a wall. Top: Complete wetting. The lower phase wets the wall. The upper phase does not contact the wall. Bottom: Incomplete wetting. The lower phase meets the wall at a finite contact angle θ. There is a line of three-phase contact.

the temperature derivative of the surface tension between the non-critical phase and the critical fluids.

The predominant chemical species in the critical phase which wets the non-critical phase will be preferentially adsorbed on the surface of the non-critical phase. In the Cahn picture, a discontinuous increase in this adsorption occurs upon crossing the prewetting line (dashed curve on fig. 1) in a direction towards the curve of two-phase coexistence. In fig. 3, we have sketched concentration profiles in the vicinity of the non-critical phase (say a wall). The profiles show an increase in adsorption as the concentration far from the wall is increased at constant temperature from a concentration to the left of the prewetting line to a concentration between the prewetting line and the coexistence curve. When the concentration far from the wall is exactly C_{pw} (i.e., on the prewetting line) two concentration profiles can coexist. These correspond to two surface phases, one of high adsorption and one of low adsorption. As the temperature is raised the difference between the high adsorption state and the low adsorption state gradually decreases. The difference vanishes at the surface critical point at the temperature T_p. The prewetting line also ends at the wetting transition where it is tangent to the coexistence curve [2, 3].

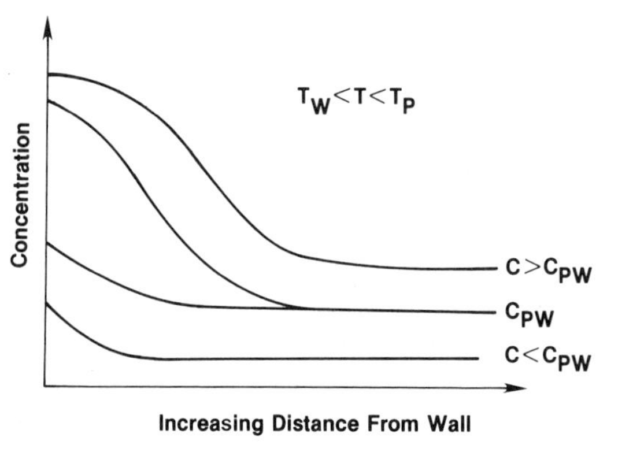

Fig. 3. Adsorption as a function of distance from the wall in the vicinity of the prewetting curve. When the bulk composition falls on the prewetting curve (i.e. $C = C_{pw}$), two profiles are possible, one of high adsorption and one of low adsorption. When the bulk composition is very close to that of two-phase coexistence ($C > C_{pw}$ on fig. 1), only a high adsorption profile occurs.

This discussion of the generic phase diagram (fig. 1) and its interpretation (figs. 2–3) has repeatedly used the phrases "critical fluids" and "non-critical phase." The vagueness of these phrases is deliberate. The same qualitative phenomena are expected either when both critical fluids are liquid mixtures near a consolute point or when one critical fluid is a liquid and the other is its vapor and they are both near their liquid–vapor critical point. (In the latter case the word "density" should be substituted for the word "concentration" throughout the discussion.) The same qualitative phenomena are expected when the non-critical phase is a solid, a fluid, or even a vacuum.

Before concluding this discussion of the generic phase diagram we note that several theories [3–8] (which are essentially generalizations of the one which Cahn used) predict that the transition from complete wetting to incomplete wetting will be a second-order phase transition when the (averaged) long-ranged part of the interatomic potential between the like species (i.e. the components which make up the two critical phases) exceeds the long-ranged potential between the unlike species (i.e. between the components of the critical phases and the non-critical phase. Under these same conditions, the prewetting curve will be absent.

3. Tests of the generic phase diagram

The generic phase diagram described just above results from the van der Waals theory of interfaces. Thus, it is expected to be quite common. Subsequent to Cahn's treatment similar phase diagrams have been published by numerous authors [3–10]. We shall not review the various theories here. Instead, we shall discuss several experiments which test features of the generic phase diagram. We discuss the following features in turn: (A) complete wetting near critical points, (B) the order of the transition from complete wetting to incomplete wetting as the distance from the critical point is increased, and (C) the search for the prewetting line.

3.1. *Complete wetting near critical points*

Heady and Cahn [11] observed the phenomenon of complete wetting of a vapor phase by one of two liquid phases of a binary mixture near an upper consolute point. Moldover and Cahn [12] showed that this phenomenon was a common occurrence in many different mixtures of organic liquids near consolute points. Perepezko et al. [13] have provided evidence of essentially the same phenomenon in alloy mixtures which separate into two partially miscible liquid phases prior to freezing. Zabel et al. [14] studied single crystals of niobium with substantial amounts of hydrogen dissolved in them. As such crystals are cooled they separate into hydrogen-rich and hydrogen-poor phases below a consolute point at 171°C. By using X-ray scattering, Zabel et al. were able to show that the hydrogen-poor phase completely surrounded the hydrogen-rich phase. Thus, even in this crystalline system one may say that near a consolute point one of two near-critical phases (the hydrogen-poor-one) completely wets a non-critical phase (the vacuum or the very thin oxide layer on the periphery of the crystal).

The complete wetting of container walls by one of the two liquid phases of binary mixtures near consolute points was also observed by Moldover and Cahn [12]. Another more quantitative study of this phenomenon has been made by Pohl and Goldburg [15].

The complete wetting of solids by a pure liquid or a liquid mixture in the presence of its vapor is hardly remarkable, even far from a liquid–vapor critical point. Indeed, the capillary rise method of measurement of surface tension is often (but not always) based on the assumption that the angle of contact between the liquid and the capillary wall is zero. In cases where non-zero contact angles are observed far from critical points, the contact angle usually decreases towards zero as the temperature is increased (towards the liquid–vapor critical point).

There are some recent measurements of non-zero contact angles between liquid–vapor interfaces and vertical solid surfaces quite close to liquid–vapor critical points. If these measurements by Stepanov et al. [16] are confirmed they would contradict the picture we are presenting here. We shall now digress to a discussion of the difficulties inherent in these measurements in order to justify our tentative decision to discount them when discussing tests of the generic phase diagram.

The measurements of contact angles near liquid–vapor critical points are subject to both optical and thermal constraints. Near typical liquid–vapor critical points the capillary length, α, vanishes as

$$\alpha = \alpha_0(-t)^{0.45} \tag{1}$$

(where t is the reduced temperature difference measured from the critical temperature: $t \equiv (T - T_c)/T_c$. The capillary length is defined by $\alpha^2 \equiv 2\sigma/(\Delta\rho g)$, where σ is the surface tension, $\Delta\rho$ is the difference between the liquid and the vapor density and $g = 9.8$ m/s^2.) The curved portion of the liquid–vapor interface extends a distance on the order of α out from a solid; thus the contact angle must be measured within a small fraction of α of the line of liquid–vapor–solid contact. If one wishes to distinguish between a contact angle of 0° and one of 10° with a smooth vertical plane wall, the standard expression for the shape of the interface [17] implies that the measurement must be made within $10^{-2}\alpha$ of the wall. Because α_0 in eq. (1) is often 2–3 mm, we see that at temperatures within 10 K of the critical temperature, the contact angle must be measured within 5 μm of the liquid–vapor–solid contact. It is a very difficult task to make accurate angle measurements so close to a wall using an image brought out through the several windows needed to maintain a high pressure, isothermal environment near a critical point. As an additional complication we mention that fluids near liquid–vapor critical points are extraordinarily sensitive to temperature gradients. The decline of the latent heat as T_c is approached means that a small temperature gradient will produce increasing mass transport via evaporation and condensation. The strong divergence of the Prandtl number near T_c leads to a divergent Marangoni number near T_c. Thus small temperature gradients will couple to the temperature dependence of the surface tension to produce strong Marangoni flows [18]. The published descriptions of the measurements in ref. 16 are not sufficiently detailed to convince us that the authors dealt adequately with these optical and thermal constraints.*

3.2. *Transition from complete to incomplete wetting*

The generic phase diagram shows a transition from zero contact angle to non-zero contact angle as the two fluid phases are taken away from the critical point. The present authors have published a detailed study of a particular example of such a transition [19]. We have observed the transition using liquid mixtures of an alcohol (i-C_3H_7OH) and a fluorocarbon (C_7F_{14}) in the presence of their saturated vapor and air. These liquid mixtures separate into two liquid phases below a consolute temperature at 90°C (see fig. 4). Between 38°C and 90°C the denser fluorocarbon-rich phase completely wets the vapor phase. It forms a layer which intrudes between the upper, alcohol-rich phase, and the vapor. Below 38°C the intruding layer contracts into a pendent drop bounded by a line of three phase contact. In fig. 5 we display the temperature dependence of one of the angles characterizing the three-phase contact. The data in fig. 5 are consistent with the expression

$$\cos\theta = 1 - 0.8(T_w - T)/T_w, \tag{2}$$

where $T_w = 311$ K. As discussed in ref. 19, these data are consistent with a discontinuity in the temperature derivative of the surface tension of the interface between the vapor and the liquid mixtures. Similar results were obtained by Moldover and Cahn [12] using mixtures of an alcohol (CH_3OH) and a hydrocarbon (C_6H_{12}). These authors measured contact angles at ambient temperature and varied the consolute temperature

Note added in proof. The recent measurements of Zheleznyi [32] appear to contradict Stepanov et al. [16] and support our view that the contact angle is zero near liquid–vapor critical points.

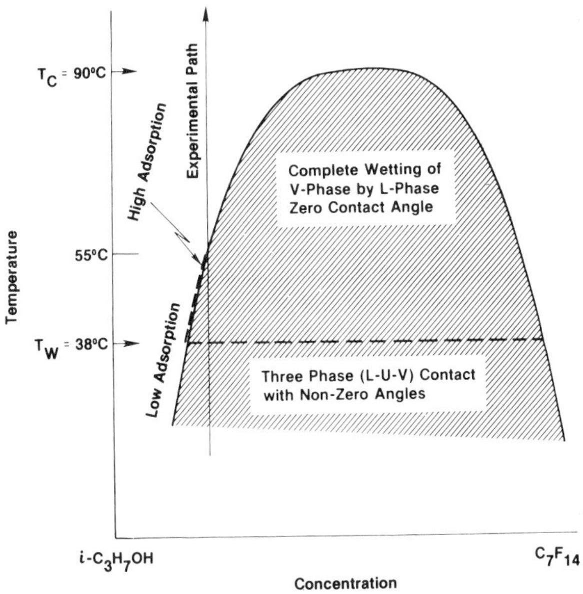

Fig. 4. Experimentally determined phase diagram for mixtures of i-C_3H_7OH with C_7F_{14}. The consolute temperature (T_c = 90°), the temperature at which the lower-phase completely wets the vapor (T_w = 38°C), and a region of high adsorption of the lower phase on the liquid–vapor interface is shown.

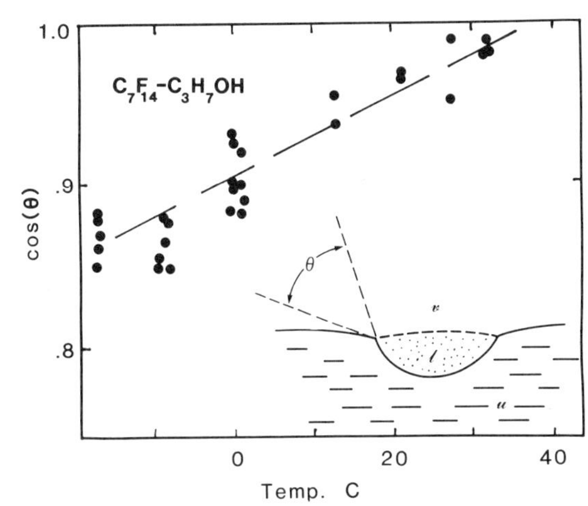

Fig. 5. The temperature dependence of cosine θ. The inset shows θ to be the complement of the dihedral angle subtended by the alcohol-rich phase at the line of three-phase contact between the alcohol-rich liquid, the fluorocarbon-rich droplet, and the vapor. The measurement techniques are described in ref. 19.

(and T_w) by adding water to the mixtures. Pohl and Goldburg [15] found a first order wetting transition near the consolute temperature of mixtures of 2,6-lutidine and water. In their experiments glass capillary tubes played the role of the non-critical phase.

3.3. *Search for the prewetting transition*

As we have indicated on fig. 1, a locus of prewetting transitions is expected to terminate at T_w. A simple thermodynamic argument leads to the conclusion that the prewetting curve should be tangent to the coexistence curve at T_w. We have searched for the prewetting curve using a mixture of i-$C_3H_7OH - C_7F_{14}$ where the vapor phase (in equilibrium with air) plays the role of the non-critical phase. As indicated in fig. 4, the mixture's overall composition was chosen such that the mixture separated into two coexisting liquid phases below 55°C and became a single homogeneous alcohol-rich liquid phase above 55°C. (The onset of phase separation could be detected with a precision of ± 0.05 K by measuring the index of refraction of the upper alcohol-rich phase as the mixture was slowly cooled.) We searched for evidence of the prewetting curve by analyzing the polarization of light reflected from the liquid–vapor surface. The experimental techniques have been described carefully elsewhere [19]. Some results are shown in fig. 6.

As the temperature is increased above 38°C, we observe that the ellipticity of the liquid–vapor interface increases abruptly. In fact the increase occurs in an interval no wider than 0.03°C. In fig. 6 this increase has been interpreted in units of length using a naive "slab" model which assumes that a uniform layer of the lower fluorocarbon-rich phase intrudes between the upper alcohol-rich phase and the vapor. At 55°C the ellipticity of the liquid–vapor interfaces abruptly returns to a value near zero. The decrease at 55°C occurs within ± 0.05 K of the temperature at which the mixture becomes a single homogeneous phase. Thus we conclude that the prewetting curve is indistinguishable from the coexistence curve (with a

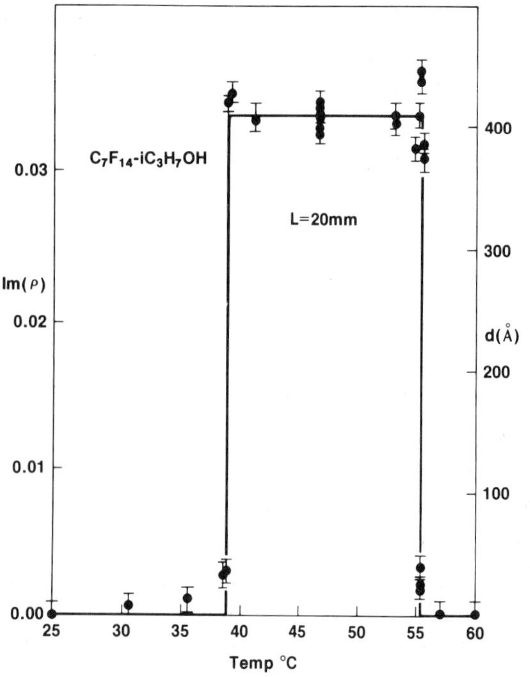

Fig. 6. The temperature dependence of the ellipticity of the liquid–vapor interface above a mixture of i–C_3H_7OH with C_7F_{14}. The ellipticity increases abruptly at the wetting transition ($T_w = 38°C$) and decreases abruptly when the mixture becomes a single liquid phase at 55°C. The prewetting transition, if it occurs, is within ±0.05°C of the coexistence curve at 55°C.

resolution of ±0.05 K) in this temperature range. We have indicated this conclusion in fig. 4 by drawing a section of the prewetting curve adjacent to the coexistence curve between 38°C and 55°C. We are now extending these measurements towards the critical temperature.

4. Multilayer adsorption

As we have indicated on fig. 1, a region of comparatively high adsorption is expected to occur at temperatures above T_w between the prewetting curve and the coexistence curve. It is possible to interpret our ellipsometry data as confirming this feature of the generic phase diagram. This interpretation relies on the rarely noted fact that the liquid–vapor interface which we have studied is very slightly outside the region of the coexistence of two liquid phases. (The conventional representation of the experimental path shown in fig. 4 implies that the interface studied is exactly in equilibrium with two liquid phases.)

The argument proceeds as follows. First, we assume that the two-phase liquid samples have reached compositional equilibrium throughout their volume. (In our experiments, the samples were brought to temperature, thoroughly stirred, and allowed to stand for 24 hours before measurements were made. Typical sample dimensions were less than 2 cm.) Next we note that the two liquid phases coexist in equilibrium separated by a plane interface at just one height within the sample cell. As one moves up in height from this unique height, the gravitational contribution to the chemical potentials increasingly favors the presence of the upper phase. At the top of the upper phase, the liquid–vapor interface forms. There, for the mixtures we have chosen to study, comparatively short-ranged surface forces favor the presence of the lower phase. These surface forces competing with gravitational forces are responsible for the intrusion of a thin layer of the lower phase between the upper phase and the vapor. We interpret the large ellipticity of the liquid–vapor interface as a manifestation of this extreme example of multilayer adsorption. The substrate for this adsorption is the upper liquid phase and near the liquid–vapor surface it is undersaturated with lower phase by an amount

$$\Delta\mu \sim \Delta\rho g L, \tag{3}$$

where L is the height spanned by the upper liquid phase. The very small chemical potential deviation from two-phase equilibrium (on the order of 1 erg/cm^3) given by eq. (3) leads to enormous adsorption even with the extremely small surface forces that exist a few hundred angstroms below the liquid–vapor interface. (We remarked above that if the ellipticity is interpreted as arising from the intrusion of a layer with the properties of the lower phase, then this layer must be several hundred angstroms thick.)

We have limited data on the L-dependence (or equivalently, the $\Delta\mu$ dependence) of the thickness of the adsorbed layer near the wetting transition of i–C_3H_7OH–C_7F_{14} mixtures. Our data show a very weak L-dependence, which is consistent with $L^{-0.06}$ behavior.

Tarazona et al.[6] have calculated the location of the prewetting curve for a theoretical model appropriate for a binary liquid mixture near a consolute point in equilibrium with its vapor. They chose parameters in their model such that $T_w/T_c \approx 0.70$. At temperatures just above T_w their model shows a prewetting transition in the thickness of adsorbed film, at $L \approx 3$ cm. Under those conditions, the thickness of the adsorbed film abruptly decreased by about 160 Å. At values of L just below 3 cm the film's thickness varied as $L^{-0.15}$. It is encouraging that the model of Tarazona et al. has three features which are consistent with our data: (1) The prewetting transition is very close to the bulk coexistence curve, at least near T_w, (2) the decrease in the film's thickness near T_w is a few hundred angstroms, a value which is much greater than the correlation length in the fluids but comparable to our measurements, and (3) the L-dependence of the thickness of the wetting layer is much weaker than $L^{-1/3}$ near the prewetting transition. It would be most interesting to repeat the calculations for model parameters which are closer to the experimental ones ($T_w/T_c \approx 0.85$).

Teletzke et al. [7] have made theoretical estimates for the location of the prewetting transition in model liquid–vapor systems where a wall plays the role of a non-critical phase. They also find that the prewetting transition is very close to the region of bulk two-phase coexistence.

We are aware of two other optical studies of multilayer adsorption on liquid substrates [20, 21]. These other studies involved three different binary mixtures. None of them are known to have a wetting transition at the liquid–vapor interface above the mixtures at temperatures higher than those at which the liquids freeze. The data for two of these mixtures obtained by Kwon et al. [20] were interpreted to show that the thickness of the adsorbed layers varied as $L^{-1/3}$ when the layers ranged from 60–400 Å thick and the temperatures were not very near the consolute temperature of the mixtures. These experiments do not provide evidence to support or to contradict the idea that the region of high adsorption is bounded by a prewetting transition. The $L^{-1/3}$-dependence of the layers' thickness can be understood with the theory of Dzyaloshinskii et al. [22] and by the theory of deGennes [23]. These theories begin with the observation that it is the longest-ranged portion of the intermolecular forces that produces the large multilayer adsorption at the smallest values of $\Delta\mu$. These forces decline as $1/r^6$ for molecules separated by distances on the order of 10–200 Å. At larger separations the decline gradually becomes more rapid, eventually approaching $1/r^7$. It is the integrated effect of all the molecules in the substrate which causes the multilayer adsorption. The arguments of Dzyaloshinskii et al. and deGennes lead to the prediction that the thickness of the adsorbed layer should vary as $L^{-1/3}$ for layers in the size range 10–200 Å and that the variation should approach $L^{-1/4}$ for much thicker layers. These arguments do not incorporate the liquid structure effects which have been used to obtain the phase diagram we are discussing; thus the fact that they are consistent with the data of Kwon et al. does not address the existence of a prewetting transition.

There are few experimental studies of multilayer adsorption on a homogeneous solid substrate where the adsorbate is very near a critical point or a consolute point. We are not aware of any adsorption study which shows evidence of a prewetting transition. Recently, Moldover and Gammon [24] have measured the thickness of layers of sulfur hexafluoride adsorbed on dielectric mirrors. Their data is close to the liquid–vapor critical point of SF_6 and was interpreted to show that the thickness of the adsorbed layer varies as $(\Delta\mu)^{-1/3}$. Their work shows that high adsorption on a solid substrate does occur on one side of a liquid–vapor critical point; however, the work does not address the issue of the existence of a prewetting transition. Recent optical studies by Beysens and Leibler [25] and by Frank and Schnatterly [26] also show that high adsorption does occur on solid substrates immersed in fluid mixtures near consolute points. These studies do not show the $\Delta\mu$-dependence of the adsorption and do not address the issue of the existence of a prewetting transition.

In summary, several experiments show that multilayer adsorption is ubiquitous near critical and consolute points. In some of the cases where it has been studied, the thickness of the adsorbed layer varies as $(\Delta\mu)^{-1/3}$, in agreement with theories which do not address the existence of wetting transitions. In our own study of adsorption near T_w we find a weaker dependence of the thickness of the adsorbed layer upon $\Delta\mu$. This should be confirmed with further experiments because it may be a precursor to the prewetting transition.

5. Consequences of the generic phase diagram

Heady and Cahn discovered the existence of wetting layers in the course of testing nucleation theory [11]. They observed that the fluorocarbon-rich layer which forms at the liquid–vapor interface above the hydrocarbon–fluorocarbon mixtures they studied acts as a site for barrierless nucleation of the fluorocarbon-rich phase when these liquid mixtures are cooled to temperatures such that two liquid phases are thermodynamically stable. Today, one would generalize their discovery to assert that any non-critical phase (such as a container wall) will act as a barrierless site for nucleation of one of the two phases which form when any fluid system is cooled below its coexistence curve near its critical point. This generalization is a direct consequence of the fact that one of the two critical phases wets any non-critical phase.

It follows that metastability against phase separation of near-critical fluid samples enclosed in a single container can be observed only on one side of the critical point. This effect has been illustrated by equation of state [27] and heat capacity [28] measurements for several different fluids near liquid–vapor critical points. In the cases studied, it is the liquid phase which wets the solid container walls and thus it can be brought into the metastable region; the vapor phase cannot be brought into the metastable region because liquid forms immediately on the walls. Analogous phenomena have been seen near consolute points. An exotic example [29] is provided by liquid ^3He–^4He mixtures below their consolute temperature at 0.9 K. Solid containers are wet by the ^4He-rich phase; thus only that phase can be taken into states which are metastable with respect to separation into two liquid phases.

Cahn predicted [30] that the presence or absence of wetting layers would be a significant factor in determining under which conditions ordered composite solids could be solidified from melts near monotectic points. Grugel and Hellawell have studied the growth of composite structures from many different alloy melts [31]. They found that ordered composites could be solidified from those binary alloy systems in which the monotectic temperature (at which the solidification occurred) was well below the consolute temperature. Under these conditions, well below T_c, we would expect that stable three-phase contact could occur between the two critical phases (the two liquid phases in the melt) and the non-critical phase (the solid emerging from the melt). Grugel and Hellawell could not grow uniform composites from copper-lead melts. In the Cu–Pb system, the monotectic temperature is quite close to T_c. Then we expect that a layer of the copper-rich liquid phase will completely wet the emerging copper-rich solid phase and this layer will tend to exclude the other liquid phase from stable contact with the solid. Surface forces will prevent the formation of the stable solidification front required to grow an ordered composite. This picture is supported by further experiments in which aluminum is added to the Cu–Pb melt. The aluminum raises T_c relative to the monotectic temperature. When the aluminum concentration exceeds a specific value, ordered composites can be grown. We expect that a first-order transition occurs from complete wetting of the copper-rich solid to incomplete wetting of the solid at that specific aluminum concentration. This result is quite analogous to the first-order wetting transition that Moldover and Cahn observed [12] when they added water to the methanol–cyclohexane system to raise T_c. The remarkable order in the composites that Grugel and Hellawell grew is documented in a beautiful series of electron micrographs in ref. 31. Such small, regular structures should have numerous technological applications.

6. Summary

The generic phase diagram proposed by Cahn provides a unified description of a wide variety of experimental results; thus we may say many aspects of it are confirmed. In particular, when two critical phases are in equilibrium with a non-critical phase, one of the critical phases completely wets the non-critical phase. (If the results of ref. 16 are confirmed an important exception to this generalization must be recognized.) Outside of the region where both critical phases coexist in equilibrium the critical phase which wets the non-critical phase is strongly adsorbed on the non-critical phase. This adsorbed layer acts as a barrierless site for nucleation.

As one moves away from the critical point, a first-order transition from complete wetting to incomplete wetting has been observed at T_w (figs. 4 and 6). These observations are also consistent with the generic phase diagram which the prewetting line to be tangent to the coexistence curve.

A search for the predicted curve of prewetting phase transitions has been inconclusive to date. The data suggest that the locus of such transitions (if they occur) remains very close to the coexistence curve through a large portion of the temperature range between T_w and T_c.

Acknowledgements

The authors thank J.W. Cahn for his continuing interest in this work. This work has been supported in part by NASA under contract H-27954B.

References

[1] J.W. Cahn, J. Chem. Phys. 66 (1977) 3667.
[2] J.W. Cahn, private communication.
[3] E.H. Hauge and M. Schick, Phys. Rev. B27 (1983) 4288.
[4] D.E. Sullivan, J. Chem. Phys. 74 (1981) 2604.
[5] H. Nakanishi and M.E. Fisher, Phys. Rev. Letters 49 (1982) 1565.
[6] P. Tarazona, M.M. Telo da Gama and R. Evans, Molecular Physics 49 (1983) 283; P. Tarazona, M.M. Telo da Gama and R. Evans, Molecular Physics 49 (1983) 301.
[7] G.F. Teletzke, L.E. Scriven and H.T. Davis, J. Colloid and Interface Science 87 (1982) 550; G.F. Teletzke, L.E. Scriven and H.T. Davis, J. Chem. Phys. 77 (1982) 5794.
[8] R. Pandit, M. Schick and M. Wortis, Phys. Rev. B26 (1982) 5112.
[9] C. Ebner and W.F. Saam, Phys. Rev. Letters 38 (1977) 1486.
[10] J.W. Rowlinson and B. Widom, Molecular Theory of Capillarity (Clarendon, Oxford, 1982) pp. 225–232.
[11] R.B. Heady and J.W. Cahn, J. Chem. Phys. 58 (1973) 896.
[12] M.R. Moldover and J.W. Cahn, Science 207 (1980) 1073.
[13] J.H. Perepezko, C. Galaup and K.P. Cooper, in: Materials Processing in the Reduced Gravity Environment, Guy E. Rindone, ed. (North-Holland, New York, 1982) pp. 491–501.
[14] H. Zabel, B. Schoenfeld and S.C. Moss, J. of Physics and Chemistry of Solids 42 (1981) 897.
[15] D.W. Pohl and W.I. Goldburg, Phys. Rev. Letters 48 (1982) 1111.
[16] V.G. Stepanov, L.D. Volyak and Yu.V. Tarlakos, InzheneroFizicheskii Zhurnal 32 (1977) 458 [English transl: J. Engineering Physics 32 (1977) 283]; L.D. Volyak, V.G. Stepanov and Yu.V. Taralakov, Zhurnal Fizicheskoi Khimii 49 (1975) 2931 [English transl: Russian J. Phys. Chem. 49 (1975) 1729].
[17] L.D. Landau and E.M. Lifshitz, Fluid Mechanics, (Addison-Wesley, Reading, Mass., USA, 1959) p. 235.
[18] S. Ostrach, Ann. Rev. Fluid Mech. 14 (1982) 313.
[19] J.W. Schmidt and M.R. Moldover, J. Chem. Phys. 79 (1983) 379.
[20] O'D Kwon, D. Beaglehole, W.W. Webb, B. Widom, J.W. Schmidt, J.W. Cahn, M.R. Moldover and B. Stephenson, Phys. Rev. Letters 48 (1982) 185.
[21] D. Beaglehole, J. Chem. Phys. 73 (1980) 3366.
[22] I.E. Dzyaloshinskii, E.M. Lifshitz and L.P. Pitaevskii, Advan. Phys. 10 (1961) 165.
[23] P.G. deGennes, J. Phys. (Paris) Lett. 42 (1981) L-377.
[24] M.R. Moldover and Robert W. Gammon, J. Chem. Phys. 80 (1984) 528.
[25] D. Beysens and S. Leibler, J. Physique Letters 43 (1982) L-133.
[26] C. Frank and S.E. Schnatterly, Phys. Rev. Letters 26 (1982) 763.
[27] J.R. Hastings, J.M.H. Levelt Sengers and F.W. Balfour, J. Chem. Thermodynamics 12 (1980) 1009.
[28] David Dahl and M.R. Moldover, Phys. Rev. Letters 27 (1971) 1421.
[29] N.R. Brubaker and M.R. Moldover, in: Thirteenth Int. Conf. on Low Temperature Physics, Boulder, Colorado (1972), vol. 1, K.D. Timmerhaus, W.J. O'Sullivan and E.F. Hammel, eds. (Plenum, New York, London, 1974) p. 612.
[30] John W. Cahn, Metallurgical Transactions 10A (1979) 119.
[31] R.N. Grugel and A. Hellawell, Metallurgical Transitions A 12A (1981) 669.
[32] V.P. Zheleznyi, Issled. Protsessov Teplo i Massoperenosa, Kiev (1979), pp. 127–130.

SECTION 3

MATHEMATICAL STRUCTURE AND COMPUTATIONAL METHODS

Physica 12D (1984) 363–368
North-Holland, Amsterdam

3.1. MATHEMATICS

SINGULARITIES AND OSCILLATIONS IN SOLUTIONS TO CONSERVATION LAWS

Ronald J. DIPERNA
Department of Mathematics, Duke University, Durham, NC 27706, USA

1. Introduction

We shall be concerned with singularities and oscillations for compressible flows modelled by hyperbolic systems of conservation laws in one space dimension,

$$u_t + f(u)_x = 0, \quad u = u(x, t) \in \mathrm{R}^m, \tag{1}$$

and incompressible flows modelled by the Euler equations in two space dimensions

$$u_t + u \cdot \nabla u + \nabla p = 0, \quad \nabla \cdot u = o. \tag{2}$$

In the second section we shall consider the fine scale features and discuss several theoretical results from [2] and [7] dealing with singularities of solutions to hyperbolic systems (1) and functions of bounded variation. In the third section we shall discuss more recent work [3, 8, 9, 13] dealing with averaged quantities and oscillations for system (1) and involving the theory of compensated compactness which originates in the work of Tartar and Murat [8, 9, 13]. We conclude by presenting a new definition of generalized solution of conservation laws relevant to the problem of describing the limits of singular perturbations.

2. Singularities

For hyperbolic systems in one space dimension the L^∞ norm and the total variation norm provide natural metrics to measure the solution amplitude and gradient respectively. Their relevance for conservation laws is established by the following constructive existence theorem due to Glimm [5] for the Cauchy problem with small initial data.

Theorem. If $\mathrm{TV}\, u_0 \ll 1$, there exists a globally defined distributional solution $u = u(x, t)$ of (1) generated by the random choice method which assumes the data u_0 at $t = 0$ and satisfies the following stability estimates,

$$|u(\cdot, t)|_\infty \leqslant \text{const.} |u_0|_\infty , \tag{3}$$

$$\mathrm{TV}\, u(\cdot, t) \leqslant \text{const.}\ \mathrm{TV}\, u_0 .$$

Thus, at any fixed time, the amplitude of the solution is bounded by a constant times the amplitude of the data, measured in the L^∞ norm, while the total amount of wave magnitude in the solution is bounded by a constant times the total amount of wave magnitude in the data, measured by the spatial total variation norm. These sharp time estimates provide the starting point of the regularity problem for systems (1).

In the setting of one space dimension, the spatial regularity at a fixed time may be easily described: for each time t, the solution has finite total variation in x and therefore one-sided pointwise limits,

$$\lim_{y \uparrow x} u(y, t) \quad \text{and} \quad \lim_{y \downarrow x} u(y, t)$$

exist for all x, with an at most countable number of

0167-2789/84/$03.00

jump discontinuities representing the shock waves. As a simple model of a BV (bounded variation) function of one variable it is useful to keep in mind a piecewise smooth function $u(x)$, for which the total variation norm registers the sum of the jumps plus the L^1 norm of the first derivative over the smooth intervals:

$$\text{TV}\, u(x) = \sum_j |[u]_j| + \int_{x_j}^{x_{j+1}} |u'(x)|\, dx,$$

where

$$[u]_j = \lim_{x \uparrow x_j} u(x) - \lim_{x \downarrow x_j} u(x),$$

and $\{x_j, 1 \leqslant j \leqslant N\}$ specifies the points of discontinuity. In the setting of conservation laws the first term represents the sum of shock strengths while the second term represents the fluctuations in the rarefaction waves and compression waves.

For hyperbolic systems (1) the main problem is concerned with regularity of the solution u jointly in x and t. For the sake of x–t considerations the notion of BV function of several variables is appropriate.

Definition. A locally integrable function u of n variables lies in the space $\text{BV}(R^n)$ if each of its first order partial derivatives in the sense of distributions is represented by a measure with finite total mass, i.e.

$$\int \frac{\partial \phi}{\partial x_j} u\, dx = \int \phi\, d\sigma_j, \quad 1 \leqslant j \leqslant n, \tag{4}$$

for all continuously differentiable functions ϕ with compact support and $\|\sigma_j\| < \infty$.

For a piecewise smooth function of one variable the distributional derivative on the left-hand side of (4) is represented by a measure σ with point masses located at the jump discontinuities and with an absolutely continuous structure over the intervals of smoothness. For a piecewise smooth function of (say) two variables the distributional derivative is represented by a measure σ with δ-functions distributed along the curves of discontinuity and with an absolutely continuous description over the intervening regions.

In several dimensions an arbitrary BV function may admit a variety of singularities. We shall first recall the standard classification of regular and singular points for an arbitrary BV function which is based on average limiting behavior and then discuss the subclass of singularities which are admissible in a BV solution to a system of conservation laws (1). It is not difficult to show that Glimm's random choice method generates solutions in BV.

Definition. A point $y \in R^n$ is a *regular point* for a measurable function $u = u(x)$ if y is either a Lebesgue point or a Lebesgue jump point. In the first case there exists a finite value $\bar{u}$ such that u is near $\bar{u}$ on the average near y,

$$\lim_{r \to 0} \frac{1}{r^n} \int_{B_r(y)} |u(x) - \bar{u}|\, dx = 0,$$

$$B_r(y) = \{x : |x - y| \leqslant r\}.$$

In the second case there exists two distinct values $\bar{u}^{\pm}$ and a distinguished direction v such that u is near $\bar{u}^{+}$ on the average locally in the positive half-space H^{+} determined by v while u is near $\bar{u}^{-}$ on the average locally in the negative half-space H^{-} determined by v,

$$\lim_{r \to 0} \frac{1}{r^n} \int_{B_r(y) \cap H^{\pm}} |u(x) - \bar{u}|\, dx = 0,$$

$$H^{\pm} = \{x : (x - y, \pm v) \geqslant 0\}.$$

Thus, at a regular point, either the function u is approximately continuous in the sense that the n-dimensional measure of the set were u deviates from $\bar{u}$ by a prescribed amount vanishes faster than the n-dimensional measure of B_r or the function u

has an approximate jump discontinuity in the sense that n-dimensional measure of the deviation sets corresponding to two fixed values and two complementary half-spaces vanishes faster than the n-dimensional measure of B_r. The first situation is reminiscent of a "continuous" flow while the second is reminescent of a shock. In both cases, we emphasize that the qualifications for a regular point involves only n-dimensional averages of a function; the pointwise behavior may be quite wild.

The subject of geometric measure theory addresses a variety of problems including the regularity of BV functions. It turns out that for an arbitrary function in $\mathrm{BV}(R^n)$ most of the space is filled with regular points while those regular points corresponding to jumps are organized inside sets with a differentiable structure. Among the available structural results dealing with BV functions of several variables we shall mention two basic properties, cf. Federer [4] and Vol'pert [10].

Theorem. If $u \in \mathrm{BV}(R^n)$ then the singular set S has vanishing $(n-1)$-dimensional Hausdorff measure, i.e.

$$H^{n-1}(S) = 0, \quad S = R^c, \quad R = \text{set of regular points}$$

while the subset J of R consisting of jump points forms at most countable union of rectifiable sets of dimension $n-1$,

$$J = \bigcup_{j=1}^{\infty} \Gamma_j, \quad \Gamma_j = \phi_j(K_j)\,;$$

$$\phi_j \in \mathrm{Lip}(R^{n-1}, R^n); \quad K_j \Subset R^{n-1}\,.$$

Remarks. We recall that in R^n a rectifiable subset Γ of dimension $n-1$ is defined to be a compact subset of a Lipschitz hypersurface, i.e. the image of a compact subset K in R^{n-1} under a Lipschitz map ϕ from R^{n-1} to R^n. Although such a set Γ is covered by a set with a differentiable structure it may itself have a complex structure. For example, in two dimensions, a Cantor subset of a Lipschitz curve qualifies as a rectifiable set.

The regularity theory for BV functions provides the starting point for the regularity theory for BV solutions to hyperbolic systems (1) dealing with the structure of the fine scale features. The basic problem concerns the classification of singularities of BV solutions to (1) which satisfy the entropy condition [6]. Are all of the singularities found in the BV category admissible in an entropy satisfying BV solution? In the setting of ideal gas dynamics, does a compressible flow respecting the second law of thermodynamics present coherent behavior only on the average or does the divergence-form structure and entropy condition convert the measure-theoretic Lebesgue limits into classical pointwise limits? A complete regularity theory of conservation laws in one space dimension has been developed which answers these and related questions. We refer the reader to [2] for a treatment of systems with non-degenerate eigenvalues such as those describing ideal isentropic gas dynamics and to [8] for a treatment of systems with linearly degenerate eigenvalues such as provided by the full equations of ideal gas dynamics. In both cases the results apply to solutions constructed by the random choice method. The uniqueness theorem which is currently available [11] is not quite strong enough to assert the equality of two BV entropy solutions.

Next we shall describe several regularity results for BV solutions to conservations (1) which are relevant to the singularity problem for the Euler eqs. (2).

Theorem. If u is a solution to a strictly hyperbolic system (1) with non-degenerate eigenvalues constructed by the random choice method then the following properties hold.

P_1: The singular set S is at most countable and each of its points can be identified as either a point of shock formation, a point of shock interaction or a focal point of a centering compression or expansion wave. In particular, S has Hausdorff dimension zero.

P_2: The singular set S together with the jump set

J forms an at most countable union of classical Lipschitz curves C_j:

$$S \cup J = \bigcup_{j=1}^{\infty} C_j .$$

Moreover, the curves C_j are continuously differentiable with the possible exception of a countable number of points corresponding to impacts from shocks and focusing compression waves. Thus, the possibility of "holes" that exists for rectifiable sets is removed and a higher degree of regularity is obtained.

P_3: All approximate limits are converted into classical pointwise limits. In particular, at each regular point on shock the solution u has classical one-sided limits which satisfy the Rankine–Hugoniot relations.

With background of this type dealing with the fine scale features of compressible 1-D flows, it is natural to entertain analogous questions for 2-D incompressible flows. In this regard we note that, in the setting of flows with concentrated vorticity, fundamental problems remain open for the 2-D Euler eqs. (2) even at the level of existence to the initial value problem. On one hand, if the data have finite kinetic energy and vorticity $\xi = \operatorname{curl} u$ forming an L^p-function, $p > 1$, then global existence may be obtained by several methods including the zero diffusion limit for the corresponding Navier–Stokes equations. On the other hand, if the initial vorticity forms a measure which is not absolutely continuous with respect to two dimensional Lebesgue measure, the existence problem is wide open. The latter case is of particular interest in treating intrinsic singularities such as vortex sheets where the vorticity is a measure assigning non-zero weight to lower dimensional sets. In this regard it is natural to ask whether or not the BV category is sufficiently broad to include all 2-D incompressible flows. From the structural results discussed above it appears that BV is too narrow, even if reference is made only to spatial regularity at a fixed time, i.e. only to the spatial velocity gradient. The BV picture may turn out to be apprpriate for a class of flows with the highest concentration of vorticity appearing on 1-dimensional sets (vortex sheets). However, for flows whose existence is conjectured, having non-zero vorticity on sets with fractional Hausdorff dimension, the BV category is too small for the following technical reason. If $u \in BV(R^n)$ then the gradient of u is absolutely continuous with respect to $(n-1)$-dimensional Hausdorff measure:

$$\nabla u(E) = 0, \quad \text{if } H^{n-1}(E) = 0 .$$

3. Oscillations

For the purpose of constructing solutions to the Euler equation it is, of course, natural to consider the zero diffusion limit of the navier–Stokes system. This singular limit falls within a general class for which new theoretical results have been obtained in one space dimension by a procedure involving a study of averaged quantities rather than the fine scale features. For background we shall recall some work dealing with the zero diffusion limit for conservation laws of the form

$$\begin{aligned} &u_t + f(u)_x = \varepsilon D(u, u_x)_x \\ &u = u(x, t, \varepsilon) . \end{aligned} \tag{5}$$

As a concrete example, one can keep in mind the compressible Navier–Stokes equations viewed as a parabolic regularization of the compressible Euler equations with the Reynolds number $(1/\epsilon)$ approaching infinity. Regarding singular limits of this type there are two view points depending on the dimension on which the process is observed in the x–t plane. From the view point of one-dimensional Hausdorff measure, i.e. the fine scale features, it is natural to investigate the dependence of both the L^∞ norm and the total variation norm of the solution u on the perturbation parameter ε. Establishing uniform control on both norms would allow one to extract a strongly convergent subsequence and pass to the limit in the nonlinear flux function f. Unfortunately, it remains an open problem to establish uniform control on the L^∞ norm and total variation norm for the classical approximation methods such as the viscosity method and conservative finite difference schemes. The only method for which such uniform control has been established is the random choice scheme in the case of initial data with small or at best "moderately large" total variation.

Although the classical strategy has not yet been carried out for proving convergence in the nonlinear singular perturbation problem (5), a new approach, which envolves the theory of compensated compactness due to Tartar and Murat [8, 9, 13] and which requires only uniform control on the amplitude of the solution, has been successful in establishing convergence. We refer the reader to [9] for a treatment of the viscosity method applied to a scalar conservation in one space dimension and to [3] for a treatment of the viscosity method and first order accurate finite difference schemes applied to non-degenerate systems of two equations. We shall state a representative result from [3] which deals with a simple form of viscosity.

Theorem. Consider a strictly hyperbolic system of two equations with non-degenerate eigenvalues to which the method of artificial viscosity is applied, diffusing each of the primitive variables at an equal rate:

$$u_t^\varepsilon + f(u^\varepsilon)_x = \varepsilon u_{xx}^\varepsilon .$$

If $\{u^\varepsilon\}$ is a solution sequence which is uniformly bounded in L^∞ then there exists a subsequence which converges pointwise almost everywhere to an (entropy satisfying) solution u of (1).

Remarks. For an arbitrary sequence of measurable functions one may only assert, in general, convergence in the weak topology, i.e.

$$\lim \iint \phi u^\varepsilon \, dx \, dt = \iint \phi u \, dx \, dt ,$$

for all test functions ϕ while nonlinear maps g are not continuous, i.e.

$$\lim \iint \phi g(u^\varepsilon) \, dx \, dt \neq \iint \phi g(u) \, dx \, dt ,$$

for all ϕ unless g is affine. As a simple example we recall that $\sin(x/\varepsilon)$ converges weakly to zero while its square converges weakly to one-half. On the other hand of u^ε converges strongly in u, i.e. in L_{loc}^1, then for all g the functions $g(u^\varepsilon)$ converge strongly to $g(u)$.

We emphasize that compactness phenomena described in the theorem above is due to the nonlinear structure of the equations. The focusing of acoustic waves along a propagating shock induces a gain of entropy and consequently a loss of information. This loss of information manifests itself in the fact that solution operator is a compact mapping from L^∞ to L_{loc}^1 uniformly in the parameter ε. It is easy to show the corresponding result does not hold for systems with linear modes of propagation, which neither enhance nor reduce propagating oscillations.

One of the tools of the proof is provided by the representing measure of L.C. Young which expresses composite weak limits in terms of expected values as follows. Suppose v_k is an arbitrary sequence of uniformly bounded functions from R^n to R^m (physical space to state space) and suppose g is a continuous real-values map on the state space R^m (for example, pressure in terms of density and temperature). Then there exists a subsequence of v_k and there exist, for almost all points y in the physical domain R^n, a probability measure ν_y on the state space R^m such that the weak limit of the composition $g(v_k)$ exists and coincides almost everywhere with the expected value of g:

$$\lim g(v_k)(y) = \int_{R^m} g(\lambda) \, d\nu_y .$$

It is not difficult to show that strong convergence of v_k corresponds tot he situation where the representing measure ν_y reduces to a point mass. The deviation between weak and strong convergence and the character of oscillations is described by the non-trivial structure of ν_y. The proof of the compactness theorem for the viscosity method proceeds by showing that the representing measure must reduce to a point mass. In [9], Tartar conjectured that the Young measure associated with a diffusion process which respects the entropy condition must reduce to a point mass. This conjecture has now been verified for general scalar eqs. [9] and for non-degenerate systems of two eqs. [3] using the theory of compensated compactness.

In order to study a general class of singular limits, we have introduced in an earlier note [12] a notion of generalized solution to a system of conservation laws which may be stated in the context of (1) as follows:

Definition. A measurable map $(x, t)\rightarrow\nu_{(x,t)}$ from the physical space R^2 to the space of probability measures over the state space R^n is a generalized solution to (1) if the expected value of the field (u,f) has zero divergence, i.e. if the time derivative of the center of mass is balanced by the space derivative of the expected value of the flux f:

$$\frac{\partial}{\partial t}\left(\int_{R^n}\lambda\,\mathrm{d}\nu_{(x,t)}\right)+\frac{\partial}{\partial x}\left(\int_{R^n}f(\lambda)\,\mathrm{d}\nu_{(x,t)}\right)=0\,. \tag{6}$$

We remark that the equality (6) is to hold in the sense of distributions. It is not difficult to show that any L^p-stable singular perturbation of (1) produces a generalized solution. The problem is to characterize those generalized solutions which correspond to diffusion processes, dispersion processes, etc. In this regard we note that the zero dispersion limit for the KdV equation generates a generalized solutions which flows as a point mass up to the time at which shocks form at the hyperbolic level and then spreads into a more complex measure whose center of mass governed by modulation equations.

We conclude with some brief remarks concerning the notion of generalized solution in the setting of the 2-D Euler equations. As we remarked above, it is an open problem to study the infinite Reynolds number of the initial value problem for the 2-D Navier–Stokes equation with initial data having finite kinetic energy and a distribution of vorticity forming a non-absolutely continuous measure. Does the Navier–Stokes solution converge strongly in L^2 to a classical distributional solution of the Euler equations? Does there exist a generalized solution in the above sense to the 2-D Euler equations which represents the Navier–Stokes limit. If so, what is its structure?

References

[1] C.M. Dafermos, Generalized characteristics and the structure of solutions to hyperbolic conservation laws, Indiana Univ. math. J. 26 (1977) 1097–1119.

[2] R.J. DiPerna, Singularities of solutions of nonlinear hyperbolic systems of conservation laws, Arch. Rat. Mech. Anal. 60 (1975) 75–100.

[3] R.J. DiPerna, Convergence of approximate solutions to conservation laws, Arch. Rat. Mech. Anal. 82 (1983) 27–70.

[4] H. Federer, Geometric Measure Theory (Springer, New York, 1969).

[5] J. Glimm, Solutions in the large for nonlinear hyperbolic systems of equations, Comm. Pure Appl. Math. 18 (1965) 697–715.

[6] P.D. Lax, Shock waves and entropy, in: Contributions to Nonlinear Functional Analysis, E.A. Zarantonello, ed. (Academic Press, New York, 1971).

[7] T.-P. Liu, Admissible solutions to systems of conservation laws, Amer. Math. Soc. Memoirs, 1982.

[8] F. Murat, Compacité par compensation, Ann. Scuola Norm. Sup. Pisa Sci. Fis. Mat. 5 (1978) 69–102.

[9] L. Tartar, Compensated compactness and applications to partial differential equations. in: Research Notes in Mathematics, Nonlinear analysis and Mechanics: Heriot–Watt Symposium, vol. 4, R.J. Knops, ed. (Pitman, London, 1979).

[10] A.I. Vol'pert, The spaces BV and quasilinear equations, math. USSR sb. 2 (1967) 257–267.

[11] R.J. DiPerna, Uniqueness of solutions to hyperbolic conservation laws, Indiana Univ. Math. J. 28 (1979) 137–188.

[12] R.J. DiPerna, Generalized solutions to conservation laws, to appear in: Systems of Nonlinear Partial Differential Equations, J.M. Ball, ed. (Reidel, Dordrecht, 1983).

[13] L. Tartar, The compensated compactness method applied to systems of conservation laws, to appear in Systems of Nonlinear Partial Differential Equations, J.M. Ball, ed. (Reidel, Dordrecht, 1983).

Physica 12D (1984) 369–374
North-Holland, Amsterdam

DIFFUSIVE FRONTS OF PENETRANTS IN GLASSY POLYMERS

Donald S. COHEN*
Department of Applied Mathematics, California Institute of Technology, Pasadena, California 91125, USA

A model for diffusion in polymers near the glass–rubber transition is developed. The model unifies many of the diverse observations made in various kinds of glassy polymers. Preliminary studies of the diffusive penetrant front are carried out.

1. Introduction

Diffusion in rubbery polymers is described by the standard Fickian diffusion equation $C_t = \operatorname{div}(D \operatorname{grad} C)$, in which the diffusivity D is at worst dependent upon the concentration C of the diffusing species; i.e., $D = D(C)$. Although problems involving this equation may be formidable nonlinear boundary value problems, the important qualitative features of diffusion in rubbery polymers can be obtained, and indeed, a rather complete theory exists.

On the other hand, there is no unifying theoretical formulation for glassy polymers. There is not even a complete classification of the various anomalous effects. Glassy polymers respond slowly to changing conditions, and most of the anomalous effects are directly related to this. For example, the changing polymer structure influences the solubility and diffusional mobility. This leads to diffusion coefficients which can depend not only nonlinearly on the state of the system, but also on the previous history of the system, and sometimes even on previous rates of change of the system [1–4]. Internal stresses can be induced due to differential swelling of different parts of the polymer [5–7]. Immobilization (and sometimes reaction) in the interstitial spaces and microvoids greatly affects the sorption of diffusing penetrants [8–10]. These and many other phenomena have given rise to a partial classification (Case II, Super-case II, Dual-mode, Anomalous non-Fickian,...) of glassy polymers according to various experimental observations.

Because of the relative ease with which they can be obtained, both experimentally and theoretically, the quantities most commonly used to define the kinetics of the sorption process are M_t, the total amount of penetrant absorbed per unit area of polymer at time t, and $x = s(t)$, the position of the glass–gel (rubber) interface at time t. Many diverse observations are reported. For example, for a given polymer and solvent different sorption behaviors are observed according to the size of the sample [17]; in large samples both M_t and $s(t)$ may vary linearly with t for small t, while for small samples $s(t)$ may vary linearly with t for small t although M_t varies as $t^{1/2}$ initially. In other situations far more general time dependences for both M_t and $s(t)$ are observed, and the following crude classification has been adopted: Look at $M_t = kt^n$ for short periods of time. Standard Fickian diffusion (i.e., a polymer in its rubbery state above the glass transition temperature) corre-

*Supported in part by the U.S. Army Research Office (Durham) under Contract DAAG29-81-K-0107 and by the National Science Foundation under grant MCS-8205407.

sponds to the usual $n = \frac{1}{2}$. Case II glassy polymers correspond to $\frac{1}{2} < n < 1$, and super-case II corresponds to $n > 1$. However, this is not a hard and fast rule, and furthermore, in most glassy polymers M_t as a function of t does not look like t^n for one constant value of n. In dual mode sorption and in anomalous non-Fickian cases the most commonly observed function for M_t is what is now universally called "sigmoid" by polymer chemists [1, 3]. That is, M_t as a function of t commonly exhibits a single point of inflection (most often at about 50% of equilibrium sorption). For example, in the penetration of cellulose acetate by acetone vapor [13], M_t versus $t^{1/2}$ is linear for both small and large time t joined for intermediate time by a curve with a single inflection point. The penetration of polystyrene by methane [14] exhibits an M_t versus t with two linear regions connected by a nonlinear region. In other contexts, for example, crystallization, the terminology explosive crystallization refers to crystallization rates like t (rather than $t^{1/2}$).

In this paper we derive and present a model which incorporates and unifies many of the diverse observations occurring in diffusive motion in glassy polymers. Our model explicitly incorporates the most important property of a glassy polymer, namely the finite relaxation time implied by the slow response to changing conditions. While this slow response property of a glassy polymer is recognized by all researchers, almost all other models do not properly account for it or do so in an artificial manner.

In section 2 we derive and propose a general model for diffusion in glassy polymers. In one dimension our model is of the form

$$\frac{\partial C}{\partial t} = \frac{\partial}{\partial x}\left(D\frac{\partial C}{\partial x}\right) + R(x,t,C,C_t,C_x), \tag{1.1}$$

$$D = \int_0^t K(t,s,C(x,s))f(C(x,s),C_t(x,s))\,\mathrm{d}s. \tag{1.2}$$

The meaning of the various terms is stated precisely in section 2. Here we note only that via the integration in eq. (2) the diffusion coefficient D at time t explicitly depends on the history of the process prior to time t, and the kernel $K(t,s,C)$ represents the history dependence (or delay or heredity) inherent in the relaxation process. The integral relation (2) is the common and natural mathematical expression describing the history dependence implied by the finite relaxation times, and the theoretical properties of eqs. (1) and (2) are very satisfying with regard to many of the observations concerning diffusion in glassy polymers.

In section 3 we employ a Karman–Pohlhausen integral averaging method to obtain the time history of the penetrant front. In section 4 the glass–rubber interface is studied by using a travelling diffusion wave connecting equilibrium states above (rubbery) and below (glassy) the interface transition. In both sections 3 and 4 we present results only; the complete mathematical details are presented in the paper of Cohen and Stanley [18].

2. The model

We initiate our derivation by appropriately generalizing Crank's [12] clear and penetrating observations of the nature of the response of a glassy polymer to a diffusing solvent. According to Crank, part of the diffusion process is instantaneous and part proceeds slowly. The instantaneous part accompanies a change in the concentration of the diffusing substance. The slow part is a drift toward an equilibrium value. Furthermore, this slow part takes place even when the concentration is not changing.

The microvoids (or holes or interstitial spaces) responsible for trapping a diffusing gas, for example, are consequences of local heterogeneities in the polymer intimately related to the slow relaxation process associated with the glass–rubber transition. Immobilization (or partial immobilization) and sometimes reaction in these interstitial spaces and microvoids greatly affects the sorption of diffusion penetrants. It is known [10] that there is relatively slow penetration of the sorbate into the holes of the polymer, and thus, the gas adsorbed into the holes is not in local equilibrium with the diffusing gas. This relatively slow hole-filling process accounts for the slow response to changing conditions or the finite (rather than instantaneous) relaxation time characteristic of the glassy state.

Let $D^{i}(C)$ denote the (concentration-dependent) instantaneous diffusion coefficient of a dry polymer, and let $D^{eq}(C)$ denote the equilibrium diffusivity corresponding to the fully saturated polymer long after all penetration is complete. We postulate that the diffusive flux J is given by

$$J = -D\frac{\partial C}{\partial x}, \quad \text{where } D = D^{i}(C) + D^{d}(C), \tag{2.1}$$

and where $D^{d}(C)$ represents the drift diffusivity corresponding to the slow part of the process. Furthermore, we postulate that

$$\frac{\partial D^{d}}{\partial t} = \alpha(C)[D^{ex} - D^{d}], \tag{2.2}$$

where

$$D^{ex} = D^{eq}(C) - D^{i}(C). \tag{2.3}$$

Thus D^{d} changes at a rate proportional to the difference of D^{ex} and D^{d}, where according to (2.3), D^{ex} is the excess diffusivity available to the drift diffusivity.

Note that $D^{d} = D^{ex}$ implies that the total diffusivity D is given by $D = D^{eq}$. Furthermore, when $D^{d} = 0$ (corresponding to a perfectly dry polymer), then $D = D^{i}$ provided that either $\alpha(0) = 0$ or $D^{i}(0) = D^{eq}(0)$. Finally, corresponding to the experimental observation that relaxation times decrease as C increases, we take the proportionality parameter α to be concentration dependent and monotonically increasing in C.

Now, we apply the basic equation of continuity (i.e. conservation of mass), namely

$$\frac{\partial C}{\partial t} = -\frac{\partial J}{\partial x}. \tag{2.4}$$

Then, (2.1), (2.2), and (2.4) immediately yield

$$\frac{\partial C}{\partial t} = \frac{\partial}{\partial x}\left(D^{i}\frac{\partial C}{\partial x}\right) + \frac{\partial}{\partial x}\left(D^{d}\frac{\partial C}{\partial x}\right), \tag{2.5}$$

$$\frac{\partial D^{d}}{\partial t} = \alpha(C)[D^{ex} - D^{d}], \tag{2.6}$$

or, equivalently, in terms of D and C,

$$\frac{\partial C}{\partial t} = \frac{\partial}{\partial x}\left(D\frac{\partial C}{\partial x}\right), \tag{2.7}$$

$$\frac{\partial D}{\partial t} = \frac{\partial D^{i}(C)}{\partial C}\frac{\partial C}{\partial t} + \alpha(C)[D^{eq}(C) - D]. \tag{2.8}$$

A diffusivity evolving like (2.8) was first proposed by Crank [12]. He noted that large α implies $D \sim D^{eq}(C)$ and small α implies $D \sim D^{i}(C)$ so that intermediate values of α should correspond to the glass–rubber transition regions and its various anomalous behaviors.

System (2.7), (2.8) is the more convenient to study analytically, but for basic theoretical and physical information it's better to look at the equivalent system (2.5), (2.6). For example, integrate eq. (2.6) to obtain

$$D^{d}(x,t) = \int_0^t e^{-\int_s^t \alpha(C(x,\tau))\,d\tau} \times \alpha(C(x,s))\,D^{ex}(C(x,s))\,ds. \tag{2.9}$$

In this form we now see explicitly that D^{d} remembers with exponentially fading memory the amount of D^{ex} it has. Furthermore, more fundamental to the physics, an integral law such as (2.9) explicitly incorporates the most important property of a glassy polymer, namely the finite relaxation time implied by the slow (rather than instantaneous) response to changing conditions.

Finally, rather than postulate the specific mechanism (2.6) we incorporate possible alternate or even more general history dependent processes and replace (2.7), (2.8) by

$$\frac{\partial C}{\partial t}=\frac{\partial}{\partial x}\left(D\frac{\partial C}{\partial x}\right)+R(x,t,C,C_t,C_x), \tag{2.10}$$

$$D=\int_0^t K(t,s,C(x,s))f(C(x,s),C_t(x,s))\,\mathrm{d}s, \tag{2.11}$$

with an appropriate (hereditary or delay) kernel $K(t,s,C)$ and an appropriate concentration dependent f. Here $R(x,t,C,C_t,C_x)$ represents the (at this stage unknown) kinetics in the process. We now briefly discuss this.

A detailed knowledge of $R(x,t,C,C_t,C_x)$ from a molecular or statistical dynamical theory would be difficult to obtain. However, for our continuum mechanics model we need only observe that $R(x,t,C,C_t,C_x)$ should provide the kinetics incorporating the change from a dry polymer to a fully saturated one. Thus, it models all the complex processes which can be written effectively as

$$\text{site + gas molecule} \underset{k_{-1}}{\overset{k_1}{\rightleftarrows}} \text{site–molecule complex.}$$

Qualitatively, there should be little difference from the simple assumption of the most commonly occurring bimolecular stationary form

$$R(x,t,C,C_t,C_x)=\mu C(k-C), \tag{2.12}$$

where μ represents the "strength" of the reaction and k represents the equilibrium concentration (or carrying capacity) towards which the reaction is driven.

3. The penetrant front

A crude but effective way to determine the behavior of the progressing penetrant front due to the type of diffusion given by (1.1), (1.2) is to employ an integral averaging method (commonly called the Karman–Pohlhausen method).

We introduce the quantity $\delta(t)$ called the penetration distance (or propagating concentration front). For $x>\delta(t)$ the polymer is at its equilibrium (i.e., initial) concentration, and there is no flux transferred beyond $x=\delta(t)$. Thus,

$$C=0, \quad \text{at } x=\delta(t), \tag{3.1}$$

$$DC_x=0, \quad \text{at } x=\delta(t). \tag{3.2}$$

It is important to note that for our problem the "signal speed" is infinite. That is, solutions have a finite value for all x for all positive t for any initial or boundary data. This is a consequence of the fact that the characteristic surfaces for parabolic equations are parallel to the spatial axes. For practical purposes and often as a basis for sophisticated approximation techniques one defines a diffusion front as the locus of points moving with a given concentration. Clearly, in our problem we proceed in the standard way to take the concentration field $C(x,t)$ to be zero for $x>\delta(t)$ *to within the accuracy of the approximation* or equivalently we define the front to be the locus of a given (presumably small) concentration.

To assess the effect of our diffusive operator we consider

$$C_t=(DC_x)_x+R(C), \quad x>0,\ t>0, \tag{3.3}$$

$$D_t=\frac{\mathrm{d}D^{\mathrm{i}}}{\mathrm{d}C}C_t+\alpha(C)\left[D^{\mathrm{eq}}(C)-D\right], \quad x>0,\ t>0, \tag{3.4}$$

$$C(0,t)=C_0, \quad t\geq 0, \tag{3.5}$$

$$C(x,0)=0, \quad x\geq 0. \tag{3.6}$$

Here C_0 is a prescribed constant.

We now assume that $C(x,t)$ can be approximated by an Nth degree polynomial in x with time dependent coefficients. This approximation is required to satisfy the boundary condition (3.5) and the conditions (3.1) and (3.2) at the front $x=\delta(t)$. Thus,

$$C(x,t)=C_0\sum_{n=2}^{N}k_n\left[1-\frac{x}{\delta(t)}\right]^n, \tag{3.7}$$

where

$$\sum_{n=2}^{N}k_n=1. \tag{3.8}$$

By differentiation in (3.3) it is possible [19] to derive $(N-2)$ conditions which the constants k_n $(n=2,\ldots,N)$ must satisfy in addition to (3.9). We thus obtain $(N-1)$ equations in $(N-1)$ unknowns for the k_n. We shall not need that information here.

We now find $\delta(t)$ by requiring (3.7) to be an approximation in the sense that

$$\int_0^{\delta(t)}[C_t-(DC_x)_x-R(C)]\,\mathrm{d}x=0. \tag{3.9}$$

As shown in [18] a first order nonlinear ordinary differential for $\delta(t)$ is easily derived. This equation involves $D(0,t)$ which can be found by an appropriate use of (3.4). The equation for $\delta(t)$ can then be studied asymptotically for both large and small t. It is found that

$$\delta^2(t)\sim\begin{cases}(\text{constant})\ D^{\mathrm{i}}(C_0)t, & \text{for small } t,\\ (\text{constant})\ D^{\mathrm{eq}}(C_0)t \\ \qquad +(\text{constant}), & \text{for large } t.\end{cases} \tag{3.10}$$

This represents so-called sigmoid sorption in which the diffusive front is Fickian (i.e. like $t^{1/2}$) for small and large times (corresponding to glassy and rubbery states well below and well above the glass–rubber transition), but in which an accelerated time behavior is observed at the glass–rubber interface.

4. Traveling waves and the glass–rubber interface

We now study the glass–rubber interface in the model

$$C_t=(DC_x)_x+\mu C(k-C), \tag{4.1}$$

$$D_t=\frac{\mathrm{d}D^{\mathrm{i}}}{\mathrm{d}C}C_t+\alpha(C)[D^{\mathrm{eq}}(C)-D], \tag{4.2}$$

where $D^{\mathrm{i}}(C)$, $D^{\mathrm{eq}}(C)$, and $\alpha(C)$ are given monotonically non-decreasing functions of C such that $D^{\mathrm{i}}(0)=D^{\mathrm{eq}}(0)$. The steady states $C\equiv 0$ and $C\equiv k$ correspond to the dry and fully saturated states, respectively. The structure of the transition between the two regimes and the role of the various terms in the equations of motion is very often nicely revealed by a study of traveling wave solutions connecting the equilibrium states. Even in more general solutions transition zones in some appropriate scaled variables are often described by traveling waves. Thus, we shall study traveling wave solutions.

We seek solutions of the form

$$C(x,t)=C(\eta),\quad D(x,t)=D(\eta), \tag{4.3}$$

where

$$\eta=x+mt. \tag{4.4}$$

Define $u=\mathrm{d}C/\mathrm{d}\eta$. Then, (4.1), (4.2) become

$$\frac{\mathrm{d}C}{\mathrm{d}\eta}=u, \tag{4.5}$$

$$\frac{\mathrm{d}}{\mathrm{d}\eta}(Du)=mu-\mu C(k-C), \tag{4.6}$$

$$\frac{\mathrm{d}D}{\mathrm{d}\eta}=\frac{\mathrm{d}D^{\mathrm{i}}}{\mathrm{d}C}\frac{\mathrm{d}C}{\mathrm{d}\eta}+\frac{\alpha(C)}{m}[D^{\mathrm{eq}}(C)-D]. \tag{4.7}$$

The system (4.5)–(4.7) has two critical points given by

$$(C,u,D)=(0,0,D^{\mathrm{eq}}(0)),(k,0,D^{\mathrm{eq}}(k)). \tag{4.8}$$

The existence of a (non-negative) traveling-wave solution of (4.1), (4.2) with $C\to 0$ as $t\to-\infty$ and $C\to 1$ as $t\to\infty$ is equivalent to the existence of a trajectory (in $C\geq 0$) of system (4.5)–(4.8) which starts at $(C,u,D)=(k,0,D^{\mathrm{eq}}(k))$ as $\eta\to-\infty$ and terminates at $(C,u,D)=(0,0,D^{\mathrm{eq}}(0))$ as $\eta\to\infty$.

Under very mild (physically acceptable) conditions on $D^{\mathrm{i}}(C)$, $D^{\mathrm{eq}}(C)$, and $\alpha(C)$, Cohen and Stanley [18] have used topological and geometrical methods to prove the existence of the desired traveling wave solution for all wave speeds $m\geq m^*(\mu,k,D^{\mathrm{eq}}(0))$, where this lower bound m^* is easily characterized. Furthermore, in order to study the analytical structure of the glass–rubber interface, they have used a perturbation expansion in the special case that the critical points can be scaled to be close. Not unexpectedly, the interface has the structure of a slightly deformed and twisted hyperbolic tangent, reminiscent of the usual weak shock in gas dynamics.

References

[1] J. Crank, The Mathematics of Diffusion, second ed. (Oxford, London, 1976).

[2] J. Crank and G.S. Park, eds., Diffusion in Polymers (Academic Press, New York, 1968).

[3] H.L. Frisch, Sorption and transport in glassy polymers – a review, Polymer Engr. and Sci. 20 (1980) 2–13.

[4] T.L. Smith and R.E. Adam, Effect of tensile deformations on gas transport in glassy polymer films, Polymer 22 (1981) 299–304.

[5] H.L. Frisch, T.T. Wang and T.K. Kwei, Diffusion in glassy polymers, II. J. Polymer Sci. 7 (1969) 879–887.

[6] T.K. Kwei and T.T. Wang, Diffusion on glassy polymers, in: Permeability of Plastic Films and Coatings, H.B. Hoferberg, ed. (Plenum, New York, 1974).

[7] C.H.M. Jacques, H.B. Hofenberg and V. Stannett, Super case II transport of organic vapors in glassy polymers, in: Permeability of Plastic Films and Coatings, H.B. Hofenberg, ed. (Plenum, New York, 1974).

[8] W.R. Vieth and K.J. Sladek, A model for diffusion in a glassy polymer, J. Colloid Sci. 20 (1965) 1014–1033.

[9] J.A. Tshudy and C. von Frankenberg, A model incorporating reversible immobilization for sorption and diffusion in glassy polymers, J. Polymer Sci. 11 (1973) 2027–2037.

[10] D.R. Paul and W.J. Koros, Effect of partially immobilizing sorption on permeability and the diffusion time lag, J. Polymer Sci. 14 (1976) 675–685.

[11] J.H. Petropoulos, Quantitative analysis of gaseous diffusion in glassy polymers, J. Polymer Sci. 8 (1970) 1797–1801.

[12] J. Crank, A theoretical investigation of the influence of molecular relaxation and internal stress on diffusion in polymers, J. Polymer Sci. 11 (1953) 151–168.

[13] E. Bagley and F.A. Long, Two-stage sorption and desorption of organic vapors in cellulose acetate, J. Amer. Chem. Soc. 77 (1955) 2172–2178.

[14] W.R. Vieth and M.A. Amini, Generalized dual sorption theory, in: Permeability of Plastic Films and Coatings, H.B. Hofenberg, ed. (Plenum, New York, 1974).

[15] D.R. Paul, Effect of immobilizing adsorption on the diffusion time lag, J. Polymer Sci. 7 (1969) 1811–1818.

[16] H.B. Hofenberg and V. Stannett, in: The Physics of Glassy Polymers, R.N. Howard, ed. (Applied Science Publishers, New York, 1973).

[17] G. Astarita and S. Joshi, Sample-dimension effects in the sorption of solvents in polymers – A mathematical model, J. Membrane Sci. 4 (1978) 165–182.

[18] D.S. Cohen and E.A. Stanley, Gaseous diffusion in glassy polymers, SIAM J. Appl. Math. (1983).

[19] T.R. Goodman, Application of integral methods to transient nonlinear heat transfer, Advances in Heat Transfer 1 (1964) 51–122.

Physica 12D (1984) 375–381
North-Holland, Amsterdam

WAITING AND PROPAGATING FRONTS IN NONLINEAR DIFFUSION

William L. KATH
Applied Mathematics, Caltech 217-50, Pasadena, California 91125, USA

A nonlinear diffusion equation is considered, where the nonlinearity arises due to the variation of the diffusion coefficient with concentration. If this diffusion coefficient vanishes at a certain value of the concentration, this produces a number of interesting interfacial effects, such as the finite propagation speed of interfaces (in contrast to the infinite propagation speed of linear diffusion) and waiting-time behavior (in which the interfaces can remain stationary for a finite time before beginning to move). A relatively simple singular perturbation analysis is given to explain these effects.

1. Introduction

The behavior of the linear diffusion equation

$$\frac{\partial u}{\partial t^*} = \frac{\partial}{\partial x}\left(D\frac{\partial u}{\partial x}\right), \quad D = \text{const.}, \tag{1}$$

is of course elementary and well known. For many problems, however, it is not sufficient to use the linear equation and nonlinear effects must be considered. In the study of reaction–diffusion problems, for example, nonlinear chemical reaction terms must be added to the right-hand sides of equations such as the above. The resulting interaction between the diffusion and reaction effects has long been known to give rise to spatial patterns (see for example Turing [14]), and these patterns can become, in fact, quite complicated. Indeed, reaction–diffusion models have been proposed for the rotating spiral waves discussed by Winfree [15] (see for example Cohen, Neu and Rosales [3]).

The above is certainly not the only type of nonlinearity that can arise in (1), however, and therefore spatial patterns should not be the only type of nonlinear effect that is expected. Another kind of nonlinearity occurs when the diffusion coefficient is allowed to depend upon the concentration, $D = D(u)$, and for the remainder of this paper the behavior of the resulting equation will be discussed. In particular, the interesting behavior will be seen to occur when the diffusion coefficient vanishes (or at least becomes very small).

Anticipating that the most interesting effects will occur near a zero of $D(u)$, it is possible to simplify the analysis by taking $D(u) = u^n$. This can be thought of as being a local approximation near the zero of D, which is taken to be at $u = 0$. With this choice the equation becomes

$$\frac{\partial u}{\partial t^*} = \frac{\partial}{\partial x}\left(u^n\frac{\partial u}{\partial x}\right). \tag{2}$$

It may be thought that the approximations made to this point are overly restrictive, but this is not the case. The above equation arises naturally in the study of a number of physical problems:

(1) Thin saturated regions in porous media, where $n = 1$ [13];

(2) percolation of gas through porous media, where $n \geq 1$ [10];

(3) thin liquid films spreading under gravity, where $n = 3$ [2];

(4) radiative heat transfer by Marshak waves, where $n = 6$ [9].

This list is definitely not exhaustive; for example, another very different appearance of equations of this type is in the study of the dynamics of stretched vortices [11].

Oleinik, Kalashnikov and Yui-Lin [12] proved the existence and uniqueness of weak solutions of

0167-2789/84/$03.00© Elsevier Science Publishers B.V.
(North-Holland Physics Publishing Division)

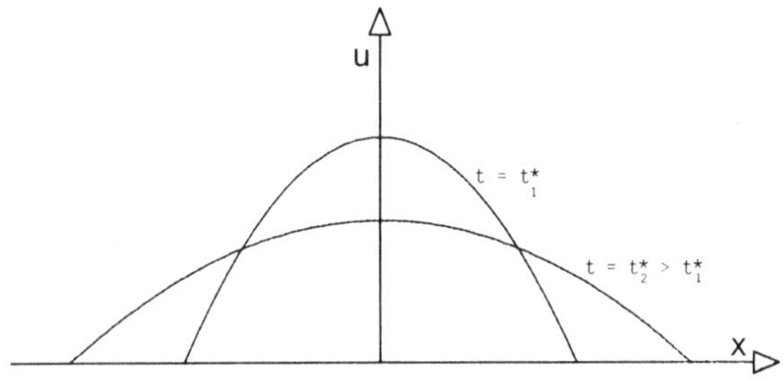

Fig. 1. The similarity solution (3).

eq. (2). They also proved, using a maximum principle, that if the initial concentration vanishes outside a finite interval, then the solution vanishes outside a finite interval at all later times for which it exists. Thus, there is a sharp boundary separating regions of zero and nonzero concentration, which is called here an *interface* or a *front*. This is shown clearly by the similarity solution plotted in fig. 1,

$$u = t^{*-1/(2+n)}\left[\left(E - \frac{n}{4+2n}\frac{x^2}{t^{*2/(2+n)}}\right)^{+}\right]^{1/n}, \quad (3)$$

where $(x)^{+} = \max(x, 0)$. Notice that the interface propagates with finite speed, in contrast with the infinite propagation speed of linear diffusion.

Aronson [1], Kamin [5] and Knerr [7] also examined eq. (2) and proved the existence of what are called here *waiting-time solutions*. These solutions occur for certain types of initial conditions, and are characterized by having interfaces which do not move immediately: they wait for a finite time before beginning to move.

2. Singular perturbation formulation

Rather than following the work of previous authors, which has typically involved the use of maximum principles to obtain rigorous bounds upon the solution and the waiting time, a different approach is used here. To begin, one first notes that when $n = 0$ eq. (2) becomes the linear diffusion equation and can be solved exactly. One might expect, therefore, that when n is nonzero but small a perturbation expansion could be used to solve it. Or said another way, if one would like to discover the qualitative difference between the behavior of linear diffusion and that of eq. (2), then the limit of small n should be studied.

Once this has been suggested, one sets $n = \varepsilon$, where $0 < \varepsilon \ll 1$, and the equation becomes

$$u_{t*} = \left(u^{\varepsilon} u_x\right)_x. \quad (4)$$

If a perturbation expansion in powers of ε is now used, immediately it is seen that the limit $\varepsilon \to 0$ is a singular one because of the diffusion coefficient u^{ε}: it is not possible to expand this function in a power series in ε that will be uniformly valid near $u = 0$.

Since the main interest is in the behavior of the solution near zero, this nonuniformity must be eliminated. To do so it is best to eliminate the occurrence of the u^{ε} term, which suggests the transformation

$$v = u^{\varepsilon}, \quad t^* = \varepsilon t. \quad (5)$$

The resulting equation for $v(x, t; \varepsilon)$ is then

$$v_t = v_x^2 + \varepsilon v v_{xx}, \quad (6)$$

which clearly does not possess such a nonuniformity near $v = 0$. In fact, from the form of the above equation it seems that the real small parameter should be εv, not just ε. Some of the following analysis, then, should hold for $\varepsilon = \mathcal{O}(1)$ and v small. It will be shown later that this is indeed the case.

Although the difficulty near zero has been removed in (6), the limit $\varepsilon \to 0$ is still a singular one, due to the appearance of regions where v_{xx} is large. In addition, there is another apparent problem. The initial conditions for v should also be affected by the transformation (5), and thus depend upon ε. The basic behavior of the solutions of (6) does not depend upon this fact, however, and it is most convenient to ignore this dependence here (for a more detailed discussion and a way to get around this difficulty, please see [6]).

Thus, it is assumed that

$$v(x,0;\varepsilon)=f(x), \tag{7}$$

where $f(x)$ is independent of ε.

3. The leading-order wave equation

Assuming at first the vv_{xx} is $\mathcal{O}(1)$ everywhere, to leading order the behavior of solutions of (6) is governed by the nonlinear wave equation

$$v_t=v_x^2. \tag{8}$$

The solution of this equation is standard [4], and is given implicitly by

$$v(\xi,t)=f(\xi)-f'^2(\xi)t, \tag{9}$$

$$x(\xi,t)=\xi-2f'(\xi)t. \tag{10}$$

Under certain conditions the subcharacteristics defined by (10) will cross, yielding multivalued solutions. This error is clearly due to the approximation made in arriving at (8), since v_{xx} becomes large when the subcharacteristics cross. Although a complete remedy involves a detailed perturbation expansion (outlined later), to leading order the difficulties can be corrected by suitable placement of corner shocks in the solution (the term corner shock is used to indicate a discontinuous jump in the first derivative). An example of this is given in fig. 2, where

$$f(\xi)=\begin{cases}\dfrac{1-\cos\pi\xi}{\pi^2}, & \text{for } |\xi|<2,\\ 0, & \text{for } |\xi|\geq 2.\end{cases}$$

From (10) the time of corner shock formation can also be ascertained. It is the time at which the subcharacteristics cross, determined by

$$\frac{\partial x}{\partial\xi}=1-2f''(\xi)t=0.$$

Any local maximum of $f''(x)$, provided it is separated sufficiently from other maxima, will generate a corner shock which forms at the time

$$t_s=\frac{1}{2f''(\xi_s)} \tag{11}$$

(here ξ_s is the position of the local maximum).

The leading-order approximation (9) and (10) also yields much information about solutions in which the interface does not begin to move immediately. The example shown in fig. 2 is such a solution, and for clarity a greatly expanded view of the interface is given in fig. 3. Note that the solution rearranges itself behind the stationary interface, until it has developed sufficiently at $t=0.5$, the time of shock formation, to start the interface moving.

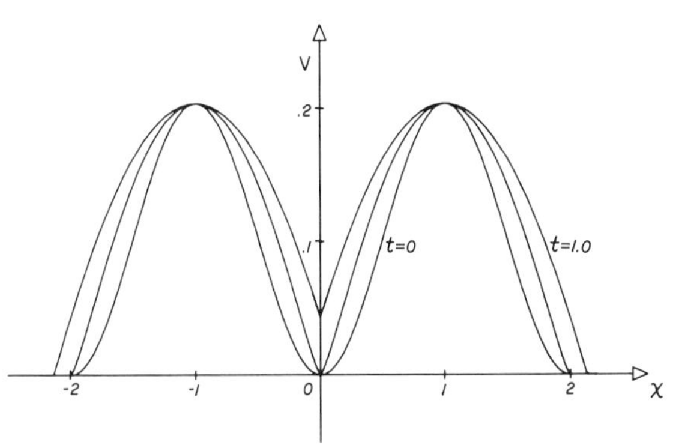

Fig. 2. The leading-order approximation for v (with corner shocks).

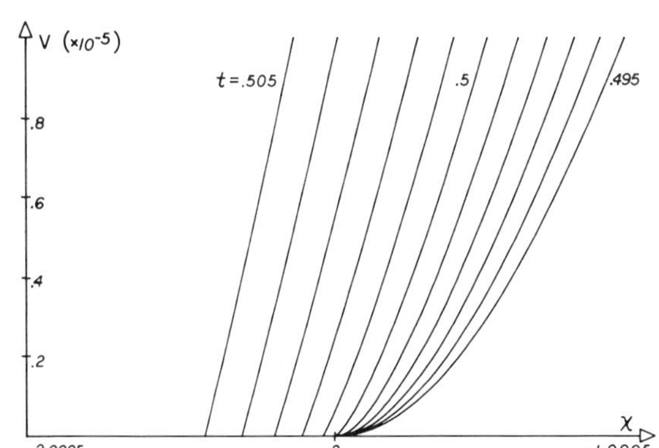

Fig. 3. Leading-order approximation near the interface: a waiting-time solution.

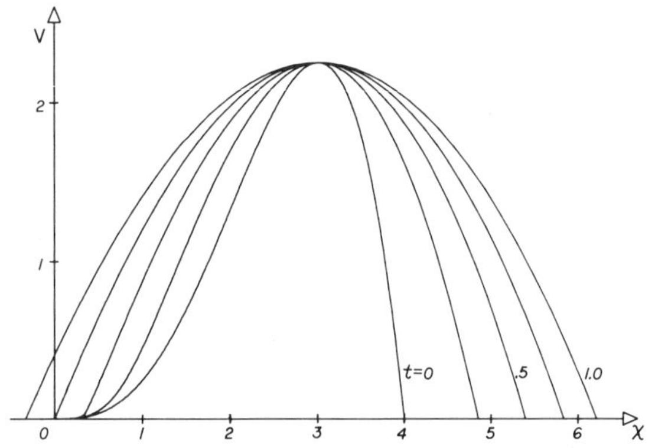

Fig. 4. A leading-order solution with other types of waiting-time behavior.

Another example of interface behavior results from taking

$$f(\xi)=\begin{cases}\frac{1}{12}(4-\xi)\xi^3, & \text{for } 0\le\xi\le 4,\\ 0, & \text{otherwise,}\end{cases}$$

and is shown in fig. 4. The interface on the right is already a corner shock, and thus this interface begins to move immediately (its waiting time is zero).

The interface on the left is more interesting, however, and an expanded plot is given in fig. 5. The time of shock formation is again 0.5, but now the corner shock forms to the right of the interface at a point where $v>0$. After forming, the corner shock moves toward the interface, and at $t=0.75$

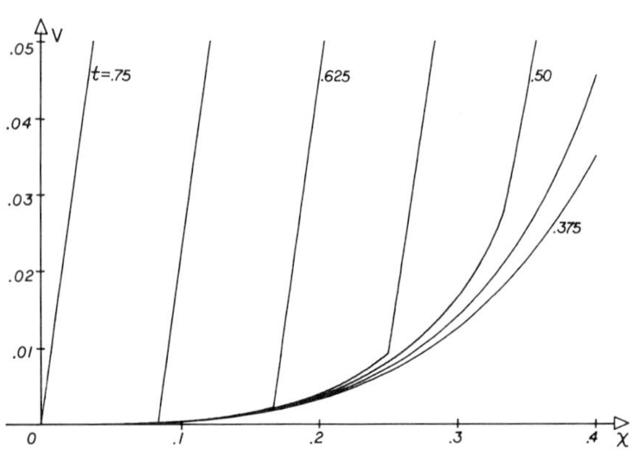

Fig. 5. A corner shock overtaking the interface.

it reaches $x=0$, $v=0$, and the interface begins to move.

Thus, it is seen that there are three possible ways in which the interface can begin to move, and in the limit $\xi\to 0$ all of them are associated intimately with corner shocks or the formation of corner shocks. It should be noted that the first two ways can be discovered using similarity solutions (and have been), while the last way, that of a corner shock overtaking an interface, cannot. This shows the usefulness of the perturbation expansion, in that it reveals this new effect, while at the same time providing a simple method for observing the other two ways in which the interface can begin to move.

The solution of (8) also gives a leading-order approximation for the waiting time, which is obtained from (9) and (10) as the time at which $v=0$ and $x=\zeta_0$ (the initial position of the interface). It is given by

$$t_c=\frac{f(\xi)}{f'^2(\xi)},\tag{12}$$

where ξ satisfies

$$f'(\xi)(\xi-\zeta_0)=2f(\xi).\tag{13}$$

If $f(\xi)$ and $f'(\xi)$ are both zero, then application of L'Hôpital's rule to (12) yields (11).

As an example, consider the use of (12) and (13) when $f(\xi)\sim K(\xi-\zeta_0)^\alpha$ as $\xi\to\zeta_0^+$, with $\alpha>0$. (Then $\xi=\zeta_0$ is a solution of (13).) A local analysis at the interface (at $\xi=\zeta_0$) shows there are three cases:

(1) If $0<\alpha<2$, then $t_c=0$ and the interface begins to move immediately, as for the right interface of fig. 4.

(2) If $\alpha=2$ then t_c is finite and nonzero, as for fig. 3.

(3) If $\alpha<2$ then $t_c=\infty$ and the interface begins to move when it is overtaken by a corner shock, as for fig. 5 (in this case the waiting time is found from other solutions of (13)).

It is possible to have more than one solution of (13). Therefore, all appropriate solutions must be examined to determine the smallest t_c, which will be the waiting time (it is also necessary to determine with which interface a solution is associated; this is relatively simple, since the subcharacteristics always move away from maxima of v). When $\alpha = 2$, however, it is not necessary to look for other solutions of (13) if $f''(x)$ is maximum at $x = \zeta_0$. In this case the corner shock will form at the interface before any others can form, so that $t_c = [2f''(\zeta_0)]^{-1}$.

4. $\mathcal{O}(\varepsilon)$ effects

Here a brief outline of the effects of the neglected term, $\varepsilon v v_{xx}$, is given. For more details, please see [6].

The most noticeable effect is upon the corner shock. To obtain a relatively quick derivation of the basic behavior, rescale in (6) using

$$\xi = \frac{x - \zeta(t)}{\varepsilon}, \quad v = v_0(t) + \varepsilon v_1, \tag{14}$$

where $\zeta(t)$ and $v_0(t)$ are the leading order position of and the value of v at the corner shock. When $v_0 \neq 0$ the result is, after neglecting terms of $\mathcal{O}(\varepsilon)$,

$$v_0 v_{1\xi\xi} = -(v_{1\xi} - w_1)(v_{1\xi} - w_2), \tag{15}$$

where $w_1 < w_2$ and

$$\zeta' = -(w_1 + w_2), \quad v_0' = -w_1 w_2. \tag{16}$$

The solution for v apart from integration constants is therefore (in terms of the original variables)

$$v \sim v_0 + \frac{w_1 + w_2}{2}(x - \zeta) + \varepsilon v_0 \ln\cosh\left[\frac{w_1 - w_2}{2\varepsilon v_0}(x - \zeta)\right]. \tag{17}$$

Thus, it is seen that a corner shock is not sharp if it occurs away from the interface ($v_0 \neq 0$), but rather is rounded into a smooth corner layer of width $\mathcal{O}(\varepsilon v_0)$. The solution is only modified to $\mathcal{O}(\varepsilon)$, however, because $v v_{xx}$ is $\mathcal{O}(1)$ away from the layer and the layer itself is narrow. Finally, we note that (17) seems to indicate that the corner layer sharpens if either $\varepsilon \to 0$ or $v_0 \to 0$.

A uniformly valid expansion can also be constructed by modifying the previous one to

$$v(x, t; \varepsilon) \sim v_0(x, t) + \varepsilon v_1(\xi, x, t) + \varepsilon^2 v_2(\xi, x, t) + \cdots, \tag{18}$$

where

$$\xi = \frac{1}{\varepsilon}\zeta(x, t). \tag{19}$$

The result is, to $\mathcal{O}(\varepsilon)$, approximately [6]

$$v \sim \frac{\tilde{v}_1 + \tilde{v}_2}{2} + \varepsilon \frac{\tilde{v}_1 + \tilde{v}_2}{2} \times \ln\cosh\left[\frac{\tilde{v}_1 - \tilde{v}_2 + \varepsilon(\alpha_1 - \alpha_2)}{\varepsilon(\tilde{v}_1 + \tilde{v}_2)}\right] + \varepsilon\frac{\alpha_1 + \alpha_2}{2} + \cdots, \tag{20}$$

where

$$\tilde{v}_{it} = \tilde{v}_{ix}^2, \tag{21}$$

$$\alpha_{it} = 2\tilde{v}_{ix}\alpha_{ix} + \tilde{v}_i \tilde{v}_{ixx}, \tag{22}$$

for $i = 1, 2$. This approximation merely consists of two regular expansions, given by $\tilde{v}_i$ and α_i, joined together smoothly through the corner layer by (20).

If one attempts to construct a similar uniformly valid expansion when the corner layer occurs at the interface (i.e. at $v = 0$), then one finds that the corner layer is not rounded, but is indeed sharp. In this instance the approximation is

$$v \sim \tfrac{1}{2}[(\tilde{v}_1 + \varepsilon\alpha_1) + |\tilde{v}_1 + \varepsilon\alpha_1|] + \cdots, \tag{23}$$

where $\tilde{v}_1$ and α_1 are the first two terms in a regular expansion, as before. This expression gives a particularly simple way of calculating v to $\mathcal{O}(\varepsilon)$ when corner layers occur only at the interface (a small modification is needed when the waiting time is nonzero, however; see [6] for the details).

5. Local results when $n = \mathcal{O}(1)$

When n is not small it is still possible to use local approximations on the equation

$$v_t = v_x^2 + nvv_{xx}. \tag{24}$$

In particular, following Lacey [8], it is possible to find solutions which are the $n = \mathcal{O}(1)$ equivalents of a shock overtaking the interface (fig. 5). In this case the small parameter is either the distance from the interface or the amount of time before it begins to move; this is why the expansion is local in nature.

Suppose that at $t = 0, v = 0$ for $x < 0$ and $v = bx^3 + \cdots$ for $0 < x \ll 1$. A local expansion in (24) near $x = 0$ then shows that $v = bx^3 + \cdots$ provided that $t < t_c$. The key point now is to look for layer-type solutions which match this cubic behavior to another region where the behavior is approximately linear in x, say $v_x = c + \cdots$. (In this region v will be something like $v = c[c(t - t_c) + x] +$ const. $+ \cdots$.)

To do so, expand for $t < t_c$ using

$$t - t_c = \delta\tau, \tag{25}$$

$$x = -\delta c\tau + \delta^3\xi, \tag{26}$$

$$v = \delta^3 w, \tag{27}$$

where $0 < \delta \ll 1$, $\tau = \mathcal{O}(1)$ (and $\tau < 0$), $\xi = \mathcal{O}(1)$ and $w = \mathcal{O}(1)$. Note that ξ will be the layer variable and the scaling for v results from the fact that $x = \mathcal{O}(\delta)$ and $v = \mathcal{O}(x^3)$. The equation for w is then

$$\delta^2 w_\tau = -cw_\xi + w_\xi^2 + nww_{\xi\xi}. \tag{28}$$

In addition, w must satisfy

$$w \sim c\xi, \quad \xi \to +\infty \tag{29}$$

(to match with the region where $v_x = c$) and

$$w \sim -bc^3\tau^3, \quad \xi \to -\infty \tag{30}$$

(to match with the region where $v = bx^3 + \cdots$).

Neglecting the $\mathcal{O}(\delta^2)$ term and using the matching conditions, it is found that

$$\frac{\mathrm{d}w}{\mathrm{d}\xi} = c - A(\tau)w^{-1/n}, \tag{31}$$

where $A(\tau) = c^{1+3/n}b^{1/n}(-\tau)^{3/n}$. This represents a solution which looks like fig. 5, but now where the corner shock has been smoothed to become a corner layer. The thickness of this layer (in x) is easily found to be $c^2b(-\delta\tau)^3$ if n is fixed as $\tau \to 0$, and $c^2nb(-\delta\tau)^3$ if n can go to zero as well. (Note that this also can be written $c^2nv + \cdots$, so that the real small parameter is nv, as mentioned previously.)

As $\tau \to 0^-$, the width of the layer goes to zero faster than its distance from the interface. Finally, when $\tau = 0$ $(t = t_c)$ the layer has reached $x = 0, v = 0$, and has become sharp. For times greater than this the interface remains sharp and moves to the left, away from $x = 0$. Note that this local analysis does not determine the waiting time t_c; a global analysis, such as when n is small, must be done to determine this.

6. Summary

Singular perturbation and local expansions have been presented to determine the behavior of solutions to a diffusion equation where the diffusion coefficient is a function of the concentration and in fact goes to zero as the concentration goes to zero. These expansions are seen to give a relatively simple understanding of the basic behavior caused by the variation of the diffusion coefficient.

This variation produces a number of interesting interfacial effects, the first of which is a finite propagation speed of the interface between zero and nonzero concentration, in contrast with the infinite propagation speed of linear diffusion. For appropriate initial conditions, these interfaces can also remain stationary for a finite amount of time before beginning to move. The perturbation expansions also easily determine the three ways in which the interface can begin to move, one of which cannot be discovered by considering similarity solutions.

References

[1] D.G. Aronson, Regularity properties of flows through porous media: A counterexample, SIAM J. Appl. Math. 19 (1970) 299–307.

[2] J. Buckmaster, Viscous sheets advancing over dry beds, J. Fluid Mech. 81 (1977) 735–756.

[3] D.S. Cohen, John C. Neu and R.R. Rosales, Rotating spiral wave solutions of Reaction–Diffusion Equations, SIAM J. Appl. Math. 35 (1978) 536–547.

[4] R. Courant and D. Hilbert, Methods of Mathematical Physics, vol. II, (Interscience, New York, 1962).

[5] S. Kamin, Continuous groups of transformations of differential equations; applications to free boundary problems, in: Free Boundary Problems, (E. Magenes, ed.) (Rome, 1980).

[6] W.L. Kath and D.S. Cohen, Waiting-time behavior in a nonlinear diffusion equation, Stud. Appl. Math. 67 (1982) 79–105.

[7] B.F. Knerr, The porous medium equation in one dimension, Trans. Amer. Math. Soc. 234 (1977) 381–415.

[8] A.A. Lacey, Initial motion of the free boundary for a nonlinear diffusion equation, IMA J. Appl. Math. 31 (1983) 113–119.

[9] E.W. Larsen and G.C. Pomraning, Asymptotic analysis of nonlinear Marshak waves, SIAM J. Appl. Math. 39 (1980) 201–212.

[10] M. Muskat, The Flow of Homogeneous Fluids through Porous Media (McGraw-Hill, New York, 1937).

[11] J. Neu, The dynamics of stretched vortices: asymptotic theory and exact solutions, Conference on Fronts, Interfaces and Patterns, Los Alamos, N.M. May 1983.

[12] O.A. Oleinik, A.S. Kalashnikov and Chzou Yui-Lin, The Cauchy problem and boundary problems for equations of the type of unsteady filtration, Izv. Akad. Nauk SSR ser. Mat. 22 (1958) 667–704.

[13] P.Y. Polubarinova-Kochina, Theory of Ground Water Movement (Princeton Univ. Press, Princeton, 1962).

[14] A.M. Turing, The chemical basis of morphogenesis, Phil. Trans. R. Soc. Lond. B 237 (1952) 37–72.

[15] A.T. Winfree, Wavefront geometry in excitable media, Physica 12D (1984) 321 (these proceedings).

Physica 12D (1984) 382–390
North-Holland, Amsterdam

ASYMPTOTIC STRUCTURES IN NONLINEAR DISSIPATIVE AND DISPERSIVE SYSTEMS

F.A. HOWES
Department of Mathematics, University of California at Davis, Davis, CA 95616, USA

Recent results in the asymptotic theory of ordinary differential equations are applied to certain scalar and vector boundary value problems that model various nonlinear dissipative and dispersive wave phenomena.

1. Introduction

We consider here some recent results on dissipative and dispersive systems of perturbed nonlinear differential equations whose solutions exhibit boundary layer behavior and shock or spike layer behavior. In the dissipative case, the equations are second-order systems, obtained for example from hyperbolic systems of conservation laws to which a viscosity term has been added; cf. [13] and [3]. The solutions represent travelling wave fronts, in which the balance between nonlinearity and dissipation leads to the formation of a shock layer connecting two stable equilibrium states. In the dispersive case, however, the equations are third-order systems, obtained for example from coupled systems of long wave equations, such as the Korteweg–de Vries equation; cf. [1], [18] and [3]. The solutions again represent travelling waves, but now the balance is between nonlinearity and dispersion. This leads to the formation of either an isolated spike layer (solitary wave) connecting the same stable equilibrium state or a periodic train of spike layers (cnoidal wave).

The asymptotic approach developed here shows very clearly the interplay of forces in both types of systems, and it provides us with a convenient way of studying the effects of dissipation and dispersion in some fairly general systems.

2. A model dissipative problem

Before turning to the general problem, let us consider the famous Lagerstrom–Cole problem

$$\begin{aligned}&\varepsilon y'' = -yy' + y, \quad 0 < x < 1,\\&y(0,\varepsilon) = A, \quad y(1,\varepsilon) = B,\end{aligned} \tag{2.1}$$

in order to illustrate some effects of the balance between nonlinearity (represented by the term $-yy'$) and dissipation (represented by the term $\varepsilon y''$, $0 < \varepsilon \ll 1$). Only three cases will be discussed. First, if $B > 1$ and $|A| < B - 1$, then the theory [12; chap. 2], [5] tells us that the solution $y = y(x,\varepsilon)$ of (2.1) satisfies

$$\lim_{\varepsilon \to 0^+} y(x,\varepsilon) = u_R(x) \equiv x + B - 1, \quad \text{in } [\delta, 1], \tag{2.2}$$

$0 < \delta < 1$, that is, there is a boundary layer (region of nonuniform convergence) at $x = 0$, since $u_R(0) \neq A$. The function u_R is the solution of the frictionless problem $uu' = u$, $u(1) = B$, which satisfies $-u_R(x) < 0$ in $[0,1]$, and the bound on $|A|$ implies that $\int_\xi^{u_R(0)}(-s)\,ds < 0$, for $A \le \xi < B - 1$. Here the sign condition on u_R guarantees that u_R is "attractive", in the sense of (2.2), while the integral condition determines the exact size of the allowable "boundary layer jump" $|A - u_R(0)|$.

Next, if $A < -1$ and $|B| < |A+1|$, then the solution of (2.1) satisfies

$$\lim_{\varepsilon \to 0^+} y(x, \varepsilon) = u_L(x) \equiv x + A, \quad \text{in } [0, 1-\delta],$$

that is, there is a boundary layer at $x = 1$, since this case is just a reflection of the previous one. The function u_L is now the solution of the problem $uu' = u$, $u(0) = A$, which satisfies $-u_L(x) > 0$ in $[0,1]$, and the bound on $|B|$ implies that $\int_{u_L(1)}^{\eta}(-s)\,ds > 0$, for $A+1 < \eta \le B$. (These sign conditions are best motivated by considering the simple linear problems $\varepsilon y'' = \pm y'$, $y(0,\varepsilon)$, $y(1,\varepsilon)$ prescribed.) Finally, suppose that $A < -1$, $B > 1$ and $B - 1 < |A| < B + 1$. Then the boundary layer jumps $|A - u_R(0)|$ and $|B - u_L(1)|$ are too large to permit the existence of boundary layers. It is known that in this case the solution of (2.1) satisfies

$$\lim_{\varepsilon \to 0^+} y(x, \varepsilon) = \begin{cases} u_L(x), & 0 \le x \le x^* - \delta, \\ u_R(x), & x^* + \delta \le x \le 1, \end{cases} \tag{2.3}$$

for $x^* = (1 - B - A)/2$, that is, there is a shock layer (interior transition) at $x = x^*$ connecting the states u_L and u_R. The location x^* of the shock is found by solving the functional equation $J(x) \equiv \int_{u_L(x)}^{u_R(x)}(-s)\,ds = 0$, which has a solution since $J(0) > 0$ and $J(1) < 0$.

In order to study system analogs of this problem we will have to proceed somewhat indirectly. Let us rederive the above shock layer result, as an illustration of this new technique. For $A < -1$ and $B > 1$ both u_L and u_R are attractive, in that $-u_R(x) < 0$ and $-u_L(x) > 0$ in $[0,1]$, and so the limiting relation (2.3) is certainly possible. It remains for us to locate x^* and to ensure that the "shock strength" $u_R(x^*) - u_L(x^*)$ is not too large. To this end, we consider the two approximate equations

$$\varepsilon y'' \approx -yy' + u_L(x) = \tfrac{1}{2}\left(u_L^2(x) - y^2\right)', \qquad 0 < x < x^*,$$

and

$$\varepsilon y'' \approx -yy' + u_R(x) = \tfrac{1}{2}\left(u_R^2(x) - y^2\right)', \qquad x^* < x < 1,$$

obtained from (2.1) by replacing the y-term with $u_L(x)$ or $u_R(x)$, and recalling that u_L and u_R satisfy $uu' = u$. Integrating once, we arrive finally at the *first-order* equation

$$\varepsilon z' = \begin{cases} \tfrac{1}{2}\left(u_L^2(x) - z^2\right), & 0 < x < x^*, \\ \tfrac{1}{2}\left(u_R^2(x) - z^2\right), & x^* < x < 1, \end{cases} \tag{2.4}$$

which is asymptotically equivalent to (2.1), in the sense that u_L, u_R are solutions of (2.4) with $\varepsilon = 0$ on $(0, x^*), (x^*, 1)$, respectively. In order to proceed further, we must have a single equation on $(0,1)$ whose right-hand side is continuous in x, and so we choose x^* such that

$$u_L^2(x^*) = u_R^2(x^*), \tag{2.5}$$

that is, $x^* = (1 - B - A)/2$, in agreement with our previous result. (The equation (2.5) is a simple form of the Rankine–Hugoniot shock condition; cf. for example [2; chap. 3] and [13].) Thus the theory [14, 6] tells us that there is a solution $z = z(x, \varepsilon)$ of (2.4) satisfying

$$\lim_{\varepsilon \to 0^+} z(x, \varepsilon) = \begin{cases} u_L(x), & 0 \le x \le x^* - \delta, \\ u_R(x), & x^* + \delta \le x \le 1, \end{cases}$$

since $(u_R(x^*) - u_L(x^*)) \cdot \tfrac{1}{2}(u_R^2(x^*) - \lambda^2) > 0$, for $u_L(x^*) < \lambda < u_R(x^*)$, that is, the shock is not too strong. In view of the relationship between (2.4) and (2.1), we conclude that the original problem (2.1) has a solution satisfying (2.3).

3. Dissipative systems

We turn now to the general dissipative system

$$\begin{aligned} &\varepsilon \boldsymbol{y}'' = \mathscr{F}(\boldsymbol{y})\boldsymbol{y}' + \boldsymbol{g}(x, \boldsymbol{y}), \quad a < x < b, \\ &\boldsymbol{y}(a, \varepsilon) = \boldsymbol{A}, \quad \boldsymbol{y}(b, \varepsilon) = \boldsymbol{B}, \end{aligned} \tag{3.1}$$

where y, g, A and B are column vectors in $\mathbb{R}^n$, and $\mathscr{F}(y) \equiv (f_1^T(y), \ldots, f_n^T(y))^T$ is an $(n \times n)$-matrix. We assume that curl $f_i^T \equiv 0$ $(1 \le i \le n)$, in order to write the differential equation in (3.1) as $\varepsilon y'' = [F(y)]' + g(x, y)$, with $F = (F_1, \ldots, F_n)^T$, for $F_i = F_i(y)$ such that $f_i = \operatorname{grad} F_i$. Problems like (3.1) arise in studying travelling wave solutions of hyperbolic systems in one space variable (cf. [13, 3]), for which the assumption that $\mathscr{F}(y)y' = [F(y)]'$ is quite natural. In this setting, the function g may represent a source term, due perhaps to chemical reaction.

To get some feeling for the asymptotic behavior of solutions of (3.1), let us begin by looking for a solution with a boundary layer at $x = a$. We assume first that the frictionless system $\mathscr{F}(u)u' + g(x, u) = 0$, $u(b) = B$, has a smooth solution $u = u_R(x)$ in $[a, b]$. Such functions are candidates for the role of limiting solution outside of the boundary layer, provided they are attractive (cf. (2.1)) in any one of a number of senses. For simplicity, we assume that the eigenvalues of the matrix $\mathscr{F}(u_R(x))$ have negative real parts in $[a, b]$. Then under these two assumptions the theory [9, 10] tells us that the problem (3.1) has a solution $y = y(x, \varepsilon)$ as $\varepsilon \to 0^+$ satisfying the norm-estimate

$$\lim_{\varepsilon \to 0^+} \| y(x, \varepsilon) - u_R(x) \| = 0, \quad \text{in } [a + \delta, b],$$

$0 < \delta < b - a$, *if* the boundary layer jump $\|A - u_R(a)\|$ is sufficiently small. (Here and below, $\|\cdot\|$ is the usual Euclidean norm in $\mathbb{R}^n$.) In order to derive a bound on $\|A - u_R(a)\|$, we proceed as in the case of (2.1), and replace (3.1) with the approximate system

$$\begin{aligned} \varepsilon y'' &\approx [F(y)]' + g(x, u_R(x)) \\ &= [F(y) - F(u_R(x))]'. \end{aligned}$$

An integration gives us finally the asymptotically equivalent *initial value problem*

$$\begin{aligned} &\varepsilon z' = F(z) - F(u_R(x)) \equiv H(x, z), \quad a < x < b, \\ &z(a, \varepsilon) = A. \end{aligned} \tag{3.2}$$

Since by construction $H(x, u_R(x)) \equiv 0$ and since by assumption the eigenvalues of the Jacobian matrix $(\partial H/\partial z)(x, u_R(x))$ $(\equiv \mathscr{F}(u_R(x)))$ have negative real parts in $[a, b]$, the theory [4, 6] implies that the solution $z = z(x, \varepsilon)$ of (3.2) satisfies

$$\lim_{\varepsilon \to 0^+} z(x, \varepsilon) = u_R(x), \quad \text{in } [a + \delta, b],$$

provided that

$$\lambda^T H(a, u_R(a) + \lambda) < 0, \tag{3.3}$$

for all λ in $\mathbb{R}^n$ such that $0 < \|\lambda\| \le \|A - u_R(a)\|$. The condition (3.3) gives us a quantitative estimate on the allowable size of the boundary layer jump; see [10] for examples and further discussion.

For systems like (3.1) it is often advantageous to seek componentwise bounds on a solution y, rather than norm bounds. Let us illustrate this approach by considering solutions of (3.1) with shock layers at points in (a, b), under the assumption that the system is weakly coupled, that is, $\mathscr{F}(y) \equiv \operatorname{diag}\{f_1(y), \ldots, f_n(y)\}$. We discuss just the simple case in which only the kth component has a shock layer at a point x^*, while the remaining components have boundary layers at $x = a$. Our basic assumption is that the frictionless system $\mathscr{F}(u)u' + g(x, u) = 0$ has two smooth solutions $u_{kL} = (u_1, \ldots, u_{k-1}, u_{kL}, u_{k+1}, \ldots, u_n)$ and $u_{kR} = (u_1, \ldots, u_{k-1}, u_{kR}, u_{k+1}, \ldots, u_n)$, which exist in $[a, b]$ and satisfy $u_i(b) = B_i$ for $i \ne k$, $u_{kL} \ne u_{kR}$, $u_{kL}(a) = A_k$ and $u_{kR}(b) = B_k$. In addition, we assume that u_{kL} and u_{kR} satisfy the frictionless system in the strong sense that for x in (a, b)

$$f_i\left(y_{u_L}^{(i)}\right) u_{iL}' + g_i\left(x, y_{u_L}^{(i)}\right) = 0$$

and

$$f_i\left(y_{u_R}^{(i)}\right) u_{iR}' + g_i\left(x, y_{u_R}^{(i)}\right) = 0,$$

where $y_{u_L}^{(i)} = (y_1, \ldots, y_{i-1}, [u_{kL}]_i, y_{i+1}, \ldots, y_n)$ and $y_{u_R}^{(i)} = (y_1, \ldots, y_{i-1}, [u_{kR}]_i, y_{i+1}, \ldots, y_n)$, for all y_j $(j \ne i)$ of interest. Here $[\cdot]_i$ denotes the ith component of the enclosed vector. We can guarantee that these solutions are attractive by assuming also that for $i \neq k$,

$$f_i\left(y_{u_L}^{(i)}\right) \leq -l_i < 0, \quad f_i\left(y_{u_R}^{(i)}\right) \leq -l_i < 0,$$

and that

$$f_k\left(y_{u_L}^{(k)}\right) \geq l > 0, \quad f_k\left(y_{u_R}^{(k)}\right) \leq -l < 0,$$

in the region of interest. Then the theory [16] implies that the problem (3.1) has a solution $y = y(x, \varepsilon)$ such that for $i \neq k$

$$\lim_{\varepsilon \to 0^+} y_i(x, \varepsilon) = u_i(x), \quad \text{in } [a + \delta, b]$$

and

$$\lim_{\varepsilon \to 0^+} y_k(x, \varepsilon) = \begin{cases} u_{kL}(x), & \text{in } [a, x^* - \delta], \\ u_{kR}(x), & \text{in } [x^* + \delta, b], \end{cases}$$

provided the boundary layer jumps $|A_i - u_i(a)|$ and the shock strength $|u_{kR}(x^*) - u_{kL}(x^*)|$ are sufficiently small, where x^* is yet to be determined. The maximum allowable sizes of the jumps are given by $(n-1)$ integral conditions of the form:

for $i \neq k$

$$(u_i(a) - A_i) \times \int_{u_i(a)}^{\xi} f_i(A_1, \ldots, A_{i-1}, s, A_{i+1}, \ldots, A_n)\,\mathrm{d}s > 0,$$

for ξ between $u_i(a)$ and A_i, $\xi \neq u_i(a)$. In the case of the kth component, the location of the shock layer is determined by a solution $x = x^*$ of the functional equation $J(x) \equiv \int_{u_{kL}(x)}^{u_{kR}(x)} f_k(\boldsymbol{u}_s(x^*))\,\mathrm{d}s = 0$, which also satisfies $J'(x^*) \neq 0$. Here $\boldsymbol{u}_s(x^*) = (u_1(x^*), \ldots, u_{k-1}(x^*), s, u_{k+1}(x^*), \ldots, u_n(x^*))$. This result can be extended in an obvious way to the case in which different components of the solution have shock layers at distinct points; we refer the reader to [16] for details and examples.

4. A model dispersive problem

It is instructive to begin our discussion of dispersive systems by considering a typical scalar problem like

$$\begin{aligned} &\varepsilon^2 y''' = [F(y)]' + g(y), \quad 0 < x < 1, \\ &y(0, \varepsilon) = A, \quad y'(0, \varepsilon) = C, \quad y(1, \varepsilon) = B, \end{aligned} \tag{4.1}$$

in which dispersion (represented by the term $\varepsilon^2 y'''$, $0 < \varepsilon \ll 1$) competes with nonlinearity (represented by the term $F_y y'$), assuming that $F_{yy} \neq 0$. The well-studied Korteweg–de Vries equation (KdV), $v_t + v_\xi + v v_\xi + \varepsilon^2 v_{\xi\xi\xi} = 0$, is transformed into such a third-order equation with $F(y) = (c-1)y - y^2/2$ and $g \equiv 0$, by introducing the variables $y(x, \varepsilon) \equiv v(\xi, t, \varepsilon)$, $x = \xi - ct$, $c > 0$. Consequently our results for (4.1) apply to (KdV) and related evolution equations; in turn, the physical implications of such equations offer a useful check on the theory.

The study of (4.1) begins with the assumption that the dispersionless problem $[F(u)]' + g(u) = 0$, $u(0) = A$, has a smooth solution $u = u(x)$. (Clearly, it is not possible for us to ask that $u'(0) = C$, in general, and on asymptotic grounds, it is likewise inappropriate to ask that u satisfy $u(1) = B$; cf. [7].) Consideration of simple linear problems like $\varepsilon^2 y''' = \pm y'$, $y(0, \varepsilon)$, $y'(0, \varepsilon)$, $y(1, \varepsilon)$ prescribed, reveals that in order for the function u to attract a solution of (4.1), in the sense that $y(x, \varepsilon) \to u(x)$ as $\varepsilon \to 0^+$, we must require that $F_y(u(x)) > 0$ for x in $[0, 1]$. Then the theory [7] implies that there is a solution $y = y(x, \varepsilon)$ of (4.1) as $\varepsilon \to 0^+$ such that

$$\lim_{\varepsilon \to 0^+} y(x, \varepsilon) = u(x), \quad \text{in } [0, 1 - \delta], \tag{4.2}$$

$0 < \delta < 1$, *provided* the boundary layer jump $|B - u(1)|$ is sufficiently small. (We note that the convergence in (4.2) is uniform at $x = 0$, since $u(0) = A$; however, $y'(x, \varepsilon) \to u'(x)$ in $[\delta, 1 - \delta]$, unless u also satisfies $u'(0) = C$.) In the context of (KdV) the restriction that $F_y(u) > 0$, for $F(y) = (c-1)y$

$-y^2/2$ and $u \equiv 0$, implies the well-known inequality $c > 1$, that is, solitary waves move supercritically against a zero background. To derive an upper bound on the allowable size of $|B - u(1)|$, let us pass from (4.1) to the approximate equation $\varepsilon^2 y''' \approx [F(y)]' + g(u(x)) = [F(y) - F(u)]'$, and integrate once. The resulting asymptotically equivalent second-order problem

$$\varepsilon^2 z'' = F(z) - F(u(x)) \equiv H(x, z), \quad 0 < x < 1,$$
$$z'(0, \varepsilon) = C, \quad z(1, \varepsilon) = B, \tag{4.3}$$

clearly displays the dispersion-nonlinearity balance within the layer at $x = 1$. Since by construction $H(x, u(x)) \equiv 0$ and since by assumption $H_z(x, u(x))\ (= F_y(u(x))) > 0$ in $[0,1]$, the theory [5] guarantees that the solution z satisfies (4.2) provided

$$\int_{u(1)}^{\eta} H(1, s)\,\mathrm{d}s > 0, \tag{4.4}$$

for η between $u(1)$ and B, $\eta \neq u(1)$. In view of the relationship between (4.1) and (4.3), the solution of (4.1) satisfies (4.2) as well, for all such values of B.

Suppose now that the function H in (4.3) is independent of x, and suppose for simplicity that $H(0) = 0$. If $H_z(0) > 0$ then $z = 0$ is a maximum point of the potential energy $\Phi(z) \equiv -\int_0^z H(s)\,\mathrm{d}s$. The theory [17] (based on phase-plane analysis) states that if there is a finite z-value z^* such that $\Phi(z^*) = \Phi(0) = 0$ and $H(z^*) \neq 0$, then the equation $\varepsilon^2 z'' = H(z)$ has a solution $z = z(x, \varepsilon)$ as $\varepsilon \to 0^+$ with a spike of amplitude $|z^*|$ at $x = 0$, that is,

$$\lim_{\varepsilon \to 0^+} z(x, \varepsilon) = 0, \quad \text{for } x \text{ in } (-L, L),\ L \gg 1,$$

except that (4.5)

$$\lim_{\varepsilon \to 0^+} z(0, \varepsilon) = z^*.$$

(If $z^* > 0$ $[z^* < 0]$ then the spike represents an elevation [depression].) As an illustration, in the case of (KdV), $H(z) = (c-1)z - z^2/2$, and so $\Phi(z^*) = 0$, for $z^* = 3(c-1)$; the theory thus reproduces the familiar amplitude-phase speed relation for a solitary wave of elevation moving to the right with speed c: $c = 1 + z^*/3$ (cf. [19]). In addition to the solution satisfying (4.5), there are solutions $z = \tilde{z}(x, \varepsilon)$ of the autonomous equation with spikes of amplitude $|z^*|$ at regularly spaced points in $(-L, L)$, that is,

$$\lim_{\varepsilon \to 0^+} \tilde{z}(x, \varepsilon) = 0, \quad \text{for } x \text{ in } (-L, L),$$

except that

$$\lim_{\varepsilon \to 0^+} \tilde{z}(x_k, \varepsilon) = z^*, \quad \text{for } x_k = L(2k/n - 1),$$

$n = 2, 3, \ldots, k = 1, 2, \ldots, n-1$; cf. [17]. In the case of (KdV) these solutions correspond to periodic cnoidal waves of elevation $z^* = 3(c-1)$, as discussed for example in [11; chap. 3].

5. Dispersive systems

Let us turn finally to the dispersive system

$$\varepsilon^2 \boldsymbol{y}''' = [\boldsymbol{F}(\boldsymbol{y})]' + \boldsymbol{g}(\boldsymbol{y}), \quad a < x < b,$$
$$\boldsymbol{y}(a, \varepsilon) = \boldsymbol{0}, \quad \boldsymbol{y}'(a, \varepsilon) = \boldsymbol{C}, \quad \boldsymbol{y}(b, \varepsilon) = \boldsymbol{B}, \tag{5.1}$$

where $\boldsymbol{y}$, $\boldsymbol{F}$, $\boldsymbol{g}$, $\boldsymbol{C}$ and $\boldsymbol{B}$ are column vectors in $\mathbb{R}^n$. Such problems arise in studying permanent wave solutions of coupled evolution equations, which describe the flow of stratified fluids [1] and the motion of large-scale atmospheric disturbances [18]. If we assume, for simplicity, that $\boldsymbol{g}(\boldsymbol{0}) = \boldsymbol{0}$, then $\boldsymbol{u} \equiv \boldsymbol{0}$ is a solution of the dispersionless problem $[\boldsymbol{F}(\boldsymbol{u})]' + \boldsymbol{g}(\boldsymbol{u}) = \boldsymbol{0}$, $\boldsymbol{u}(a) = \boldsymbol{0}$, and we can pass directly to the asymptotically equivalent second-order problem

$$\varepsilon^2 z'' = F(z) - F(0) \equiv H(z), \quad a < x < b,$$
$$z'(a, \varepsilon) = C, \quad z(b, \varepsilon) = B. \tag{5.2}$$

Let us now look for componentwise bounds on a solution of (5.2) (and hence, of (5.1)) having a boundary layer at $x = b$. Conditions which enable us to do this are given in [15, 8]. The first one is the condition that for $i = 1, \ldots, n$

$$H_i(z_1, \ldots, z_{i-1}, 0, z_{i+1}, \ldots, z_n) \equiv 0 \tag{5.3}$$

for each z_j, $j \neq i$, in the region of interest. This condition is an obvious strengthening of our basic assumption that $H(0) = 0$, and it allows us essentially to "decouple" the system and study each of the component equations separately. The second condition is the "attractiveness" assumption that for $i = 1, \ldots, n$ and all such z_j

$$H_{i,z_i}(z_1, \ldots, z_{i-1}, 0, z_{i+1}, \ldots, z_n) > 0. \tag{5.4}$$

Then the theory guarantees the existence of a solution $z = z(x, \varepsilon)$ of (5.2) as $\varepsilon \to 0^+$ such that

$$\lim_{\varepsilon \to 0^+} z_i(x, \varepsilon) = 0, \quad \text{in } [a, b - \delta], \tag{5.5}$$

$0 < \delta < b - a$, provided the boundary layer jumps $|B_i|$ are sufficiently small. A precise estimate of the allowable size of each $|B_i|$ is given by an integral condition of the form (cf. (4.4))

$$\int_0^{\eta} H_i(B_1, \ldots, B_{i-1}, s, B_{i+1}, \ldots, B_n)\, ds > 0, \tag{5.6}$$

for η between 0 and B_i, $\eta \neq 0$. Thus, for such values of B_i, the original problem (5.1) has a solution $y = y(x, \varepsilon)$ as $\varepsilon \to 0^+$ satisfying

$$\lim_{\varepsilon \to 0^+} y(x, \varepsilon) = 0, \quad \text{in } [a, b - \delta].$$

As an illustration, consider the following second-order problem on $(0, 1)$:

$$\begin{aligned} &\varepsilon^2 z_1'' = z_1 - z_1^2/2 - z_1 z_2 \equiv H_1(z_1, z_2), \\ &z_1'(0, \varepsilon) = C_1, \quad z_1(1, \varepsilon) = B_1, \\ &\varepsilon^2 z_2'' = z_2 - z_2^2/2 - z_1 z_2 \equiv H_2(z_1, z_2), \\ &z_2'(0, \varepsilon) = C_2, \quad z_2(1, \varepsilon) = B_2; \end{aligned} \tag{5.7}$$

it arises through the coupling of two (KdV) equations. Here $H_1(0, z_2) \equiv 0$, $H_2(z_1, 0) \equiv 0$, for all z_1, z_2, and $H_{1,z_1}(0, z_2) = 1 - z_2 > 0$, $H_{2,z_2}(z_1, 0) = 1 - z_1 > 0$, for all $z_1, z_2 < 1$. The integral conditions $\int_0^{B_1} H_1(s, B_2)\, ds > 0$ and $\int_0^{B_2} H_2(B_1, s)\, ds > 0$ imply the inequalities $B_1 < 3(1 - B_2)$ and $B_2 < 3(1 - B_1)$, respectively. Consequently, for all $B_1, B_2 < \frac{3}{4}$, the problem (5.7) has a solution satisfying the limiting relation (5.5).

The system (5.1) also possesses solutions which are the analogs of the solitary and cnoidal wave solutions of the scalar problem (4.1). To see this, let us assume that conditions (5.3) and (5.4) obtain again for the asymptotically equivalent system (5.2). Guided by the scalar theory we define the "potential energies"

$$\Phi_i(z_1, \ldots, z_n) \equiv -\int_0^{z_i} H_i(z_1, \ldots, z_{i-1}, s, z_{i+1}, \ldots, z_n)\, ds,$$

for $i = 1, \ldots, n$. It follows from (5.3) and (5.4) that $z = 0$ is a maximum point of each Φ_i. We consider only two cases. First, suppose we do not allow two components of a solution to have spikes at the same x-value. In this case we assume that for each $i = 1, \ldots, n$ there is a finite number z_i^* such that $\Phi_i(0, \ldots, 0, z_i^*, 0, \ldots, 0) = \Phi_i(0) = 0$ and $H_i(z_1^*, \ldots, z_n^*) \neq 0$. Then, provided the boundary values B_i satisfy (5.6), the theory [8] guarantees the existence of solutions $z = z(x, \varepsilon)$ of (5.2) as $\varepsilon \to 0^+$

such that

$$\lim_{\varepsilon \to 0^+} z_i(x, \varepsilon) = 0, \quad \text{in } [a, b-\delta],$$

except that (5.8)

$$\lim_{\varepsilon \to 0^+} z_i(x_{i,k}, \varepsilon) = z_i^*,$$

at points $x_{i,k}$ in (a, b). In the second case, suppose we want each of the components of a solution to have a spike at the same x-value. We assume now that there are n finite numbers $z_1^{**}, \dots, z_n^{**}$ such that $\Phi_i(z_1^{**}, \dots, z_n^{**}) = \Phi_i(\mathbf{0}) = 0$ and $H_i(z_1^{**}, \dots, z_n^{**}) \neq 0$, for $i = 1, \dots, n$. The same theory implies the existence of a solution of (5.2) as $\varepsilon \to 0^+$ satisfying the limiting relations (5.8) (with z_i^* replaced by z_i^{**}), provided $\boldsymbol{B}$ satisfies (5.6).

These ideas are best explained by looking at an example, and so let us reconsider the two-component system (5.7). In any subinterval of (0, 1) where z_2 (z_1) is asymptotically zero, the z_1-equation (z_2-equation) reduces to the uncoupled equation $\varepsilon^2 z'' = z - z^2/2$. Consequently, the scalar theory [17, 7] implies that for each integer $n \geq 2$, the problem

$$\begin{aligned} &\varepsilon^2 z'' = z - z^2/2, \quad 0 < x < 1, \\ &z'(0, \varepsilon) = C, \quad z(1, \varepsilon) = B < \tfrac{3}{4}, \end{aligned} \tag{5.9}$$

has two solutions $z = z(x, \varepsilon)$ as $\varepsilon \to 0^+$ such that

$$\lim_{\varepsilon \to 0^+} z(x, \varepsilon) = 0, \quad \text{in } [0, 1-\delta],$$

except that

$$\lim_{\varepsilon \to 0^+} z(x_k, \varepsilon) = 3, \quad \text{for } x_k = k/n, \ k = 1, \dots, n-1.$$

We conclude, for example by taking solutions of (5.9) for $n = 2$ and $n = 3$, that the problem (5.7) has the four solutions pictured below, along with the four solutions obtained from these by interchanging z_1 and z_2; see figs. 1–4. Using the solutions of (5.9) for $n = 2, 3, \dots, p$, p a prime, we see that there are solutions of (5.7) with spikes of height 3 at the x-values $1/2$, $1/3$, $2/3$, $1/5$, $2/5$, $3/5$, $4/5, \dots, 1/p$, $2/p, \dots, (p-1)/p$. At a point in (0, 1) where *both* components z_1, z_2 have a spike, the behavior of a solution is determined by the energies $\Phi_i(z_1, z_2) \equiv -\int_0^{z_i}(s - s^2/2 - s z_j)\,\mathrm{d}s$, for $i, j = 1, 2$ $(i \neq j)$. Clearly, Φ_1 and Φ_2 have maxima at (0, 0), and since $\Phi_i(3/4, 3/4) = 0$ with $H_i(3/4, 3/4) \neq 0$, for $i = 1, 2$, we conclude that the height of a spike in both components at the same x-value is only $3/4$. Thus the nonlinear coupling of the components, due to the interaction term $z_1 z_2$, decreases the height of the spikes from 3 to $3/4$.

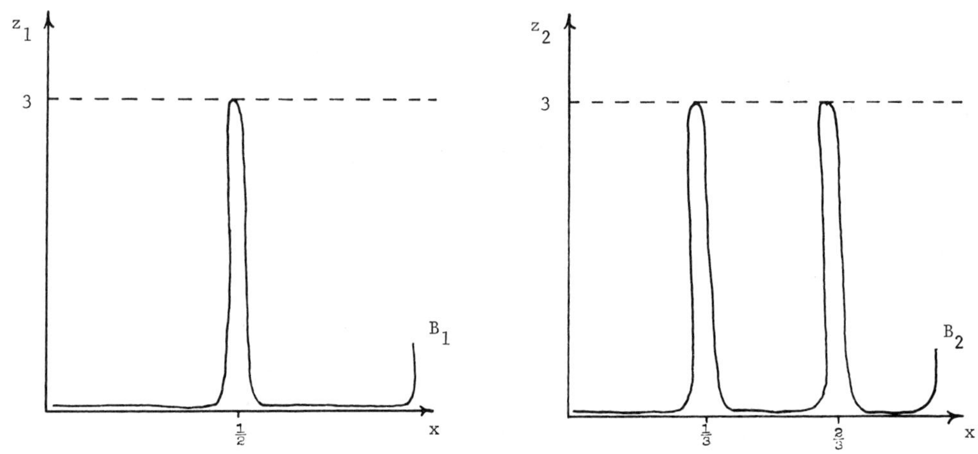

Fig. 1. One solution of (5.7).

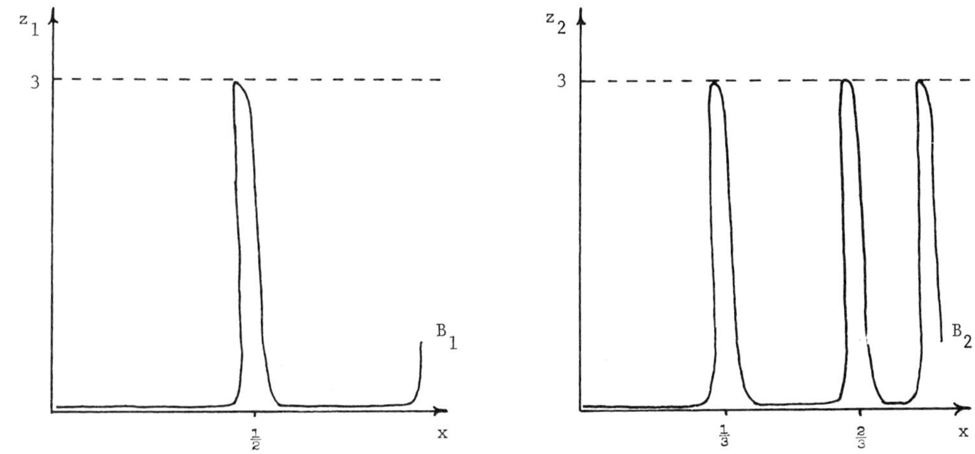

Fig. 2. A second solution of (5.7).

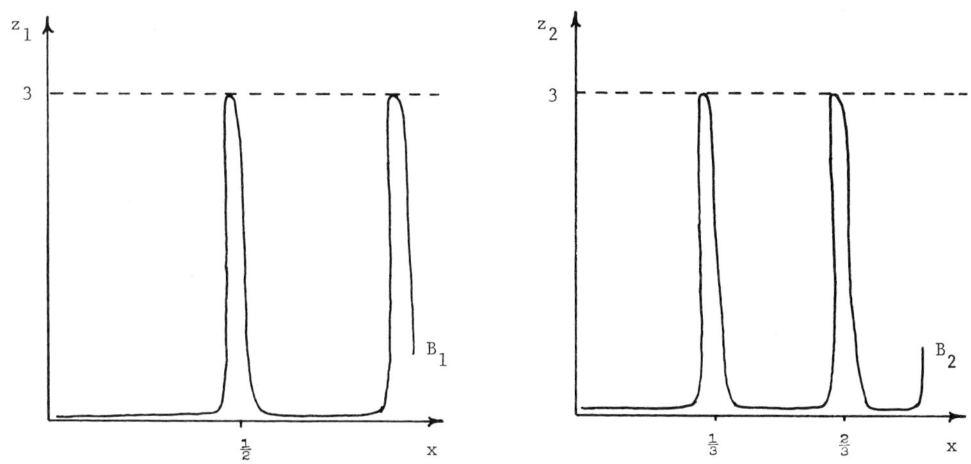

Fig. 3. A third solution of (5.7).

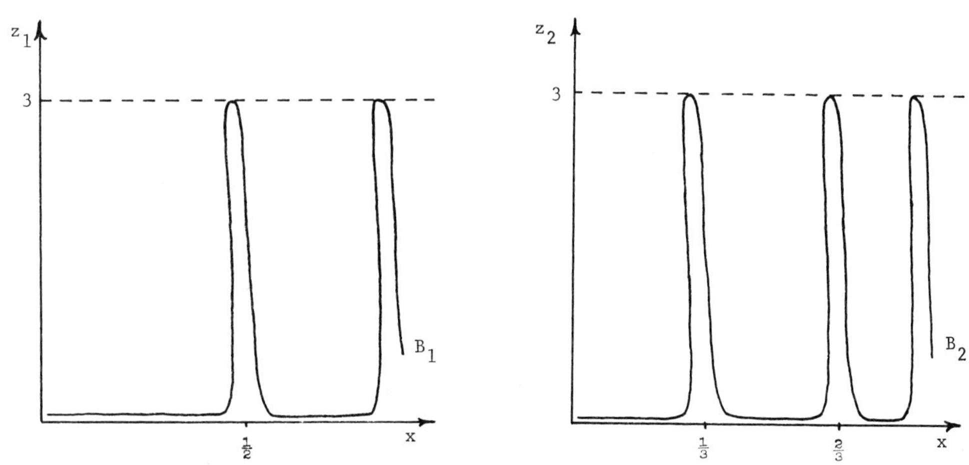

Fig. 4. A fourth solution of (5.7).

Acknowledgements

The author wishes to thank the typist, Mrs. Ida Zalac, for her prompt and accurate typing of the original manuscript.

References

[1] D.J. Benney, J. Math. and Phys. 45 (1966) 52.
[2] R. Courant and K.O. Friedrichs, Supersonic Flow and Shock Waves (Springer, New York, 1976).
[3] R.J. DiPerna, Physica 12D (1984) 363 (these proceedings).
[4] F. Hoppensteadt, Comm. Pure Appl. Math. 24 (1971) 807.
[5] F.A. Howes, Memoirs Amer. Math. Soc. no. 203 (1978).
[6] F.A. Howes, Bull. Acad. Polonaise des Sci. 30 (1982) 117.
[7] F.A. Howes, SIAM J. Appl. Math. 43 (1983) 993.
[8] F.A. Howes, Studies in Appl. Math. 69 (1983) 75.
[9] F.A. Howes, J. Math. Anal. Appl., in press.
[10] F.A. Howes and R.E. O'Malley Jr., in: Lecture Notes in Math., vol. 827 (Springer, New York, 1980) p. 131.
[11] V.I. Karpman, Nonlinear Waves in Dispersive Media (Pergamon, Oxford, 1975).
[12] J. Kevorkian and J.D. Cole, Perturbation Methods in Applied Mathematics (Springer, New York, 1981).
[13] P.D. Lax, Hyperbolic Systems of Conservation Laws and the Mathematical Theory of Shock Waves, Regional Conf. Series in Appl. Math., vol. 11 (SIAM, Philadelphia, 1973).
[14] J.J. Levin and N. Levinson, J. Rational Mech. Anal. 3 (1954) 247.
[15] M.A. O'Donnell, SIAM J. Math. Anal. 15 (1984) 317.
[16] M.A. O'Donnell, Ph.D. Thesis, Univ. of California at Davis, 1983.
[17] R.E. O'Malley Jr., J. Math. Anal. Appl. 54 (1976) 449.
[18] L.G. Redekopp and P. Weidman, J. Atmos. Sci. 35 (1978) 790.
[19] N.J. Zabusky, in: Proc. Symp. on Nonlinear Partial Differential Equations, W.F. Ames, ed. (Academic Press, New York, 1967) p. 223.

Physica 12D (1984) 391–395
North-Holland, Amsterdam

REMARKS ON THE KURAMOTO–SIVASHINSKY EQUATION

Basil NICOLAENKO and Bruno SCHEURER†
Center for Nonlinear Studies, MS B258, Los Alamos National Laboratory, Los Alamos, NM 87545, USA

We report some preliminary results on the Kuramoto–Sivashinsky equation which models instabilities for both flame front cellular patterns and chemical reaction–diffusion waves. We present existence and uniqueness results and a fractional step method of numerical resolution.

1. Introduction

We report here a joint work in progress on the Kuramoto–Sivashinsky equation. The question we address is the analytical study of the following fourth order nonlinear evolution equation:

$$\frac{\partial u}{\partial t} + \Delta^2 u + \Delta u + \tfrac{1}{2}|\nabla u|^2 = 0. \tag{1.1}$$

This equation has been obtained by Sivashinsky [8] in the context of combustion and independently by Kuramoto [3] in the context of reaction diffusion-systems. Both were motivated by (nonlinear) stability of travelling waves. Numerical calculations have been done on this equation: we should mention the work of Michelson and Sivashinsky [6], Aimar [2], Manneville [5], and Hyman [7]. All the results seem to indicate a "chaotic" behavior of the solution. Therefore, the analytical study is of interest in analogy with the Burger's and Navier–Stokes equations. Here we give some existence and uniqueness results for eq. (1.1), in space dimension one (section 2), and we also study a fractional step method of numerical resolution (section 3). In a forthcoming joint paper with R. Temam, we will study the asymptotic behavior, as $t \to +\infty$, of the solution of (1.1) and give an estimate on the number of "determining modes" (see [9]).

2. Existence and uniqueness in space dimension one

We are considering the following initial-value problem (for $0 < T < \infty$)

$$\frac{\partial u}{\partial t} + \frac{\partial^4 u}{\partial x^4} + \frac{\partial^2 u}{\partial x^2} + \tfrac{1}{2}\left|\frac{\partial u}{\partial x}\right|^2 = 0, \quad \text{in } [0,1]\times[0,T], \tag{2.1}$$

$$u(0) = \vartheta, \tag{2.2}$$

†On leave, Centre d'Etudes de Limeil and Université Paris-Sud (Orsay), France.

with *periodic boundary conditions* (the unit interval can be therefore identified with the one dimensional torus). The case of the boundary conditions

$$\frac{\partial u}{\partial x}=\frac{\partial^3 u}{\partial x^3}=0, \quad \text{on } x=0 \text{ and } x=1, \tag{2.3}$$

could be handled in the same way. Rewriting the problem (2.1) as an integral equation,

$$u(t)=S(t)\vartheta+\tfrac{1}{2}\int_0^t S(t-s)\left|\frac{\partial u}{\partial x}(s)\right|^2 \mathrm{d}s,$$

where $S(t)$ denote the semigroup associated to $\partial^4/\partial x^4+\partial^2/\partial x^2$, we easily obtain local in time existence and uniqueness results. It is sufficient to apply the standard Picard iteration scheme. We therefore need a priori estimates on u in order to get existence on $[0,T]$ (instead of $[0,T_*]$, $T_*<T$). Indeed we will prove ($\|\cdot\|$ is the usual L^2 norm):

Theorem 1. Let ϑ satisfy $\partial^k\vartheta/\partial x^k\in \mathrm{L}^2$, $0\le k\le 2$. Then, for any solution of (2.1), (2.2), one has:

$$\sup_{0\le t\le T}\left\|\frac{\partial^k U}{\partial x^k}(t)\right\|, \int_0^T\left\|\frac{\partial^{k+2} U}{\partial x^{k+2}}(t)\right\|^2 \mathrm{d}t < \text{constant}, \tag{2.4}$$

where $0\le k\le 2$, the constant depends only on ϑ and T and $U\equiv u(x,t)-\int_0^1 u(x,t)\,\mathrm{d}x$.

Corollary 1. Under the hypothesis of theorem 1, the problem (2.1), (2.2) admits a unique solution u such that: $u\in \mathrm{L}^2(0,T;H^4)\subset \mathrm{L}^\infty(0,T;H^2)$, $\mathrm{d}u/\mathrm{d}t\in \mathrm{L}^2(0,T;\mathrm{L}^2)$. (Here H^k is the Sobolev's space of functions u such that $\partial^j u/\partial x^j\in \mathrm{L}^2$, $0\le j\le k$.)

Proof of theorem 1. If the space dimension is one, we make the following remark. Set $v\equiv\partial u/\partial x$, then v satisfies a "Burger's like" equation,

$$\frac{\partial v}{\partial t}+\frac{\partial^4 v}{\partial x^4}+\frac{\partial^2 v}{\partial x^2}+v\frac{\partial v}{\partial x}=0. \tag{2.5}$$

In other words, we multiply (2.1) by $\partial^2 u/\partial x^2$; after integrating by parts we get

$$\tfrac{1}{2}\frac{\mathrm{d}}{\mathrm{d}t}\left\|\frac{\partial u}{\partial x}(t)\right\|^2+\left\|\frac{\partial^3 u}{\partial x^3}(t)\right\|^2-\left\|\frac{\partial^2 u}{\partial x^2}(t)\right\|^2=0 \tag{2.6}$$

(here and in what follows, we denote by $\|\cdot\|$ the usual L^2 norm). This identity reflects the fact that, in (2.1) or (2.3), the term $\partial^4 u/\partial x^4$ (resp. $\partial^2 u/\partial x^2$) is a sink (resp. source) of energy. We now distinguish two cases.

a) The sink term is dominant. In other words, by Poincaré's inequality:

$$\left\|\frac{\partial^2 u}{\partial x^2}(t)\right\|^2\le\Lambda_1\left\|\frac{\partial^3 u}{\partial x^3}(t)\right\|^2, \tag{2.7}$$

where Λ_1 is the constant of Poincaré (notice that $\partial^2 u/\partial x^2$ has 0 mean value). From (2.6) and (2.7) we conclude, by Gronwall's lemma

$$\sup_{0\le t\le T}\left\|\frac{\partial u}{\partial x}(t)\right\|, \int_0^T\left\|\frac{\partial^3 u}{\partial x^3}(t)\right\|^2 \mathrm{d}t < C\left(\left\|\frac{\partial\vartheta}{\partial x}\right\|\right), \tag{2.8}$$

where the bound is uniform in T.

b) The sink term is not dominant. We use the interpolation inequality

$$\left\|\frac{\partial^2 u}{\partial x^2}(t)\right\|^2 \le \tfrac{1}{2}\left\|\frac{\partial^3 u}{\partial x^3}(t)\right\|^2 + C\left\|\frac{\partial u}{\partial x}(t)\right\|^2 \tag{2.9}$$

and again we obtain, from (2.6) and Gronwall's lemma, the bound (2.8). But now the constant depends on T.

Coming back to (2.1), we note that $\bar{u}(t) \equiv \int_0^1 u(x,t)\,\mathrm{d}x$ satisfy

$$\frac{\mathrm{d}}{\mathrm{d}t}\bar{u}(t) + \tfrac{1}{2}\int_0^1\left|\frac{\partial u}{\partial x}\right|^2 \mathrm{d}x = 0, \tag{2.10}$$

therefore $U(x,t) \equiv u(x,t) - \bar{u}(t)$ satisfy

$$\frac{\partial}{\partial t}U + \frac{\partial^4}{\partial x^4}U + \frac{\partial^2 U}{\partial x^2} + \tfrac{1}{2}\left|\frac{\partial U}{\partial x}\right|^2 - \tfrac{1}{2}\int_0^1\left|\frac{\partial U}{\partial x}\right|^2 \mathrm{d}x = 0, \tag{2.11}$$

where now, $\int_0^1 U(x,t)\,\mathrm{d}x = 0$. Multiplying (2.11) by U and integrating, we get

$$\tfrac{1}{2}\frac{\mathrm{d}}{\mathrm{d}t}\|U(t)\|^2 + \left\|\frac{\partial^2 U}{\partial x^2}(t)\right\|^2 = \left\|\frac{\partial U}{\partial x}(t)\right\|^2 - \tfrac{1}{2}\int_0^1\left|\frac{\partial U}{\partial x}\right|^2 U\,\mathrm{d}x. \tag{2.12}$$

The first term in the right-hand side of (2.12) is bounded using an inequality similar to (2.9). For the second term we use Hölder inequality:

$$\int_0^1\left|\frac{\partial U}{\partial x}\right|^2 U\,\mathrm{d}x \le \left\|\frac{\partial U}{\partial x}(t)\right\| \left\|\frac{\partial U}{\partial x}(t)\right\|_{\mathrm{L}^4} \|U(t)\|_{\mathrm{L}^4}. \tag{2.13}$$

If the space dimension is one, by the Sobolev imbedding theorem and interpolation (see for instance [1, 4]), we know that:

$$[H^2, \mathrm{L}^2]_{7/8} = H^{1/4} \subset \mathrm{L}^4, \tag{2.14}$$

$$[H^2, \mathrm{L}^2]_{3/8} = H^{5/4}. \tag{2.15}$$

Using the norm inequalities associated with (2.14), (2.15) we deduce finally from (2.13)

$$\int_0^1\left|\frac{\partial U}{\partial x}\right|^2 U\,\mathrm{d}x \le \left\|\frac{\partial U}{\partial x}(t)\right\|\left\{C_\varepsilon\|U(t)\|^2 + \varepsilon\left\|\frac{\partial^2 U}{\partial x^2}(t)\right\|^2\right\}, \tag{2.16}$$

for $\varepsilon > 0$ small. Thanks to (2.8), we can control $\|\partial U(t)/\partial x\| = \|\partial u(t)/\partial x\| \le \sup_{0\le t\le T}\|\partial u(t)/\partial x\|$, and obtain from (2.12) and (2.16)

$$\tfrac{1}{2}\frac{\mathrm{d}}{\mathrm{d}t}\|U(t)\|^2 + C'_\varepsilon\left\|\frac{\partial^2 U}{\partial x^2}(t)\right\|^2 \le C''_\varepsilon\|U(t)\|^2. \tag{2.17}$$

By Gronwall's lemma, we get

$$\sup_{0\le t\le T}\|u(x,t)-\bar{u}(t)\|, \int_0^T\left\|\frac{\partial^2 u}{\partial x^2}(t)\right\|^2 \mathrm{d}t < \underset{T}{C}\left(\|\vartheta-\bar{\vartheta}\|, \left\|\frac{\partial\vartheta}{\partial x}\right\|\right). \tag{2.18}$$

The remaining needed estimates are proved by multiplying (2.1) by $\partial^4 u/\partial x^4$; after integration by parts we get

$$\tfrac{1}{2}\frac{\mathrm{d}}{\mathrm{d}t}\left\|\frac{\partial^2 u}{\partial x^2}(t)\right\|^2 + \left\|\frac{\partial^4 u}{\partial x^4}(t)\right\|^2 = \left\|\frac{\partial^3 u}{\partial x^3}(t)\right\|^2 - \tfrac{1}{2}\int_0^1\left|\frac{\partial^2 u}{\partial x^2}\right|^2\frac{\partial^2 u}{\partial x^2}\,\mathrm{d}x. \tag{2.19}$$

The right-hand side is bounded using (2.8), the Hölder inequality and the following version of Poincaré's inequality $\|\partial^2 u(t)/\partial x^2\|_{L^\infty} \le 2\Lambda_1\|\partial^3 u(t)/\partial x^3\|$. We conclude then

$$\sup_{0\le t\le T}\left\|\frac{\partial^2 u}{\partial x^2}(t)\right\|, \int_0^T\left\|\frac{\partial^4 u}{\partial x^4}(t)\right\|^2 \mathrm{d}t \le \underset{T}{C}\left(\|\vartheta\|, \left\|\frac{\partial\vartheta}{\partial x}\right\|, \left\|\frac{\partial^2\vartheta}{\partial x^2}\right\|\right). \tag{2.20}$$

This completes the proof of (2.4). □

3. A fractional step method

It is natural to decompose eq. (2.1) in two parts corresponding to the nonlinear term and the linear terms. Precisely, we can split (2.1) in the following way:

$$\frac{\partial u}{\partial t} - \frac{\partial^2 u}{\partial x^2} + \tfrac{1}{2}\left|\frac{\partial u}{\partial x}\right|^2 = 0, \tag{3.1}$$

$$\frac{\partial u}{\partial t} + \frac{\partial^4 u}{\partial x^4} + 2\frac{\partial^2 u}{\partial x^2} = 0. \tag{3.2}$$

According to (3.1), (3.2) we define the following scheme. For N given large set $\tau = T/N$ and $u^n = u(n\tau)$, $0 \le n \le N$. For u^n given $(u^0 = \vartheta)u^{n+1}$ is obtained via

$$u^{n+1/2} - \tau\frac{\partial^2 u^{n+1/2}}{\partial x^2} + \frac{\tau}{2}\left|\frac{\partial u^{n+1/2}}{\partial x}\right|^2 = u^n, \quad \text{in } [0,1], \tag{3.3}$$

$$u^{n+1/2} + \tau\frac{\partial^4 u^{n+1}}{\partial x^4} + 2\frac{\partial^2 u^{n+1}}{\partial x^2} = u^{n+1/2}, \quad \text{in } [0,1]. \tag{3.4}$$

A convenient equivalent form of (3.4) is

$$u^{n+1} + \tau\left(\frac{\partial^2}{\partial x^2} + 1\right)^2 u^{n+1} - \tau u^{n+1} = u^{n+1/2}. \tag{3.5}$$

We still suppose periodic boundary conditions. From (3.3), multiplying by $-\partial^2 u^{n+1/2}/\partial x^2$ and integrating by parts, we get

$$\left\|\frac{\partial u^{n+1/2}}{\partial x}\right\|^2 - \left\|\frac{\partial u^{n}}{\partial x}\right\|^2 + \left\|\frac{\partial u^{n+1/2}}{\partial x} - \frac{\partial u^{n}}{\partial x}\right\|^2 + 2\tau\left\|\frac{\partial^2 u^{n+1/2}}{\partial x^2}\right\|^2 = 0. \tag{3.6}$$

Similarly from (3.5) we get

$$(1-2\tau)\left\|\frac{\partial u^{n+1}}{\partial x}\right\|^2 - \left\|\frac{\partial u^{n+1/2}}{\partial x}\right\|^2 + \left\|\frac{\partial u^{n+1}}{\partial x} - \frac{\partial u^{n+1/2}}{\partial x}\right\|^2 + 2\tau\left\|\left(\frac{\partial^2}{\partial x^2}+1\right)\frac{\partial u^{n+1}}{\partial x}\right\| = 0. \tag{3.7}$$

Therefore, combining (3.6) and (3.7):

$$\left\|\frac{\partial u^{n+1}}{\partial x}\right\|^2 \le (1-2\tau)^{-1}\left\|\frac{\partial u^{n}}{\partial x}\right\|^2. \tag{3.8}$$

Consequently, for τ small enough,

$$\left\|\frac{\partial u^{n+1/2}}{\partial x}\right\|^2 \le \left\|\frac{\partial \vartheta}{\partial x}\right\|^2, \tag{3.9}$$

$$\left\|\frac{\partial u^{n+1}}{\partial x}\right\|^2 \le \mathrm{e}^{2\tau}\left\|\frac{\partial \vartheta}{\partial x}\right\|^2. \tag{3.10}$$

Coming back to (3.3), (3.4) we can find bounds, as in section 2, for $\|u^{n+1/2}\|$, $\|u^{n+1}\|$, $\|\partial^2 u^{n+1}/\partial x^2\|$. Then, thanks to the maximum principle we can estimate in L^∞, $u^{n+1/2}$ and $\partial u^{n+1/2}/\partial x$; as a consequence we can pass to the limit in the non-linearity. So are the main estimates in order to prove stability and convergence results for the scheme defined by (3.3)–(3.4). Details will be given elsewhere.

Acknowledgements

This research was supported by the Center for Nonlinear Studies, Los Alamos National Laboratory. Work also performed under the auspices of the U.S. Department of Energy under contract W-7405-ENG-36 and contract KC-04-02-01, Division of Basic and Engineering Sciences.

References

[1] A. Adams, Sobolev Spaces (Academic Press, New York, 1979).
[2] M.T. Aimar, Thèse 3° Cycle, 1982.
[3] Y. Kuramoto, Prog. of Theor. Phys. 63 (1980) 1885–1903.
[4] J.L. Lions and E. Magenes, Problèmes aux limites non homogenes, (Dunod, Paris, 1968).
[5] P. Manneville, unpublished.
[6] D. Michelson and G.I. Sivashinsky, Acta Astronautica 4 (1977) 1207–1221.
[7] J.M. Hyman, private communication.
[8] G.I. Sivashinsky, Acta Astronautica 4 (1977) 1117–1206.
[9] C. Foias, O. Manley, R. Temam and Y. Treve, Physica 9D (1983) 157.

Physica 12D (1984) 396–407
North-Holland, Amsterdam

3.2. COMPUTATIONAL METHODS

NUMERICAL METHODS FOR TRACKING INTERFACES*

James M. HYMAN
Center for Nonlinear Studies, Theoretical Division, MS B284, Los Alamos National Laboratory, Los Alamos, NM 87545, USA

An overview is given of special numerical methods for tracking discontinuous fronts and interfaces. These methods include surface tracking methods based on connected marker points along the interface, volume tracking methods that track the volume occupied by the solution regions bounded by the interfaces, and moving-mesh methods where the underlying mesh is aligned and moved with the interface. The pros and cons of the current methods are discussed and a new method is proposed that overcomes some of the difficulties encountered in approximating equations with multiple interacting interfaces.

1. Introduction

Interface tracking methods are often necessary to efficiently compute accurate numerical approximations to partial differential equations with moving discontinuous interfaces in the solution. There are a few well-established algorithms that account for these discontinuities, but most are numerical schemes† still in their early developmental stages. No simple rules exist for choosing the best method for the more difficult problems. In this paper I will give an overview of the current methods in order of their ability to handle problems of increasing difficulty. I will then introduce a new adaptive moving-mesh scheme and speculate on what will be the more significant future developments.

Much of our understanding of the laws of nature is based on integral equations and constitutive relationships that hold across discontinuous interfaces. Away from these discontinuities, where the solution is smooth, these equations can be well approximated by partial differential equations (PDEs). Most numerical predictions of the laws of nature are based on discrete approximations to these PDEs. For these numerical methods to be accurate near discontinuities, where the PDEs fail to approximate the integral equations, they must treat the discontinuity as a special case. Otherwise, the method may not accurately approximate the physically relevant solution.

Two commonly used approaches are to smear the interface by adding artificial dissipation or viscosity to the PDEs and solving this nearby problem, or to treat the discontinuity as an internal boundary, solving the PDEs away from the discontinuity and imposing the appropriate jump conditions across this boundary.

The better artificial dissipation methods are extremely easy to implement, perform excellently for a restricted class of problems, but are not well understood theoretically. One of the more important problems they cannot treat adequately is tracking a moving internal material interface or slip line when the equation-of-state for the two materials is radically different. For example, the water/air interface of a bubble should not be artificially smeared, or the method will not accurately account for the effect of surface tension on its motion.

Another difficulty with the artificial dissipation method occurs when it is used in conjunction with an adaptive static rezone method [2, 18]. These

*This work was supported by the U.S. Department of Energy under contract W-7405-ENG-36 and the Office of Basic Energy Sciences, Department of Applied Mathematics under contract number KC-04-02-01-0.

†Scheme – (def): an especially sly or devious plan of action. The Random House Dictionary of the English Language, (Random House, New York, 1967).

0167-2789/84/$03.00© Elsevier Science Publishers B.V.
(North-Holland Physics Publishing Division)

adaptive methods automatically adjust the mesh so that it is dense in regions with sharp transitions and sparse where the solution is smooth. The artificially smeared discontinuities are likely to have the large gradients causing the adaptive mesh algorithm to introduce mesh points that resolve the structure of the transition layer. This structure is often an artifact of the artificial dissipation and does not significantly affect the behavior of the solution. (If it did, then the artificial dissipation method would not have been appropriate.) These extra mesh points can greatly increase the computational expense without significantly enhancing the accuracy of the calculation.

The interface tracking methods described in this paper were developed to overcome the deficiencies of the artificial dissipation approach. The tracking methods have little or no artificial dissipation near the interface since the singularity is directly computed and treated explicitly as a discontinuity. These methods are more difficult to implement, perform excellently for a large class of problems, but, like artificial dissipation methods, are not well understood theoretically for the more difficult problems.

Interface tracking methods can be divided into three catagories. In order of increasing flexibility and computational complexity, these are surface tracking methods, volume tracking methods, and moving mesh methods.

Surface tracking methods track the location of the interface by interpolating between marker particles along the interface. Because this is a lower dimensional problem, the additional effort to accurately resolve small-subgrid-scale structure in the interface is usually small compared with the overall solution time. The surface tracking methods are the simplest to implement, until interactions occur that change the topology of the interface during the computation.

Volume tracking methods overcome the changing topology problems by dividing the domain into a union of disjoint solution regions. The boundary between these regions is the interface location. The regions are identified by marker points, or alternatively, the fractional volume of each solution region located in each computational cell is calculated and advanced during the computation by solving an auxiliary evolutionary PDE. These fractional volumes can be used to reconstruct an approximate interface location at any time. Unlike the surface tracking methods, very little subgrid scale structure is retained during the calculation.

Moving-mesh methods can be used to track the location, account for changes in the interface topology, and resolve small-scale structures in the interfaces. Here a multivalued solution is defined at mesh points located on the interface. The interface mesh points move with the interface in a Lagrangian manner. To prevent the mesh from tangling, a dynamic data structure is used so the moving mesh points can change their nearest neighbors during the calculation. Also, new points are added when the mesh becomes sparse or a new interface appears, and mesh points are removed when they are more dense than necessary for an accurate calculation.

In all the above methods, the location of the interface is advanced by solving a lower dimensional PDE derived from an appropriate constitutive jump condition. The effect of the interface movement is transferred to the solution to the original PDEs on either side where the interface is treated as a moving boundary.

2. Surface tracking methods

In surface tracking methods the interface is specified by an ordered set of marker points located on the interface [11, 12, 22, 25]. Between these points its position is approximated by an interpolant, usually a piecewise polynomial. These time-dependent interfaces divide the problem domain into connected regions. The solution defined at the marker points and along the interpolated interface may be multivalued to account for discontinuities.

2.1. *Surface representation*

The marker points may be represented by the distance from some reference surface such as the chain-of-line segments defined by a height function (fig. 1a) or by a parametric interpolant as shown (fig. 1b).

The distance function is simpler to implement, but the interface deformation is severely limited because this representation breaks down if the curve becomes multivalued with respect to the reference surface. The parameter representation does not have these limitations and is only slightly more complicated to implement.

Both of these methods can provide the fine resolution and detail needed to track small-sub-grid-scale structures in the interface in two and three space dimensions. This can be especially important when tracking an unstable interface. The underlying computational grid resolution is usually chosen to resolve the structure of the smooth solution away from the interface and is rarely sufficient to resolve the fine-scale interface detail, such as the onset of a slip-line Helmholtz instability rollup.

The surface tracking methods are hybrid numerical schemes – splitting the solution process into two parts: the interface tracking and the smooth solution algorithm. Both the smooth solution and the interface are treated as separate computational objects. The interface position is stored and dynamically updated along with the smooth solution away from the interface. The numerical method may be implemented to do this simultaneously, or operator splitting may be used, first advancing one and then the other.

2.2. *Surface evolution*

The evolution equations for the interface are lower dimensional differential equations derived from constitutive relationships, usually obtained by applying the divergence theorem to an integral formulation of the physical model. Similar methods are used to define an accurate approximation to the smooth solution near an interface. This approach, sometimes called the finite volume method, results in equations for the interface–not for the marker points. In practice, however, it is the marker points that are evolved. Because of this, extra care is needed to maintain the appropriate relationships, such as the conservation laws, near the interface.

When the underlying PDE system is hyperbolic, the equations for the interface can be derived by solving one-dimensional Riemann problems nor-

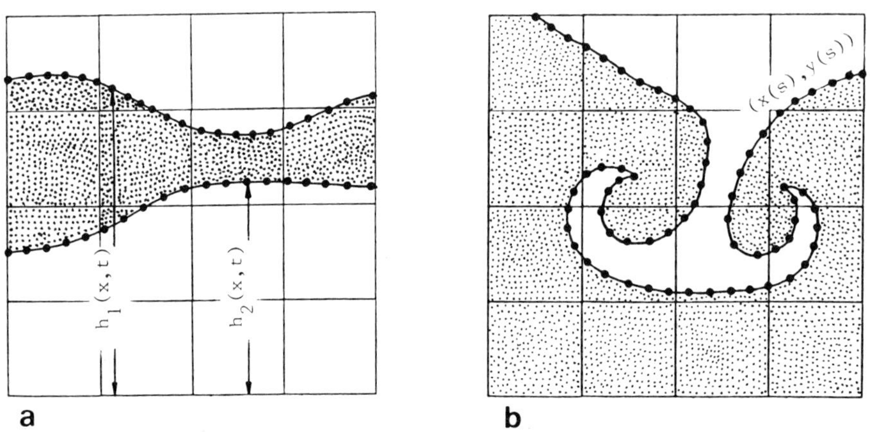

Fig. 1. a) Multiple distance functions designate the locations of the interfaces. b) A parametric interpolant designates the locations of the interfaces.

mal to the interface [11–13]. A Riemann problem is a type of a nonlinear normal mode expansion for a one-dimensional Cauchy problem with initial data that is constant everywhere except for a single jump discontinuity. The solution evolves into nonlinear waves that propagate coherently in time. For example, a material interface discontinuity moves with the underlying fluid velocity, and a shock wave moves with the speed given by the Rankine–Hugoniot jump conditions. When used to advance a 2-D interface in the normal direction, curvature effects can be incorporated as source terms.

When multiple waves are emitted in the Riemann problem, then a single wave, such as the contact discontinuity corresponding to a material interface, must be selected and tracked at all the points along the same interface. Alternatively, additional interfaces can be inserted to track the new discontinuities that arise.

The constant states used for the Riemann problem are the bounding states on either side of the interface. A slightly improved version could be implemented by using a linear or quadratic variation in the states normal to the front, extending the ideas of van Leer [28]. The velocities for the marker points are taken to be the tracked wave velocities normal to the interface. The tangential velocities of the underlying solution also contribute to the interface movement, but are less important, since when the marker points are displaced tangentially, they remain on the interface.

In addition to the Riemann problem equations, the evolution equations may contain additional terms to approximate the effects of surface tension [15, 21], flame propagation [1, 3, 26] or phase changes. These all influence the boundary conditions imposed on the smooth solution at the interface. For example, if the interface were a flame front that converts the fuel ahead of the interface to burned material behind it, the interface boundary conditions would be conservative outflow or inflow conditions, respectively, with the appropriate heat source term to account for the energy released at the interface.

The above procedure is simplified if a local curvilinear coordinate system orthogonal to the interface is employed. The orthogonal coordinate system also simplifies the interface boundary conditions that connect the smooth solution to the interface. Additionally, when approximating the spatial derivatives, we can smoothly map this coordinate system onto a fixed regular computational grid [19, 24].

Sometimes the evolution of the interface is sensitive to small amounts of noise in the computed solution on either side, and it may be necessary to regularize or smooth the solution states in the tangential direction. Glimm, Marchesin, and McBryan [11, 12] average in a circle about each marker point on the interface to filter out short wavelength fluctuations. Another way to reduce the effects of small errors on the motion of the interface is to modify the interface evolution equations slightly by adding artificial surface tension along the interface.

Instead of solving Riemann problems, the vortex methods [3, 26] introduce small lines on surfaces of vorticity along the interface. The vortex motion is defined by a Hamiltonian system of ordinary differential equations with a Coulomb-type interaction term. For viscous PDEs, these equations also contain a diffusive term.

The contour dynamics method of Overman and Zabusky [25] follows the motion of point vortices along a closed contour line bounding a region of constant density or vorticity. The contour velocity is obtained from boundary integral equations derived by applying Green's theorem to the area integral in the Green's function solution of a Poisson equation.

2.3. *Interpolants*

The accuracy of the surface tracking methods depends strongly on stability and accuracy of the interpolation method approximating the interface location between the marker points. The shape-preserving Hermite piecewise-polynomial interpolants [16] remain consistently well behaved and smooth by retaining the convexity and monotonicity properties of the original data. Therefore, extraneous bumps or wiggles are not introduced between the data points. This is particularly important for unstable interfaces such as a Rayleigh–Taylor or Buckely–Leverett interface.

The piecewise-linear interpolant is the simplest shape- and area-preserving interpolant, but introduces fictitious corners in the interface location. In fluid flows, unless the solution is artificially smoothed, these corners can create unrealistic velocities in the smooth solution that eventually destroy the global accuracy of the calculation. The higher order Hermite piecewise-polynomial interpolants are smoother but can still easily accommodate necessary corners in the interface by retaining left and right derivatives at the marker points. Usually, cubic piecewise polynomials are sufficient.

As the interface grows or shrinks, the distribution of the interface marker points must change to continually resolve the interface. The static rezone methods [18] use a mesh function or performance index based on the arc length and curvature of the interface as a guide when removing extraneous mesh points or adding new ones when the existing points are too dense or sparse. Alternatively, a fixed number of marker points can be continuously redistributed so as to best resolve the interface. The first approach is usually more efficient in 2-D calculations where the marker points are easily reordered as points are added or deleted. The data structure is simpler in the second approach and it is more commonly used in 3-D calculations.

2.4. *Interactions*

In many calculations the topology of the interfaces is constantly changing. The interfaces can interact with each other, spontaneously disappear, or new ones can be created when an existing interface bifurcates or is formed by a compression wave, a flame ignition, or a phase change. Multiple interaction points where three or more interfaces meet are common occurrences and also require special data structures.

Fortunately, the interactions are usually a lower dimensional event; in 2-D calculations the 1-D interfaces intersect at points; in 3-D, the 2-D surfaces interact along 1-D lines. Therefore, it takes little computer time to identify and track the interactions. Unfortunately, the data structure and algorithms needed by surface tracking methods to account for interactions greatly increase the program complexity. Also, near the interaction many of the lower dimensional interface evolution equations based on 1-D Riemann problems are no longer valid. That is, near an interaction point, the solution often is intrinsically 2-D or 3-D and the interface motion cannot be well approximated by a 1-D Riemann approximation. The reason for this is that near the interface the components of the solution normal to the interface interact strongly with the interface through the interface boundary conditions, whereas the tangential components have a weak, if any, interaction with the interface. Therefore, away from interactions, applying the 1-D jump conditions normal to the interface is a valid approximation. Near an interaction, however, the solution cannot be split into components that are normal and tangential to both interfaces simultaneously and the splitting algorithm is no longer valid.

Simple material interfaces which move with the underlying fluid velocity are not a problem, but for more complicated interactions, such as an oblique shock reflection, then a local grid refinement may be necessary to reduce the splitting errors.

These hybrid methods, combining and interface tracking algorithm with local grid refinement, are also excellent when there is a boundary layer in the solution adjacent to the interface. This might be caused by the energy generated in a flame front or governed by an internal mixing length. If each region is being separately solved on an interface-fitted curvilinear coordinate system, then standard adaptive methods can be used to resolve the boundary layer [2, 24].

Because of the complexity of handling interactions and adaptively refining the interface in 3-D, to my knowledge, the only surface tracking codes that resolve multiple interactions are in 1- and 2-D. In most 2-D codes with interface interactions, and in all 3-D codes I know of, the interfaces are tracked by a volume tracking method.

3. Volume tracking methods

The volume tracking methods are less capable than the surface tracking methods in providing subgrid-scale resolution, but they can simply and accurately account for the interactions of many different smoothly varying interfaces. Here, the interface tracking equations have the same dimension as the underlying PDEs and, therefore, are potentially more expensive than the surface tracking methods. In practice, however, the interface data need only be stored and the equations solved in the cells along or near the interface. This tactic increases the computational complexity but improves the efficiency when the interfaces are well separated.

3.1. *Marker and cell*

One of the earliest volume tracking methods for material interfaces is the marker and cell (MAC) method [29]. Marker particles are scattered initially to identify each material region in the calculation. These particles are transported in a Lagrangian manner along with the materials. Their presence in a computational cell indicates the presence of the marked material. The material boundary is reconstructed using the marker particle densities in the mixed cells with marker particles of two or more materials. The interface reconstruction scheme may also use the density of particles in the surrounding cells to reconstruct a more accurate interface location. A moderate number of particles must be in the mixed cells to reconstruct an accurate interface. If there are only a few marker particles, the interface location will be poorly defined and sensitive to small errors. This often happens in expansion regions unless new particles are continually added and deleted during the calculation.

To improve the efficiency of the MAC method, we can scatter initial marker particles more densely (or only) near the interfaces or, if there are only two materials, only use marker particles to identify one of the materials.

The MAC methods do not require special logic for colliding surfaces but need many marker particles per computational cell to get a well-defined interface. Also, numerical errors in transporting the marker particles can cause an artificial numerical diffusive mixing near the interface resulting in a fuzzy interface. The fuzzy interface makes it harder to generalize the MAC methods for complicated interfaces such as detonations, flame fronts or multiphase flow problems. To reduce the fuzziness far more marker particles than computational cells are needed, increasing the computational cost. The fractional marker volume methods were developed to overcome these drawbacks.

3.2. *Fractional marker volumes*

The fractional marker volume methods (sometimes called the volume of fluid, VOF [14, 15, 21, 22], or simple line interface calculation, SLIC [23], methods) define the surface by calculating the fractional volume of each material occupied in each computational cell. These numbers range from zero (no material) to one (completely filled with material). The interfaces occur in the cells with fractional volumes (fig. 2).

The volume fractions are updated during the calculation according to the appropriate advection equations. On each time step, the interface position is reconstructed cellwise using the fractional volume of a cell and its nearest neighbors. This localness is especially good for long thin interfaces or fingers. When interfaces collide, the fraction volumes are added, and the interface intersections are simply accounted for in the reconstruction step.

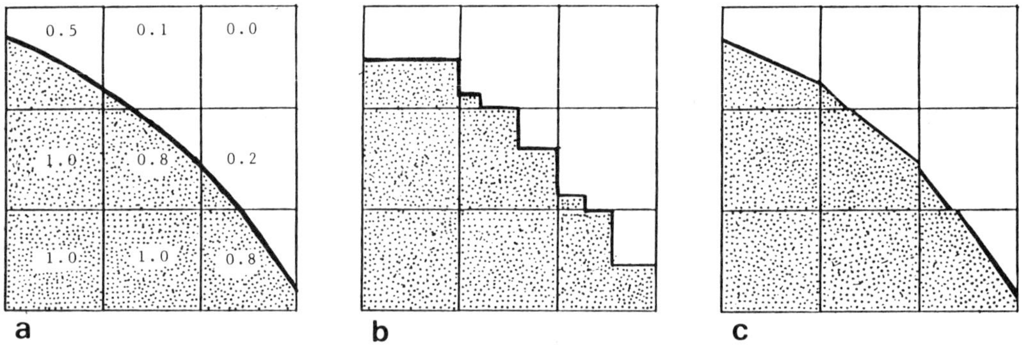

Fig. 2. The original interface separating two regions and the associated volume fractions in each computational cell are shown (2a). In figs. 2b and 2c are two possible reconstructed interfaces using the rectangular and piecewise-linear fractional volume methods, respectively.

3.3. *Interface reconstruction*

Even though the volume fraction in the interface cells are between zero and one, the reconstructed discontinuity is sharp. Within each cell, the volume regions can be represented by unions of rectangles, triangles, or regions bounded by piecewise-polynomial surfaces. Thin rectangles can be used to accommodate fingers. The curvature of the interface can be estimated using finite-difference approximates based on the neighboring fractional volumes. Although the more complicated methods yield a better approximation to the position of the interface, they are more prone to numerical diffusion and nonphysical mixing caused by small pieces shedding off the corners of the reconstructed interface. Imposing monotonicity and convexity constraints [16, 28] on the reconstructed interface greatly reduces the shedding problems.

There are as many fractional volume reconstruction schemes as there are practitioners [1, 3, 4, 14, 15, 20, 21, 23, 26]. The best reconstruction algorithm depends upon the application and the importance of subgrid scale structure. For example, Chorin [3] modified the original SLIC algorithm by allowing for multiple rectangles in a cell and developed a SLICer one for flame propagation. Barr and Ashurst [1] then combined the ideas of slope determination in the VOF method with Chorin's modifications and have one of the SLICest methods in use for turbulent flame propagation.

Another variation of the fractional volume method is to maintain and evolve a point on the interface in each fractional volume cell to aid the reconstruction algorithm. These marker points are moved with the interface during the evolution step. After each reconstruction, the multiple marker points within a single cell are combined to form a single new point on the reconstructed interface, and new marker points are added along the interface in cells with no marker particles. This approach is similar to the reconnecting dual mesh method discussed in the next section.

If the underlying mesh is not a tensor product mesh, then the mapping method can be used to reconstruct the interface using fractional volumes in the uniform logical reference mesh. The curvilinear mesh in the original domain could be chosen adaptively to accurately resolve the solution [2, 18]. The fractional volumes of the reconstructed interface are easily transferred from one mesh to another after a static rezone.

3.4. *Interface evolution*

Once constructed, the interface can be advanced using the same techniques as described for the surface tracking methods. More often, however, they are advanced using a fractional step method. That is, the interface is reconstructed and evolved in each spatial dimension separately. The frac-

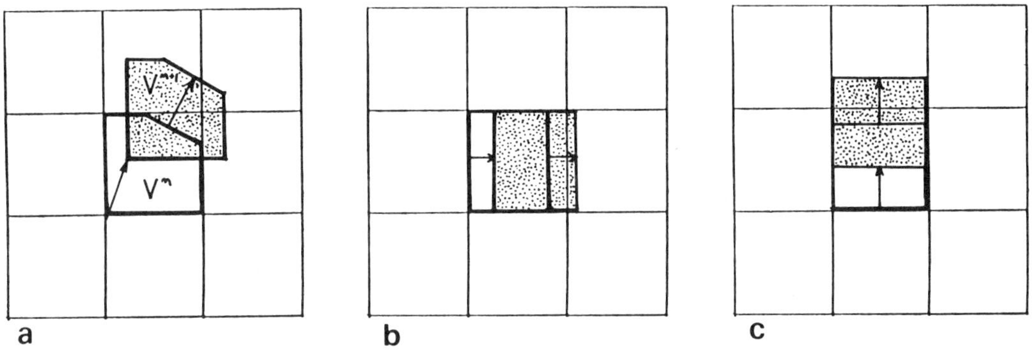

Fig. 3. The reconstructed interfaces for an unsplit algorithm (a), in the x-sweep (b), and in the y-sweep (c).

tional step algorithms are more attractive than the surface tracking methods for almost all interacting surfaces in 2-D and, more so, in 3-D calculations. The unsplit fractional volume method is more difficult to implement because of the odd shaped regions that must be accounted for, such as the one shown in fig. 3a.

In a fractional step method, the interface reconstructed in the x-sweep may be different from what it is in the y-sweep, (figs. 3b, c). By alternating the sweep direction, averaging the results, or using conservative limiters that preserve the symmetry, much of the sweep dependence can be reduced.

4. Moving mesh methods

4.1. *Local adjustment methods*

When using either a surface or volume tracking method, if the reconstructed interface is continuous, then on each time step a local adjustment in the underlying mesh location can be used to approximate the interface. Each mesh point within one half mesh spacing of the interface is moved to the nearest location horizontally or vertically where the interface intersects a mesh line. The interface is now well approximated by the cell edges and diagonals on the new mildly distorted mesh, as seen in the example in fig. 4.

The PDEs are then solved on the new grid treating the mesh points on the interface as a moving boundary; the boundary conditions are determined by the appropriate jump conditions. The correct solution values for the possibly multivalued solution along the interface are easily selected. Also, the discrete approximations to the equations at the few irregular mesh points near the interface are easily derived on each time step [19]. If the finite volume method is being used, the solution jumps across the interface are taken into account in the surface integrals. In many PDEs, accurate and efficient global solution algorithms are available if the discontinuities occur only along mesh boundaries and diagonals [5].

4.2. *Lagrangian methods*

The next logical step is not to have a separate algorithm to track the interface, but to subdivide the initial regions into discrete elements with the cell boundaries aligned with the interfaces. The PDEs are then solved cell by cell, and the cell edges are all treated like interfaces and, therefore, will continue to track the interface. As new interfaces are created, new cells are added (or existing ones readjusted); as the volume of existing cells goes to zero, they are combined or deleted. In this manner, numerical mixing between the different regions is avoided, and varying size structures that move with the solution are easily tracked.

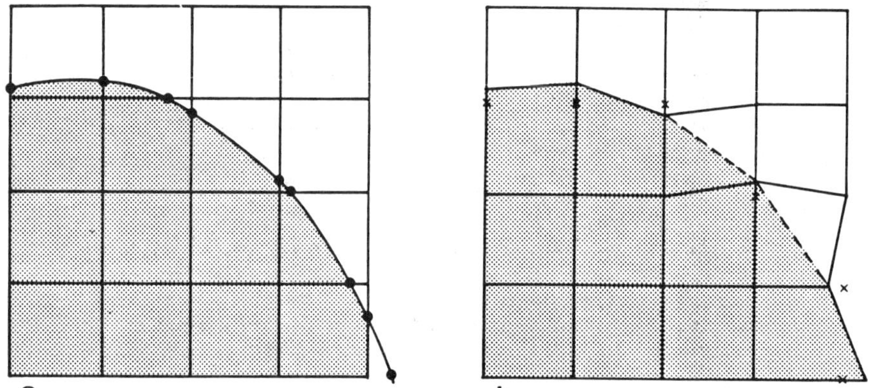

Fig. 4. A simple local adjustment of the mesh aligns it with the front, so the interface is located only along cell edges or diagonals. a) The original underlying mesh and interface; b) the locally adjusted mesh is aligned with the interface.

Depending upon the goal of the calculation, any one of several moving-mesh methods can be used away from the interface [6–10, 13, 17]. Unfortunately, unless the moving cells are allowed to change their nearest neighbors in a calculation, then continual rezoning is necessary to prevent the mesh distortion, in even the simplest flows, from destroying the accuracy of the calculation. This is shown for the very early stages of a Lagrangian Rayleigh–Taylor calculation on a logically rectangular grid in fig. 5. Soon after the time of this plot, numerical errors caused the grid lines to cross and the calculation to fail.

The grid distortion can be somewhat alleviated if only the mesh points along specified interfaces are required to move with the interface. The grid points away from the interface are distributed to prevent mesh tangling or to better resolve the structure of the smooth solution. In addition, the mesh points along the interface can be redistributed along the interface using the surface tracking rezone methods described earlier.

Fig. 5. Distorted Lagrangian mesh in the early stages of a Rayleigh–Taylor instability calculation.

The mesh tangling is not caused by the Lagrangian equations but is an artifact of the mesh data structure. A more flexible data structure that prevents mesh tangling is the neighborhood grid, where pointers are kept to all the nearest neighbors of each mesh point. These are used to approximate the equations locally using a finite element or finite volume discrete approximation. As the mesh moves, the nearest neighbor pointers are continually updated. Usually these reconnections alone will not completely solve the problem of grid resolution and regularity, and some form of rezoning is till necessary. But here, adding and deleting points is much easier than on a logically rectangular or cuboid mesh.

There has been considerable success in using the reconnecting moving neighborhood mesh in fluid flows with material interfaces [7–9, 27]. These free Lagrangian codes, as they are called, could easily be adapted for more complicated flows where the interface moves with the shock or flame velocity rather than with just the underlying fluid velocity.

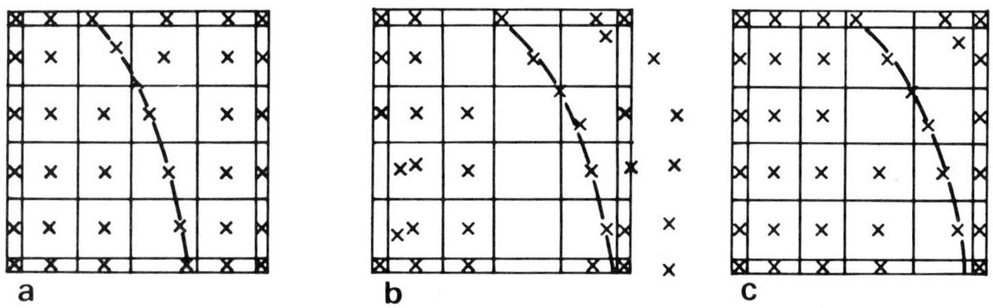

Fig. 6. The dual computing mesh is advanced according to the appropriate moving mesh equations. At the end of each time step it is regularized as in 6c. a) The reference mesh and the dual computing mesh; b) the new dual mesh after one time step; c) the regularized dual mesh at the new time.

The major disadvantage of neighborhood meshes is the data structure. Most of the fast and accurate numerical methods (and multidimensional plotting packages) are for logically rectangular and cuboid grids. The accuracy of local discrete approximations of second derivative operators is particularly poor on neighborhood grids.

One of the more promising reconnecting mesh algorithms retains a logically rectangular or cuboid data structure by approximating the solution on the dual mesh of a logically rectangular or cuboid reference mesh [13, 17]. The dual mesh consists of one mesh point within each cell of the reference mesh (fig. 6a).

At the beginning of every time step, the mesh is regular (one grid point per reference cell). The dual mesh is advanced (fig. 6b) according to the appropriate moving-mesh equations for the interface depicted by a solid line in fig. 6. If desired, an underlying reference mesh can now be chosen to resolve the solution. The mesh is then regularized (fig. 6c) by adding new dual mesh points to empty cells or combining them in reference cells with more than one dual mesh point. In this way, the mesh points where the solution is computed can continue changing their nearest neighbors while maintaining a logically rectangular data structure.

In addition, the dual mesh computing points can be tagged as special interface points (fig. 7a). In fig. 7b the dual mesh points, separating materials □, × and ○, have been advanced forming some new mixed cells in the center. In the regularization stage (fig. 7c) these are combined conservatively to form two new triple points.

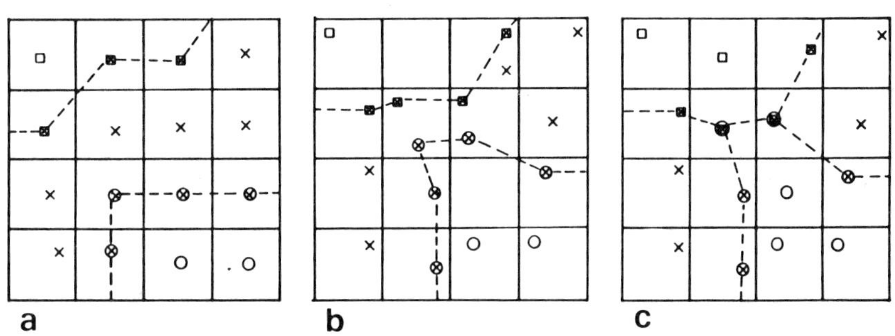

Fig. 7. Interface tracking on the dual mesh. a) Original dual mesh marker points and interface; b) predicted new dual mesh points; c) regularized dual mesh and interface.

5. Summary and conclusions

Most numerical methods for tracking interfaces are based on either following the surface of the interface using marker points, calculating the fractional volume of each region separated by the interfaces as they pass over a reference grid, or moving control volumes aligned with the interface.

The interface is advanced by solving a compatibility equation, usually derived from an integral formulation of the weak form of the PDEs. The smooth solution away from the interface is treated with standard finite-difference or finite-element methods. The interaction of the smooth solution with the interface is accounted for by treating the interfaces as a moving boundary with boundary conditions determined by the compatibility equations.

Reliable methods are available that approximate multiple interactions in 1-D, simple interactions in 1- and 2-D, and isolated interfaces in 1-, 2-, and 3-D. Better methods are still needed for interacting interfaces in 2- and 3-D. It is not yet clear which of the existing methods will prove to be the most effective in these complicated situations.

Acknowledgements

I thank Andy White for pointing out the distinction between a numerical method and a numerical scheme.

References

[1] P.K. Barr and Wm.T. Ashurst, An Interface Scheme for Turbulent Flame Propagation, Sandia National Laboratories report SAND82-8773 (1982).

[2] J.U. Brackbill and J.S. Saltzman, Adaptive Zoning for Singular Problems in Two Dimensions, J. Comp. Phys. 46 (1982) 342–368.

[3] A.J. Chorin, Flame Advection and Propagation Algorithms, J. Comp. Phys. 35 (1980) 1–11.

[4] P. Colella, P. Concus and J.A. Sethian, Some Numerical Methods for Discontinuous Flows in Porous Media, Lawrence Berkeley Laboratory report LBL-15932, April 1983.

[5] J.E. Dendy and J.M. Hyman, Multigrid and ICCG for Problems with Interfaces, Elliptic Problem Solvers, Martin Schultz, ed. (Academic Press, New York, 1981), pp. 247–254.

[6] J.K. Dukowicz, A Simplified Adaptive Mesh Technique Derived from the Moving Finite Element Method, Los Alamos National Laboratory report LA-UR-82-3664.

[7] J.K. Dukowicz, Lagrangian Fluid Dynamics using the Voronoi–Delaunay Mesh, Los Alamos Scientific Laboratory report LA-UR-81-1421.

[8] M.J. Fritts, Lagrangian Simulations of the Kelvin–Helmholtz Instability, Science Applications Inc. report SAI-76-632-WA, September 1976.

[9] M.J. Fritts and J.P. Boris, The Lagrangian Solution of Transient Problems in Hydrodynamics using a Triangular Mesh, J. Comp. Phys. 31 (1979) 173–215.

[10] R.J. Gelinas and S.K. Doss, The Moving Finite Element Method: Applications to General Partial Differential Equations with Multiple Large Gradients, J. Comp. Phys. 40 (1981).

[11] J. Glimm, E. Isaacson, D. Marchesin and O. McBryan, Front Tracking for Hyperbolic Systems, Adv. Appl. Math. 2 (1981) 91–119.

[12] J. Glimm, D. Marchesin and O. McBryan, A Numerical Method for Two Phase Flow with an Unstable Interface, J. Comp. Phys. 39 (1981) 179–200.

[13] A. Harten and J.M. Hyman, A Self-Adjusting Grid for the Computation of Weak Solutions of Hyperbolic Conservation Laws, J. Comp. Phys. 50 (1983) 235–269.

[14] C.W. Hirt and B.D. Nichols, A Computational Method for Free Surface Hydrodynamics, American Society of Mechanical Engineers 80-C2/PVP-144, April 1980.

[15] C.W. Hirt and B.D. Nichols, Volume of Fluid (VOF) Method for the Dynamics of Free Boundaries, J. Comp. Phys. 39 (1981) 201–225.

[16] J.M. Hyman, Accurate Monotonicity Preserving Cubic Interpolation, Los Alamos Report No. LA-8796-MS, to be published in SIAM J. Sci. and Stat. Comp.

[17] J.M. Hyman, Adaptive Moving Mesh Methods for Partial Differential Equations, Advances in Reactor Computations (American Nucl. Soc., La Grange Park, IL, 1983) pp. 24–43.

[18] J.M. Hyman, Adaptive Static Rezoning Methods, to be published in the proceedings AMS-SIAM Conference on Large Scale computations in Fluid Mechanics (1983).

[19] J.M. Hyman and B. Larrouturou, The Numerical Differentiation of Discrete Functions Using Polynomial Interpolation Methods, Appl. Math. and Comp., vols. 10–11 (1982) pp. 487–506.

[20] P. Lötstedt, A Front Tracking Method Applied to Burgers' Equation and Two-Phase Porous Flow, J. Comp. Phys. 47 (1982) 211–228.

[21] B.D. Nichols, C.W. Hirt and R.S. Hotchkiss, SOLA-VOF: A Solution Algorithm for Transient Fluid Flow with Multiple Free Boundaries, Los Alamos Scientific Laboratory Report LA-8355, August 1980.

[22] B.D. Nichols and C.W. Hirt, Methods for Calculating Multi-Dimensional, Transient Free Surface Flows Past Bodies, Los Alamos Scientific Laboratory report LA-UR-

75-1932; Proc. First Int. Conf. on Numerical Ship Hydrodynamics, Gaithersburg, MD, October 1975.

[23] W.F. Noh and P. Woodward, SLIC (Simple Line Interface Calculation), in Lecture Notes in Physics, vol. 59: Proc. Fifth Int. Conf. on Numerical Methods in Fluid Dynamics, A.I. van de Vooren and P.J. Zandbergen, eds. (Springer-Berlin, 1976).

[24] J.F. Thompson, ed., Numerical Grid Generation (North-Holland, Amsterdam, 1982).

[25] E.A. Overman II and N.J. Zabusky, Evolution and Merger of Isolated Vortex Structures, Inst. for Computational Mathematics and Applications Report ICMA-81-31 (1981).

[26] J. Sethian, Turbulent Combustion in Open and Closed Vessels, Lawrence Berkeley National Laboratory report LBL-15744, Feb. 1983.

[27] H.E. Trease, A Two-Dimensional Free Lagrangian Hydrodynamics Model, Ph. D. Thesis, Sept. 1981.

[28] Bram van Leer, Towards the Ultimate Conservative Difference Scheme. V. A Second-Order Sequel to Godunov's Method, J. Comp. Phys. 32 (1979) 101–136.

[29] J.E. Welch, F.W. Harlow, J.P. Shannon and B.J. Daly, The MAC Method: A Computing Technique for Solving Viscous, Incompressible, Transient Fluid Flow Problems Involving Free Surfaces, Los Alamos Scientific Laboratory report LA-3425, 1966.

Physica 12D (1984) 408–425
North-Holland, Amsterdam

ADAPTIVE MESH TECHNIQUES FOR FRONTS IN STAR FORMATION

Karl-Heinz A. WINKLER* and Michael L. NORMAN
Max-Planck-Institut für Physik und Astrophysik, Institut für Astrophysik, Garching bei München, Fed. Rep. Germany

and

Michael J. NEWMAN#
X-Division, Los Alamos National Laboratory, University of California, Los Alamos, USA

We present an implicit, adaptive-grid, finite-difference technique specifically designed to locate and track arbitrary fronts, interfaces and structures in a radiation hydro flow. The adaptive mesh is constructed in such a way that the "average" change of the flow-variables per grid zone is the same throughout the entire grid. A local grid refinement of up to a factor of 10^6 compared to an equidistant mesh has been obtained on a machine with 14 decimal digits per word. Crucial to this type of approach is the use of a stiffness operator, which prevents tangling of the mesh. The elliptic equation for the grid motion is solved implicitly coupled with the equations for the flow variables. Per timestep very steep and narrow flow features can travel many times their own width in true space. However, with respect to the adaptive mesh the motion of such flow features is always less than one zone. Thus, the Courant–Friedrichs–Lewy [1] (CFL)-condition's accuracy limitation is always fulfilled although its stability criterion does not pose any timestep limitations for our implicit numerical method. In fact, the choice of timesteps is based purely on physical and accuracy considerations. Typical applications run on Courant numbers ranging from 0.1 to 10^{12}. The performance of this numerical scheme is demonstrated on a planar 1-D shocktube and the formation of a one solar mass protostar in spherical symmetry.

1. Introduction

Within the context of ideal compressible fluid dynamics, shock fronts and contact discontinuities appear as true flow discontinuities. At these discontinuities the differential form of the Euler equations mathematically breaks down. However, the integral form of the equations still holds and can be used to derive the Rankine–Hugoniot jump conditions which connect both sides of a fluid discontinuity. Accordingly, one numerical approach for computing discontinuous fluid flows is to use interface following and/or front tracking algorithms. Such methods were used e.g. by McBryan [2], Youngs [3], and Smarr, Norman, and Winkler [4], this conference, and are authoritatively summarized by Hyman [5] in his conference paper "Moving Mesh Methods for Shocks and Interfaces".

However, front tracking is not always feasible or desirable. For example, in flows in which a large number of fluid discontinuities are generated and/or interact with each other, this numerical approach tends to become progressively complicated and cumbersome. In addition to these complications, typical astrophysical flows are also governed by a variety of external forces and microphysical processes which impose complicated physical structures onto the flow with time and length scales alien to pure ideal fluid dynamics. In particular, the formation of stars out of collapsing interstellar gas clouds develops unique flow patterns due to self-gravity, non-local radiation effects, dissociation and ionization, and finally nuclear burning. These features require special numerical techniques for their proper treatment, some of which are the main topic of this paper. But first,

*X-Division Consultant, Los Alamos National Laboratory.
#Visitor, Max-Planck-Institut für Astrophysik.

0167-2789/84/$03.00 © Elsevier Science Publishers B.V.
(North-Holland Physics Publishing Division)

we shall describe the additional complicating physical processes that operate in protostellar collapse which led us to the development of our implicit 1-dimensional adaptive-mesh method.

For example, self-gravity can force a diffuse interstellar cloud to collapse into a highly condensed state, a star, which typically has an extremely sharp transition zone at its edge, an atmosphere, where over a very small relative length dr/r density ρ falls off by many orders of magnitude. A density variation of up to $\rho \sim r^{-5000}$ is not unusual for accreting atmospheres. Moreover, the accretion time scale, to be covered by the numerical simulation, in which the original circumstellar cloud slowly rains onto the newly formed embryo star, is up to a factor of 10^9 longer than a typical dynamical time scale of the central object.

Radiation effects can change dramatically the structure of shock fronts, Heaslet and Baldwin [6]. They either produce a very narrow relaxation zone just downstream of the mechanical shock, or in the case of a strong radiation field coupled with a weak shock, a discontinuous transition in some or all of the fluid quantities through the mechanical shock front. Further complications arise when light fronts diffuse through opaque material. This phenomena is called a Marshak wave. In contrast to pure fluid dynamics, radiation fields are generally non-local and must be described by the integro-differential transport equation. Also, whereas spherically symmetric hydrodynamics constitutes a truly one-dimensional computation, spherically symmetric radiation hydrodynamics is a genuine three-dimensional phenomena, because the radiation quantities are functions of radius, angle of sight, and frequency.

Microphysical processes considered include realistic equations of state as well as detailed opacities for the gas and dust mixtures out of which stars are formed. Dissociation and ionization processes typically occur in very narrow regions of space – in dissociation zones or ionization fronts – and likewise leave their imprint on the fluid flow. Initially, a star is heated alone by gravitational contraction in the Kelvin–Helmholtz contraction phase. Later, nuclear energy is the dominant energy source. The nuclear burning is governed by a large reaction network of different isotopes and occurs predominantly in shells. In these burning shells the relative abundance of the nuclear species changes dramatically with respect to the unburned areas. Again, the need for a high resolution numerical method arises to properly describe the sharp transition of specific nuclear abundances from the burning to the unburned regions.

As many of the narrow physical structures (exponential atmospheres, relaxation and dissociation zones, ionization and other reactive fronts) encountered here cannot be described by jump conditions, we abandon explicit interface following and front tracking methods and treat instead all flow features as smooth flow. We use a truly adaptive mesh algorithm which allows us to completely resolve all flow structures. Shock fronts are artificially smeared out to a tiny fraction of a normal zone size and can then be treated in the same way as the other fronts. The usual penalty for narrow grid zones in explicit methods in the form of timestep limitations due to the Courant–Friedrichs–Lewy [1] condition is avoided by using an implicit numerical scheme. The implicit equations are solved iteratively with a second order converging Newton–Raphson type method, in which particular advantage is taken of the block-diagonal structure of the resulting system of linear equations.

One objection sometimes raised against this numerical procedure is the following. In principle, all narrow flow features can be described by some kind of generalized "jump conditions". It is assumed in this argument that all narrow flow features can be described by steady state equations. The resulting set of ordinary differential equations is then solved in a black box type fashion and used to connect both sides of the flow feature. Thus, the black box ordinary differential equation solver assumes the role of a jump condition that has to be called every time the usual partial differential equation solver encounters a special flow feature on its equidistant mesh. The computational advantage with respect to our implicit adaptive mesh method seems to consist of combining well-known numerical methods for ordinary differential equations with an otherwise standard explicit difference method. However, the price being paid is the assumption of steady state equations.

We are not willing to pay the price for this unnecessary assumption. Indeed, one could use ordinary differential equation techniques throughout without any further assumptions. In the method of lines, Hyman [7], a system of partial differential equations is discretized in the spatial direction, then considered as a coupled system of ordinary differential equations, with time as the independent variable, and solved with standard ODE routines. It is used in connection with an adaptive grid method by Paleologou and Mouschovias [8], and offers us an alternative solution procedure for a system of adaptive mesh equations. However, the basic character (i.e. adaptive grid technique, spatial discretization, artificial viscosity, etc.) of the numerical procedures is the same. Only future work will show which of the two approaches is more cost effective on a given machine.

In this paper we will be concerned exclusively with the adaptive mesh aspects of our numerical method. The additional numerical techniques such as conservative radiation hydro equations, artificial tensor viscosity with a fixed length or a fixed relative length, and a second-order accurate, implicit, monotone, upwind centered, interface flux-differencing algorithm for advection terms are described in Tscharnuter and Winkler [9], Winkler and Norman [10], and Winkler, Norman and Mihalas [11].

In section 2 an example of the performance of the adaptive mesh technique in ideal fluid dynamics is given for Sod's [12] shock tube. After developing the adaptive mesh equation in section 3, we apply it within a radiation hydro context on a test problem with an exponential atmosphere in section 4. Finally, in section 5 the space-time evolution of a realistic protostar computation is discussed.

2. Hydrodynamics in a flat background

We begin by showing the performance of our numerical method in a pure gas dynamical flow unaffected by the other physical processes mentioned in the introduction, so that we can later sort out flow features due to these additional effects. In particular, we avoid here the operation of hydrodynamics in an extreme density stratification of 20 or more orders of magnitude, characteristic of gravitational flow problems. This is the reason for calling the chapter hydrodynamics in a flat background.

For our first example we apply our numerical method to Sod's shock tube [12]. This test problem allows us to compare the numerical solution with the analytical one for the early phases of the evolution, Chorin [13], so that we directly can see the "error bars" of the numerical method. Later, the evolution leads to a number of interactions of non-linear waves, which we will compare qualitatively with text book examples of Courant and Friedrichs [14]. Finally, the numerical solution is carried through to late times of the evolution where no analytical solution exists.

Fig. 1 shows the initial configuration, space-time evolution, and analytical distribution of pressure P, density ρ, specific internal energy e, and velocity u at the evolutionary time $t = 0.25$ of the shock tube. Initially two constant, but different states of gas, $P_1 = 1.0$, $\rho_1 = 1.0$, $e_1 = 2.5$, $u_1 = 0.0$ at one side, and $P_0 = 0.1$, $\rho_0 = 0.125$, $e_0 = 2.0$, $u_0 = 0.0$ at the other are separated by a diaphragm at position $x = 0.5$. Furthermore, the ideal equation of state

$$P = (\gamma - 1) e\rho, \tag{2.1}$$

with ratio of specific heats $\gamma = 1.4$, and reflecting boundary conditions at position $x = 0$ and $x = 1$ are assumed. At time $t = 0$ the diaphragm is re-

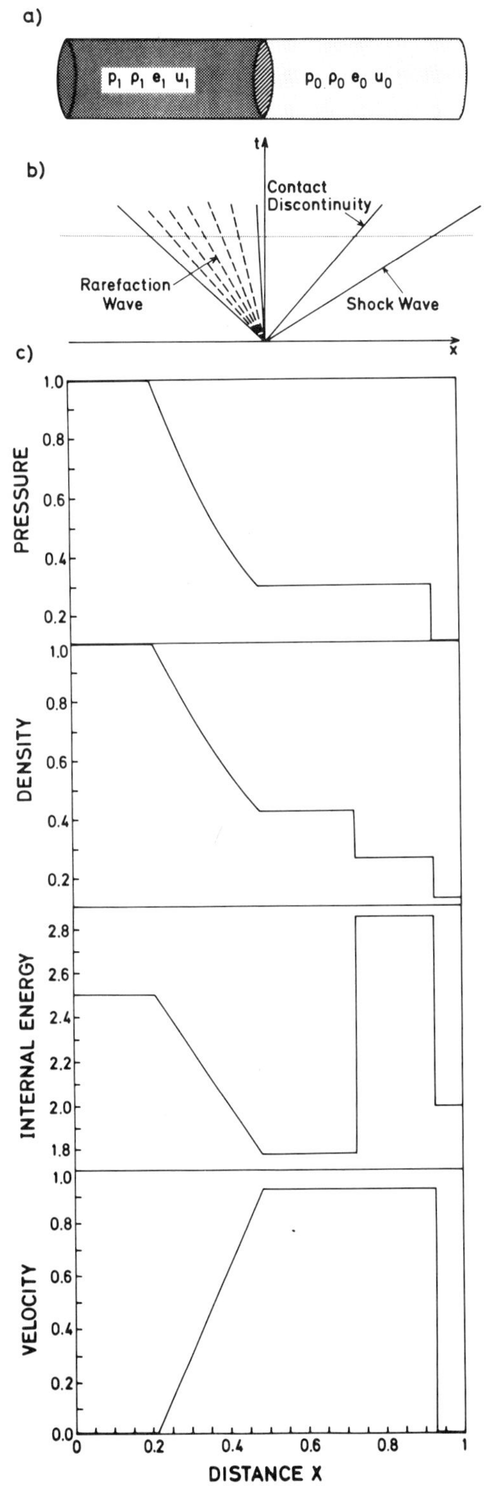

Fig. 1. Shock tube. a) schematic setup; b) space–time evolution; c) analytic solution showing well separated waves.

moved. The subsequent evolution of the gas is assumed to be described by the equations of ideal gas dynamics. The resolution of the initial discontinuity between the 2 different constant states constitutes a Riemann problem and leads to the formation of 2 waves, breaking away at different sides from the contact discontinuity, which is left behind in pressure and velocity equilibrium. Because of the pressure ratio of $P_1/P_0 = 10$ the wave, going to the right, is a shock front, and the wave, going to the left, a rarefaction wave. Specific internal energy and velocity vary linearly through the rarefaction wave, whereas the pressure and density profiles are curved. For detailed numbers of the different physical states we refer to table II of Sod [12].

In fig. 2 four different snapshots of our numerical evolution are presented. Each of the 292 spatial gridpoints used in the computation is shown individually in the distributions of P, ρ, e, and u. Columns from left to right refer to evolutionary times of $t = 0.0108$, 0.228, 0.364, and 1.0. For this calculation the weight factors of the adaptive grid equation (3.2) were chosen to be $Wx = 1$, $Wxz = 0.5$, $Wel = 2$, $W\rho l = 4$, and $WPl = 1$. The fixed length l, used in the artificial viscosity for treating shocks, was set to 0.005 of a normal (equidistant) zone width of $\mathrm{d}x = 0.0034$, i.e. $l = 0.005 * \mathrm{d}x = 1.7 \times 10^{-5}$. This also means that the quadratic term of the artificial viscosity is reduced by a factor of 4×10^4 compared to an equidistant run. The widths of the artificially broadened shock fronts typically are 4 to 5 l in our calculation, only a tiny fraction of a normal zone size. A prerequisite for this level of resolution is that the grid equation (3.1) always manages to pull "enough" gridpoints into these flow features, which guarantees a smooth and stable numerical solution.

Column 1 of fig. 2 shows the flow structure immediately after breakup of the initial discontinuity. The right facing shock front and the left facing rarefaction wave, visible in all four variables, as well as the central contact discontinuity, only seen in ρ and e, are well developed and accurately resolved by the numerical method. In

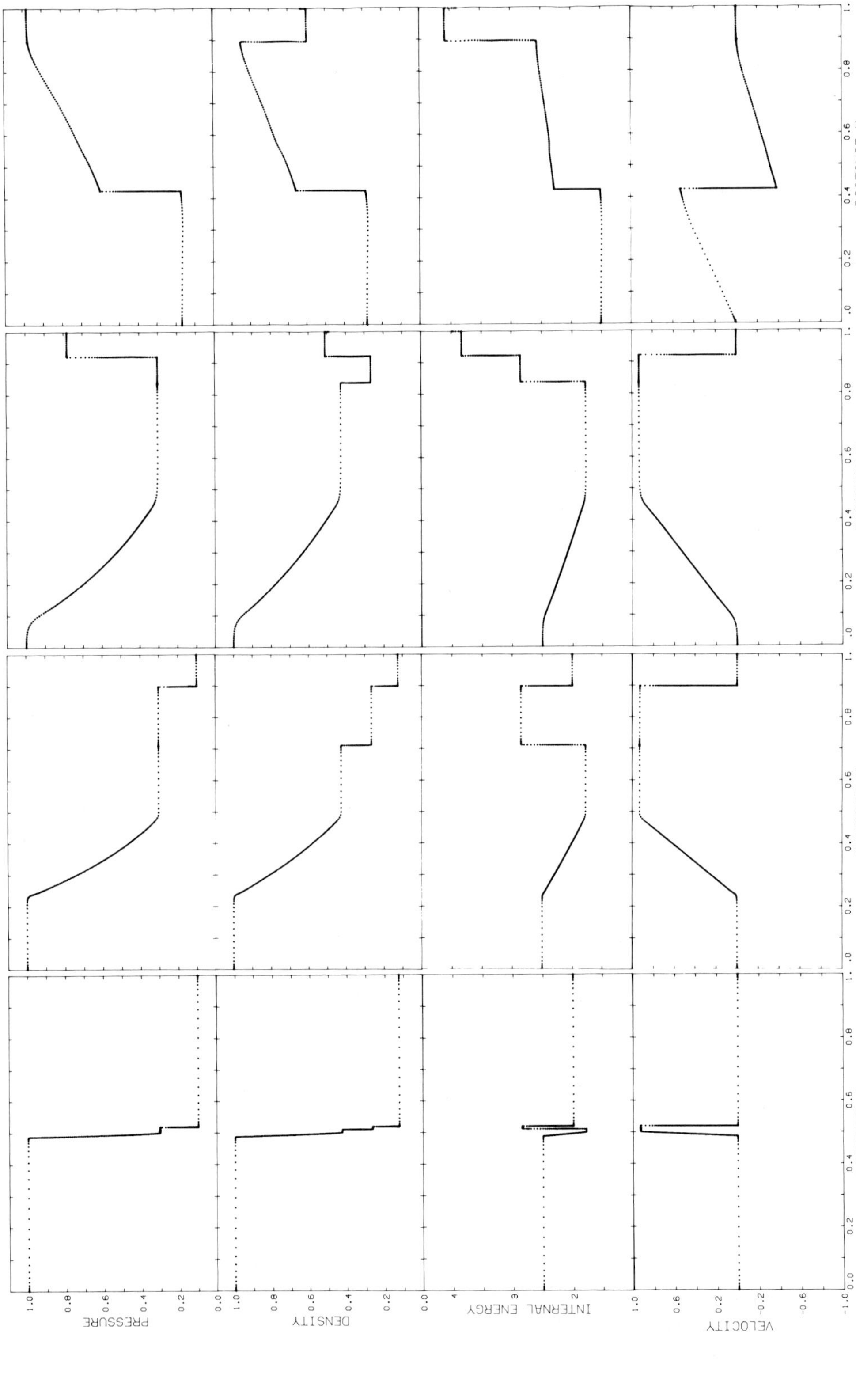

Fig. 2. Numerical solution of Sod's shock tube test problem at 4 different times. Gridpoints are shown individually.

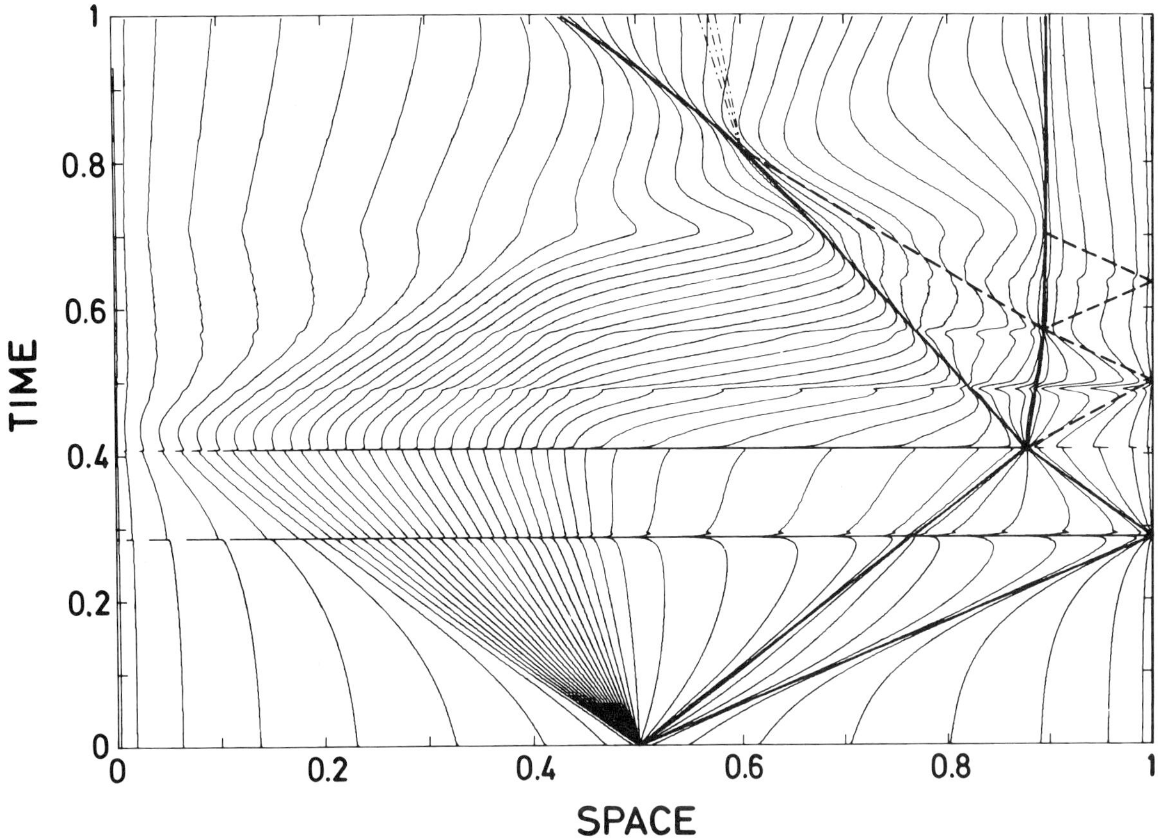

Fig. 3. Numerical space–time evolution of the adaptive mesh for Sod's shock tube problem. The trajectory of every 4th gridpoint is shown. Strong shocks and contact discontinuity are made visible by the dense concentration of gridpoints. Weak waves are indicated by broken lines.

column 2 all three flow features are fairly well separated. Here, shock front and contact discontinuity are still moving to the right. All significant flow features of the numerical solution agree to better than 0.01 of a percent with the mathematically correct solution of fig. 1. In columns 3 and 4 of fig. 2 waves have already reflected off the boundaries. In column 3 the shock has reflected off the right hand boundary and is now approaching the contact discontinuity. When the shock interacts with the contact discontinuity the contact discontinuity is brought to rest by multiple shock reflections. In column 4 the transmitted shocks have merged into a single left-facing strong shock, thereby sending off a new rarefaction wave, clearly seen in the numerical solution around position $x = 0.57$.

All these wave interactions can be understood a lot easier with the help of the space-time evolution of the adaptive mesh, shown in fig. 3. Here, the spatial position of every fourth gridpoint of the computation is shown as a function of time. Because the adaptive mesh is constructed in such a way that it automatically finds flow features and resolves them with a large number of gridpoints, as we will see in section 3, fig. 3 not only shows us the motion of the computational mesh but also the evolution and interaction of nonlinear waves. Five different types of wave interactions can be distinguished. In comparison with similar interac-

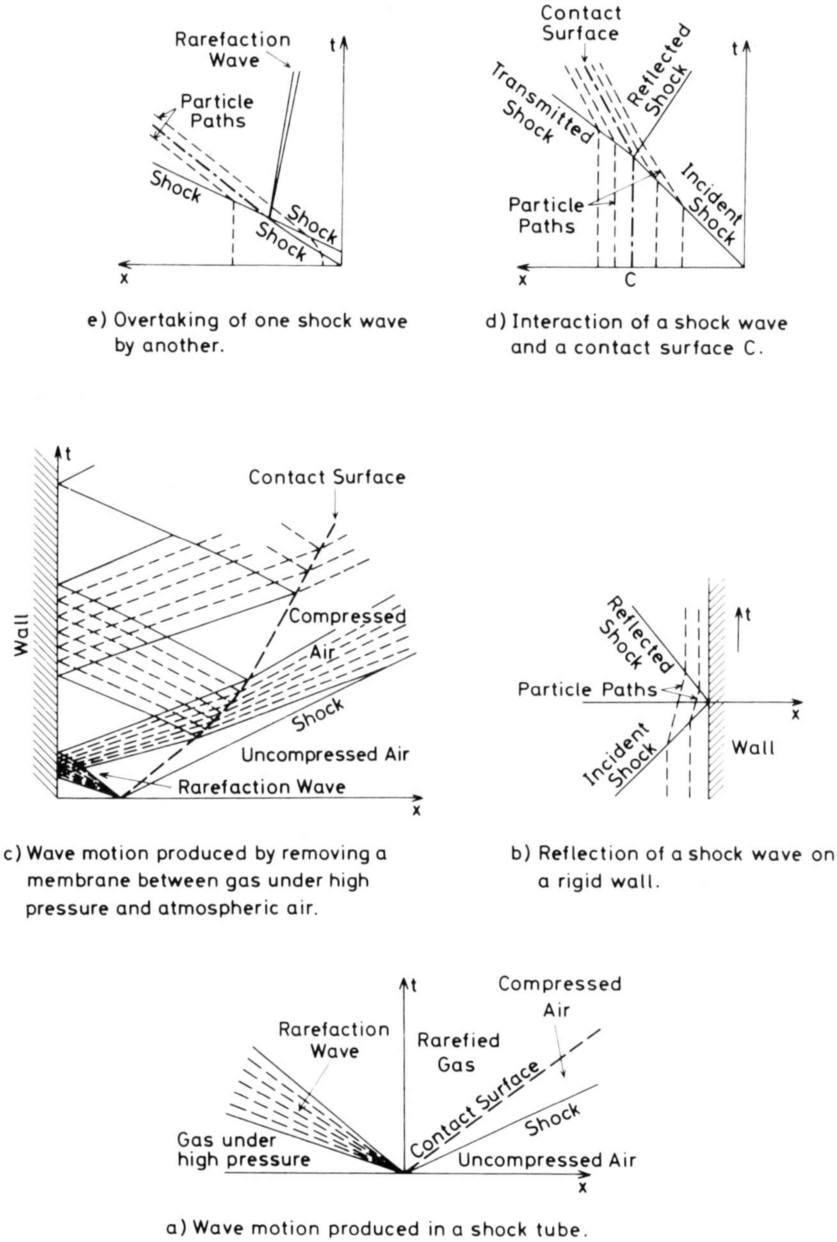

Fig. 4. Interactions of nonlinear waves, adapted from Courant and Friedrichs [14].

tions, adopted from Courant and Friedrichs [14] and shown in fig. 4, we identify the following ones:

a) break up of a shock tube;

b) shock reflecting off boundary;

c) rarefaction wave reflecting off boundary;

d) interaction of a shock wave with a contact discontinuity, resulting in a transmitted and a reflected shock;

e) one shock overtaking another, thereby forming a new contact discontinuity and a new rarefaction wave.

One final comment concerns the almost horizontal grid motion in fig. 3. Whenever nonlinear waves interact with each other the flow variables are restructured, i.e. gradients disappear and show up again. Consequently, gridpoints are first being

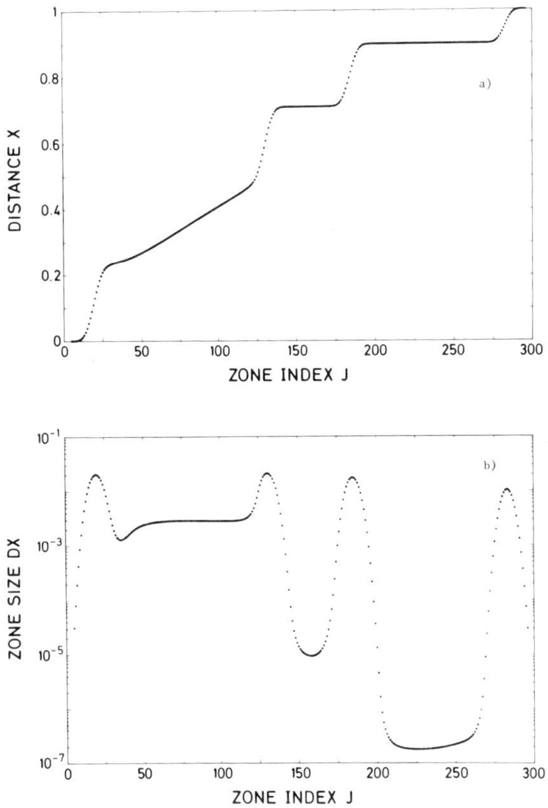

Fig. 5. Adaptive mesh distribution. a) position; and b) spacing of gridpoints at $t = 0.228$ of shock tube test problem.

pulled out of, and then put back into those flow features. Because the grid motion is free of inertia, the redistribution of gridpoints is affected everywhere.

In fig. 5 the metric of our numerical simulation is shown at time $t = 0.228$. By metric we mean the dependence of the dependent spatial variable x and the grid spacing dx as a function of our independent zone index j. In comparison with column 2 of fig. 2 we see that shock front and contact discontinuity cover quite a number of gridpoints. However, in true space these features are extremely narrow. Overall, a grid spacing varying by 5 orders of magnitude was used in this example of ideal gas dynamics, which was run on a Courant–Friedrichs–Lewy number of up to 50000.

3. Adaptive mesh techniques

In many flow situations of interest complicated features with steep gradients develop, which cannot adequately be resolved with standard Eulerian or Lagrangean finite-difference techniques. In general, there is an urgent need for the use of some kind of adaptive coordinate system in the numerical methods employed. The so called "rezoning" or "moving grid" methods were developed, see e.g. refs. 5, 16–34. In these methods, the computational grid is occasionally or more often rearranged according to some prescription. In any case, rearranging of the numerical grid is done only in between the time steps of the physical equations.

We would like to distinguish between these historically older methods and the truly "adaptive mesh" techniques of refs. 8, 9, 15, 35–42. In these new methods, an extra numerical equation, the grid distribution equation, is added to the appropriate set of physical equations. This new set of coupled grid and physical equations is then solved simultaneously, (i.e. implicitly) with either finite difference or finite element techniques, or with the method of lines. Only in this way can one avoid Courant-like conditions, which would otherwise limit a reasonable mesh refinement to order of the square of the fractional zone size of a typical equidistant mesh [43].

In the remainder of this chapter we follow the basic philosophy of Winkler [35], Tscharnuter and Winkler [9], and Winkler and Newman [36, 37] in deriving the adaptive mesh technology. However, our new grid equation now contains an essential new operator, the stiffness operator, which automatically stabilizes the grid distribution completely and prevents tangling of the mesh.

The starting point for any type of adaptive mesh technique is the formulation of the physical equations in an arbitrarily moving grid system. Hereby, we have to distinguish between the coordinate- and the grid-dependence of the equations. We mean coordinate systems such as Cartesian, spherical, or axial coordinates, whereas grid stands for the discretized description of a particularly chosen coordinate system. For our purposes we choose either a 1-dimensional Cartesian coordinate system or spherical coordinates. The corresponding grid of the 5-point difference molecule we use is shown schematically in the following graph:

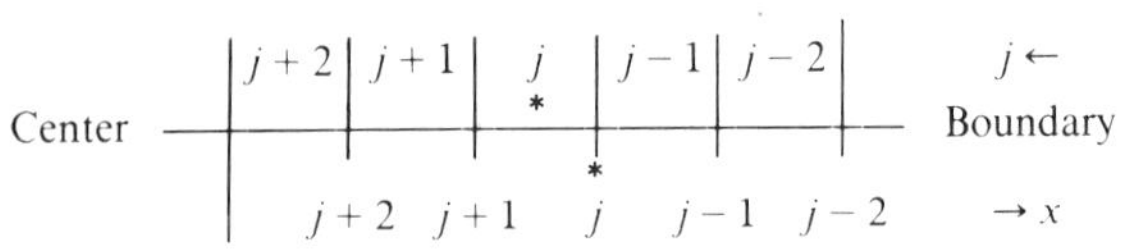

Index j varies inversely with the spatial variable x, which is either the position x in Cartesian coordinates or the radius r in spherical coordinates. We use a staggered mesh and define vector quantities or tensors of odd rank at zone interfaces and scalars or tensors of even rank at zone centers. For later use we also define that the index of the sum/difference of 2 neighboring interface quantities gets the index of the enclosed zone centered quantity, and vice versa, the index of the sum/difference of 2 neighboring zone centered quantities the index of the enclosed interface quantity.

We have almost complete freedom in choosing the motion of individual grid points within a once and for all preselected type of coordinate system. For example, we could choose spherical coordinates and move all grid points in the radial direction according to some prescription. In that case, the physical equations are written in spherical coordinates and would have to take explicitly into account the advection effects which are caused by the radial motion of the gridpoints. Thus, we could choose a different motion for any of the grid points as a function of time, as long as we keep the grid points monotonically ordered.

In practice, we determine directly the new spatial position of an interior gridpoint j by solving the elliptical equation

$$\mathrm{d}f(j)=f(j-1)-f(j)=0, \tag{3.1}$$

at the new time level with function f defined by

$$\begin{aligned}
f(j)= {} & Wx \;*\mathrm{d}x(j) && (3.2)\ \mathrm{T1}\\
&+Wxz \;*\mathrm{d}x(j)**2/(\mathrm{d}x(j-1)*\mathrm{d}x(j+1)) && \mathrm{T2}\\
&+Wxl \;*\mathrm{d}xl(j) && \mathrm{T3}\\
&+Wxzl*\mathrm{d}xl(j)**2/(\mathrm{d}xl(j-1)*\mathrm{d}xl(j+1)) && \mathrm{T4}\\
&+Wm \;*\mathrm{d}m(j) && \mathrm{T5}\\
&+Wml*\mathrm{d}ml(j) && \mathrm{T6}\\
&+Wql \;*\mathrm{d}ql(j+1) && \mathrm{T7}\\
&+Wel \;*\mathrm{d}el(j+1) && \mathrm{T8}\\
&+W\rho l \;*\mathrm{d}\rho l(j+1) && \mathrm{T9}\\
&+WPl \;*\mathrm{d}Pl(j+1) && \mathrm{T10}\\
&+W\mu \;*\mathrm{d}\mu(j+1) && \mathrm{T11}\\
&+WEl \;*\mathrm{d}El(j+1) && \mathrm{T12}\\
&+WNl*\mathrm{d}Nl(j+1), && \mathrm{T13}
\end{aligned}$$

where $\mathrm{d}x, \ldots, \mathrm{d}Nl$ are given by

$$\mathrm{d}x(j)=(x(j)-x(j+1))/X\text{ scale}, \tag{3.3}$$

$$\mathrm{d}xl(j)=(x(j)-x(j+1))/(x(j)+x(j+1)), \tag{3.4}$$

$$\mathrm{d}m(j)=(m(j)-m(j+1))/M\text{ scale}, \tag{3.5}$$

$$\mathrm{d}ml(j)=(m(j)-m(j+1))/(m(j)+m(j+1)), \tag{3.6}$$

$$q(j)=(u(j)+u(j+1))**2/e(j)+q\text{ floor}, \tag{3.7}$$

$$\mathrm{d}ql(j)=|q(j-1)-q(j)|/(q(j-1)+q(j)), \tag{3.8}$$

$$\mathrm{d}el(j)=|e(j-1)-e(j)|/(e(j-1)+e(j)), \tag{3.9}$$

$$\mathrm{d}\rho l(j)=|\rho(j-1)-\rho(j)|/(\rho(j-1)+\rho(j)), \tag{3.10}$$

$$\mathrm{d}Pl(j)=|P(j-1)-P(j)|/(P(j-1)+P(j)), \tag{3.11}$$

$$\mathrm{d}\mu(j) = |\mu(j-1) - \mu(j)|, \tag{3.12}$$

$$\mathrm{d}El(j) = |E(j-1) - E(j)|/(E(j-1) + E(j)), \tag{3.13}$$

$$\mathrm{d}Nl(j) = |N(j-1) - N(j)|/(N(j-1) + N(j) + N\ \text{floor}). \tag{3.14}$$

Here, x is the spatial variable, m the Lagrangean mass variable, e the specific internal energy of the gas, E the energy density of the radiation field, ρ the gas density, P the gas pressure, μ the mean molecular weight of the gas, N the nuclear energy generation per gram and per second, and u the Eulerian velocity of the gas. Quantity q is, aside from a constant, the ratio of gas kinetic to gas internal energy. Constant weight factors are denoted by $Wx, \ldots, WNl$. Other constants are X scale, M scale, q floor, and N floor.

Central and boundary points cannot be computed from eq. (3.1). They are set to either fixed Eulerian or fixed Lagrangean coordinate values, or they are computed as a free boundary value problem with a given outside pressure.

As a first consequence of eq. (3.1) we note that the spatial variable x and the mass variable m are no longer independent variables. Instead, the spatial index j and the time index n are the only independent quantities. As the grid distribution equation (3.1) is solved simultaneously with our set of physical (i.e. self-gravitating radiation hydrodynamics) equations in an iterative fashion, we can say that the grid system actually used is only known after convergence of the iterations.

Secondly, we note that the grid distribution is entirely determined by the solution of the dependent variables at the new time level. The intention behind eq. (3.1) is to construct the grid in such a way that it automatically recognizes characteristic features in the flow and moves along with them. In places where the physical variables of interest vary sharply over small distances, the grid points should run closely together. But in places where the variables hardly vary, the grid points should be spread apart. As a further constraint, our accordian-type grid distribution should be numerically stable, e.g. one grid point should never overtake another one. How eq. (3.1) fulfills indeed all these conditions will be shown next in a detailed discussion of the individual terms of the equation.

Consider the first term T1 of eq. (3.2). Inserted into equation (3.1) we get

$$Wx(\mathrm{d}x(j-1) - \mathrm{d}x(j)) = 0. \tag{3.15}$$

The constant X scale (see eq. (3.3)) is used to scale the spatial variables and typically set to the width of the computational domain. With appropriate boundary conditions and say $Wx = 1$ we find the solution of (3.15) to be an equidistant mesh, i.e.

$$\cdots = \mathrm{d}x(j+1) = \mathrm{d}x(j) = \mathrm{d}x(j-1) = \mathrm{d}x(j-2) = \cdots . \tag{3.16}$$

If we use instead only term T3 we have

$$Wxl(\mathrm{d}xl(j-1) - \mathrm{d}xl(j)) = 0. \tag{3.17}$$

Noticing that eq. (3.4) can be rewritten as

$$\mathrm{d}xl(j) = (x(j) - x(j+1))/(x(j) + x(j+1)) = 0.5 * \mathrm{d}\ln x(j), \tag{3.18}$$

we see that the solution of eq. (3.17) with appropriate boundary conditions is a mesh which is equally spaced in the natural logarithm ln of x, i.e.

$$\cdots = \mathrm{d}xl(j+1) = \mathrm{d}xl(j) = \mathrm{d}xl(j-1) = \mathrm{d}xl(j-2) = \cdots . \tag{3.19}$$

This grid distribution is advantageous when the spatial variable x covers many orders of magnitude in a problem. We also see that the logarithm automatically scales the equation.

If we insert only term T8 (or T9, T10, or T12) into eq. (3.1) we find as the corresponding grid solution a mesh which is equidistant in the logarithm of e (resp. ρ, P, or E), provided these variables are not constants. In that case the solution is undetermined. In general, we are interested only in grid distributions in which all variables are resolved equally well. This can be achieved by inserting terms T1 and/or T3 with the terms T9, T10, and T12 into eq. (3.1). Now the solution of the grid equation will describe an adaptive mesh in which the "average" change or the "average" relative change of the considered flow-variables is the same per grid zone throughout the entire mesh. To the list of variables x, e, ρ, P, or E we add all those quantities, which are essential to our problem, mass m (T5, T6), ratio of kinetic to internal energy (T7, important for pulling grid points into shock fronts), mean molecular weight (T11), nuclear energy generation rate (T13), or other new variables, e.g. fractional isotope abundances, optical depth, etc. Except for x or m (used either linearly or logarithmically) and μ all other variables are used only logarithmically. Our experience is that this procedure gives excellent results for widely different problems. A typical value for the constant M scale is the total mass of the problem, q floor and N floor are typically of order 1. The weight factors $Wx, \ldots, WNl$ are likewise of order 1 and reflect approximately the inverse ratios of the order of magnitudes by which the flow variables change with respect to x.

For many applications, the terms described so far give an excellent grid distribution. However, in some cases it turns out that this reduced grid distribution equation leads to numerically unstable results. Characteristic of this instability is a grid distribution with alternating narrow and wide zones as shown below. This difficulty can be overcome by use of the stiffness operators T2 and T4. Let us consider the following grid distribution and study the effects of T2, T4 on it.

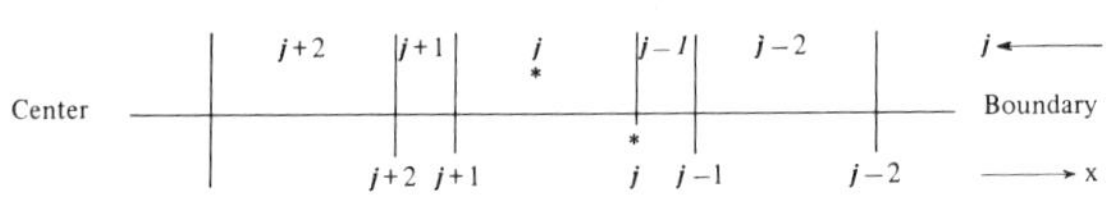

Adding term T2 or T4 to eq. (3.1) with adequate weight factor Wxz or $Wxzl$ will result in a widening of the narrow zones and a narrowing of the wide zones, thus stabilizing the mesh. This property of the stiffness operators T2 and T4 can be deduced by inserting only term T2 (or T4) into eq. (3.1),

$$Wxz*(\mathrm{d}x(j-1)**2/(\mathrm{d}x(j-2)*\mathrm{d}x(j)) - \mathrm{d}x(j)**2/(\mathrm{d}x(j-1)*\mathrm{d}x(j+1)))=0. \tag{3.20}$$

Consider now all positions x fixed, except for $x(j)$, and solve eq. (3.20) for $x(j)$. One then finds the above properties.

Another property of the stiffness operator T2, T4 is desired, namely that they do not change the basic underlying mesh. Depending on the boundary conditions used for the solution of eq. (3.20) (resp. the corresponding equation for T4), one also finds the following 2 grid distributions as possible solutions.

equidistant distribution, $\cdots = \mathrm{d}x(j+1) = \mathrm{d}x(j) = \mathrm{d}x(j-1) = \cdots$

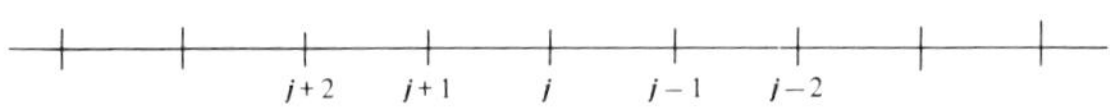

power law distribution, $\mathrm{d}x(j+n) = \mathrm{d}x(j)^n$

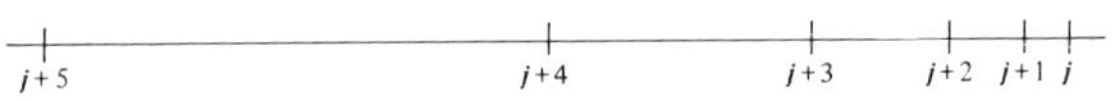

These results can be verified by direct substitution. The 2 possible solutions imply that adding terms T2 or T4 to an equidistant or logarithmically equidistant grid distribution will not change that distribution. In regions with large spatial gradients of the flow variables the stabilizing operators influence the grid distribution in such a way that a smooth regular mesh refinement results. Additionally, the grid in smooth regions anticipates steep flow features and becomes gradually narrower. Consequently, the mesh is already quite narrow at the onset of the actual flow feature and is further refined in a smooth way.

Eqs. (3.1) and (3.2) are designed to concentrate grid points at discontinuities of the flow variables. Additional terms based on second differences of the flow variables can be introduced that focus on discontinuities in the first derivative. This is desirable for example at the head and tail of a rarefaction fan. However, these terms have been omitted in the present discussion because we feel that they are special purpose tools. Eqs. (3.1) and (3.2) constitute a minimum set of terms for a stable and robust adaptive mesh.

So far, we have described adaptive mesh techniques for only the spatial coordinate in planar or spherical geometry. However, spherically symmetric radiation hydrodynamics is a 3-dimensional problem in radius, angle, and frequency and ultimately requires a 3-dimensional adaptive mesh treatment. Such a technique is presently being developed by Winkler, Norman and Mihalas [44] and Mihalas, Winkler and Norman [45] and will be completely described in future papers of that series.

4. Radiation hydrodynamics in an exponential background

Gravitational flows differ from ordinary laboratory-type flows in two respects. First, their motion is driven by an external gravity force in addition to internal pressure forces. Second, gravity imposes a length scale l of order $l = GM/a**2$ onto the flow, so that gravitational flows are no longer scale invariant, in contrast to normal scale free hydrodynamical flows. Here, G is the gravitational constant, M the mass of the system, and a the local speed of sound. Since gravity is an attractive force, a nearly homogeneous medium will tend to become clumpy on a length scale l. As it does so, however, gravitational compression generally heats the gas, leading to a stratification of sound speeds and hence length scales. Thus, gravitational flows are in general characterized by stratifications in all flow variables over a variety of length scales set by local conditions and some nonlocal influences (e.g. radiation), and by a wide range of dynamical time scales.

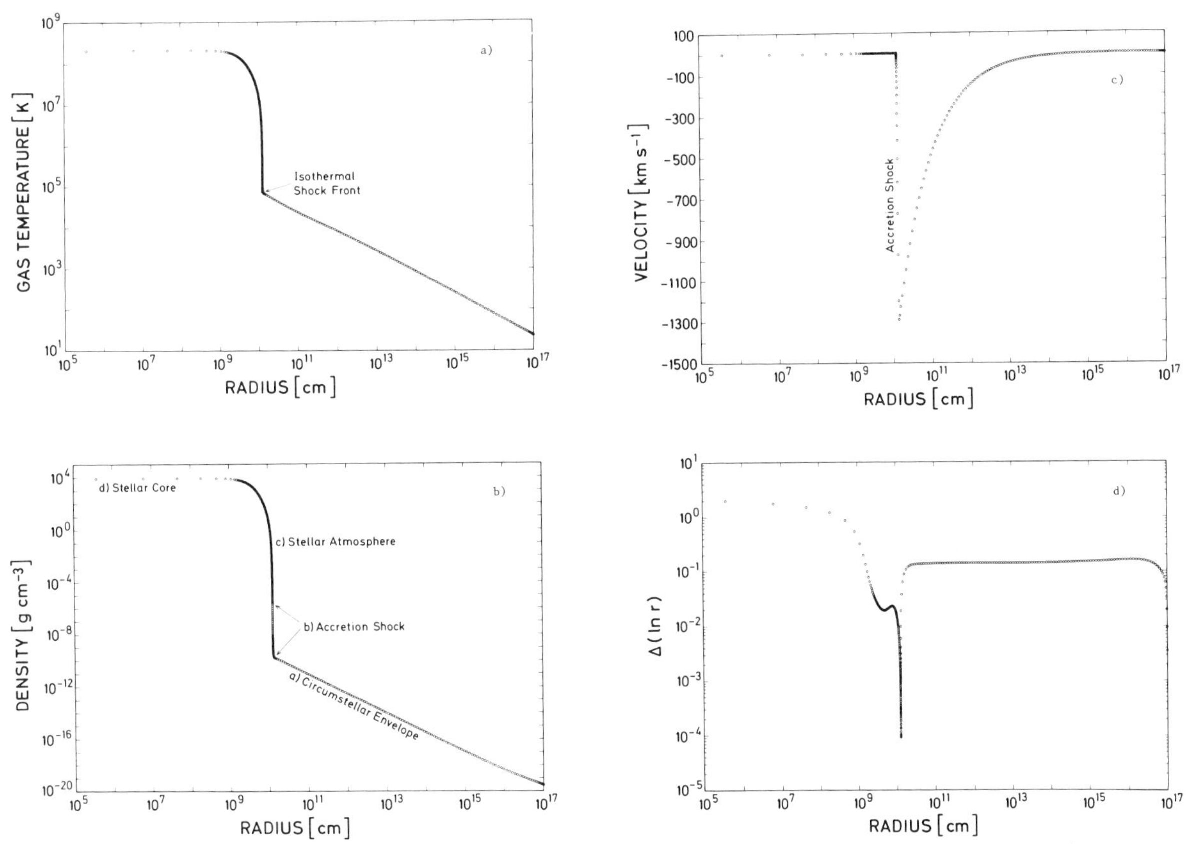

Fig. 6. Protostar test problem showing stellar core, atmosphere, accretion shock and circumstellar envelope when 90 percent of the total mass is assembled in the core.

An example of a gravitational flow which develops an extreme density stratification is a protostellar accretion flow. In order to show the influence of gravity alone onto the flow we have eliminated the complicated realistic equation of state and opacity. Instead, we assume for this test problem an ideal equation of state with $\gamma = 5/3$ and a constant opacity of $\kappa = 1\ \mathrm{cm}^2/\mathrm{g}$. As initial conditions we assume a homogeneous cloud of gas of 1 solar mass $= 1.989 \times 10^{33}$ g with constant temperature of 10 K filling a volume of radius $R = 1.0 \times 10^{17}$ cm. As the boundary condition we assume a constant volume, embedded in a 10 K background radiation field. These parameters are chosen in such a way that the gas cloud is gravitationally unstable, i.e. the gas cloud will collapse under its own gravity because it cannot be stabilized by pressure gradients. Later, a tiny core forms onto which the remaining cloud slowly rains down in the accretion phase.

Fig. 6 shows a snapshot of gas temperature, density, velocity, and grid spacing as a function of radius which is typical for an accretion flow situation. In fig. 6b we have indicated the essential physical regimes of this problem, i.e.

a) the circumstellar freely falling envelope, which provides the accreting material;

b) the accretion shock, in which the incoming material is brought to rest; and

c) the stellar atmosphere, which acts as a transition zone to

d) the stellar core.

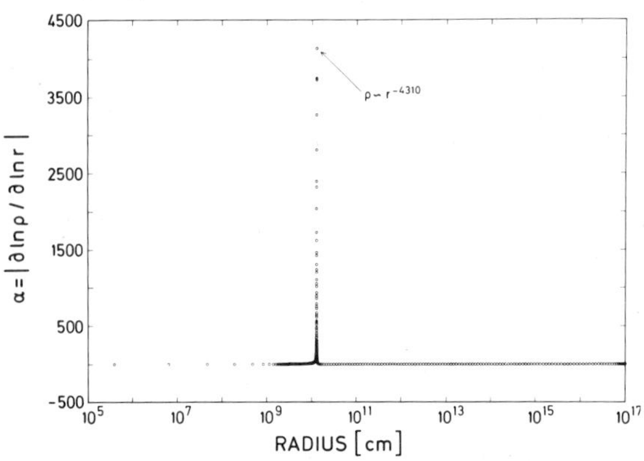

Fig. 7. Extremely large density gradient $\alpha = |\mathrm{d} \ln \rho / \mathrm{d} \ln r|$ of the stellar atmosphere as resolved by the adaptive mesh.

In the stellar atmosphere, temperature and density vary by many orders of magnitude over a short distance in true space, requiring the high resolution of our adaptive mesh technique. That the achieved grid refinement of approximately 4 orders of magnitude is indeed sufficient for this challenging problem can be seen in fig. 6d, where the grid spacing is shown. All grid points are shown individually. The grid distribution was determined from eq. (3.1) with the parameters set to $Wxl = 1$, $Wxzl = 3$, $Wm = 50$, $Wml = 0.01$, $Wql = 3$, $Wel = 1$, $W\rho l = 5$, $WPl = 1$, $WEl = 1$, q floor $= 1$; the remaining parameters were set to zero.

As a consequence of our simplified assumptions the test problem develops an isothermal accretion shock, indicated in fig. 6a, in contrast to real accretion shocks, which have a complicated relaxa-

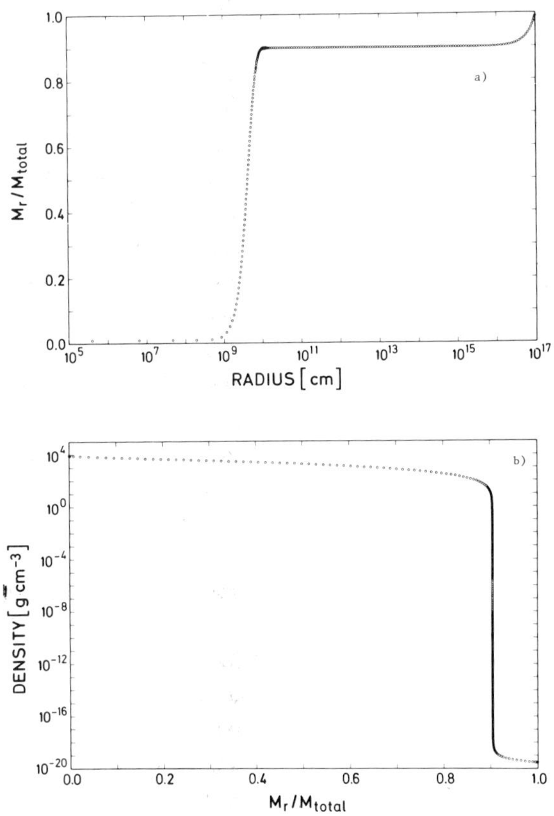

Fig. 8. a) Radial distribution of the integrated mass; b) density distribution as a function of the normalized Lagrangean mass variable. Plateau in fig. 8a and steep fall-off in fig. 8b pose insurmountable problems for a Lagrangean calculation.

tion zone. This will allow us to check the accuracy of our numerical solution against the analytical one for this highly nonlinear phenomenon. As $\mathrm{d}r/r$ for the isothermal shock front in this spherical problem is very small, and can therefore be considered as planar, we can compare the density variation across our accretion shock of $\rho_1/\rho_0 = 8.824 \times 10^3$ directly with the analytical result of $\rho_1/\rho_0 = 1.25\ M_0 ** 2$ for an isothermal planar shock front with constant $\gamma = 5/3$, where M_0 is the preshock Mach number of the flow. With M_0 measured to be $M_0 = 84.31$ in our numerical solution, we find an agreement in the density jump across the shock of 0.7 per cent.

Next, we would like to address the question of whether this problem could be solved successfully with either an Eulerian or Lagrangean numerical hydro method. The answer to this question has to be no. All attempts in this direction have had to rely on additional simplifying physical assumptions or ad hoc numerical procedures which modified the character of the original physical equations. The inadequacy of an Eulerian or Lagrangean grid method originates in the extraordinary steep structures of the flow which are neither fixed in space nor mass, and therefore cannot be resolved adequately by those methods.

This is illustrated in figs. 7 and 8. Fig. 7 shows that the exponent $\alpha = |\mathrm{d}\ln\rho/\mathrm{d}\ln r|$ of the piecewise constant power law of $\rho \sim r^{-\alpha}$ defined locally for the density distribution in fig. 6b varies between 1.5 and 4310. This extremely steep and narrow feature can travel many times its own width per computational time step. In one time step 30 to 40 gridpoints will typically move across a fixed position in radius, whereas the density at the same position may vary by up to 6 orders of magnitude. However, if we ask ourselves by how much does the density maximally change for a given mesh index, we find a change of only 3 to 5 per cent. If we used instead an Eulerian mesh and tried to obtain the same accuracy, a very fine zoning in space and time would have to be used for most of the computational mesh, resulting in a substantial (many orders of magnitude) cut back of our time step of 10^{12} Courant–Friedrichs–Lewy numbers during the accretion phase.

For the accretion flow problem, Lagrangean coordinates are even worse than Eulerian coordinates. Fig. 8a shows the normalized mass within a given volume of radius r as a function of radius. The plot shows the situation when about 90 per cent of the total mass has been accreted onto the stellar core of radius 10^{10} cm. The remaining 10 per cent are still at radii of 10^{16} to 10^{17} cm. In between the outer mass reservoir and the inner core is the freefall region, where up to 10^{12} times more mass per unit time is being transported through than actually resides there at any instant in time. If we used an equidistant zoning in mass, a mass element would first be stretched during its accretion by 4 or more orders of magnitude, and then pushed back together. Such a deformation of a mass element necessarily degrades any Lagrangean grid solution.

That a Lagrangean difference scheme is indeed more poorly suited than an Eulerian scheme is demonstrated in fig. 8b. Here density, shown as a function of normalized mass, drops suddenly by almost 20 orders of magnitude, whereas in fig. 6b density dropped only by about 10 orders of magnitude in a short distance of radius. The reason for this difference in behavior is the moderate power law fall off of density as a function of radius in the more or less mass-free free-fall region of the circumstellar envelope, which adds 9 more orders of magnitude of sudden fall off in density as a function of mass.

5. Space–time evolution of a 1 solar mass protostar

Now let us consider the evolution of a realistic protostar and describe the additional complicating physical effects due to equation of state, opacity, and nuclear burning, which leave their imprint on the self-gravitating radiation hydro flow. The general features of the evolution, adapted from Winkler and Newman [37], are contained in fig. 9. It presents a summary of the results of numerous

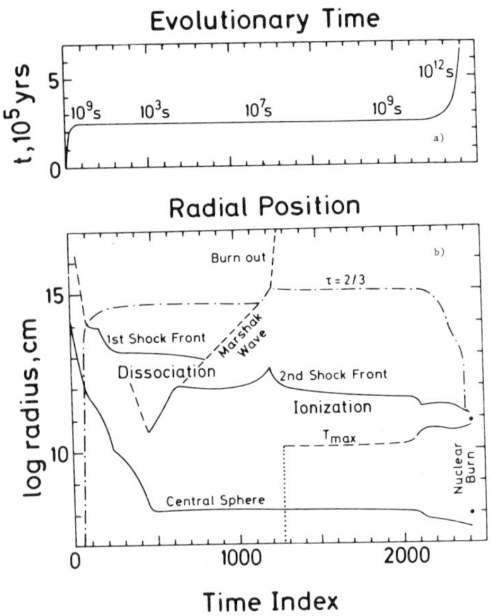

Fig. 9. Space–time evolution of a 1 solar mass protostar as described in the text. a) Evolutionary time as a function of time index; b) radial position of various fronts and significant features as a function of time index.

investigations and review papers [36, 37, 46–58].

As the isothermal collapse begins, a rarefaction wave travels inward from the outer boundary stratifying the outer layers of the cloud. The position of the head of the wave is plotted in fig. 9b. As the collapse continues the density increases and the material begins to become optically thick to radiation; the point where the optical depth $\tau = 2/3$ (measured from the fixed outer boundary) begins to move outward in radius, as shown in fig. 9b. When the point $\tau = 2/3$ crosses the head of the rarefaction wave, the gravitational energy released by the collapse can no longer be radiated away directly, and begins to contribute to the heating of the gas. The collapse is halted, the first stellar core is formed at a radius near 10^{14} cm, and the rarefaction wave is reflected as a standing accretion shock. At this phase the central density has already increased by 6 orders of magnitude from initially 10^{-19} g cm^{-3} to 10^{-13} g cm^{-3}.

Following the end of the first collapse and the formation of the first hydrostatic core, the time scale of the evolution increases. The core contracts quasi-hydrostatically and nearly adiabatically, the temperature increasing until it reaches values near 3000 K in the center, when hydrogen molecules begin to dissociate. The gravitational energy released then goes into dissociation energy of the molecules, rather than into thermal energy. The adiabatic exponent γ becomes smaller than 4/3, and the inner regions of the core again become gravitationally unstable; a second collapse ensues, involving material far below the outer shock front. The second collapse continues until H_2 molecules are completely dissociated in the central region. When the heating resumes, the resultant pressure increase arrests the collapse, and the second (final) stellar core with an accretion shock is formed at a radius of about 4.2×10^{10} cm and a central density of 10^{-3} g cm^{-3} before rebounding to about 2.8×10^{12} cm. For a time a double shock front structure persists. The outer material is not aware of the crisis below, and the quasi-static contraction of the first (outer) core continues, the outer shock eventually simply disappearing as the material below it falls onto the second shock front. The material at the point $\tau = 2/3$ is not disturbed by the transition, and its radial position does not change appreciably.

The time scale of the second collapse is short, approximately 10^3 s. The following expansion of the newly formed core is at first rapid (time index numbers 450–600) and then slow (600–1200). Typical time scales for various phases of the evolution are indicated in fig. 9a, where the true evolutionary time is shown as a function of the computational cycle number.

A third collapse, which might be expected to be produced by the ionization of atomic hydrogen does not occur. In contrast to the dissociation process ionization of the gas procedes rather slowly and does not lead to a gravitationally unstable situation.

The gravitational energy released during the second collapse is primarily transformed into radiation by the second shock front. As the time scale for the production of the radiation is very short compared to the transport of the radiation through the opaque envelope, a luminosity spike builds up. The light wave then slowly diffuses through the envelope in about 10^7 s. This Marshak wave heats the freely falling material of the envelope somewhat, increasing the optical thickness of the outer regions. The point $\tau = 2/3$ moves sharply outward in radius shortly before the wave has burned completely through the envelope and reaches the outer boundary of the protostellar cloud. The flare-up in luminosity is the first sign to the outside world that a new star is in the process of being born.

The first and second collapse can also be recognized in the contraction of the central (innermost) sphere of our adaptive mesh system. In order to be able to resolve all newly developing structures deep inside the gas cloud, which by now increased its central density by 16 orders of magnitude, we prescribe a Lagrangean grid for the central sphere with a mass of 10^{-10} times the total mass of the system.

During the slow contraction of the stellar core from about time index 1250 to 2100 the dynamical flow pattern is transformed into a typical accretion flow: a supersonic infall is stopped by a strongly radiating shock front and converted to a hydrostatic equilibrium in the central core. For details of the shock structure including its relaxation zone we refer to Winkler and Newman [36, 37]. Now the use of an adaptive mesh becomes completely indispensible for the further accurate simulation of the evolution. As accretion begins in earnest and the envelope is depleted of material, the point $\tau = 2/3$ moves inward, at first slowly, then more rapidly, until the radiation coming from the core and the shock front is able to shine through the envelope without degradation, and the initial infrared spectrum characteristic of the envelope is replaced by a normal stellar appearance.

Towards the end of the accretion process temperatures and densities in the stellar core reach high enough values for the onset of nuclear burning. The additional energy release contributes to the heating of the gas and slows down the stellar contraction, which so far was the primary energy source of the core. The birth process is over, the adult life of the star begins. Nuclear burning alone will now supply the energy, which lets the star shine for billions of years.

6. Summary

The formation of a protostar is a complicated physical process involving the fluid dynamics of a radiating, self-gravitating gas wherein multiple and interacting fronts of various sorts (see fig. 9) are of primary dynamical importance and interest. Our numerical method was designed with these complexities in mind, in particular the use of an implicit adaptive mesh. The advantages of our method for constructing numerical models of star formation are: a) the enormous variation in flow variables and length- and time-scales are easily handled, b) fronts are automatically detected and tracked as they arise in the flow, and c) a local grid refinement of up to a factor of 10^6 (limited by machine roundoff!) can be used to resolve the structure of fronts which cannot be approximated as simple discontinuities (important for radiating shock fronts, dissociation and ionization fronts, nuclear burning fronts, Marshak waves etc.). However, as the shock tube examples show, our method can also be applied to more standard dynamical gas flows with good results. We would therefore conclude that the utility of adaptive mesh methods extends beyond problems in astrophysics, and that adaptive mesh techniques offer a promising new approach to a wide class of pure and applied problems where the above-mentioned capabilities are required.

Acknowledgements

K.-H.A.W. thanks X-Division of the Los Alamos National Laboratory and M.J.N. thanks the Max-Planck-Institut für Physik und Astrophysik for their hospitality while portions of this work were done.

References

[1] R. Courant, K. Friedrichs and H. Lewy, Math. Ann. 100 (1928) 32.
[2] O.A. McBryan, this conference.
[3] D.L. Youngs, Physica 12D (1984) 32 (these proceedings).

[4] L.L. Smarr, M.L. Norman and K.-H.A. Winkler, Physica 12D (1984) 83 (these proceedings).
[5] J.M. Hyman, Physica 12D (1984) 396 (these proceedings).
[6] M.A. Heaslet and B.S. Baldwin, Phys. Fluids 6 (1963) 781.
[7] J.M. Hyman, The Method of Lines Solution of Partial Differential Equations, technical report (1976) C00-3077-139, Courant Institute of Mathematical Sciences, New York University.
[8] E.V. Paleologou and T.Ch. Mouschovias, The Magnetic Flux Problem and Ambipolar Diffusion during Star Formation: One-Dimensional Collapse. I. Formulation of the Problem and Method of Solution, (1983), preprint.
[9] W.M. Tscharnuter and K.-H.A. Winkler, Computer Phys. Commun. 18 (1979) 171.
[10] K.-H.A. Winkler and M.L. Norman, in: Astrophysical Radiation Hydrodynamics, Proc. of NATO ARW (Garching, August 2–13, 1982), K.-H.A. Winkler and M.L. Norman, eds. (Reidel, Dordrecht, 1984), in preparation.
[11] K.-H.A. Winkler, M.L. Norman and D. Mihalas, An Implicit Adaptive-Grid Radiation Hydrodynamics, in: Computational Techniques (1984), in press.
[12] G.A. Sod, J. Comp. Phys. 27 (1978) 1.
[13] A.J. Chorin, J. Comp. Phys. 22 (1976) 517.
[14] R. Courant and K.O. Friedrichs, Supersonic Flow and Shock Waves (Interscience, New York, 1948).
[15] P. Eggleton, Monthly Notices Roy. Astron. Soc. 151 (1971) 351.
[16] D.K. Davison, Dynamical Zoning, an Efficient Technique for Finite-Difference Solution of Hydrodynamic Initial-Boundary Value Problems with Steep Gradients and Local Condensations, Ph.D. Thesis (1977) University of California, Santa Cruz.
[17] J.I. Castor, C.G. Davis and D.K. Davison, Dynamical Zoning within a Lagrangian Mesh by use of DYN, a Stellar Pulsation Code, technical report (1977) LA-6664, Los Alamos National Laboratory.
[18] H.A. Dwyer, R.J. Kee and B.R. Sanders, AIAA Journal 18 (1980) 1205.
[19] H.A. Dwyer, M.D. Smooke and R.J. Kee, in: Numerical Grid Generation, J.F. Thompson, ed. (Elsevier, New York, 1982).
[20] J.H. Bolstad, An Adaptive Finite Difference Method for Hyperbolic Systems in One Space Dimension, Ph.D. Thesis (1982) Stanford University.
[21] S.F. Davis and J.E. Flaherty, SIAM J. Sci. Stat. Comput. 3 (1982) 6.
[22] J.M. Hyman, Adaptive Mesh Strategies for the Numerical Solution of Differential Equations, technical report (1981) LA-UR-81-2594, Los Alamos National Laboratory.
[23] J.M. Hyman, Adaptive Moving Mesh Methods for Partial Differential Equations, technical report (1982) LA-UR-82-3690, Los Alamos National Laboratory.
[24] A. Harten and J.M. Hyman, A Self-Adjusted Grid for the Computation of Weak Solutions of Hyperbolic Conservation Laws: I. One-Dimensional Problems, (1982), preprint.
[25] J.K. Dukowicz, Lagrangian Fluid Dynamics using the Voronoi-Delaunay Mesh, technical report (1981) LA-UR-81-1421, Los Alamos National Laboratory.
[26] A.J. Scannapieco, An Automatic Rezoner for Lagrangian Quadrilateral Hydrocodes, (1983), preprint, Los Alamos National Laboratory.
[27] A.M. Winslow, J. Comp. Phys. 1 (1966) 149.
[28] S.K. Godunov and G.P. Prokopov, USSR Comp. Math. Phys. 12 (1972) 182.
[29] J.F. Thompson, F.C. Thames and C.W. Masten, J. Comp. Phys. 15 (1974) 299.
[30] M.M. Rai and D.A. Anderson, Numerical Grid Generation Techniques, NASA Conf. Publ. 2166 (1980) 409.
[31] J.U. Brackbill, in: Numerical Grid Generation, J.F. Thompson, ed. (Elsevier Science, New York, 1982).
[32] J. Saltzman, A Variational Method for Generating Multidimensional Adaptive Grids, technical report (1982) DOE/ER/03077-174, Courant Mathematics and Computing Laboratory, New York University.
[33] J.U. Brackbill and J.S. Saltzman, J. Comp. Phys. 46 (1982a) 342.
[34] J.U. Brackbill and J.S. Saltzman, Variational Formulation of Adaptive Mesh Generators, technical report (1982) LA-UR-82-1635, Los Alamos National Laboratory.
[35] K.-H.A. Winkler, A Numerical Procedure for the Calculation of Nonsteady Spherical Shock Fronts with Radiation, Ph.D. Thesis, (1976), University of Göttingen, (issued as technical report MPI-PAE-Astro 90 by the Max-Planck-Institut für Physik und Astrophysik, Munich; (1977), English transl. UCRL-Trans-11206, Lawrence Livermore National Laboratory.
[36] K.-H.A. Winkler and M.J. Newman, Ap. J. 236 (1980a) 201.
[37] K.-H.A. Winkler and M.J. Newman, Ap. J. 238 (1980b) 311.
[38] K. Miller and R. Miller, SIAM J. Numer. Anal. 18 (1981) 1019.
[39] R.J. Gelinas and S.K. Doss, The Moving Finite Element Method: A Semi-Automatic Solver for Diverse PDE Applications, (1981) preprint.
[40] R.J. Gelinas, S.K. Doss and K. Miller, J. Comp. Phys. 40 (1981) 202.
[41] A.B. White, SIAM J. Numer. Anal. 19 (1982) 683.
[42] M.D. Smooke and M.L. Koszykowski, Fully Adaptive Solutions of One-Dimensional Mixed Initial-Boundary Value Problems with Applications to Unstable Problems in Combustion, Sandia report (1983) SAND83-8219, Sandia National Laboratories.
[43] B. Swartz, Courant-Like Conditions Limit Reasonable Mesh Refinement to Order H^2, technical report (1981) LA-UR-81-2037, Los Alamos National Laboratory.
[44] K.-H.A. Winkler, M.L. Norman and D. Mihalas, Adaptive-Mesh Radiation Hydrodynamics. I. The Radiation Transport Equation in a Completely Adaptive Coordinate System, (1984) J.Q.S.R.T, in press.
[45] D. Mihalas, K.-H.A. Winkler and M.L. Norman, Adaptive-Mesh Radiation Hydrodynamics. II. The Radiation and Fluid Equations in Relativistic Flows, (1984) J.Q.S.R.T, in press.
[46] R.B. Larson, Dynamics of a Collapsing Protostar, Ph.D. Thesis, 1968), Calif. Inst. Tech., Pasadena (University Microfilms order number 68-13, 070).
[47] R.B. Larson, Fund. Cosmic Physics 1 (1973) 1.

[48] I. Appenzeller and W. Tscharnuter, Astron. Astrophys. 40 (1975) 397.
[49] C.K. Westbrook and C.B. Tarter, Ap. J. 200 (1975) 48.
[50] P. Woodward, Ann. Rev. Astr. Ap. 16 (1978) 555.
[51] K.-H.A. Winkler, Moon and Planets 18 (1978) 472.
[52] K.-H.A. Winkler and M.J. Newman, Space Science Rev. 27 (1980) 261.
[53] M.J. Newman and K.-H.A. Winkler, The Pre-Main Sequence Sun: A Dynamic Approach, Proc. Conf. Ancient Sun, Boulder, Colorado, 1979, R.O. Pepin, J.A. Eddy and R.B. Merrill, eds. (Pergamon, Oxford, 1980), p. 551.
[54] S.W. Stahler, F.H. Shu and R.E. Taam, Ap. J. 241 (1980a) 637.
[55] S.W. Stahler, F.H. Shu and R.E. Taam, Ap. J. 242 (1980b) 226.
[56] S.W. Stahler, F.H. Shu and R.E. Taam, Ap. J. 248 (1981) 727.
[57] I. Appenzeller, Fund. Cosmic Physics 7 (1982) 313.
[58] J.E. Tohline, Fund. Cosmic Physics 8 (1982) 1.

Physica 12D (1984) 426–433
North-Holland, Amsterdam

THE EFFECTS OF TRUNCATION ERROR IN NUMERICAL MODELLING OF FLOOD-FRONTS IN OIL RESERVOIRS

C.P. JACKSON and J. RAE
Theoretical Physics Division, AERE Harwell, Oxon. OX11 ORA, USA

In oil reservoir simulations shock fronts or discontinuous solutions of the two-phase flow equations can form when the capillary pressure is constant. The well-known Buckley–Leverett solution is one example. Attempts to solve these equations by standard finite difference or element techniques lead to shock-like solutions with rapid changes in saturation over small distances. However, many authors have reported numerical shock speeds and heights in the Buckley–Leverett case which are wrong by as much as 20%. This paper shows that this is due to the truncation error of different schemes used and presents details of an analytic technique for studying the shock-like solutions to the discretised equations.

1. Introduction

Many authors have reported difficulties in numerical modelling of two-phase flow in oil reservoirs when flood fronts form. The problems encountered (not necessarily all in the same simulation) are excesssive spreading of the fronts by numerical dispersion, ripples before or after the shock, and incorrect front speeds. This last is the most serious problem since it may lead to estimates of the ultimate oil recovery which are considerably in error.

We show in this paper that these phenomena are consequences of the truncation error of the particular numerical schemes being used.

The paper illustrates the general principle that

"The phenomena that occur numerical modelling of partial differential equations can usually be best understood from the first few terms of the appropriate truncation-error equation".

(There are some phenomena that only appear when all the terms of the truncation-error equation are included, and these are often best understood directly in terms of the difference equations.) Orszag and Jayne [1], and Chin [2] have shown the utility of the approach via the truncation-error equation for a linear problem. This paper shows that it can be even more useful for a non-linear problem.

The results show that great care should be taken when choosing a numerical scheme for solving highly non-linear hyperbolic equations in order to obtain correct shock speeds. However, a suitable scheme is inherently dissipative and liable to suffer from excessive spreading of shocks. This is a currently unsolved problem.

2. Flood fronts

The equations governing two phase incompressible flow in oil reservoirs can be written as a parabolic equation for the saturation of one phase, and an elliptic equation relating the pressure to the saturation. Under certain circumstances the parabolic equation becomes hyperbolic, of the general form in one dimension,

$$\frac{\partial u}{\partial t} + \frac{\partial}{\partial x} g(u) = 0\,, \tag{1}$$

where x and t are the space and time variables. The functions g arising in oil reservoir simulation are highly non-linear. In the simplest case they are S-shaped as shown in fig.1. As we shall show later

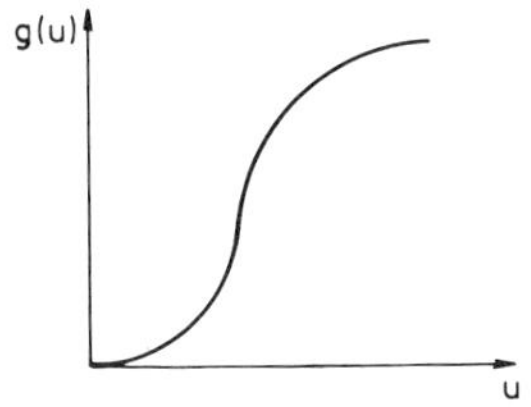

Fig. 1. Typical g in oil reservoir simulation.

this leads to effects that cannot occur unless g is purely convex (or concave) which is the case in simple gas dynamics problems for example.

Flood fronts correspond to discontinuous weak solutions of (1). At a discontinuity where u tends to v from the left and w from the right, the shock jump condition

$$c(w-v)=g(w)-g(v) \tag{2}$$

must hold where c is the shock speed. (This is a generalisation of the Rankine–Hugoniot shock condition). If weak solutions are allowed then eq. (1) together with given initial and boundary conditions do not define a unique solution. An additional constraint has to be imposed [3] such as:

"The physical solutions should be the limit of the smooth solution to eq. (1) with an added dissipation term which is taken to zero".

(This is the approach often adopted in gas dynamics where viscosity provides the dissipation mechanism.)

3. Truncation error in finite element and difference discretisations

A typical finite difference (or element) scheme for solving equation (1) on a uniform grid with space and time mesh spacings Δx and Δt is

$$\frac{u_i^{j+1}-u_i^j}{\Delta t}=-\theta\frac{(g_{i+1}^{j+1}-g_{i-1}^{j+1})}{2\Delta x}-(1-\theta)\frac{(g_{i+1}^{j}-g_{i-1}^{j})}{2\Delta x}, \tag{3}$$

where

$$u_i^j=u(i\Delta x, j\Delta t),$$
$$g_i^j=g(u_i^j),$$

and θ is a parameter which can be used to vary the implicitness of the method. Four values of θ are of particular interest, namely 0.0, 0.5, 0.878, 1.0. They give respectively an explicit forward Euler scheme, a Crank–Nicholson scheme, the Liniger scheme [4], and a fully implicit backward Euler scheme. (The first of these is unstable when applied to the linear case $g(u)=ku$ with constant k.) For all values of θ the scheme is conservative. (The scheme (3) is only used as an example to illustrate the technique of this paper. As will be shown later it is not the most sensible scheme to use for (1), although it has actually been used by some authors.)

The truncation error equation (TEE) is obtained by expanding (3) in a Taylor series about u_i^j. We shall concentrate on the case of small Δt when the temporal truncation error can be neglected. The TEE is then

$$u_t+g_x+\frac{1}{6}(\Delta x)^2 g_{xxx}+\frac{1}{124}(\Delta x)^4 g_{xxxxx}+\cdots=0, \tag{4}$$

where

$$u_t=\frac{\partial u}{\partial t},\quad g_x=\frac{\partial g}{\partial x},\quad \text{etc.}$$

It should be noticed that the truncation error is dispersive in character rather than dissipative.

In a linear problem Orszag and Jayne [1], have shown the usefulness of the TEE in determining a priori the error of a numerical scheme. Chin [2] has shown that the significant features of their work can be obtained from the first truncation term. This paper shows that the TEE is even more useful in non-linear problems.

In the linear case the shock speeds obtained by numerical methods are correct but there is an error in the form of the shock front, a decaying ripple behind the shock. In the non-linear case incorrect shock speeds are obtained from various methods and this is more serious than the localised error in the linear case. It can lead to predictions in waterflood breakthrough times in error by more than 10% independent of the grid and time step refinement, which is obviously undesirable. The theory developed below, based on an analysis of the TEE, enables one to predict such anomalous results.

That the difficulties are due to the truncation error can be seen as follows. One is interested in how well a solution of (3) solves (1) at a given x, t in the limit $\Delta x \to 0$, $\Delta t \to 0$. The solutions of (3) can be classified broadly into two groups although many solutions have features of both. Firstly, there are solutions with an intrinsic length scale determined by the physical problem, so significant changes in the solution take place over some length independent of Δx. Secondly, there are solutions with a length scale of order Δx, including of course shock-like solutions. The accuracy of the first class of solutions is expected to improve as Δx is decreased while that of the second is almost certain not to as can be seen from (4).

In particular, although the limit of the solution of (3) as Δx and Δt tend to zero is a weak solution of (1), it is not necessarily the physical weak solution, and we shall show in the next section some examples of this.

Alternative discretizations such as upwind differences may be used, or an explicit dissipation term may be added. These both lead to a TEE of the form

$$u_t + g_x - \psi \Delta x g_{xx} + \tfrac{1}{6}(\Delta x)^2 g_{xxx} + \cdots = 0\,, \qquad (5)$$

where $\psi = \frac{1}{2}$ corresponds to upwind differences.

4. Shock-like solutions of the truncated SDPDE

It is found that when (3) is applied in cases when shocks are expected to develop that the solutions are "shock fronts", that is they have a stable shape which propagates at a constant, albeit wrong, speed. This section shows by explicit construction of approximate solutions to (4) that the anomalous speeds can be explained simply in terms of the TEE (4) truncated after a finite number of terms. The construction is by the method of matched asymptotic expansions. The approximate solutions are solutions of (1) away from the front, while in its neighbourhood the truncated TEE is reduced to an ordinary differential equation from which the speed, height and shape of the shock can be determined.

Thus consider a shock-like solution of (4) for which there is a narrow region of the order of a few mesh intervals over which the solution changes rapidly from v on the left to w on the right, while outside this region the solution is slowly varying as illustrated in fig. 2.

Away from the front the derivatives of g are independent of Δx and so, from (4),

$$u_t + g_x = \mathcal{O}[(\Delta x)^2].$$

Near the front additional terms in (4) cannot be neglected and the solution is a solution of (4) truncated after a certain number of terms, which tends to v and w on the left and right. As the truncated equation (4) is a higher order partial differential equation more boundary conditions have to be imposed. For the moment these will be obtained by requiring that at least all the derivatives of u and g appearing in the truncated equation tend to zero away from the shock. This is motivated by the shape of calculated shocks but its ultimate justification comes from the very good agreement between shocks predicted in this way and those obtained in numerical experiments.

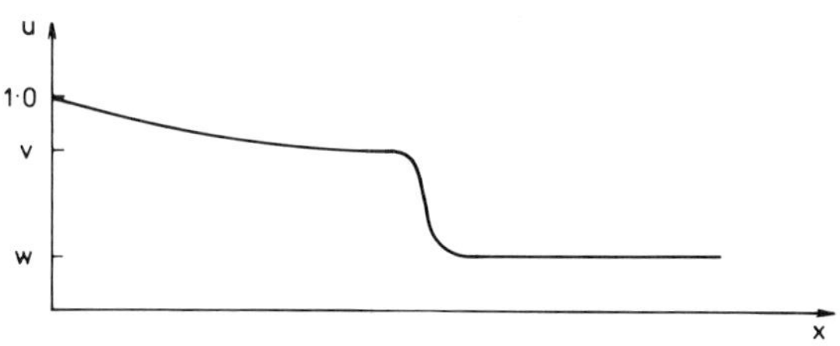

Fig. 2. Shock-like solution.

We now consider the equation

$$u_t + g_x + \tfrac{1}{6}(\Delta x)^2 g_{xxx} = 0 \qquad (6)$$

obtained by truncating (4) after the first error term. We are looking for a solution with the form of a shock front, thus a solution of the form

$$u = U(\xi)\,, \qquad (7)$$

where

$$\xi = \frac{x - ct}{\Delta x}\,, \qquad (8)$$

with constant speed c. This will be a solution provided

$$-cU_\xi + [g(U)]_\xi + \tfrac{1}{6}[g(U)]_{\xi\xi\xi} = 0\,, \qquad (9)$$

and

$$\begin{aligned} U &\to v \quad \text{as } \xi \to \xi_0\,, \\ U &\to w \quad \text{as } \xi \to \xi_1\,, \end{aligned} \qquad (10)$$

where $\xi_0 < \xi_1$ (and one or both may be infinite) and all derivatives appearing in (16) tend to zero in these limits. Eq. (9) may be integrated to give u as the solution of

$$\frac{du}{d\xi} = \left(12 \int_w^u \{(cu - g(u)) - (cw - g(u))\} \frac{dg}{du}\, du\right)^{1/2} \bigg/ \frac{dg}{du}\,, \qquad (11)$$

provided firstly that

$$cv - g(v) = cw - g(u)\,, \qquad (12)$$

which is of course just the Rankine–Hugoniot condition, and also

$$\int_w^u [(cu - g(u)) - (cw - g(w))] \frac{dg}{du}\, du = 0\,, \qquad (13)$$

which can only be satisfied for certain values of c.

Inspection of a graph of g, as in fig. 3, reveals the fundamental difference between the standard case of concave (or convex) g and the S-shaped curve of the present study.

In the standard case the quantity

$$F(u) = (cu - g) - (cw - g(w)) \qquad (14)$$

is always positive (or negative) and so (26) cannot be satisfied for any c.

It can be seen by retracing the argument that if w, v, c exist which satisfy (12) and (13) then the function u determined by integrating (11) satisfies all requirements unless either w is zero or v is one (assuming that the only points where $dg/du = 0$ is $u = 0$ and $u = 1$). If $w = 0$ (or $v = 1$) more care is needed. If $gu \simeq ku^p$ as $u \to 0$ then, if $p \leqslant 1$ the solution can be constructed as above, but if $p > 1$ then it is necessary to include more terms of the TEE up to the $(2p + 1)$ derivative.

Thus we can construct a shock-like solution of the truncated TEE. We have not shown that this is the only one, nor that it is stable, nor that it will develop from given initial conditions. However, the numerical results of the next section provide an empirical proof of these.

We postulate that

"Numerical solutions of (3) will lead to functions which approximate very closely the shock-like solution described above rather than the physical shock solutions".

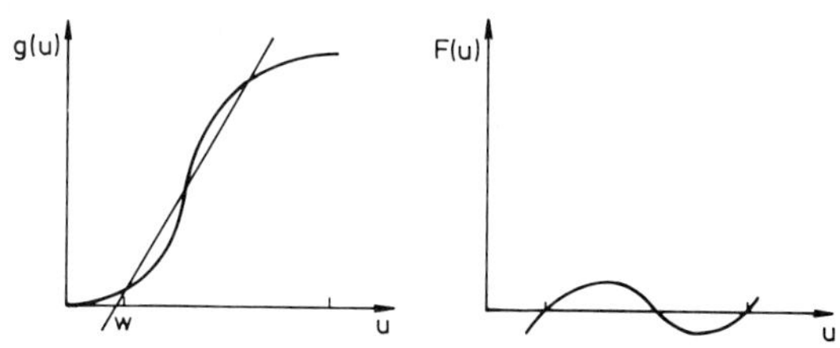

Fig. 3. Typical g and F from oil reservoir simulation.

5. Results

Eq. (11) can be integrated analytically in the case when

$$g(u)=\tfrac{1}{2}u^2-\tfrac{1}{3}u^3 .$$

The function

$$u=\begin{cases}1 & \text{for } x-\frac{1}{6}t\le\frac{3\pi\Delta x}{2\sqrt{6}},\\ \frac{1}{2}-\frac{1}{2}\sin\frac{\sqrt{6}}{3\Delta x}\left(x-\frac{1}{6}t\right) & \\ & \text{for } -\frac{3\pi\Delta x}{2\sqrt{6}}<x-\frac{1}{6}t<\frac{3\pi\Delta x}{2\sqrt{6}},\\ 0 & \text{for } x-\frac{1}{6}t\ge\frac{3\pi\Delta x}{2\sqrt{6}},\end{cases} \qquad (15)$$

derived by solving (11) satisfies all the requirements with $w=0$, $v=1$, and a speed of 1/6. (It corresponds to a water flood front moving into an initially oil filled reservoir.) In contrast the physical Buckley–Leverett solution has shock height 3/4 and speed 3/16. Fig. 4 shows the comparison of (15) with the results obtained from numerical experiments using (3). The small discrepancy is due to the neglected higher derivatives.

Tables I and II compare the front height and speed respectively obtained from numerical experiment using (3), predicted by the theory of (11), (12) and (13), and the physical Buckley–Leverett values for various functions g and $w=0$. (All calculations were performed on an IBM 3033 with considerable care taken to minimise roundoff.) The agreement between the predicted values, and the results of numerical experiment is extremely good. (The effects of time truncation error neglected in the

Table I
Shock heights for method (3) predicted and experiment. The Buckley–Leverett values are included for comparison

	Experiment				
	$\theta=1.0$		$\theta=0.5$		
$g(u)$	$\Delta t=0.1\Delta x$	$\Delta=0.01\Delta x$	$\Delta t=0.1\Delta x$	Prediction	Buckley–Leverett
$\frac{1}{2}u_2-\frac{1}{3}u^3$	0.99329	0.99934	1.0000	1.0000	0.75000
$\frac{1}{2}u^2-\frac{2}{3}u^3+\frac{1}{4}u^4$	0.75486	0.759	0.75975	0.75975	0.53752
$\frac{1}{3}u^3-\frac{1}{2}u^4+\frac{1}{5}u^5$	0.99847	0.999	1.0000	1.0000	0.72402
$g(u)=\dfrac{u^2}{u^2+\alpha(1-u)^2}$					
α					
1.0	0.95335	0.99542	1.0000	1.0000	0.70711
0.9	0.92881	0.97217	0.97689	0.97689	0.68825
0.8	0.90037	0.94509	0.94997	0.94997	0.66667
0.7	0.86703	0.91319	0.91824	0.91824	0.64167
0.6	0.82740	0.87508	0.88030	0.88030	0.61237
0.5	0.77943	0.82870	0.83411	0.83411	0.57735
0.4	0.72000	0.77093	0.77652	0.77652	0.53452
0.3	0.64389	0.69664	0.70226	0.70226	0.48038
0.2	0.54111	0.59522	0.60119	0.60119	0.40825
0.1	0.38700	0.44196	0.44810	0.44810	0.30151

above analysis can be seen in the large time step $\theta = 1$ results. As is predicted by analysis the time truncation error acts to increase the speed and decrease the height of the shock.)

The effect of upwinding or including explicit dissipation can be analysed similarly. Although it is not possible to write down a closed form for the front analogous to (11), (12) and (13) considerable

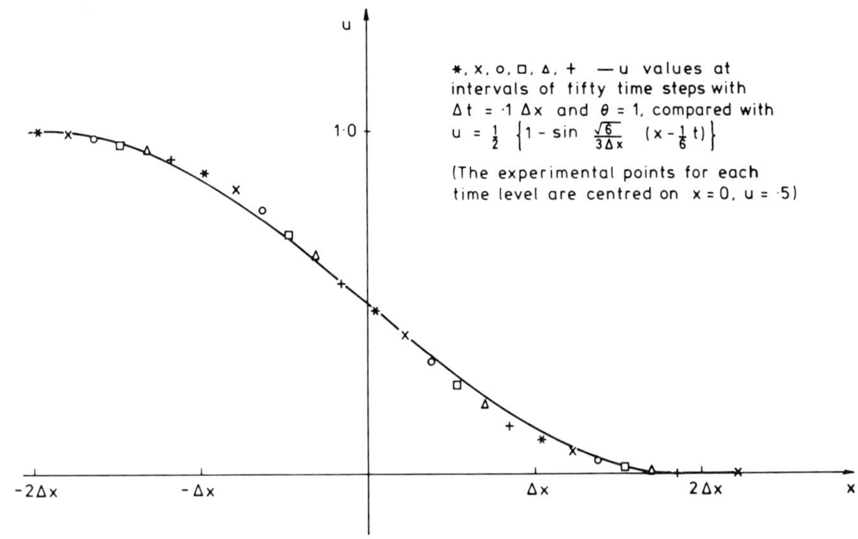

Fig. 4. Comparison of experiment with (15).

Table II
Shock speeds for method (3) predicted and experiment. The experimental values were obtained from the results at $\theta = 0.5$, $\Delta t = 0.1\Delta x$, by measuring the displacement of a particular u value (determined by linear interpolation) over many time steps

$g(u)$	Experiment	Prediction	Buckley–Leverett
$\frac{1}{2}u - \frac{1}{3}u^2$	0.16667	0.16667	0.18750
$\frac{1}{2}u^2 - \frac{2}{3}u^3 + \frac{1}{4}u^4$	0.10491	0.10470	0.11497
$\frac{1}{3}u^3 - \frac{1}{2}u^4 + \frac{1}{5}u^5$	0.033333	0.033333	0.039926
$g(u) = \dfrac{u^2}{u^2 + \alpha(1-u)^2}$			
α			
1.0	1.0000	1.0000	1.2071
0.9	1.0222	1.0231	1.2265
0.8	1.0507	1.0503	1.2500
0.7	1.0841	1.0830	1.2792
0.6	1.1243	1.1235	1.3165
0.5	1.1748	1.1756	1.3660
0.4	1.2488	1.2465	1.4354
0.3	1.3524	1.3511	1.5408
0.2	1.5242	1.5288	1.7247
0.1	1.9399	1.9377	2.1583

qualitative information can be obtained from a careful phase plane analysis. The general effect of the dissipation is to increase the shock speed and decrease the shock jump. The predicted shock shape and speed can be obtained numerically. Table III illustrates the effect of varying the amount of dissipation ψ, comparing the predictions from the TEE with numerical calculations using (3).

6. Remarks

The detailed numerical comparisons presented here substantiate the claim that the TEE is a very useful tool for analysing the error of numerical schemes for solving partial differential equations. It is almost the only method capable of providing qualitative a priori information about the error of a given numerical scheme applied to a non-linear problem.

One point emerges very strongly from the analysis, namely that any method with an $o(1)$ truncation error will probably give reasonable results if the physical problem under discussion is expected to behave smoothly without shocks or sharp changes, but if shocks are expected then the method used must be considered very carefully in the context of the problems to which it is to be applied.

In the context of oil reservoir simulation various comments can be made about the methods discussed in this paper. Thus method (3) using central spatial differences is not very useful unless there is considerable physical dissipation, which is large compared to the mesh spacing. Although for any given non-zero physical dissipation this can be ensured by the use of a sufficiently refined mesh, this would almost certainly be prohibitively expensive.

In the case when the physical dissipation is small compared to the mesh spacing, when a physical shock front is expected to occur over a length much less than a mesh spacing, then the only way to obtain the correct shock speed without refining the mesh is to add on a large artificial dissipation either directly or indirectly by using upstream differences. This of course spreads the shock front over a distance which is observed experimentally to be of the order of 9 or 10 mesh intervals. It is possible that shock sharpening techniques could reduce this, maybe even to the ultimate of one mesh spacing, and this is being researched currently. It should be emphasized that the case of small dissipation is very important physically.

There are various possible alternatives to method

Table III
Comparison of prediction and experiment with dissipation or upwinding

$g=\frac{1}{2}u^2-\frac{1}{3}u^3$	Shock height		Shock speed	
ψ	Experiment	Prediction	Experiment	Prediction
10^{-3}	0.99911	0.999	0.16690	0.167
10^{-2}	0.99104	0.991	0.16836	0.168
5×10^{-2}	0.95480	0.954	0.17335	0.174
10^{-1}	0.90828	0.908	0.17928	0.179
2×10^{-1}	0.81	0.811	0.18647	0.186
$g=\frac{1}{2}u^2-\frac{2}{3}u^3+\frac{1}{4}u^4$				
ψ				
0.001	0.75873	0.757	0.10471	0.105
0.01	0.74961	0.749	0.10562	0.106
0.05	0.7100	0.709	0.10833	0.108
0.1	0.662	0.659	0.11155	0.111

(3) which may lead to more accurate methods. For example, there is some evidence [1, 5] that higher order schemes might be better, perhaps requiring a much smaller dissipation. (Higher order schemes can be derived from quadratic or cubic finite elements, or higher order difference schemes.) Another alternative is to use some form of shock fitting. This may well be best, as it is probably unreasonable to expect a method using a mesh spacing Δx to predict accurately phenomena on a scale much less than Δx.

In conclusion, whatever method is used, the TEE can and should be used to determine the accuracy of the method.

Acknowledgements

This work has been supported by the United Kingdom Department of Energy, the British Gas Corporation and the Britoil P.L.C..

References

[1] S.A. Orszag and L.W. Jayne, Local errors of difference approximations to hyperbolic equations, J. Comp. Phys. 14 (1974) 93–104.
[2] R.C.Y. Chin, Dispersion and Gibbs Phenomena associated with difference approximations to initial boundary value problems for hyperbolic equations, J. Comp. Phys. 18 (1975) 233–247; Notes on the article "Local errors of difference approximations to hyperbolic equations", J. Comp. Phys. 16 (1974) 311–313.
[3] P.D. Lax, Hyperbolic systems of conservation laws II, Comm. Pure. Appl. Math. 10 (1957) 627–646.
[4] Liniger, Optimization of a numerical integration method for still systems of ordinary differential equations IBM Research Report RC2198 (1968).
[5] P.M. Gresho, R.L. Lee and R.L. Sani, Advection dominated flow, with emphasis on the consequence of mass lumping, Finite Elements in Fluids, vol. 3, Gallagher, R.H. et al., eds. (Wiley, New York).

LIST OF CONTRIBUTORS

ANALYTIC SUBJECT INDEX